X.media.publishing

For further volumes:
http://www.springer.com/series/5175

Christoph Meinel · Harald Sack

Internetworking

Technological Foundations and Applications

 Springer

Christoph Meinel
Hasso Plattner-Institute for IT Systems
  Engineering
Potsdam
Germany

Harald Sack
Hasso Plattner-Institute for IT Systems
  Engineering
Potsdam
Germany

ISSN 1612-1449
ISBN 978-3-662-51227-2   ISBN 978-3-642-35392-5   (eBook)
DOI 10.1007/978-3-642-35392-5
Springer Heidelberg New York Dordrecht London

Translation from the German language edition: Internetworking. Technische Grundlagen und Anwendungen. © Springer Berlin Heidelberg 2012.

Printed on acid-free paper

Springer is part of Springer Science+Business Media (www.springer.com)

Preface

What should still be something pretty amazing seems just like a part of everyday life to a lot of people. In recent decades, the old dream driving human development toward a mobility spanning time and space has become unprecedented reality. And this without a single physical law being broken. Instead we have learned to deal with a surprising number of things in life in a dematerialized, digitalized form. Dematerialized in the sense that instead of confronting the things themselves directly we interact with their digital "shadows." In essence, their descriptions, which are coded in the form of zeros and ones and then transported at the speed of light via electromagnetic signals to be processed at any computer. Two technological developments have made this possible. First, computers in all their forms provide the cosmos to give these digital shadows complete expression. Here, they can be realized anew, processed, linked and stored. Second, the Internet offers the possibility of transporting the digital shadows almost anywhere in the world at the speed of light where they achieve their full effect at another computer, even if it's at the other end of the world.

In fact, the computer and Internet rank among those very few technological developments in human history that have intrinsically changed people's lives and actions. The industrial revolution of the 19th and 20th centuries expanded our physical mobility in way that had been unparalleled up to that time. Just as cars, airplanes and space ships have dramatically increased the radius of human physical activity, the drivers of the digital revolution - computers and Internet technology - have extended our intellectual mobility in a way that was previously unthinkable. Our range of mental activity has been freed from (almost) every physical limitation. While it is likely that even the most modern physical transportation medium will still need some hours to bring a person from one continent to another, he or she is now able to bridge this distance almost immediately with the help of the Internet. Feelings, thoughts and instructions can be sent within seconds as we respond to the wishes and needs of those far away. And in contrast to physical transportation this can be done without significant cost.

The Internet is not even 50 and the WWW a good 20 years old - a young and ongoing history, with the rapid developments in computer and network technology continuing unabated. This makes it all the more interesting to look behind the scenes and gain an understanding of the technical basics of how the Internet and the WWW really work. This is just what "Internetworking" has set out to do. Together with this trilogy's other volumes, "Digital Communication" and "Web Technologies," we aim to offer the reader an understandable, comprehensive, trustworthy, informative and detailed guide.

The specific concepts presented in volume 1 of the trilogy, "Foundations of Digital Communication," (computer networks, media and their coding, communication protocols and security in computer networks) form the basis for the book at hand, volume 2, "Internetworking." Against the background of the developmental history of the Internet and a short guide through the Internet with its various actors, the actual functioning of Internet technology - the TCP/IP protocol stack - is introduced. We will look in detail at the physical layer required for every digital communication, as well as at the network access layer with its numerous technologies - wireless LAN, wired LAN, WAN. Attention will be given to the Internet layer with the protocols that support the Internet: IPv4, IPv6 and Mobile IP and the transport layer with the second protocol of the Internet - the eponymous TCP, as well as the application layer with its multi-faceted Internet services that have helped the Internet achieve its revolutionary significance for today's society. Only the World Wide Web has been excluded here and will be examined on its own in the third volume, "Web Technologies." Underlying Web technologies, such as URL, HTTP, HTML, CSS, XML, web programming, search engines, Web2.0 and the Semantic Web are presented in detail there.

The multi-dimensional material, containing understandable descriptions complemented by numerous, technically detailed excursus and glossaries offering chapter-related commentaries, indexes and bibliographic references, is arranged in such a way as to be an invitation for further research and reading. The reader is assisted in gaining the easiest access possible to the fullness of available material and also guided in making an interest or topic-based selection.

We have gone to great lengths in the hope that those of you who are interested laypeople will be infected by the fascination of the new digital world offered in this book. We also aim to provide hard-working students - who don't shy away from a bit of extra effort - with a useful and comprehensive textbook. Furthermore, we would like to present readers who are seasoned professionals with a reliable, handy reference book that can serve to classify areas of specialization easily and safely within the context of the huge complex of digital communication.

Many thanks to our colleagues at the Hasso Plattner Institute, "Department of Internet Systems and Technologies" for every imaginable support in research and teaching, as well as to Springer Verlag in the person of Hermann Engesser and Dorothea Glaunsinger for their trust in the success of this book project and their patience in its realization, and to Ivana and Anja for the forbearance and tolerance shown when we each disappeared into the office on countless weekends and holidays and for your love which accompanied us there.

Potsdam, August 2013   *Christoph Meinel*
                                                                          Harald Sack

Table of Contents

Chapter 1
Prologue

> *"Read quickly, because nothing is more certain than change on the Internet!"*
>
> – Anita Berres (German journalist)

Digital communication has developed into one of the driving technical and cultural forces in the 21st century. More than ever has it become impossible to imagine living in today's world without the Internet as our universal communication medium. In the first volume of this series the focus was on the general foundation of digital communication, i.e., the basics of computer networking, media encoding technology and digital security. In this volume, the Internet, with its various contributors, technological foundation and numerous protocols and technologies, will be covered.

The computer and the Internet belong to the group of rare technological developments in the history of humankind that have changed the life and actions of people in a fundamental way. As a driving force of the digital revolution it has expanded our intellectual mobility to a degree that would have been inconceivable in the past, and freed our mental radius of action from (nearly) every physical limitation. Together with the other volumes of this series, "Fundamentals of Digital Communication" and "Web Technologies", this book serves as a comprehensive and instructive guide through the world of modern, digital communication. This present volume could be considered the heart of the series, in which the basic technological and functional principles of the communication infrastructure of the Internet are described in detail.

A short historical look back at almost 50 years of Internet history starts off the prologue, introducing the key participants and their tasks in the global Internet.

The communication tasks taken on by the Internet are enormously complex. Various computer architectures, based on local need, connect networks of different dimensions and technologies into a virtual communication network. The user is, at the same time, always given the impression that the resulting Internet is in fact a unified and homogeneous entity. In order to master this monumental complexity, a hierarchical, modular approach was chosen in the form of a communication layer model. This has the added benefit that the tasks to be carried out at each layer are self-contained and the interaction between theses layers proceeds over a fixed interface. Chapter 2

C. Meinel and H. Sack, *Internetworking*, X.media.publishing,
DOI: 10.1007/978-3-642-35392-5_1, © Springer-Verlag Berlin Heidelberg 2013

is devoted to the communication layer model known as the **TCP/IP Reference Model** and examines in detail the general tasks and functions of the individual protocol layers.

Subsequently, the following chapters each reflect on layers of the TCP/IP reference model and present the communication protocols located at these layers. Chapter 3 starts with the **physical layer**, which is officially not a component of the TCP/IP reference model, but forms the basis upon which it is set up. If we compare simple situational parameters, such as distances to be bridged, mobility, technical effort or cost, different situations demand a variety of technologies each based on different physical communication media and infrastructures. Accordingly, chapter 3 covers the theoretical fundamentals of communication with electromagnetic signals and presents wired and wireless technological variations.

Building on this, follows a presentation of the lowest, first layer of the TCP/IP reference model in chapter 4, the so-called **data link layer**, where the local area networks (LANs) and the basic wide area network, with various technologies, are found. We will first look at the **wired LAN technologies** and the most important technologies in this group, for example, Ethernet, token ring, FDDI and ATM.

Wireless LAN technologies will then be examined. Their popularity continues to increase and any differences to their wired competitors pertaining to capability are almost non-existent today. Yet in contrast to cabled medium, wireless displays distinctive differences in terms of its demands on network communication in the areas of range, reliability or – particularly – in security. Chapter 5 explains the foundation of wireless and mobile network technologies, presenting the most important technological representatives, such as WLAN, as well as those technologies limited to close range: Bluetooth and ZigBee.

As the number of devices connected to a network increases as well as the distance between communication partners, alternative application must be implemented. These are the **wide area networks** (WANs) discussed in chapter 6. WANs can be put into use to connect local networks in different locations. Of great importance in this regard are special pathfinding procedures – so-called routing algorithms. The most important WAN technologies are introduced from the historical ARPANET to the wideband radio network standard WiMAX. The chapter closes with a presentation of various technologies by which an end user gains access to a wide area network. This span ranges from the (historical) analog telephone networks to the LTE, a mobile communication technology of the 4th generation.

In order to be able to communicate beyond the network technologies in a unified way to communicate, as it were, in one network, the Internet protocol (IP) makes a simple, but comprehensive communication service available on the **Internet layer** of the TCP/IP reference model. In the version **IPv4** it has been the heart of Internet as we have known it for for 30 years. Its successor **IPv6** has been waiting in the wings for several years now and increasingly asserting itself in offering further growth for the Internet. Together with other communication protocols from the Internet layer, IPv4 and IPv6 will be discussed in detail in chapter 7.

In the **transport layer**, located above the Internet layer, protocol functions are made available that make it possible to offer a connection-oriented and reliable transport layer on the basis of the simple, wireless and unreliable IP.

This task is taken over by the **TCP** protocol, offering secure end-to-end communication between two communication partners, or services, in Internet. It will be presented together with further protocols of the transport layer in chapter 8.

The numerous network services, whose applications are part of our daily communication repertoire, are based on the TCP/IP reference model. For the most part they follow the so-called client/server communication scheme. A client requests information or a service from a server. The server provides this information or service and delivers it to the requesting client, who with the proper authorization receives it. In accordance with their tasks, a distinction is made between name and directory services, electronic mail, data transfer services, network management, and real-time transport services. All follow this scheme and their protocol is located in the **application layer** of the TCP/IP reference model These are presented in detail in chapter 9.

The epilogue at the conclusion of this book offers a brief look at the two other volumes of the trilogy. Volume 1 is devoted to the theme "digital communication" and introduces the fundamentals of networking, encoding and digital security. Volume 3, the conclusion of the trilogy, deals with "web technologies." It sums up the technological foundation of the World Wide Webs, the most important Web applications and current developments in the World Wide Web.

1.1 Computer Networks and the Internet – A Historical Perspective

Presumably nothing else has advanced human development as much as the capability of communicating and exchanging information. The Internet, with its development having begun a good 50 years ago, has extended this capability into nearly the bounds of endlessness. Every kind of information imaginable is only a few mouse clicks away and the user can access it irrespective of time and place.

1.1.1 ARPANET

The origin of the Internet can be traced back to the days of the Cold War. With the purported intention of ensuring fail-safe and reliable command and communication connections, even capable of withstanding a nuclear attack, the idea of a packet-switched communication service was developed that could bridge different computer networks. The idea of packet switching was already developed at the beginning of the 60s. Responsible for this fundamental pillar of Internet technology, enabling secure communication in an unsecure, error-prone network, were Paul Baran at the American RAND Corporation, Donald Davies at the British National Physical Laboratory (NLP) and Leonard Kleinrock at the Massachusetts Institute of Technology (MIT).

At a meeting of the ARPA research directors in spring 1967, the **Information Processing Techniques Office** (IPTO or only IPT) under the leadership of *Joseph C. R. Licklider* (1915–1990) and *Lawrence Roberts* (*1937) presented for the first time the topic of bridging heterogeneous networks. That is, connecting noncompatible computer networks. The first specification could already be discussed in October 1967 for the **Interface Message Processor** (IMP). Special minicomputers, similar to Internet routers used today, were intended to be installed via telephone lines and coupled to the computer. The decision of using standardized connection nodes for the coupling of proprietary hardware to a communication subnet simplified the development of the necessary network protocol.

Software development for communication via IMP and the proprietary computers could be left to the respective communication partner. In 60's and 70's there was no need to be concerned about the implemented computers following a standardized architecture. Neither the operational system nor the employed hardware shared a common interface. This would have meant having to develop an individualized interface for every communication between two computers.

At the end of 1968, the final specifications of the IMPs could be laid out, based on the work of the **Stanford Research Institutes** (SRI). In order to connect with the communication subnetwork, the respective host computer communicated via a bit serial high-speed interface with the upstream IMPs. The IMPs themselves communicated with each other via modems connected over permanently switched telephone lines to provide intermediate storage for data packets and to forward them as well (store-and-forward packet switching). The first four connected network nodes of the **ARPANET** – named after its sponsor the US government agency ARPA (Advanced Research Project Agency) – belonged to the university research centers at the Universities of Los Angeles (UCLA, Sigma-7), Santa Barbara (UCSB, IBM-360/75), Stanford (SRI, SDS-940) and Utah (DEC PDP-10). On October 29, 1969 the day had finally come and the first IMPs were successfully connected to each other as well as to their host computer. The era of the Internet had begun.

In March 1970 the expansion of the new ARPANET first reached the east coast of the USA. By April 1971 there were already 23 hosts and 15 nodes connected with each other. The first "popular" application of the new network was software for the transfer of text messages. This was the first email program, developed in 1971 by *Ray Tomlinson* (*1941) from BBN. In January 1973 the number of computers in the ARPANET increased to 35 nodes. Starting from the middle of 1973 computers in England and Norway were also joined adding the first international nodes. The first data transfer application, the File Transfer Protocol (FTP), was employed the same year. Network nodes outside of the USA were connected via satellite beginning in 1975. The number of computers in the network grew from 111 connected host computers in 1977 to over 500 hosts in 1983. One of the first highly successful public demonstrations of Internetworking took place in November 1977. Via special gateway computers, the ARPANET was interconnected with one of the first wireless data networks, the Packet Radio Network and a satellite network, the Atlantic Packet Satellite Network.

1.1.2 Internet

The great turning point in ARPANET history took place in 1983. The communication software of all connected computing systems of the old **Network Control Protocol** (NCP) was converted to the **TCP/IP**. The communication protocol suite was developed under the direction of *Vinton Cerf* (*1943) (University Stanford) and *Robert Kahn* (*1938) (DARPA). Initiated by the **Department of Defense**, this conversion to the TCP/IP protocol became necessary as NCP enabled only limited communication over heterogeneous networks. The conversion was a decisive condition for the worldwide expansion, which this "network of networks" finally attained. The ARPANET was also split up into a military (MILNET) and a civilian area in 1983. Two different networks – administrative and operational – were now in place. But as they were linked by a gateway the user remained unaware of this division. The ARPANET had become a full-fledged internet Integrated in the ARPANET at the beginning of the 1980's, the CSNET (Computer Science Network) of the American **National Science Foundation** (NSF) connected increasingly more American universities. It successor, the NSFNET, finally made it possible for all universities to be connected via a high-speed backbone created especially for this purpose. Every college student could now become an Internet user. Thus NSFNET quickly became the actual cornerstone of Internet, and that was not only because its transmission performance was more than 25 times faster than that of the old ARPANET's. In addition to scientific utilization, economic use also established an foothold in the NSFNET. This was an area that had been strictly forbidden in the original ARPANET. At the beginning of the 90's, the number of computers linked worldwide over NSFNET exceeded those in ARPANET by a large number. In the meantime the ARPA had become the **Defense Advanced Research Project Agency**, and on the occasion of their 20 year founding in August 1989, the DARPA management decided that it was now time to shut down the ARPANET.

The NSFNET, and the regional networks that had evolved from it, now became the central backbone of the Internet as we know it today. The actual date of birth of the **Internet** is often placed on January 1, 1983. This is the date of the conversion of the until then valid network protocol NCP to the new **protocol family TCP/IP**, with its basis protocols IP (Internet Protocol), TCP (Transmission Control Protocol), including ICMP (Internet Control Message Protocol). These had already been established since 1981 when they were determined and released as Internet standards. With the use of the TCP/IP protocol family a shared interconnection of different network technologies was possible for the first time in a simple and efficient way.

On the evening of November 2, 1988 the first Internet worm, a self-reproducing program, paralyzed an amazing 10% of the then 60,000 computers connected to the Internet. It was an incident that received a great deal of public attention. The important role that data networks such as the Internet had in the meantime achieved, combined with an increasing dependency on them, gave rise to the feeling that such attacks posed a direct threat to public life. It was a fear echoed in the assumption that in an extreme case an entire country and its economy could be plunged into an information chaos.

1.1.3 World Wide Web

The **World Wide Web** (**WWW**) and its easy to use graphic user interface, the browser, finally helped the Internet to achieve its permanent success and worldwide propagation. The fact that the browser is capable of not only requesting and presenting websites, but also of acting as an integrative interface to unify the access to many different type of Internet services, such as email or file transfer, simplified the use of the new medium to such a degree to enable its development into a revolutionary means of mass communication. The foundation for the World Wide Web is the linking of individual documents over so-called **hyperlinks**. A hyperlink is nothing more than an explicit reference to another document on the Web, or to another place within the same document. **Hypertext** documents refers to linked, text-based documents. *Tim Berners Lee* (*1955) formulated a proposal at the Swiss nuclear research institute CERN in 1989, "*Information Management: A Proposal*," in which he suggested a divided hypertext based document management system This was intended as a way to manage the documents and research data that flooded CERN. The following year he got the okay to put his ideas into practice together with *Robert Cailliau* (*1947) on the NeXT computer system. The first WWW server was already operational in November 1990. Tim Berners Lee gave it the name **World Wide Web**. In March 1991 the first WWW browser followed.

The American physicist *Paul Kunz* from Stanford Linear Acceleration Center (SLAC) visited CERN a few months later in September 1991 and was introduced to the WWW. Excited about the idea, he took a copy of the program back with him and in December 1991 the first WWW server outside of the CERN appeared on the network at SLAC. The development of a new server can indeed be credited to the independent initiative of the university members. During 1992 only 26 WWW servers existed, whereas at the beginning of 1993 the number of worldwide operated WWW servers had doubled to almost 50. The first WWW browser with a graphic user surface, the NCSA mosaic from *Marc Andreesen* (*1971) for the X-Windows system, made it possible for the layperson to use the WWW from the end of 1993 onward. This was particularly the case for the versions released shortly thereafter by NCSA for IBM PC and Apple Macintosh. The number of WWW servers had now increased to 500, with the WWW responsible for approximately 1% of the worldwide Internet data traffic. However, the year of the WWW would actually turn out to be 1994. The first international World Wide Web conference was held in May 1994 at CERN. While 400 researchers and developers registered for the conference, a much greater number wished to take part. The available space was simply too small for the huge interest in the WWW. Reports about the WWW were gaining increasing prominence in the media, and in October a second conference was held in the USA attended by 1300 people. Thanks to the popularity of the further developed Netscape Navigator's Mosaic Browser and its market competitor the Microsoft Internet Explorer, which had been in every Microsoft operating system sold since 1995, the growth of the World Wide Web continued unabated. If the growth before this time had meant a yearly doubling of connected computers, now this number doubled itself every three months. The WWW spread like wildfire across the entire globe as it entered the workplace and the home.

Starting in 1995 the term **e-commerce** came into use. Business and trade discovered the WWW and the possibilities it had to offer. The first Internet shopping systems were set up and companies such as *Amazon.com* or *Google.com* suddenly emerged, becoming giants on the stock market almost overnight. Fees became required for the registration of Internet addresses and names, and large companies often paid huge sums to ensure the legal protection of their names on the WWW. The hype was unrestrained and the entire economy was anxious to jump on the bandwagon. The media celebrated the Internet-based economic model, described euphorically as the "New Economy." The breeding ground for the **dot-coms**, so-called because of the suffix in their WWW address **.com**, was the American Silicon Valley. What mostly started with a simple business idea for a Web-based service was built up with the help of venture capital and investors in only a few months and, in the case of success, then bought up by a bigger competitor with a resulting astronomical profit. Yet the real profit these companies boldly forecast often did not materialize. Consumers also remained cautious about on-line shopping, at least as long as there was no standardized and secure transaction mechanism available. In the middle of 2000 the market suddenly collapsed: the so-called "dot-com bubble" had burst.

1.1.4 Web 2.0 and the Semantic Web

The **WWW** has evolved greatly in terms of content since its inception in 1990. Initially, a networked document management system accessible over a hyperlink, it was only available to a small number of insiders. In the years that followed, it developed into the largest information distribution system of all time. When e-commerce arrived, the focus of the WWW shifted from being a medium for personal and public communication by specialists to a medium of mass communication. Information production and information consumption remained strictly separate. Only specialists were in a position to place their own contents on-line in the WWW. The public at large consumed the information offered, which took the role of a commercial information provider in a traditional broadcast medium. User interaction was limited solely to reading websites, ordering goods on-line or clicking on advertising banners. But the WWW was still changing.

New technologies were subsequently developed allowing laypeople to release their own information contents in a simple way. Weblogs, chat rooms, peer-to-peer exchanges, tagging systems, and wikis conquered the Web and opened to the user on a broader basis the way to real interaction and participation in the digital world. Media entrepreneur and network pioneer *Tim OReilly* (*1954) presented this changing face of the WWW to an audience exclusively made up of industry experts under the name **Web 2.0** in October 2004. At that time it could not yet be anticipated that this "renaissance of the WWW" would grow to such a huge extent. The Internet had transformed itself from a pure broadcast medium into a genuine interactive marketplace, with the user functioning as both information consumer and information producer. The new interactivity provided impetus for the creation of **social networks** such as Facebook, which are used in every country by millions of people.

In addition to the evolutionary development of the WWW, the information offering itself has continued to grow by leaps and bounds. In order to find the way through the maze of the WWW information universe, **search engines** such as Google were created. Google administers a gigantic index that with the input of a search term offers comprehensive access to the relevant web document in a matter of seconds. However, it should be noted that only documents appear in the results list containing the term verbatim. It is not possible to find descriptions and synonyms. The completeness and exactness of the search result can never even come close to natural language, seen alone from the standpoint of the problematic interpretation. To do this a systematic supplement to the web document with corresponding additional significant data – so-called **metadata** – would be necessary.

This type of web document, supplemented with metadata, would need to contain along with every term relevant to this document a reference to the concepts describing this term. These conceptual descriptions, also called **ontologies**, could be stored in a machine-readable, standardized form and additionally be analyzed by a search engine, therefore raising the qualitative hit ratio of the present search result. The WWW Consortium (W3C), responsible for the standardization of the WWW, has already created the necessary foundation in the form of ontology descriptive languages such as: RDF, RDFS or OWL. Semantically annotated websites make it possible for autonomous agents to collect goal-oriented information. Based on this, they are then entitled to make independent decisions in the interest of their client and to initiate transactions via the WWW. This semantic network (**Semantic Web**) presents the next evolutionary level of the WWW and is intended to become a reality in the near future.

1.2 A Guide Through the World of the Internet

The Internet as the "network of networks" is not controlled by a central body but rather conceived and organized as an almost completely decentralized entity. There exists neither a headquarters nor a central instance for monitoring the organization or seeing to it that standards are held. Yet because its international ties include the most diverse of communication infrastructures, internal steering and standardization are indispensable. Internationally linked nonprofit organizations along with the entire Internet community regulate technical organization in the form of an open standardization process . The base component of this process is the so-called **Request for Comments (RFC)**, where the specifications for new standards are proposed. In the ratification process, the proposed standard proceeds through different stages. The progression through these stages is the responsibility of intervening organizations. The glue holding the ultimate network together is composed of a group of, in part, old standards addressing the exchange of data and files extending beyond the borders of physical networks. The most important norms of the Internet are the **Internet Protocol** (IP) and the pair **Transmission Control Protocol/Internet Protocol** (TCP/IP). These ensure the transport of data from one endpoint in the network to another in a (possibly different) network. Further standards regulate the exchange of electronic mail, the preparation of WWW sites, as well as the structure and functionality of Internet addresses. All of these standards are the foundation that makes it possible for millions of people to be able to communicate with each other daily via the Internet – despite different hardware requirements. In spite of its decentralized structure and distributed responsibilities, seen from a technical point of view the organization of Internet is based on a strict and largely hierarchical foundation. But who sets these standards? And who makes sure that they are followed as well as bearing responsibility for the further development and operation of the Internet? For this purpose, there are a group of organizations in charge of matters pertaining to the Internet as well as its further development. The most important of these organizations as well as their special tasks are introduced here briefly.

1.2.1 Internet Architecture Board – IAB

The **Internet Architecture Board (IAB)** – formerly the Internet Activities Board or Internet Advisory Board – evolved from the ARPA reorganized Internet Control and Configuration Board (ICCB) in 1983. The primary concern of the IAB is to guide the development of the Internet. This means, among other things, that the IAB is responsible for determining which new protocols are necessary and which official policies should be followed regarding their introduction and the further path of the Internet. The original idea was to bring together the main parties responsible for the development of the Internet technology. It was hoped to encourage the exchange of thought among them, to fix mutual guidelines and research goals and bring them to fruition. Until the first big IAB reorganization in 1989, an autonomously acting institution had grown out of the initially ARPA-centralized research group. The Internet and related TCP/IP technologies had developed far beyond the original research project in 1989. Hundreds of companies working on products and new standards connected to TCP/IP found that these could no longer be put into practice overnight. The commercial success of Internet technology made the reorganization of the steering committee IAB necessary in order to respond to changing policies and commercial requirements. The role of the chair was defined in a new way. Scientists moved from the actual **Board** into assigned support groups, and the new IAB Board was made up of representatives from the new Internet community. The IAB is composed of ten so-called **Internet Task Forces (ITF)**. Each of them focuses on particular problems in the area of the Internet. The two most important task forces are the **Internet Engineering Task Force (IETF)** and the **Internet Research Task Force (IRTF)**. The IAB organizes general meetings where each of the ITFs presents its status report. Technical specifications are checked, improved, and the respective policies laid down.

The head of the IAB, the so-called **chairman**, has the task of creating technical directives from the proposals, as well as organizing the work of the various ITFs. At the suggestion of the other IAB members, the chairman sets up new ITFs and represents the IAB externally. What may come as a surprise is that the IAB has never had substantial financial resources at its disposal. Members of the IAB are volunteers, as a rule, who recruit further volunteers for each of the ITFs. Most of them come from the university or industrial sector of Internet research. Volunteer membership is worthwhile for those involved. On one hand, it offers an opportunity to constantly keep abreast of the newest trends and technologies. On the other hand, it provides a chance to take an active part in shaping the Internet.

Among the most important tasks of IAB is monitoring the standardization process. A special **RFC editor** is named for this responsibility. Furthermore, the IAB is concerned with managing protocol parameter assignment through the **Internet Assigned Numbers Authority** (IANA).

Internet Engineering Task Force – IETF

The IETF is one of the two most important organizations that work under the auspices of the IAB. The main task of the IETF involves solving short and medium term problems affecting the further technical development of the Internet in order to improve its functionality. In contrast to the more research-oriented Internet Research Task Force (IRTF), the IETF is concerned with immediate problems of the Internet, in particular standardizing the implemented communication protocols. This involves the task of working out highly-qualitative, relevant, technical documentation for Internet protocols. The IETF is an open international organization of volunteers where network technicians, manufacturers, network operators and researchers are all acting contributors. Participation is open to everyone who is interested, and there is no formal membership process or special membership condition. As a loose organization the IETF does not have a legal structure and is formally active under the auspices of the **Internet Society (ISOC)**.

The IETF had already existed as one of the task forces of the ICCB, the predecessor of the IAB. The success of its work was one of the determining factors that led to the reorganization of the IAB. The other task forces had mostly consisted of just a few specialists working together on a specific problem. In contrast, the IETF was large from the beginning, and its numerous members worked on various problems at the same time. Because of this, it was subdivided into over 20 working groups

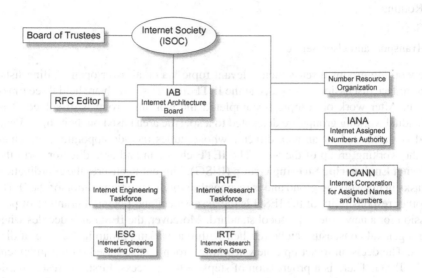

Fig. 1.1 Organigramm of the Internet administrative body.

Table 1.1 RFCs for the organization and working method of the IETF.

RFC 3233 „Defining the IETF"
RFC 3935 „A Mission Statement for the IETF"
RFC 4677 „The Tao of IETF
A Novice's Guide to the Internet Engineering Task Force".

each working on just one particular problem. The working groups held their own meetings to formulate appropriate solutions to problems. They then presented these at the regular meetings of the IETF, where an important point of discussion was their inclusion in the Internet standard work. Often hundreds of people attended the IETF meetings. The task force eventually became too large to be successfully led by one chairman. Following reorganization of the IAB, the IETF continues to play an important role separated into individual working groups. Each working group is led by an appointed director and has its own charter. This lays out the goals of the working group and indicates when which documents should be produced. At this time, the IETF working groups are dedicated to the following eight areas:

- Applications
- General
- Internet
- Operations and Management

- Real-Time Applications and Infrastructure
- Routing
- Security
- Transport and User Services

The working groups discuss their relevant topic via email over open mailing lists. Face-to face discussions take place at the IETF meetings, which are held three times a year. After work on a topic is completed the working group is dissolved. The individual working groups are delegated to a specific area based on their topic. Each field is supervised by an area director whose duties include appointing the chair for the working group of the area. The IETF chairman and area director form the **Internet Engineering Steering Group** (IESG). This entity ensures the coordination of tasks in the working group and is responsible for the entire operation of the IETF. Another responsibility of the IESG is the assessment and potential granting of permission for a new official protocol standard. Moreover, the IESG also decides whether a general consensus can be reached within a working group in the case of disputes. The decision to set up a new working group is also based on the judgment of the IESG. There is a progression of steps in this process. First, interested participants discuss a new theme with an area director. Such a first meeting of similarly interested parties is called a **Birds of a Feather** meeting (BOF). It can be held during a IETF meeting. In the framework of a BOF, problems are discussed that the newly set up working group is expected to solve. The first proposals for a charter for this new working group are also created. It is entirely possible for BOF meetings to take place more often until it is clear whether enough volunteers are available for the founding of a new working group. The organization and working methods of the IETF are defined in several standard documents (RFCs) (cf. Fig. 1.1).

Internet Research Task Force – IRTF

In a further step of reorganization by the IAB, the **Internet Research Task Force** (**IRTF**) was founded in 1998. This is the research counterpart of the IETF, an independent organization under the auspices of the IAB. The IRTF coordinates long-term research activities involving the TCP/IP protocol suite and is generally responsible for the development of Internet architecture. Just as the IETF, the IRTF is led by a chairman who is selected every two years by the IAB. There exists a steering group within the IRTF, the **Internet Research Steering Group** (IRSG). It consists of the IRTF chairman, the leaders of each research group and, additionally, especially chosen individuals. The tasks of the IRSG consist of establishing each of the current research priorities as well as coordinating and leading the corresponding research activities.

In addition to its leadership task, the IRSG puts on various workshops, with a focus on research themes of importance for the current development of the Internet, or at which research priorities are discussed from an Internet perspective. In contrast to the IETF, the IRTF is much smaller at this time and less active. Many research activities continue to be carried out by IETF. The following IRTF research groups are currently working:

- Anti-Spam Research Group
- Crypto Forum Research Group
- Delay-Tolerant Networking Research Group
- End-to-End Research Group
- Host Identity Protocol (HIP)
- Internet Measurement Research Group
- IP Mobility Optimizations Research Group
- Network Management Research Group
- Peer-to-Peer Research Group
- Routing Research Group Charter
- Transport Modeling Research Group
- Internet Congestion Control Research Group
- Scalable Adaptive Multicast Research Group
- End Middle End Research Group

Further information regarding the organization and working methods of the IRTF may be found in RFC 2014 *"IRTF Research Group Guidelines and Procedures."*

1.2.2 Internet Society – ISOC

The Internet had increasingly distanced itself from its roots in U.S. government agencies by 1992. An international society was founded with the intent of taking on the task of encouraging participation in the Internet and guaranteeing its further development.
The **Internet Society**, called **ISOC** for short, is an international, non-governmental and non-profit organization (Non-Governmental Organization, NGO) headquartered near Washington, D.C. Made up of more than 150 national and regional sub-chapters, it has over 16,000 individual members from 170 countries. The members of the ISOC have dedicated themselves to ensuring the worldwide propagation of the Internet and its continued existence (*"...to assure the open development, evolution and use of the Internet for the benefit of all people throughout the world."*).
The ISOC **Board of Trustees** consists of 15 members who are elected by all ISOC members worldwide. In matters of organization, ISOC is oriented on the "National Geographic Society," founded in the 19th century. Whereas its role model aims to support the field of geography, ISOC concentrates its activities on the Internet and its continued development.
The formation of ISOC was based on the initiative of several long-year members from the Internet Engineering Task Force (IETF) who had occupied themselves with

standardization questions affecting the Internet. Their objective was to see that these questions were in line with the organizations: Internet Architecture Board (IAB), Internet Research Task Force (IRTF) and Internet Assigned Number Authority (IANA) and also that secure financing was guaranteed. The IETF was financed primarily by the US government until 1990, with funding supplied by institutions such as DARPA, NASA or NSF. Yet there were long-term issues concerning the security of this financing. This combined with the goal of international support inspired the idea of starting a society that could obtain alternate means of support, particularly from leaders in global industry.

The ISOC was officially founded in January 1992 at the INET Conference in Kobe, Japan. In 1993, the responsibilities of the ISOC were summarized in a special Request for Comments, the **RFC 1602** *"The Internet Standards Process."* It was the first time that the ISOC had participated in this process. The ISOC maintains offices in Virginia (USA) and Geneva (Switzerland). Until 2004 the ISOC organized a yearly INET conference where plenary sessions, tutorials and lectures took place. These INET meetings are usually regional now and conducted in connection with other conferences.

The tasks of the ISOC in creating new Internet standards concentrate on the independent supervision and coordination of the standardization process. Therefore all RFC documents are copyrighted by the ISOC even when these documents are freely available to everyone. All of the organizations named in Fig. 1.2 continue to play a role in the Internet standardization process.

Table 1.2 WWW addresses of the most important Internet Organizations.

IAB	Internet Authority Board	http://www.iab.org/
IETF	Internet Engineering Task Force	http://www.ietf.org/
IRTF	Internet Research Task Force	http://www.irtf.org/
ISOC	Internet Society	http://www.isoc.org/
IANA	Internet Assigned Number Authority	http://www.iana.org/
ICANN	Internet Corporation for Assigned Names and Numbers	http://www.icann.org/

1.2.3 IANA and ICANN

Besides the determination and further development of Internet protocol standards, regulating the assignment of worldwide unique addresses and names on the Internet is of particular importance. This task was carried out by the **Internet Assigned Numbers Authority** (IANA) until 1998. This later became an independent department within the **Internet Corporation for Assigned Names and Numbers** (ICANN). The IANA originally consisted of only one person, *Jon Postel* (1943–

IAB – Internet Architecture Board
The IAB functions as the technical advisory group of the ISOC. Among other things it is responsible for maintaining an overview of the Internet and its development. It must confirm nominations to the IESG, which are made by the nominating committee of the IETF.

IETF – Internet Engineering Task Force
The IETF is a loosely connected, self-organized group of experts that provide technical and other contributions to the further development of the Internet and its related technologies. While acting as the main participant in the development of new standards, the IETF is, however, not a component of the ISOC. It is made up of individual working groups, divided into task fields, each led by a working group manager. Nominations to the IAB or to the IESG are made by a nominating committee. This is comprised of volunteer participants who are arbitrarily chosen for this duty at the IETF meetings.

IRTF – Internet Research Task Force
The IRTF is not directly involved in the standardization process Rather, the IRTF is involved in the long-term development of the Internet and the handling of topic areas considered too vague, too advanced, or not well enough understood too-date to be suitable for standardization. As soon as the IRTF releases a specification it has worked on – seen as as stable enough for a standardization process – this specification is then processed further by the IETF according to the guidelines.

ISOC – Internet Society
The main duty of the ISOC lies in the determination of new standards. The ISOC is an international society concerned with ensuring the further Internet growth and development. A part of this is dealing with questions regarding the use of the Internet or the social, political or technical consequences resulting from Internet development. The ISOC **Board of Trustees** confirms nominations for the IAB, which are made by the IETF nominating committee.

IESG – Internet Engineering Steering Group

The IESG supervises the technical management of the IETF activities and the Internet standardization process. It is also a part of the ISOC. The main responsibility of the IESG involves all activities concerning the entrance of new standardization proposals as well as the progress of those already in the processing phase. The IESG is the final authority before ratification of an official Internet standard. It is made up of the working group mangers from the IETF.

IANA – Internet Assigned Number Authority

Originally the IANA had control over the organization, distribution and assignment of Internet addresses In reality, however, actual control of Internet addresses at the IANA until 1989 was held by Jon Postel, one of the founders of the Internet. From the beginning Postel acted as RFC editor. After his death, control of address assignment went to the **ICANN (Internet Corporation for Assigned Names and Numbers)**.

Fig. 1.2 The most important Internet organizations

1998) from the Information Sciences Institute at the University of Southern California. He took over the task of address assignment in a contractual agreement with the US Department of Defense. After his death in October 1988, this assignment was transferred to the ICANN.

The main task of the IANA, as has already been noted, is the provision of global unique names and addresses as well as the determination of unique numbers for Internet protocol standards. To enable this, a close collaboration with the IETF and the RFC editor was necessary. IANA delegates the job of assigning IP addresses and domain names to regional organizations (**Regional Internet Registry, RIR**). Each RIR administers the assignment of a certain number of addresses for a certain geographic area. IANA gave each a IPv4 address bundle (as a rule 224 single addresses or more), which were subsequently given out again by the RIR in smaller numbers to the local Internet Service Provider (ISP). Following the exhaustion of free IPv4 addresses in spring 2011, IANA took over the task of assigning IPv6 addresses. Their bundling and assignment is less critical due to the large size of the IPv6 address space. The IANA also administer the data for the root name server, located on the top place of the DNS address space hierarchy. A further area of responsibility is the administration of those protocol parameters from the Internet protocol standards for which a central registration is necessary. Among these are e.g., the names of URI schemes or character encoding for use in Internet.

The work of the IANA is monitored by the IAB and laid down in the RFC 2860 *"Memorandum of Understanding Concerning the Technical Work of the Internet Assigned Numbers Authority."* In agreement with the US Department of Commerce, the IANA has worked as a subdividion of the ICANN since 1998. Similar to the ISOC, the ICANN is a private, non-profit organization under U.S. jurisdiction headquartered in Marina del Rey (USA). The ICANN was founded in October 1998 as a fusion of different interest groups that came from the industry as well as the business sector. Included in ICANN's areas of responsibility are a series of technical guidelines which had previously been taken care of by IANA and various other groups. The most important tasks are the administration of names and addresses in the Internet as well as the determination of technical standard procedures. In this way the ICANN coordinates many technical aspects of the Internet, without them becoming binding by law. ICANN is ofter portrayed as a kind of "world government of the Internet" in the media.

Under the auspices of the US Department of Commerce until October 2009, the ICANN had thus been controlled by the US government. Since then a mutual "Affirmation of Commitments" has replaced the previous agreement (Joint Project Agreement, JPA). It is intended that representatives from the government and relevant interest groups regularly check whether or not the ICANN organization is fulfilling its duties in accordance with the statutes set. The Governmental Advisory Committee (GAC) of the ICANN is composed of worldwide government representatives and has its own seat at the EU Commission in Brussels. As the ICANN headquarters is located in the USA, the network administration organization continues to be under US jurisdiction. This special position held by the US government by way of its state supervision and contract with ICANN is often criticized and the topic of numerous and controversial discussions.

The **Board of Directors** – the executive committee of ICANN – totals 21 international members. Of these only 15 have voting rights. This means 8 voting members are elected by the nominating committee, 2 by the Address Supporting Organization (ASO, part of ICANN and responsible for the assignment of domain names), 2 by the **country-code Name Supporting Organization** (ccNSO, part of the ICANN and responsible for global guidelines in connection with country codes of the DNS Top-Level Domains), 2 by the **Generic Name Supporting Organization** (GNSO, successor of the Domain Name Support Organization, DNSO) and the chair. The 6 members without voting rights are nominated by the advisory organization. In 2000, 5 members were publicly elected for three years as representatives of the Internet user. This public election was, however, abolished again in 2003.

1.2.4 World Wide Web Consortium

The **World Wide Web Consortium**, or **W3C** for short, is the international committee for the standardization of technologies and languages in the World Wide Web. Even though the W3C has authored numerous de facto standards, it is not an intergovernmentally recognized organization and therefore technically not authorized to determine standards. For this reason, the final standards ratified by the W3C are called "Recommendations." The W3C has committed itself to only using technologies that are free from patent fees. The organizational form of the W3C is that of a consortium whose membership organization assigns its own personnel to develop WWW standards. Currently there are 322 member organizations functioning in W3C (as of February 2011). Tim Berners-Lee holds the chair of the W3C. He created the foundation for the Internet with the conception of the HTTP protocol: the URI and HTM. According to its motto: *"To lead the World Wide Web to its full potential by developing protocols and guidelines that ensure long-term growth for the Web,"* the W3C sees its primary role in the worldwide expansion and ongoing development of WWW technology.

Tim Berners-Lee left the European nuclear research center CERN in October 1994 and founded the World Wide Web Consortium at the Laboratory of Computer Science at Massachusetts Institute of Technology (MIT/LCS), supported by the DARPA and the European Commission. The main objective was to guarantee compatibility with the incorporation and development of new standards affecting WWW. For example, before the founding of W3C the inconsistency of different WWW documents was a great problem. Different companies offered different HTML dialects and extensions. Under the single roof of the consortium all of these providers were to agree on common principles and components, which would then be supported by all. Since 2006, the W3C maintains 16 locations around the world. The consortium is mutually administered by the Computer Science and Artificial Intelligence Laboratory at MIT (CSAIL), the European Research Consortium for Informatics and Mathematics (ERCIM) in France and Keio University in Japan.

The development of a W3C Recommendation proceeds in a similar way to the Internet standardization process described in detail in the next chapter. Before a to-be-ratified standard achieves the status of a W3C Recommendation it must proceed through the levels: "working draft," " last call," "candidate recommendation," and "proposed recommendation." Along with the recommendation, an "errata" may be released as well as the issue of a new "edition" of a recommendation. Recommendations that have already been issued may also be withdrawn in case a reworking is necessary (for example in cases of RDF). Moreover, the W3C issues so-called "notes," without normative requirements. The W3C does not explicitly dictate to the manufacturer a necessity to follow the recommendations. Many recommendations, however, define, so-called "Levels of Conformance. " These have to be fulfilled by the manufacturer in order that its product may be identified as "W3C Compliant." The recommendations themselves are without a sovereign patent. This means that anyone can implement them without having to pay a license fee.

1.2.5 Open Standards in the Internet – Regulated Anarchy

Most of the organizations mentioned in section 1.4 are involved in the development of new Internet standards. They have dedicated themselves to the mutual goal of shaping the Internet standardization process according to the following criteria:

- technical excellence,
- early implementation and practical tests,
- clear, precise and easy to understand documentation,
- openness, balance and
- the highest possible timeliness.

Request for Comments (RFC) – Point of Departure for the Standardization Process

The TCP/IP technology does not represent a property in the sense of the word, and no manufacturer may claim this privilege for itself. Accordingly, the documentation of the protocol standard cannot be obtained from a manufacturer. The protocol documentation is available on-line to everyone without a fee. The specification of every later Internet standard, proposals for the establishment of new or the revision of existing standards are first released by IESG or IAB in the form of a technical report, a so-called **Request for Comments** (RFC). It is just as likely that RFCs are extremely detailed as it is that they are brief. They may contain a completed standard or merely suggestions for the form a new standard could take. Although not assessed in the same way as scientific works of research, they are edited. For many years, until 1998, their publication was the responsibility of one single person who held the position of RFC Editor, Jon Postel. Today this task is assumed by the IETF working group managers. RFCs serve the IESG as well as the IAB as official publications for the communication with the Internet community. There exist a number of servers with public access to all of the RFCs via WWW, FTP or other document retrieval systems available. The name alone is intended to indicate that the submitted specification proposal is a matter of public debate. The long series of released RFCs had already begun in 1969 within the framework of the original ARPANET project. Besides the Internet standard, many research themes and discussions as well as status reports about the Internet are covered. The release of new RFCs is the responsibility of the **RFC editors** and follows the general guidelines of the IAB. The series of the RFCs is consecutively and chronologically numbered. Every new RFC, or the revision of one, has its own number. It is therefore up to the reader to find the respective most current number to a certain thematic area from the series of the RFCs. An RFC index offers help in this regard.

The Standardization Process

During the time that a new Internet standard is being developed, there are provisional versions, so-called **drafts**, available. These can be accessed over the IETF **Internet Drafts Directory**, and are made available to certain computers in the network as an outlet for discussion and opinion. Every working document is thus presented to the entire Internet Community for assessment and, if applicable, for revision. If such a draft remains for more than 6 months in the Internet Drafts Directory without being suggested for publication by the IESG, it is simply removed from the directory. However, an Internet draft may be replaced at any time by a reworked version. In this case the 6 month revision period then begins anew. An Internet draft does not count as a publication, meaning that it has neither a formal status nor can it be quoted. The IETF has a cornerstone function in the Internet standardization process. It initiates a significant number of all technical contributions and it functions as an integration point for other standards that are defined outside of the Internet standardization process. (cf. Fig. 1.3).

A specification that is to be released as an Internet standard runs through a fixed process of developmental steps, with differing levels of growth. Starting with the **Proposed Standard** and on to the status of the **Draft Standard**, the specification can finally come to fruition in the fully developed **Internet Official Protocol Standard**.

Proposed Standard

A new specification enters the standardization process as a **Proposed Standard**. The IESG of the IETF is responsible for a proposed specification entering the standardization process. A proposed standard is already considered stable and necessary design decisions have been made in advance. Generally, the problem addressed in a proposed standard has already been throughly investigated beforehand. It has also been evaluated by the Internet community and deemed worthy of further attention. An implementation or other operational application is not necessary to establish a specification as a proposed standard. Nevertheless, for fundamental Internet protocols or core components, the IESG normally requires implementation and experience in the area of operation before the status of a proposed standard is granted. The

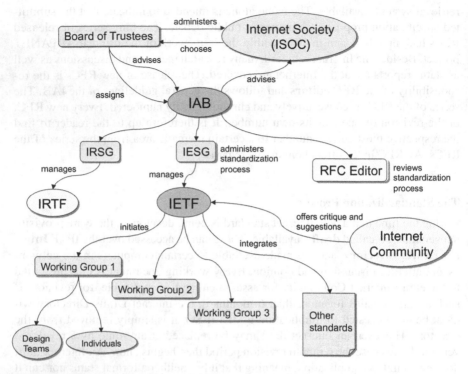

Fig. 1.3 The IETF as the cornerstone of the Internet standardization process.

proposed standard presents a still undeveloped specification for the implementer .
Therefore a timely implementation is desirable in order to be able to incorporate the
experiences gained in further forms of the specifications.

Draft Standard

When at least two independent implementations have been developed for a proposed standard and enough operational experiences collected, a proposal can reach
the level of a **Draft Standard** after the end of a 6 month period at the earliest It
is necessary that a draft standard already be generally recognized as well as stable.
Further application experiences are, however, still necessary in particular, for example, when the large-scale employment in a production environment is involved.
Usually a draft standard is already considered a final specification and will only be
changed in the event that this is necessitated by unforeseen problems.

Internet Official Protocol Standard

When sufficient operational experience has been gleaned with the help of various
implementations based on the specifications for the draft standard, it is possible to

achieve the status of an official **Internet Standard**. An Internet standard may be distinguished by its high level of technical sophistication. There is a considerable degree of confidence in the significant contribution that the specification is making to the Internet and its users. A specification that reaches the level of an Internet standard retains its original RFC number. Fig. 1.4 presents a brief summary of the standardization process.

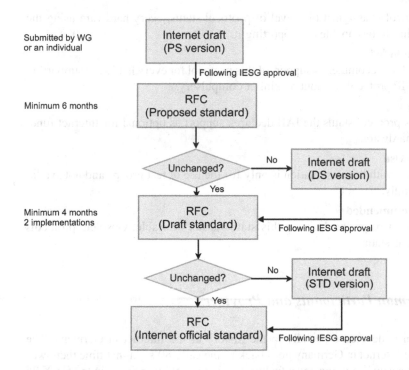

Fig. 1.4 The Internet standardization process.

There are two additional subtypes of RFC standards. Although they do not have a practical significance, they are mentioned here for the sake of completeness:

- **Experimental Standard**
  This type of standard has been implemented on a trial basis for testing purposes to evaluate new methods and technologies. It has, however, never gone through the prescribed standardization process. This kind of standard has been principally deemed to be of little interest for practical application.

- **Historic Standard**
  While this standard has gone through the prescribed standardization process, it is already considered outdated and has been, e.g., replaced by a new standard. As every Internet standard can suffer this fate, RFCs are periodically released to

provide a better overview of the current standards. These are listed there officially.

Along with the level of a RFCs, the IAB determines a standard a level of significance
(**Protocol Status**) for every protocol released. This fixes the conditions necessary
for the implementation of this protocol:

- **Required**
  If a protocol is assigned this level of protocol status, every hardware using the
  TCP/IP has to be capable of supporting it.

- **Recommended**
  Here the IAB recommends support of this protocol for every hardware supporting
  the TCP/IP protocol, without making it compulsory.

- **Elective**
  With this protocol status the IAB declares support as optional for Internet functioning hardware.

- **Limited Use**
  A protocol with this designation is only for the use of test groups and not available generally.

- **Not Recommended**
  The support of protocols given this status is not desirable. Obsolete protocols
  receive this status.

1.2.6 German Participants and Providers

This section is devoted especially to the history of the Internet in Germany. The
origin of the Internet in Germany goes back to the early 80's. At that time there was
the sense of a new beginning and a feeling of excitement hung heavy in the air. X.25
offered worldwide as a reliable public data communication service was beginning
to take off. Btx promised consistent information services for citizens throughout
Europe. ISDN, the all-encompassing digital telecommunications service, was on
the horizon.
With the founding of computer science departments at German universities (1971-
75) the wish for Internet-like Infrastructures begin to grow, spurred on by what
was taking place in the USA. Nevertheless, the starting signal for the launch of the
Internet in Germany came relatively late. At the University of Karlsruhe the first
German node was connected to the CSNet in August 1984. This enabled Germany,
for the first time, to communicate with research centers in the USA and with the
other connected countries of Canada, Sweden and Israel. Internet communication at
this time mostly took the form of emails. On August 3, 1984 at 10:14 CET Michael
Rotert at the University of Karlsruhe received the first Internet email in Germany at
his address `rotert@germany`. It was copied to the director of the project, Prof.

Werner Zorn (now professor emeritus at Hasso Plattner Institute, University of Potsdam), and thereby confirmed Germany's membership in the network system of the
CSNet. Gradually, more research institutes in Germany connected to the CSNet and
the first commercial clients, such as Siemens or BASF in Ludwigshafen, followed.
The University of Dortmund took a different path. In 1983, the IBM company had
already proposed to several German universities, which used IBM hardware, to link
up with each other on the basis of dedicated lines financed by IBM. This **European Academic and Research Network** (EARN) was then started in Paris in 1984,
using the same protocols and namespaces which were also used by the more famous BITNET. Via an email link, EARN was directly connected with the USA and
from there connected over a special gateway to the Internet in 1984. In the same
year, the University of Dortmund was connected to EARN nodes over a dedicated
line to the already connected Open University of Hagen. The history of the German
Internet is closely connected to the **EUnet** (European Unix Network). It established the first international, dial-up connection-based UUCP network between the
countries Great Britain, the Netherlands, Denmark, and Sweden in 1982. From the
initially informal, collaborative work of individual locations, under the supervision
of the European UNIX Users Group (EUUG, later EurOpen), there quickly grew a
commercial network provider called "EUNet International Ltd." In April 1988, the
computer science operations group at the University of Dortmund put on an EUNet
workshop. For the first time, plans for the "InterEUNet" were introduced. This was
a European-wide IP network with a direct connection to the Internet. It could already be tested in the fall of 1988 over the North-Rhine-Westphalian NRW computer
network. At the beginning of 1989, when the old Datex-P-connection Dortmund–
Amsterdam was replaced by a dedicated line, it was then possible to speak of the
University of Dortmund's first, genuine and direct Internet connection. In 1991, likewise in Dortmund, the first voluntary nameserver service for Germany and the
domain . de were set up – the **DENIC**. The University of Dortmund remained the
central hub for email traffic in Germany up to 1992.
The wish to create a high-performance network infrastructure for science and to increase the competitiveness of German industry were the focus from the beginning
when the **Deutsche Forschungsnetz** (DFN) ("German Research Network") emerged in 1984. Instead of the sovereign state, a registered association of experts – the
DFN – were to take over the steering and formation of a new scientific network with
the support of the Federal Ministry of Education and Research. However, the venture
turned out to be more difficult than expected due to differences between the parties
involving the implementation of norms and standards. Not until spring of 1990, after
an investment of 100 million German Marks, could the first narrow band scientific
network (WiN) with 64kbps be put into operation. It suffered the slings and arrows
directed at it from various sides and was hampered by the year-long preferences of
the monopolist German Telekom in the sector of voice-based communication vs.
data communication. It was necessary to traverse a long, rough road in order to arrive at the modern GWiN – the gigabit Wissenschaftsnetz (scientific network) – that
German universities provide with a core network bandwidth of between 2.5 Gbps
and 10 Gbps. Also in 1990, the DFN came to the realization that it was necessary to

have an IP service of its own, as development, also European-wide, was changing more and more from X.25 to the Internet technology (TCP/IP). A WiN-IP planning group – still without an official okay – was formed. In November 1990, it was then officially decided to also test and introduce TCP/IP services in WiN. This was preceded by intense discussions between the defenders of the still-being-developed ISO/OSI reference model and those in support of the TCP/IP. While the development of ISO/OSI continued, the decision by the DFN was the determining factor for the establishment of TCP/IP in WiN. Nevertheless, even after this decision was made it still took another year until the IP services of the WiN were also connected to the international Internet.

The DFN, headquartered in Berlin, is made up of 370 members from industry, universities and research. Just as in the past, its tasks still include the creation and guaranteeing of **open net structures**, i.e., manufacture-independent network technologies and transmission protocols with connectivity to the commercial Internet world, as well as the expansion and demand-oriented further development of the DFN. Along with the DFN are numerous commercial providers–in Germany there are regionally and nationally over 200 – such as **XLink** (eXtended Lokales Informatik Netz), which originated in a university project that connected the University of Karlsruhe with the USA (NYSERNet).

XLink first came to life in 1989 at the University of Karlsruhe, initially as a third-party funded project. In 1993, it was privatized and became a commercial Internet Service Provider.

The relevant namespace for Germany – the **.de domain** – was officially assigned in November 1986 by the IANA to the local administration organization" **DENIC**" at the data processing center of the University of Dortmund. The technical administration by the primary nameserver of the .de domain as well as the registration of domain names within the .de domain was carried out there. In this way, the DENIC also took over the allocation of IP addresses to German companies and institutions. However, as the registration of domain names and IP addresses was free of charge the operators soon found themselves facing financial difficulties. In reaction, as well as to avoid the possibility of government control of this Internet resource, the **Deutsche Interessengemeinschaft Internet** (DIGI e.V.) ("German Community of Internet Interests") was founded at the same time as the ISOC in 1992. The defined tasks of the DIGI e.V. include the coordination of Internet activities by individual participants as well as the securing the administration of domain and address designation. In the course of 1992, the operation of the DENIC was advertised for bids by the DIGI e.V. This was done to enable the possibility of a new regulation for managing the .de domain and assigning IP addresses in 1993. This decision had been made because the University of Dortmund was not able to guarantee its operation beyond 1992. The administration was the targeted area, as it had been able to support itself through revenues made from the registration of .de domains and IP addresses. The University of Karlsruhe took over this service in the end, after it had been performed in the interim by the EUNet GmbH and the XLink-Projekt. The interest group DENIC (IV-DENIC) was founded to handle ongoing financing. In 1997, DENIC eG was formed, as a registered cooperative, in Frankfurt/Main.

It took over the technical administration of the .de domain from the University of Karlsruhe in 1999.

The DIGI e.V. was recognized as the **German Chapter** of the Internet Society by ISOC in 1995. Since this time, DIGI e.V. has operated under the abbreviation and Internet domain: **ISOC.DE e.V.** Just as the global ISOC, the ISOC.DE has positioned itself to deal with organizational, technical, and political topics related to Internet. The ISOC.DE has taken clear positions on topics such as securing the DENIC, the cryptography debate in Germany, Internet and censorship, as well as structural questions on the Internet in Germany.

The Internet was initially not an issue in the former GDR up to the time of Germany's political transformation. Several local computer networks existed, e.g., at the University of Jena or at the Charite in Berlin, however these networks were not connected to each other. At the end of the 1980's, a dedicated line was installed between the Free University in West Berlin and the Humboldt University in East Berlin. For a long time it was intended to be the only dedicated line between the Federal Republic of Germany and the GDR, used exclusively for civil purposes. The namespace reserved for the GDR, .dd, was, however, never activated. After the reunification of Germany, eastern German universities and research institutes were connected to the **ErWiN**, the " Erweiterte Wissenschaftsnetz" (Expanded Scientific Network) of the DFN.

1.3 Glossary

ARPANET (Advanced Research Projects Agency Net)   The first packet switching data network and forerunner of the Internet. It was founded by the DARPA, a research initiative of the US Department of Defense. The first network node (Interface Message Processor, IMP) was ready for employment on August 30, 1969 and the ARPANET began operation in December 1969 with 4 IMPs in Stanford, Santa Barbara, Los Angeles and Utah. It was comprised of multiple satellites at its highpoint, among them were those from the West to the East Coast of the USA to Hawaii, Great Britain, Norway, Korea and Germany. The ARPANET ended its operation in July 1990.

Client   This designates a program that contacts a server from which it requests information services. The browser used in WWW is in this sense a client. There are also other clients in WWW that contact the WWW server in order to download information, e.g. search engines or agents.

Client/server architecture: An application is carried out based on a network of multiple computers connected to each other. The server provides specific services, the client, on the other hand requests services. Except for allocation and answering in connection with the task at hand, the components are independent of each other. Interface and the type of communication necessary for order submissions and responses are clearly defined.

Communication   Communication can be understood as the process of individually or mutually supplying, transmitting, and processing information by people or technical systems.

Communication protocol   A communication protocol (or simply protocol) is a collection of rules and requirements that define the data format of information to be transmitted as well as the way of transmission. They contain agreements about the data packets to be sent, the establishment and termination of the connection between the communication partners as well as how the data is to be sent.

Digital communication: Digital communication is a term used to describe the exchange of digital information over specialized, digital communication channels intended for this purpose. The type of media (text, image, audio, video, etc.) determines the data format of the information. The information is transmitted according to the requirements of the implemented communication protocols over the communication channel (e.g., Internet or WWW).

Dot-com bubble: The expression dot-com-bubble was a word originating in the media that came to be used for a worldwide phenomenon. It primarily gained use in connection with the collapse of stock market speculation in March 2000 – when "the bubble burst." The so-called dot-com companies were particularly affected by this turn of events. Particularly in industrial countries it led to major losses for small investors. Dot-com companies are characteristically technological companies whose field of business involves Internet services. The name comes from the ".com" ending of the domain names of these companies. It was first coined in the jargon of the stock exchange and then adopted in the media.

Electronic commerce (also e-commerce):   Electronic commerce describes that part of electronic business involving Internet-based agreements and legally-binding business transactions. E-commerce normally encompasses the three transaction phases: information, agreement, transaction.

HTML: Hypertext Markup Language; the standardized document format for hypermedia documents in WWW. Documents that can be transmitted and represented by the browser in the WWW by means of HTTP are encoded in HTML.

HTTP: Hypertext Transfer Protocol; the protocol regulating the WWW communication of browsers and WWW servers. If a browsers requests a document from the WWW server or answers a request from the WWW server this communication must obey the conventions of the HTTP protocol.

Hyperlink: Reference to another hypermedia document or text passage in the same hypertext. This enables a (non-linear) linking of information in different documents.

Internet: The Internet is the largest computer network in the world. It made up of many networks connected to each other as well as individual resources. Among the most important services of the Internet are: electronic mail (email), hyper-media documents (WWW), file transfer (FTP) and discussion forums (Usenet/Newsgroups). The global network gained its popularity primarily through the introduction of the World Wide Web (WWW). While this is often spoken of as synonymous with the Internet, it is in fact just a subset and represents one of the many Internet services.

Internetworking: A technique for bridging multiple different and possibly incompatible networks (LANs, WANs) into one Internet. To enable this, suitable switching computers (routers) are necessary, which mediate the path of a data packet through the network and ensure a secure delivery. The network appears as a homogeneous, virtual network (Internet) to the user.

Internet standard: Because many companies and organizations were involved in the development of the Internet it was necessary to create standardized protocols and interfaces to simplify the development effort. These were adopted in the form of Internet standards in a public standardization process. It thus became possible for every user to take a position on a new proposal for a future standard (Request for Comment, RFC) and thereby play a role in steering the course of Internet development.

Internet Protocol (IP):   Protocol on the network layer of the TCP/IP reference model, more precisely called **IPv4**. As one of the pillars of the Internet, IP ensures that the global Internet, consisting of many heterogeneous, individual networks appears as a unified homogeneous network. A standardized addressing scheme (**IP addresses**) enables word-wide unambiguous computer identification. IP additionally provides a **connectionless, packet switching datagram service** that cannot fulfill quality of service guarantees, but always works according to the **best effort** principle. For the communication of control information and error notification, the **ICMP** protocol is an integral component of the IP.

Medium: The expression of a transportation channel for information transmission between a sender and a receiver. In order to transport information it must be be exchanged over a carrier medium set up between the sender and receiver.

Multimedia: A multimedia information presentation is referred to when the information presented involves multiple and varied media, for example, text, image, and sound.

Network application: An application program whose operation includes access to resources that are not locally available from the executing computer but attainable at a remote computer that then must be accessible over a network.

Request for Comments (RFC): New technologies in the Internet developing from expert discussions are recorded in so-called RFCs. In the course of the Internet standardization processes a consecutively and chronologically numbered collection of documents was begun, in which technologies, standards and further Internet-related information have been documented and standardized.

Semantic Web: The semantic web represents an expansion of the existing World Wide Web. Every information content represented in the semantic web is given a well-defined and machine-readable meaning. This should enable a program, acting autonomously, to interpret the information and to be capable of making decisions based on it. The concept of the semantic web originated in a proposal by the WWW founder Tim Berners-Lee.

Server: Designates a process which is contacted by clients in order to send back requested information or to make resources available. A computer on which a server process is carried out is often referred to as a server.

TCP/IP reference model (also TCP/IP protocol suite, TCP/IP communication model): This term refers to the communication layer model for the Internet. The TCP/IP reference model is divided into 4 protocol layers. These are: the data link layer, the Internet layer, the transport layer, and the application layer. In this way it is possible for different computers and protocol worlds to be able to communicate over unified interfaces on the Internet.

Uniform Resource Identifier (URI): Serves as the world-wide unique identification of information resources in the WWW. Implemented as **Uniform Resource Locator (URL)** at this time in the form of a unique address. As the address of an information resource in WWW is subject to change, work is being done on the creation of a **Uniform Resource Name (URN)** that identifies an information resource explicitly by its name and not by its address.

Web 2.0: Web 2.0 identifies a seemingly new generation of web-based services that are characterized by the simple possibility they also offer to lay people to participate and interact in the WWW. Typical examples of these services are wikis, weblogs, image and video portals, or peer-to-peer exchanges.

World Wide Web: Term for the "worldwide data net" (also WWW, 3W, W3, Web). The most successful service on the Internet distinguishes itself by its high degree of user-friendliness as well as its multimedia elements. WWW in fact designates a technology that implements a distributed, Internet-based hypermedia document model. Internet and the World Wide Web (WWW) are often used as synonyms today, however the WWW is only a special service in the Internet using the HTTP protocol.

Chapter 2
The Foundation of the Internet : TCP/IP Reference Model

> *"The borders of my language define*
> *the borders of my world."*
> *– Ludwig Wittgenstein, (1889 – 1951)*

Spanning the world with its almighty presence, today's Internet connects computers, telephones, entertainment electronics and, in a short time also the household devices and the goods we need for daily life. More and more it penetrates the surface of our lives. To enable these different devices to communicate smoothly and efficiently with each other their communication must follow defined rules – so-called communication protocols. These mold the individual layers of Internet communication determining the tasks, level of abstraction complexity and respective range of functions. By what means and way these specifications are put into practice is, however, not defined by the model but depends on the specific implementation. The TCP/IP reference model thus assumed a concrete form through practical application and builds today, as well as in the foreseeable future, a solid foundation for all of the communication tasks on the Internet.

2.1 Communication Protocol and the Layer Model

Let us first take a closer look at the basis of computer communication. The hardware of a computer network is made up of components that have the task of transmitting information, encoded in the form of bits, from one computer to another. If one wanted to organize computer communication solely on this level it would be like programming a computer in a rudimentary machine language, i.e., only using zeros and ones. It would be virtually impossible to control the required effort and complexity needed to carry out this task. As in computer programing, complex software systems – called network operating systems – were therefore created for the control and use of computer networks. With their help, computer networks can be controlled and implemented in a comfortable way from a higher level of abstraction. These network operating system are based on the idea of handling communication tasks and functions in different degrees of abstraction and complexity. Tasks and

C. Meinel and H. Sack, *Internetworking*, X.media.publishing,
DOI: 10.1007/978-3-642-35392-5_2, © Springer-Verlag Berlin Heidelberg 2013

functionalities at the same level of abstraction are bundled together into "layers." Structured one on top of the other, different layers are defined in such a way that with the increasing level of abstraction, communication tasks with different complexity are handled. They are available to the user or computer application via a suitable interface. This type of an approach is also called a **layer model of communication**. The protocols acting on different layers are interlocked via the interface and together form a family of communication protocols (protocol family, protocol suite). The user, but also the majority of application programs communicating over the network to exchange data and offer services, only come into contact with this network operating system. It is only an extremely rare case that contact occurs with the network hardware hidden underneath.

2.1.1 Protocol Families

All of the parties involved must agree to follow common, fixed rules concerning the exchange of information to enable communication – and that not only in the case of digital communication in computer networks. This applies to the language used for communication, as well as to all the codes of behavior, that first make an efficient communication possible. In technical language these codes of behavior are described with the term **communication protocol** or simply, **protocol**

In addition to laying down the format of the information to be exchanged by communication partners, a communication protocol specifies a variety of actions necessary for the transmission of this information. With the development of the first computer networks, hardware was the primary focus and protocol software was viewed as secondary. This strategy has changed radically and today protocol software is highly structured. Instead of providing immense, highly complex and universal network protocols that regulate the entire task range in network communication, the problem of network communication plays out according to the principle of "divide and conquer" (divide et impera). There is a breakdown into a multiple number of manageable sub-problems. (Sub-)protocols focusing on a specific problem are provided in each case in order to deal with these.

These special protocols must work together smoothly and seamlessly. This poses a particular problem that is not to be underestimated in its complexity. In order to ensure this interaction, the development of protocol software is seen as a comprehensive task which is to be accomplished through the availability of an accompanying **family of protocols** (protocol stack, protocol suite). All of the individual protocols are efficiently integrated with each other to solve the overall problem of network communication.

While the various protocol families do in fact share many concepts, they are developed independently from one another as a rule and therefore not compatible. It is, however, possible to implement different protocol families simultaneously and parallel on the computers in a network, allowing them to use the same physical network interface without any resulting interference.

The term "protocol" is normally used here with two different meanings. On one hand, protocol is used to define an abstract interface. Included are all functions and operations that are made available over this interface. On the other hand, the term protocol sums up all of the information formats and their meaning. The definition of the **protocol specification** proceeds most of the time as a combination of specifying texts, images, status transition diagrams, and algorithms in pseudocode. It is necessary that the specifications be precise enough to enable the interoperability of different protocol implementations. Two of the different implementations can then successfully exchange information.

2.1.2 The Layer Model

To support protocol designers in their work, tools and models were developed that finely break down the entire process of network communication, ordering it hierarchically. In this way, clear interfaces are established between the individual levels of the hierarchy. These facilitate the largely independent development and improvement of the individual network protocol located on each layer, simplifying this process as much as possible.The best known variation of this model is the **layer model** (protocol stack) (cf. Fig. 2.3). The entire network communication process is separated into individual layers that are arranged one on top of each other. Each layer addresses a sub-problem of the network communication, with the addition of a new abstraction level of communication. The top layer provides the interface for application programs wishing to exchange information with applications on other computers. On the basis of such a layer model, the protocol designer constructs a complete protocol family, the so-called **protocol stack**, in which the individual protocol solves exactly the tasks addressed on each layer. Principally, in such a layer model the transmission of information from the application program of one computer to the application program of another computer follows a specific organization. The information from the source computer is passed along, processed in parts, from the top to the bottom through the different protocol layers. It is then physically transmitted over the transmission medium to the destination computer over the same protocol layers in reverse order – passed from the bottom to the top and finally transferred to the receiving application (cf. Fig. 2.1).

In the layer model, every layer is responsible for solving a specific part of the numerous tasks that come up within network communication. So that these tasks are carried out correctly, command and control information is created on the side of the transmitting computer, and used on each individual layer of the protocol stack. This is added to the transmitted data (cf. Fig. 2.2). At the receiving computer this supplementary information is read by the protocol software corresponding to each layer and processed further. In this way, the transmitted data can be correctly received at the end.

In accordance with the layer model of network communication, the protocol software of a certain layer k at the receiving computer must receive exactly the information

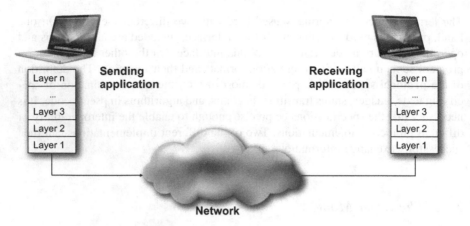

Fig. 2.1 Data transmission via a protocol stack.

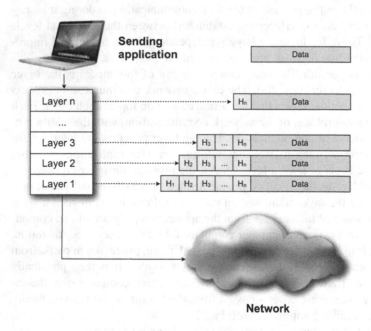

Fig. 2.2 Every layer of the protocol stack adds its own command and control information ($H_n...H_1$) to the data to be transmitted.

General Information about the Layer Model

Layer models play an important role in communication technology as well as in other areas of information science. The **shell model** is a modified form. Rather than being made up of stacked, hierarchically-ordered layers, it is comprised of individual shells.

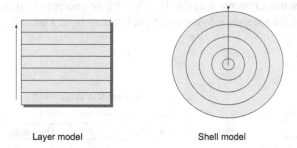

Layer model   Shell model

An application of such a model is useful for the following reasons:

- **Divide and Conquer (Divide et Impera)**
  According to this strategy a complex problem is broken down into individual sub-problems. Each one can be handled separately and is thus easier to manage and solve. This might be the only way that a problem can be solved as a whole.

- **Independence**
  The individual layers collaborate with each other, every layer only using the interface specifications of its direct predecessor. With fixed, predetermined interface specifications, the internal structure of a layer does not play a role for the other layers. Therefore, implementations on a layer could be exchanged without additional effort in the case of improved implementations. They would only have to be oriented on the respective interface specifications. In this way, implementations at individual layers are **independent** from those at the other layers and a **modular** (building-block) construction of the whole system is made possible.

- **Shielding**
  Each individual layer communicates only with its two directly neighboring layers. An **encapsulation** of the single layers is achieved. The challenge posed by the level of complexity to be overcome sinks drastically.

- **Standardization**
  The breakdown of the overall problem into individual layers also makes the development of standards easier. An individual layer and its interface allows a faster and easier standardization with the neighboring layers than with the complex system in its entirety.

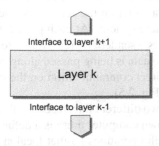

Interface to layer k+1

Layer k

Interface to layer k-1

Fig. 2.3 General information about the layer model.

that is sent it by the protocol software of level k at the transmitting computer. This means that every change that the protocol of a specific layer makes to the data being transmitted must be completely reversed by the receiver. If layer *k* adds additional command and control information to the data to be transmitted, then layer *k* at the receiving computer must remove it again. If in layer *k* an encryption of the data has taken place, then at the receiver side the encrypted data in layer *k* must be decrypted (cf. Fig. 2.4).

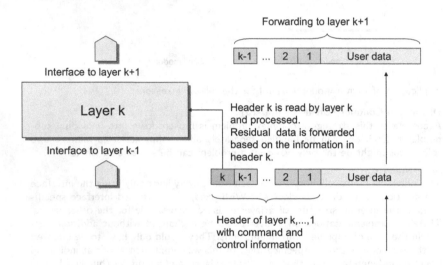

Fig. 2.4 Every layer of the protocol stack reads the data received in the layer-relevant header with the command and control information necessary for processing at this layer.

The actual communication in the protocol stack is always carried out in a vertical direction. When sending data, each protocol layer adds its own command and control information to this data. Typically, this information is passed on to the layer above it in a header prepended to the data packet. In this way the data packet is "encapsulated." The protocol software receives the necessary command and control information from this additional data, which ensures the correct and reliable further processing of transmitted data either on the receiver's side or in the corresponding protocol layer of an intermediate system . At each protocol layer it appears as if the protocol software on both sides – sender and receiver – are in direct communication with each other. But in fact the data is being passed along vertically through the protocol stack. The seemingly direct communication on the individual layers is called **virtual communication** (cf. Fig. 2.5).

Every protocol layer defines two different interfaces. So that applications can use the services of a protocol on its own computer there is a defined **service interface**. The service interface establishes all operations so that local applications can be carried out based on the protocol. Additionally, in every protocol layer there is also a defined

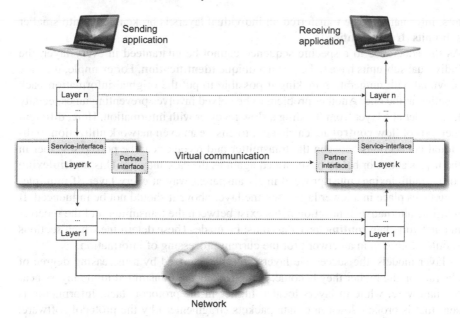

Fig. 2.5 On each level it appears that the individual layers of the protocol stack communicate directly with each other horizontally (**virtual communication**), actually the communication is always carried out vertically.

interface, a counterpart to the corresponding protocol layer on the other computer, the so-called **partner interface** (peer interface). The partner interface specifies the format of information that is exchanged between the neighboring protocol layers on different computers and establishes their meaning. However, communication over the partner interface proceeds in an indirect way. This means that every protocol layer communicates with its counterpart through the delivery of information to a lower, or higher protocol layer, which then sends this information in the same way it to its counterpart at the remote computer.

If a protocol family is implemented in the form of a layer model, several basic areas have to be observed in designing the protocol involved. These can apply to several or to all of the protocol layers So that information can in fact be exchanged between a transmitter and a receiver, a specific form of **addressing** is necessary on every layer so that the correct receiver can be identified among the many that are possible. Furthermore, rules for **data transfer** have to be established at every layer. Does data flow in both directions (bi-directional, duplex operation) or is data traffic only possible in one direction (uni-directional, simplex operation)? Can several (logical) channels be established and used within a communication connection, e.g., a channel for regular data, a channel to control communication, and a further channel for data with high priority? Any error occurring in transmission must also be detected and corrected (**error control**). This task is relevant for all layers and is carried out at different layers with different procedures. Based on technical and logical parameters, information to be transferred on individual layers is broken down into smaller sub-units (**fragmentation**).

As the adherence to a specific sequence cannot be guaranteed in every layer, the individual sub-units have to be given a unique **identification**. For example, they are provided with numeration making it possible to put the original information back together at the end. Another problem to be solved involves preventing an especially fast sender at a layer from flooding a slow receiver with information. Here, different methods of **flow control** are carried out to ensure an even network utilization. Individual connections between the transmitter and receiver can be pooled together in the layers above or below or separated again as the case may be. This **multiplexing** (or demultiplexing) must proceed in a transparent way at every layer. If multiplexing takes place in a lower layer then the layer above it should not be influenced. If multiple alternative connection paths exist between the transmitter and the receiver in a network then **routing** decisions must be made. These determine which sections should be chosen in a network for the current processing of information.

In layer models, the successive layers are characterized by a increasing degree of abstraction the higher they become. Data packets are transmitted in the layers near the hardware, while in layers located higher in the protocol stack, information is sent that is broken down into data packets (fragmented) by the protocol software. These higher located layers conceal unnecessary communication details from users and provide comfortable **services** for communication and data transmission.

A general distinction in service is made between **connectionless service** and **connection-oriented service**. Connectionless service works in a way similar to the traditional postal system. Every piece of information is like a letter or package affixed with a complete receiver's address and sent through the network, independent of any other message. As connectionless services do not take prescribed paths through the network, the sequence of the received information packets to the receiver can deviate from that of the sender's. A **reliable service** always confirms the successful delivery of a message by way of an acknowledgment. The transmitter is thus provided certainty as to whether the receiver has in fact gotten the transmitted information. In the case of a **non-reliable service** an acknowledgment of receipt is not provided.

A **connection-oriented service**, in contrast, works in a similar way to the telephone. This means that before a message can be transmitted, a connection to the receiver must be set up. All of the information is then sent along this connection, until the connection is terminated by both communication partners again. Reliable connection-oriented services can transmit data as **message sequences**. In this case, strict attention is paid that message borders are retained upon delivery. As an alternative, reliable, connection-oriented services can also send messages as **byte streams**. Here, message borders are ignored. A further variation is non-reliable, connection-oriented services. Here, a connection is set up prior to data transfer, but transmitter and receiver do not acknowledge confirmation of a message that has been received. This variation is chosen, e.g., for the transmission of audio or video data, as transmission delay caused by the absence of a confirmation receipt would not be acceptable. Transmission errors that occur are perceived as interference or noise and are sooner tolerated in, e.g., live transmission than delay would be. A **connectionless service** is known as a **datagram service**. In an analogy to the traditional postal service, a datagram service parallels a telegram or a postcard: upon successful delivery the sender receives no acknowledgement confirming this fact. If the (connectionless) communication between the transmitter and the receiver is limited to the exchange of a single message, one speaks of a **request-reply service** (request-reply service). A reliable, connectionless service (**reliable datagram service**), is advantageous when only a short message is to be sent, without the wish to establish an explicit connection. This variation can be compared to a registered letter with receipt. Here, with the receipt of a delivery confirmation the transmitter can be certain that the recipient has in fact really gotten the letter. Table 2.1 presents an overview of these different types of services.

Table 2.1 Service types

	Service	Example
	reliable message stream	sequence of individual images
Connection-oriented	reliable byte stream	terminal login
	unreliable connection	video stream
	unreliable datagram	unconfirmed email message
Connectionless	confirmed datagram	confirmed email message
	request/answer	client/server handshake

The difference between **protocols** and **services** has a special significance in this context. While protocols determine the rules and data formats governing the data that is to be exchanged within a given layer, the service designates a collection of operations (service primitives) that one layer makes available to the layer above it (cf. Fig. 2.5). Service primitives are understood as individual operations that prompt a certain action or report on its status (cf. Fig. 2.6). The specification of a service is summarized in the **service primitive**. However, not the actual manner of how this service is to be carried out is specified, but only the interface description between two neighboring layers in the protocol stack. Protocols, in contrast, carry out the services available in each layer.

Based on the layer model, two important **reference models** have been developed: the ISO/OSI reference model and the TCP/IP reference model. A reference model is a term used to describe an abstract model that serves as the basis upon which concrete implementation can be created. Specific protocols are linked with the two reference models named. These are located on the individual layers of the model

Service Primitives for the Implementation of a Connection-Oriented Service

To implement a connection-oriented service it is necessary that service primitives are available for the following operations:

- **Connection setup**:
  In order to establish a connection with a communication partner (on the same layer in the protocol stack), an operation is necessary that takes as parameter the address of the communication partner to which it sends a login (CONNECT).

- **Waiting for connection**:
  When one communication partner is ready to connect with another communication partner the latter is put into a special state of waiting for the establishment of the connection (LISTEN).

- **Transmission of messages**:
  When the connection has been established the active communication partner can send its partner a message (SEND).

- **Receiving of messages**:
  With the establishment of a connection each communication partner alternatively is put in a special waiting state for the message of its counterpart (RECEIVE).

- **Connection termination**:
  After the communication is finished a corresponding operation is necessary to actively end the connection to a communication partner (DISCONNECT).

A distinction is made between the active communication partner, who begins a communication process (client) and the passive communication partner (server), who waits for the establishment of communication to it, in order, e.g., to ask for services or information. A server is initially in the LISTEN state and waits for a connection setup. The clients starts its request with a CONNECT to a specific server and waits for an answer. If the server receives the connection request, it confirms this and executes RECEIVE to wait for data sent by the client. After receiving the confirmation of a connection establishment, the client executes SEND to send out further requests for services or data. In subsequently executing RECEIVE it waits for the answer from the server. A dialog exchange thereby arises between the client and the server. It is actively ended by the client with the execution of DISCONNECT. The server then also implements DISCONNECT and moves into the LISTEN status again (cf. also the section 8.1.4).

Fig. 2.6 Example for an application by service primitives

and enable those services to be carried out that are contained within that layer. The ISO/OSI model and accompanying protocol were conceived as a theoretical model and had an important didactic significance demonstrating the tasks and services of each layer. Conversely, the TCP/IP reference model evolved from the practical development of the Internet. It was based on a protocol model with practical application.

Excursus 1: ISO/OSI Reference Model

To enable the development of network protocol families, the International Standards Organization (ISO) made the **ISO/OSI Reference Model** available starting in 1977 for communication in open networks (Open Systems Interconnection). It broke down the entire process of network communication into seven individual layers and is designed as a conceptual tool for the development of protocol families (cf. Fig. 2.7).

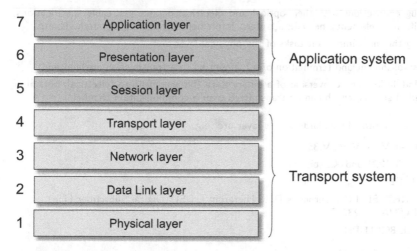

Fig. 2.7 The individual layers of the ISO/OSI Reference Model.

The network protocols existing before the ISO/OSI initiative were mainly of a proprietary nature and developed by the individual manufactures of the network devices themselves. Among these pre-ISO/OSI network protocol standards were, for example, IBM SNA, AppleTalk, Novell Netware and DECnet – all of which were not compatible with each other. While the standardization efforts for the ISO/OSI were still in progress the Internet -based protocol family TCP/IP was rapidly gaining in importance in heterogeneous networks that were comprised of components from different manufacturers. It gained a leading position before the ISO/OS succeeded in achieving standardization.

The ISO/OSI reference model was given the name **Open Systems Interconnection**, as it was intended for the connection of open systems, i.e., systems that are open for communication with other systems. The primary idea behind the design of the ISO/OSI Reference Model was that every single layer was to implement an exactly defined function. A new, higher layer would always be added if a new degree of abstraction was necessary to carry out the tasks at hand. The ISO/OSI model does not itself offer a network architecture. Only the tasks of the individual layers are determined in it and no decisions are made about the implementation of the functionality of the services and protocols.

In the ISO/OSI reference model the bottom layer in the protocol stack corresponds to the actual network hardware (physical level). The layers building on it each comprise the firmware and software implemented on this network hardware. The highest layer (layer seven) is finally the application layer providing an interface between the communication system and the various applications wishing to use the communication system for their own purpose. Layers 1-4 are designated in general as the **transport system** and layers 5-7 as the **application system**. These provide the increasingly general functionalities of the communication process. Although the name is the same, they should not be confused with the actual application programs located outside of the layer model. The tasks of the individual layers of the ISO/OSI reference model are described as follows.

- **Layer 1: Physical layer**
  The physical layer defines the physical and technical properties of the transmission medium (transmission channel). In particular, the relation between the network hardware and the physical transmission medium is regulated. For example, layout and assignment

of plug connections with their optical and electrical parameters, cable specifications, amplification elements, network adapters, implemented transmission methods, etc.

Among the most important tasks of the physical layer are:

– Establishment and termination of a connection to a transmission medium and
– Modulation, i.e., conversion of a binary data (bit stream) into (electrical, optical or radio) signals, which can be transmitted over a communication channel.

Important protocol standards of this layer are, e.g.,

– ITU-T V.24, V.34, V.35
– ITU-T X.21 and X.21bis
– T1, E1
– SONET, SDH (Synchronous Data Hierarchy), DSL (Digital Subscriber Line)
– EIA/TIA RS-232-C
– IEEE 802.11 PHY

● **Layer 2: Data link layer**
In contrast to the physical layer, whose main task is regulating the communication between a single network component and the transmission medium, the data link layer is concerned with the interaction of multiple (at least two) network components. The data link layer ensures that a reliable transmission can take place at a point-to-point-connection – despite potential periodic errors on the physical layer. This point-to-point-connection can be carried out either as a direct connection or via a **diffusion network** as in the case of e.g., Ethernet or WLAN. All connected computers in a diffusion network can receive the transmitted the data of all other connected computers without the need for an intermediate system.

The tasks to be carried out on the data link layer include

– the organization of data into logical units, called **frames** on the data link layer,
– the transmission of frames between network components,
– bit-stuffing, i.e., the padding of a frame that is not filled completely with special fill data, and
– the reliable transmission of frames through simple error detection methods, such as checksum calculation.

Among the most well known protocol standards of this layer are:

– BSC (Bit Synchronous Communication) und DDCMP (Digital Data Communications Message Protocol), PPP (Point-to-Point Protocol)
– IEEE 802.3 (Ethernet)
– HDLC (High-level Data Link Protocol)
– X.25 LAPB (Link Access Procedure for Balanced Mode) und LAPD (Link Access Procedure for D-Channels)
– IEEE 802.11 MAC (Medium Access Control)/LLC (Logical Link Control)
– ATM (Asynchronous Transfer Mode), FDDI (Fiber Distributed Data Interface), Frame Relay

- **Layer 3: Network layer**
  The network layer provides the functional and procedural means to enable the transfer of data sequences of variable lengths (**data packets**) from a transmitter to a receiver over one or more networks.

  Numbering among the tasks of the network layer are:

  - the assigning of addresses to end and intermediate systems,
  - the targeted forwarding of data packets from one end of the network to the other (**routing**) and thereby
  - linking individual networks (**internetworking**),
  - the fragmentation and reassembly of data packets, as different networks are determined by different transport parameters, and
  - the forwarding of error and status notification, related to the successful delivery of data packets.

  Some of the most important protocol standards at this layer are:

  - ITU-T X.25 PLP (Packet Layer Protocol)
  - ISO/IEC 8208, ISO/IEC 8878
  - Novell IPX (Internetwork Packet Exchange)
  - IP (Internet Protocol)

- **Layer 4: Transport layer**
  The transport layer facilitates a transparent data transfer between end users, providing the layers above it with a reliable transport service. The details necessary to ensure a reliable and secure data transmission are defined by the transport layer. Here it is assured that a sequence of error-free and complete data packets reach the receiver from the transmitter. On the transport layer the mapping of network addresses to logical names take place. The transport layer provides the participating end systems an end-to-end connection. Because it conceals the the details of the network infrastructure in between it is described as **transparent** The protocols located on this layer number among the most complex in network communication.

  Among the most important protocol standards in layer 4 are:

  - ISO/IEC 8072 (Transport Service Definition)
  - ISO/IEC 8073 (Connection Oriented Transport Protocol)
  - ITU-T T.80 (Network-Independent Basic Transport Service for Telematic Services)
  - TCP (Transmission Control Protocol), UDP (User Datagram Protocol), RTP (Real-time Transport Protocol)

- **Layer 5: Session layer**
  The session layer is also called the communication control layer because it controls the dialog between two computers connected over the network.

  The main duties of the session layer include:

  - the establishment, management and termination of all connections between local and distance services,
  - the controlling of full-duplex or simplex data transport and
  - the establishment of security mechanisms, for example authentification via the password procedure.

  Important protocols of this layer are:

- SAP (Session Announcement Protocol), SIP (Session Initiation Protocol)
- NetBIOS (Network Basic Input/Output System)
- ISO 8326 (Basic Connection Oriented Session Service Definition)
- ISO 8327 (Basic Connection Oriented Session Protocol Definition)
- ITU-T T.62 (Control Procedures for Teletex and Group 4 Facsimile Services)

- **Layer 6: Presentation layer**
  The presentation layer creates a context between two entities (applications) of the application layer above it, so that the two applications are able to use different syntax (e.g., data formats and encoding) and semantics. The presentation layer is therefore responsible for a correct interpretation of the transmitted data. Additionally, the respective local encoding of the data is transcribed in a special, standardized transfer encoding and transformed back at the receiver into the locally valid encoding .
  Additionally, data compression and encryption belong to the tasks on this layer.

  Among the most important protocol standards of the presentation layer are:

  - ISO 8322 (Connection Oriented Session Service Definition)
  - ISO 8323 (Connection Oriented Session Protocol Definition)
  - ITU-T T.73 (Document Interchange Protocol for Telematic Services), ITU-T X.409 (Presentation Syntax and Notation)
  - MIME (Multipurpose Internet Mail Extensions), XDR (External Data Representation)
  - SSL (Secure Socket Layer), TLS (Transport Layer Security)

- **Layer 7: Application layer**
  The application layer provides an interface for application programs wishing to use the network for their own purposes. Application programs themselves do not belong in this layer but only use its services. The application layer provides simple and easy-to-manage service primitives that conceal network internal details from the user or the programmer of the application program and therefore enable a simple use of the communication system. Some of the most important functions of the application layer are:

  - identification of the communication partner,
  - determination of the availability of resources and the
  - synchronization of communication.

Numbering among the most important protocol standards located on this layer are:

- ISO 8571 (FTAM, File Transfer, Access and Management)
- ISO 8831 (JTM, Job Transfer and Manipulation)
- ISO 9040 und 9041 (VT, Virtual Terminal Protocol)
- ISO 10021 (MOTIS, Message Oriented Text Interchange System)
- FTP (File Transfer Protocol), SMTP (Simple Mail Transfer Protocol), HTTP (Hypertext Transfer Protocol), etc.
- ITU-T X.400 (Data Communication for Message Handling Systems). ITU-T X.500 (Electronic Directory Services)

Since the development of the ISO/OSI Reference Model, concepts for protocol families in different locations have changed and many of the newly developed protocols no longer fit into this scheme. Nevertheless, a large part of the terminology, especially designations and numeration of the individual layers, has remained the same until today.

Further Reading:

U. Black: OSI − A Model for Computer Communications Standards, Upper Saddle River, NJ, USA (1991)
H. Zimmermann: OSI Reference Model − The ISO Model of Architecture for Open Systems Interconnection, in IEEE Transactions on Communications, vol. 28, no. 4, pp. 425–432 (1980)

2.2 The Physical Layer as the Basis for Computer Communication

The protocols in the lowest layer of the TCP/IP Reference Model (the network access layer) are based on the physical transmission medium (transmission channel). This transmission medium is also called the **physical layer**, but it is normally not included in the protocol stack of the TCP/IP reference model (cf. Fig. 2.10). The physical layer together with the four layers of the TCP/IP reference model comprise the so-called hybrid TCP/IP reference model. In contrast, the physical layer forms a layer of its own with the same name in the ISO/OSI reference model .

2.2.1 Physical Transmission Media

In general, the physical layer defines the physical and technical properties of a physical or analog transmission medium, used for data transmission. The relations between the network hardware and the physical transmission medium are particularly regulated, such as the layout and assignment of plug connections with their optical/electrical parameters, cable specifications, amplification elements, network adapters, implemented transmission methods, etc. The actual task of the physical layer is to translate a series of bits (bit stream) into a sequence of physical signals, which is then forwarded from the sender to the receiver with the help of the transmission medium.
Depending on the nature of the transmission medium, numerous methods and processes can be used. Bit sequences are transformed securely and reliably into physical signals in different ways to be sent over the transmission medium. On the side of the receiver they are reassembled again into the output information. This procedure is known as **modulation**, while in the opposite direction it is called **demodulation**. Media implemented for data transmission can generally be separated into the groups of **wired** (guided) transmission media and **wireless** (unguided) transmission media. With wired (guided) transmission media, electromagnetic waves n are forwarded along a solid medium. There are many different variations of this medium, from copper cable, such as twisted pairs or coaxial cable (wirepath, conductor), to different fiber optical cable variations (fiber optics) (light path, waveguide). To gain

access to a network based on a wired transmission medium, a direct, physical contact must first be created. The high transmission speed results from the low rate of error, which can be achieved thanks to good shielding possibilities. However, wired network architectures also involve substantial costs as cables must be bought and laid.

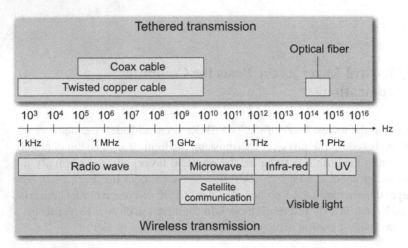

Fig. 2.8 Wired and wireless transmission media in the electromagnetic spectrum.

In the case of wireless (unguided) transmission media, electromagnetic waves are transmitted over different frequency ranges of the electromagnetic spectrum into space. These include radio transmission via short or ultra-short wave, microwave transmission, infrared or laser light. A distinction is made between directed transmission, such as with a laser beam, directional radio or satellite direct radio link, and undirected (isotropic) transmission, such as mobile communication, terrestrial or satellite broadcast. In comparison to wired transmission media, a wireless network architecture is flexible and ideal for mobile implementation. There are no costs for complex cabling. But, on the other hand, it is possible to penetrate a wireless network without direct physical contact being necessary. For this reason, the use of complex software technologies are a necessary security measure, for example encrypted data transmission. Furthermore, transmission speed is decreased due to signal transmission disturbances caused by reflections from objects or atmospheric interference.

2.2.2 Characteristic Properties of Physical Transmission Media

All physical transmission media are restricted by their individual limitations. This affects the maximum information (bandwidth) transported per time unit or the speed at which a signal can spread over the transmission medium. Generally, every signal that spreads along a physical medium is subject to a signal damping. As the distance to the transmitter increases, the signal weakens accordingly. In contrast to a perfect transmission medium, actual transmission media is constantly at the mercy of interference (noise). If signal damping weakens the signal to such an extent that it can no longer be distinguished from noise, then the signal on the side of the receiver cannot be reconstructed or correctly interpreted. For this reason, as a signal moves along the transmission medium it must be refreshed in certain places, i.e., strengthened, in order that it can be received with the greatest reliability and in the purest form possible.

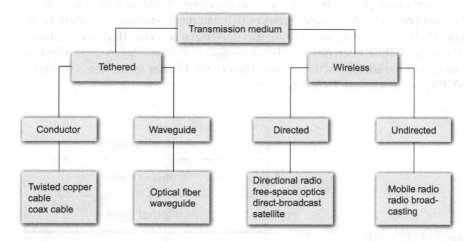

Fig. 2.9 A classification of physical transmission media.

Independent of the physical transmission medium, different modulation procedures are implemented to ensure that the encoding and transmission of binary (digital) information over a physical (analog) medium is as efficient as possible. The various physical transmission media, their limitations, characteristic properties as well as their employment in the Internet will be looked at in detail in Chap. 3.

2.3 The TCP/IP Reference Model

The complex tasks that are necessary for computer communication on the Internet are regulated with the help of different hierarchically linked protocols. Their functionality is best described with a **layer model**, as was presented in the last chapter. A group of tasks is defined in each layer that must be handled by the protocols assigned to that layer. The protocols of any given layer must take into account the neighboring layer above and below it. The interface to the adjacent layers on the same terminal is called the **service interface**. Over the service interface, a protocol provides a specific service to a layer higher in the protocol stack and in carrying out its task can fall back on the services of the layers below. In addition, a protocol defines a further interface to its counterpart on a remote terminal in the form and meaning of the information exchanged between the terminals. This interface is therefore also called a **partner interface**. Communication within the layer model proceeds in a vertical direction between neighboring layers, with the actual data exchange only occurring on the lowest level, the physical layer. In contrast, via the partner interface a seemingly virtual connection is implemented over which data and control information is exchanged. In each case only a specific protocol layer is affected. At the time of the Internet's development the focus was on making seamless communication possible over a multiple number of network architectures. Based on both of the primary Internet protocols: the Internet Protocol (IP) and the Transmission Control Protocol (TCP), the resulting architecture was designated the TCP-IP reference model. The actual TCP/IP reference model encompasses four layers (layers 2–5). Those together with the addition of the physical layer (layer 1) make up the five-layer hybrid TCP/IP reference model (cf. Fig. 2.10).

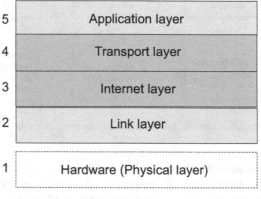

Fig. 2.10 The TCP/IP reference Model is comprised of four layers (2-5), together with the physical layer (1). This model is also called the hybrid TCP/IP reference model.

The **link layer** of the TCP/IP reference model corresponds to the first two layers of the ISO/OSI reference model (physical layer and data link layer). Its main task focuses on the secure transmission of data packets of pooled bit sequences. It is followed by the **internet layer**, which corresponds to the network layer of the ISO/OSI reference model. Its main responsibility is to enable the data communication of two end systems at a given location in the heterogeneous communication network. The **transport layer** above it corresponds to the layer of the same name in the ISO/OSI reference model. It enables two user programs on different computers in the communication network to exchange reliable and connection-oriented data. The **application layer** of the TCP/IP reference model includes the three upper layers of the ISO/OSI reference model and serves as an interface for the actual application programs that wish to communicate with each other over the network.

The **TCP/IP reference model** stands in marked contrast to the ISO/OSI reference model. Unlike the ISO/OSI reference model, it was not conceived and planned theoretically but was derived from the protocols that had been put into practice on the Internet.

The ISO/OSI protocols, on the other hand, were planned theoretically and adopted before protocols could be invented that implemented the different function for the layers of the ISO/OSI reference model. Today these protocols are no longer used and the TCP/IP reference model protocols, which had been developed from practical application, dominate the Internet. After a short excursion into the historical development of the TCP/IP protocols, a look will be taken at the similarities and differences between the TCP/IP reference model and the ISO/OSI reference model. Finally, the individual layers of the TCP/IP reference model will be presented.

2.3.1 Historical Background and Distinction from the ISO/OSI Reference Model

The Internet, and its successor the ARPANET, came into use purely as research networks starting in 1969.

The initial four computers connected to the network increased to several hundred in just a short time, linking universities, research institutes and military installations with each other over leased telephone lines. As technologically different networks, such as satellite networks and wireless networks, were connected to the Internet the originally implemented protocols were soon overburdened by the data traffic translation required from one network to the other.

In 1972, *Robert E. Kahn* (*1938) was working on data transmission in satellite networks and wireless networks at the DARPA Information Processing Technology Office (IPTO), which was responsible for the further development of the Internet at that time. There, he soon realized the advantage of enabling data traffic via different network technologies. A new, flexible network paradigm had to be developed with the initial focus on linking heterogeneous networks based on different technologies. *Vinton Cerf* (*1943) joined Kahn's team in 1973. He had been one of the developers of the **Network Control Program**, at that time still implemented as a network protocol in ARPANET, and he joined Kahn to work on protocols for an open network architecture

In the summer of 1973, Kahn and Cerf introduced a fundamentally new network architecture. Its main characteristic was to virtually unite the different network technologies over a common "Internetworking protocol," via the respective protocols of the actual network technologies. In contrast to the existing ARPANET, where the network itself was responsible for reliable data transport, from now on end systems (hosts) connected with networks should be responsible for reliable data transmission. The functionality of the network itself would be limited from this time on to the simplest possible data packet transport. In making this move, Kahn and Cerf also succeeded in connecting the most different network technologies with one another.

The connection of different networks was to proceed via special computers, so-called **package brokers** (routers), which are solely responsible for the forwarding of data packets between different networks. Cerf's own research group at Stanford University worked on the first specification of the **Transmission Control Protocol** (TCP, RFC 675) until 1974.

They were strongly influenced by the network research group of the Xerox PARC (Palo Alto Research Center), and the work being done there on the PARC Universal Packet Protocol Suite (PARC UPPS).

The DARPA commissioned the company BN Technologies, Stanford University and University College London with the technical implementation. Their task was to employ the new protocol standard on different hardware platforms. Following the versions TCP v1 and TCP v2 , the protocol was broken down and further developed into TCP v3 and IPv3. The resulting network architecture was later given a desi-gnation based on the two most important protocols, TCP and IP (Internet Protocol), as TCP/IP reference model (RFC 1122). This led to the 1978 development of an operational version **TCP/IP v4** (Version 4), which is still used on the Internet today. proof of the operational readiness of the TCP/IP could be demonstrated with the linking of two different networks between Stanford University and the University College London via TCP/IP. A test with three different network architectures follo-wed between the USA, Great Britain and Norway in 1977. The final transition of the complete Internet to TCP/IP v4 took place on January 1, 1983.

Robert E. Kahn and Vinton Cerf received the highest award in information tech-nology for their achievements in 2004 – the Turing Award. A year later both were recipients of the the Presidential Medal of Freedom, the highest civil order of merit in the USA. At the 2nd German IPv6 Summit in 2009 at Hasso Plattner Institute in Potsdam, Robert E. Kahn was named a HPI Fellow, an honor he shares with German Chancellor Dr. Angela Merkel and, since 2011, with his former colleague, Vinton Cerf.

Both the ARPANET as well as the Internet already existed when the ISO tackled the development and standardization of the ISO/OS reference model. The TCP/IP reference model manifested on the Internet, therefore had a decisive influence on the development of the ISO/OSI reference model. The seven layers of the ISO/OSI reference model allow themselves to be transferred to the protocol architecture of the Internet, whereby the TCP/IP reference model is only made up of four layers (or five layers in the hybrid TCP/IP reference model, with the inclusion of the physical layer). Implementation possibilities and application areas (network access, Internet, transport, application) are specified in the individual layers of the TCP/IP reference model. In the ISO/OSI reference model, concrete rules are given for operation, se-mantics of the data and network technologies. The TCP/IP reference model does not contain concrete hardware specifications and neither does it standardize the physical data transmission as such, but rather binds these aspects to the implementation of individual layers.

The most important protocol family today, the **TCP/IP** protocol suite, is not based on the specifications of a standardization committee, but grew out of requirements and experience from the developing Internet. The ISO/OSI reference model is adaptable to the extent that it can also serve to describe the TCP/IP protocol stack, but both express different principles.

The **TCP/IP reference model** was in fact first defined completely after the protocols described in it were implemented and being used successfully. This had the advantage that the described layer specifications corresponded perfectly with the protocol implementation. However, an application of this model in other protocol families could not be carried out easily. The first description of the TCP/IP reference model (RFC 1122) can already be found as early as 1974, even before the first specifications of the ISO/OSI model were carried out.

Principally, the TCP/IP protocol family can be divided into four single layers, which are organized around the core layers TCP and IP (cf. Fig. 2.10). In fact descriptions of the TCP/IP reference model as comprised of five different layers can also be found in technological literature. The communications hardware descriptive layer (physical layer, hardware) is included in the original four layer TCP/IP reference model. This five layer model is often called the **hybrid TCP/IP Reference Model**. The designation of the single layers correspond to the underlying RFC 1122 and will be used throughout this book.

The four layers of the TCP/IP Reference Model can be compared in the following way with the seven layers of the ISO/OSI Reference Model (cf. also Fig. 2.11):

- Layer 2 of the TCP/IP reference model (link layer) is often designated as the data link layer in technological literature, or is also called the network access layer or host-to-network layer. It corresponds to the first two layers of the ISO/OSI reference model (physical layer, data link layer).

- Layer 3 of the TCP/IP reference model (Internet layer) is also called the network layer or Internetwork layer and corresponds to layer 3 of the ISO/OSI reference model (network layer).

- Layer 4 of the TCP/I reference model (transport layer) is also designated the host-to-host layer and corresponds to layer 4 of the ISO/OSI reference model (transport layer).

- Layer 5 of the TCP/IP reference model (application layer) corresponds to layers 5 – 7 of the ISO/OSI reference model (session layer, presentation layer, application layer).

In the following sections, the tasks and protocols of the individual layers of the TCP/IP reference model will be looked at in more detail.

2.3.2 Link Layer

The link layer of the TCP/IP reference Model combines the first two layers of the ISO/ISO reference model: layer 1 – the physical layer and layer 2 – the data link layer. However, the link layer does not contain the aspects of the physical layer that are part of the ISO/OSI reference model. The link layer is therefore the lowest

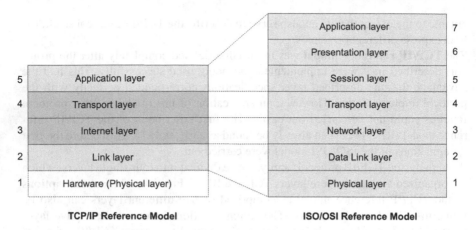

Fig. 2.11 A comparison of the TCP/IP Reference Model and the ISO/OSI-Reference Model.

layer of the TCP/IP reference model. The primary task of the link layer is the secure transmission of individual data packets between two adjacent end systems. The bit sequences to be transmitted are pooled together into fixed units and provided with the additional information necessary for transmission, e.g., checksums for simple error detection. The adjacent end systems can be either directly connected with each other by a transmission medium or by a so-called bus (diffusion network), which connects multiple end systems directly, therefore without intermediate systems.

On this layer a distinction is made between **secured** and **unsecured services**. With unsecured services, data packets recognized as defective are eliminated. The request for the necessary re-transmission follows, but first on a higher layer of the protocol stack. In contrast, a secured service takes over the request for a retransmission itself.

In local networks (LANs), layer 2 of the TCP/IP reference model is normally sub-divided into two further layers:

- **Medium Access Control (MAC)**
  This sublayer regulates access to the shared (with many other) computer systems transmission medium. As these are in competition with each other to gain access to the transmission medium, it is necessary that protocol mechanisms are provided allowing fair and efficient access to all participants (**multiple access protocols**). This includes methods for the discovery of collisions or their avoidance, as many participants wish to transmit data at the same time (**collision detection**, **collision avoidance**). Additionally, every participant at this layer must have an individual and unique address as a means of identification (**MAC addressing**). On the MAC sublayer different (but homogeneous) subnets are already connected with each other over a so-called **switch** (**LAN switching**). Thereby data packets are each forwarded only within the subnet where the respective target computer is located. Here, the switch takes over the task of filtering the data traffic (**MAC filtering**). Two types of switches may be identified. The **store-and-forward** switch

always save the data packets to be filtered before they are analyzed and finally forwarded. The **cut-through** switch carries out the forwarding without prior caching. Additionally, in this sublayer there are also tasks of administering (**data packet queuing** and **scheduling**) if data packets cannot be forwarded quickly enough before new data packets are delivered and it is necessary to determine priority.

- **Logical Link Control (LLC)**
  This sublayer forms the so-called data link layer of the LAN. The tasks regulated there are on a higher abstraction level than the MAC sublayer below upon which it is located. Problems and tasks of the LLC sublayer are determined in the **IEEE 802.2** standard. Among its tasks are avoiding overload situations by potential receivers of transmitted data through targeted interventions in the data flow (**flow control**) and control of data transmission (**link management**). Over the LLC sublayer a first quality control of the transmitted data also takes place. Data transmission errors must be recognized and if possible corrected. For this purpose the protocol located on the LLC sublayer carries out various **error detection and error correction procedures**. Additionally, the LLC sublayer synchronizes the transmitting and receiving of data units (data packets). To enable this, the data must be subdivided into size-limited data packets (**fragmentation**) in accordance with the physical and logical conditions of the respectively chosen transmission form. Following transmission it is essential that the beginning and ending of a data packet always be correctly recognized (**data packet synchronization**). Besides this, the LLC sublayer ensures so-called **multi-protocol capability**. This is the capability of using different communication protocols at the same time.

Included with the most important protocols of the link layer of the TCP/IP reference model are those from the IEEE, based on the **IEEE 802** LAN standard of the standardized LAN protocols These are technologies such as **Ethernet** (IEEE 802.3), **token ring** and **FDDI** (IEEE 802.5), as well as different wireless **WLAN** technologies (IEEE 802.11). We will take a closer look at these in Chap. 4.

The most important protocols of the link layer of the TCP/IP protocol family are:

- **ATM (Asynchronous Transfer Mode)**
  ATM is a packet-switching network protocol that breaks down and forwards the data to be transported in cells of a fixed size (cell relay). Behind this design principle was the idea of providing time-critical, real-time data, such as video or audio information, together with regular data over a standardized protocol. Attention was paid to keeping switching and transfer delay as small as possible. ATM is connection-oriented and additionally establishes a virtual connection between two endpoints in the network before the actual data transfer begins. ATM is employed in LAN as well as in wide area networks, so called WANs.

- **ARP (Address Resolution Protocol) and RARP (Reverse Address Resolution Protocol)**
  With the help of the ARP protocol described in RFC 826 the MAC address of a host can be determined from the IP address in the Internet protocol of the layer

above it. This is important if a data packet from the Internet is to be delivered
to a local network whose MAC address must be determined from the stored IP
address of the receiver for forwarding in the LAN. ARP is only implemented in
LANs or point-to- point connections. The reverse service is delivered by the standardized RARP protocol in RFC 903. For each MAC address a corresponding IP
address can be determined .

- **NDP (Neighbor Discovery Protocol)**
  The functions of the NDP protocols are very similar to the ARP protocol and
  serve to explore and discover further hosts in the local network. In contrast to
  ARP, NDP was developed for the next generation of the Internet protocol IPv6,
  while ARP works under the current version IPv4.

- **LLTD (Link Layer Topology Discovery)**
  The proprietary LLTD protocol was developed by the Microsoft company for
  exploration of the present network topology and verification of the guaranteed
  quality of service in a network.

- **SLIP (Serial Line Interface Protocol) and PLIP (Parallel Line Interface Protocol**
  SLIP, established in RFC 1055, and PLIP are simple point-to-point network protocols that serve in the transport of encapsulated IP data packets between personal
  computers over a serial (SLIP) or parallel interfaces, i.e., these IP data packets
  are packed in data packets of the SLIP or PLIP protocols. For the most part, SLIP
  and PLIP have been replaced by modern PPP protocol.

- **PPP (Point to Point Protocol)**
  PPP is a simple point-to-point network protocol that serves the connection between two network nodes. PPP is used by the majority of Internet providers (ISPs)
  to offer their customers a dial-up connection over a standard telephone line in the
  Internet. Modern access via DSL (Digital Subscriber Line) is implemented by
  the ISPs over the encapsulated protocol variations PPPoE (PPP over Ethernet,
  RFC 2516) and PPPoA (PPP over ATM, RFC 2364).

- **STP (Spanning Tree Protocol)**
  STP was established in the IEEE 802.1D standard, and describes a protocol that is
  intended to ensure cycle freedom within a LAN architecture consisting of multiple network segments. As indicated in the name of the protocol, a so-called
  spanning tree is created from the network graph that is present. Over the spanning tree it is guaranteed that the LAN does not contain any closed loops, where
  the data packets would then be traveling for an unlimited time.

2.3.3 Internet Layer

The main task of the TCP/IP reference model Internet layer involves enabling data
communication between two end systems on different ends of the communication
network, i.e., over different network architectures.

The methods described in the Internet layer for bridging and uniting different network architectures with the help of special intermediate systems (routers) is also called **Internetworking**. The necessary tasks to be solved are described in RFC 1122. In Internetworking there needs to be a clear addressing scheme that exceeds the given network borders (**IP addressing**) . Each data packet to be sent has to be provided with the addresses from the transmitter and receiver in order to be delivered correctly. As the communication proceeds over one or more independent operating networks, the computers at the connection and switching centers (intermediate systems, routers) must be in a position to choose the correct path to follow when forwarding data packets (**routing**). In data packet switching across different network types, often different rules are valid in connection with the maximum size of data to be transported in a single data packet (Maximum Transmission Unit, MTU). The transmitting intermediate systems therefore have to break down data packets (**fragmentation**)) to be transmitted in a network with stricter limitations. At the receiver they will then be put back together. [1] Furthermore, further technical differences can occur between the bridged networks. These have to be compensated by appropriate transfer and calculation methods, for example the switching between encrypted networks and unencrypted networks or different time or volume-based calculation procedures.

The three basic tasks of the Internet layer are composed of the following functions:

- Outgoing data packets must be forwarded to the next switching location or to the receiving end system. To do this, the responsible communication protocol must choose the next (direct) receiver (**next hop**) along the mediated path and send the data packet to it through transfer to the respective, responsible protocol in the link layer.

- Incoming data packets have to be unpacked, the control information read from the header of the data packet and, if relevant, the transported user data passed on to the active transport protocol in the above layer.

- Additionally, diagnostic tasks are taken over and a simple error handling implemented. However, on the Internet layer there are just **unreliable services** offered, i.e. there is no guarantee that a transmitted data packet actually reaches its receiver. The transport can therefore take place only "as well as possible" (**best effort**). Control of a reliable communication is the responsibility of the two endpoints of communication (transmitter and receiver). It is carried out on a higher layer of the TCP/IP reference model to unburden the network on this layer from this difficult task. The scalability and error tolerance of Internet technology is first made possible through this "best effort" strategy. Only in this way has it been possible for the Internet to grow to its present-day size.

The central protocol of the Internet layer is the **Internet Protocol (IP)**. IP offers an unreliable, data packet-oriented, end-to-end information transmission. It is responsible for fragmentation and defragmentation into so-called **IP datagrams** and has

---

[1] This task is omitted in the new version of the Internet protocol IPv6. The end systems involved in the communication carry out a pre-fragmentation themselves in order to facilitate a faster transmission of data.

protocol mechanisms for forwarding via intermediate systems to the designated receiver of the information. IP exists today in two versions, IPv4 (RFC 791) and IPv6
(RFC 2460). It takes its place among the most important protocols in the Internet.
In addition, the protocol **ICMP (Internet Control Message Protocol)** is implemented in the Internet layer. It is responsible for the notification of specific errors occurring during IP transmission, as well as for further diagnostic tasks, such as sending
echo requests to test the availability of a computer and the necessary transmission
time. ICMP is a protocol that sits directly on the IP. Two different variations of the
ICMP protocols exist, one for IPv4 (RFC 792) and another for IPv6 (RFC 4443).

Besides IP and ICMP there are further protocols that number among those in the
Internet layer of the TCP/IP protocol stack, for example:

- **IPsec (Internet Protocol Security)**
  IPsec is comprised of a protocol suite to ensure the secure execution of IP data
  traffic. Within a data stream IP datagrams can be authenticated (Authentication
  Header, AH) and encrypted (Encapsulating Security Payload, ESP) (RFC 4835).
  Additionally, included in IPsec are protocols for the handling, establishment and
  exchange of secure cryptographic keys (Internet Key Exchange Protocol, IKE,
  RFC 2409).

- **IGMP (Internet Group Management Protocol)**
  The IGMP protocol (RFC 1112, RFC 2236, RFC 3376) carries out the administration of IP multicast groups of end systems within a TCP/IP network. Special
  multicast routers administer address lists of end systems that can be addressed
  commonly via one multicast address. Through the use of multicast addresses the
  burden on the transmitter and on the entire network is reduced. IGMP exists only
  in one version for IPv4, as IPv6 implements multicasting differently.

- **OSPF (Open Shortest Path First)**
  The OSPF protocol (RFC 2328) is a so-called link-state routing protocol, that
  transmits IP datagrams within a single routing domain (autonomous system). It
  belongs to the group of the Interior Gateway Protocols (IGP). OSPF is the most
  widely used routing protocol on the Internet.

- **ST 2+ (Internet Stream Protocol, Version 2)**
  The Internet stream protocol (ST, RFC 1190, und ST 2+, RFC 1819) is an experimental protocol of the Internet layer. As a supplement to the Internet protocol,
  it is intended to provide a connection-oriented transport of real-time data with
  the guarantee of a constant quality of service.

2.3.4 Transport Layer

The transport layer in the TCP/IP reference model corresponds roughly to layer 4
of the ISO/OSI reference model. Its primary task entails the establishment and use
of a communication connection between two application programs residing on different computers in the network. The protocol of the transport layer establishes a

direct, virtual end-to-end communication connection. To allow multiple application programs on the same computer to have parallel communication, a statistical multiplexing is among the tasks of the transport layer. Every application program is assigned a so-called **port number** to provide unique identification. On the transport layer, every unit of data sent must contain the port number of the sender and the receiver in order to be transmitted correctly. Together with the IP address, the port number defines what is commonly called a **network socket**, a unique connection endpoint in the network. A complex flow control is likewise implemented on the transport layer. This ensures that overload situations are avoided to the greatest extent possible (congestion avoidance). Finally, measures are taken to ensure that the transmitted data arrives at the receiver error-free and in the correct order (sequence numbers). An acknowledgment mechanism is provided over which the receiver can confirm correctly transmitted data packets, or send a new request in the case of defective data packets.

Unlike the Internet layer, the transport layer is not under the control of the network operator. It therefore offers the user or application program of the communicating end systems the possibility to influence problems in data transmission that are not handled at the Internet layer. These include the bridging of failures on the Internet layer and the subsequent delivery of data packets that have gotten lost on the Internet layer.

The **Transport Control Protcol (TCP)** is a further core element of the Internet protocol architecture and the most popular protocol of the transport layer in the TCP/IP reference model. Standardized as RFC 793, it carries out a reliable, connection-oriented, bi-directional data exchange between two end systems, which in the TCP/IP reference model is based on an unreliable, connectionless datagram service of the Internet layer. TCP enables the establishment of what are called **virtual networks** (virtual circuits). After a virtual connection is set up, a data stream (byte stream) is transmitted that hides the packet-oriented transmission of information from the application layer above it. A reliable service is thereby initiated by an acknowledgment mechanism (Automatic Repeat Request, ARQ), over which the transmission of lost data is initiated.

Next to TCP, the **Universal Datagram Protocol (UDP)** is the second most prominent protocol of the transport layer. Standardized as RFC 768, it transmits independent data units, so-called datagrams, between application programs that reside on different computers in networks. However, the transmission is unreliable, i.e., possibly combined with data loss, proliferation of datagrams and changes in sequence. The datagrams recognized as false are discarded by UDP and do not even reach the receiver. In comparison to TCP, UDP is clearly less complex,which is reflected in its increased data throughput. Yet this is compensated by a dramatic loss of reliability and security. The application program above has to take care of this itself.

Other important protocols of the transport layer are:

- **DCCP (Datagram Congestion Control Protocol)**
  DDCP (RFC 4340) is a message-oriented protocol of the transport layer. In addition to reliably establishing and terminating connections, it also distributes overload notifications (Explicit Congestion Notification, ECN). It provides overload

control functions (congestion control) and can be used for the negotiation of transmission parameters.

- **RSVP (Resource Reservation Protocol)**
  The RSVP (RFC 2205) protocol is used for the request and reservation of network resources using IP to transmit data streams. It is not intended for the actual data transport and bears a similarity to the ICMP and IGMP protocols at the Internet layer. RSVP can be implemented by end systems as well as routers without having to reserve and maintain specified service qualities.

- **TLS (Transport Layer Security)**
  As predecessor of the **Secure Socket Layer Protocol (SSL)**, TLS supplies (RFC 2246, 4346 and RFC 5246) cryptographic protocols for secure data transport in Internet. Individual TCP segments are encrypted by TLS and SSL.
  TLS provides protocols for the negotiation of transmission parameters (peer negotiation), for the exchange of cryptographic keys and authentication as well as for encrypting and digital signature.

- **SCTP (Stream Control Transmission Protocol)**
  SCTP (RFC 4960) is a proposal for a highly scalable and performant version of the original TCP protocol and is specialized in transmitting large amounts of data.

2.3.5 Application Layer

The functions on the application layer in the TCP/IP reference model encompass the tasks of layers 5-7 in the ISO/OSI reference model. The application layer primarily functions as an interface for the actual application programs wishing to communicate over the network (process-to-process communication). The applications themselves are located outside of this layer and even outside of the TCP/IP reference model.

The services offered and programming interfaces (Application Programming Interface, API) of the application layer have a high level of abstraction. The user or communicating application is for the most part shielded from the details of communication, which is regulated on the lower protocol layers. Protocols and services of the application layer normally carry out translations and transformation of data between application programs at the semantic level. Among the services offered are: naming services, which translate IP addresses into readable names and visa versa, redirect services, which reroute requests that cannot be filled to another host, as well as directory services and network management services. The protocols of the application layer work mainly according to the client/server communication principle. An active client contacts a passive, waiting server and transmits a service request to it. The server accepts the request of the client, processes it and sends back an answer to the client, i.e., in a positive case the requested service. Communication between the client and server therefore does not proceed symmetrically.

A few of the many important protocols of the application layer located in the TCP/IP protocol family are:

- **TELNET (TELetype NETwork)**
  TELNET (RFC 854) allows the setup of an interactive, bidirectional communication connection to a remote computer, additionally providing a command-line interface. This enables a virtual terminal to be established at the remote computer via TCP, on which commands and actions can be initiated.

- **FTP (File Transfer Protocol)**
  FTP (RFC 959) facilitates the transmission and manipulation of data between two computers connected over a TCP/IP network. FTP functions according to the client/server paradigm. A client initiates the connection and requests a service. The server takes the connection request and answers the service request. The actual data transfer of the control command transmission proceeds over two different TCP ports at FTP.

- **SMTP (Simple Mail Transfer Protocol)**
  SMTP (RFC 821) is a simple, structured protocol for the transmission of electronic mail on the Internet. Today, as a rule, ESMPT (Extended SMTP, RFC 5321) is used as it allows a transparent transmission of messages in different formats. SMTP is used by the Message Handling Systems (MHS) of the email service for sending and receiving messages. End systems, i.e., systems on which the end user works, use SMTP exclusively for sending email messages forwarded from a mail server.

- **HTTP (Hypertext Transport Protocol)**
  The HTTP protocol (RFC 2616 among others) is used for data transmission in the World Wide Web. Just as many other protocols of the application layer, it works according to the client/server paradigm and is based on the reliable transport protocol TCP.

- **RPC (Remote Procedure Call)**
  The RPC protocol (RFC 1057 and RFC 5531) is used for inter-process communication. That means, it allows a computer program to call an external, subroutine located in another addressing area. This is carried out externally and only the result transmitted to the requesting computer where it is further processed.

- **DNS (Domain Name System)**
  The DNS service establishes a name and directory service that delivers the assignment between readable end system names (strings) to IP addresses for all participating systems on the Internet. The name space for end systems administered over DNS is hierarchically organized and works with local intermediate storage and proxies to ensure an efficient implementation. DNS is standardized in RFC 1123 and in numerous other RFCs.

- **SNMP (Simple Network Management Protocol)**
  With the help of the SNMP protocol, network management systems can monitor, administer and control individual systems connected to the network. SNMP is standardized in RFC 3411 and in many other RFCs.

- **RTP (Real-time Transport Protocol)**
  With the help of the RTP protocol (RFC 1889) real-time audio and video data can be transmitted over the Internet. For this purpose, the RTP protocol defines a separate data format for the efficient transport of a media data stream. Normally, the transport and the service quality achieved is monitored with the help of the RTCP (RTP Control Protocol). Although the protocol standard intends the TCP protocol for actual data transport, in practical application most of the time the unreliable but faster, UDP protocol is implemented. This is done to avoid inherent waiting times during connection management and error correction.

2.4 Glossary

Authentication: Serves to give proof of a user's identity. Certificates of a trustworthy authority are used to check identity in carrying out authentication. To verify the integrity of a message, digital signatures are created and transmitted with it.

Broadcast: A broadcast transmission is a transmission conducted simultaneously from one point to all participants. Classical broadcast applications are radio and television.

Circuit switching: Method of information exchange via a network. At the beginning of the information exchange an exclusive connection between the communicating terminals is established and remains in effect for the duration of the communication. Analog telephone networks, for example, work according to this principle.

Client: Designates a program that contacts a server and requests information from it. The browser employed in the WWW is in this sense a client. But there are also other clients in the WWW that contact WWW servers and download information from it, e.g., search engines or agents.

Client/Server-Architecture: An application is carried out as a collaborative effort over a network of connected, multiple computers. The server provides specific services, the client conversely requests services. Except for the actions of placing an order and responding to it, the components are independent of each other. Interfaces and the way of communicating in placing orders and responding are explicitly defined.

Communication protocol: A communication protocol (also simply protocol) is a collection of rules and regulations that determines the data format of the information to be transferred as well as the mechanisms and procedures concerning their transmission. Protocols contain agreements about the establishment and termination of a connection between the communication partners and methods of data transmission.

Computer network: A computer network (**network**) offers autonomic computing systems, which each have their own storage, periphery and computational ability, the infrastructure for data exchange. All of the subscribers are linked with each other via the computer network, therefore each has the possibility to get into contact with every other network participant.

Connection-oriented/connectionless service: A basic distinction is made between **connection-oriented** and **connectionless** services on the Internet. Before the start of the actual data transmission, connection-oriented services must establish a connection over predetermined switching locations in the network. This specified connection route is used for the duration of the entire communication. Connectionless services do not choose a fixed connection route in advance. The transmitted data packets are each, independent of one another, transmitted in potentially different ways via the Internet.

Cryptography: A branch of information technology and mathematics that is involved with the construction and evaluation of encryption. The objective of cryptography is to prevent unauthorized third parties from gaining access to confidential information.

Diffusion network (broadcast network):   In a diffusion network the signal of a transmitter is received by all the computers connected in the network, with the respective time delay taken into account. Every receiver must itself determine if the message is intended for it and whether or to process it.

Flow control: Method to ensure an even and as continuous as possible data transmission between network terminals that do not work synchronously. Flow control intercedes to regulate the transmission sequence of network terminals and to slow down transmission power. This takes effect when congestion situations occur along the path to the receiver in order to avoid potential data loss.

Fragmentation/defragmentation: Because of technical restrictions, the length of the data packet sending a communication protocol in a packet-sending network is always limited below the application layer. If the message to be sent is larger than the respectively prescribed data packet length, then the message is broken down into single submessages (fragments) corresponding to the required length restrictions. To enable the original message to be put back together again correctly at the receiver after transmission (defragmentation), the fragments are provided with **sequence numbers**. This is necessary as the transmission sequence in the Internet cannot always be guaranteed.

Internetworking: The bridging of multiple, different networks that are separated from each other (LANs, WANs) into one internet. The appropriate switching computers (routers) are needed to do this. They mediate the path of a data packet through the network and ensure a secure delivery. The network appears as a homogeneous, virtual network (internet) to the user.

Internet standard: Because there were many companies and organizations involved in the development of the Internet, it was necessary to create unified protocols and interfaces to simplify the development effort. These took the form of Internet standards and were ratified in a public standardization process. Every user was principally allowed to make suggestion for future standards (Request for Comment, RFC), and thereby to steer the course of the Internet.

Internet Protocol (IP):   Protocol on the network layer of the TCP/IP reference model, more precisely called **IPv4**. As one of the pillars of the Internet, IP ensures that the global Internet, consisting of many heterogeneous, individual networks appears as a unified homogeneous network. A standardized addressing scheme (**IP addresses**) enables word-wide unambiguous computer identification. IP additionally provides a **connectionless, packet switching datagram service** that cannot fulfill quality of service guarantees, but always works according to the **best effort** principle. For the communication of control information and error notification, the **ICMP** protocol is an integral component of the IP.

ISO/OSI Reference Model: A specification of the ISO that was designed and made public as the basis for the development of communication standards. It is an international reference model for data transmission consisting of seven layers. The ISO/OSI reference model has the goal of enabling different computer and protocol worlds to communicate with each other. In contrast to the TCP/IP reference model, the protocol standard underlying the Internet, the ISO/OSI reference model has become increasingly less important.

Layer model: Complex problems allow themselves be broken down hierarchically into subproblems, built one on top of the other. The resulting layering of individual subproblems makes the modeling of the problem as a whole easier. The abstraction level increases on each individual layer. Therefore, a layer located higher in the layer model is from detail

problems handled on a lower layer. Layer models play an important role in communication technology, but also in other areas of information technology. A further representation
may be seen in the corresponding **shell model**. Instead of hierarchical layers its structure
is composed of individual shells.

Local Area Network (LAN):   A spatially limited computer network that can only accommodate a limited number of terminals (computers). A LAN enables an efficient and equal
communication for all the connected end systems. As a rule, the connected computers
share a common transmission medium.

Multicasting: One source transmits simultaneously to a group of receivers in a multicast
transmission. This is a 1:n-communication. Multicast is often used for the transmission
of multimedia data.

Network application: An application program whose process includes the access to resources that are not available locally on the exporting computer, but rather on a remote
computer across the network.

Overload (congestion):   With its means of operation (transmission media, router, and
other intermediate systems) a network is able to manage a specific load (communication,
data transmission). If the load created in the network nears 100% of the available
capacity, an overload (congestion) occurs. The network must react in an appropriate
way to avoid data loss and the breakdown of communication.

Packet header: In a packet switching network, the communication protocols implemented require the fragmentation of the information to be transmitted into individual data
packets. In order to ensure that the data packets reach the designated receiver in the
correct form, and can be reassembled into their original information, command and
control information is added to the data packet in a so-called data packet header.

Packet switching   The primary communication method in digital networks. The message
is broken down into individual data packets of a fixed sized. The packets are then sent
individually and independently of each other from the transmitter, over any existing switching centers, to the receiver. A distinction is made between **connection-oriented** and
connectionless (datagram network) packet switching networks. In connection-oriented
packet switching networks, a connection is established over a fixed packet switching
center established in the network before the start of the actual data transmission. Conversely, in connectionless networks there is no predetermined connection path.

Protocol stack: The various subproblems of network communication are each handled by
special protocols. These must all work together smoothly to solve the problem of network
communication as a whole. In order to guarantee the functioning of this interplay, the
development of network protocol software is seen as a comprehensive task. To solve it,
an accompanying **family of protocols** (protocol suite) was developed that addresses each
subtask and integrates them efficiently with each other. The entire problem of network
communication may be represented with the help of a **layer model**. As the individual
protocols of a protocol family are each assigned to a specific layer, the term **protocol
stack** is used. The most well-known protocol stacks are the TCP/IP protocol suites
of the Internet and the ISO/OSI layer model, which often serves as an instructional
example.

Quality of service: Quantifies the performance of a service offered by a communication
system. It is described by means of the performance of service, quality attributes, performance fluctuation, reliability and security, which in each case are specified via individual,
quantifiable service quality parameters.

Reference model: An abstract model that serves as the basis for deriving more specialized
models or concrete implementations. Reference models are often used as general objects
of comparison with other models describing the same technical concept. In the area of
computer networks there exist two well-known reference models. The ISO/OSI reference

model, which is primarily used today for didactic purposes, and the model actually implemented in the Internet: the TCP/IP reference model.

Request for Comments (RFC): New technologies pertaining to the Internet under discussion by experts are recorded in so-called RFCs. In the course of the Internet standardization process there evolved a collection of consecutively numbered documents where technologies, standards and miscellaneous information connected to the Internet were documented and standardized.

Router: A switching computer that is capable of connecting two or more subnets with each other. Routers work in the transport layer (IP layer) of the network and are able to forward arriving data packets along the shortest route through the network based on their destination address.

Routing: Along the path of a WAN there are often multiple switching elements between transmitter and receiver. These carry out mediation of the transmitted data to the respective receiver. The determination of the correct path from transmitter to receiver is called routing. The dedicated switching centers (**routers**) receive a transmitted data packet, evaluate its address information and forward it correspondingly to the designated receiver.

Server: Describes a process that clients request in order to receive information or be provided with resources. The computer on which a server process runs is often known as the server.

Security: In network technology the term security encompasses different security objectives (quality of service parameters), describing the degree of integrity and authenticity of the transmitted data. Among the most important goals of security are: **confidentiality** (no unauthorized third party being able to eavesdrop on data communication between the transmitter and the receiver), **integrity** (ensuring accuracy of the received data), **authentication** (guarantee of the identity of the communication partner), **liability** (legally binding proof of a completed communication) and **availability** (guarantee that an offer of service is in fact available).

Transmission Control Protocol (TCP): Protocol standard on the transport layer of the TCP/IP Reference Model. TCP provides a reliable, connection-oriented transport service upon which many Internet applications are based.

TCP/IP Reference Model (also TCP/IP protocol suite, TCP/IP communications model): Designates a communication layer model for the Internet. The TCP/IP reference model is divided into 5 protocol layers and enables different computers and protocol worlds to communicate with each other via standardized interfaces on the Internet.

Topology: The topology of a computer network is understood as the geometric form of the distribution of individual computer nodes in the network. Widespread topologies for computer networks are the **bus topology**, **ring topology** and **star topology**.

Wide Area Network (WAN): A freely scalable computer network that is not limited by spatial or capacity restrictions. Individual subnets are connected with each other by switching systems (routers), which coordinate data transfer in the WAN. The WAN technology supplies the foundation for **internetworking**.

Chapter 3
Physical Layer

"The visible comprises the basis for the knowledge of that which is invisible."

– Anaxagoras, (499 – 427 B.C.)

Every output, transfer and reception of information takes place by way of a physical medium, meaning a carrier of this information. Also on the Internet, the transfer of information is conducted via a medium of physical communication. These range from traditional electric conductors, simple cables and fiber-optic cables (fiberglass cables) to conductor-independent forms, whereby electromagnetic waves in different frequencies serve as information carriers. But how does the information to be transferred, digitally coded as a series of "zeros" and "ones," actually arrive at the physical communications medium? Before taking a detailed look in this chapter at the different wired and wireless media of transfer, the theoretical foundation of physical data transfer will be explained, whereby so-called modulation procedures and multiplex procedures stand at the forefront of how this binary information is put into a physically transferable "form," thus enabling an efficient data transfer, also over shared media.

The protocols on the single layers of the TCP/IP reference model of the Internet are always based on a physical transfer medium. For this reason, there is often talk of the so-called physical layer, which is in itself not a component of the actual TCP/IP-reference model. Together with the 4 layers of the TCP/IP reference model it forms the hybrid TCP/IP reference model.

Generally, all of the physical and technical characteristics of the physical medium to be used for data transfer are defined in the physical layer. In the forefront stands the interchange between the network hardware and the physical transmission medium. It is necessary to determine the layout of the connectors with their respective electrical or optical parameters, the specifications of the physical properties of the cables (electrical and optical), as well as the specifications of the amplification elements, network adapters and the data transmission procedures implemented.

The basis of every communication is **signal transmission**, i.e., the transport of signals via an appropriate medium of transmission that further conveys this signal

C. Meinel and H. Sack, *Internetworking*, X.media.publishing,
DOI: 10.1007/978-3-642-35392-5_3, © Springer-Verlag Berlin Heidelberg 2013

over a spatial distance. The sender uses a signal source for the activation of a transmission channel, in the medium of transmission, and to transport the message that follows, while the receiver at the other end of the transmission medium has a corresponding signal sink for the reception of the transmitted signal (cf. Fig.3.1).

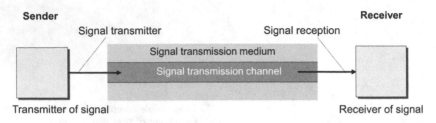

Fig. 3.1 Signal transmission over a spatial distance using a suitable medium.

The physical layer receives a binary data stream of "zeros" and "ones" from the layer above it, which must then be translated (cf. also Fig. 3.3). This conversion takes place with the aid of different **modulation processes** through which the variations of single or multiple signal parameter information is "imprinted" (cf. Fig.3.2). The processes are distinguished by the variety of variable signal parameters, the respective strength in the level of variation and their combinations with each other, whereby different amounts of information may be transferred at the same time. On the receiver's side these signals have to be translated back again by the physical transmission medium into a stream of binary information with the help of a **demodulation process**.

Therefore, many users share a common communication medium on the Internet, and processes for an efficient delegation of resources must be employed. The logical part of solving this task rests on the upper protocol layers, while conditions for the common use of the physical communication medium have to be created on the physical layer. Additionally, specific areas of the varying parameters, such as time or frequency, or their respective combination, are assigned to the individual users so that during transmission there is no mutual interference. Depending on the physical communication medium and data transmission process selected, different processes of **multiplexing** and **demultiplexing** come into play.

We distinguish between **wired** and **wireless media** as the physical transmission media for data traffic. In wired transmission, media signals sent out with the help of electromagnetic waves are dispersed along a solid medium. One of the simplest examples of this is a copper cable used for the transmission of electrical signals.

Here, the material property and construction principle of the cable are determining factors for the desired transmission. In the following sections, a distinction is made between variations of so-called twisted pair copper cables and coaxial cables. With light wave conductors we similarly distinguish between monomodal and multimodal fiber optic cables. Monomodal conductors, in contrast to multimodal, are limited to the use of a single frequency.

Modulation

The range of the signal transported over a special transmission medium does not only depend on the respective physical properties of the medium, but is frequently subject to regulatory constraints as well. Therefore, the frequency range used for transmission may often only be used in a limited way. With the help of a line code a digital signal for transmission may be directly adapted to a transmission channel. An even more efficient utilization of an available transmission channel can be achieved if one uses a signal especially adapted for this transmission channel – the so-called **carrier signal** – and imprints or "modulates" it to the information to be transmitted – the **original signal** – using variations of the signal parameters in the framework of the chosen transmission channel.

In the most elementary case, modulation involves a simple shift in the frequency of one original signal to another. Otherwise, with the help of a complex and modern modulation process it is often possible to achieve a nearly optimal adaption of the spectrum of an original signal to the available transmissions channel. In the process, the original signal always increases the frequency bandwidth of the carrier signal. In order for the receiver to recover the original information from the transmitted signal, a **demodulation** takes place.

A distinction is made between a time and value continuous **analog modulation process** and a time and value discrete **digital modulation process** (called *shift keying*).

Further reading:

Haykin, S., Moher, M.: An Introduction to Digital and Analog Communications, 2nd. ed., John Wiley and Sons, USA (2006)

Fig. 3.2 Modulation

The signals of non-wireless media are transmitted through space by way of different frequency ranges within the electromagnetic spectrum. Data transmission via mobile communication or wireless LAN is especially popular today. In relation to the frequency range of the electromagnetic spectrum used, very different processes are employed. In the following, we shall discuss radio wave transmission via short and ultra-short wave, microwaves and infrared, as well as data transmission via laser.

3.1 Theoretical Principles

With wired as well as wireless transmission media, the transportation of information takes place with the help of **electromagnetic waves**, which are dispersed either along a cable (wired) or freely into space (wireless). Thereby, a connection exists between the amount of information that can be transmitted with the help of electromagnetic waves and the respective transmission range. Depending on the transmission medium, different frequency ranges of the electromagnetic spectrum can be used for data transmission. These are always limited by a **maximum frequency**, which in turn normally determines the so-called **bandwidth** of the signal used for transmission.

Physical Signals

A **signal** is understood as the information-carrying temporal process of measurable quantity within a physical system. Thereby, one distinguishes between signals whose value takes a continuous course (**time-continuous signals**) and those whose value only changes in discrete intervals (**time-discrete signals**). If the measurable quantity can only take on a finite number of values, then we speak of a **value discrete signal**, otherwise of a **value continuous signal**. If there are only two possible measures then a **binary signal** is evident.

A signal that is both time and value discrete is referred to as a **digital signal**. A signal whose information-bearing measurable quantity is able to take on any arbitrary value is called an **analog signal**.

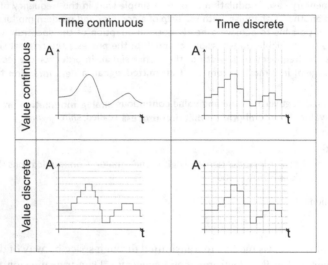

A distinction is made between **spatial-variant signals**, whose signal parameters can be represented as a function of space (e.g., data storage), and **time-varying signals**, whose signal parameters can change value over time (represented thus as a function over time) and are used for communication. Every spatial-variant signal can be transferred into a time-varying signal and vice versa. For example, stored data (spatial-variant signals) can be read and sent through a network as a time-varying signal to a receiver.

Signals can be understood as a physical representation of data by way of characteristic spatial and/or time variable values of physical measurements. With them it is possible to represent abstract data, e.g., logical values, in the physical (real) world.

With regard to their ability to represent data, signals can be divided into information-bearing **original signals** or **interference signals**. Interference signals originate from either natural sources such as atmospheric or static disturbances (noise), or have a technical origin, e.g., the crosstalk of signals or impulse-like disturbances through technical devices.

Further reading:

Shannon, C. E., Weaver, W.: The Mathematical Theory of Communication, University of Illinois Press, Urbana, Illinois (1949)

Shu, H. P.: Schaum's Theory and Problems: Signals and Systems. Schaum's Outline Series, McGraw-Hill, Inc., USA (1995)

Fig. 3.3 Physical signals and signal classes.

Fundamental Terms from the Theory of Signal Transmission

- **Bandwidth**
  The term bandwidth defines a given physical measurement in hertz (1 Hz = 1/s), and is used in physics, telecommunications, and in computer science in different meanings. Seen physically, bandwidth B designates the difference between two frequencies: f_1 (the lower cutoff frequency) and f_2 (the upper cutoff frequency), which form a continuously connected frequency range (frequency band), $B=f_2-f_1$.

  In analog telecommunications, bandwidth designates the frequency range in which electrical signals with an amplitude decrease of up to 3 dB can be transmitted. The larger the bandwidth, the more information can theoretically be transmitted in one unit of time. The term bandwidth is also used in computer science, whereby here the **transmission rate** (also called the data rate) by which digital signals are transmitted is meant. It is used as a measurement for the speed at which data, in the form of bits per time unit, can be sent via a transmission medium. There is, however, a direct connection between the bandwidth and the transmission rate, meaning that in data transfer the transmission speed reached depends directly on the bandwidth of the transmission medium. The maximum bandwidth utilization for binary signal is 2 bits pre hertz.

- **Dynamic**
  The dynamic defines the range between the maximum and minimum signal level (signal value). It is often measured as a logarithmic value and is of importance in determining the susceptibility of signals to interference.

- **Modulation**
  In telecommunication, modulation describes a process by which an original signal (data) to be transmitted changes a carrier signal and thereby the making possible the transmission of the original signal via the normally higher frequency carrier signal.

- **Multiplexing**
  Methods of the signal and telecommunication transmission are identified as a multiplex process when several signals are combined (bundled) and simultaneously transmitted over a medium.

- **Signal parameters**
  Identified as signal parameters are those physical parameters of a signal, whose value or value parameters, respectively, represent the actual data itself that is transmitted with the help of a signal. Typical signal parameters are the **frequency** (the number of oscillations per time unit), the **amplitude** (signal strength) and the **phase** (temporal shift) of the signal .

- **Signal level**
  The signal level is the ratio from a measured signal value and a reference value within a transmission system. Levels are often represented in a logarithmic standard, in order that the representation of extremely large dynamic areas can be represented in a manageable number range.

Further reading:

Ibbotson, L.: The Fundamentals of Signal Transmission: Optical Fibre, Waveguides and Free Space, Butterworth-Heinemann (1998)

Fig. 3.4 Selected fundamental terms from the theory of signal transmission.

The easiest way to transmit binary information via an electrical conductor consists of coding "ones" and "zeros" in regular intervals as "power on" and "power off" impulses. (cf. Fig. 3.5). The voltage fluctuations occurring in the specified interval pattern results in a **square-wave signal**, which can be represented as a superimposition of an infinite number of many single fluctuations of different frequency and amplitude. (cf. Chap. 3.1.2).

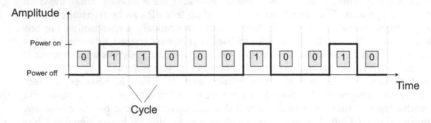

Fig. 3.5 A simulated digital transmission on an electrical conductor via two different amplitudes (power on / power off) results in a square-wave oscillation.

But even this simple operation reveals several problems. No signal can be transported without the loss of energy. Based on the physical characteristics of each transmission medium, this loss of energy appears in different frequency ranges in different levels of strength. This means that independent of the material properties of the conductor, as the signal path increases the signal becomes continually more distorted until the receiver is no longer able to recognize (decode) it. Additionally, the signal can be distorted through outside influences (signal noise). But even a perfect, noise-free transmission channel has only a limited transmission capacity (cf. Section 3.1.3). This means that not all of the frequencies necessary to correctly send a square-wave signal can be transmitted.
A solution to this problem is found in the use of **narrow band signals**, that is, signals represented by the lowest possible (ideally only a single) frequency. In this way, the bandwidth limitation and signal distortion have only a minimum interference on the signal transmission. In order to still be able to transmit information with the help of this narrow band signal, other signal parameters are "modulated," and signal strength, signal phase or signal frequency varied with the help of an appropriate encoding method.

3.1.1 Electromagnetic Spectrum and Signal Transmission

The transmission of information from one point to another is carried out via a communication medium. Normally, the information to be transmitted must be encoded, this means brought into a form that allows itself to be transported by way of the communication medium. This can occur in completely different ways, e.g., the transport of a letter with the postal service, the transport of sound waves (audio information) through the air, as well as the transport of electromagnetic waves along a conductor or through space.

Attempts to transmit information (data) with the help of electromagnetic waves were conducted as early as the 18th century (cf. Fig.3.6). This method distinguishes itself by its virtually instantaneous information transmission. This means that the information-bearing signals are dispersed in space at the speed of light, and in electrical conductors nearly at the speed of light. That this transmission is, however, not completely instantaneous can be evidenced in satellite-based communication, where perceptible delays occur as correspondingly large distances (often covering several relay points) must be bridged.

Electromagnetic waves are identified as those waves made up of coupled electric and magnetic fields. A time-varying electric field always creates a magnetic field, just as a time-varying magnetic field simultaneously creates an electric field. The one does not occur without the other, even if there is no carrier. (cf. Fig. 3.7).

Among the types of electromagnetic waves are: radio waves, microwaves, infrared radiation, visible light, UV radiation, as well as x-rays and gamma rays. These are components of the continuous **electromagnetic spectrum**. The noted manifestations are the sole result of the designated frequency changing properties of the radiation, respective of its origin and use. The only type of electromagnetic waves that can be perceived by humans, without relying on outside aid, is visible light, whose rays possess a wave length of between 380 nm and 780 nm. Fig. 3.8 shows the electromagnetic spectrum, the assigned types of electromagnetic waves with frequency and wave length, as well as their application in wired and wireless media.

The electromagnetic spectrum begins with **static electromagnetic fields**. This includes, e.g., the terrestrial magnetic field, which is virtually constant apart from small fluctuations in the 24-hour rhythm and long-term changes.

Directly connected to this is the **low frequency range** (Very Low Frequency, VLF), which ranges from the lowest frequency to approx. 30 kHz. This includes the electrical railway network (16.7 Hz) and the European alternating currency (50 Hz). The range of radio waves begins at 10 kHz. For example, those in the range of 10 kHz to 30 kHz are used by navy radio installations (VLF signals can penetrate several meters under water and are used in the communication of submerged submarines). Additionally, VLF signals are used for radio communication in mines, radio navigation and for the transmission of radio signals. Very low frequency transmitters are often huge installations, made up of multiple masts towering more than 100 meters high and covering an area of several square kilometers.

The **high frequency range** extends from over 30 kHz to 300 GHz. It is used particularly in communication and broadcasting (low frequency, medium frequency, high frequency, very high frequency, and ultra high frequency), as well as in tracking (radar). The frequency range between 30 kHz and 300 MHz, designated radio waves, is used for the radio broadcasting of audio signals as well as video signals. The characteristics of radio waves depend on their respective frequencies. When the frequency is low, radio waves can easily penetrate obstacles, however, with increasing distance the radiation energy drops considerably.

Electricity and Information Transmission

The idea of using electricity for telecommunication dates back to the 18th century. In the past, electrical phenomena were often dismissed as curiosities or parlor games, despite the fact that the Greek philosopher *Thales of Miletus* (approx. 640 – 546 B.C.) had already described the attractive force of static electricity.

In 1730 British physicist *Stephen Gray* (1666-1736) succeeded in proving that electricity can be propagated along a metallic wire. Thus, the idea of electric information transmission was born. In 1800 Italian physicist *Alessandro Volta* (1745–1827) developed the voltaic pile, the first constant source of electricity, while in 1820 French mathematician and physicist *André Marie Ampère* developed the principle of the electromagnetic needle telegraph, based on the work of Danish physicist and chemist *Hans Christian Oersteds* (1777–1851). It was first put into practical application as the pointer telegraph in 1833 by *Carl Friedrich Gauss* (1777–1855) and *Wilhelm Weber* (1804–1891). But it was not until 1837 that *Samuel Morse* (1791–1872) achieved a commercial breakthrough with the writing telegraph, thus signaling the beginning of the era of the electrical telegraph.

Different from the electric telegraph, in which a coded letter is made up of a sequence of short and long (although otherwise homogeneous) signal impulses, the transmission of complex acoustical information, e.g., the human voice, was considerably more difficult. First the problem of transforming sound waves into electrical voltage fluctuations would have to be solved. Building on the work of English physicist *Michael Faraday* (1791-1867) in the area of **electromagnetic induction**, *Alexander Graham Bell* (1848–1922) succeeded a half century later, when he developed the telephone in 1876. For the first time, the human voice could be transmitted over a large distance almost instantaneously. Unlike the telegraph, now a whole spectrum of different electric frequencies had to be transmitted via an electric conductor. Different frequencies are, however, damped to varying degrees, and as the cable length becomes progressively longer, the output signal becomes more and more distorted until it can no longer be recognized. For this reason, a corresponding robust signal amplifier had to first be developed for long-rang communication, and therefore the telephone was only first able to see its breakthrough in the 20th century.

Electromagnetic waves are not only able to disperse along a conductor, but also in space. In 1865, the Scottish physicist *James Clerk Maxwell* (1831–1879) was the first to postulate the existence of **electromagnetic waves**, which could then be proven experimentally in 1885 by *Heinrich Hertz* (1857–1895). The Italian engineer *Guglielmo Marconi* (1874–1937) combined the results of Heinrich Hertz (high frequency generator, sender), *Alexander Popow* (1858–1906, antenna and relays) and *Eduard Branly* (1846–1940, coherer for the transformation of electromagnetic waves into electric impulses, receiver) into a complete **wireless telegraphy system**, which he had patented in 1896.

Further reading:

Meinel, Ch., Sack, H.: Digital Communication – Networking, Multimedia, Security, Springer, (2013)

Fig. 3.6 Electricity and information transmission.

Electromagnetic Waves

Scottish physicist James Clerk Maxwell unified all essential laws of electricity and magnetism in 1860 in the so-called **Maxwell's equation**, together with those laws in which the existence of electromagnetic waves could be evidenced. An electric field that changes over time always creates a magnetic field, just as, conversely, a magnetic field that changes over time always creates an electric field. Through the periodically changing fields, a progressive electromagnetic wave is created. Unlike, e.g., sound waves, these waves need no transmission medium to propagate. They enter into free space as **transverse waves** and propagate in a vacuum at the speed of light. Thereby both vector fields of the electrical field (E) and the magnetic field (B) stand perpendicular to one another and have a fixed dimension (wave impedance).

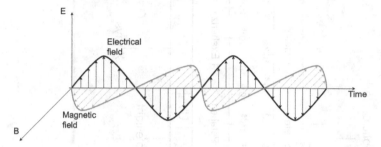

The **wavelength** λ of the electromagnetic wave is measured as a mechanical wave, from one wave tip to the next, between two points t_1, t_2 of the same phase. Thereby, two points have the same phase if, in their temporal sequence, they have the same deflection (amplitude) and the same direction of motion. The wave length λ therefore corresponds to the temporal equivalent of the cycle duration of the wave $\delta t = t_2 - -t_1$. The reciprocal value of the length of the period $1/\delta t$ is designated as frequency f. Thus, between the wave length λ, the velocity of the propagation of the wave c and its frequency f exists the following connection:

$$\lambda = \frac{c}{f}.$$

The propagation velocity c of electromagnetic waves is expressed as:

$$c = \frac{1}{\sqrt{\mu_0 \cdot \varepsilon_0} \cdot \sqrt{\mu_r \cdot \varepsilon_r}},$$

whereby ε_0 designates the electrical field constants and μ_0 the magnetic field constants; the dielectric constant ε_r and the permeability μ_r depend on the material of the conductor (medium) in which the wave propagates. For the vacuum, i.e., the propagation of the electromagnetic waves in space, applies: $\varepsilon_r = \mu_r = 1$. The electromagnetic waves are refracted in the transmission medium depending on their frequency and (contingent on the medium) on their polarization and propagation velocity.

Further reading:

Tipler, P., Mosca, G.: Physics for Scientists and Engineers, 6th ed., Palgrave Macmillan (2007)

Fig. 3.7 Electromagnetic waves.

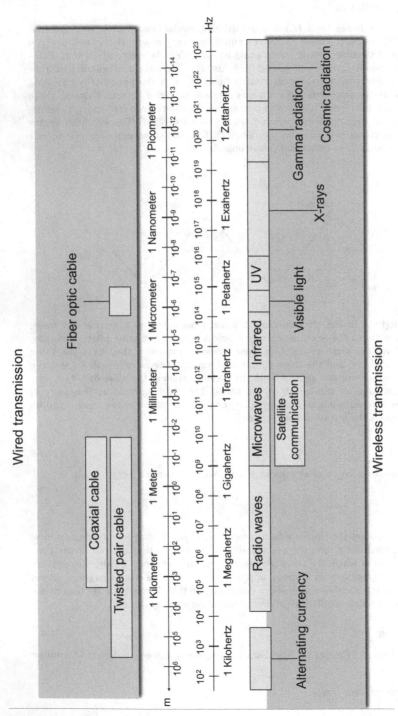

Fig. 3.8 The electromagnetic spectrum.

The low frequency range (LF) encompasses wave lengths in the range of 1,000 – 10,000 meters, i.e., frequencies between 30 kHz and 300 kHz. This frequency range is used by broadcasting stations, radio beacons (Non Directional Beacon, NDB) and radio navigation systems.

Following this is the medium frequency range (MF) with wave lengths of 100 – 1,000 meters, i.e., frequencies between 300 kHz and 3,000 kHz. In contrast to the low frequency transmissions – which spread out as a linear ground wave following the curvature of the earth – they are reflected in the ionosphere (skywave) and are therefore especially well suited for the transmission of radio signals. Application fields for medium frequency signals are in particular radio, amateur radio and maritime radio.

The high frequency range of the short wave (HF) encompasses wave lengths between 10 and 100 meters, i.e., frequencies between 3 MHz and 30 MHz. Because of their good reflection properties, high frequency signals can be received worldwide. While the ground wave has a scope of between only 30 km and 100 km, the skywave is reflected in the various layers of the ionosphere (that is, in approx. 70 km to 400 km altitude) and returns to the earth's surface where – if conditions are favorable – it is reflected again. In this way, distances of several thousand kilometers can be bridged. High frequency plays an important role primarily in the area of broadcasting and worldwide communication networks, e.g., in amateur radio.

The very high frequency range (VHF) encompasses frequencies between 30 MHz and 300 MHz. In contrast to high frequency, the scope of very high frequency is limited due to the so-called radio horizon. Rather than being reflected in the ionosphere, the VHF signal propagates directly in a straight line. This means that a line of sight must exist between the sender and the receiver. Depending on the location of the sender and receiver, the transmission range of ultra-short wave radiation is between 10 and approx. 200 km. VHF has a great importance for radio and television signals but also for aircraft radio and radio navigation.

Above the VHF range in the frequency spectrum are microwaves. These cover the range between 300 MHz and 300 GHz. The lowest frequency range is held by decimeter waves (Ultra High Frequency, UHF), whose frequencies lie between 300 MHz and 3 GHz. Decimeter waves are used for terrestrial television broadcasting, mobile communication, as well as for wireless LAN (WLAN) and radar. In addition, decimeter waves are also utilized in microwave ovens. In the area over 1 GHz, waves proceed in straight lines and can therefore be bundled closely together, enabling a higher signal-to-noise ratio to be reached. Here, sender and receiver antennas must be aligned exactly to one another.

Centimeter waves (Super High Frequency, SHF) have a wave length in the range between 1 cm – 10 cm, this means that the frequency band lies in the area between 3 GHz and 30 GHz. Centimeter waves also belong to the category of microwaves and are used in radio relay, radar, and television broadcasting.

Millimeter waves (Extreme High Frequency, EHF) are microwaves with wave lengths in the millimeter range, thus between 1 mm and 10 mm, which is equivalent to a frequency band of 30 GHz to 300 GHz. Millimeter waves are utilized in military radar, building surveillance and in cruise control systems to enable distance control between vehicles.

As the frequency rises, terahertz radiation is evident, and, finally, infrared radiation, which is located slightly below the spectrum range visible to the human eye. This range of the spectrum, extending to the beginning of the ultraviolet-range (UV), is also known as **non-ionizing radiation**. This is because, though numbering among electromagnetic radiation waves, it does not have the capacity to force electrons out of atoms or molecules or, in other words, to ionize them.

Terahertz radiation (submillimeter waves) exhibits wave lengths of between 1 mm and 100 µm, and therefore covers the frequency range between 300 GHz and 3 THz. Terahertz radiation – which is sometimes classified with the more distant infrared – was not used, or barely used, in the past, thus it was common to speak of a terahertz gap in the electromagnetic spectrum. While being able to penetrate many materials as well as biological tissues, terahertz radiation, in contrast to x-rays or gamma rays, does not have an ionizing effect. Currently, terahertz radiation is utilized in spectroscopy, material testing, and safety engineering.

Infrared radiation (IR) follows terahertz radiation with wave lengths of between 100 µm and 780 nm. Here, subdivisions are made further into: near IR (NIR), short-wavelength IR (SWIR), mid-wavelength IR (MWIR), long-wavelength IR (LWIR), and far IR (FIR). Infrared radiation is often popularly equated with heat radiation, even though it is actually the complete frequency range between microwave radiation and visible light that contributes to heat radiation. Infrared radiation is utilized in heat generation, infrared remote control or in the infrared interface of computers. In signal transmission, the use of NIR is preferred with fiber optic cables because of its low absorption and dispersion.

The frequency range for light visible to the human eye covers wave lengths of between 750 nm and 380 nm, representing only a tiny fraction of the entire electromagnetic spectrum.

Next in the spectrum of visible light is the area of ultraviolet radiation (UV), in which wave lengths span from 380 nm to 1 nm. According to DIN 5031, Part 7, the radiation in the ultraviolet range is subdivided into 3 frequency bands UV-A, UV-B and UV-C. Ultraviolet radiation is utilized in disinfection, spectroscopy, fluorescence testing, and lithography.

Ionizing radiation first begins the group of energy-rich ultraviolet radiation (UV-C), which includes x-rays, gamma rays, high-altitude radiation, and cosmic radiation. The frequency range of x-rays begins below extreme ultraviolet radiation, extending over wave lengths from 10 nm to 1 pm. Overlapping occurs in the areas of hard x-rays and the adjoining gamma rays. The difference between gamma and x-rays in the same frequency range is related to their origin. Whereas gamma rays originate in the processes of the atomic nucleus, x-rays evolve from high-energy processes in the electron shell of the atom. X-rays are used in medicine, material physics, as well as in chemistry and biochemistry. Gamma radiation evolves from the radio-active decay within the atomic nucleus and encompasses the entire frequency range beyond x-radiation. Gamma radiation is used in medicine as well, and also in sensors and material testing. High-energy gamma radiation is identified as high-altitude radiation or cosmic radiation, which originates in space.

3.1.2 Bandwidth-restricted Signals

With the help of electromagnetic waves, signals can be transported along a conductor or in space. These signals are able to carry encoded information that is "imprinted" on the output signal through variations of specific signal parameters. One of these parameters is voltage, or electrical current, the variations of which determine the signal characteristics. By way of a simple single-valued function of time $f(t)$, the characteristics of the signal can be modeled and mathematically analyzed (cf. Fig. 3.9).

Fourier analysis

Any periodic signal may be decomposed into a (potentially infinite) sum of simpler sine and cosine functions. This decomposition is called **Fourier analysis**, after the French mathematician and physicist *Jean-Baptiste Fourier* (1768–1830). If g(t) is an arbitrary periodic function with a period T, then g(t) can be decomposed into

$$g(t) = \frac{1}{2}c + \sum_{i=1}^{\infty} a_i \sin(2\pi i f t) + \sum_{i=1}^{\infty} b_i \cos(2\pi i f t),$$

with f=1/T as fundamental frequency. All summands (frequency components) are integer multiples (harmonic numbers) of this fundamental frequency, and a_i and b_i determine the amplitude of the frequency components. This decomposition is also designated **Fourier series**. If the period T and all the amplitudes a_i and b_i are known, then it is possible to reconstruct the original function.

Normally, data signals are limited in time and therefore not true periodic signals. This problem lets itself be remedied easily by repeating, as often as desired, the time-limited signal pattern at hand.

The amplitude values of the sine and cosine terms a_i and b_i let themselves be calculated for every function g(t) in the same way as the constant c in the above equation as

$$a_i = \frac{2}{T}\int_0^T g(t)\sin(2\pi i f t)dt, \quad b_i = \frac{2}{T}\int_0^T g(t)\cos(2\pi i f t)dt, \quad c = \frac{2}{T}\int_0^T g(t)dt \; .$$

Further reading:

Tanenbaum, A. S.: Computer Networks, Prentice-Hall, Inc., Upper Saddle River, NJ, USA (1996).

Fig. 3.9 Fourier analysis

If a bit string encoded message, as shown in Fig. 3.5, is to be transmitted as a series of synchronized voltage fluctuations (square-wave signals), the single components that make up the resulting function can be determined with the help of the Fourier coefficients. An example of this is seen in Fig. 3.10.

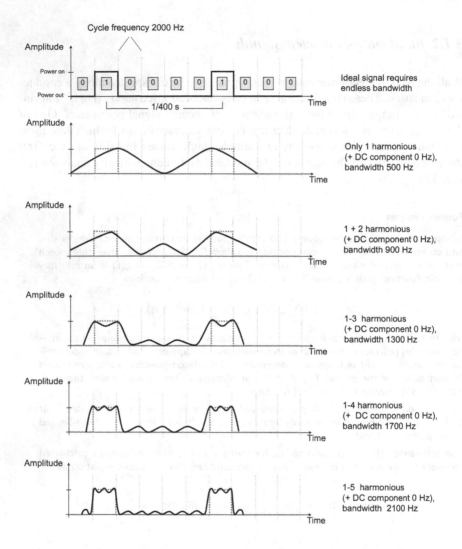

Fig. 3.10 A minimum bandwidth is necessary for the transmission of a bit string with a predetermined step frequency.

A simple bit string (0100001000) should be encoded with the help of signals in a basic cycle of 2000 pulses per second (2000 Hz). This means both voltage peaks of the bit string have a distance of 5/2000 seconds. To give an exact representation of this function as a Fourier series we would need an infinite number of a_i, b_i coefficients, which with a rising index i, represent i-times the fundamental frequency (so-called *i*. harmonics or overtones). Every physical transmission medium has only a limited bandwidth; therefore, a transmission media is only able to transmit a limited frequency band. For this reason, it is not possible to ever transmit all the necessary harmonics. The highest transferable frequency is determined by the material properties (bandwidth) of the transmission medium. The signals to be transferred must therefore be adapted to the transmission characteristics of the medium.

With every signal transmission, energy is lost as the distance to the transmitter increases. In the case of cable-bound media, there exists a simple logarithmic dependency between signal attenuation and conductor length. However, in the case of wireless media a more complex relation of dependency exists between signal attenuation and conductor length. If all harmonics of the signal were to be effected in the same way, signal strength would simply diminish with increased distance to the transmitter (signal damping, attenuation), meaning that the amplitude would become weaker. The frequency range in which a signal can be transmitted over a transmission medium without significant damping is also designated as bandwidth. However, within a transmission medium different portions of frequency are damped in different strengths, i.e., the single harmonics are damped in differing degrees of strength depending on the material properties of the transmission media (e.g., the construction of the conductor, including thickness and length). Thereby, with increased distance from the sender, the signal becomes distorted beyond recognition (**signal distortion**). A differentiation is made between the just described **damping distortion**, which damps different harmonics of a signal based on the material properties, and **delay distortion**, which influences the output signal based on the signal transmission time to the receiver. Further, the propagation velocity of a signal via a cable-bound transmission medium varies depending on the respective frequency (i.e., transmission velocity is higher in the middle of the spectrum than on the edges). This means that the receiver gets the individual signal components at different points in time. Delay distortion is especially critical in the transmission of discrete (digital) signals, as parts of one signal may overlap another. In order to compensate for damping effects, the signal to be transmitted must have sufficient output (signal strength), in order for the receiver to receive the signal and correctly interpret it. A countermeasure against signal damping when transmitting signals over large distances is found in periodic signal amplification by way of suitable technical signal amplifiers.

In addition to these systematic signal disturbances, further disruptions can appear in the form of **noise** and impulse disturbances. that originate in transient and stochastic processes. A distinction is made between **thermal noise**, released in the transmission medium through the thermal oscillations of molecules and **Intermodulation noise**, resulting from a signal being influenced by other signals in the medium. Disturbances resulting from an unwanted coupling of signal pathways, for example, in the case of two unshielded electrical conductors running directly parallel to each other, are known as **crosstalk**. When they are triggered by irregular impulses, e.g,. lightning or other system irregularities, such disturbances are called **impulse noise** or impulse disturbance. They distinguish themselves by their short duration and high amplitude, respectively.

These disturbances effect signals to such a degree that there is a loss of quality in the transmitted signal. In continuous signals this leads to arbitrary changes in the signal parameters, while in the transmission of discrete or digital signals it may result in transmission errors. For example, signal values to be interpreted as 0 can be transformed into a 1 and vice versa. In order to guarantee correct signal transmission, the strength of the signal must be higher than the strength of the noise. This can also be achieved through suitable signal amplification.

Let us look again at the example showing the transmission of a simple bit string. Fig. 3.10 shows what the signal would look like if the bandwidth could only accommodate the transmission of the lowest frequency (i.e., only the first Fourier coefficient) and how the signal can ever more closely approximate the output signal when the implemented bandwidth is increased.

Despite having a **perfect channel**, meaning a transmission medium that is ideal and disturbance-free, the capacity of how much information can be transmitted is limited. The Sweden-born, American physicist, *Harry Nyquist* (1889–1978), who worked as an engineer at AT&T, already recognized this in 1928 and calculated the maximum data transmission rate of a noise-free channel with a limited bandwidth. Any bandwidth-limited signal with maximum frequency f_{max}=H may be reconstructed exactly through 2H signals (samples) per second (**sampling theorem**, cf. Fig. 3.11).

From the Nyquist sampling theorem it can be determined that the data rate, in other words, the maximum amount of data to be transferred in bits per second (bps), for an ideal and error-free channel, is only limited by the bandwidth of the channel. The maximum data transfer rate for a noise-free channel with a bandwidth H for binary signals is:

$$\text{maximum data rate} = 2H \text{ bps.}$$

If instead of a binary signal, a signal with M discrete steps is chosen (cf. Fig. 3.12), then the capacity increases accordingly :

$$\text{maximum data rate} = 2H \log_2 M \text{ bps.}$$

The higher the data transmission rate – meaning the higher the frequency of cycles and the more signals chosen for encoding the information – the "shorter" or lower the respective voltage differences in representing the transmitted bits. For this reason, disturbances in higher data rates have a greater effect than those in lower rates. Important in this connection is the relation between signal strength and strength of noise, the so-called **signal-to-noise ratio** (SNR). The signal-to-noise-ratio between signal strength S and noise strength N is represented on a logarithmic scale in order to keep the resulting values in easily accessible terms and to show them as **decibels** (dB):

$$\text{Signal-to-noise-ratio} = 10 \log_{10} \frac{S}{N}.$$

Taking this standard into account in the case of disturbance gives the result found in the **Shannon-Hartley theorem** by *Claude E. Shannon* (1916–2001) and *Ralph*

Sampling Theorem

(According to Nyquist (1928), Whittaker (1929), Kotelnikow (1933), Raabe (1939) and Shannon (1949))

A signal function that only contains frequencies in a limited frequency band (band-limited signal), whereby f_{max} is simultaneously the highest occurring signal frequency, is completely defined by its discrete amplitude in the time interval

$$T_0 \leq \frac{1}{2 \cdot f_{max}}$$

This means that the sampling frequency f_A has to be twice as high as the highest frequency that appears in the signal to be sampled f_{max} (Nyquist criterion or Raabe condition):

$$f_A \geq 2 \cdot f_{max}$$

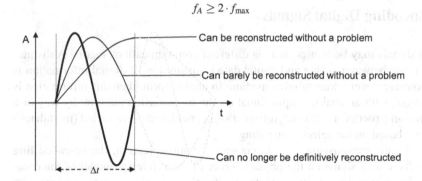

Further reading:

Shannon, C. E.: Communication in the presence of noise. Proceedings of the IRE 37(1), pp. 10-21 (1949), reprint in Proc. IEEE 86(2) (1998)

Fig. 3.11 Sampling theorem.

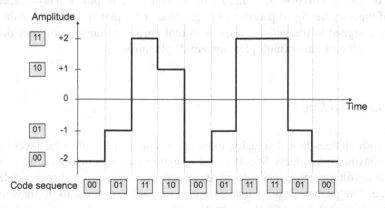

Fig. 3.12 An example of a multiple value digital signal.

Hartley (1888–1970). This concept presents the theoretical maximum limit of data transmission rate for a transmission channel that is dependent on bandwidth and signal-to-noise ratio.

$$\text{maximum data rate} = H \log_2 \left(1 + \frac{S}{N}\right).$$

The Shannon-Hartley theorem provides a theoretical maximum, achieved with a hypothetically optimal channel encoding, without having to provide information on the procedure by which this optimum is reached.

3.2 Encoding Digital Signals

Digital signals may be transported via different communications media. A distinction is made between analog and digital data transmission. If digital information is to be conveyed over a transmission medium in analog form, then this information is "imprinted" with an analog output signal on the transmission medium by way of a **modulation process**. To this end, different signal parameters are varied (modulated) over time based on the selected encoding.

In digital data transmission, e.g., in computer communication, the so-called **line code** is fixed as a signal on the physical layer of the reference model of the communication protocol and translated into specific signal level sequences, to enable its efficient transmission. The physical system here is designated as a "line." In order to transmit information, its state is changed by the sender, just as the receiver is capable of assessing this state in order to get information. To facilitate this, the sender transforms a character into a physical signal, which the receiver assesses and then transforms back again into a character. In the same way, the conditions for the shared use of the transmission medium have to be created on the physical layer. Specific areas of the varying signal parameters (e.g., time or frequency, or their combinations) are assigned to the individual users so that during a transmission they do not interfere with each other (multiplexing/demultiplexing).

3.2.1 Line Coding

A **line code** defines how a signal is to be transmitted on the physical layer in the digital transmission process. Specific level sequences, such as light intensity in fiber optics and voltage or currency on an electrical line, are assigned to a single bit sequence. The primary task of line coding is to form the signal to be transmitted in such a way that it can fit the respective transmission medium as optimally as possible

Line codes display the following important properties:

- **Clock recovery**
  The line code enables recovery of the time cycles underlying data transmission from the signal values. Clock recovery is first necessary if there is no separate clock line available for data transmission. The clock of a line code should always be independent of the contents of the information to be transmitted. Clock recovery is carried out by a permanent, periodical level change of the signal values (self-clocked signal). Thereby, every signal is divided into two phases: a status change taking place at the beginning of the first phase, followed by a further status change at the beginning of the second phase. The receiver can reconstruct the clock from the edges of the signal in one of the two phases. However, twice as many status changes – meaning the doubled bandwidth – as signals are necessary.

- **Direct current component**
  The line code should be as DC free as possible when electrical transmissions take place, as a transmission of DC voltage (signals with low frequency) over long lines is not possible. In most cases, this challenge cannot be filled in an absolute sense, but only on a statistical average.

- **Error dectection**
  Line codes should already enable the detection of transmission errors on the physical signal layer. This is made possible by additional channel encoding and redundancy. [1]

- **Transmission range**
  The transmission range is causally related to signal attenuation. For a metal cable it approximately applies that the damping constant is proportional to the root of the frequency to be transmitted, i.e., higher frequencies are damped more strongly than lower ones. Therefore, in the case of some line codes, the line bandwidth decreases in order to make better use of the transmission network when reduced signal damping occurs. While having the lowest possible frequency, a line code should have the maximum power density, and at the same time possess a low spectral bandwidth.

- **Number of encoded signs**
  More bits can be encoded simultaneously in one signal value (group encoding). This can be achieved, for example, if instead of two signal levels (binary digital signal, binary codes) more signal levels are allowed (multiple digital code). However, as the number of steps increases, the more prone a signal is to disruption.

- **Resynchronization**
  Because there is not a separate clock signal, several bits are combined together for the purpose of synchronization in so-called **frames**. A pure bit synchronization, as is possible with the help of clock recovery, is not sufficient for synchronization and thus for a correct transmission execution. A specific sequence of

---

[1] A detailed presentation of error detection codes may be found in Excursus 2: "Error Detection and Error Correction Codes," in the first volume of this series: Meinel, Ch., Sack, H.: Digital Communication – Networking, Multimedia, Security, Springer (2013).

different signal levels corresponds to a synchronization signal and can be recognized by the receiver. A frame always consists of a sequence of bits, which is limited on both ends by reserved synchronization and control characters.

- **Modulation rate**
  The average number of signal transitions during a bit duration – the so-called modulation rate – sets the necessary bandwidth and because of this should be kept as low as possible.

- **Error recovery**
  It is rare that all transmission errors can be avoided. Therefore, any errors which may occur should have as little effect as possible (ideally none) on the transmission of subsequent communication. Through the generation of frames for synchronization, the beginning of a new transmission can always be identified. Thus, the use of frames enables error recovery, as any data blocks recognized as defective (via a protocol layer on a higher abstract level) can be requested again.

Line codes

Line codes can be divided into the following groups:

- **Binary line codes**
  Binary line codes are distinguished by the use of only two different signal levels, which also directly determine the respective signal value.

- **Biphase line codes**
  In contrast to binary line codes, the signal values in biphase line codes are encoded using (two) phase jumps.

- **Ternary line codes**
  In ternary line codes, both of the logical values 0 and 1 are in three signal values {-1, 0, +1}.

- **Block codes**
  Block codes are combined into line code groups with respective m bits and encoded to a new block with the length n. The general designation of block codes is „mBnX", whereby m designates the number of bits in a binary word that are combined into a block with the length m in the desired form (e.g. ternary encoding, quaternary encoding). The advantage of this type of coding is based on the fact that the clock cycle is reduced by the factor m, whereby the damping coefficient is likewise also reduced, enabling a greater transmission range.

Further reading:

Anttalainen, T.: Introduction to Telecommunication Network Engineering, 2nd ed., Artech House, USA (2003)

Fig. 3.13 Types of line codes.

The following gives a brief overview of the most important techniques of line coding,

- **Non-Return-to-Zero (NRZ)**
  The NRZ code (1B1B coding) is a simple line code for the transmission of bi-

nary codes over a transmission medium, in which the sender varies the condition of the medium between two levels (0,1). The NRZ codes are the simplest line codes in digital transmission technology. The transmitted bits are transferred directly as a voltage level onto the transmission medium. One of the two signal levels can thereby also be represented with the value of 0V. The designation Non-Return-to-Zero refers to the fact that each of the two possible voltage levels carries information and, in contrast to RZ coding, there is no signal level that is not assigned an information value. In order to correctly receive data, NRZ codes need a separate clock signal, which is either transported parallel or obtained from a superordinate channel coding/framework creation, respectively. Generally speaking, NRZ coding always has a direct current component.

The **NRZ-I coding** (Non- Return-to-Zero Inverted) is, in contrast to NRZ coding, a so-called differential line code, meaning that the coding is stateful and the respective signal level is determined in part by previous states. A distinction is made between **NRZ-M** coding (Non-Return-to-Zero Mark), in which the incoming data sequence is not inverted, and **NRZ-S** coding (Non-Return-to-Zero Space) in which inverting is present,

$$\text{NRZ-M: } out_i = in_i \oplus out_{i-1}, \text{ NRZ-S: } out_i = \overline{in_i} \oplus out_{i-1},$$

whereby in_i represents the binary input data sequence, out_i the binary NRZ output data sequence and the operator $\oplus$ the logical XOR operation (cf. Abb. 3.14).

- **Return-to-Zero (RZ)**
  The RZ code is a simple line code for the sending of binary codes over a transmission medium and a further development of the NRZ code. Here, the sender varies the status of the medium between three levels (transmitting symbols most of the time designated as +1, 0 and -1). In the transmission of a logical zero with the level +1, the RZ code returns to level 0 following a half pulse. In the transmission of a logical zero, level -1 is also transmitted for half a pulse in order to consequently return to level 0. Thus, with the transmission of every bit there is always a level change, which the receiver uses to facilitate clock recovery (synchronization) (cf. Fig. 3.15).
  In contrast to the NRZ code, the doubled bandwidth is necessary for the RZ code. Unless additional measures are undertaken, the RZ code is not free of direct current components. **Unipolar RZ coding**, is a special form of RZ coding, in which only 2 signal levels (0, +1) are used for coding. While this variation is easier to implement, it has a disadvantage in that clock recovery is not possible due to the zero sequences, which also precludes a synchronization. Additionally, the unipolar RZ coding always has a direct current component.

- **Manchester encoding**
  Like RZ encoding, Manchester encoding (1B2B encoding) is a simple line code and enables the recovery of the clock signal from the transmitted signal. In fact, Manchester encoding belongs to the group of phase modulated codes (biphase line codes, phase shift keying, PSK). The actual information is obtained from the rising or falling edge of each signal. Thereby, a change of a signal level from a

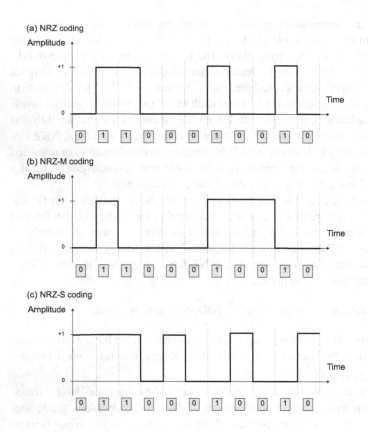

Fig. 3.14 Examples of (a) NRZ coding, (b) NRZ-M coding and (c) NRZ-S coding.

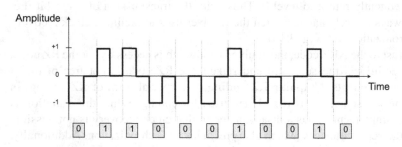

Fig. 3.15 Example of a RZ encoding.

low to a high level (rising edge) indicates a logical zero, and the change from a one to a zero (falling edge) indicates a logical one. In this way it is ensured that at least one signal change per clock step (bit interval) always occurs in the middle of clock interval and, hence, that the system clock can be recovered.[2] The signal of the Manchester code does not have a DC component. A single transmission error can be identified easily by the absence of an expected signal. A simple variation on generating a Manchester encoded signal can be achieved through the logical XOR operation of a clock signal with a NRZ coded signal. However, at the same time a doubling of the number of necessary signal changes also occurs and therefore also the needed bandwidth of the signal to be transmitted.

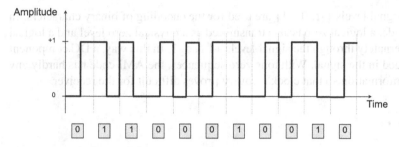

Fig. 3.16 Example of a Manchester encoded bit stream.

- **Differential Manchester encoding**[3]
  The differential Manchester code (1B2B coding, conditional dephase encoding, CDP) is a simple line code based on Manchester encoding. Here, the clock signal can be regained from the coded signal, just as in the case of Manchester encoding. In differential Manchester encoding a bit sequence modulates the change of the phase angle of a clock signal and, hence, presents a form of the differential, digital phase modulation (differential phase shift keying, DPSK). Just as with Manchester encoding, there is always a change of the signal level in the middle of a clock interval in differential Manchester encoding. In the transmission of a logical zero, the clock signal is inverted (180 degrees) in relation to the output from the previous phase angle. However, in the transmission of a logical zero, a signal value change also takes place at the beginning of the clock. The major advantage of differential Manchester encoding is that the polarity of the encoded signal plays no role in the correct reception and subsequent decoding.
- **Alternate Mark Inversion (AMI)**[4]
  The AMI code (1B1T coding) is a pseudo-ternary line code in which three dif-

---

[2] This variation of the Manchester code is also designated as biphase-L or Manchester-II. A variation with reversed edge polarity is found in 10MB Ethernet based on IEEE 802.3.

[3] Differential Manchester encoding is applied in the IEEE 802.5 token ring. In technical literature, Manchester can also be found with reverse polarity for logical zero and one.

[4] The ISDN multiplex-net uses this coding under the name AMI-NRZ. There also exists an AMI-RZ-variation, in which the signal level change takes place in the middle of the clock interval.

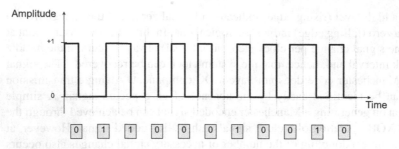

Fig. 3.17 Example of a differential Manchester encoding.

ferent signal levels {-1, 0, +1} are used for the encoding of binary characters. In AMI code, a logical zero is also transmitted as a physical zero level and a logical one alternately through the signal levels +1 and -1. In this way, a DC component is avoided in the signal. With long zero sequences, the AMI code has hardly any clock information so that clock recovery proves difficult for the receiver.

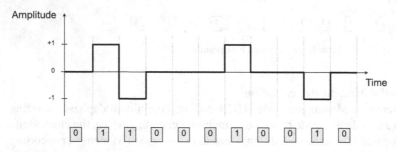

Fig. 3.18 Example of an AMI coding.

In addition to the examples mentioned, there exist a large number of other line codes, a few of which shall be looked at here:

- **MLT-3 Code** Multi-level transmit coding, implements binary sequences on three voltage levels, has a low DC component, implemented in FDDI and Ethernet 100BASE-[TF]X.

- **Biphase Mark Code** Two-phase marking code, comparable to differential Manchester code, is used for digital audio transmission in AES-3 and S/PDIF.

- **4B3T Code** 4 bits signs are mapped on 3 ternary signal values, used in ISDN-basic connections.

- **4B5B Code** 4 bits are mapped onto 5 bits, normally with a DC component, used in FDDI und Ethernet 100BASE-TX.

- **8B10B Code** 8 bit symbols are mapped onto 10 bits, application in, e.g., Gigabit Ethernet, Fibre Channel, USB 3.0 and Serial ATA.

- **64B66B coding** 64 bit symbols are mapped onto 66 bits, application in, e.g., 10GB Ethernet and Fibre Channel.

3.2.2 Analog Modulation Methods

Special modulation processes (cf. also Fig. 3.2) designated **Analog Spectrum Modulation** (ASM) transmit a time and value continuous signal, that is to say, an analog information signal. These processes find application in speech transmission (audio information) as well as music, image or video information, which are not digitalized before transmission. Analog transmission channels are also used in digital technology. Here as well, the digital original signal to be transmitted must first be modulated onto an analog carrier signal before the modulated information can be transmitted onto an analog path. On the side of the receiver, a demodulation of the received analog signal from the carrier signal takes place to recover the digital original signal (cf. Fig. 3.19). Because the data transmission equipment used in signal conversion consists primarily of a *mo*dulator and a *dem*odulator, the name *modem* was coined. A modem is necessary, e.g., to transmit data via an analog telephone network.

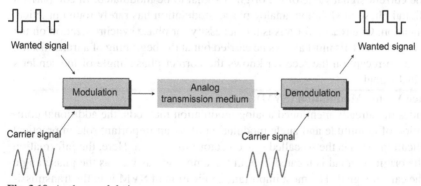

Fig. 3.19 Analog modulation.

For analog signal transmission, the modem uses continuous carrier signals in the form of sinusoidal oscillations, while in digital transmission discrete pulse carriers are normally used.

The signal parameters, amplitude, frequency and phase, are varied for information transmission in analog modulation. Generally, a distinction is made between **amplitude modulation** und **angle modulation**, whereas the latter of the two methods combines frequency and phase modulation.

- **Amplitude Modulation (AM)**
  In amplitude modulation, the amplitude of a high frequency carrier wave is changed, based on the low frequency signal (to be modulated) of the transmitted original signal. Amplitude modulation was used from the very beginning in radio

engineering, because such modulated signals let themselves be easily generated and demodulated. Theoretically, in amplitude modulation, the signal value of the carrier signal is multiplied with the signal value of the original signal. However, this results in a phase shift of 180 degrees as soon as the amplitude of the original signal becomes negative. However in practical application both signals are simply superimposed (added) and then subsequently distorted and filtered.

- **Frequency Modulation (FM)**
  In contrast to amplitude modulation, in frequency modulation the frequency is changed depending on the current value of the original signal to be modulated. Unlike amplitude modulation, frequency modulation also enables a better utilization of the information signal's bandwidth and is less prone to disruption. The reduced susceptibility of frequency modulation is based on the fact that the amplitude value of the modulated signal plays no significant role in the demodulation process. Frequency modulation is used in analog radio, in transmitting audio signals, in analog televisions and video recorders.

- **Phase Modulation (PM)**
  Just as frequency modulation, phase modulation belongs to the family of angle modulation. It is not the current frequency that changes in phase modulation but rather the phase angle of the carrier signal. The phase angle changes in proportion to the current signal value of the original signal to be modulated. In comparison to digital phase modulation, analog phase modulation has rarely found practical application. One reason for this is the necessity for phase synchronization on the receiver's side. This must always be carried out at the beginning of a transmission so that on reception the receiver knows the correct phase angle of the sender's original signal.

- **Space Vector Modulation (SVM)**
  Besides the already mentioned analog modulation methods, the additional combination of amplitude and angle modulation plays an important role in practical application. This is the so-called space vector modulation. Here, the information of the original signal is located in both the amplitude as well as the phase angle of the carrier signal. The most important application of SVM is in the transmission of color information in PAL or NTSC color picture signals (FBAS) in analog color televisions. The phase angle of the carrier signal determines the color saturation of the transmitted image, the amplitude and respective hue.

A further variation of the classic modulation method is the **pulse modulation method**. The primary difference to classic modulation is found in the signal form of the pulse modulation method. Its shape is namely pulsed or rectangular as opposed to sinusoidal. Because of this, the frequency spectrum of a pulse modulation also contains numerous harmonic components of the carrier frequency. The modulation of a pulse-shaped carrier signal proceeds in two separate steps: sampling the original signal and the actual pulse modulation itself. Pulse modulation can take place in the amplitude, pulse width or duration and pulse phase.

- **Pulse Amplitude Modulation (PAM)**
  Pulse amplitude modulation is an analog modulation method in which the ampli-

tude of the original signal is sampled in specific time intervals. The information content of the PAM signal corresponds to the height of the respective impulse, which at the time of the sampling correlates to the existing amplitude of the signal voltage. PAM is well suited to transmission in a time multiplexing process. In the time between the individual PAM impulses of one communication channel, PAM impulses from other channels can be transmitted. Because of a high susceptibility to disturbance, PAM is not suitable as a transmission process over large distances. In this case, the respective impulse heights are effected too strongly by characteristics of the transmission path or are subject to distortion.

- **Pulse Width Modulation (PWM)**
  Pulse Width Modulation (also called Pulse Duration Modulation, PDM) is present based on the sampling of the impulse width of the signal in proportion to the respective sampling value of the amplitude of the original signal. The amplitude of the pulse signal can therefore take only two values. PWM is used in control engineering and power electronics as well as in digital-to-analog conversion and in the sound generation of synthesizers.

- **Pulse Position Modulation (PPM)**
  Pulse position modulation is based on the pulse angle (phase angle) proportional to the respective signal sample of the amplitude of the phase-shifted original signal. Here, a fixed impulse width and impulse amplitude are used and the carrier frequency remains constant. If the carrier is not modulated, the result is a series of square-wave signals with the same time interval (reference clock). PPM is rarely used in practical applications. When, then in model-making, for example, as pulse interval modulation or in ultra-wideband applications.

- **Pulse Code Modulation (PCM)**
  Pulse code modulation is a pulse modulation method that transforms a time and value continuous analog signal into a time and value discrete digital signal. The sampling cuts the time-continuous sequence of an original signal into discrete time points and records its value continuous amplitude value in these discrete time points. The exact sampled values are rounded to the subsequent binary coding, set within predefined quantization intervals. PCM, which was developed in 1938 by the English engineer *Alec A. Reeves* (1902–1971), is mainly utilized in analog-to-digital conversion.

3.2.3 Digital Modulation Methods

Digital modulation methods are of central importance for the data transmission of time and value discrete information. Digital modulation methods (shift keying, Digital Spectrum Modulation, DSM) transmit symbols that are clearly defined for both sender and receiver. For the transmission of analog signals, such as speech or music, the information must first undergo a digitalization before digital modulation can take place. The time span of the actual modulation signal is, in contrast, time and value continuous. Several digital modulation methods are based on corresponding analog modulation methods, or are derived from them. However, there also exist a large number of digital modulation methods without an analog counterpart.

- **Amplitude Modulation (Amplitude Shift Keying, ASK)**
  In digital (binary) amplitude modulation, the value of a digital symbol is represented by two different amplitude values A_0 and A_1 of a sinusoidal signal. This signal is thereby sent for each previously defined symbol length T_s. Usually, one of the two amplitude values selected is equal to zero, i.e., the carrier signal will either be switched on or off by the modulation. Thus, digital amplitude modulation is also known as on-off-keying (OOK) (cf. Fig. 3.20).

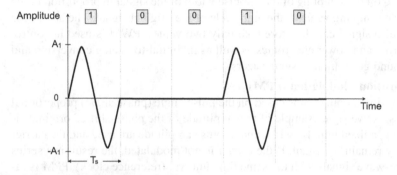

Fig. 3.20 Amplitude shift keying.

A disadvantage of this simple variation is that it it is not possible to definitively determine whether a binary zero sequence has been transmitted or if a transmitter failed or is malfunctioning. Because of this, it is often not the binary character sequence which is transmitted but only the bit transition. A simple method to increase the bandwidth consists of using not just two, but rather several different amplitude values for the encoding of bit sequences, for example, four amplitude values to represent the bit sequences 00, 01, 10 and 11. Amplitude shift keying is currently used, for example, in Central Europe for the DCF77 signal to transmit time signals for radio clocks. It is particularly well suited for synchronization tasks due to the non-changing carrier frequency of its signal.

- **Frequency Modulation (Frequency Shift Keying, FSK)**
  In digital (binary) frequency modulation, the value of a binary symbol is assigned two different frequencies f_1 and f_2 of a sinus-shaped signal. This signal is sent each time for a previously defined symbol duration T_s.
  The example illustrated in Fig. 3.21 presents a special case: no phase jumps occur because the selected symbol duration T_s is a multiple of the period of the two signals. A special case occurs if and only if the difference of both implemented frequency values correlates to half of the symbol rate, i.e. $f_2-f_1=1/(2 \cdot T_s)$. Accordingly, the frequency values of the associated signals distinguish themselves by

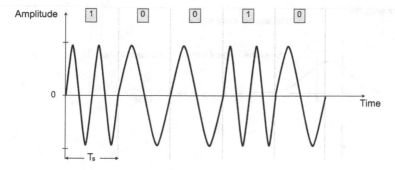

Fig. 3.21 Frequency shift keying.

exactly half a period, i.e., the signals are orthogonal. This form of digital frequency modulation is called Minimum Shift Keying (MSK). In order to increase
the bandwidth of transmission in digital frequency modulation, several discrete
frequency values can be used for the encoding of bit sequences. If multiple frequency values are used, this is also called M-FSK, whereby M stands for the
number of symbols or the various frequency values. The oldest use of this method – a frequently used modulation technique in telecommunications – is the
wireless telegraph.

- **Gaussian Minimum Shift Keying (GMSK)**
  This modulation process is a frequency modulation method (as is Minimum Shift
  Keying, MSK) with an upstream Gauss filter. By way of a Gauss filter (low-
  pass filter), the steep edges of the digital signal to be transmitted are flattened
  before the actual modulation takes place (cf. Fig. 3.22). As a result, the high-
  frequency components of the signal drop off. This phenomenon occurs as a result
  of the hard transition between two frequencies. These hard transitions take the
  form of crosstalk disturbances in adjacent channels, which can occur in radio
  transmission. Less bandwidth is necessary in signal transmission. The signal re-
  shaping that occurs in the filtering of digital square-wave signals is generally
  known as pulse shaping.
  Over a longer time the original square wave input signal becomes "smeared,"
  due to the implementation of Gauss filtering. This leads to an overlapping (malfunction) in the signals of the individual transmission symbols. A so-called intersymbol interference arises, which is, however, calculable, and can be offset with
  the help of a special error correction algorithm (Viterbi Algorithm). Among its
  other applications, GMSK is used in the still wide-spread mobile communications standard: Global System for Mobile Communication (GSM).

- **Phase Modulation (Phase Shift Keying, PSK)**
  In digital (binary) phase modulation, the value of a binary symbol is associated
  with two different phase angles of a sinusoidal carrier signal. Typically, the phase
  angles 0° and 180° are equivalent to the binary symbols 0 and 1 (cf. Fig. 3.23).
  The two signals differ only in sign, thus the designation: Phase Reversal Shift

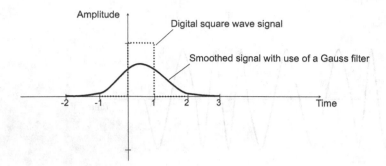

Fig. 3.22 Application of a Gauss filter.

Keying (PRK). Here, the signal is sent for a previously defined symbol duration T_s.

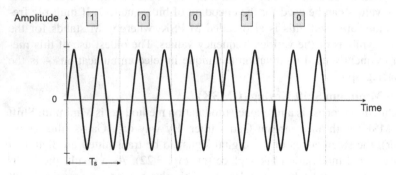

Fig. 3.23 Phase shift keying.

In phase differential modulation (Differential Phase Shift Keying, DPSK) the symbols to be transmitted are encoded by a change in the phase angle, thus, e.g., changing the phase angle by $0°$ represents the symbol 0 and changing the phase by $180°$ represents a 1.

- **Quadrature Amplitude Modulation (QAM)**
  In this modulation process, amplitude modulation and phase modulation are combined. Thereby, a sinusoidal carrier wave is imprinted with two signals that are independent of one another. In principle, this means that both signals – each with an amplitude modulation on a carrier of the same frequency yet with different phase angles – are modulated and finally added to each other. To illustrate the chosen modulation variation, amplitude and phase angle are represented as polar coordinates on a two-dimensional level (constellation diagram) (cf. Fig. 3.24). The most basic variation is 4-QAM (also QPSK, quadrature phase shift keying), which at a constant amount of amplitude uses the 4 phase angle $45°$, $135°$, $225°$ and $315°$, so that 2 bits per time step T_s may be encoded. At the same velocity

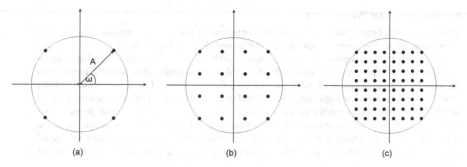

Fig. 3.24 Constellation Diagram for 4-QAM (a), 16-QAM (b) and 64-QAM (c).

$V_s=1/T_s$ twice as much information can be transferred with 4-QAM as with the basic phase modulation PSK. In addition, there exist further processes, among them the methods 16-QAM (4 bits per symbol period), 64-QAM (6 bits per symbol period), 256-QAM (8 bits per symbol period), up to 4096-QAM (12 bits per symbol period). 256-QAM is used in the transmission of cable television signals (digital video broadcast – cable, DVB-C).

The modulation methods we have looked at so far use only one carrier signal, yet it is also possible to divide the data stream to be modulated onto several different carrier signals. In particular, **Multi-Carrier Modulation Methods (MCM)** are used with multiple carrier signals to guarantee the best possible adaptation to the technical parameters and general conditions of specific transmission channels. If because of a malfunction individual carrier signals (sub-channels) are not available, only the data throughput rate as a whole is reduced as the remaining carrier signals can still be used. In the simplest case, each of these sub-channels is assigned the same modulation scheme and consequently the same transmission rate. It is, however, more efficient to set the bit rate of each sub-channel according to its interference-free functioning. In sub-channels with a lower rate of malfunction a more complex, multi-stage modulation method is used, while in sub-channels with a higher rate of malfunction a more robust, simpler method is implemented. In this way, a sub-channel can find nearly optimal utilization. Typically, in the application of MCM a time safety interval is added between the individual symbols or groups of symbols to be sent. This is done to ensure the possibility of a quicker reaction from the receiver to problems concerning multi-path reception (multi-path, multiple reception of an individual output signal through reflection, see also Fig. 3.25).

The modulation methods in this category are closely connected with multiplex techniques. The main task of such techniques is to transmit several different signals at the same time without mutual interference via a shared communication channel. Among the digital modulation methods with a plurality of carrier signals are:

- **Discrete Multitone (DMT)**
  The fundamental idea of the discrete multi-tone method is to divide the frequency band available for signal transmission into an array of individual sub-

Multi-Path Signal Propagation

The **multi-path propagation** of signals presents one of the greatest problems in wireless communication. Electromagnetic waves can travel directly from sender to receiver, but can also reach the receiver in indirect ways, such as being reflected off obstacles or scattered. In the process, the transmitted signal covers distances of varying lengths on its way to the receiver, meaning it arrives at different times. The effect which results from this multi-path propagation is called **delay spread**. It is a typical characteristic of wireless communication media because there is no single direction of signal propagation as predefined by a transmission line. The delay spread in city centers averages 3 µs. The GSM -system, which is used in mobile communication, can compensate for signal time differences of up to 16 µs. This corresponds to a signal length difference of about 5 km.

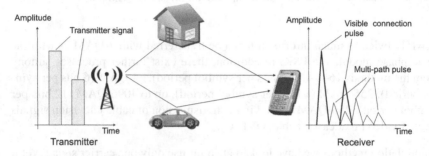

The delay spread causes a **pulse propagation** of the transmitted signal, i.e., a short transmitter impulse is "smeared" over a long period of time. In the process, the individual partial signals are also subject to variable signal damping on their different propagation paths; meaning, the signals are received in different strengths. While on the sender's side consecutive pulses, which can either represent an individual symbol or a symbol group depending on the modulation method, are clearly present as time separated, the delay spread ensures time-overlapping consecutive signals. This effect is called **intersymbol interference**. The higher the transmitted symbol rate, the closer together the individual symbols are situated and the more seriously they are affected by interference. This can even lead to the deletion of consecutive symbols by interference-affected signals. One way to get around the negative effects of multi-path propagation consists of "training" the characteristics of the transmission channel and the accompanying signal propagation paths, then equalizing them at the receiver's end with the help of this training data. Additionally, the sender must transmit regular training sequences, whose signal pattern is already known to the receiver. With the help of actually received signals, an equalizer can be programed accordingly, based on a comparison with the known training sequence to achieve a balance of the interference caused by multi-path propagation.

Further reading:

Pahlavan, K., Krishnamurthy, P.: Principles of Wireless Networks: A Unified Approach, Prentice Hall PTR, Upper Saddle River, NJ, USA (2001)
Stallings, W:: Wireless Communications and Networks, Prentice Hall Professional Technical Reference (2001)
Wesel, E. K.: Wireless Multimedia Communications: Networking Video, Voice and Data, Addison-Wesley Longman Publishing Co. Inc., Boston, MA, USA (1997)

Fig. 3.25 Multi-path signal propagation.

channels. The data stream to be transmitted is subdivided into symbol groups. These are simultaneously modulated and transmitted as sum signals via the sub-channels. DMT is used in ADSL (Asymmetric Digital Subscriber Line) broad-band connections. Here, the carrier signal for transmission is subdivided into 255 sub-channels via the telephone line. The data to be sent is modulated with different transmission rates via QAM and then transmitted. Further examples of application are ADSL2+ and VDSL2 (very high bit rate DSL).

- **Orthogonal Frequency-Division Multiplexing (OFDM)**
  In contrast to the DMT method, the OFDM method, as a multi-carrier modula-tion method, uses orthogonal carrier signals for data transmission. OFMD is a special variation of the frequency division multiplex method (cf. Section 3.27). Through the orthogonality of the carrier, crosstalk of the transmitted signals on the adjacent carriers is reduced. In the same way as DMT, the data stream to be transmitted is first divided into several partial data streams with a low bit transmission rate. With the help of a conventional modulation method, it is mo-dulated with a low bandwidth. Finally, the individual partial signals are added together to a complete signal again. In order to distinguish the individual partial data streams when demodulation takes place at the receiver end, it is necessary that the respective carrier signals are positioned orthogonally to each other. As with DMT, in this way it is possible to achieve an especially good adaptation to the physical characteristic of a transmission channel. OFDM is used in digital au-dio broadcast (DAB), in which a signal can be split from 192 to 1536 individual sub-carriers. Other application examples are terrestrial digital television (Digital Video Broadcast – Terrestrial, DVB-T), Wireless LAN based on IEEE 802.11a and IEEE 802.11g standard, Worldwide Interoperability for Microwave Access (WiMAX) based on IEEE 802.16e and the technologies for mobile communica-tion networks of the 4th generation (4G, B3G, Long Term Evolution, LTE).

- **Coded Orthogonal Frequency Division Multiplex (COFDM)**
  The COFDM method expands OFDM to include forward error correction. In this way, higher stability is offered for multi-path reception and a safeguard against its associated frequency selective deletion (fading – the fluctuations in reception field strength caused by interference or shading) and burst errors (errors occur-ring in blocks). COFDM is used in digital broadcasting (DVB) and in terrestrial digital television (DVB-T).

3.2.4 Multiplex Methods with a Constant Bandwidth

Modulation methods and multiplex technology are closely linked. Multiplex me-thods serve to transmit multiple signals simultaneously. Ideally, this should be done without mutual interference via a shared channel in bundled form, for example, over a cable or a single radio spectrum (cf. Fig. 3.26). The bundling (**multiplexing**) normally occurs after the data signals are modulated onto a carrier signal. At the

receiver a corresponding unbundling (demultiplexing) takes place initially, before
the data signals are demodulated and recovered.

Multiplexing

Multiplex methods are processes of signal and communication transmission whereby
several signals are combined (bundled), and at the same time transmitted via a medium.
At the receiver, first a demultiplexing of the bundled signals takes place.

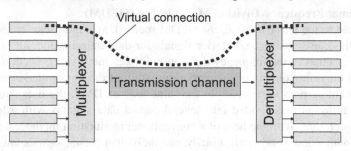

Multiplexing provides an efficient utilization of a physical communication connection
through the mutual compensation of logical connections.

Further reading:

Hsu, H.P.: Schaum's Outline of Theory and Problems of Signals and Systems, Schaum's
Outline Series, McGraw-Hill Co, Inc., USA (1995)

Fig. 3.26 Multiplexing.

Practically-speaking, a distinction is made between four different dimensions in the
multiplex method: space, time, frequency, and code (cf. also Fig. 3.27). Based on
the chosen procedure, a data channel is assigned a certain space at a certain time on
a set frequency with a specific coding. A distinction is often made between "multiplexing" and "multiple access." With **multiplexing**, a hardware solution is normally
involved in which at the beginning of a signal transmission path a multiplexer bundles different signals and a demultiplexer separates them again at the end. One speaks
of **multiple access** if several pairs of senders and receivers independently share a
transmission medium, for example, in the mobile communication between basis stations as the central instance and the subscriber terminals. The technical procedures
applied in multiplexing and multiple access are identical.

- **Space Division Multiple Access (SDMA)**
  Space Division Multiple Access is understood as the transmission or relaying of
  several communication signals via parallel transmission paths. These paths are
  made available to the connected senders and receivers for their exclusive use. A
  differentiation is made between wireless and wired space division multiple access
  methods. Wired space division multiple access methods were already available
  in the early days of telecommunication in the form of cable bundles (trunks)
  or, as today, in the crossbar switch (a switching matrix composed of conductors

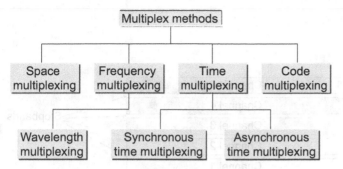

Fig. 3.27 Overview of the various multiplex methods.

and switching elements, through which the desired connections between sender and receiver can be controlled). Different directional radio links, each covering a separate area, are used in wireless space division multiple access. Normally used in wireless data transmission are either time or frequency multiplex methods, or a combination of both. Space multiplexing is generally first implemented when the number of communication networks increases and the available frequencies become scarce. Then the same frequency is used repeatedly in corresponding spatial distances. A minimum distance between the areas is necessary to avoid the risk of interference.

- **Frequency Division Multiple Access (FDMA)**
  Frequency multiplexing comprises all multiplexing methods that divide the frequency spectrum into multiple, non-overlapping frequency bands for shared transmission. Each transmission channel receives its own frequency band. Each sender can use this frequency band exclusively and continuously. Two adjacent frequency bands must be separated from each other by a guard interval so there is no crosstalk occurring on the signals to be transmitted. This procedure parallels the well-known frequency allocation to radio stations in a broadcasting area. The necessary coordination between sender and receiver lies in the correct selection of the frequency band required to receive a certain station. Frequency multiplexing can be used in wired or wireless form. The first frequency-based method for the multiple use of transmission lines was proposed in 1886 by the co-inventor of the telephone, *Elisha Gray* (1835–1901), for the electrical telegraph.
  In contrast to the use of frequency modulation in radio broadcasting, its application in mobile communication is problematic. While a radio station broadcasts permanently on an assigned channel, mobile communication usually takes place for only a short time so that a fixed frequency allocation for certain transmitters is very inefficient. For this reason, today's norm is the application of several different multiplexing methods. A special variation of frequency multiplexing called Wavelength Division Multiple Access (WDMA) is used in the transmission of data signals on optical fibers. For the transmission on optical fibers, via laser or on light-emitting diodes (LED), light signals are generated in different spectral

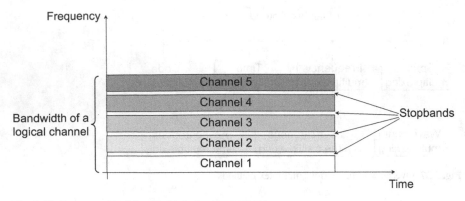

Fig. 3.28 Frequency Division Multiple Access (FDMA).

colors. Each spectral color signifies a specific transmission channel on which the data of the sender is modulated.

- **Time Division Multiple Access (TDMA)**
  As space and frequency multiplexing are particularly useful in analog transmission technologies, these methods were already implemented early on in telecommunication. Time multiplexing, in contrast, can be mainly used just in digital transmission technologies. While in frequency multiplexing each channel claims a certain frequency band, in time multiplexing all channels share the frequency band available, but at different times. This means that every channel has the entire bandwidth of the transmission medium at its disposal for the length of the time frame assigned to it. In time multiplexing, (time) guard intervals must be observed between each channel otherwise overlapping and reciprocal disturbances with adjacent channels could occur. Historically, the time multiplexing method goes back to the Italian physicist *Giovanni Abbate Caselli* (1815–1891) and the French engineer *Jean-Maurice-Émile Baudot* (1845–1903). In 1865, Caselli developed the pantelegraph, an early electromechanical facsimile machine that transmitted two pictures using a a line-by-line multiplex method. Baudot was the inventor of a printing telegraph system and the code for it that was named after him. In 1874, he developed a telegraph system on the basis of synchronous time multiplexing. This made it possible to transmit 4-6 telegraph signals over a shared line.
  Synchronous time multiplexing methods allocate separate transmission times (time frame, time slot), divided into fixed intervals, to each channel. These repeat periodically at defined intervals. Sender and receiver facilitate this procedure by way of a precise, synchronized clock pulse. In demultiplexing, a transmission channel can be identified by the position of its time slot. If permanent transmission does not take place, many time slots remain unused in this procedure. Capacity utilization then becomes inefficient. Synchronous time multiplexing is used in the technology of the backbone computer networks: SONET (Synchronous Opti-

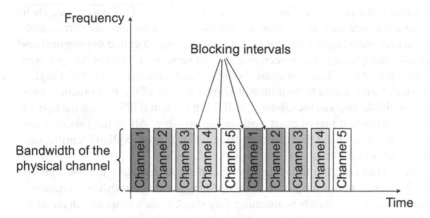

Fig. 3.29 Time Division Multiple Access (TDMA).

cal Network), SDH (Synchronous Digital Hierarchy) und PDH (Plesiochronous Digital Hierarchy).

In **asynchronous time multiplexing methods**, in contrast, the individual channels are assigned a time slot based on need. Asynchronous time multiplexing thus improves channel utilization in many cases. Logical channels only occupy separate time slots when necessary so that no empty time slot can be transmitted. To guarantee correct channel allocation on the receiver side after demultiplexing, the time frames must be clearly assigned a channel identification number. Asynchronous time multiplexing can be found in the technologies of the computer networks Frame Relay or ATM (Asynchronous Transfer Mode).

Today, combinations of frequency and time multiplex methods are often implemented. Here, a specific channel remains active for a fixed period of time within a specific frequency range and has the possibility to be occupied. At the same time, barriers are needed in the temporal dimension as well as between the required frequency bands. A channel can successively, in fixed time intervals, always occupy the same frequency range, but it's just as likely that each time it could be assigned to a different one. This makes synchronization between sender and receiver more difficult, yet at the same time also ensures an increased robustness with respect to narrow band disturbances. If the order of sequential channel occupancy is not known to the receiver, this method already provides some measure of protection against unwanted listeners.

- **Code Division Multiple Access (CDMA)**

  In the code multiplexing method every sender and every channel is assigned its own special code to identify it unambiguously and separate it from other channels. All channels transmit at the same time and on the same frequency band. The coded signals of all the senders are superimposed to form a sum signal. The receivers are able to filter out of the sum signal the data intended for them, in each case based on prearranged codes. To ensure that there is clearance between the channels, the codes used must have an appropriate code distance, i.e., their Hamming distance must be as large as possible in order to minimize error, malfunction and interference. Ideal for this purpose are the so-called **orthogonal codes**. Code multiplexing can be seen as a special variation of the spread spectrum technique (cf. section 3.2.5), because the modulated, encoded signal has a higher band width than the data to be transmitted over it. One of the first practical uses of code multiplexing was the Global Positioning System (GPS). With the help of multiple satellites, it makes exact positioning possible. All of the satellites use the same frequency band, but different codes. This allows explicit identification from the side of GPS receivers.

The principle of code multiplexing can be compared to a party to which guests from different countries have been invited. Many guests establish communication channels among themselves, meaning they speak to each other simultaneously using the same frequency band at the same time. If all of the party guests were to converse in the same language, then the individual groups communicating together would have to distribute themselves appropriately around the room as not to disturb each other, i.e,. space multiplexing would need to take place. The precondition for this would be that the party guests restrain from speaking too loudly to avoid disturbing each other. But if each group communicates in a different language then each listener only will concentrate on his own language, markedly reducing the danger of a mutual interference. The respective sender, therefore, each uses an individual code that allows itself to be filtered out of the sum signal of all of the conversations taking place simultaneously. Conversations in different languages are then only registered as background noise. Furthermore, if each group converses in a language which is not understood by the other guests it is also possible to increase the privacy sphere at the party. This highlights a further advantage of code multiplexing: if the code used by the sender is unknown then although the signal can still be registered by an unwanted listener, it cannot be decrypted and understood. But if, on the other hand, guests in two of the neighboring groups are having a conversation in a very similar language (i.e., the code distance chosen is too small) then this can lead to unwanted interference and mutual disturbance. To blank out conversations heard in a foreign language there must be a "minimum distance" between the foreign language and one's own – in the best case scenario this should be orthogonal. Indeed, the main problem with the code multiplexing method is finding just the right codes that effectively demarcate borders between different users of the same communication medium and at the same time enable an adequate distinction to be made between the user signal and background noise (cf. Fig. 3.30).

Code multiplexing methods are used in particular in wireless communication. Here, their robustness and security are especially valuable as protection against eavesdroppers. In comparison to time multiplexing and frequency multiplexing the available code space here is huge. However, a receiver must know the applied code in advance and the sender and receiver have to be perfectly synchronized to filter out the code used from the total signal and participate in communication. This increases the complexity of the communication process and the infra-

Code Multiplexing and Orthogonal Codes

In code multiplexing several signals are bonded to a total signal with the help of different codes which, accordingly, can be regained by the receiver with knowledge of the exact codes. Which codes can be used for this application? It is generally true that the correct codes must have a good autocorrelation and be as orthogonal as possible in relation to the other codes used. In signal processing **autocorrelation** is understood as a signal sequence that correlates with itself (=is brought into a correspondent relationship). If, additionally, we see a code as vector $a=(a_1,\ldots,a_1)$, then there is a good autocorrelation if the amount of its scalar product (inner product, $\sum_{i=1}^n a_i a_i$) is as large as possible with itself. In contrast, if two different codes are compared $a=(a_1,\ldots,a_1)$, $b=(b_1,\ldots,b_1)$ with each other, then these codes are **orthogonal** in relation to each other if their scalar product equals zero $(a \cdot b = \sum_{i=1}^n a_i b_i = 0)$, therefore

$$a \cdot b = \begin{cases} 1 & \text{if } a = b \\ 0 & \text{otherwise} \end{cases}$$

is valid. In signal transmission, original codes often no longer correspond to their exact standard due to disturbance and noise. For this reason, after transmission these codes are often only "nearly" orthogonal.

Example: Let us assume the following orthogonal codes are given for the encoding of the two channels A and B: $A=(1,-1)$ and $B=(1,1)$. In practical application the orthogonal codes for the coding of channels are considerably longer. To keep the example simple channel coding with a length of only 2 bit was chosen. Different data should be transmitted on the two channels A and B: $data_A=(1,0,1,1)$ and $data_B=(0,1,1,1)$.

On both channels a spread of the data to be transmitted takes place with the help of an assigned code sequence: $code_A = A \cdot data_A = (1,-1),(0,0),(1,-1),(1,-1)$ and $code_B = B \cdot data_B = (0,0),(1,1),(1,1),(1,1)$.

Both signals are now superimposed, i.e., added to a total signal $(1,-1),(1,1),(2,0),(2,0)$ that is finally transmitted.

In order to be able to reconstruct the data of channel A at the receiver, only the channel code $A=(1,-1)$ is used for the spreading of the signal. For this purpose, the scalar product between the channel code and the received total signal is formed thus: $(1 \cdot 1)+(-1 \cdot -1)=2$, $(1 \cdot 1)+(1 \cdot -1)=0$, $(2 \cdot 1)+(0 \cdot -1)=2$, $(2 \cdot 1)+(0 \cdot -1)=2$. In the obtained signal sequence $(2,0,2,2)$ all the values above zero are interpreted as a logical 1 and therefore result in $(1,0,1,1)=data_A$. In the same way, the data on channel B may be regained through a despreading of the total signal with the channel code $B=(1,1)$.

Because in practice disturbance and noise can arise during transmission, in despreading the individual channels it is no longer possible to ascertain clear-cut numerical values as in the above presented best case scenario. In addition, an exact synchronized superimposition of the signals is assumed here and a possible signal damping of individual frequency ranges is not taken into account.

Further reading:

Viterbi, A. J.: CDMA: Principles of Spread Spectrum Communication, Addison Wesley Longman Publishing Co., Inc., Redwood City, CA, USA (1995).

Fig. 3.30 Code multiplexing and orthogonal codes.

structure necessary for communication. Originally, CDMA was developed for the military and first employed in World War II. Due to its high insensitivity to interference, it was used as a countermeasure by the British allies against German attempts to disrupt allied radio transmission (cf. also Fig. 3.34).

Besides the presented synchronous multiplex methods, which share an even (static) distribution of their entire channel capacity on several connections, there are also **asynchronous multiplexing methods**, which carry out a dynamically defined distribution of the available capacity over several connections. Asynchronous channel allocations are made solely on the basis of time multiplexing, i.e., a channel is allocated variously long time intervals as required, without a fixed multiplex grid. Asynchronous multiplexing adapts flexibly to the respective communication load and uses the total capacity optimally for data transmission. Since no fixed time grid is given, the data to be transmitted must be given special header and trailer information on a higher protocol level so that it can be clearly assigned to a channel, which means significantly more effort. Asynchronous multiplexing is employed in the data transmission method ATM (Asynchronous Transfer Mode).

3.2.5 Spread Spectrum Methods

Spread spectrum describes a method in which a narrow band signal is converted into a signal for information transmission with a wider than necessary bandwidth. In the process, the electromagnetic transmission energy, which had until then been concentrated within a small frequency range, is now distributed over a larger frequency range. In this way, the power density of the signal can be lower than that of a narrow band signal without data getting lost during information transmission. An advantage of what at first glance appears to be an inefficient method is found in a considerable robustness, which forms a contrast to narrow band disturbances. Fig. ?? shows how with the help of the spread spectrum technique interference and disruptions can effectively be reduced. While at the sender the narrow ban signal is spread for transmission (a), at the receiver narrow and broad band disturbances are reduced in their strength by the opposite process of despreading (b). Afterwards, the signal also passes through a band-pass filter which cuts off the frequencies above and below the narrow band signal (c). The receiver can reconstruct the original data from the signal that remains. This is because the payload data has a higher output than the disturbances still in the signal.

A further advantage is that the output density of the spread spectrum signals can be selected at such a low level as to be even less than the background noise. In this way, potential unwanted listeners are not able to recognize that communication is even taking place.

Spread spectrum methods are often employed together with multiplexing techniques. If, for example, several signals are to be transmitted parallel over a common frequency band with the help of a FDMA method, each individual signal is initially assigned a narrow frequency band with a sufficient signal-to-noise-ratio in relati-

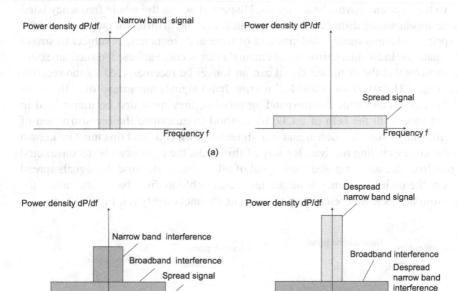

(a)

(b)

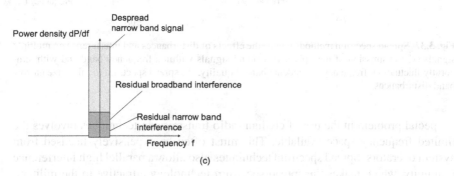

(c)

Fig. 3.31 Spread spectrum methods reduce the effect of disturbances and interference in the signal: (a) spreading of the signal at the transmitter, (b) despreading of the signal at the receiver, (c) a band-pass filter removes frequency components above and below the signal.

on to the adjacent narrow band signals. Dispersed across the whole frequency band there appear strong disturbances at different times and in different ranges so that the respective channel quality, independent of time and frequency, is subject to strong fluctuations. Individual narrow band channels can become affected to such an extent in sections that the disturbed signal can no longer be reconstructed by the receiver (cf. Fig. 3.31 (a)). If the individual narrow band signals are spread over the whole frequency band with a corresponding process, they must first be transmitted in coded form with the help of a CDMA method to guarantee the reconstruction of the individual signals. Each signal is assigned its own code and this must be known to the corresponding receiver. By way of this code, the signal can be reconstructed again from the superimposed total signal of all channels. Because the signals spread across the entire frequency band are less susceptible to disturbance and noise, the transmission is less dependent on the current channel quality (cf. Fig. 3.32 (b)).

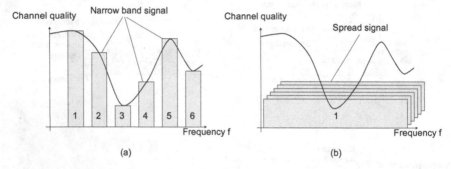

Fig. 3.32 Spread spectrum methods reduce the effects of disturbances and interference on multiple signals (a) transmission of multiple narrow band signals within a frequency band and with temporally fluctuating, frequency dependent channel quality, (b) spread spectrum to minimize narrow band disturbances.

A special problem in the use of civilian radio transmission techniques involves the limited frequency space available. This must often be expensively licensed from system operators. Spread spectrum techniques also allow a parallel high interference immunity, which makes the spread spectrum technology attractive to the military. At the same time, it is a technology that is also finding increased use in civilian radio systems. Use of the available frequencies, also for new radio transmission systems, is possible without disturbing the operation of existing systems that occupy the same frequency range. In addition to the transmission of military communication, spread spectrum techniques are employed in civilian wireless data transmission. Some examples of this are wireless LAN or Bluetooth, as well the mobile communication area standard UMTS.

The higher bandwidth required by the spread spectrum signal creates an initial problem. In code multiplexing, the signals not intended for the receiver are perceived as background noise. Yet this artificially created noise raises the existing noise level. If a certain level is exceeded, the receiver can no longer reconstruct the original signal, i.e., reception fails. Although recovering the original signal on the receiver's side requires more effort, it can also proceed on a large scale for multiple signals parallel without a complex infrastructure. This can be done with the help of modern signal processors.

The two best known band-spreading methods are Frequency Hopping Spread Spectrum (FHSS) and Direct Sequence Spread Spectrum (DSSS):

- **Frequency Hopping Spread Spectrum (FHSS)**

  This method is based on the principle of the "frequency jump," which draws on a combination of FDMA and TDMA. Here, an available frequency band is divided into a series of narrow band sections separated from each other by a sufficiently large signal-to-noise ratio (FDMA). These individual narrow band channels are each allocated to a sender for only a short time (dwell time) before changing to another channel (TDMA). The sequence of the different channel allocations is determined by the sender and receiver with the help of a pseudo-random number generator. A sequence of the sender's channel occupation is called a "hopping sequence." A distinction is made between a short dwell time (fast hopping) and a long dwell time (slow hopping).

  Whereas in fast hopping, the frequency of a single bit changes several times already when sending, in slow hopping several bits are always being transmitted on a frequency during the dwell time. Because of its longer dwell time, slow hopping is easier to implement and it is also is able to compensate for fluctuations within a certain range of the dwell time. In fast hopping, both sender and receiver have to be exactly synchronized with one another. However, in view of narrow band disturbances the slow hopping method is considerably more sensitive than fast hopping, which occupies a potentially disturbed frequency range for a significantly shorter time. In Fig. 3.33 the slow hopping and fast hopping FHSS methods are illustrated with a short data sequence; slow hopping is given a frequency of 1 hop/bit and fast hopping 3 hop/bit. The binary data to be transmitted is presented in the middle of the two procedures.

  Each jump sequence of the FHSS method is determined with the help of a pseudo-random number generator. This sequence is either mutually agreed upon in advance by the two communicating parties – as would normally be the case when this method is put into application by the military – or the parties agree on the starting value of the random numbers sequence at the beginning of communication, which could proceed as follows:

  - The transmitter, i.e., the party who initiates the communication, sends a request via a predetermined frequency or over a separate control channel.
  - The receiver responds with a randomly chosen number, the so-called seed value.
  - The sender uses this seed value as the starting value for a predefined pseudo-random number algorithm. From this, it generates a sequence of quasi-random frequencies. It acknowledges to the receiver the correct reception of the seed value via the control channel with the help of a synchronous signal, based on

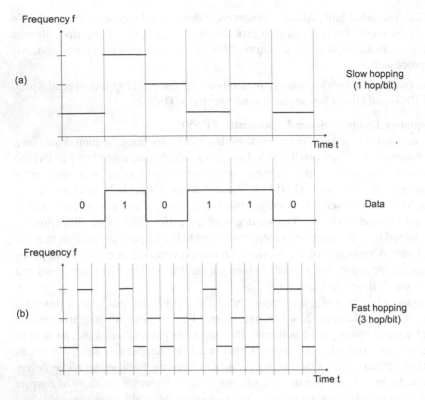

Fig. 3.33 Example for application of the Frequency Hopping Spread Spectrum (FHSS) with (a) slow hopping and (b) fast hopping.

  the derived random sequence of frequencies. The correct calculation is thereby confirmed.

– At this point the actual communication between sender and receiver begins, based on the calculated frequency sequence.

Among other applications, the FHSS spread spectrum method is used in wireless LAN IEEE 802.11, with its fixed 79 single channels in the 2,4 GHz frequency range with a minimum jump rate of 20 frequency jumps per second and a jump interval of at least 6 channels.

- **Direct Sequence Spread Spectrum (DSSS)**
  Direct sequence spread spectrum involves a spread spectrum method that is based on the idea of spreading the output signal with the help of a predetermined bit sequence. This bit sequence is also called a **chipping sequence**. As in other spread spectrum methods, in DSS the signal energy is also spread over a greater bandwidth. To do this, the user data to be transmitted is linked with the chipping sequence, which has a higher data rate, via a logical exclusive OR operation (XOR) (cf. Fig. 3.35). Each bit of the user data is linked with the entire chip-

The Historical Background of Band-Spreading Methods

The first description of a frequency switching method can be found in 1900 when *Nikola Tesla* (1856–1943) submitted a patent for a radio-controlled submarine Therein, he described a method that was particularly resistant to radio interference. It was approved by the United States Patent Office in 1903. His patents outline the first frequency hopping method as well as frequency multiplexing and – also worthy of mention – contain the first description of a logical circuit based on an electronic and switching-based element.

Frequency hopping was also mentioned in a book about radiotelegraphy by the German radio pioneer *Jonathan Zenneck* (1871–1959) in 1908. Zenneck's ideas were said to have resulted in the Telefunken company trying out these techniques several years previously. The German military made limited use of the frequency changing technology in World War I, employing it for interception-proof communication between individual commandos. British military forces were unable to eavesdrop on these connections because of the then unavailable technology for tracking frequency. In the 1920's and 1930's further patents for frequency switching methods were issued.

The most well known of these was the 1942 patent granted to Hollywood actress *Hedy Lamarr* (1914–2000) and composer *George Antheil* (1900–1959) for a "secret communication system." Before her film career in the USA began, Hedy Lamarr had been married to the Austrian arms merchant Friedrich Mandl. Besides producing weapons and munitions, Mandl was also involved in aircraft construction and in this field focused on the problems of radio remote control, among other things. Through him, Lamarr likely first gained insight into the technology of frequency switching.

The system developed by Antheil and Lamarr was based on piano rolls, containing musical pieces for a player piano. Their invention materialized when they wanted to synchronize one of Antheils compositions written for 16 player pianos. To make this possible, sender and receiver were given identical piano rolls (punched cards). Corresponding to the 88 keys on a piano, the device could switch between 88 different frequencies. It was intended to be used for radio controlled torpedoes, making it more difficult for the enemy to find and disrupt them. But the US military did not implement the patent and thus it never found practical application. In the 1950's it reappeared as ITT Corporation and other companies began to look at the technique of CDMA for civilian use. The frequency hopping method developed by Lamarr and Antheil was in fact first used in 1962 during the Cuban Missile Crisis – after the patent had already expired. In 1997, Hedy Lamarr was awarded the EFF Pioneer Award by the Electronic Frontier Foundation in honor of her and Antheil's invention, which was to become the basis of modern mobile technology.

Further reading:

Scholtz, R. A.: The Origins of Spread-Spectrum Communications, IEEE Transactions on Communications, Vol. 30, No. 5, p. 822 (1982).

Price, R.: Further Notes and Anecdotes on Spread-Spectrum Origins, IEEE Transactions on Communications, Vol. 31, No. 1, p. 85 (1983).

Fig. 3.34 The historical background of band-spreading methods.

ping sequence XOR, so that a binary 0 is represented by the chipping sequence itself and a logical 1 by the complementary sequence. These code series are called **chips** or **pseudo-static codes**. Depending on the method used to create the chipping sequence, the effect can be similar to random noise. Thus, it is often identified as a pseudo-noise sequence. On the receiver side, the user data stream can be reconstructed again by means of a new XOR linking operation with the correct chipping sequence.

The **spreading factor** s is the designation for the relationship between the duration of a user data bit t_N and one of the bits in the chipping sequence t_s, $s = t_N/t_s$. The spreading factor s determines the bandwidth of the resulting signal, by multiplying the original bandwidth b with the spreading factor s. In contrast to the example in Fig. 3.35, in civilian applications spreading factors are between 10 and 100 and in military use even up to 10,000. DSSS is used in the WLAN standard IEEE 802.11 as well as in GPS, UMTS, UWB, ZigBee, WirelessUSB and in model-making remote radio control applications.

DSSS is unaffected by narrow band disturbances. This because an interference signal at the receiver is also linked with the chipping sequence XOR and thus despread again. The data signal is then linked for a second time with the chipping sequence XOR and despread. The power density of the interference signal is reduced by the spreading factor and can no longer interfere with the despread data signal. The interference signal disappears into the background noise.

In addition to DSSS and FHSS, there are further modulation methods based on the fundamental principle of the spreading spectrum. Numbering among them is the so-called **time hopping spread spectrum**. (THSS) This is a time hopping method in which the user data bits to be sent by one party are only transmitted in short increments of time, similar to a TCMA procedure. The distance between the individual time increments can vary within a transmission period.

The **chirp spread spectrum** method (CSS) uses what is called a **chirp** pulse for the spread spectrum. A chirp pulse has a sinusoidal signal behavior whose frequency linearly increases or decreases over time. This signal behavior is used in the context of CSS as an elementary transmitting pulse for the transmission of binary user data. A data transmission based on the CSS method is made up of a string of increasing and decreasing chirp pulses. A special characteristic of this method is found in its robustness in conjunction with interference caused by the Doppler effect. These are interferences occurring as a result of movement, or acceleration, of transmitter and receiver in relation to each other. This is because only the direction of a frequency change must be recognized when decoding a chirp pulse, irrespective of its actual frequency. CSS is used in Wireless Personal Area Network (WPAN) and standardized in IEEE 802.15.4 .

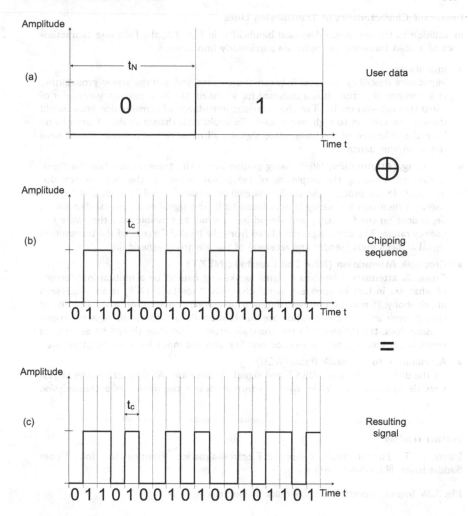

Fig. 3.35 Example for the application of Direct Sequence Spread Spectrum (DSSS) with (a) data signal, (b) chipping sequence and (c) the resulting signal from the XOR combination of (a) and (b).

3.3 Guided Transmission Media

In **guided transmission** (also called bounded transmission), media signals are guided along a solid medium with the help of electromagnetic waves. The various forms of guided transmission media each possess certain physical properties that have their own advantages and disadvantages in terms of bandwidth, delay, cost, installation, and maintenance. The fundamental characteristics of wired transmission media are briefly summarized in Fig. 3.36. The most important variations are presented as follows.

Important Characteristics of Transmission Lines

In addition to the previously discussed **bandwidth** in Fig. 3.4, the following characteristics of guided transmission media are particularly important.

- **Impedance**
  Impedance describes the ratio between the current and voltage waves propagating in a common direction. It is a standard for assessing the high-frequency behavior of wired transmission media. The characteristic impedance of transmission lines should remain constant up to high frequencies. To avoid disturbance in data transmission from the reflection of the transmitted signals, all network components should have the same impedance.

- **Damping (Attenuation)** While being guided along the transmission line the signal is damped, meaning the amplitude of the signal lessens as the distance traveled increases. In the process, the oscillation energy is transformed into another form of energy. When no new energy is introduced to it, the signal ebbs. Signal damping is dependent on the frequency and should be as small as possible over the entire frequency range. The damping is calculated from the level difference of the transmitted signal at both ends (sender and receiver) of the electro-magnetic line.

- **Crosstalk Attenuation (Near End Crosstalk, NEXT)**
  Crosstalk attenuation involves a signal weakening caused by a mutual interference of what are in fact independent signal channels. Crosstalk (XT) is an effect seen in telephony. It may be observed during a telephone conversation when the bits of speech from another conversation are unintentionally picked up. In adjacent transmission lines, the difference in the crosstalk attenuation level should be as large as possible between the output level on one line and the input level on the other line.

- **Attenuation to Crosstalk Ratio (ACR)**
  As the difference between NEXT and signal attenuation, ACR describes the largest possible signal-to-noise ratio, which in turn expresses the quality of a transmission line.

Further reading:

Ulaby, F. T.: Fundamentals of Applied Electromagnetics. Prentice-Hall, Inc., Upper Saddle River, NJ, USA (1997).

Fig. 3.36 Important characteristics of transmission lines.

3.3.1 Coaxial Cables

Coaxial cables are bipolar cables with a concentric structure consisting of a core of solid copper wire (inner conductor), surrounded by an insulation layer (dielectrium, in the norm composed of synthetic materials or gases). Flexible coaxial cables can also have an inner conductor made of thin braided or stranded copper wire. The insulation jacket itself is encased in a hollow cylindrical outer conductor, which mostly takes the form of woven metallic braid (cf. Fig. 3.37). This outer conductor is encased again in a protective jacket that is normally made of a synthetic material and has insulating properties as well as being noncorrosive and waterproof. Conventional coaxial cables have an outer diameter of 2 to 15 mm and special variations of 1 to 100 mm. In coaxial cables, signal transmission proceeds over the inner conductor.

The outer conductor serves as a reference ground and at the same time is used for signal feedback.

Because of its construction and good insulation (shielding) coaxial cable has a high immunity to noise and is thus well suited for application in the high bandwidth range (today mostly up to 1 GHz). For this reason, coaxial cables are often called high frequency cables. Coaxial cables were often used in the past in long-distant telecommunications, but have been largely replaced today by fiber optic cables. Throughout the 1990's coaxial cables were used in computer networks in Ethernet-based technologies. Today, a common application of coaxial cables can be found in television technology (cable networks).

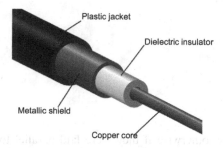

Fig. 3.37 Construction of a coaxial cable

At present, there are two types of the coaxial cable in widespread use: one for digital transmission applications, the 50-ohm coaxial cable (characteristic impedance 50 ohm) and one for analogous transmission and cable television application, the 75-ohm coaxial cable. This distinction has a historical background. The first bipolar antennas had an impedance of 300 ohm and the manufacturer's corresponding 4:1 transformer was a less expensive model. The first patent was filed for the coaxial cable in 1880 in England by the British electrical engineer, inventor and physicist *Oliver Heaviside* (1850–1925).

The baseband[5]-coaxial cables are used for the so-called "thin Ethernet" (10Base-2) network wiring and have the designation RG-58. RG-58 coaxial cables are connected with each other by a so-called **BNC connectors** (Bayonet Neill-Concelman connector) with a bayonet fastening device.

3.3.2 Twisted Pair Cables

Wiring types where both wires of a wire pair are twisted together are known as **twisted pair cables (TP)** (twisted pair). Normally, a simple twisted pair cable consists

---

[5] The term **base band** (also basis band), designates the frequency range where the signal is located. Through modulation, a base band signal can be shifted by means of a carrier frequency into another frequency range (bandpass position), which could potentially be better suited for transmission.

of two insulated, approx. 1 mm thick, copper wires that are twisted spirally around each other. It is often the case, however, that different wire pairs consisting of various degrees of twisting – the so-called lay length – are twisted into a cable (cf. Fig. 3.38).

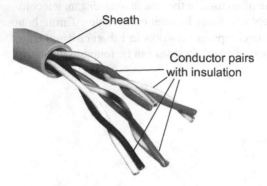

Fig. 3.38 The construction of a twisted pair cable.

The twisting of the cables is a necessary, otherwise if they were laid parallel to each other they would function as an antenna and their transmitted signals would cause interference in data transmission (crosstalk). Through twisting, the resulting waves are broken and most of the time cancel each other out. In turn, the cable is less susceptible to interference. For the same reason, twisting also protects against external interference caused by alternating magnetic fields and other electrostatic influences. Additionally, twisted pair cables often have an electrically conductive of aluminum sheet or copper braiding for protection against outer, interfering electromagnetic fields. Twisted pair cables are employed in analog as well as digital transmission. Thereby, symmetrical signals are transmitted on both strands, so that on the other end of the cable path the original signal can be reconstructed as well as possible by calculation of the difference. Each strand consists of a plastic insulated copper conductor that normally has a diameter of 0,40 mm or 0,6 mm. The standard designation of typical twisted pairs cables are 4x2x0,4 (4 strand pairs, strands running in pairs, strand diameter 0,4 mm) or 4x2x0,6.

A basic distinction is made between the following types of twisted pair cables:

- **Unshielded Twisted Pair (UTP)**
  In accordance with ISO/IEC-11801, a standard cable with unshielded twisted pairs and without an overall shield. In German-speaking countries, UTP cables are rare, despite being the most used cable in the world. More than 90% of all Ethernet LANs worldwide use UTP. In comparison to shielded cables, it is possible to just achieve a lower bandwidth and range with UTP. UTP is, however, sufficient for application in Gigahertz Ethernet (bandwidth 1 GHz). The advantage of UTP cables compared to shielded cables is their small outer diameter. Owing to the absence of shielding, UTP cables are easier and cheaper to use.

- **Foiled Twisted Pair (FTP)**
  In accordance with ISO/IEC-11801, a standard cable with strand pairs encased in a metallic shield (normally an aluminum-laminated synthetic sheet). Because of this shield, the FTP cable has a slightly larger outer diameter and is therefore more difficult to lay than UTP cable, due to its lower bending radius. Compared with UTP cables, FTP cables exhibit a higher electromagnetic compatibility. This means that the crosstalk between the signals on the wire pairs is reduced by the shield.

- **Screened Foiled Twisted Pair (S/FTP)**
  In accordance with ISO/IEC-11801, a standard cable that follows the same basic construction as FTP cable. There is, however, an additional metallic shield around the entire conductor bundle. The shield can be designed as sheet or as a wire braid, or as a combination of both.

- Besides these variations, there also exist the **S/UTP (Screened Unshielded Twisted Pair)** cable, where unshielded wire pairs are surrounded by a overall shield, and the **ITP (Industrial Twisted Pair)**, a cable variation used in industry, where instead of the usual 4 wire pairs only 2 pairs are twisted.

Furthermore, there are seven more cable categories. They are defined based on the capacity of built-in individual components. Categories 1 and 2 are only informally defined, categories 3 and 4 are in the meantime obsolete.

- **Category 1**
  Cat.1 cables are designed for maximum operating frequencies of up to 100 kHz. This cable variation has found application particularly in speech transmission via telephony applications and the wiring of doorbells as UTP cable.

- **Category 2**
  Cat. 2 cables are designed for maximum operating frequencies of up to 1,5 MHz and have been used for, e.g., house wiring for ISDN primary multiplex connections as a UTP cable or also in the basic version of token ring networks.

- **Category 3**
  Cat. 3 cables are non-shielded twisted pair cables designed for maximum operating frequencies of 16 MHz and used for a transmission capacity of up to 16 Mbps. Cat. 3 cables are used in 10 MBit Ethernet (10BASE-T).

- **Category 4**
  Cat. 4 cables are designed for maximum data transmission rates of up to 20 Mbps and were used in the 16 Mbps variation of the token ring network.

- **Category 5**
  Cat. 5 cables are now the most widely used type of twisted pair cable used today and are designed for operating frequencies of up to 100 MHz. Cables in category 5 are often used in the structured cabling of computer networks, such as in Fast (100BASE-T) or Gigabit Ethernet (1000BASE-T).

- **Category 6**
  Cat. 6 cables are designed for operating frequencies of up to 250 MHz and are

used in speech as well as data transmission. Cables of the category **Cat-6a** (augmented cable) have an even higher capacity with an operating frequency of up to 500 MHz, used in 10 gigahertz Ethernet (10GBASE-T).

- **Category 7**
  Cat.7 cables (class F and FA) are designed for operating frequencies of up to 1000 MHz and have four individually shielded pairs (Screened/Foiled Shielded Twisted Pair S/FTP) within a single shield. Cat.7 cable is also suitable for use in 10 gigahertz Ethernet (10GBASE-T) networks.

3.3.3 Fiber Optic Cables

Fiber optic cables or **optical fiber cables** (OFC) are fibers, or a body, made of a transparent, light permeable material (most of the time glass or synthetics), which serve to facilitate the transport of visible or infrared light (cf. Fig. 3.40). Fiber optic cables are comparable to coaxial cables as both contain a glass core where light used to transmit information is spread. This glass core can have different diameters. **Multi-mode optical fibers** have a core diameter of 50 μm (approximately as wide as a human hair), whereas, in comparison, **mono-modal optical fibers** have a much smaller core diameter of approx. 8-10 μm.

A simple (unidirectional) optical transmission path consists of three components: a light source – today normally semiconductor lasers (for data transmission over large distances) or light-emitting diodes (LED, for data transmission over a short distance) are employed – as well as an optical transmission medium (glass fibers) and a detector. The detector generates an electrical impulse as soon as light is cast upon it. The light source is controlled by an electrical signal that sets off light impulses (light impulse = 1 bit, no light impulse = 0 bit) that are transmitted over the optical fiber cables. When the light impulses reach the detector at the other end of the optical fiber cable, they are transformed by it back into electrical signals.

In this way the modulation for data transmission is implemented with the light source itself or with the help of external modulators (normally a Mach-Zehnder modulator or an electro-absorption modulator). At the receiver, the amplitude modulated light stream is transformed back into an electrical signal by a light sensitive photodiode. If transmission proceeds via phase shift keying then it has to be transformed beforehand into an amplitude shift keyed signal with a suitable demodulator, before the signal of the photodiode can be transformed into an electrical signal.

Light waves on the interface of optical fibers overlap each other through constructive interference as a consequence of the total reflection of transmitted light into so-called **modes**. The possible modes of a light conductor are numbered consecutively from the flattest to the steepest reflection angle. In extremely thin light wave conductors, which only exhibit the thickness of a few wave lengths, just the first mode (fundamental mode) is possible. These are therefore called **monomodal** optical fibers (single-mode fiber). With monomodal optical fibers, light waves travel

Cable Types

A distinction is made between cable types according to their intended purpose. In addition to permanently installed cables, which run directly into a wall socket, so-called **connection cables** are required for connection to the endpoints on these sockets.
To connect network devices that are spatially close to each other, short **patch cables** are used, for example, to connect a network switch to a router housed inside a network cabinet in a data processing center.

For TP cables, the connection and patch cables normally end in so-called **RJ45 connectors** (registered jack connectors), each with 8 contacts, which are standardized according to TIA/EIA 568A and 568B. Depending on their application, the connection contacts are assigned differently (pin assignment), e.g., in an Ethernet network the strand pair 2 and 3, or in a token ring network the strand pair 1 and 3.

For communication between network-enabled devices a distinction is made in the connection between two similar devices, for example the direct connection between two computers and the connection between different network components. An example for the latter is the connection between computer and network switch. If two computers are directly connected to each other then a **crossover cable** is required. Here, the receiving plug contacts of one side (RX) are connected cross-wise to the sending plug contacts of the opposite side (TX). To connect a computer to another network component, e.g., a switch or router, the respective plug contacts must be connected or "looped through" the identical plug contacts of the opposite side. Therefore this cable variation is referred to as a **straight-through cable**.

Another type is the so-called **rollover cable**, in which all of the pins are crossed in pairs. By way of illustration, plug contact 1 is connected with plug contact 8 of the opposite side, and 2 with 7, 3 with 6 and so on. Rollover cables are employed, for example, in the first configuration of routers via serial interface over an external console (terminal), when the router is not yet accessible via Internet Protocol.

Further reading:

Vacca, J. R.: The Cabling Handbook, Prentice Hall Professional Technical Reference, Upper Saddle River, NJ, USA (1998).

Fig. 3.39 Cable types and their application.

Light Transmission Via Fiber Optic Cables

The material of glass was already known to ancient Egyptians. But before glass fibers can be used for interference-free transmission, the glass used must demonstrate a high degree of purity. It was only first possible in the Renaissance to achieve the level of purity necessary for the manufacture of window glass. Glass fibers have their origin in the 18th century in the work of glass blowers in the Thuringian Forest, who made so-called fairy or angel hair out of spun glass for decorative purposes.

British physicist *John Tyndall* (1820–1893), came up with the idea of directing light along an optically transparent conductor when he attempted to guide a light beam along a jet of water. But it was only in the 1950's that optical conductors were first employed in the area of medicine. The loss of light still proved so great that their use was only possible over short conductor paths. With the development of the first lasers in 1960 by *Theodore Maiman* (1927–2007), a strong enough concentrated light source was available so that optical conductors could also be used for data transmission. For the discovery of the the high loss of transmission power resulting from glass impurities as well as his pioneering work in the field of fiber optics, *Charles Kuen Kao* (*1933) was awarded the Nobel prize in physics in 2009.

Light conduction in optical data transmission proceeds on the basis of **total internal reflection**. Normally, a cylindrical glass fiber is surrounded by a medium with a lower refractive index, which is shielded by a protective covering. In the interface between both of the transparent media, which have a different refractive index, light is reflected virtually without loss when it is cast at a certain angle. However, if a light beam from an optically transparent medium crosses over into another, it is refracted on the interface between the two media. The strength of the refraction depends on the two materials (refractive index). If the angle of incidence is above a critical angle (acceptance angle), it is reflected completely and a total reflection takes place. The light is captured within the optical wave conductor and can spread out nearly without loss and be guided in a desired direction. Fiber optic cables can transport information in the form of light for more than 20,000 meters, without the need for amplification.

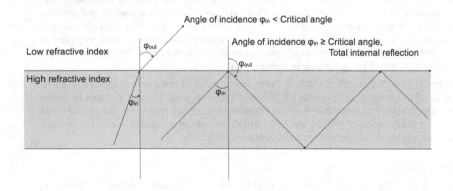

Further reading:

Mitschke, F: Fiber Optics : Physics and Technology, Springer, Berlin, Heidelberg (2010).

Fig. 3.40 Transmission of light over fiber optic cables.

over a single possible light path. Because of this, with a set frequency there are no time differences evidenced between different light paths. Monomodal optical fiber cables are more costly and are mainly utilized for long-distance transmission.

With light conductors having a larger diameter many modes appear, as many light beams in various angles meet above the acceptance angle on the interface and are reflected differently. These thicker light wave conductors are called **multi-modal** optical fibers (multi-mode fibers).

Besides the other physical characteristics mineral glass – the raw material used for light wave conductors – possesses, damping has a critical influence on an efficient conduction of light. The damping of light through glass depends on the wavelength of the light. Thus, for optical data transmission in optical fibers, the band widths are used where the least damping occurs. While the visible spectrum of the wave length range encompasses between 400 nm and 700 nm, the damping in longer wave lengths in the infrared range is significantly lower. The loss of power caused by damping in the last two bands amounts to only 5% per km. The single bands each have a bandwidth of between 25.000 GHz and 30.000 GHz.

If a light impulse is sent on a light wave conductor it stretches in length as a result of transmission. This effect is known as **chromatic dispersion**, and is dependent on the wavelength of the light impulse. This broadening can lead to the succeeding light impulses overlapping and thus interfering with each other (intersymbol interference). An enlargement of the pulse intervals would, however, mean a reduction of the data transmission rate. Through a special type of pulse shaping, the so-called **soliton**, the effect of chromatic dispersion can be prevented for the most part. A soliton is a light pulse that maintains its shape when spread over a distance (cf. Fig. 3.41).

Optical fiber cables can be connected to each other in a variety of ways. In all variations, reflections appear on the joints of two optical fiber strands that can potentially disrupt the signal to be transmitted:

- The simplest way of connecting optical fiber components is to use plug connectors, i.e., an optical fiber terminates in a connector that is inserted into a corresponding counterpart – an optical fiber connector. Although this type of connection simplifies the configuration of system components, as with every plug connection there also occurs a power loss of between 10% and 20%.

- A further possibility is the exact, mechanical splicing of two optical fiber ends. The ends are cut clean, laid next to each other in a special jacket and clamped in place. Often the mutual alignment of both optical fiber ends can be optimized during the act of splicing. Light is sent through the conductor while small corrections continue to be carried out. The splicing of optical fibers requires a trained specialist and causes an approx.10% power loss at the interface.

- The cleanest – while at the same time most complicated – connection possible is achieved by fusing two optical fiber ends to form a single fiber glass strand. This type of connection is also called "fusion splicing." Minimal signal damping occurs here as well, however this method exhibits the lowest amount of loss.

Data transmission using optical fibers has an advantage over electrical transmission due to its higher bandwidth (gigabit up to terabit range). The signal transmitted via

Soliton

Optical fiber cables that transmit light pulses are used for data transmission. This involves wave packets made up of different individual frequencies. The propagation velocity of a wave in a dispersive, non-linear medium depends on its frequency. Therefore, a transmitted wave packet which changes with increasing distance to the sender, becomes broader and "smears." This effect is known as **chromatic dispersion**.

Utilizing nonlinear effects it can be achieved that the individual frequencies, which make up a wave packet, convert into each other. Thus higher proportions change into lower ones and vice versa. In this way, a dynamic balance is created. Such a phenomenon is called a **soliton**. A soliton can spread without changing its form and has the following three basic characteristics:

1. Solitons possess a permanent form and do not change their shape.
2. Solitons are a localized phenomenon.
3. Solitons can interact with other solitons without their form being influenced or changed (except in the case of a phase shift).

The soliton phenomenon was first described in 1834 by the ship-building engineer *John Scott Russell* (1808–1882). While attempting to find the most efficient design possible for a canal boat, Russell observed that a wave in a canal may spread out over several kilometers without changing its form. In further experiments he was able to determine that the wave, which he called a „translation wave" did not only not merge with other waves but that a larger wave can even pass up a smaller one. The theoretical explanation of the soliton effect was first supplied by the Korteweg-de Vries equation in 1895.

The significance of solitons in practical application was only first realized at the end of the 1960's. In 1973, the existence of solitons in optical fiber cables and their utilization for communication and data transmission in optical fiber networks was predicted by the British physicist *Robin K. Bullough* (1929–2008), and in 1980 experimentally proven. In 1991, a research group at Bell Labs succeeded in error-free soliton data transmission at the rate of 2.5 gbit/s over a distance of more than 14,000 km.

Further reading:

Scott, A. C., Chu, F. Y. F. , McLaughlin, D. W. : Soliton: A new concept in applied science. in Proc. IEEE 61, Nr. 10, pp. 1443–1482 (1973)

Fig. 3.41 Soliton

optical fibers is resistant to electromagnetic interference fields and offers increased protection against electronic eavesdropping. The laying of fiber optic cables requires a greater effort than simple copper cables because material-wise a minimal bending radius may not be exceeded. If the minimum bending radius is exceeded this results in a coupling of light. If this happens, the optical fiber cables can even be destroyed as the consequent absorption in the shell leads to the development of excessive heat. Moreover, the bending radius is limited by the mechanical breaking strength. In order to prevent breakage of fiber optic cables they are normally provided with a shell or a reinforcement that prevents dropping below the minimum bending radius. Optical fiber cables are significantly lighter than copper cables and also evidence a smaller diameter. They do not need a grounding and can also be used in explosive atmospheres, as sparking due to electrical short circuits cannot occur. Corroding substances cannot affect optical fiber cables, thus making their application in critical industrial areas possible. In contrast to the limited availability of the raw material copper, optical fiber cables are made of mineral glass and the basic component, silicon dioxide, is virtually unlimited.

A further advantage of optical fiber cables over copper cables is their reduced signal damping. With an optical fiber wide-area network, signal amplification using a repeater needs to take place every 50 km, while with copper wire this is already necessary at approximately 5 km. In contrast to copper cables, optical fiber cables have a higher bandwidth at a considerably lower weight. This means that mechanical support systems and infrastructures can be constructed with far less effort and are cheaper in comparison.

3.4 Wireless Transmission Media

Wireless communication has been gaining rapidly in importance since the end of the 1990's, freeing us from"the leash" of wired transmission media. This has led to a whole new mobility in speech and data communication. While on the go we can communicate and exchange data via mobile, or another wireless data transmission technology, from nearly every place and at any time. Wireless transmission media has been made possible by the utilization of **electromagnetice waves**, which spread themselves either directed or undirected in space and have the capability to transmit data via the previously discussed modulation and multiplex methods. Principally, a distinction can be made between **terrestrial radio networks**, whose signals spread along the earth's surface (transmitter and receiver are both on the earth's surface), and **satellite radio networks**, whose signals are received from the earth's surface by a satellite encircling the orbit of the earth and sent back to the earth's surface. Depending on the type of wave propagation, radio transmission can be divided into the following groups:

- broadcast (radio and television networks),
- cellular radio (mobile communication networks) and
- radio link.

Radio and television networks and cellular radio networks are designed as broadcast networks, which means that a transmission is received by all radio members. Mobile communication networks are made up of so-called **radio cells**, in the center of which a base station communicates with the mobile members of the cells or with other base stations. Adjacent cells use different frequencies so as not to interfere with each other. To facilitate this in GSM mobile radio networks (Global System for Mobile Communications), seven cells each with different frequencies (F1–F7) are fused together into a macrocell. This means that in adjacent macrocells individual radio cells with the same radio frequency are always as far away from each other as possible (cf. Fig. 3.42).

With radio link networks, point-to point, directed connections can normally be installed as optical directional radio links or microwave directional radio links. Count-

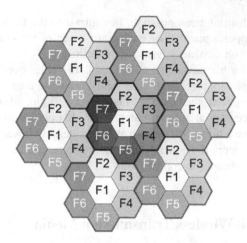

Fig. 3.42 Distribution of cells
with different frequencies
(F1–F7) in a cellular radio
network

less areas of the electromagnetic spectrum are suitable for the wireless transmission
of data, beginning with radio waves through the area of visible light and subsequent
ultra-violet light. As previously described, these waves possess different characteristics, which has led to the the development of completely diverse transmission
technologies. Their use is described briefly in the following.

3.4.1 Radio Transmission via Short Waves and Ultra-short Waves

The frequency range for short waves and ultra-short waves was originally used for
radio broadcasting, as were middle and long waves. In radio broadcasting, electromagnetic waves are generated over high frequency alternating currents with an
electrical oscillating circuit. They are modulated onto the signal with the help of a
suitable modulation process and then radiated by an antenna. The actual conversion
of high frequency alternating current fields into electromagnetic waves is carried
out in the transmitting antenna, which normally consists of metal rods and metal
reflectors.

Short waves and ultra-short waves are **omnidirectional**, and used especially for
the transmission of radio and television signals. In this case, the transmitter and
receiver need not be aligned exactly to each other. The transmission rate reached
for digital transmission is currently in the 100 Mbps range. Thus, large distances
can be bridged and obstacles can be penetrated effectively with lower frequencies.
However, short waves and ultra-short waves are prone to interference caused by rain
and stray radiation from electrical devices or motors. With increasing distance, the
radiated transmission power drops quadratically. At higher frequencies, radio waves
run linearly and are absorbed or reflected by obstacles.

Radio waves can cover very large distances, which means that overlapping and interference may result from other users. Accordingly, the use and allocation of radio

frequency bands for radio broadcasting is strictly controlled worldwide. With lower frequencies in the VLF, LF and MF areas, radio waves spread out as ground waves and follow the profile of the earth's surface (cf. Fig. 3.43). Their transmission range can measure up to 1000 km (the higher the frequency in this range, the shorter the scope) and they penetrate buildings and other obstacles easily. Because of their extremely limited bandwidth these low frequency bands are not used for digital communication.

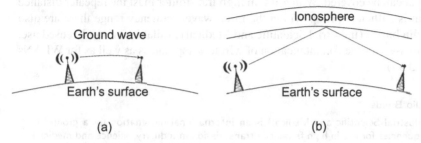

Fig. 3.43 In low frequency ranges (VLF, LF and MF bands) radio waves follow the curve of the earth's surface (a) and in high frequency ranges (HF bands) are reflected in the ionosphere (b).

In high frequency transmissions in the HF and VHF ranges, the ground waves can only achieve a short reach and are reflected back to the ground by the ionosphere, a layer of the upper atmosphere at the height of 100 – 500 km containing ionized electric particles. They can then be reflected once again under ideal conditions (zigzag reflection, multi-hop) and in this way cover a great distance.

3.4.2 Data Transmission via Microwaves

Starting from a frequency of approximately 100 MHz, radio waves spread out in virtually a linear form and let themselves be bundled with the help of parabolic reflectors, horn antennas, or horn radiators. These antennas are normally placed in raised locations in order to avoid possible obstructions. In this way a point-to-point communication is made possible, with superior signal-to-noise ratio when compared to short and ultra-short wave communication. Obstacles along the line of vision cause interference in the frequency range utilized and damp the transmitted signal. Furthermore, other disturbances can arise as a result of selective damping and the refraction of individual frequency areas on different atmospheric layers. The phase shift caused by the delayed arrival of the affected frequency range can lead to signal deletion. Selective fading (multipath fading) always depends on weather and on the selected transmission frequency.

Most of the time nowadays radio carrier frequencies are used in the microwave range between 1 and 40 GHz. Above 4 GHz it is possible for signal damping to occur as a result of raindrops. At this frequency range an absorption of electromagnetic waves sets in through the water. Before the beginning of optical fiber networks in the 1990's, radio link transmission formed the foundation of the long-distance telephone network. By way of radio link networks it is possible to bridge distances of up to 100 km. In order to accomplish greater ranges, amplifying **repeaters** are set up in appropriately large distances to one another. These repeaters receive the radio link signal transmitted from the respective remote station and pass it on in amplified form to their nearest neighbor. The higher the transmitter mast the greater the distance that can be covered. With a 100 m high transmitter mast the repeater distance of 80 km is sufficient [261]. Within the microwave frequency range there are also **ISM radio bands** (Industrial Scientific and Medical) available for non-licensed use. These are used for the communication of wireless appliances as well as for WLANs (cf. Fig. 3.44).

ISM Radio Bands

ISM (Industrial Scientific and Medical) is an international designation for a group of radio frequencies for use in high frequency transmission in industry, science and medicine, as well as in consumer electronics. No special license must be acquired for their use. Devices that use this frequency only need a general permit. Among them: microwave ovens (2,4 GHz), medical applications, such as high-frequency magnetic field therapy (150 MHz), cordless DECT telephones (1,88 − 1,93 GHz, not yet internationally regulated), radio frequency identification tags (Smart Tags, 13,56 MHz), model-making remote control (27 MHz, 35 MHz, 40,6 MHz und 2,6 GHz), and data transmission via WLAN (2,4 GHz und 5,8 GHz) and Bluetooth (2,4 GHz).

As transmitters in the same or an overlapping frequency area cause mutual interference, the existing frequency space is traded as a precious commodity. This was particularly evident at the UMTS auction of licenses for mobile communications systems of the 3rd Generation in 2000, which unlike ISM are not free. License fees received from the auction in Germany alone totaled 50.8 billion euros, and in Great Britain 38 billion euros The regulation of the use of frequency space is based on national and international agreements. On the international level the ITU (International Telecommunication Union) coordinates the worldwide use of the frequency range so that mobile and device applications can be carried out across national borders. These countries, however, are not strictly bound by ITU policies.

In the ISM frequency range, radio applications are pooled together. Because of their susceptibility only short distances can be bridged or short-term operational disturbances tolerated. Devices that implement frequencies from the ISM frequency range can be used by everyone without a special frequency allocation or a fee.

Further reading:

Code of Federal Regulations, Title 47: Telecommunication, Chap. 1, Federal Communications Division; Part 18, Industrial Scientific and Medical Equipment (1985).

Fig. 3.44 ISM radio bands.

3.4.3 Infrared, Millimeter Waves and Optical Transmission

Undirected **infrared and millimeter waves** are used primarily today in short-range transmission. In contrast to long wave radio waves, infrared and millimeter waves already have nearly all the characteristics of optical light i.e., that which is perceptible for humans. They spread out linearly along an optical connection and are not capable of penetrating solid obstacles. This makes them ideal for short distance application within rooms. Because of this restriction, devices used in areas implementing infrared and millimeter waves exhibit only minimal malfunction properties. Also the risk of eavesdropping by unauthorized third parties is relatively small due to spatial restrictions. In addition to the operation of remote controls, optocouplers, and light barriers, infrared transmission is used particularly in the desktop area, for example in the data transmission between computer and peripheral devices in a Personal Area Network (PAN). The importance of infrared transmission in the computer network area has waned over time and been replaced, to a great extent, by more powerful, radio-based technology, such as e.g., Bluetooth.
It is also possible to carry out optical transmission over greater distances along an undisturbed optical connection. Signal transmission using visible light is therefore also known as optical free-space transmission. Here, lasers are used in conjunction with light-sensitive photo detectors. Optical data transmission via lasers achieves a high data transmission, but is strongly dependent on the prevailing weather conditions. In the case of rain or fog the laser beam is scattered and it is necessary to fall back on alternative communication paths. Furthermore, the airflow resulting from convective currents in the atmosphere can divert and disturb the beam.

3.4.4 Satellite Communication

Communication satellites have been orbiting the earth since the 1960's. A communication satellite principally acts as an amplifier, or a relay station, to disseminate a signal transmitted from the earth over an extensive geographic area. The "optical-dissemination spreading characteristics" of high frequency signals are put to use for full-coverage transmission along the earth's surface, where otherwise a multiple number of repeaters would be necessary. The communication satellite houses one or more transponders for various frequency ranges over which the incoming signal is amplified and sent back again on an alternate frequency.
Satellite communication is commonly applied in point-to-point communication, e.g., in satellite telephony, for the worldwide dissemination of radio and television, as well as in data communication. For the end user of a data connection, normally only the link from provider to end user is conveyed via satellite. The return channel from the end user to the provider often proceeds via a traditional, wired communication connection as providing a return channel for a large number of participating member is often too laborious.

A principle distinction is made between two types of satellites dependent on their orbital altitude above the earth's surface. Based on Kepler's law of the orbital motion of celestial bodies, the revolution period of a satellite is dependent on its altitude. The higher a satellite the more time needed for its orbit around the earth. Thus, there is a difference between satellites that move synchronous to the earth's rotation and those that revolve in a lower orbit and have a shorter revolution period.

- **Geostationary Satellites**:
  A satellite that circles the earth at a height of 35,790 km above the earth's surface around the equator with an orbital inclination of 0 degrees moves synchronous to the earth's rotation. It thus always remains locally fixed over the same place on the earth's surface and can therefore also be used as a communication relay for relatively high data transmission rates, without it being necessary to track the satellite's movement. Satellites located in this altitude are called **GEO satellites** (Geostationary Earth Orbit). The idea of worldwide radio communication with the help of geostationary satellites was proposed for the first time in 1945 by British science-fiction writer *Arthur C. Clarke* (1917–2008). Just 19 years later his idea would become reality with the geosynchronous satellites Syncom 2 (1963) and Syncom 3 (1964). At the same time, geostationary orbit is very much coveted and the available places around the equator scarce. Modern technology allows the positioning of a geostationary satellite in 2 degree intervals within the 360 degree equatorial plane. A maximum of 180 satellites can be placed thusly, whereby the coordinates of the revolution position is designated by the ITU.
  A modern communication satellite can carry up to 40 different transponders and be run within different frequency ranges based on the principles of frequency and time multiplexing. ITU reserves 5 frequency bands for satellite communication (L, S, C, Ku und Ka) in areas between 1,5 GHz and 30 GHz. In order to guarantee a global (extending up beyond the polar regions at 82 degrees latitude) coverage of communication over geostationary satellites, 3 satellites are needed.

- **LEO/MEO Satellites (Low Earth/Medium Earth Orbiting)**:
  Satellites with an altitude of between 200 km and 1,500 km are called LEO satellites (Low Earth Orbiting). The time they need to make one revolution around the earth is a relatively short 90 minutes to 2 hours. With this type of satellite the problem exists that they invariably tend to move quickly out of the receiving range of the end user and must then be tracked. LEO orbits are the energy-lowest orbits and can consequently be reached with the lowest effort. Therefore, for a long time they only played a secondary role as a communication satellite. In order to ensure coverage for global communication with LEO satellites about 50-70 satellites would be necessary.
  Only in the 1990's did the idea of a worldwide satellite mobile communication system again become a subject of public focus when the **Iridium** project was started at the initiative of the Motorola company. It proposed a network of 66 (originally 77) active LEO satellites at an orbit altitude of 750 km, in order to ensure a full-coverage mobile communication network. However, the necessary investment costs proved to be too high and acceptance was still so low that the

project was threatened with being shelved in 2000. Since March 30, 2001 the Iridium network has been commercially operated.

Altitudes between 1,500 km and the geostationary orbit below 36,000 km are covered by MEO satellites (Medium Earth Orbiting). It should be noted that the areas between 1,500 km and 3,500 km as well as between 15,000 km and 20,000 km are avoided because of their high radiation level. The Van Allen Belt, named after the American astrophysicist *James van Allen*, (1914–2006) is located there. This is a torus of high-energy cosmic particles that are trapped in the magnetic field of the earth. 10-12 satellites are required to assure global network coverage over MEO satellites. The time slot available for communication is 5-12 hours. Similar to the just mentioned Iriduim network, the satellite network, **Globalstar** has 48 MEO satellites which can be found in an orbital height of approximately 1,400 km. The Globalstar network is not able to cover the polar areas which can only be reached by the Iridium network. In contrast to Iridium satellites, Globalstar satellites do not maintain direct contact with one another, but are always transmitted over a gateway on the ground within the footprint of the active satellite via a conventional communication system. This is the reason that areas on the high seas as well as in parts of Africa and Southern Asia cannot yet be reached over the Globalstar network.

In contrast to terrestrial connections, satellite connections have a higher transmission rate caused by the enormous distance involved in reaching the e.g., geostationary satellites. Although the signals disseminate at the speed of light, the transmission time from endpoint to endpoint can be from 250 ms to 300 ms. In comparison, terrestrial microwave connections have a signal dissemination delay of approx. 3 μs/km and optical fiber cables of approx. 5 μs/km.

3.5 Glossary

Bandwidth: A given measurement in hertz (1 Hz = 1/s), used in physics, telecommunications and computer science with different meanings. Seen physically, bandwidth B designates the difference between two frequencies (lower and upper cutoff frequency), which form a continuously connected frequency range. In analog telecommunication technology, bandwidth designates the frequency range in which electrical signals with an amplitude decrease of up to 3 dB are transmitted. The larger the bandwidth, the more information can theoretically be transmitted in one unit of time. In computer science the term "bandwidth" means the data transmission rate (also called transmission rate or data rate) of digital signals, used as a measurement for speed in which data can be transmitted over a transmission medium in the form of bits per time unit.

Channel capacity: Channel capacity defines the upper limit of an error-free transmitted information flow (number of transmitted bits) in a specific transmission medium, based on the signal-to-noise ratio and in relation to the frequency spectrum to be transmitted.

Chirp Impulse: A signal whose sinusoidal signal waveform continuously rises, or falls, over time in frequency is called a chirp impulse. Among other technical applications, chirp impulses are used in spread spectrum modulation methods, e.g., the Chirp Spread

Spectrum (CSS). This method has proven resilient to interference caused by the Doppler effect as only the frequency change of a chirp impulse over time is of importance. In nature the chirp impulse is used by bats for tracking.

Coaxial cable: A bipolar cable with a concentric structure, normally containing a core of solid copper wire (inner conductor), surrounded by an insulation layer (dielectrium, that is usually made up of synthetic materials or gases). Flexible coaxial cables can also have an inner conductor of thin, braided or stranded copper wire. The isolation jacket itself is encased in a hollow cylindrical outer conductor that mostly takes the form of a woven metallic braid. This outer conductor is once again encased in a protective shield that is normally made of synthetic material and has insulating properties, as well as being noncorrosive and waterproof. Signal transmission over coaxial cables proceeds via the inner conductor, while the outer conductor serves as the reference ground and is used at the same time for signal feedback.

Communication: Communication is the process of a one-way or reciprocal exchange, conveyance and reception of information by humans or technical systems.

Communications protocol: A communications protocol (also simply protocol) is a collection of rules and regulations that determines the data format of the information to be transmitted as well as establishing the guidelines for its transmission. Protocols contain agreements about the to-be-transmitted data packets, the establishing and terminating of a connection between communication partners, as well as the manner of data transmission.

Digital: (digitus=[lat.]finger), The designation for a technology/method that only uses discrete, discontinuous, i.e. cascading mathematical values. The basis for digital technology is the binary (two value) number system, that can only distinguish between the conditions "true" and "false," or the number values "1" and "0." These binary numerical values are called **bits** (**binary digit**) and represent the smallest units of information.

Digital communication   Digital communication refers to the exchange of digital information over specialized digital communication channels. The data format of the information determines the respective media (text, image, audio, video, etc.). The information is transmitted according to the conditions of the implemented communication protocols via a digital communication channel (e.g. Internet or WWW).

Direct wave propagation: Electromagnetic waves in the frequency range of 30 MHz and 30 GHz are no longer reflected in the ionosphere and propagate as direct waves. Ground waves in this frequency are damped so strongly that in order for even a small area to be covered, line of sight propagation between the antennas is necessary.

Doppler effect: A phenomenon named after the Austrian physicist *Christian Doppler* (1803–1853), the Doppler effect describes the change in perceived or measured frequencies of all wave types in which the source and the observer approach each other or recede from one another. As the observer and source move closer to each other the frequency perceived by the observer increases, when they move farther apart the frequency decreases. The best-known example of this is the siren of a passing ambulance. The Doppler effect plays an important role in wireless telecommunication with mobile terminals and is necessary to take into account when modulation and multiplexing methods are applied.

Electromagnetic wave: The designation for a wave that consists of interlinked electrical and magnetic fields. A time-varying electrical field always generates a magnetic field, while at the same time a time-varying magnetic field always generates an electrical field. One does not occur without the other, even if there is no carrier. Electromagnetic waves are used for the transmission of analog and digital signals.

Fiber optic cable: The term for fibers made out of a transparent, light-permeable, material (usually glass or plastics) that serve to transport light or infrared radiation. Fiber optic cables are comparable to coaxial cables as both have a glass core (coaxial cables a copper core) where the light used for information transmission propagates. Based on their diameter, a distinction is made between multi-modal optical fibers (thick fiber optic cable, in which during light transmission runtime interference occurs) and mono-modal optical fibers.

Fourier analysis: Any periodic signal g(t) can be decomposed into a seemingly endless number of simple sine und cosine functions. This decomposition is called Fourier analysis after the French mathematician and physicist *Jean-Baptiste Fourier* (1768–1830).

Ground wave propagation: Electomagnetic waves in the frequency range of 3 kHz up to approx. 30 kHz propagate as ground waves along the earth's surface. The range of a ground wave is dependent on both the frequency and the conditions of the ground.

Internet: The Internet is the world's largest virtual computer network, made up of computer networks linked over the standard Internet Protocol Suite (TCP/IP). The most important offers of the Internet – also called "services"– include electronic mail (email), hypermedia documents (WWW), file transfer (FTP) and discussion forums (usenet/-newsgroup). Its huge popularity has been achieved through the introduction of the World Wide Web (WWW). Often thought of as synonymous with the Internet, the WWW is in fact only one of the many services the Internet offers.

Internetworking: The implementation of several different individual networks to an internet of connected networks. To enable this, suitable linking network devices (routers) are necessary. They direct the path of a data package through the network system and, with the help of the Internet protocol, enable a secure delivery. The network system appears to the user as a homogeneous, virtual network (Internet).

Intersymbol interference: In intersymbol interference (also called crosstalk) there disturbances appear between temporal, successive, transmission symbols in a digital data transmission on a single transmission channel.

Ionosphere: The layer of the earth's atmosphere, where a significant amount of charged particles (electrons and ions) are found. In radio transmission this atmospheric layer is used for the transfer of shortwave signals (HF frequency range). Here is where signals in this frequency range are reflected back to the earth's surface and in this way can be transmitted worldwide. If a radio wave strikes the ionosphere the charged particles are induced to oscillate. The energy thus spent in transmitting energy is not lost as the oscillating electrons are also transmitters, and through their vibrations emit the same frequency in phase-shifted form again.

ISO/OSI Reference Model: A specification of the ISO that was designed and published as the basis for the development of communication standards. The core of this standard is an international reference model consisting of seven layers. The goal of the ISO/OSI Reference Model is to enable the communication of different computer and protocol worlds over uniform interfaces. In contrast to the TCP/IP Reference Model, which forms the very fabric of the Internet, the ISO/OSI Reference Model has decreased significantly in importance.

Line coding: In digital telecommunications, a line code determines how a signal is to be transmitted to the physical layer. Specific level sequences, such as light intensity in fiber optics and voltage or currency on electrical lines are assigned bit sequences. A distinction is made between line codes and other forms of coding, such as channel coding or source coding. With the help of supplementary redundancy, channel coding enables the recognition, or correction of data transmission and memory errors. Source coding, by contrast, removes redundancy from the information of a data source and is used for data compression.

Medium: The form of a transportation channel for the transmission of information between sender and receiver. In order to transmit information there must be an appropriate transmission medium for exchange between sender and receiver. A distinction is made between wired and wireless media.

Modulation: In information technology, modulation is described as a process in which the to-be-transmitted (data) signal changes (modulates) a carrier signal, thereby enabling transmission of the original signal over the normally higher frequency of the carrier signal.

Multiplexing: A designation for methods of signal and information transmission in which multiple signals are combined (bundled), and at the same time transmitted over a medium.

Noise ratio: This describes the relation between the amplitude of the output signal and the amplitude of interference signals. The noise ratio is also called the signal-to-noise ratio (SNR) and is indicated in decibels (dB).

Physical layer: Generally speaking, all of the physical and technical characteristics of a physical medium used for data transmission are defined in the physical layer. The protocol in the individual layers of the TCP/IP Reference Model of the Internet are all directed toward the physical transmission medium. The physical layer, which is itself not a part of the TCP/IP Reference Model, defines the reciprocal action between the network hardware and the physical transmission medium. Determined therein are: the layout of the plug connections and their electrical or optical parameters, the specifications of the physical nature of cables (electrical and optical), as well as the specifications of amplifying elements, network adapters and, additionally the data transmission method implemented.

Refractive index: The refractive index (also index of refraction, refraction) is a fundamental parameter of geometric optics and describes the refraction (change of direction) and the reflection behavior (reflection and total reflection) of electromagnetic waves upon impacting an interface from one medium to the other.

Skywave propagation: So-called skywave propagation occurs in electromagnetic waves within the frequency range of 300 kHz and 30 MHz. Here, electromagnetic waves from the sender propagate in space and through reflection in the ionosphere return to earth to be reflected again. In this way, skywaves achieve an extremely long range and, in the ideal case, can even be spread around the entire globe.

Signal: A signal is understood as the information-bearing temporal process of a measurable quantity within a physical system.

Signal noise: No signal can be transmitted without a loss of energy. The distortion (interference) of a signal through outer influences is called signal noise.

Signal-to-noise ratio: This describes the relation between the amplitude of the output signal and the amplitude of interference signals. The signal-to-noise ratio is given in decibels (dB). It is also designated as the dynamic and is an important standard for the quality of a signal.

Soliton: A soliton is a wave packet that can spread out over long distances without being disturbed by other waves. A wave packet is made up of different individual elements all having different (frequency dependent) propagation velocities in a propagation medium (dispersive, nonlinear). Due to non-linear effects, it is possible that the individual frequencies that make up a wave packet can be transformed into one another thus achieving a dynamic balance. This results in a **soliton**, which propagates without changing its form.

Spread spectrum: With spread spectrum or frequency spreading a technique is defined in which a narrow band signal is converted into a signal with a larger bandwidth than is necessary for information transmission. In the process, an electromagnetic transmission energy, which had previously been concentrated in a smaller frequency range, is now distributed over a larger frequency area. The power density of the signal can thereby be lower than in a narrow band without the data getting lost during data transmission. An advantage of this method is the greater robustness of the signal to narrow band disturbances.

Total reflection: Total reflection is a wave phenomenon of light and manifests itself on
the interface between two media, e.g, air and water. For example, here light is not
refracted on the interface but reflected completely, i.e., it is reflected in the output
medium. The transmission of optical signals in fiber optic cables is facilitated by total
reflection.

Twisted pair cable: Designation for a type of copper cable in which both strands of a
wire pair are twisted together. Normally, a simple twisted pair consists of two insulated
approx. 1 mm thick copper wires, which are twisted spirally around each other. Often
different stranded pairs with different strengths of twisting (lay lengths) are stranded
into a cable.

Chapter 4
Network Access Layer (1): Wired LAN Technologies

> *„Only he who treats his neighbors with honor,*
> *can become an honorable man."*
> *– Proverb*

Local networks – so-called LANs (Local Area Networks) – connect computers that are in close proximity to each other. The triumphant march of LANs remains unbroken, extending from a simple point-to- point connection between two computers in the same room, to company/campus networks linking several hundred or even thousands of computers all communicating with each other over a common transmission medium. In respect to their geographical scope and the number of computers that can be connected over them, LANs, however, do have technical limitations. Different operational criteria, such as cost, throughput rate, spatial expansion and arrangement have led to the development of widely varying LAN technologies. These all follow their own protocol mechanisms and are each suited for application in different scenarios – whether stationary or mobile.

For the manufacturer as well as for the user, the necessity of a standardization was an urgent matter. Under the umbrella of the organization IEEE, the IEEE 802 chapter has been formed with its respective working groups. This chapter has created standards for widespread LAN technologies such as Ethernet, token ring, or the indispensable WLAN, and continues to develop them further. We will look at the development of various technologies and describe standards in the area of the local network. The topics covered range from thematic classification in the TCP/IP reference model to protocol mechanisms, address management in the LAN, to a description of various LAN technologies. The focus of here will be on wired LAN technologies. Due to its growing importance, a separate chapter has been dedicated to wireless LAN.

4.1 Network Access Layer

While on the physical layer only electrical, electromagnetic or optical signals are exchanged between two or more of the involved communication partners, on the **link layer** of the TCP/IP reference model, communication is already taking place

C. Meinel and H. Sack, *Internetworking*, X.media.publishing,
DOI: 10.1007/978-3-642-35392-5_4, © Springer-Verlag Berlin Heidelberg 2013

on an abstract logical level, detached from the physical level. Sequences of bits –
logical zeros and ones – are combined into larger units and sent out. The network
access layer is implemented in various network technologies beginning with so-
called personal area networks (PANs), linking only personal areas directly located
within a few meters of each other, and on to the widely used local networks (local
area networks, LANs) connecting buildings. A wider reach is achieved with the
metropolitan area networks (MANs), linking entire cities, and still further with the
global wide area networks (WANs). The following chapter provides an outline of
the general tasks of the network access layer. Within this layer a distinction is made
between the **medium access layer** (MAC, access control layer), on which the actual
access to the physical network is organized and the **logical link control** (LLC), on
the MAC layer, which allows communication to take place on a higher abstract level.
Afterwards, the general principles of LAN technologies as the most important group
of technological examples of the network access layer will be explained. The most
popular implementations of stationary (wired) technologies are introduced: Ethernet, token ring, FDDI and ATM. This is followed by a discussion of the expansion
possibilities of local networks. Wireless LAN technologies, among which are included personal area networks (PANs) as well as today's widespread WLAN technologies, will then be looked at in the next chapter. The LAN-expanded MAN and WAN
technologies, which are also classified in the network access layer, will be examined
in a separate chapter.

4.1.1 Elementary Tasks and Protocols

The network access layer (link layer, data link layer) is the lowest layer of the
TCP/IP reference model. It lies directly on the physical transmission medium (physical layer), which is itself not a part of the TCP/IP reference model. The physical
layer translates the bit sequences sent along a physical transmission medium into
physical signals. The main task of the network access layer involves bringing these bit sequences together into structured units (data packets, datagrams, frames)
(framing) and exchanging them between two (adjacent) end systems along a communication channel. In the process, the data packets receive additional information
that support a correct transmission, for example checksums for simple error recognition (error handling). The data packets used on the network access layer are
commonly referred to as **frames**. The communicating end systems can either be directly linked with each other through a transmission medium or be connected to a
so-called bus (diffusion network, broadcast network), which directly connects multiple end systems without a intermediate system. Protocols enabling communication
on the network access layer define the format of the frames to be exchanged between
the communicating end systems. They also determine the necessary action for the
transmitting and receiving of the frames. In addition, the network access layer is
responsible for making sure that slow receivers are not flooded with data from fast
transmitters. This regulating of the data stream is also called **flow control**.

The most important tasks of the network access layer therefore consist of:

1. providing a defined service interface for higher protocol layers,
2. regulating the multiple access of a shared communication medium,
3. handling of transmission errors (error recognition, new transmission, error correction, etc.), and
4. flow control.

Most of the time a multiple number of end systems are connected with each other over a shared medium as a diffusion network (also a broadcast network) in close proximity to the local networks, while in wide networks usually two respective end systems are linked over a transmission medium exclusively used by them. In this way, completely different demands are placed on the applied network technology in near range use, as the number of the communicating end systems here and their spatial distancing to each other have technological and practical limitations. In order to ensure equal access and efficient communication in a shared transmission medium for a multiple number of connected end systems, complex arbitrator algorithms are applied that regulate entry to the transmission medium. If in wide area networks a transmission medium is exclusively used by the two end systems connected to it (a point to point connection), much simpler communication algorithms can be used to guarantee the highest transmission capability possible.

In doing so, attention must be paid that the implemented physical transmission media is always given a limited transmission capacity. As the transmitted signals cannot be sent at an unlimited speed, delay plays an important role. This means the time between the transmission of a bit and its reception. Depending on the technology applied, the protocols used for communication take these factors into consideration to ensure efficient communication. A further distinction is made in the network access layer between **secured** and **unsecured services**. A data packet that is recognized as faulty is thereby eliminated by an unsecured service, however a wider ranging error correction, or a new request for the faulty data packet remains reserved for the protocols of a higher layer in the protocol stack. A secured service is, on the other hand, is in a position to initiate a possible necessary repeat transmission itself.

Among the most important protocols of the TCP/IP reference model network access layer are those from the IEEE, according to the **IEEE 802** LAN standard, specified LAN protocols. The most important representatives of the IEEE 802 LAN protocol standard are:

- Ethernet – IEEE 802.3
- Token Ring – IEEE 802.5
- Fiber Distributed Data Interface (FDDI) – derived from IEEE 802.4 (token bus)
- Wireless LAN – IEEE 802.11
- Wireless PAN – IEEE 802.15

Data in the global Internet that is transmitted from end system to another normally passes through a variety of different network technologies on the network access layer. In this process local networks, wide area networks and local area networks are used, each of which are optimally suited to its purpose (cf. Fig. 4.1). Connecting these to each other, or communicating between them is the task of the **Internet layer**, which lies above the network access layer.

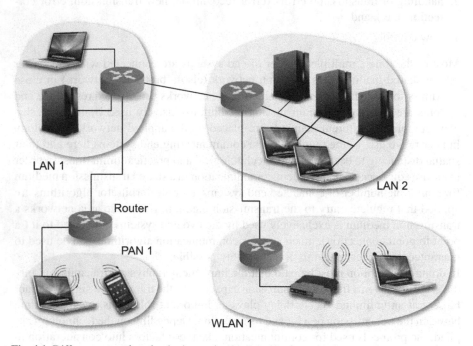

Fig. 4.1 Different network technologies are implemented in the network access layer.

An important task of the network access layer involves structuring the bit stream delivered from the physical layer into data packets of fixed or variable lengths, which are then transferred further to the higher protocol layers.

This **framing** divides the arriving bit stream into discrete units. As errors can occur in data transmission, a checksum is calculated for every frame. This serves to help identify transmission errors or, as the case may be, even correct them. For this purpose, different methods may come into play, however the general procedure remains the same. A checksum is determined from the payload, which is transmitted along with the user data in the frame. At his end, the receiver of the frame calculates the checksum from the received payload and compares it with the transmitted checksum. If the transmitted and the newly calculated checksums agree, then the degree of certainty is high that no transmission error has occurred. If, however, both do not agree then the transmission was faulty. Depending on the service type, the

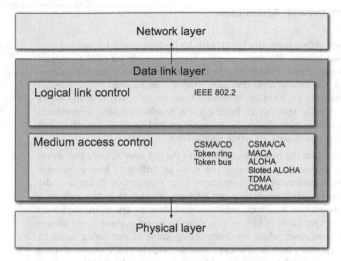

Fig. 4.2 The network access layer in the TCP/IP reference model and several protocols located therein

appropriate measures for error handling are triggered. These range from erasing the faulty frames to requesting a new transmission.[1]

In addition to checksum information, frames contain extra information that take care of the correct transport and successful delivery of the transmitted frame to the receiver. This extra information is summed up in a **header**, which normally precedes the payload. But it can also follow the payload in the form of a **trailer**. Included in the additional information contained in the header/trailer is:

- address information identifying the transmitter and receiver and necessary for correct data delivery,
- frame identification to uniquely identify a sent frame and to enable a correct decoding at the receiver,
- further data fields with control information.

So that the network access layer can correctly identify the transmitted frame of the communication partner upon receiving a bit stream from the physical layer, both the beginning and end of the frame have to be made identifiable. An intuitive method would be the addition of time intervals between the sending of separate frames. However, many network technologies can provide no guarantee of keeping to certain timed processes during data transmission, meaning the time intervals between the individual frames are subject to change or can even disappear completely. Various protocols in the network access layer solve the problem of **frame identification** in various ways. A distinction is made between **bit oriented protocols** and **byte**

---

[1] on this topic see also Volume 1 of the trilogy, Ch. Meinel, H. Sack: Digital Communication: Networking, Multimedia, Security, Springer (2013), Excursus 2: „Error-detecting and error-correcting codes"

Fundamental Tasks of the Network Access Layer

- **Framing:** In the network access layer the bit stream to be transmitted is pooled into coherent units, so-called frames. Normally, such a frame is made up of the payload to be transported and additional control information to ensure the correct transport of the user information. Control information can "frame" the payload in the form of a header, or, as the case may be, a trailer. It should however be noted that the entire control information is mostly referred to by the term **header**. Header information contains, e.g., address information about the transmitter and receiver, length values of the user contents and checksums for error recognition.

- **Medium access:** Protocols of the network access layer do not only specify the structure of the frame to be transmitted, but also the rules according to how a frame is to be transmitted over a communications medium (e.g. connection cable or radio connection). These access circuit protocols can be very simple, as in the case of a basic point-to point connection in which only two end devices are connected with each other over a connecting cable. Thus, in this case each end system could always transmit exactly at the moment when the connecting cable is not being used by the end system at the opposite end. In the case where multiple end systems share a connection cable, then the **multiple access problem** has to be solved with the help of special arbitrator algorithms.

- **Flow control:** The end systems taking part in communication have a buffer on the network access layer for the intermediate storage of incoming frames. If these are, however, processed and forwarded at a slower rate relative to their arrival at the processing end system then the buffer storage overflows and data loss can result. The protocol of the network access layer therefore offers the possibility to prevent a transmitting end system from overloading a communication partner with sent frames.

- **Error recognition:** If data is sent over a transmission medium, interference and signal noise can cause so-called **bit errors**, i.e., a bit originally sent as a logical "1" is as a result of the disturbance received as a "0" or vice versa. The protocols on the network access layer have reliable error recognition mechanisms that work on the basis of checksums.

- **Error correction:** Error correction, just as error detection, can be implemented with the help of redundant codes. In this case, the redundancy is chosen on such a large scale that the original data may be reconstructed from the distorted data that has been transmitted. In the case of the ATM protocol, error correction on the network access layer is only carried out for the data header of the transmitted frame.

Fig. 4.3 Fundamental tasks of the network access layer (Part 1)

oriented protocols. In byte oriented protocols the received bit stream is interpreted as an 8 bit character, in bit oriented protocol, on the other hand, byte boundaries do not occur:

- **Byte counting:** The length of a frame is simply given at the beginning (e.g., in bytes) for identification. The protocol of the network access layer reads the length values upon receipt and thereby obtains both the beginning and ending position of a frame of the bit stream received. Problematic in this procedure is that the length value can become distorted as a result of transmission errors. In this case, the end of one frame and, subsequently, the beginning of the next

Fundamental tasks of the network access layer (Part 2)

- **Secured transmission:** Besides unsecured services that discard frames identified as faulty, secured services are also possible on the network access layer. Over a mechanism responsible for confirmation and new transmission they ensure reliable data transmission. This applies particularly in the case of communication connections via unreliable media, meaning media with a high rate of errors, for example radio transmissions. Here, a special attempt is made to correct errors that appear locally, i.e., in the network access layer, rather than transferring the faulty error to a protocol instance on a higher layer of the TCP/IP reference model for error correction and waiting for new transmission requirements. In media with a very low rate of errors, for example with fiber optic cables or other cable bound media, error correction on the network access layer is viewed as a unnecessary ballast. For this reason, unsecured services are employed on the network access layer.

- **Bi-directional transmission:** Some of the protocols on the network access layer allow for a simultaneously bi-directional operation (**full-duplex**), meaning that transmitting and receiving end systems may transmit data at the same time, while other systems are only designed for an alternating operation of transmitter and receiver (**half-duplex**).

Further literature:

J. F. Kurose, K. W. Ross: Computer Networking: A Top-Down Approach, 5th ed., Addison-Wesley Publishing Company, USA (2009)

Fig. 4.4 Fundamental tasks of the networkaccess layer (Part 2)

cannot be correctly identified. This leads to a synchronization loss between the transmitter and receiver. Although an erroneously transmitted length value can be recognized with the help of checksums, a resynchronization of the transmission is problematic as long as no additional mechanisms for the identification of the start frame have been agreed upon.

- **Flag bytes and byte stuffing in byte-oriented protocols:**
  To solve the problem of lost synchronization between transmitter and receiver, the frame beginning and ending is marked with special character patterns called flag bytes (sentinels). Countless protocols use identical flag bytes to mark the frame beginning and ending. In the event that synchronization between the transmitter and receiver is lost due to a transmission error, the receiver waits until receiving another flag byte to re-determine the frame end. Two flag bytes following each other mark the last frame end and the start of the next frame. This flag byte identification becomes problematic when the bit pattern used also appears in the payload. In order to distinguish flag bytes from identical bit patterns within the payload, these bit patterns are prepended with a special control character (escape byte, ESC). The protocol of the network access layer additionally verifies whether a received flag byte has been prepended with an escape byte. If this is not the case, a frame has been correctly identified. If the escape byte is identified before a flag byte then it is clear that the received bit pattern is solely

payload. Before the user date is transferred to a higher protocol layer, the escape byte is usually removed. The additional insertion of escape bytes is also called **byte stuffing**. If the bit pattern of an escape byte appears in the payload to be transmitted, then it will likewise be prepended with an additional escape byte. In this way, a single escape-byte bit sequence will always be identified as an escape byte. The double escape bit is then identified as the bit pattern of the payload.

- **Start and end flags in conjunction with bit stuffing**:
Bit-oriented protocols are not dependent on the use of 8-bit long codes, i.e., the implemented frames may consist of any number of bits. Character codes of any bit length may also be used. Here as well, the principle of identifying the frame start and end through special bit sequences (start flag and end flag) is possible. In the case where a bit sequence specified as a start or end flag appears in the transmitted payload, then this data string is simply marked by the appropriate extra **filling bits** (bit stuffing). Upon receiving a bit sequence, the protocol on the network access layer corresponding to the start or end flag can verify whether this sequence follows a filling bit. If this is the case, the bit pattern is interpreted as payload and the filling bit removed from the payload before it is transmitted to a higher protocol layer. If synchronization between the transmitter and receiver is lost, the receiver searches the arriving bit stream for the bit pattern of the start and end flag. These can only appear at the respective frame borders without the subsequent filling bits and in this way make a clear frame identification possible. In practical application, many of the protocols located on the network access layer use a combination of byte counting and flag byte (start or end flags) procedures to provide additional security. This means that following the arrival of a frame, the length values at the beginning of the frame are also read in order to determine the end of the frame. Only if the correct end flag is found and a correct checksum can be calculated is the frame accepted as valid. If not, the input data stream for the next limitation field (i.e., a start or end flag) is searched.

- **Cycle-based frame generation**:
This variation of frame designation also uses special bit patterns to mark the beginning and end of a frame. Yet it is important to note that no additional filling bits or escape bytes are used. How is it then possible to distinguish between a bit pattern in the payload and a special pattern at the beginning or end of a frame? Cycle-based frame generation is founded on the principle of a frame being composed of periodically repeating patterns. This technique is applied, for example, in the SONET network technology for wide area networks. For instance, the STS-1 SONET Standard defines frames that are made up of 9 lines with a length of 90 bytes. The first 3 bytes of each line receive a special bit pattern (section overhead) that marks the beginning of a new line (cf. Fig 4.5). Although this bit pattern can also appear in the transmitted payload data, the receiver searches the received bit stream to see whether or not specific bit patterns repeat in the correct places. If this is the case then the transmitter and receiver are working synchronously.

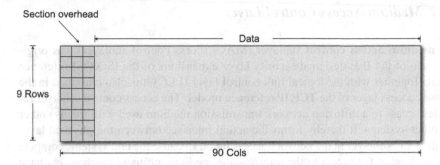

Fig. 4.5 Cycle based frame generation in the STS-1 frame of the SONET standard

A further problem that has to be solved on the network access layer as well as on the protocol stack is that of a transmitter sending frames at a higher transmission speed than the receiver is in a position to correctly receive them and process them securely. This problem can occur between network users with different hardware capacities or different current processing loads. If the fast sender does not slow down transmission speed or interrupt data transmission, then frames will be delivered in such a rapid succession to the slow receiver that it will eventually not be able to process them correctly. If this happens, the incoming frames can be lost, something that should be fundamentally avoided. The dynamics necessary to regulate transmission velocity to facilitate a perfect functioning of the entire system is known as **flow control**.

In principle, the procedure of feedback-based flow control can be distinguished from the simpler velocity-based procedures. **Feedback-based flow control** uses an extra information exchange between the transmitter and receiver. The transmitter is informed about the state of the receiver and a deceleration or acceleration then depends on the current state or requests of the receiver. Normally, most of the implementations of feedback-based flow control are based on well-defined rules regarding whether or not (and when) a transmitter can send the next frame to the receiver. These rule depend on the implicit or explicit agreement of the receiving computer. In contrast, **velocity-based flow control** functions without this reciprocal information flow and limits the respective transmission velocity of the sending computer. Feedback-based flow control procedures are exclusively employed on the network access layer.

In local networks (LANs), the network access layer, as the lowest layer of the TCP/IP reference model, divides the different tasks to be manged into two sublayers:

- Medium Access Control Layer (MAC, access control sublayer), whose main task consists of solving the multiple acess problems of the the network access layer, and

- Logical Link Control (LLC), in which framing, flow control, error recognition and error correction and secure transmission is facilitated.

4.1.2 Medium Access Control Layer

The **medium access control** sublayer (MAC, access control sublayer) was originally one of the IEE-designed security layer expansions of the ISO/OSI-reference model. Together with the logical link control layer (LLC) it is also classified in the network access layer of the TCP/IP reference model. The access control sublayer regulates access to a diffusion network transmission medium used with (many) other computer systems. It thereby forms the actual interface between the physical layer and the LLC sublayer in the network access layer. Because the end systems compete with each other for access to the transmission medium, protocol mechanisms have to be provided which allow a fair and efficient access to all users (**multiple access protocols**). Also necessary are methods for the discovery, or avoidance, of collisions as two or more users who are sharing a transmission medium may wish to transmit data simultaneously (**collision detection, collision avoidance**). In the access control sublayer a distinction is made between methods with **controlled access** (collision avoidance) and methods with **competing access** (collision resolution). Alternatively, transmission resources are exclusively reserved and logical (virtual) transmission channels set up (**channel access control**).

Included among the most popular multiple access protocols are:

- In wired networks:
  - CSMA/CD (Carrier Sense Multiple Access / Collision Detect), implemented in Ethernet (IEEE 802.3)
  - Token Bus (IEEE 802.4)
  - Token Ring (IEEE 802.5)

- In wireless networks:
  - CSMA/CA (Carrier Sense Multiple Access / Collision Avoidance), implemented in WLAN (IEEE 802.11)
  - ALOHA, simple stochastic access method
  - slotted ALOHA, stochastic access method with predetermined time intervals (slots)
  - TDMA (Time Division Multiple Access), time multiplexing method
  - CDMA (Code Division Multiple Access), Code multiplexing method

In the MAC sublayer every end system participating in the communication must have an individual and unique address by which it can be identified (**MAC addressing**). MAC addresses are unique serial numbers that can be activated by the network hardware and are also known as physical addresses. They are usually assigned to the network hardware by the manufacturer. In this way, it can be made certain that no two network elements have the same MAC address. Specialized network hardware, on which the protocols of the MAC sublayer are implemented, is also known as the **medium access controller**. In contrast to diffusion networks, no MAC sublayer is necessary in point-to-point connections. Nevertheless, some point-to-point protocols also use MAC addresses for reasons of compatibility. On the MAC sublayer the borders of the local networks may be crossed and different (albeit homogeneous) subnets connected with each other over a so-called **switch** (**LAN switching**). At the same time, frames (data packets) are each forwarded only to the subnet in which the corresponding target computer is located. The switch takes over the task of filtering the data traffic (**MAC filtering**) here. A general distinction can be made between two general types:

- the **store-and-forward** switch always saves the data packets to be filtered before an analysis and subsequent forwarding takes place.
- the **cut-through** switch carries out the forwarding without prior intermediate storage.

Additionally in the MAC sublayer, tasks of queue management (**data packet queuing** and **scheduling**) are performed when frames in the network cannot be forwarded quickly enough before new frames arrive, and decisions have to be made about the priority of forwarding.

4.1.3 Logical Link Control

The **logical link control** sublayer forms the so-called security layer of the LAN. The tasks regulated here are on a higher abstraction level than the lower MAC sublayer, upon which it is located. Primarily, the LLC sublayer functions as the mediation between the above Internet layer in the TCP/IP reference model and the MAC sublayer. Its conception and implementation is independent of the network technologies used. LAN technologies such as Ethernet, token ring, or WLAN can be operated together with the same LLC sublayer. In this context, the logical link control also has its own standardized network protocol of its own as IEEE 802.2.

Among the tasks of the LLC sublayer is the avoidance of overload situations during data transmissions to potential receiving systems by way of targeted intervention in the data flow (**flow control**) as well as the control of data transmission (**link management**). The first quality control of the transmitted data takes place in the LLC sublayer . Data transmission errors must be recognized and – if possible – corrected. For this purpose the protocols located on the LLC sublayer implement various **error recognition** and **correction methods**. Additionally, the LLC sublayer synchronizes the transmission and reception of data units (frames). Thereby, the data has to be subdivided into data packets that are limited in length to correspond to the physical and logical conditions of the respectively chosen transmission form (**fragmentation**). Also following transmission the beginning and ending must be correctly identified (**data packet synchronization**). Furthermore, the LLC sublayer ensures so-called **multi-protocol ability**, that is the ability to use different communication protocols at the same time. The data transmission services in the LLC sublayer may be divided into connection-oriented and connectionless services.

The following protocols belong to the LLC sublayer:

- IEEE 802.2 Logical Link Control
- PPP (Point-to-Point Protocol)
- SLIP (Serial Line Internet Protocol)
- SNAP (Subnetwork Access Protocol)
- HDLC (High-level Data Link Protocol)
- LAPB (Link Access Procedure, Balanced)
- LLDP (Link Layer Discovery Protocol).

4.2 LANs – Local Area Networks

The connection of computers to each other for the purpose of data transmission can proceed in different ways and in a variety of arrangements. Normally networks are classified based on their spatial dimensions and user number, as well as their topology, meaning the spatial layout, distribution and connection of the individual participating computers. The most basic type of networking is the direct **point-to point connection** between two computers. But if more computers share a common transmission network then the computers are indirectly connected to each other and – depending on the distance bridged – one speaks of a **local network** (**local area network, LAN**) or a **wide area network** (**Wide Area Network, WAN**) (cf. also Table 4.1).

Table 4.1 LANs and the classification of computer networks based on their spatial dimensions.

Distance	Unit of order	Example
0.1 m	Circuit board	Multi-processor system
1 m	System	Multi-processor cluster
10 m	Room	Personal Area Network
0.1 km – 1 km	**Building complex**	**Local Area Network**
10 km – 100 km	City	Metropolitan Area Network
100 km – 1,000 km	Continent	Wide Area Network
>10,000 km	Planet	Internet

LANs are distinguished by the fact that the participating computers share a common transmission medium. This means that there are special requirements regarding the implementation of communication protocols as well as limitations concerning the scalability of geographic dimensions and total number of participating computers involved. Of particular importance in the LAN area are the pooled technologies of the standard for wired and wireless local networks under IEEE 802 .

4.2.1 Use of Shared Communication Channels

The first basic computer networks were based on the principle of the point-to-point connection. The communication between two computer systems takes place over an individual separate channel that creates a permanent connection between both systems and is used exclusively for mutual data exchange. If, however, many computers are connected over separate point-to-point connections then the number of needed connections increases dramatically. In in a point-to-point connected network the number increases quadratically with the number of connected computer systems. In order, for, e.g.,10 separate computer systems to completely link to each other in pairs, 45 cable connections are necessary, and with 100 computers 4,950 cable connections are needed. For this reason, it would appear to make sense for one and the same communication connection to be shared by many stations, to be used interchangeably by the transmitter and receiver. Naturally, a shared transmission medium also demands increased coordination and administrative efforts as well as unavoidably higher transfer times for the sending of data.

A network with a transmission medium shared by all of the connected subscribers is called a **broadcast network** (or a diffusion network). Data that is sent to one of the computers connected to the communication network is received by all of the other computers connected to the network. In order to be able to explicitly identify the transmitter and sender of the data, the sent frames in this networks must have appropriate address fields available. As soon as a computer in the broadcast network receives a data frame it will first verify the receiver address field. It the frame is intended for it, will be processed, otherwise the frame is discarded.

Broadcast networks also offer the possibility of addressing a data frame simultaneously to all connected computers. Such a data frame is received and then processed by all the connected computers – a procedure known as **broadcasting**. Moreover, some systems allow the transmission to a previously fixed group of computers connected to the communication network, so-called **multicasting**. A possibility of implementing multi-casting involves reserving one of the n possible address bits that indicates the address used as a multicasting transmission. The remaining $n-1$ address bits are used to specify a user group to whom the packet should be sent.

Despite there being many exceptions, a general rule of thumb applies: small, spatially limited communication systems are often realized as broadcasting networks, while networks that are widely distance from each other geographically are commonly linked via point-to-point connections. This is because the effort to coordinate a multiple number of computers, shared by a network for data communication, is strongly dependent on the respective transfer and waiting time in this network. In local area networks there usually occurs only a minimal transfer and waiting time in the short distances to be bridged. Thus broadcasting network technologies have gained predominance in LAN. With connections spanning large distances, and with a correspondingly long transfer time, point-to-point connections are usually preferable.

4.2.2 The Importance of LANs

The most widespread form of computer networks today are local area networks
(LANs). Worldwide most computers are linked over LANs. Through the shared use
of available network resources for all connected computers, LANs have an extremely high rate of efficiency. One of the reasons computer networking in the form of
LANs is so efficient has been known for a long time in the field of computer architecture – this is the principle of **locality of reference**. It means that the probability
of memory access taking place at one of the currently neighboring memory cells is
greater than the probability that the content will be requested by the program from a
distant cell. In particular, fast-paced intermediate storage – so-called caches – make
use of this feature for increased storage efficiency.

If the locality of reference principle is transferred to the realm of computer networks,
it reveals the communication there as not completely arbitrary. In the following
instances, locality of reference is the governing principle in temporal as well as in
spatial terms:

- **Temporal locality of reference**
  If two computers communicate with each other, there is a great likelihood that
  they will subsequently do so again.

- **Spatial locality of reference**
  The likelihood of two neighboring computers in the network communicating
  with each other is greater than the communication with other distant computers
  in the network.

Usually LANs involve private networks that can essentially be installed and operated without special regulations or fixed users fees by everyone. Geographically
speaking, LANs are initially limited to the property of the individual owner, however, it is also possible for wireless LAN islands to be distributed over different
properties. Widely spread networks, on the other hand (WANs, MANs), are dependent on the network operator (**carrier**), who support the network and usually makes
them available for a fee. Network operators are normally private or public providers
who operates within the framework of legal regulations. But it is also possible for a
company to operate its own WAN, without the service being made available to outsiders (**corporate network**). The necessary infrastructure (line) is often rented from
a network operator and an apparent proprietary network is established. The company itself is then responsible for the operation and management of the company-own
network.

4.2.3 IEEE 802 Local Area Networks

The **IEEE (Institute of Electrical and Electronics Engineers)** is responsible for
the **standardization** of the individual local area networks and metropolitan area
networks used in Internet. Of particular importance is the IEEE-802 LAN/MAN

Definition: Local Area Network (LAN)

(Definition based on ISO/IEC DTR 8802-1): A local network serves the bit-serial information transmission between independent devices that are connected to each other.
It is entirely within the bounds of the user's legal area of decision and limited to his
property.

(Definition based on „Kauffels: Dictionary of Data Communication"):
Local networks are systems for the high performance transfer of information. They
allow a number of users, with equal rights, to conduct a partnership-based information
exchange in a spatially limited area over a fast transmission medium with a high level
of quality.

Fig. 4.6 Definition of a LAN.

Standards Committee (LMSC), with its various working groups, which began operation in February 1980. Strictly speaking, the standard networks defined as IEEE
802 are described in terms of variable data packet sizes. An overview of the working groups of the IEEE 802 LMSC is presented in table 4.2. The working groups
are divided as follows:

- **802.1 – High Level Interface (Internetworking)**
  Generally, the higher level working group 802.1 focuses on the technology overlapping networking of IEEE 802 LANs, MANs and wide area networks (WANs).
  The areas of concentration include: addressing, network management and Internetworking. There is an emphasis on link security, as well as the IEEE 802
  complete network management affecting the protocol layers above the MAC and
  LLC sublayer. The most important standards adopted or currently being developed include:

    - 802.1D – Spanning Tree Protocol, MAC Bridges
    - 802.1H – Ethernet MAC Bridging
    - 802.1P – General Registration Protocol
    - 802.1pQ – Quality of Service
    - 802.1Q – Virtual Bridged LANs
    - 802.1S – Multiple Spanning Tree Protocol
    - 802.1W – Rapid Spanning Tree Protocol
    - 802.1X – Port Based Network Access Control
    - 802.1AB – Link Layer Discovery Protocol

- **802.2 – Logical Link Control**
  In this working group the protocol mechanisms for the upper sublayer (Logical
  Link Control) of the network access layer is standardized. Today it is no longer
  active.

- **802.3 – Ethernet**
  This working group prepared a series of standards that concern the physical layer
  and the lower MAC sublayer of the network access layer for Ethernet technology. Beginning with the first experimental Ethernet implementation in 1972 until

today, progressively more powerful variations have been developed and standardized, for example IEEE 802.3bx for data transmission with a bandwidth of 40–100 Gbps. The IEEE 802.3 Ethernet technology will be looked at more closely in section 4.3.2. Among the most important standards of the IEEE 802.3 that have been adopted or are currently under development are:

- 802.3 (1983) – 10BASE5 with 10 Mbps over coaxial cables (baseband)
- 802.3a (1985) – 10BASE2 with 10 Mbps over thin coaxial cables
- 802.3c (1985) – specifications for10 Mbps repeater
- 802.3i (1990) – 10BASE-T with 10 Mbps over copper cables (twisted pair)
- 802.3j (1993) – 10BASE-F with 10 Mbps over fiber optical cables
- 802.3u (1995) – 100BASE-TX/-T4/-FX fast Ethernet with 100 Mbps
- 802.3y (1998) – 100BASE-T2 100 Mbps over copper cablesl (twisted pair)
- 802.3z (1998) – 1000BASE-X over fiber optical cables with 1 Gbps
- 802.3ab (1999) – 1000BASE-T over copper cables (twisted pair) with 1 Gbps
- 802.3ae (2003) – 10GBASE-SR/-LR/-ER/-SW/-LW/-EW over fiber opticcal cables with 10 Gbps
- 802.3ak (2004) – 10GBASE-CX4 over coaxial cables with 10 Gbps
- 802.3an (2006) – 10GBASE-T over copper cables (unshielded twisted pair) with 10 Gbps
- 802.3ba (2010) – 40 Gbps and 100 Gbps Ethernet

- **802.4 – Token Bus**
  This working group was responsible for the development of the standard for the implementation of the token bus technology and has since been disbanded. Token bus is the implementation of a virtual token ring (cf. IEEE 802.5) on one of the ends of an unconnected coaxial cable (bus). This technology was used mainly in industrial applications, e.g., at the General Motors company for the Manufacturing Automation Protocol (MAP).

- **802.5 – Token Ring**
  In the framework of this working group a series of standards have been developed that apply to the physical layer and the lower MAC sublayer of the network access layer for token ring technology. In contrast to Ethernet, which is founded on a bus or star topology, the token ring technology is based on a ring topology. This means that the data is transported along one of its ends to the interconnected medium. The IEEE 802.5 token ring technology is covered in more detail in section 4.3.3. This working group 802.5 no longer exists today. Included among the most important standards adopted in IEEE 802.5 are:

- 802.5 (1998) – Token ring access method and physical layer specification
- 802.5c (1991) – Supplements and best practices for dual ring operation
- 802.5r/j (1998) – Supplements for operation of token ring over fiber optical media
- 802.5t (2000) – Token ring with 100 Mbps bandwidth
- 802.5v (2001) – Gigabit token ring with 1 Gpbs bandwidth
- 802.5w (2000) – Token ring maintenance and servicing

- **802.6 – Metropolitan Area Networks**

  This working group developed standards for specifications of the technologies in the MAN area and was conducted under the guidance of the ANSI (American National Standards Institute). For the operation of MANs, IEEE 802.6 focuses on the less expensive and less complicated distributed queue dual bus (DQDB) technology, which supports bandwidths from 150 Mbps with a distance of to 160 km, in contrast to the ANSI proposed FDDI technology (Fiber Distributed Data Interface). The suggested standard was, however, unsuccessful for the same reason as the FDDI, so that today in the MAN range SONET (Synchronous Optical Network) or ATM (Asynchronous Transfer Mode) technologies are employed, which are not standardized by IEEE 802. The IEEE 802.6 working group has been disbanded in the meantime.

- **802.7 – Broadband LAN via Coaxial Cables**

  This working group developed standardization proposals for the operation of local broadband networks. These were published as a proposal in 1989 but then later recalled. The IEEE 802.7 working group is no longer active today.

- **802.8 – Fiber Optic Technical Advisory Group**

  This working group was entrusted with the task of developing standardization proposals for high speed optical fiber networks. These were intended to be based on a token transferal mechanism similar to the FDDI standard. The IEEE 802.8 working group has since been disbanded.

- **802.9 – Integrated Services LAN**

  The 802.9 working group of the IEEE 802 LAN/MAN Standards Committee developed standardization proposals for the integration of telephone traffic (voice transmission) and simultaneous data transmission of the already existing customary telephone networks (category 3 twisted pair cabling). The most important standard proposal in this area became known under the name "ISO Internet," which combines a 10 Mbps Ethernet data transmission with 96 ISDN channels, each with 64 kbps. Because of the success of the more powerful fast Ethernet (100 Mbps), this proposal was abandoned and the working group disbanded.

- **802.10 – Interoperable LAN Security**

  This working group developed standardization proposals for the implementation of security standards for LANs and MANs based on IEEE 802 technologies. This includes security relevant tasks such as key management, access control, confidentiality or data integrity.

  The IEEE 802.10 standard proposals were recalled in 2004 and the working group is no longer active. The standardization for LAN security included the following sections:

  - 802.10a – Security model and security management
  - 802.10b – Secure Data Exchange Protocol (SDE)
  - 802.10c – Key management
  - 802.10e – SDE via Ethernet 2.0
  - 802.10f – SDE sublayer management

- **802.11 – Wireless LAN**
The working groups in this area have developed standardization proposals for the operation of local wireless networks in the frequency bands 2.4 GHz, 3.6 GHz and 5 GHz. With advancing technological developments, standardization has evolved, starting with IEEE 802.11 and data transmission rates of 1–2 Mbps to the current IEEE 802.11n with rates up to 150 Mbps. Among the most important adopted or currently being developed IEEE 802.11 standards are:

  – 802.11 (1997) – an already obsolete WLAN standard today with data transmission rates from 1–2 Mbps
  – 802.11a (1999) – WLAN standard with data transmission rates up to 54 Mbps
  – 802.11b (1999) – simpler WLAN standard with data transmission rates from 1–11 Mbps
  – 802.11g (2003) – improved WLAN standard (over 802.b) with data transmission rates from 1–54 Mbps
  – 802.11n (2009) – currently the latest expansion of the IEEE 802.11 standard, with data transmission rates of 7–150 Mbps

- **802.12 – Demand Priority**
This working group focuses on a number of standardization proposals from network technologies that compete with fast Ethernet. These include the 100Base-VG Ethernet technology developed by the Hewlett Packard (HP) company, which is based on the network access procedure of the demand priority access method (DPAM). This method works on a star topology. This means that if a computer wishes to gain access to a network a prior request for permission must be made to the responsible hub for allocation of a free slot. These adopted standards encompass the general demand priority access methods as well as the technical specifications for the physical layer and the network repeater. The IEEE 802.12 working group has since been disbanded.

- **802.13 – Fast Ethernet**
The IEEE 802.13 standard is not used. Originally the IEEE 802.13 working group was intended to develop standardization proposals for the 100BASE-X fast Ethernet technology, operated with 100 Mbps transmission data rate. IEEE 802.13 Ethernet is, however, virtually identical with the standardized 10Base-T Ethernet in IEEE 802.3i, and is thus often also referred to as 100Base-T. It uses the prescribed Ethernet bus topology and the CSMA/CD network access mechanism.

- **802.14 – Cable modems**
Within this working group, standardization proposals for the specification and operation of cable modems and associated protocols of the MAC sublayer were developed. Cable modems serve as access interface for the implementation of broadband Internet access via traditional television cable networks. The IEEE 802.14 working group has been disbanded.

- **802.15 – Wireless Personal Area Networks**
This working group focuses on the development and adoption of standardization

proposals for wireless personal area networks (wireless PANs), allowing wireless network access between devices that are located only a few meters from each other. It includes the following sub-working groups:

- 802.15.1 – WPAN/Bluetooth published in 2002 the Bluetooth v1.1 standard and in 2005 Bluetooth v1.2.
- 802.15.2 – co-existence of various WPAN technologies with other wireless devices working in the unlicensed ISM frequency band, e.g., WLANs.
- 802.15.3 – High Rate WPAN, wireless PANs with transmission speeds of 11–55 Mbps.
- 802.15.4 – Low Rate WPAN, wireless PANs with low transmission speeds for the most energy efficient data transmission possible.
- 802.15.5 – Meshed networks on the basis of WPAN technologies.
- 802.15.6 – Body Area Networks (BAN) with the goal of achieving low energy wireless data transmission at close range.
- 802.15.7 – Visible Light Communication (VLC), wireless data transmission in the area of visible light.

- **802.16 – Broadband Wireless Access (WiMAX)**
  This working group is occupied with the development of standardization proposals for different wireless broadband technologies. While the official designation for the standard developed here is "WirelessMAN," it has become known by the name "WiMAX," (Worldwide Interoperability for Microwave Access). The most popular implementation under the IEEE 802.16 standard is 802.16e (2005) Mobile Wireless MAN, available in more than 140 countries throughout the world). Among the most important adopted or currently being developed IEEE 802.16 standards are:

  - 802.16 (2009) – Air interface for stationary and mobile broadband wireless access
  - 802.16.2 (2004) – Recommendations for the co-existence of different wireless network standards
  - 802.16e (2005) – Mobile Broadband Wireless Access
  - 802.16f (2005) – Management Information Base (MIB) for IEEE 802.16
  - 802.16k (2007) – Bridging of IEEE 802.16 networks
  - 802.16m – Advanced air interface with data rates from 100 Mbps in mobile operation and 1 Gbps in stationary operation

- **802.17 – Resilient Packet Ring**
  This working group is concerned with the optimal method for data transport via optical fiber networks in ring topologies. Robust technologies should be developed and standardized, for example as implemented in the circuit switched SONET networks, with the advantages of packet-switched networks, such as Ethernet, for an optimal transmission of IP-based data traffic.

- **802.18 – Radio Regulatory Technical Advisory Group**
  This working group was set up as an advisory board for the purpose of balan-

cing the different interests of all IEEE 802 projects involving wireless systems. National as well as international interests are taken into consideration and the appropriate regulatory agencies are advised. The following IEEE 802 projects are represented:

- IEEE 802.11 – WLAN
- IEEE 802.15 – Wireless PAN
- IEEE 802.16 – Wireless MAN
- IEEE 802.20 – Wireless Mobility
- IEEE 802.21 – Interoperability Between Networks
- IEEE 802.22 – Wireless Regional Area Network (WRAN)

- **802.19 – Wireless Coexistence Technical Advisory Group**
  This working group was formed to ensure a trouble-free co-existence between the various wireless data transmission technologies in the unlicensed area of the frequency spectrum (ISM band). This is necessary because different devices can operate within the same frequency spectrum at the same place and it is necessary to avoid reciprocal interference.

- **802.20 – Mobile Broadband Wireless Access**
  Strategies and access interfaces are developed in this working group for the worldwide application of interoperable wireless mobile network technologies. The goal is to create reasonably-priced mobile broadband networks that ensure permanent network connectivity ("always on"). The network client is in fact mobile, also across network borders (MobileFi). An architecture should be created consistent with the existing IEEE 802 standards. The physical layer, the MAC sublayer, and the LLC sublayer also need to be taken into consideration. Thus the goal of the 802.20 working group also corresponds to that of the 802.16e working group, which is to create the specification for a mobile WiMAX.

- **802.21 – Media Independent Handoff**
  This working group took up its work in 2004 with the primary intention of developing and standardizing algorithms that are intended to ensure a seamless transition (handover) between different network technologies (Media Independent Handover, MIH), e.g., between mobile communication standards GSM and GPRS over WLAN IEEE 802.11 or Bluetooth IEEE 802.15, which each require different handover mechanisms.

- **802.22 – Wireless Regional Area Network (WRAN)**
  Within this working group, standards should be developed based on utilizing regional and, until now unused, frequency areas of the TV frequency spectrum for wireless data communication. The intention is to also provide a broadband network connection to hard-to-reach rural areas with a low population density. Of primary interest is the development of technologies that enable an inference-free operation together with other devices (digital and analog televisions, digital and analog radios, as well as licensed devices with low power requirements, e.g., a wireless microphone).

- **802.23 – Emergency Services Working Group**
  Set up in 2010, the intention of this working group is to create a standard for a media-independent framework within the IEEE 802 technologies. This standard is to comply with the requirements set up by the civil authorities for communication systems. A particular focus is on accessing IP-based data communication for emergency and rescue services.

The standards developed here were also adopted as international standards ISO 8802 by the International Standards Organisation (ISO).

The following working groups no longer operate actively or have been disbanded:

- IEEE 802.2 – Logical Link Control
- IEEE 802.4 – Token Bus
- IEEE 802.5 – Token Ring
- IEEE 802.6 – MANs
- IEEE 802.7 – Broadband TAG
- IEEE 802.8 – Fiber Optic TAG
- IEEE 802.9 – Integrated Services
- IEEE 802.10 – Interoperable LAN Security
- IEEE 802.12 – 100 Base VG ANyLAN
- IEEE 802.14 – Cable Data Modems

4.2.4 Local Address Management

As previously discussed, LANs are typically based on the principle of a commonly used network infrastructure. Data packets (frames) are simultaneously sent to all computers in the network (broadcast network). This means that with every LAN data transmission all other computers linked in the LAN receive the data packet. The following section examines the explicit identification of a computer within a LAN. In this respect, a special role is played by addressing, which takes place on the MAC sublayer (Medium Access Control). So that data can be correctly sent from the transmitter to the receiver in a LAN, the correct address details must be contained in every data packet transmitted. In order to "speak to" a certain computer uniquely in a LAN, it is necessary that each computer be assigned a so-called **address**. These addresses are numerical values in different formats that clearly identify a computer within a LAN. Addresses with different formats may be used on each layer of the TCP/IP reference model. The simplest analogy can be made to a telephone number used to communicate with another participant in the telephone network. If the participant is located in the same area network (the same LAN in data communication), a short number is already enough to be able to identify this party uniquely. In the other case, the area code for this location must also be provided for the telephone network. Finally, the telephone network allows an additional country code for international phone calls. This hierarchical breakdown of address space is again found in

Table 4.2 An overview of IEEE 802 working groups.

Nr.	Designation	Content
802.1	Bridging (Networking) and Network Management	Deals with common aspects of all LANs according to the IEEE 802 standard
802.2	Logical Link Control	Defines the LLC protocol (TCP/IP layer 2)
802.3	CSMA/CD	Ethernet protocol standard
802.4	Token Passing Bus	Token bus protocol
802.5	Token Passing Ring	Token ring protocol (MAC-Schicht)
802.6	MAN	Deals with Metropolitan Area Network standard
802.7	Broadband Technical Advisory Group	Advises other area IEEE 802 groups in broadband technology
802.8	Fiber Optical Technical Advisory Group	Advises other IEEE 802 groups in optical fiber technology
802.9	Integrated Voice and Data Networks	Deals with LAN variations handling both data and language (Isochronous Ethernet)
802.10	Network Security	Deals with security in LANs
802.11	Wireless LAN	Deals with wireless local networks
802.12	100 Base VG AnyLAN	Higher speed LAN standard in competition with fast Ethernet
802.13	100 Base-X Ethernet	Deals with fast Ethernet
802.14	Cable Data Modem	Deals with the application of cable networks for data communication
802.15	Wireless Personal Area Networks (WPAN)	Deals with wireless networks over short distances
802.16	Broadband Wireless Access WiMAX	Deals with wireless high speed networks
802.17	Resilient Packet Ring	Development of high speed networks with self-repair mechanisms
802.18	Radio Regulatory Technical Advisory Group	deals with questions regarding frequency allocation
802.19	Coexistence Wireless Technical Advisory Group	Deals with questions regarding the coexistence of different wireless network standards
802.20	Mobile Broadband Wireless Access	Development of universal mobile broadband interfaces
802.21	Media Independent Handoff	Deals with the seamless transitions between various networks
802.22	Wireless Regional Area Network	Deals with the utilization of until now unused regional TV frequency areas for data transmission

data communication, albeit with different addresses in the different address formats of the single TCP/IP reference model layers.

Every transmitted packet within a data network contains the address of the sender (source) and the receiver (destination), as well as additional information necessary for a correct and error-free data transmission. The **LAN communication interface** of a computer connected to the network filters incoming data packets based on the addresses they contain as follows:

- If the address of the receiver contained in the data packet agrees with the computer's own address the data packet will be transfered to the operating system for further processing.
- If the address in the data packet does not agree with the computer's own address it will be discarded.

It is in fact not the computer itself that has its own LAN address (also called a physical address), but rather the computer's LAN communication interface (LAN adapter). In transmission, the CPU of the computer wishing to send data passes the data on to the LAN communication interface. All details of the data exchange are taken over by the LAN communication interface. Since the LAN communication interface works without placing any demands on the computer's CPU, the normal operation of the computer is not disturbed during data transmission (cf. Fig. 4.7).

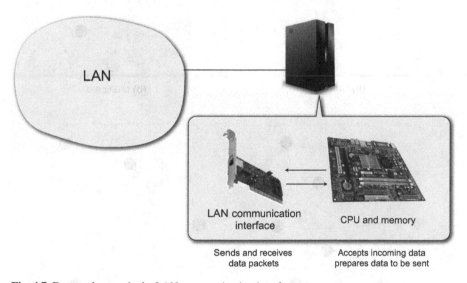

Fig. 4.7 Data exchange via the LAN communication interface.

As a rule, the addresses within a network neighborhood are **individual addresses**, i.e., addresses that identify a specific single computer (or, more precisely, it's network interface). Besides these, there are special **multicast addresses**, which identify and speak to a certain group of computers within a network. There are also so-called **broadcast addresses**. These address all computers connected to a network and can

facilitate "mass mailings." Network technologies with shared media utilization enable an efficient broadcast procedure as a data packet must be sent in any case to
all of the linked computers sharing the transmission medium that is being used.
Not yet mentioned is a special variation of addressing in computer networks called
anycast. Similar to multicast addresses, anycast addresses identify a group of individual computers that are, however, normally distributed over several networks for
reasons of burden sharing and greater availability. The computers brought together
in an anycast group all render the same service. Each requesting computer speaks to
the computer geographically closest to it along the shortest connection route. If this
computer is not available or subject to a high processing load, the request is sent on
to the next computer in the anycast group and processed there (cf. Fig. 4.8).

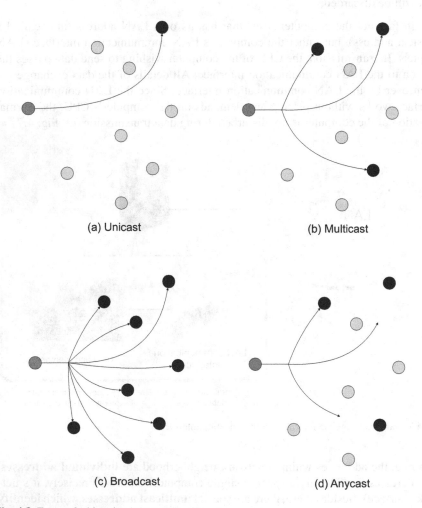

Fig. 4.8 Types of addressing in computer networks.

The so-called **MAC addresses** (Medium Access Layer Address), found on the level of the LAN and its responsible protocol layer, are of particular importance. MAC addresses are hardware addresses assigned to the LAN communication interface on the computer linked to LAN. Network devices generally only then need a MAC address if they are to be explicitly addressed on the network access layer to provide services to higher levels of the protocol stack. Network intermediate systems, such as repeaters or hubs, only forward data in LANs and are themselves not visible in this process (transparent). It is therefore not necessary that repeaters and hubs have their own MAC address. It is the same with bridges and switches. These are network intermediate systems and carry out tasks on the network access layer in the LAN. Bridges and switches investigate the data packets in the LAN and forward the corresponding varying criteria to different LAN segments. To accomplish these basic tasks the network intermediate systems do not need their own MAC address. However, an individual MAC address is necessary for the network intermediate system if these are to be managed and monitored via the network itself. Bridges and switches in networks that are redundantly designed are a special case. A special algorithm (spanning tree algorithm), that needs its own MAC address, is implemented here to avoid looping.

Within a LAN, a MAC address must be unique in each case. Based on the validity of assignment, a distinction is made between three categories of MAC address formats:

- **Static addresses** are given to every LAN interface by the manufacturer as an unique worldwide hardware address. This assignment is permanent and only changes when hardware is exchanged. This is why these addresses are also known as burned-in addresses (BID). As the hardware addresses are already uniquely configured in prior arrangement between the different hardware manufacturers on delivery, hardware from various manufacturers can be connected to the LAN without address conflicts occurring.

- **Configurable addresses** can be freely set by the network operator. This often takes place manually using the switch on the LAN interface card intended for this purpose or through the software solution of setting the address in a non-volatile memory (EPROM). As a rule, this configuration only takes place the one time when the hardware is installed by the operator. Because address clarity must only be maintained within the LAN, short addresses may be used here.

- **Dynamic addresses** offer the highest degree of flexibility. Here, the network interface is automatically assigned a new hardware address when first connected to a computer network. This process is initiated by the LAN interface itself and can include a number of failed attempts, in which the LAN interface "unknowingly" assigns a number of addresses that have already been given out in the LAN. This procedure can be repeated every time the computer is restarted and every computer can thereby be assigned a new hardware address. Because only the uniqueness of the address must be maintained within the LAN, the addresses used here can be very short as well.

Configurable and dynamic MAC addresses are administered locally. But most of the time MAC addresses are globally unique and statically assigned (Organizationally

Unique Identifier, OUI). This is particularly the case when the **IEEE 802 addressing scheme** is used. The IEEE 802 MAC address organization (MAC-48) was derived from the original Ethernet standard of the Xerox company. The 48-bit MAC addresses are uniquely assigned globally. For this purpose, the IEEE assigns every manufacturer of LAN communication interfaces (network adapters) a block of manufacturer-specific address sections (OUI). These are fixed with the first 3 bytes (in the transmission sequence). The manufacturer completes this partial address for every manufactured network adapter with a sequential number (Network Interface Controller Specific Address, NIC), whose allocation mode can be determined by the manufacturer himself. The NIC also contains 3 bytes (cf. Fig. 4.9).

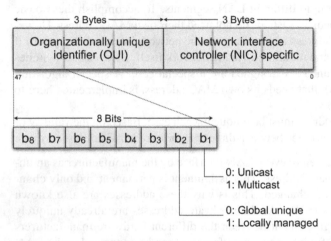

Fig. 4.9 Addressing scheme according to IEEE 802.

To make MAC-48 addresses easier for people to read, they are each grouped into 8-bit hexadecimal numbers, separated by a dash or a colon (e.g., 01-23-45-67-89-ab). Locally administered addresses are normally determined by the network administrator himself and do not need an OUI. In the most significant bytes of the MAC address the second least significant bit (b_2) determines whether or not it is a globally unique MAC address (b_2=0) or a locally administered MAC address (b_2=0). The least significant bit in the most significant byte of the MAC address (b_1) determines whether it is a unicast (b_1=0, data is only sent to one NIC) or a multicast addressing (b_1=1, data is only sent once, but can received by several NICs). If all bits of the MAC address are set to 1 (ff-ff-ff-ff-ff-ff), the data sent this way is received by stations in the network (broadcast).

So that a sufficient number of explicit hardware addresses is available in the future, an additional 64-bit addressing scheme has been devised, designated as EUI-64 . EUI-64 addresses follow the MAC-48 standard in their form, only the manufacturer-dependent NIC is expanded to 5 bytes. While MAC-48 is used by almost all the protocols defined in the IEEE 802 standard, the use of EUI-64 is intended for the following protocols:

- FireWire (IEEE 1394 series bus interface)
- IPv6 (as least significant 64 bit of a unicast address, or a link-local address, if stateless auto-configuration is used)
- ZigBee / IEEE 802.15.4 / 6LoWPAN (wireless PANs).

4.2.5 Local Data Administration

MAC addresses identify the transmitter and the receiver in the LAN. In addition to these address details, a data packet (frame) also contains information about the type of transported payload that could be relevant for further transmission. Every LAN technology exactly defines its own data packet format, which typically consists of the actual payload (body) and the accompanying metainformation (header). In the hierarchical structure of the protocol layers within a network technology, each layer includes its own header on the data to be transmitted. User data and the accompanying header become the payload information of the protocol layer above, which in turn attaches its own header.

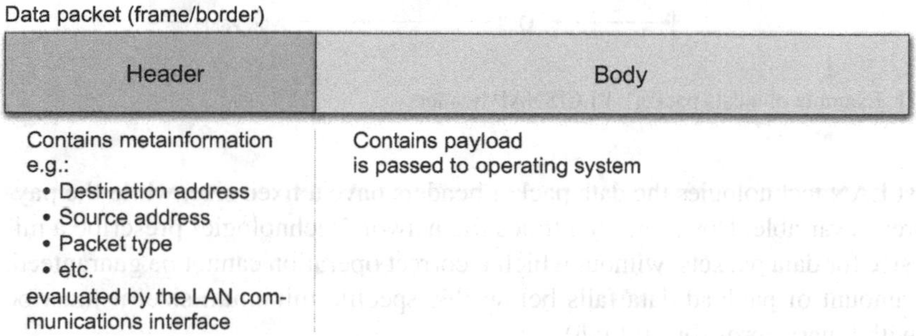

Data packet (frame/border)

Header	Body

Contains metainformation
e.g.:
- Destination address
- Source address
- Packet type
- etc.

evaluated by the LAN com-
munications interface

Contains payload
is passed to operating system

Fig. 4.10 Data packet – principle structure

An identification of the payload in the transmitted data packets is often necessary in order to be able to influence the further processing of the transmitted data. This can be done in the following way:

- **Explicit packet types**: An explicit type field is inserted in the header of the data to be transferred. Such packet types are also designated "self-identifying."
- **Implicit packet types**: Here, the data packets contain only payload data. In this procedure, transmitter and receiver must be in agreement regarding the type and the contents of the exchanged data packets, i.e., they must have previously exchanged all information necessary for correct transmission. Otherwise transmitter and receiver can also agree on characteristics or a part of the payload as type identification.

The data formats of the different network technologies described in the following chapter all correspond to the explicit packet type. To ensure that the packet types identified as explicit are also able to correctly identify the specific type, different values for type fields have been standardized.

It should also be noted that varying standards do exist. These have been determined by various standardization boards. The solution from IEEE provides that in addition to a type value, the responsible standardization board is identified. Thus the **Logical Link Control** (**LLC**) marked part of the of IEEE 802.2 standards for LANs sets the specification of a **SubNetwork Attachment Point** (**SNAP**). Fig. 4.11 shows a LLC/SNAP header that contains 8 bytes. The first 3 bytes contain the LLC component, which indicate that a type field – the SNAP part – follows. The SNAP part of the header contains a 3-byte long field identifying the standardization board (**Organisational Unique Identifier**, **OUI**). This is followed by a 2-byte long type value defined by this board.

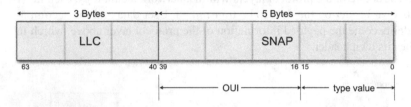

Fig. 4.11 Example of a data packet's LLC/SNAP header.

In most LAN technologies the data packet headers have a fixed size, while the payload area is variable. However, oftentimes the network technologies prescribe a minimal size for data packets, without which a correct operation cannot be guaranteed. If the amount of payload data falls below this specific minimum size, it must be filled with binary zeros (bit stuffing).

4.2.6 Special Network Hardware

The execution of network communication on the network access layer of the LAN proceeds over special network hardware located on the LAN interface. This hardware has the task of handling the communication details and keeping this processing effort at a distance from the central computing unit to not burden it additionally. So-called network analyzers play a special role in the network hardware because of their ability to monitor and control the entire data traffic in a LAN in real-time.

The access of a computer to a local network – and via this usually to Internet – proceeds over the **LAN interface** (also network adapter, LAN communication interface or network card, network interface controler). LAN interfaces work according to the principle of the typical input/output device. They are laid out for a specific network technology and take over details of the data transmission in the network without needing access to the CPU of the computer. The LAN interface can correctly interpret the electric or optical signals in a LAN, the velocity in which the data has to be transmitted, as well as the details of the data packet format. Principally, the LAN interface of a computer is responsible for seeing to it that only those data packets reach the computer that are in fact intended for it. This is because in a LAN, data packets are initially sent to all of the LAN-connected computers via a shared transmission medium, and consequently received by all of them. For this purpose the LAN interface has a **filter function** to filter out non-relevant data packets and and in this way to counteract the needless use of computing power. Because of this, the computer can meet its primary responsibilities. Originally, network adapters were mostly designed for use as expansion cards (separate network adapter) and connected over a bus interface to the actual computer. However, current computers usually have a LAN interface that is directly integrated in the computer's motherboard. Separate network adapters are generally only necessary today if, for reasons of efficiency and performance, it is desired that the computer be connected to the network via several interfaces simultaneously.

A device that determines the performance of a LAN is designated as a **network analyzer** (network analyzer, network sniffer, packet sniffer, LAN analyzer). This means that it monitors and analyzes all of the data sent over the shared transmission medium. Independent of the structural design of the LAN (network topology), the entire data traffic of the LAN can be monitored from a single access point. This normally involves a portable device or a specialized software. After the network analyzer is connected and activated on the LAN, it monitors specific characteristics and collects data in order to glean statistics about network capacity. Network errors may be identified based on these statistics. In an Ethernet LAN, the average number of collisions of data packets, for example, can be determined. In a token ring network the average dwelling time of a token data packet (cf. section 4.3.3), or the token cycle time, is established this way. Network analyzers are also put to use when it comes to controlling the data traffic of a specific computer or accounting for only certain types of data packets. While classic network analyzers work in real-time while connected to the transmission medium, newer implementations focus on processing the measured data that has been recorded off- line.

So that a network analyzer is able to evaluate all of the data packets that pass through it, the conventional address recognition of the LAN interface is circumvented and it switches into an operational mode known as a **promiscuous mode** (mixed mode). Because nearly every LAN interface card is able to function in this operational mode – with the operator's sufficient knowledge – no guarantee of confidentiality can be given regarding the data packets sent within the LAN. This is only possible if they are transmitted in encrypted form. Functioning in the promiscuous mode, the LAN interface card takes over every incoming data packet and then passes it on to the analysis software of the network analyzer. A check is subsequently made for errors in the data packet header. Using the appropriate configuration parameters, the user can determine for herself the area of concentration. parameters. Wireless LANs (WLANs) present a special case. Even if the network analyzer in a WLAN operates in the promiscuous mode, its network adapter ignores all data packets that do not belong in its so-called service set, i.e., any data packets coming from foreign WLAN networks. To still be able to receive and process data packets from foreign WLANs, the network adapter must be operated in the **monitor mode**.

Among the typical user scenarios for the network analyzer are:

- Analyzing network problems
- Discovering intrusion attempts in the network
- Acquiring information about methods of network intrusion
- Monitoring network usage
- Collecting statistical data regarding network data traffic and network use
- Filtering suspicious content in network data traffic
- Exposing sensitive user information (passwords, encrypting methods, etc.)
- Encrypting proprietary network protocols (reverse engineering)
- Monitoring and error analysis in data traffic of client/server communication (debugging)
- Monitoring and error analysis of network protocol implementation.

4.3 Important Examples of LAN Technologies

Many different technologies are employed in the field of local networks. These are based on different types of transmission media, e.g., electrical and optical conductors, or electromagnetic waves, and also distinguish themselves in the geometric arrangement of the connected computers (topology). The various LAN topologies and their respective advantages and disadvantages shall be discussed first. Afterwards, the most important historical and current wired LAN technologies will be looked at in detail with an emphasis on Ethernet, the most widely-used LAN standard in the area of wired transmission media. This will be followed by a presentation of further examples of important LAN technologies: token ring, FDDI and ATM. A separate chapter is devoted to wireless LAN technologies, which continue to gain in importance today.

4.3.1 LAN Topologies

The topology of a computer network describes the geometric arrangement of individual network nodes and their distribution. Different LAN topologies imply different properties. Knowledge of the respective advantages and disadvantages of each is therefore the foundation for selecting the suitable LAN topology for a planned application. Principally, a distinction is made between the **physical topology** of a computer network, which concerns the actual physical installation of a network in-cluding its end devices and their location and cabling, and the **logical topology**. The latter concerns the ways that data packets are actually sent to a network, and in fact often forms a contrast to physical installation. Topologies can be subdivided according to their **dimensions**. A n-dimensional topology is distinguished by the fact that it can only be laid out without intersecting in a n-dimensional space. One-dimensional topologies have gained dominance in LANs. The network topology is determined by the mapping of the physical or logical connections of the individual network nodes on a graph. Therefore, an analysis of different network topologies is also based on a graph-theoretical foundation. Over time the following topologies have been established:

- Bus topology
- Ring topology,
- Star topology and
- Tree topology.

Other two-dimensional topologies, such as grid or systolic arrays, as well as multiple dimensional topologies (meshes) are primarily of importance in parallel computing. Fig. 4.12 shows a graphic representation of basic LAN topologies.

The following criteria can be used to compare different network topologies:

- **Cabling effort**
  Which cable length is necessary for the prescribed geographic arrangement of the computer? If an additional computer is to be admitted into the network, what effort must be carried out in view of its cabling?

- **Total Bandwidth**
  What is the bandwidth of the networks for a given number of sections?

- **Efficiency**
  How large is the number of intermediate nodes along a given connection from one computer to another? The mediation effort rises as the number of switching computers in a network increases and makes an optimal throughput more diffi-cult.

- **Robustness**
  What effect does the failure of one or more computers or sections have on the network?

In the following, the advantages and disadvantages of the basic topologies will be discussed. Several examples of technologies and their implementations will then be looked at in more detail.

Bus Topology

In bus topology all of the computers are arranged along a lineal transmission medi-um (e.g., a long cable) – the bus – so that a loop does not form. Care must be taken so that at the end of the connection signals are not reflected and rebound from the

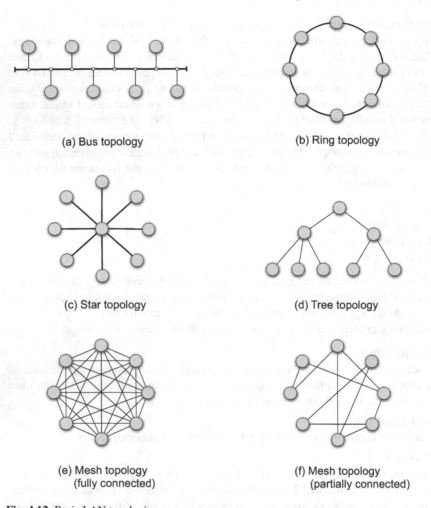

(a) Bus topology   (b) Ring topology

(c) Star topology   (d) Tree topology

(e) Mesh topology   (f) Mesh topology
(fully connected)   (partially connected)

Fig. 4.12 Basic LAN topologies

medium. This would cause interference. At any given time only one of the connected computers is free to carry out transmission, all of the others must hold off with their sending activity until the transmission procedure of the engaged computer has been completed. An additional arbitrator mechanism is implemented to ensure that all computers receive equal access to the bus. It is also able to solve conflicts when two or more of the connected computers wish to transmit at the same time. This arbitrator mechanism can be implemented centrally or decentrally. If a computer connected to a bus transmits information, it will be routed in both directions – depending on the basis of the terminal position of the computer – until it reaches the ends of the bus. All other computers connected to the bus receive the data packet and verify whether or not it is intended for them, i.e., if the MAC address of the receiver contained in the data packet is in agreement with its own MAC address. If this is not the case then the data packet is ignored.

- The **advantages of the bus topology** are primarily founded on its simple extensibility: if a new computer is to be linked to the bus, a new tap is simply installed on the bus and joined with the connected computer. When tapping new computers it is necessary that restrictions are considered, e.g., maximum length of expansion, minimum distance between both connectors, etc. The connecting and disconnecting of individual computers to the bus during normal operation is usually possible without the networks on the bus having to be switched off or without transmission errors occurring. Additionally, the failure of one computer does not have any consequences for the functionality of the network as a whole. The cabling of the bus topology is for the most part modular. A bus topology is also relatively inexpensive in view of the required cabling as only a single bus cable is needed.

- The main **disadvantage of the bus topology** surfaces if one of the bus connectors fails. In this case, the entire system is taken down. An expansion of the network is limited based on technological restrictions. At any one time only a single computer can transmit in the network and during this time the remaining computers are blocked. In the event of data collision a difficult to estimate waiting time can occur, during which the arbitrator mechanism attempts to mediate the collision conflict. The higher the data traffic on the bus the more likely it is of collisions happening, and thus the lower utilization of the transmission medium.

The most important example of a bus topology today is the widespread **Ethernet**.

Ring Topology

In a ring topology all of the computers are arranged in the form of a ring in a closed loop. The connection of two adjacent computers within the ring proceeds via a point-to-point connection. Every computer in the ring has exactly one predecessor and one certain successor. The data packet to be transmitted is sent from one computer to its successor. The successor verifies whether the data packet is meant for it. If this is not the case, it sends on the packet to its successor. This continues as long as is necessary for the data packet to reach receiver it is intended for. Because each computer in the ring resends the information to be transmitted, the transmission signal to the receiver is continuously increased and with this method kilometer-long distances can easily be bridged.

- The **advantages of the ring topology** are found in its simple extensiblity. If a new computer is added to the existing ring, the cabling effort is minimal. All of the computers in the ring act as amplifiers, thus great distances can be bridged. The delay connected with the information transmission in the ring is proportional to the number of linked computers as information must pass every single connected computer. In contrast to bus topology, collisions cannot occur in the ring topology transmission medium. Therefore it is possible to guarantee a deterministically identifiable transmission delay and transmission bandwidth.

- On the other hand, even a single cable break between two neighboring computers can lead to the failure of the entire ring because data packets can no longer be forwarded. This is undoubtedly the most critical **disadvantage of the ring topology**. This problem can be remedied by redundant cabling, or bidirectional utilization of the ring. As transmission delay is proportional to the number of linked computers in ring topology, high latency may appear when communicating with distant computers.

Ring topology in a pure form rarely exists in practical application, due to the described disadvantages concerning reliability. While practical technologies may be based on principles of the ring topology, they do not form a true ring as far as cabling is concerned. The most important examples of ring topologies are IBM **token ring** or **FDDI**.

Star Topology

The star topology is the oldest form of cabling used for the creation of networks. Here, the individual computers are linked with each over, arranged in a star form around a central point called the **hub**. Computer and hub are each connected with one another via point-to-point connections. Classical mainframe computer systems are often cabled according to this scheme: the mainframe as host in the center and the I/O systems linked to it in the periphery. The central hub has the responsibility for the entire network communication and controls the traffic. When one of the connected computers wishes to send information to another, this message is transmitted to the hub in the middle. There are different possibilities to coordinate communication in the star. One is that all linked computers send their requests to the hub in the middle and wait for its answer. The request as well as the answer are transmitted via the hub. So that no requests are lost, the hub must have a large enough cache and a high switching capacity. Another possibility is the implementation of a polling procedure, in which the central computer sends a query to each of the connected computers asking whether it has information to forward. If the hub encounters a computer that wants to transmit at this time then it will be served immediately. Here as well an arbiter routine must decide how long the sending computer may occupy the hub so that the other connected computers are also served fairly and not forced to wait an unnecessarily long time for transmission.

- In a star-formed network the hub is subject to a higher load. If it fails the entire network fails. For this reason, in practical application, the central hub is often arranged redundantly. Besides this, a further **disadvantage of the star topology** is presented in the relatively high cabling effort: from each of the connected computers a cable must be laid to the hub and then back again.

- However – and this is a considerable **advantage of the star topology** – this structure is relatively disturbance-free in case one of the connected computers fails, or a cable breaks between one of the networks members and the hub. Additionally, the star topology can be extended easily.

The **ATM** (Asynchronous Transfer Mode) is an important example of the star topology.

Table 4.3 A comparison of three network typologies.

	Bus	Ring	Star
Extensibility	simple, modular	simple	simple, dependent on hub
Speed	fast	slow dependent on number of computers in network	fast dependent on number of computers in network
Quality of service	not guaranteed	guaranteed	guaranteed
Waiting time	cannot be predicted	constant	dependent on hub
Total failure	failure of the bus	simple cable break or computer failure	failure of hub
Computer failure	network functions	network does not function	network functions
Cabling effort	small	minimal	large

Tree Topologies and Meshed Networks (Mesh Topology)

Tree topologies are organized hierarchically. On the highest level of the hierarchy they have a root node that can be connected with a multiple number of computers on the second hierarchical level. This can in turn be connected with more computers. Connections between adjacent computers are implemented in a tree topology over a point-to-point connection as well. Seen technically, a tree topology can be considered a network topology in which multiple star topologies are hierarchically linked with each other. As in the case of the star topology, the failure of an end node does not have an influence on the functionality of the total network. However, if one of the distributor nodes from a higher hierarchical level fails then the entire section of the tree below the disturbed node cannot be reached. As the number increases on the hierarchical level, the distance that must be bridged during communication between two end nodes in farther tree sections of the network increases as well. This brings with it the possibility of higher latency times.

In a **meshed network** (**mesh topology**) every computer is linked with one or more other computers over a point-to-point connection. When every computer is directly connected with every other computer in the network it is then described as a completely meshed network. In case a computer or line fails it is still usually possible to communicate in a meshed network through data detours (routing).

With the just described characteristics of various network topologies as our starting point, the most important wired LAN technologies will now be looked at in detail. Wireless LAN technologies will be discussed in the following chapter.

4.3.2 Ethernet – IEEE 802.3

In the meantime, Ethernet has become the most important technology representative in the LAN segment of the (wired) network market. Although in the 1980s, and at the beginning of the 1990s, competitors such as token ring, FDDI, or ATM presented a great challenge, none of these technologies succeeded in overtaking the leading market position held by Ethernet since its introduction in the 1970s. There are a number of possible reasons why Ethernet was able to assert its position on such a wide front. Thanks to its practical application, network administrators developed a great deal of trust in the technology and its various features and were thus more likely to take a sceptical view of the new LAN technologies that later arrived. Further, token ring and ATM are considerably more complicated in view of their infrastructure and administration. They are also more expensive in comparison to Ethernet, another reason preventing network administrators from abandoning Ethernet technologies. One reason that the alternative LAN technologies seemed initially more attractive was due to their higher bandwidth. However, Ethernet technologies were repeatedly successful in catching up with their competitors and even in some cases overtaking them. Because of the widespread use of Ethernet, the hardware equipment necessary is correspondingly inexpensive. The reasonable operating cost ratio results from Ethernet's own multiple-access protocol, which runs fully decentralized. This allows a simple design of hardware components.

Historical background

The original Ethernet was developed as a bus topology at the beginning of the 1970s by *Robert Metcalfe* (*1946) and *David Boggs* (*1950) at the research center of Xerox PARC, located in Palo Alto. During his studies at Massachusetts Institute of Technology (MIT), Robert Metcalfe was part of the group working to develop the ARPANET, a precursor to the present-day Internet. As a doctoral candidate he became acquainted with the ALOHAnet, the first radio LAN, which connected the islands of Hawaii with each other. He also became familiar with the Random Access Communication protocol developed there (cf. Fig. 4.13). When he began working at Xerox PARC in 1972 he was introduced to the then highly advanced ALTO computer, which already had many of the features of the modern PC, for example a window-based graphic user interface, a mouse as a simple input device, and the use of a desk top metaphor, i.e., the arrangement of the currently used resources on the computer screen modeled after a desk. Metcalfe immediately saw the necessity of linking these computers with each other as efficiently and inexpensively as possible and together with his colleague David Boggs began to develop the Ethernet.

The first of the two developed Ethernets had a bandwidth of 2.94 Mbps and was able to link up to 256 computers with a maximal cable length of 1,000 meters. The bandwidth of 2.94 Mbps was necessitated by the use of the system clock of the ALTO computer, which supplied this frequency. The first experimental network at Xerox PARC operated under the name "Alto Aloha Network." In 1975, Xerox registered a patent application for the **Ethernet** computer network naming as the inventors Robert Metcalfe, David Boggs, *Charles P. Thacker* (*1943) and *Butler Lampson* (*1943) .

With this new name, Metcalfe and Boggs wanted to make clear that this technology they had developed was not only intended for use on the ALTO computer, but could be implemented everywhere. Furthermore, the technology was not only oriented on the ALOHAnet. The choice of the word "Ether" was meant to describe a fundamental characteristic of the technology. The physical medium (the cable) transmits the data to all of the connected stations, such as the historical ether in physics passes electromagnetic waves in space.

Metcalfe finally succeeded in bringing the companies Xerox, Digital and Intel together to form an alliance, laying down the 10 Mbps standard, the so-called "DIX standard" (after Digital/Intel/Xerox). It was ratified by the IEEE as a first draft standard on September 30,1980. In 1979, Metcalfe founded his own company 3COM. Producing Ethernet cards for PCs, Metcalfe's company profited greatly from the PC boom of the 1980s.

The improved **Ethernet V1.0** was developed by Xerox, Digital and Intel in a common project and culminated in the so-called DEC Intel Xerox standard (DIX). Its almost unchanged specification was introduced to the network standards committee of the IEEE in 1979, as IEEE 802 draft B. It was adopted as the IEEE 802.3 standard in 1985.

Published under the name "*IEEE 802.3 Carrier Sense Multiple Access with Collision Detection (CSMA/CD) Access Method and Physical Layer Specifications,*" the Ethernet standard was subsequently adopted by the ISO as a world-wide network standard. Since it's introduction, the 802.3 standard has continually been adapted to fit the developing network technologies. Starting in 1985 with a bandwidth of 10 Mbps, rapid technological advancement brought an increase in the Ethernet bandwidth to 100 Mbps, then 1 Gbps and 10 Gbps.

All of these methods share the structure of the Ethernet data packets, so-called frames, as well as the implemented arbiter algorithm **CSMA/CD (Carrier Sense Multiple Access / Collision Detect**), to handle competition for access on the bus. The network topology, in contrast, has changed from the initial pure bus topology with coaxial cables, to a star topology with twisted pair cables and multi-port repeaters and then further to a star topology with a bidirectional, switched point-to-point connection.

Among the fundamental, characteristic features of the original Ethernet standards are:

The origin of the Ethernet.

Robert Metcalfe began his career at Massachussetts Institute of Technology (MIT) where he earned two Bachelor degrees. In the course of his subsequent graduate studies at Harvard he focused on the problem of computer networks. Even before earning his doctorate he worked at Xerox PARC, which had commissioned him with developing the first PC network.

Xerox PARC was to be equipped with the first modern proprietary computer, the Xerox Alto, as well as with the first laser printer, also developed there. This computer was so small and reasonably priced that it would now be possible for the first time to equip every office in a building with one or more of them. In order to do this a network was necessary that had the capability of bearing the load of a new computer or network without the necessity of switching off and newly configuring the entire system. The network would also have to be powerful enough to allow for the operation of a fast laser printer.

The idea that finally lead to the development of the present-day Ethernet came to Metcalfe in 1970 while he was reading a conference article by *Norman Abramson* (*1932). Abramson, from the University of Hawaii, wrote about the packet-switched wireless network ALOHAnet, that connected the Hawaiian Islands with each other via a simple and inexpensive amateur radio transmitter and receiver. Each node in ALOHAnet transmitted its information in streams of individual, separated data packets. If it happened that the transmission of a packet was not confirmed, e.g., in a situation where two transmitters are sending at the same time, the transmitted packet were seen as "lost in the ether." If a packet got lost in the ether, the transmitting computer waited a random time period before beginning a new transmission attempt. Because of this random principle, it became possible even under a heavy traffic load, to quickly resolve the unavoidable data packet collisions that occur in bus topologies. Generally speaking, it was seldom the case that a transmitter had to dispatch its packets more than once or twice before they could be registered by the designated receiver – a certainly more efficient method than preventing collisions using an elaborate collision-resolution algorithm.

Although ALOHAnet was successfully implemented, Abramson showed that it already reached its maximum capacity at only 18% of the theoretically possible transmission utilization. This was because the number of collisions rose disproportionately with an increase in network capacity. Metcalfe addressed this problem in his thesis and finally showed that with the application and exploitation of the mathematical queuing theory an efficiency of up to 90% of the theoretical maximum capacity can be reached without the system becoming blocked due to packet collision.

Further literature:

Metcalfe, R., Boggs, D.: Ethernet: Distributed Packet Switching for Local Computer Networks, in Communications of the ACM, 19(7), pp. 395–404, ACM, New York, NY, USA (1976)

Fig. 4.13 The origin of the Ethernet.

- relatively high rate of data,
- minimal delay as a result of dispensing with storage and transport logic in the network,
- network diameter up to max. approx. 1 km,
- support of hundreds of independent computers in the network,
- high reliability, no central control,
- very simple algorithm for access to the communication medium and addressing,
- efficient use of the shared communication medium,
- fair access distribution to all stations,
- high stability also under load,
- low cost.

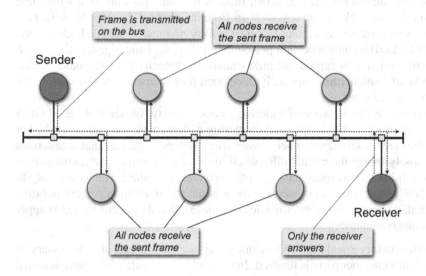

Fig. 4.14 The fundamental principle of the Ethernet.

Basic components of the Ethernet

There are three basic components that are defined by the Ethernet standard and which determine the respective Ethernet technology:

- the **physical medium** of the Ethernet channel, over which the signals between the connected computers are transmitted,

- the **framework of rules for access** to the Ethernet channel, which allows the large number of connected computers fair and equal access to the shared Ethernet channel, and

- the **Ethernet data packet** (frame), that fixes the structure of bits in a data set sent over an Ethernet channel.

Before taking a closer look at the physical medium, and accordingly different types of Ethernet, the fundamental algorithm and structure of the Ethernet data packet will be discussed.

The Ethernet multiple access algorithm – CSMA/CD

The nodes in an Ethernet LAN are all connected to one shared transmission medium, the so-called Ethernet broadcast channel (Ethernet bus). As soon as one of the connected nodes carries out a transmission all of the nodes connected on the LAN receive this data packet (cf. Fig. 4.14). Indeed, it is only possible at any one time for a single data packet to be sent over an Ethernet channel. In order to make sure this happens correctly an accumulation of collisions must be avoided. Unlike the so-called ALOHA communication protocols (cf. Fig. 4.15 and Fig. 4.16), in which a data transmission is carried out independent of the activity of the other nodes, the **CSMA/CD algorithm**, especially developed for Ethernet, does indeed take this activity into account.

Let us compared the network operation to a cocktail party. We see that the ALOHA protocol allows guests to simply start chattering without paying attention to whether or not they are disturbing any other conversation. However, the Ethernet access rules more closely follow the normal rules of etiquette at our cocktail party. One person has to wait his turn to speak until the other person is finished. This is not only to avoid "collisions," but also to increase the amount of information content per time unit. In this scenario there are two rules of conversational etiquette that also apply to Ethernet communication:

- **"Listen before speaking."** When one party wants to speak, it is necessary to wait until the other party is finished. In networking this rule of etiquette is called **carrier sensing**. A computer node listens to the transmission medium before it begins transmitting information. If the data packet of another computer is in the transmission medium at that point, the other computer waits a random period of time (**backoff**), before it checks again whether or not the communication channel is free. Then after the expiration of an additional waiting period (**interframe gap**) the computer begins transmitting its data packet.

- **"If someone starts talking when you're speaking then stop speaking."** Transferred to a network situation, this procedure is called (**collision detection**). A computer that is momentarily occupied with a data transmission listens at the same time to the communication medium. As soon as it detects that another computer is sending information that collides with its own, both computers break off their transmission (**jam signal**). The time span (backoff), in which the nodes wait until a new transmission begins, is determined by an appropriate algorithm.

The ALOHA multiple access protocols.

At the beginning of the 1970s, at the University of Hawaii, Norman Abramson developed a simple and efficient multiple access method allowing multiple stations to communicate over one, mutually used communication channel. The ALOHA system developed by Abramson was implemented in the newly developed radio-based network communication between the main islands of Hawaii. The ALOHA protocols can however be used on any wired transmission media. Two distinctive varieties exist: Pure ALOHA, which does not require global time synchronization of the stations, and Slotted ALOHA in which all stations are synchronized in time and the transmitted frames are always required to fit in the predetermined time slots.

Pure ALOHA – Every station can send a data packet (always of the same length) at any point in time. If more stations send data packets at the same time, these collide and are destroyed (unreadable). This is even the case if only the first bit of a sent data packet collides with the last bit of a data packet just sent before. As every station can transmit and receive at the same time, it is possible for each to determine if a collision has occurred. An acknowledgment mechanism must be used if transmitting and receiving is not possible. When a data packet is destroyed in a collision it is then re-sent following an randomly determined waiting period. The maximum throughput with this method is 18%. It can be calculated as follows: t is the time period needed to transmit a data packet of a fixed predetermined length. In this calculation an (unrealistically) endless number of users is assumed. These create in the mean N data packets per time span t, whereby it is assumed that N is subject to a Poisson distribution. In order that N remain constant in the event of collisions and consequent blockage the number of users must be assumed as unlimited. So that a collision does not inevitably occur in every time slot, $0<N<1$ must apply. Besides the N new data packets sent per timespan t, also the additional new transmissions caused by collisions must be calculated in. The probability that a total of k transmission attempts occur during t is also Poisson-distributed with the mean $G \geq N$. The data throughput S results from the mean value of G multiplied by the probability of a a successful (collision free) transmission p_0, d.h. $S=Gp_0$.

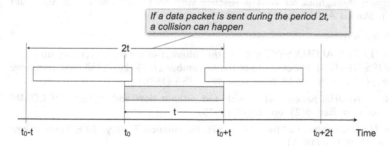

A collision occurs if a new data packet is sent within the time span of an already begun transmission. Because prior to the transmission of a data packet at the time period kt t_0 it is not verified in ALOHA if another party is already sending, the conditions for a collision-free transmission is that in the time period t_0-t to t_0 no data packet was sent. Additionally, during the transmission time t of the current frame, up to t_0+t no further transmission may take place.

Fig. 4.15 The ALOHA multiple access protocols.

The ALOHA multiple access protocols – Part 2 The probability that within the time span t exactly k data packets will be sent is determined by Poisson distribution

$$p[k] = \frac{G^k e^{-G}}{k!}.$$

The probability that within the time span for a successful transmission 2t no further data packet (k=0) will be sent, is determined by $p_0 = e^{-2G}$, as during the time span 2t on the average 2G data packets will be sent. For the data throughput S results

$$S = Ge^{-2G}.$$

The maximum throughput for G=0,5 is reached with S=1/2e≈0,184, i.e., channel usage in the best case amounts to just 18%.

Slotted ALOHA -In contrast to Pure ALOHA the stations here work synchronously, meaning they must maintain predetermined time slices (slots) of exactly the length of the data packet to be transmitted, which is clocked by a central timer. Every station can transmit in one of these slots at any time. If two stations transmit at the same time then a collision occurs. Unlike Pure ALOHA, in Slotted ALOHA it is possible that two data packets can completely overlap each other when a collision takes place due to the predetermined slots. Partial overlapping need not be taken into account. Because when sending a data packet it is always necessary to wait for the beginning of the next time slice, the time interval for transmission in which a collision can take place is cut in half, i.e., of t, in contrast to Pure ALOHA,. The probability that within the necessary time span for a successful transmission t no further data packet (k=0) will be sent, results from $p_0 = e^{-G}$, as during the time span t in the mean G data packets are sent. Thus, for the data throughput S the result is

$$S = Ge^{-G}.$$

The maximum throughput for G=1 is reached with S=1/e≈0,368, i.e., channel use improves in Slotted ALOHA by 37%.

Further literature:

Abramson, N.: THE ALOHA SYSTEM: another alternative for computer communicati-ons. in: AFIPS '70 (Fall): Proceedings of the November 17-19, 1970, fall joint computer conference, pp. 281–285. ACM, New York, NY, USA (1970)

Roberts, L.G.: ALOHA packet system with and without slots and capture. SIGCOMM Comput. Commun. Rev. 5(2), pp. 28–42 (1975)

Abramson, N.: Development of the ALOHAnet. Information Theory, IEEE Transactions on 31(2), pp. 119–123 (1985)

Fig. 4.16 The ALOHA multiple access protocols, Part 2.

This prevents both computers from starting their transmission again at the same time following completion of the time period – and the possibility of a renewed collision.

These two simple rules are observed by the families of **CSMA-** (**Carrier Sense Multiple Access**) and **CSMA/CD** protocols (**CSMA with collision detection**). Many variations of these two algorithms have been proposed. The main difference between them is the variation in the type of backoff chosen.

There are different possibilities of how a transmission-ready computer can react in a CSMA algorithm. The different cases are categorized as follows:

- **Nonpersistent CSMA**:
  If the transmission medium is free, the computer transmits immediately, but if it is occupied a waiting time is given, the length of which is determined by a random number. A check is made again afterwards to see if the medium is free. If so, transmission takes place, if not, a random time span for waiting is provided.

- **1-persistent CSMA**:
  If the transmission medium is free, the computer transmits immediately, otherwise the computer monitors the medium. As soon as the computer determines that the transmission medium is no longer occupied, transmission is carried out. 1-persistent corresponds to the subsequent p-persistent for p=1.

- **p-persistent CSMA**:
  If the transmission medium is free, then in all probability p, $0 \leq p \leq 1$, will be sent with the probability of (1-p) expected for a time unit. In the event that the transmission medium is occupied, it will then be monitored. As soon as the medium is free transmission will then be carried out with probability p, or there will be a period of waiting for a random time span with probability (1-p).

The value of p must of course be chosen appropriately. If n computers are connected to the network and waiting for access to the communication channel, then the product n·p must be smaller than 1. While a value chosen for p that is too small leads to waiting periods that are too long.

Fig. 4.17 Variations of the CSMA algorithm.

Before looking at the Ethernet variety of the CSMA/CD protocol, several fundamental characteristics of this method will be discussed. One of the first questions that arises in connection with CSMA is why collisions occur at all given the fact that all of the connected computers are equipped with carrier sensing, capable of monitoring if the communication medium is free.

With the help of a space-time diagram it is possible to show the reasons for this situation very simply. In Fig. 4.18, 4 computers (A,B,C and D) are seen in a space-time diagram all connected to a linear bus. The vertical axis shows the spatial distribution of the computers on the bus. The horizontal axis represents the time line. At time t_0 computer C recognizes that the bus is free at this moment and therefore begins transmission of its data packet, which is to be sent in both direction on the bus. At time point t_1, $t_1 > t_0$ computer A also wishes to send a packet. It monitors the system bus and cannot determine traffic as the data packet from computer C has not yet reached computer A. Thus computer A begins transmission of its own data packet in compliance with the CSMA protocol. But a short time later there is interference between the two transmitted data packets on the bus and this leads to a collision.

Here it can be clearly seen that the delay caused by signal runtime on the transmission channel (**Channel Propagation Delay**) plays an important role when describing the characteristics of the transmission medium's capacity. The greater the delay caused by the signal runtime in a transmission channel, the more likely it is that a computer wishing to transmit cannot determine if another computer in the network has already begun its own transmission.

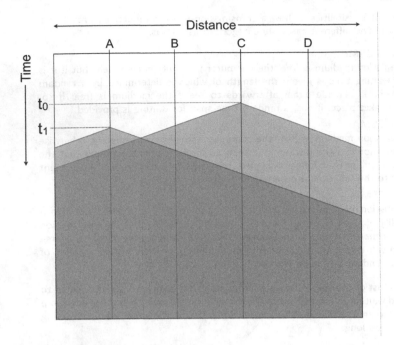

Fig. 4.18 CSMA without collision detection.

Incidentally, the two computers in Fig. 4.18 are working without collision detection; C and A continue their transmission although a collision has occurred. When a computer works with collision detection it stops its transmission immediately upon discovering a collision (cf. Fig. 4.19). It should also be noted that collision detection raises transmission capacity in a network, because the transfer of data packets damaged by collision is avoided. The Ethernet protocol uses the CSMA algorithm along with its collision detection, CSMA/CD.

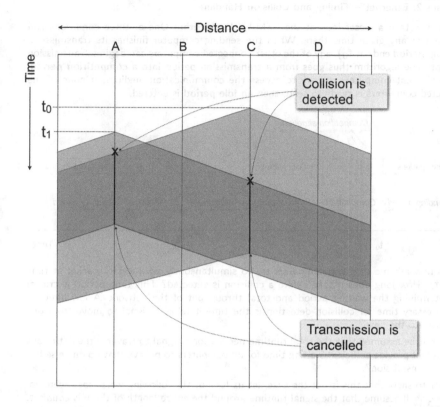

Fig. 4.19 CSMA with collision detection.

The Ethernet arbiter algorithm for competing access to a shared transmission medium: Carrier Sense Multiple Access with Collision Detection (CSMA/CD).

1. A computer can begin its transmission in a network at any time. There are no fixed time slots for transmission.

2. A computer never begins transmission when it determines that another computer has already started sending a data packet (**carrier sensing**).

3. A computer that has just started sending interrupts its transmission as soon as it determines that another computer has already begun transmission (**Collision detection**).

4. Before attempting a new transmission, the computer wishing to transmit waits a random time span the length of which is determined by network capacity (**Random backoff**).

Fig. 4.20 The CSMA/CD algorithm.

Excursus 2: Ethernet – Timing and Collision Handling

Let us now take a closer look at the CSMA/CD algorithm One of three conditions can be found at any given time there. When the sending computer finishes its transmission, (**sending period** ends at t_0), any of the other connected computers can begin a transmission attempt. The algorithm thus goes from a transmission period into a **competition period**, where different computers can try to access the communication medium. If none of the connected computers is ready to send, then an **idle period** is entered.

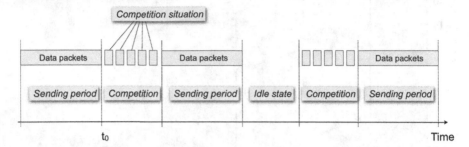

Let us now assume that two computers try to simultaneously send a data packet at time point t_0. How long does it take before a collision is detected? This time period is crucial for determining the waiting period and total throughput of the network. A low limit for the necessary time for collision detection is the time it takes a signal to move from one computer to the next.

But can it be assumed that the total runtime needed for a signal to traverse the entire bus does in fact provide sufficient waiting time for all computers to believe that no one else has started transmission?

In order to show that this is not the case, let us look at the following worst case scenario. Here we shall assume that the signal runtime around the entire length of the bus equals τ. At time point t_0 computer A begins its transmission. At time point $t_0+\tau-\varepsilon$ – a fraction less than the time the signal needs to traverse the entire length of the bus – computer B, the computer farthest away from A, starts its transmission. The collision is detected by B almost immediately and thus B ends its transmission. Yet on the other side, computer A needs the time $t_0+2\tau-\varepsilon$ to detect the collision. Therefore, in this worst case scenario, computer A can first be sure that its transmission was really successful if a collision notification has not arrived after expiration of the time 2τ. The borderline case where a collision can only just barely be detected by the transmitting computer, is at a transmitting period length τ_r with $\tau_r \geq 2\tau$. This is the so-called **runtime condition**, which must be fulfilled for the correct functioning of the Ethernet.

Collision detection

To be able to detect collisions on the Ethernet bus (coaxial cable) it is necessary to recognize a DC-offset caused by the overlapping of two or more signals. While in an undisturbed data transmission the transmitted signal has a voltage of up to -1293 mV, collision detection is evident from a voltage of between -1448 mV and -1590 mV. For Ethernet variations that function on the basis of twisted pairs or fiber optic cables, alternative methods must be used for collision detection. This is because there is a strict separation between the transmission and reception channels. If during transmission the reception of a data packet is registered there by a participant, it will be considered a collision.

Collision Handling

If a computer discover a collision, it first sends a special bit sequence (**jam signal**). This signals the collision to the entire network. When collisions occur, a fixed transmission time and, hence, a maximum response time can only be maintained within a certain degree of probability. Thus, CSMA/CD is not a deterministic procedure.

In the event of a collision, those on the Ethernet connected computer will have to wait for 0 or 1 time periods (**slot**), dependent on a randomly determined number. Perhaps a collision occurs and both computers choose the same time period – the result is another collision. If a second collision takes place, the computers involved wait for 0, 1, 2 or 3 time periods, whereby the exact number of waiting periods is again determined randomly. In the case of another collision, the number of possible waiting periods grows to $2^3=8$. Therefore it generally applies that after i collisions a random number determines how many time periods between 0 and 2^i-1 must be waited out until a new transmission can be started. After the tenth collision the maximum waiting time interval is frozen at $2^{10}=1.023$ time slots, after 16 collisions the computer wishing to transmit gives up and the transmission broken off.

This algorithm is called the **binary exponential backoff**. The exponential algorithm type is chosen purposely because of its dynamic adaption to the number of computers connected in the network. If one were to fix the random interval from the beginning at 1.023, then, although the probability of a second collision occurring two computers would be very low, the average waiting time following a collision would be several hundred time periods and thus unnecessarily extended. However, if on the other hand the waiting interval would be limited to only 0 and 1, then the network would immediately be paralyzed if 100 computers simultaneously wished to begin a transmission. The waiting time would only first end when 99 computers chose the 0 and one the 1 or vice versa. However, since the algorithm dynamically adapts the waiting time intervals to the number of occurring collisions, it provides a guarantee of short waiting times when only a few computers cause collisions. Furthermore, it sees to it that collisions on the other side can be quickly remedied when a large number of computers are involved. With every new data packet transmission, the CSMA/CD algorithm starts again from the beginning again without being preoccupied with previously caused delays. Therefore it is possible for a computer to successfully send a new data packet while others are still stuck in exponential backoff.

But CSMA/CD alone is not able to guarantee safe transmission. An **acknowledgement** mechanism is also necessary to confirm that data packets have been received. Furthermore, the non-occurrence of collisions in no way guarantees that data packets will not be damaged in other ways by transmission errors on the physical medium. The receiver must verify the checksums in a transmitted data packet and in the case of success send back an acknowledgment.

Further literature:

Tanenbaum, A. S.: Computer Networks, 4th ed., Prentice-Hall, Inc., Upper Saddle River, NJ, USA (2002)

Institute of Electrical and Electronics Engineers: 802.3: Carrier Sense Multiple Access with Collision Detection, New York: IEEE (1985)

All of the various Ethernet implementations provide a **connectionless service**, i.e., if the Ethernet network adapter from computer A communicates with the Ethernet network adapter from computer B, A simply begins sending data packets without previously having set up a connection via some kind of handshake protocol. Furthermore, all Ethernet technologies only provide a so-called **non-reliable service**. While the Ethernet data packet has a checksum in order to verify correct transmission, after reception of a data packet no explicit acknowledgment is sent back. This is the case both in the event of success and failure. If a checksum error in a received data packet is identified, then this data packet is simply abandoned. A protocol from a higher layer then bears the responsibility for error correction or retransmission.

Implementation of the CSMA/CD algorithm

The Ethernet network interface in a computer connected to Ethernet must determine if another computer is in the process of transmission at that moment and, at the same time, be able to detect collisions while being in the transmission process itself. The Ethernet network interface – often called the **Ethernet network adapter** – solves this task by measuring the voltage in the transmission medium before and during transmission. Every Ethernet network adapter thus has the capability of implementing the CSMA/CD algorithm independently, without central coordination. This proceeds as follows:

1. As soon as a data packet is ready for dispatch, the Ethernet network adapter forwards it to an output buffer.

2. If the Ethernet network adapter detects that the channel is free for transmission (i.e., it does not determine any measurable signal voltage in the cable), it begins to transmit the data packet. If, on the other hand, the Ethernet network adapter determines that the cable is occupied, it waits until no further signal voltage is measured and then begins transmission.

3. During transmission, the Ethernet network adapter monitors the channel to determine if signal energy is measured that originates from a different Ethernet network adapter. If the Ethernet network adapter succeeds in sending the complete packet without a foreign signal energy being detected in the measured channel, then transmission of the data packet is complete.

4. If the Ethernet network adapter detects foreign signal energy on the channel during transmission it terminates the transmission, instead sending an 48 bit long collision enforcement jam signal.

5. Following transmission termination, the Ethernet network adapter enters the exponential backoff phase. After n collisions occurring one after the other, the adapter randomly chooses a number $k \in \{0,1,2,\ldots,2^m - 1\}$ with $m=\min(n,10)$ and waits for exactly this time span – equaling a transmission time of $k \cdot 512$ bits. It then returns to step 2.

Fig. 4.21 Implementation of the CSMA/CD algorithm for Ethernet.

The Ethernet data packet format

The various Ethernet technologies on the market today have something more in common than just the CSMA/CD algorithm. That something is the Ethernet data packet format. This format is often called the Ethernet frame format. All of the technological variations, e.g., coaxial cable or copper cable, from 10 Mbps to 10 Gbps, use the same basic structure for dispatched data packets. Fig. 4.22 shows the Ethernet data packet format.

There are two variations of the Ethernet data packet format. These can be distinguished by the field after the source address. In the **Ethernet format** (also known as Ethernet II, DIX or Blue Book) this field is called a type field. It specifies the network protocol corresponding to the transported data. The minimum allowed value of the type is 1,501. The IEEE 802.3 data format, on the other hand, describes this 2-byte long field based on the length of the transported payload. The maximum value allowed is 1,500. Every IEEE 802.3 Ethernet data packet starts with a **preamble** of 8 bytes, each containing the bit pattern 10101010. In this preamble the last byte

Ethernet II

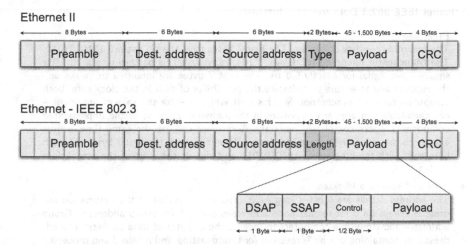

Ethernet - IEEE 802.3

Fig. 4.22 The Ethernet data packet format.

serves as a marker for the start of the data packet itself (Start-of-Frame Delimiter, SFD). It ends with the pattern 10101011, which synchronizes the system clocks of transmitter and receiver. The receiving computer then knows that the coming 6 bytes denote the **destination address**.

The subsequent **type field** in the Ethernet format characterizes the data packet as associated with the Ethernet protocol. As a protocol, Ethernet is arranged in the data link sublayer of the protocol stack. Because different types of protocols exist here capable of taking over transport (e.g. Novell IPX, Appletalk, etc.), this field has an Ethernet-typical identifier. In the IEEE 802.3 data packet, there is a length field in this location. It gives information about the length of the transported payload.

In the following **data field** the first two bytes, **DSAP** (Destination Service Access Point) and **SSAP** (Source Service Access Point), are followed by a half control byte, whereby DSAP and SSAP identify respective service access point addresses based on the TCP/IP reference model. In the data field the IP datagram is encapsulated in the Internet layer of the TCP/IP reference model lying above the network access layer. The maximum length (Maximum Transmission Unit, MTU) of the data field is 1,500 bytes. If the length of the IP datagrams that are to be sent exceeds a MTU of 1,500 bytes, the IP datagram must first be divided into smaller sections of maximum 1,500 bytes long (fragmentation) before they can be forwarded via Ethernet. The minimum length of a data field is 46 bytes. If shorter payload is to be transported, the remaining bytes of the data field must be filled up to this size (padding). The following 4 bytes receive a checksum for the datapacket. In Fig. 4.23 the individual data fields of the Ethernet format and of the IEEE 802.3 data format will be explained in more detail.

Ethernet IEEE 802.3 Data format – Internals

- **Preamble**
  The preamble consists of 8 bytes. The first 7 bytes have the bit pattern 1010101010. The **Manchester encoding** (cf. Chap. 3.16) transforms this pattern into a 10 MHz square wave signal for exactly 5.6 μs. The first 7 bytes are intended to "wake up" the receiver and to ensure that despite the possibility of drift in the clock rate, both computers function synchronously. This drift which is – for the sake of argument – not calculable can appear everywhere in the network, as none of the implemented components works completely perfectly. A receiver can, at the same time, adjust exactly to the respective transmitter by synchronizing with the first 7 bytes. The 8th byte ends with the bit pattern 1010101011 and signals the start of the actual data packet.

- **Goal and source addresses**
  The address details are 6 bytes long each. The high-level bit of the destination address is always an 0 for regular node addresses and a 1 for group addresses. Group addresses allow multiple computers to receive the dispatched data packets. The address only containing one bit is reserved for **broadcasting** and received and processed by all computers in the network. Moreover, bit 46 (directly adjacent to the high-level bit) is used to distinguish between local and global addresses. **Local addresses** are assigned by network administrators and have no significance for the world outside the internal network. **Global addresses** are assigned by the IEEE and should ensure that worldwide no two identical Ethernet addresses exist. Every computer should be capable of communicating worldwide with one of the 2^{46} possible global addresses. Ethernet does not stipulate how these addresses are to be localized. This is the task of protocol layers located on higher levels.

- **Length field**
  The 2-byte long length field describes the length of the user date in the data packet. The maximum value allowed is 1,500. If the length field is used as a **type field**, the minimum value allowed is 1,501. Although a payload data field with the length 0 would be allowed, it creates significant problems. If the Ethernet adapter discovers a collision, it terminates the transmission of the current data packet immediately. The consequence of this action is that fragments of the dispatched packet might at any time be found on the bus. In order to facilitate the distinction between junk data and permissible data packets, IEEE 802.3 requires that every data packet be at least 64 bytes long (cf. also Fig. 4.24).

- **Checksums (Frame Check Sequence)**
  The FCS uses a CRC check word for the fields of the addresses, the type/length field, and the payload.
  The generator polynomial used is:

$$G(x) = x^{32} + x^{26} + x^{23} + x^{22} + x^{16} + x^{12} + x^{11} + x^{10} + x^8 + x^7 + x^5 + x^4 + x^2 + x^1 + 1$$

  It is possible for bit errors to appear. These are caused by signal damping and electromagnetic environmental interference along the transmission medium or at the network adapter. Bit errors can be recognized with the help of a checksum procedure. The transmitting computer additionally calculates this checksum out of all of the remaining bits of the transmitted data packet, without the preamble, and records this in the FCS field. The receiving computer calculates the checksum out of all of the received bits from the data packets and compares the result with the contents of the transmitted FCS field (CRC test). If both values do not match, there is transmission error.

Fig. 4.23 Ethernet data packet internals.

Collision detection and data packet minimum length

By fixing a minimum data packet length it is ensured that the transmission of a data packet has not already ended before it reaches the spaced ends of the transmission medium and there may possibly be a collision.

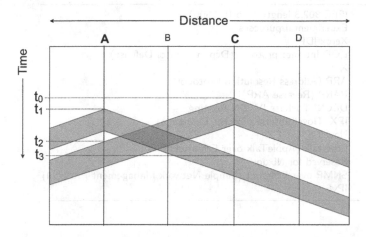

Let us assume that the time necessary for transmission from computer C to computer A equals τ_1. Computer C starts its transmission at the time point t_0. At time point $t_1 < t_0 + \tau_1$ computer A begins its transmission. If only very short data packets are sent and their transmission is already finished before they have reached the end of the transmission medium, it is possible that computer A will likewise send a data packet (at time period t_1). This means that the packet will already have been sent in its entirety at the point of a possible collision detection (t_3). The sender then falsely concludes that the transmission has been carried out successfully.

To prevent this from happening, every transmission must cover at least 2τ transmission time, whereby τ designates the signal run time from one end of the transmission medium to the other. Accordingly, a simple 10 Mbps Ethernet LAN with a maximum length of 2,500 meters, for example, requires the minimum allowed data packet length broadcasting time of at least 51,2 µ. This equals the signal runtime of a data packet of 64 bytes. Data packets that contain shorter payload information must consequently be filled up.

With a growing bandwidth, the minimum data packet length needs to be increased or the cable length reduced. For a 1 Gbps Ethernet LAN this means either the minimum packet length is fixed at 6,400 bytes or the cable length of the transmission medium is reduced to 250 meters, whereby the packet length is simultaneously increased to 640 bytes.

Fig. 4.24 Collision detection and frame minimum length.

Table 4.4 Ethernet type field values.

Hexadecimal-Value	Description
000-05DC	IEEE 802.3 length field (0-1500)
0101-01ff	Experimental purposes
0600	Xerox IDP
0800	DOD Internet protocol (Department of Defense)
0805	X.25
0806	ARP (Address Resolution Protocol)
0835	RARP (Reverse ARP)
6003	DECNET Phase IV, DN Routing
8037	IPX (Novell Internet Packet Exchange)
8038	Novell IPX
809B	Ethertalk (AppleTalk over Ethernet)
80A3	Reserved for Nixdorf
814C	SNMP over Ethernet (Simple Network Management Protocol)
86DD	IPv6

Ethernet hardware components

Before the different variations of Ethernet technologies and cabling are addressed, several terms will be addressed. In today's network construction a distinction is made between three variations of Ethernet topology (cf. Fig. 4.25).

- One variation is the true **bus topology** where broadcast messages are transported. The signal of a transmitting computer is received directly by all connected computers, depending on the distance it is time-delayed. Only one station can transmit at any one time over this true broadcast network.

- The application of a **hub** within an Ethernet LAN results in a star topology. Logically-speaking, however, it is still considered a bus topology. The hub is also designated a multi-port repeater as the signals it receives on the input port can again be passed on to all other ports. A with pure bus topology, one can speak here of broadcast networks.

- With the application of a **switch** (also switching hub oder LAN switch) a true star topology evolves from the original bus topology, which also no longer requires a common medium. The hub connects these computers directly with each other for the duration of communication. A distinction is made between a **store-and-forward switch**, which receives the complete data packet, analyses it and forwards it to its receiver and a **cut-through switch**. A cut-through switch reads the destination address at the beginning of the data packet and then immediately forwards the remaining part of the data packet to this address. While the store-and-forward switch is capable of recognizing data packets with errors and only transmitting error-free packets, a cut-through-switch is not able to recognize faulty data packets as it does not evaluate their checksum field. Many Ethernet products are in a position to offer both types of switching.

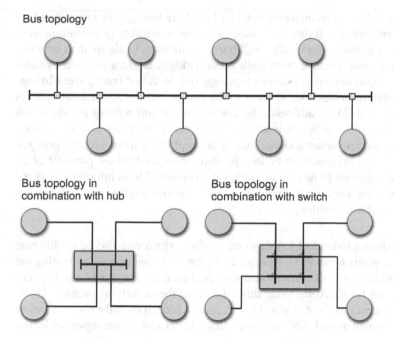

Fig. 4.25 Ethernet topology variations and their network construction.

An Ethernet LAN can be made up of different segments that are connected to each other by intermediate systems (repeater, bridges and also routers). In this context, a group of computers connected together to form a single transmission medium is called a **segment**. The simplest of Ethernet configurations – a single bus – is represented simultaneously as a complete segment as well as a **collision domain** (area through which collisions are forwarded). Based on this configuration, in Ethernet LANs the already mentioned intermediate systems are implemented with the following functions (cf. Fig. 4.26):

- **Repeaters**
  Repeaters interconnect individual segments of a Ethernet LAN into a larger network. This then serves as a single large collision domain. Repeaters enable the spatial expansion of Ethernet LANs and are logically completely transparent. The number of connected computers on an Ethernet LAN expanded this way can exceed the number of those allowed in a single segment. Because of the increased collision danger that results, this can lead to throughput reduction as well as a high network load.

- **Bridges**
  Bridges serve to separate segments logically and physically. It is their task to forward data packets whose destination addresses are in another network segment. A collision domain ends at a bridge and cannot extend beyond this point. Because of the physical separation, the network load sinks in the individual segments

allowing parallel communication to take place there locally. The principle of **locality of reference** is followed. It states that the probability of communication taking place between physically neighboring computers is higher than between those at a greater distance from each other. Bridges are in a position to independently adjust to a given network topology (the so-called **transparent bridge** or **self-learning bridge**). In addition to the source address A_Q of a data packet to be forwarded (**MAC address**), the corresponding port number p_Q via which the packet reaches the bridge is also saved. If the bridge receives a data packet with A_Q as its destination address later, it is simply forwarded over the port p_Q. For this reason, bridges must be in a position to understand the protocol of all connected segments to be able to extract the correct address information. If the connected computers use different protocols, a corresponding translation takes places (**translation bridge**).

- **Routers**
  Routers connect individual LANs to each other, which can also be of different types (in contrast to bridges). Routers carry out their function by evaluating the network addresses of the various sent data packets (e.g., the IP address, network layer/Internet layer). At the same time, details of the underlying protocol layers (network access layer and physical layer) remain hidden from the router. Routers are implemented in the LAN to connect different network technologies with each other. The gateway to a WAN is also achieved via a router.

In an Ethernet LAN, all of the introduced components are employed in different combinations.

Ethernet technologies

Ethernet network technologies exist in many different forms today. While all share the same fundamental principles, a variety of physical transmission media with differing performance indicators are used. Therefore, a variety of Ethernet wiring options are available. These begin with the original 10 Mbps 10base5 coaxial cable and continue up to the newest 10 Gbps 10GBase EX optical fiber variations. All have differing parameter limits in respect to their physical complexity, the number of computers that can be connected or the available bandwidth. Generally-speaking, the amount of hardware necessary for Ethernet wiring is minimal. Every connected computer must be fitted with transmitting and receiving hardware (Ethernet adapter) to control access to the transmission medium used as well as to monitor network traffic.

Ethernet variations with 10 Mbps

Depending on the various types of physical transmission media, there are also different topological structures. The originally used coaxial cables required a pure bus topology, contrasting with variations based on twisted wire pair cables, for which a bus topology is less attractive. These are mostly operated in the form of a star topology with a central hub, from which pairs of laid twisted pair cables lead to

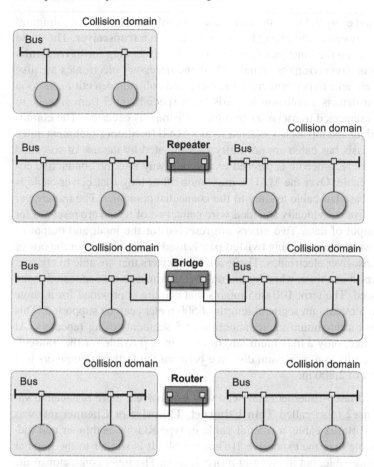

Fig. 4.26 Ethernet intermediate systems.

the connected computers. Fiber optic systems may likewise be operated in this to-pological arrangement. Double running cabling and a hub allows the simultaneous transmission and reception of data (**full-duplex operation**).

The oldest Ethernet variety is called **10Base5** (also **thick Ethernet, thick wire** or **yellow cable**). A pure bus topology is supported with 10Base5. The ends of the coaxial cable (type RG 8) used as the bus are normally insulated with a yellow shielding (the IEEE 802.3 standard does not require that this color be used but it is suggested), and thus the name "yellow cable." They are provided with 50 Ω termi-nating resistors. The inner conductor of this type of coaxial cable has a diameter of 2,17 mm and is covered with an outer conductor of between 6,16 mm and 8,28 mm. It is therefore a relatively stiff and heavy cable type with a minimal bending radius of 25 cm, making it difficult to lay. For this reason this cable has also been given the name "thick wire."

A marking is found every 2.5 m on the RG 8 coaxial cable. It specifies the minimum distance of taps between participating hosts connected via a **transceiver**. The name transceiver is based on the function of the tap, namely the sending forth (**transmission**) and reception (**receiving**) of signals. Here, the necessary electronics are also installed, which are able to perform carrier sensing and collision detection. As soon as the transceiver detects a collision it sends out a special signal (jam signal), to inform all other connected transceivers that the collision has occurred. The combination of transceiver and tap is also referred to as a **MAU** (medium attachment unit). The taps on the basis bus cable are normally implemented by the use of so-called (**vampire taps**). Here, a needle is pierced exactly halfway into the conductive core of the coaxial cable. Over the MAU a maximum 50 m long transceiver cable is connected to the base bus cable leading to the connected computer. The transceiver cable consists of five individually shielded wire pairs, two of which are reserved for the input and output of data. Two others are reserved for the input and output of control signals, while the remaining twisted pair is used in most cases for the power supply of the transceiver electronics. There are transceivers that are able to connect up to 8 computers at one time, whereby the number of transceivers connected to the bus can be reduced. The term 10Base5 means that cabling is provided for a range of 10 Mbps, and a maximum segment length of 500 meters can be supported. This technology allows a maximum interconnection of 5 segments using repeaters. At this expansion stage, only a maximum length of 30 m is possible for the transceiver cable. This results in a maximum distance between the farthest computers in a 10Base5 Ethernet of 2,800 m.

As the original 10Base5-Ethernet cable was very expensive a more reasonable variation, the **10Base2** (also called **Thin Ethernet**, **Thinwire** or **Cheapernet**) was developed. The 10Base2 cable, a coaxial cable of type RG-58, is thinner and therefore more flexible than the expensive 10Base5 cable. It is similar to the popular analog TV antenna cable, but lighter and more flexible. The inner conductor of the RG-58 coaxial cable has a diameter of approximately 0.89 mm, the outer conductor of 2.95 mm. Including insulation the RG-58 cable thus has a total diameter of about 5 mm and thanks to its bending radius of only 8 cm is considerably easier to lay than the 10Base5 cable. However, the reduced shielding of the RG-58 cable also means a higher signal damping, which leads to limitations of the maximum segment length at 185 m, or a maximum of 30 connected computers.

In contrast to 10Base5, the MAU of thin wire Ethernet is integrated into the Ethernet network adapter of the connected computer. Thin wire Ethernet does not take the conventional form of a coaxial cable. Individual sections are joined to each other and to the computer via BNC connectors and so-called t pieces. This connection technique is less complex, easier to install, cheaper and more reliable than the vampire taps in 10Base5. The minimum distance between two active components in a 10Base2 network is 0.5 m. Altogether 10Base2 allows the interconnection of a maximum of 5 segments, whereby active network nodes can only be connected on 3 segments. This results in a maximum distance of 925 between two computers in a 10Base2 Ethernet network. For transmission of a data packet the connected computer sends this packet over its t piece via the thin wire cable. The data packet is forwarded from the t piece in both direction to the different ends of the bus, where it is absorbed by the terminating resistors. The coupling of multiple Ethernet segments proceeds over **repeaters**, whereby with suitable repeaters 10Base2 as well as 10Base5 segments can be interconnected.

The repeater can be a separate network component over which both of the segments are centrally coupled with each other. Two segments that are spatially separated from each other can be coupled over a so-called **remote repeater**. With remote repeaters the two repeater halves are components of the respective segments. A 4-wire conductor is stretched between both of these halves on which no other computers can be connected. It can measure up to 1,000 meters. Repeaters must never form loops. Because of runtime conditions the number of repeaters in Ethernet is limited to a maximum of three.

Determining and localizing a cable break, a defective clip, or loose connectors in 10Base2 and 10Base5 can present a big problem. For this reason, suitable techniques were developed to localize these defects exactly. Basically, a known impulse is sent to the cable. If this impulse encounters a hindrance along the way, or if the cable end is suddenly reached, a signal echo is produced that returns to the starting point. Through an exact measurement of the transit time, the site of damage can be determined with a sufficient degree of accuracy.

The last mentioned problems led to the development of a further technological variation in which cabling emanates from a central hub in a star form. Through the use of a star topology, safeguarding measures against network failure resulting from a cable break on the Ethernet bus can be considerably reduced. A simple twisted pair telephone cable (**unshielded twisted pair**) suffices as a transmission medium. This is usually a component of the cabling in most buildings and has a diameter of between 0.4 mm and 0.6 mm. This form of cabling, which is relatively inexpensive, is known as **10BaseT** (or simply as twisted pair). The central hub is also called a star coupling and presents a multiport repeater that passes on the signal it receives as input to all the outputs. Hubs can be cascaded by switching, which is topologically seen as a star made up of stars. Viewed logically, however, the network does not change with the implementation of hubs. All of the connected computers are located as before in a broadcast domain and are linked to each other via the CSMA/CD procedure. The star topology therefore exists only in appearance. Physically and logically the bus structure remains in place. A hub can also allow the connection of various media. Between any two computers connected to the network on the Ethernet LAN, a maximum of five segments or four repeaters are allowed. Hubs can often have an extended functionality and then act as bridges or routers on a higher protocol layer. They are then known as **intelligent hubs**. The ability of hubs to offer network management functions has been a major factor in their widespread popularity. If one of the connected Ethernet adapters has a malfunction and is permanently sending data packets over the network (so-called **jabbering**), this would mean the end of all data communication for a 10Base2 Ethernet. None of the other connected computers would be able to communicate anymore. The hub recognizes this malfunction in 10BaseT Ethernet and excludes the faulty adapter from the network. This saves the network administrator from having to be wake up in the middle of the night to fix the problem manually, which can then be taken care of the next morning. Moreover, many hubs provide a monitoring function which allows statistical information to be determined such as bandwidth, number of collisions, or average data packet lengths. This is then passed on via a separate computer that is directly connected to the hub. Not only can this information be used for remedying errors and localizing problems, but it is also very helpful when it comes to planning the expansion of an Ethernet LAN landscape. The disadvantage of the 10BaseT variation is the limited cable length between the hub and the connected computer, which only reaches a maximum of 100 m (150 m using special, shielded high-quality cables). A high-performance hub can be a very expensive investment.

The fourth cabling variation for the 10 Mbps Ethernet standard, **10BaseF**, is based on fiber optical cables. This variation is considerably more expensive than the others presented so far due to the costs involved for connectors and terminators. However, its extremely low susceptibility to disturbance makes it the first choice for wide-range cabling between buildings or for hubs located at a distance from each other. A true bus topology cannot, however, be easily implemented using optical fiber cables. The necessary coupling between the bus fibers and the output of the computer is problematic. In order to carry out this implementation, so-called directional couplers are used – each with two inputs and two outputs. The amount of light reaching an input must be forwarded equally to both outputs. However, a coupling of both inputs is not desirable. Therefore when optical fibers are implemented the alternative star topology is usually preferred.

Here, the star coupler can be implemented purely **passively** as an optical star coupler. No transformation of optical into electrical signals occurs. At the same time, the possibility exists of implementing the star coupler **actively** as a coupler that works with electrical signals. Each of these are transformed into optical signals on the interface. Although the electro-optical signal transformation of signals is more time intensive and necessitates the complicated technical layout of a star coupler, it allows the connection of significantly more participating computers compared to the passive solution. In a passive star coupler, optical fibers (8 as a rule) are brought into contact in parallel form and fused together to a specific length. In this way, the arriving light energy is distributed equally over all of the connected threads. At the same time this even distribution causes high signal damping therefore this solution can only be taken into account when fewer stations are to be interconnected. All glass fiber variations use glass fiber pairs, in each of which one fiber is responsible for one transmission direction. The Manchester encoding used is simply converted into optical signals: light flow is represented as high voltage (HIGH) and the non-sending of light flow as low voltage (LOW).

100 Mbps Ethernet
With more and more users working in an Ethernet-LAN and an increase in network-based multimedia applications, the 10 Mbps bandwidth is no longer sufficient to ensure smooth operation.

The 100Base-T with its available bandwidth of 100 Mbps – **Fast Ethernet** – introduced in 1995, was undoubtedly not always exploited fully. At the same time

10 Mbps Ethernet Technologies in Comparison

The copper-based varieties: 10Base5, 10Base2 and 10BaseT as well as various optical fiber varieties (10Base-FB, 10Base-FL and 10Base-FP) are implemented based on the type of intermediate system used.

	10Base5	10Base2	10BaseT	10Base-FP
Description	Thick wire	Thin wire	Twisted pair	Optical fiber
Max. segment length	500 m	185 m	100 m	500 m
Nodes per segment	100	30	w/o spec.	33
Max. number repeaters	2	4	4	2
Min. node distance	2,5 m	0,5 m	w/o spec.	w/o. spec.
Connector	DB15	BNC	RJ-45	ST1
Cable diameter	10 mm	5 mm	0,4-0,6 mm	62,5/125 μm
Topology	Bus	Bus	Star	Star
Medium	Coaxial cable 50 Ω	Coaxial cable 50 Ω	UTP 100 Ω	Multimode optical fiber

For 10 Mbps Ethernet the following optical fiber variations are implemented in accordance with IEEE 802.3j:

- **10Base-FB (fiber backbone)**:
  An optical fiber point-to-point connection between two neighboring repeaters that can be a maximum distance of 2 km from each other. The transmission of a repeater proceeds synchronously, meaning an arriving optical signal is regenerated and sent out again with the local clock of the repeater.

- **10Base-FL (fiber link)**:
  An optical fiber point-to point- connection between individual computers, or repeaters, in which the maximum distance that can be bridged is also 2 km. In contrast to 10Base-FB, a repeater here transmits asynchronously.

- **10Base-FP (fiber passive)**:
  Optical fiber connection for a star typology with a passive star coupler, which can only bridge a maximum distance of 1 km.

Fig. 4.27 A comparison of 10 Mbps Ethernet types.

the increased bandwidth made it possible to overcome so-called bursts. These appear when multiple users simultaneously send multimedia contents over the network. Besides the high transmission velocity, fast Ethernet, as a logical development of 10Base-T, offered the advantage of being able to fall back on a standardized technology. This included the development of hubs, repeaters, Ethernet adapters and other Ethernet components. It also meant that the migration of the existing 10Base-T environment to the new **100Base-T** could be done as a relatively inexpensive

10 Mbps Ethernet – Technical Limitations

The standard Ethernet specification defines various limitations regarding maximum cable length. These restrictions arise from the maximal signal runtime and the clocking.

Segment lengths

Twisted pair telephone cables (UTP) and coaxial cables each have type-specific damping and must be provided with appropriate terminating resistors: coaxial cables with 50 Ω and UTP with 100 Ω. The length of an Ethernet segment may not exceed 500 m.

Network extension

Between two Ethernet segments a signal can be refreshed and forwarded with the help of a repeater. To better note the 10 Mbps Ethernet limitations, these are summed up in the "5–4–3-rule." The maximum number of segments in an Ethernet is 5, this means that between these segments up to 4 repeaters can be interconnected. Of the interconnected segments there can be computers connected to Ethernet in only 3 segments. 10Base5 allows segment lengths of up to 500 m, transceiver cable lengths of up to 30 m between baseband and repeater, or 50 m between baseband and end device. This results in a 10Base5 Ethernet maximum network expansion of 2,800 m.

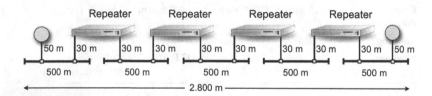

Point-to-point connection The maximum distance in which a point-to-point connection can be bridged is also 1500 m. Such a connection can be implemented, e.g., between computers that are located in different buildings.

Transceiver distances

Transceivers should never have less of a distance between each other than 2.5 m. In addition, there should never be more than 100 transceivers connected on a single segment. Transceivers that are too close together can cause interference, therefore raising the risk of collisions. Every connected transceiver leads to a reduction of impedance in the network and damps the transmitted signal. Too many transceivers can lower the electrical characteristics of the network to such an extent that it is no longer possible to guarantee a dependable functionality.

Fig. 4.28 Standard Ethernet limitations.

process. It was possible for the most part to retain the existing 10Base-T cabling as the new 100Base-T media specification (100Base-TX, 100Base-T2, 100Base-   and 100Base-FX) on all of the twisted pair cabling (UTP category 3, 4, and 5), shielded twisted pair cables (STP) or fiber optical cables can be implemented. 10 Mbps-Ethernet and 100 Mbps-Ethernet can be implemented in a mixed form be-

---

100Base-T2 and 100Base-T4 both use UTP Cat-3/4/5 cabling, however both never succeeded in gaining a niche in the market.

cause switches with ports for both speeds are available. This makes a step-by-step migration possible.

Cabling variations for fast Ethernet

All variations share the transmission rate of 100 Mbps. The segment lengths are each 100 m, network extension can be 200 m with 100Base-FX as well as 412 m (half-duplex operation) or 2,000 m (full-duplex operation). In applications of a single mode optical fiber (type 9/125 μm), 100Base-FX technology offers the possibility of bridging distances up to 15 km.

Type	Medium
100Base-T4	Cable with 4 twisted pairs (3 pairs for data transmission, 1 pair for collision reports), UTP-(3/4/5), 100 Ω impedance, Line code 8B6T, no full-duplex
100Base-T2	Cable with 2 twisted pairs (1 pair output, 1 pair input), UTP-(3/4/5), line code PAM5, full-duplex
100Base-TX	Cable with 2 twisted pairs, UTP-5, alternatively 2 twisted pairs STP (1 pair output, 1 pair input), 150 Ω impedance, line code MLT-3, full-duplex
100Base-FX	2 multimode optical fibers (62,5/125 μm), line code 4B5B, NRZI, full-duplex

Full-duplex Transmission

With the original bus topology for Ethernet only one computer was allowed to transmit at any one give time. From the perspective of the transmitting computer this was understood as a half-duplex transmission. A full-duplex can ideally increase the data rate to double the time. With the introduction of the star topology for 10Base-T, separated twisted pairs – one each for transmitting and receiving – were now available. This meant that it was no longer necessary to worry about a mutually shared medium. The resulting method became known as **switched Ethernet** and could function without collision detection and collision resolution. Every computer maintains its own collision domain within this star for itself. The IEEE 802.3 specification can be retained in view of the implemented data format and the CSMA/CD algorithm, even when no more collisions occur. An extra flow control is installed for the star coupler. This should prevent the buffer storage in the star coupler from overflowing. So-called pause packets are also sent by the hub for this purpose. These prompt the transmitting computer to stop sending data packets for a fixed period of time. The full-duplex transmission method for Ethernet was adopted as IEEE802.3x-Full-Duplex/Flow-Control Standard in 1996.

Further literature:

Spurgeon, C. E. Ethernet: the Definitive Guide. O'Reilly & Associates, Inc. (2000)

Fig. 4.29 100Mbps Ethernet variations and full-duplex operation.

All variations of fast Ethernet use the star topology. Those variations with "T" use access methods and data packet formats based on the IEEE 802.3 specification (cf. Fig. 4.29). Those variations with "X" use the physical layer originally specified for FDDI. In contrast to most of the earlier Ethernet variations, fast Ethernet technologies support a full-duplex operation, which allows a data rate per station of maximum 200 Mbps. The 100Base-TX variety is chiefly used for the wiring between individual levels (tertiary cabling). 100Base-FX is increasingly being used in secondary wiring . In contrast, there has been no significant increase in the use of 100Base-T4 and 100Base-T2 in practical application.

Although both use the same data packet format, Ethernet and fast Ethernet identify the start of their data packets in different ways. With 10 Mbps Ethernet no data signal is transmitted between the sending of a frame and the start of a new frame. The reception of a data signal serves here to identify a carrier (carrier detect), while the beginning of a frame is identified solely by the start-of-frame delimiter (SFD). 100 Mbps fast Ethernet permanently sends idle signals between the transmission of single frames. On the one hand, the possibility for link detection is created in this way (link integrity test) while, on the other hand, synchronization of the individual network nodes takes place via the idle signals. Additionally, because of this continuous transmission the complex transient procedure is limited to the start-up phase on the receiver's side, when the segment is activated. In order to make clear the beginning and end of a frame, fast Ethernet uses special group of symbols (start-of-stream delimiter and end-of-stream delimiter, SSD und ESD), at the beginning and before the end of every frame to encapsulate it (cf. Fig. 4.30).

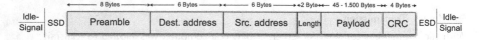

Fig. 4.30 Fast Ethernet data packet format, encapsulated with start-of-stream delimiter, end-of-stream-delimiter and idle signal.

To be able to reduce the transmission frequency necessary for sending 100 Mbps and make it possible to implement the cheapest and simplest cable media, a multilevel threshold signaling method with 3 signal values (-1, 0,+1) is used for the encoding of the data to be sent (MLT-3) by fast Ethernet. In this way, the necessary maximum transmission frequency is reduced to 31,25 MHz. In addition to the MLT-3 signaling procedure, a scrambling process is carried out. It distributes (smooths) the peak values in energy and frequency response over the entire frequency spectrum and thus should result in less electromagnetic radiation.

An alternative approach by Hewlett Packard company followed the strategy of developing a 100 Mbps Ethernet with a completely new, demand-based MAC control mechanism which is able to transport token ring data packets. This technology was released under the name **100VG-AnyLAN** and standardized by the IEEE in 1995 as IEEE 802.12. Because of the high costs of the 100VG-AnyLAN network components and their limited availability, in comparison to the IEEE 802.3 fast Ethernet network technology, the standard was never completely accepted. Already by 1998 the 100VG-AnyLAN technology had disappeared from the market. Fig. 4.31 gives a short overview of the functionality of the 100VG-AnyLAN technology.

100VG-AnyLAN Technology: An Overview

The 100VG-AnyLAN technology was developed by the Hewlett Packard and AT&T companies to succeed the 10 Mbps Ethernet standard. It was standardized in 1995 and designated IEEE 802.12. Unable to compete with the cheaper and more widespread 100Base-X technology, it already disappeared from the market in 1998.

The 100VG-AnyLAN technology follows severals principles:

- Unlike the classic Ethernet, a deterministic access method is used, whereby collisions on the shared medium can be avoided.

- To this end, each participating computer communicates with a central hub and reports its clear to send status and transmission priority.

- The hub determines the time point when a participating computer is clear to send. In doing this, the hub cyclically contacts all participating computers informing them of their clear to send status (round-robin process).

- In a cyclical run the hub gives preference to the transmission requests in order of priority.

- Every participating computer has its own timer that starts as soon as a new, initially low, priority transmission request is made. If the timer runs out before the computer gets a clear to send, its transmission request is then classified with a high priority.

Further literature:

Costa, J. F., Planning and Designing High Speed Networks Using 100VG-anyLAN, 2nd ed., Prentice Hall, (1995)

Fig. 4.31 An overview of the 100VG-AnyLAN technology.

Excursus 3: An Ethernet Performance Analysis

To be able to assess the performance of a traditional Ethernet system (half-duplex, with CSMA/CD access algorithm), we can look at an Ethernet LAN under load. In this situation, k computers are prepared to begin transmitting at any time. In their authoritative work, Metcalfe and Boggs assumed the likelihood of repeated transmission in each time slot as constant. This means that when every competing computer in a time slot with probability p begins transmission, the result is probability P(A) that another computer will occupy the transmission channel during this time (A = channel acquisition),

$$P(A) = kp(1-p)^{k-1}.$$

P(A) is the maximum, with $P(A) \rightarrow 1/e$ for $k \rightarrow \infty$, if p=1/k. The probability that during a competition situation exactly j time slots elapse is $P(A)(1-P(A))^{j-1}$. This means that in a competition situation the average number of resulting time slots is

$$\sum_{j=0}^{\infty} jP(A)(1-P(A))^{j-1} = \frac{1}{P(A)}.$$

Every time slot has a duration of exactly 2τ, the average competition interval w is $w=2\tau/P(A)$.

If we choose p as the optimum, the average number of competing time slots never exceeds e, i.e., $w \leq 2\tau e$ applies. If a data packet needs exactly t seconds to be transmitted on the average and multiple computers also have data packets waiting to be sent, then the results of channel efficiency CE is

$$CE = \frac{t}{t + 2\tau/P(A)} \ .$$

The longer now that the transmission medium is, the longer will be the competition interval. With a maximum of 2.5 km cable length and at the most 4 repeaters between two computers, the maximum length of transmission (roundtrip time) is 51.2 µs. This corresponds to the of 10 Mbps Ethernet minimum data packet length, which is 64 bytes or 512 bits.

We now express the channel efficiency over the data packet length F, the network bandwidth B, the cable length L and the velocity of the signal spread c for the optimal case of exactly e competition slots per data packet. With t=F/B the above equation becomes

$$CE = \frac{1}{1 + 2BLe/cF} \ .$$

If bandwidth B or the expansion of network L is continually increased with a constant data packet size, the network loses efficiency. Naturally, research aims to find ways to continually increase this size. While high bandwidth in wide-spread networks are desirable in WAN applications, the IEEE 802.3 specification is however not the right choice for this.

The following rough observations can help to ascertain how many computers k under high load conditions wish to transmit on the average. Every data packet blocks the transmission channel for a competition interval plus the transmission time, i.e., t+w seconds. The amount of transmitted packets per second is $r_1=1/(t+w)$. If the average transmitting rate of a computer is λ data packets per second, then the total rate of all k computers is $r_2=\lambda k$ data packets per second. Because $r_1=r_2$, the equation for k can be solved, whereby w is also a function of k.

To increase the efficiency of data transmission in Ethernet networks, numerous theoretical analyses have been carried out. Most all of them assumed that the resulting load corresponds to a Poisson distribution. Yet in reality this is seldom the case. This means that an average over a longer time period does not smooth out data traffic. The average number of data packets for every minute in an hour has the same variance as the average number of data packets per second in a minute. Because of this, theoretical Ethernet performance analyses often do not reflect the true facts.

Further literature:

Bertsekas, D., Gallagher, R.: Data Networks, 2nd ed., Prentice Hall, Englewood Cliffs, NJ, USA (1991)

Paxson, V., Floyd, S.: Wide Area Traffic: The Failure of Poisson Modeling, Proc. SIG-COMM'94 Conf., ACM, pp.257-268 (1994)

Willinger, W.,Taqqu, M. S., Sherman, R., Wilson, D. V.: Self Similarity through High Variability: Statistical Analysis of Ethernet LAN Traffic at the Source Level, SIGCOMM5'9 Conf., ACM, pp.100-113 (1995)

Spurgeon, C.: Ethernet − The Definitive Guide, O'Reilly, Sebastopol CA, USA (2000)

1 Gbps Ethernet

The amount of data being transmitted between computers is constantly increasing. With the introduction of switched Ethernet it was still possible for users to have the full bandwidth of the network available at their workplace. Now this has become increasingly difficult with the rising number of network users. The more users take advantage of broadband network applications, the more likely it is that congestion occurs. That's why even as fast Ethernet technology was being introduced, research was being done on how to increase transmission speed. This led to the development of **Gigabit Ethernet** technology. Gigabit Ethernet increases the bandwidth to 1,000 Mbps, or 1 Gbps. The data packet format and the CSMA/CD of the IEEE 802.3 specification were retained (cf. Fig. 4.32). Gigabit Ethernet is often used as a backbone technology to connect subnets operating in different bandwidths. They are structured hierarchically over central gigabit switches.

One of the goals in developing the Gigabit Ethernet standard was to ensure a high as possible compatibility with the existing Ethernet components. It was crucial to ensure a smooth migration to the new standard. Therefore, the existing data formats as well as the access mechanism used should ideally remain as unchanged as possible. If the CSMA/CD algorithm also remains in force at 1,000 Mbps, the maximum possible expansion of the network is reduced. This is because in spite of the high data rate the electromagnetic signals still spread with the same speed. To guarantee a reliable collision detection it must be ensured that the first bit of the smallest possible data packet length can reach the farthest computer in the network (more specifically, the collision domain). A data rate of 1,000 Mbps results in a bit time of 1 ns. Based on the minimum data packet length prescribed for Ethernet of 64 bytes (512 bit), the maximum expansion of a collision domain can be only 20 m, in compliance with the existing rule. To raise the collision domain to that of the prevailing fast Ethernet 200 m, the minimum packet length for Gigabit Ethernet is simply increased to 512 Bytes (4.096 Bit). Table 4.5 gives an overview of timing and frame size requirements for the different Ethernet standards.

Table 4.5 Timing and frame sizes for different Ethernet standards.

Parameter	Ethernet	FastEthernet	Gigabit- Ethernet
Data rate	10 Mbps	100 Mbps	1.000 Mbps
Bit time	100 ns	10 ns	1 ns
Minimum length	512 Bit	512 Bit	4.096 Bit
Maximum length	1518 Byte	1518 Byte	1518 Byte
Interframe gap	9,6 µs	0,96 µs	0,096 µs

If a smaller payload is sent than can fulfill the minimum packet length, it is filled with special extension characters to fit the required minimum length (padding). To still achieve an efficient network utilization even for small payload sizes, the Gigabit Ethernet standard provides for the grouping of multiple smaller data packets (frame bursting). The first smaller data packet is filled up with extension characters; further small packets are then attached to this one without the need for additional extension characters. The distance between the individual data packets in this frame group is given in the Interframe Gap (IFG). To prevent a computer from blocking the transmission medium for an undefined time period as a result of frame bursting, the total

Gigabit Ethernet Standards

The **IEEE 802.3z** Gigabit Ethernet standard specifies two variations for optical fiber transmission: 1000Base-SX (short wavelength laser) and 1000Base-LX (long wavelength laser). Long wavelength can be used with both monomode (single-mode) as well as multimode fiber optical conductors. Short wavelength is limited to multimode optical fibers.

Variation	Medium	Max. distance
1000Base-T	4 pairs UTP-5, line code 4D-PAM5	100 m
1000Base-CX	twinax copper cable, 150 Ω, line code 8B10B one pair each per direction	25 m
1000Base-SX	multimode optical fiber 770–860 nm	
	core diameter 62,5 μm	260 m
	core diameter 50 μm	550 m
1000Base-LX	multimode optical fiber 1.270–1.355 nm	
	core diameter 62.5 μm	550 m
	core diameter 50 μm	550 m
	core diameter 9 μm	5.000 m

Extension of the CSMA/CD Algorithm

There is a further sublayer provided in the layer model between the MAC layer (media access controller) and the line code (physical layer controller, PHY). This is the GMII (gigabit media-independent interface). It can be optionally inserted (except in 1000Base-T). Like the MII interface for fast Ethernet, the GMII interface serves beyond connecting different Ethernet media of 10–1,000 Mbps. It also handles automatic detection of the medium and data exchange on the state and characteristics of the current connection.

For half-duplex connections with hubs there exist two extensions of the CSMA/CD procedure. The so-called **carrier extension** adds as many symbols as necessary so that the short MAC data packet is at least 4,096 "bit times" (i.e., the time in which the data can spread out completely on the transmission medium) – therefore longer than the total runtime in the network.

The procedure of **frame bursting** makes it possible to send multiple short data packets one after another without the carrier extension. In this way, the load on the network added by carrier extension can be somewhat reduced. This extension is not needed for switched Ethernet. Here, there is no shared medium and consequently no collision can occur.

Further literature:

Spurgeon, C. E. Ethernet: the Definitive Guide. O'Reilly & Associates, Inc., (2000)

Fig. 4.32 1 Gbps Ethernet Variations.

amount of data to be sent in frame bursting is limited to 65,536 bits (8,192 bytes) (burst limit). However, this collision detection procedure is only implemented if Gigabit Ethernet is operated in half-duplex. The CSMA/CD algorithm is not used in the full-duplex procedure.

The Gigabit Ethernet standard has four further variations: 1000Base-SX, 1000Base-LX and 1000Base-CX. These were drawn up by the Gigabit Ethernet alliance and adopted by the IEEE in June 1998 as IEEE 802.3z. A year later followed the 1000Base-T as IEEE 802.3ab.

10 Gbps Ethernet

10 Gigabit Ethernet (10GE) raises the bandwidth of fast Ethernet again by a factor of 10 to 10 Gbps. Compared to the previous Ethernet technologies, with 10GE it is possible for the first time to bridge great distances. This allows a joint deployment in MAN or WAN. Additionally, when determining the 10GE standard, half-duplex transmission and the CSMA/CD access method were dispensed with for the first time. The Ethernet data packet format used up to now was retained and its exclusive use in full-duplex codified. 10GE includes a total of 10 different variations. Eight of these are optical fiber transmission based and summarized in the standard IEEE 802.3ae (June 2002). Two are copper cable technologies in IEEE 802.3ak (10GBASE-CX4, 2004) and IEEE 802.3an (10GBASE-T, 2006). The bandwidth10 Gbps (or 9,58464 Gbps) has been specified, the latter in terms of compatibility with STM-64 or SONET OC-19. These are used in WAN applications; therefore the attempt is made to achieve greater market acceptance. For communication between the MAC sublayer and the physical layer of the network a new interface with the designation XΛUI (10 gigabit attachment unit) was defined along with an extra sublayer XGXS[3] (XGMII extended sublayer). In comparison to the GMII (gigabit medium independent interface) of gigabit Ethernet, the XGMII has an expanded function capacity together with a higher flexibility for the connection of different types of media. The variations of optical fiber transmission are distinguished by single mode optical fiber transmission at 1.310 nm (10GBASE-LR and 10GBASE-LW4 for application in the LAN up to 10 km) and 1.550 nm (10GBASE-ER and 10GBASE-LR for application in LAN and WAN up to 40 km) used for greater distances. The multimode optical fiber transmission at 850 nm (10GBASE-SR) is used for distances between 300 m and 1.310 nm (10GBASE-LRM) with a range of max. 220 m. A special multimode variation, 10GBASE-LX4, is provided for 4 different wave lengths (1.275 nm, 1.300 nm, 1.325 nm and 1.350 nm) (wavelength-multiplexing, wavelength division multiplexing). It is suitable for applications in the range between 200 and 300 m.

One variation of the copper cable-based 10GE is 10GBASE-CX4. It uses double twinaxial copper cable and has a maximum length of 15 m. But compared to the popular backward-compatible standard 10GBASE-T it is becoming less important. Like its predecessor, 1000BASE-T, 10GBASE-T uses four pairs of twisted pairs (TIA-568A/B, ISO/IEC 11081) for data transmission. The range attained with 10GBASE-

---

[3] In both designations " X" stands for the roman numeral 10 and thus specifies the achieved data transmission rate of 10 Gbps.

T is dependent on the type of cable implemented. In order to reach a range of 100 m, CAT6a/7 cable must be used. In comparison, the range achieved in 1000BASE-T with Cat 5e cable is only 50 m. Data is distributed independent of each other along the four twisted pairs. In each case there is 5 Gbps for transmitting and receiving in full- duplex mode. The data is then re-assembled at the other end.

Through the high data transmission rate of 10 Gbps new disturbance factors are evident in the 10GE standard due to the closely aligned different signal levels. These have to be damped using special shielding measures. Disturbances on the cable caused by crosstalk from neighboring twisted pairs are lessened passively by a crossbar in the cable. This provides distance between the twisted pairs. In addition to the active components of the physical layer of the 10GE copper variation, digital signal processors are used to calculate disturbances. These interferences could be cross talk, signal reflection or feedback. It is not possible to compensate disturbances caused by adjacent cables – so-called alien cross talk – with digital signal filters. For this reason, the use of high-quality, shielded cables is indispensable in 10GBASE-T. The defined minimum intervals between the laid cables and connectors must also be strictly adhered to.

40 Gbps/100 Gbps Ethernet

With the establishment of the 10GE Ethernet standard, the development of this technology field is far from over. The IEEE 802 High Speed Study Group had been working on a 100 Gbps Ethernet standard as IEEE 802.3ba since 2007. It was adopted in June 2010 offering for the first time an Ethernet standard with two different data transmission speeds: 40 Gbps and 100 Gbps. In the 40 Gbps area there is also a standard for transmitting so-called "backplanes," , which are implemented for distances spanning meters (40GBase-KR4). Here, 4 cable pairs each are deployed together. Each piece of data is transmitted on them separately at 10 Gbps. Twinax copper cables may be used for distances up to 10 m. These also transmit data in 4 pairs each (40GBase-CR4) and reach a reach a data transmission rate of 40 Gbps, or with 10 pairs (100GBase-CR10) 100 Gbps. The use of fiber optic cables is exclusively planned from 100 m. On multimode optical fibers (OM3 and OM4) 4 or 10 wire pairs with 40 Gbps /100 Gbps may also be implemented. Single mode optical fiber cables bridge larger distances. 40GBase-LR4 specifies 4 wave lengths in a SMF optical fiber cable with 10 Gbps each and a distance of up to 10 km. 100GBase-LR4 specifies 4 wave lengths each with 25 Gbps and a distance of up to 10 km. While 00GBase-ER4 expands the distance to be bridged up to 40 km. Just as with the 10GE standard, implementation of 100 Gbps Ethernet is possible in both LAN and WAN applications. Whether or not a cheaper version also becomes standardized based on low-cost twisted pair copper cables for the 100 Gbps Ethernet (as was the case with the slower IEEE 802.3 standards) is not clear at this time. In comparison to the optical variety, it displays a considerably less favorable energy balance. A technical implementation is therefore not significant economically.

In Fig. 4.33 further Ethernet variations will be examined.

Table 4.6 Ethernet variations and additions.

Designation	IEEE	Year	Data rate	Medium
10Base5	802.3	1983	10 Mbps	Coaxial, RG-8 A/U
10Base2	802.3a	1988	10 Mbps	Coaxial, RG-58
1Base5	802.3e	1988	1 Mbps	StarLAN: TP, cat 3
10Base-T	802.3i	1990	10 Mbps	2 UTP, cat 3/4/5
10BROAD36	802.3b	1988	10 Mbps	Coaxial, 75 Ω
FOIRL	802.3d	1987	10 Mbps	2 multi-mode (62,5/125 µm)
10Base-FB	802.3j	1992	10 Mbps	Fiber backbone
10Base-FL	802.3j	1992	10 Mbps	Fiber link
10Base-FP	802.3j	1992	10 Mbps	Fiber passive
100Base-TX	802.3u	1995	100 Mbps	2 pair UTP-5 / STP
100Base-T4			100 Mbps	4 pair UTP-3/4/5
100Base-FX			100 Mbps	2 optical fibers
FDX	802.3x	1997	100 Mbps	Full-duplex Ethernet with flow control
100Base-T2	802.3y	1997	100 Mbps	2 pair UTP-3
1000Base-CX	802.3z	1998	1 Gbps	Twinax, 150 Ω
1000Base-LX	802.3z	1998	1 Gbps	Multi-/monomode fiber 1.300 nm
1000Base-SX	802.3z	1998	1 Gbps	Multi-mode fiber 850 nm
1000Base-T	802.3ab	1999	1 Gbps	4 pair UTP-5
Link Aggregation	802.3ad	1999		parallel links between switches increased bandwidth
10GBase-SR	802.3ae	2002	10 Gbps	Optical fiber 850 nm without WAN
10GBase-SW	802.3ae	2002	10 Gbps	Optical fiber 850 nm with WAN
10GBase-LR	802.3ae	2002	10 Gbps	Optical fiber 1.310 nm without WAN
10GBase-LW	802.3ae	2002	10 Gbps	Optical fiber 1.310 nm with WAN
10GBase-ER	802.3ae	2002	10 Gbps	Optical fiber 1.550 nm without WAN
10GBase-EW	802.3ae	2002	10 Gbps	Optical fiber 1.550 nm with WAN
10GBase-LX4	802.3ae	2002	10 Gbps	Optical fiber 1.310 nm WDM for LAN
10GBase-CX4	802.3ak	2004	10 Gbps	IB4X cable
10GBase-T	802.3an	2006	10 Gbps	4 pair UTP-6a/7, STP-5e/6a/7
40GBase-KR4	802.3ba	2010	40 Gbps	4 x back plane conductor
40GBase-CR4	802.3ba	2010	40 Gbps	4 x twinax copper cable
40GBase-SR4	802.3ba	2010	40 Gbps	4 x OM3/OM4 multi-mode glass fiber
40GBase-LR4	802.3ba	2010	40 Gbps	4 x single mode glass fiber
100GBase-CR10	802.3ba	2010	100 Gbps	10 x copper cable
100GBase-SR10	802.3ba	2010	100 Gbps	10 x OM3/OM4 multi-mode glass fiber
100GBase-LR10	802.3ba	2010	100 Gbps	10 x single mode glass fiber
100GBase-LR4	802.3ba	2010	100 Gbps	4 x single mode glass fiber
100GBase-ER4	802.3ba	2010	100 Gbps	4 x single mode glass fiber

Broadband Ethernet

In contrast to the standard 10 Mbps Ethernet variations, the already obsolete broadband Ethernet does not transmit a digital baseband signals. It modulates them instead over carriers of a specific frequency. An analog data transmission takes place. Standard coaxial cables with an impedance of 75 Ω are employed. Using a frequency multiplex procedure it is possible to operate multiple Ethernet systems over a single coaxial cable. With the use of repeaters great distances could be bridged (up to 3,600 m) – even more so than with digital Ethernet. Yet it must be noted that with frequency modulation only a unidirectional transmission can take place. This means that a separate return channel is necessary.

The broadband Ethernet in the standard IEEE 802.3b is specified as 10BROAD36. Because of its high infrastructure costs, 10BROAD36 technology was not able to make headway against the cheaper digital Ethernet technologies.

Power over Ethernet (PoE)

The procedure designated Power over Ethernet (PoE) provides network-capable devices a supply of electrical power over the cabling of an Ethernet network. This procedure standardized in 2003 as IEEE 802.3af is employed primarily if it is not possible to put an extra cable supplying power to network-compatible devices to use. Particularly network-compatible devices with reduced power input are supplied with electricity in this way, e.g., IP telephones, cameras, or hubs. During use, actual data transmission and the function of network devices with their own external energy supply should remain unaffected and the maximum feasible segment length not be reduced.

The PoE standard differentiates between devices functioning as sources of electricity (power sourcing equipment, PSE), and those supplied with electricity over Ethernet cable (powered devices, PD). Switches or hubs are typical examples of PSE (Endpoint PSE), or so-called midspans, which expand the available intermediate network systems with a PoE functionality.

Further literature:

802.3b-1985 – Supplement to 802.3: Broadband Medium Attachment Unit and Broadband Medium Specifications, Type 10BROAD36 (Section 11), IEEE Standards Association, 1985)

Roebuck, K.: 802.3at Power Over Ethernet: High-impact Technology – What You Need to Know, Emereo Pty Limited (2011)

Fig. 4.33 Further Ethernet variations.

4.3.3 Token Ring – IEEE 802.5

For a long time the main competitor of Ethernet in area of LAN applications was **token ring**, which was developed for the market by IBM. The use of ring topology in fact goes back a long way in the history of computer networks. The original concept, called the Newhall-Ring, was first introduced in 1969. It sparked interest in the fields of WAN and LAN areas since it offered the great advantage of consisting solely of a chain of point-to-point connections of single computers. The administrative and access algorithms could therefore be created easily as the ring formed did not – in a pure sense – use a shared transmission medium. The development of marketable products with token ring technology first succeeded in 1985. This happened after

Ethernet Sources of Error

Besides cabling problems, errors can appear in a LAN that come about as a direct result of the LAN technology used. The errors and disturbance caused by Ethernet may be separated into the following categories:

- **Local Collisions**
  Collisions occur when multiple computers transmit at the same time. If the collision rate is unusually large the cause of the problem can often be found in a cabling error.

- **Late Collisions**
  These are collisions that occur outside of the 512 bit window (time slot). They can result from a defective computer that no longer keeps the CSMA/CD conventions. It could also be that installation conventions regarding the maximum cable length are exceeded. The latter circumstance results in the maximum signal runtime being exceeded with collisions recognized too late.

- **Short Frame**
  Faulty Ethernet network adapters may transmit data packets under the 64 byte minimal length.

- **Jabber**
  Designation for extremely long packets that exceed the maximum length of 1,526 bytes. This condition also indicates a defective Ethernet network adapter.

- **Negative Frame Check Sequence**
  The checksum (FCS) of the original packet does not correspond with that of the transmitted packet. The cause of this problem is also often found in the cabling.

- **Ghosts**
  An incorrectly working Ethernet network adapter may transmit packet fragments over the transmission medium.

Ethernet Sources of Error and Causes

Cause	Collisions	Short Frames	Jabber	FCS	Ghosts
CSMA/CD	×				
software driver		×	×	×	
defective adapter	×	×	×	×	
defective transceiver	×		×	×	×
too many repeaters	×				
cable to long	×		×		
cable defective	×		×	×	×
termination	×		×	×	×

Fig. 4.34 Ethernet sources of error

Year	Event
1973	Xerox develops prototype of a communication controller with a transmission rate of 3 Mbps
1976	Work of Metcalfe and Boggs published, Ethernet is first introduced publicly
1979	Metcalf (Xerox) and Bell (DEC) develop Ethernet with a LAN standard
1980	DIX group, a fusion of the DEC, Intel and Xerox companies publish the first Ethernet specification: Ethernet V1
1982	IEEE adopts Ethernet in the newly founded working group 802.3 and establishes a new specification for 10Base5: Ethernet V2
1983	Work on 10Base2 begins
1985	Ethernet becomes the ISO standard ISO/DIS 8802/3 RFC 948 enables support of the TCP/IP communication protocol on IEEE 802.3 networks
1986	Ethernet standard 10Base2 adopted
1990	IEEE 802.3 specification adopted as ISO standard
1991	Ethernet standard 10Base-T published
1992	Ethernet together with 10Base-F enables operation over Fiber optic links
1993	Two variations of a 100 Mbps Ethernet are standardized: fast Ethernet and 100VG-AnyLAN
1996	IEEE founds Gigabit Ethernet task force
1998	Gigabit Ethernet standard ratified
1999	Founding of the 10 gigabit Ethernet Alliance
2002	Adoption of the 10 gigabit Ethernet standard
2007	IEEE 802.3 Higher Speed Study Group begins work on 40 Gbps/100 Gbps Ethernet
2010	Adoption of the 100 Gbps Ethernet standard

Table 4.7 The history of Ethernet.

IBM had adopted the concept and introduced a token ring network environment to the market based on the IEEE 802.5 standard.

Like Ethernet, token ring is also standardized by ISO as ISO 8802.5. Token ring is able to to provide a reliable response time behavior, also in cases of high network load. The ring provides fair network access to all of the stations. For possible network access, there is, however, an upper limit that defines the number of participating computers. Token Ring is therefore not recommended for application in very large networks or in networks with nodes that are physically located at a great distance from each other. The token ring technology is implemented on a wholly digital basis, while in Ethernet analog components are needed, e.g., for collision detection. On the other hand, maintenance of token ring is considerably higher than with Ethernet, especially when there are network errors, or an existing network is expanded. Following its introduction by IBM, token ring enjoyed great popularity for a long time until Ethernet passed it up and it was finally squeezed out of the market. The IEEE 802.5 standard specified a bandwidth of 4 Mbps (or 16 Mbps) for token ring. This was supplemented by High Speed Token Ring with 100 Mbps and Gigabit Token Ring with 1 Gbps.

Token Ring Access Method

The CSMA/CD procedure in Ethernet can be compared to a polite conversation at a cocktail party, and similarly the token ring method can be better understood using an analogy. To look at how token ring works let us picture a group of American Indians sitting around a campfire smoking a peace pipe. Only the one who holds the peace pipe has the right to speak. When his speaking times runs out or when he has finished speaking he must pass the pipe on to the next person. During the time when no one speaks the peace pipe is simply being passed around the circle without anyone asking for the privilege of speaking. In this way everyone has the chance to take part in the discussion as soon as he receives the pipe.

From a logical perspective, all of the computers connected to the token ring are sequentially linked to each other in ring form. This does not always have to be the case physically, e.g., if ring concentrators are used. In a logical ring the data is always passed on from one computer to the next and only in one direction (counterclockwise). Each participating computer receives data from its predecessor in the ring (downstream) and passes it on to its successor in the ring (upstream). Like the bus cable in Ethernet, the ring is the shared transmission medium for all connected computers. When a data packet is sent from one computer to another it passes by all of the computers connected in the ring. However, it must also be noted that only the receiver specified in the data packet header actually evaluates the data packet and saves a copy of it locally.

The word token ring originated from the type of access administration in the ring. A **token** (comparable to the peace pipe described in the previous image) – i.e., a specifically distinguished bit sequence – is passed on in a ring and regulates the transmission authorization in the case of competing access. This technique is also called **token passing** and is based on the following rules that all computers in the ring are obliged to follow:

- Before a ready-to-send computer begins its transmission it needs an authorization.

- All computers connected to the ring work as repeaters, i.e., they always pass on the received data packets to their successor.

- When a transmitted data packet returns from its way around the ring to the source computer again, this computer will take the data packet out of the ring.

A ready-to-send computer receives authorization to transmit in the form of a token. Most of the procedures work with a single token. This token continually circles the ring. In the case of a new initialization, or token loss a **monitor station** is responsible for putting a new, free token back in the ring. The token passing procedure defines the order of the computers connected to the ring. This corresponds to the

sequence in which the token passes the connected stations. The sequence of the stations in the ring is called the **physical order**. It is an order independent of whether
the individual stations are active or not. In contrast, the **active order** only considers
those stations actively involved in the communication process. The stations in the
ring are checked every seven seconds by a so-called **ring poll process**. This process verifies whether or not they are ready to transmit data. If not prepared for data
transmission they are re-classified as passive. The whole procedure is controlled by
the token rotation timer, which gives the time alloted for maximum rotation of a
token. It also determines the maximum time that a connected computer is given for
attaching the payload to be transmitted on a token. In a LAN environment these
standard values change continually because terminals are constantly switching on
or off and being incorporated into the ring or dropping out of it. It is not possible
to predict exactly when a connected computer will receive transmission authorization. The actual process of data transmission in a token ring proceeds as follows: (cf.
Fig. 4.35):

1. The token designated as free travels around the ring. The ready-to-send computer waits until the token passes by it. This computer checks whether or not the
   captured token is marked as free or occupied (cf. a).
2. If the token is recognized as free then the data packet to be transmitted is simply
   attached to it and the token is sent into the ring again. A free token always circu-
   lates alone in the ring, i.e., if after receipt of the token no further data is received
   the token is considered free. In the other case, a packet is attached to the token
   with its address and user information.
3. The combination of token and information passes by every connected computer.
   Each computer checks if the data packet is intended for it based on the destination
   address (cf. b). To do so the computer takes the message itself from the ring,
   and then following verification places it back again. Each connected computer
   therefore acts as a repeater.
4. If one of the connected computers identifies itself as the receiver of the transmitted information, it copies it into its memory, marks the information as copied and
   places the token back on the ring again (cf. c).
5. The original transmitter now has the task of removing the packet, which it initiated and has likely marked as a copy, out of the ring again and replacing it
   with a new free token (cf. d). In this simple way the transmitting computer is in
   a position to determine if the data it has sent reached the receiver or not.

The access problem of the shared transmission channel is therefore solved.
The original 4 Mbps and 16 Mbps token ring variations distinguish themselves in
how they pass on the token. In the 4 Mbps variation the release of the token only
occurs after the transmitter of last packet has received this last packet in its entirety
again. A sender is allowed to transmit for the maximum length of the so-called **token
holding time** (THT), which typically corresponds to a time period of approximately
10 ms. If there is still enough time to send further data packets after completion of
the transmission process of the first data packet, these can also be sent within the

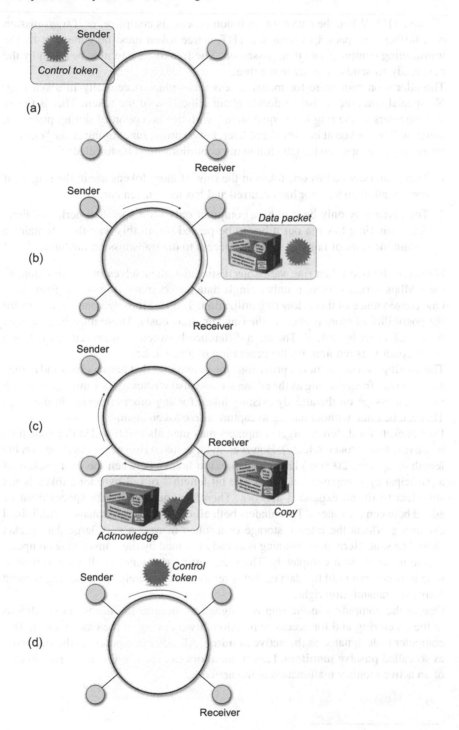

Fig. 4.35 Principles of data transmission in the token ring protocol.

allowed THT. When the entire transmission process is complete, or if transmission of a further data packet exceeds the THT, a free token must be generated by the transmitting computer and then passed on. The free token can now be taken by the next ready-to-send computer in the ring.

The allocation procedure for media access takes place decentrally in token ring. No special computer exists to decide about allocation of the token. That is to say all computers in the ring have equal status, with the exception of setting priorities, which will be looked at in more detail later. For a correct functioning of the "classic" token ring (4 Mbps version) the following conditions must be fulfilled:

1. There can only be just one token in the ring. If more tokens are in the ring, or if none at all, then an error has occurred and has to be taken care of.

2. The token may only be kept by a computer only for a specified period of time. After this time has run out it has to be passed on. In this way there remains a certain measure of fairness regarding access to the transmission medium.

However, the two token ring variations distinguish themselves in this function. In the 4 Mbps variation there is only a single data packet in the ring at any given time. One consequence of this is low ring utilization. The 16 Mbps, in contrast, allows for the possibility of more packets in the ring simultaneously. These must be separated from each other by spaces. The main difference between both procedures is found in the conditions required for the generation of a new token.

The 16 Mbps variation makes provisions for a computer that has received and copied a data packet from the ring as the addressee and also wishes to transmit. It can attach its own message on the already existing token for any other computer in the ring . This can be done without having to capture a free token again

Data packets for different target computers may then also be found at the same time in the ring (early token release). However, in large token ring networks (from a cable length of approx. 2000 m) larger transmission breaks between the data packets of a participating computer can occur. The bit length [4] of a 3 byte long token is not sufficient to fill an expanded network. Therefore multiple empty spaces must be added between the tokens. This hinders both efficiency and performance. Additional expansion effects the interim storage of a token in case a very large data packet should be sent. Here, the beginning is already retained by the transmitting computer before it can be sent completely. The transmitting computer is allowed to wait as long as necessary until the data packet is removed completely from the ring without losing its transmission right.

One of the computers in the ring is assigned to monitor the necessary conditions in the token ring and for access or functionality recovery in the case of error. This computer is designated as the **active monitor**. All other computers in the ring work as so-called **passive monitors**. Passive monitors can take on the active role in case of an active monitor malfunction in the network.

---

[4] Physical space that occupies a unit of information on the transmission medium.

Token Ring Administration and Maintenance

Monitoring an error-free function of the network is not done by a central instance with token ring, but rather by different computers belonging to the ring. In doing so the following situations must be monitored and in the case of error corrected appropriately.

- Ensuring an error-free functioning of the token passing mechanism.
- The management of computers that are additionally added to the ring or have to leave it.
- The detection and correction of errors in software or on the hardware side.

In addition to the previously mentioned active monitors, there are also the functions of the ring error monitor, the ring parameter monitor and the network manager. A token ring always has an active monitor, which is identified by a specific functional address (rather than the hardware address). All other computers are identified as passive monitors via their hardware address. The equipment (hardware and software) necessary for operation of an active monitor is a component of every regular token ring network adapter. It serves as interface between the computer and the network. Additional software is usually necessary for operation of a ring error monitor, a ring parameter monitor, or a network manager in the token ring.

The tasks of an active monitor includes the following four points:

1. Preserving ring functionality. This involves:

   - Detecting data packets and tokens with errors.
   - Ensuring that time restrictions are kept in the ring (timer settings),
   - Adding a new token in the case of loss or damage,
   - Introducing and monitoring the ring poll process,
   - Preventing the multiple passage of a data packet through the ring and
   - Ensuring the minimum storage capacity of the ring.

2. Identification and isolation of errors in the transmission medium, or in the network interface of the connected computer.
3. Detecting hardware and software errors of the network adapter and the connected computer.
4. Collecting status information from the individual computers connected to the token ring.

Token ring network construction

In a token ring network all of the connected computers are active network components. They are all mutually responsible for the correct functioning of the ring. The connected computers receive and regenerate data packets, evaluate them and pass them on to the succeeding computer in the ring. If one of the computers breaks down, for the time being the ring is interrupted.

To avoid a complete breakdown of the network, the participating computers are not directly connected to the ring, but to so-called **ring concentrators** (trunk coupling units – TCU or multistation access units – MSAU), which are connected to multiple computers simultaneously, added in the ring. This means that the individual computers are connected in a star form to the ring concentrator (star shaped ring). Seen logically, the topology remains a ring topology. The connection of the computer to the ring concentrator proceeds via a lobe cable. The connections of the individual ring concentrators to each other is called a trunk connection. On a single ring concentrator typically four to eight computers equipped with token ring network adapters are connected. The ring concentrators can be cascaded as desired and connected to each other via the so-called Ring In (RI) and Ring Out (RO). This ensures a flexible network construction (cf. Fig. 4.36).

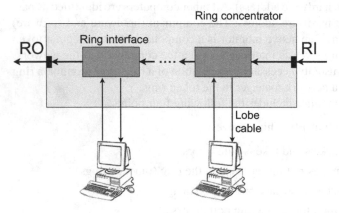

Fig. 4.36 Token ring: ring concentrator.

Ring concentrators can be divided into passive and active types. A passive ring concentrator does not need its own power supply and is therefore also not capable of regenerating or increasing incoming signals. Therefore the length of the lobe cable is limited with the use of passive ring concentrators. If a new computer is added to the token ring, or if an already integrated computer is taken out, the necessary administrative function are performed by the connected computers. For this purpose, in coupling on the token ring network, a computer creates so-called phantom voltage. This serves to activate an electromechanical relay to the junction box of the ring concentrator. Through the phantom voltage the relay is opened and the computer to be connected can be added to the ring. If errors appear in the connected computer or in the cable link, the junction box in the ring connector of this computer is immediately shorted. The defective computer is then removed from the token ring. To raise the stability of the token ring further a **secondary ring** is often added to the primary ring. If a connected computer goes down or in the event of a cable break the ring concentrator refaces the location of error using the secondary cabling.

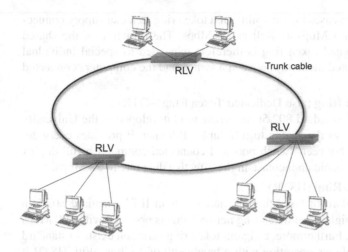

Trunk cable

RLV

RLV

RLV

Fig. 4.37 An interconnected token ring cascade via ring concentrators.

Token Ring Implementation and Development

In a token ring network a maximum of 260 computers can be connected, e.g., via 33 ring concentrators with 8 connections (4 of which remain unused). The maximum expansion of a token ring depends of the type and number of implemented ring concentrators. The type of cable used and the chosen token ring variation (4 Mbps oder 16 Mbps) also play a role. For example, the implementation of 260 computers in token ring lobe cabling lengths of maximum 100 m are required, whereas with fewer than 9 computers the token ring lobe cable lengths of up to 390 m are allowed. The transmission medium can be chosen from various cable types: coaxial cables, unshielded twisted pairs (UTP), shielded twisted pairs (STP) or fiber optics. In the IBM token ring variation these cables have the names IBM type 1 to IBM type 9.

If a passive ring concentrator is implemented, meaning a ring concentrator without its own power supply, the cable length must be shortened. One reason for this is the possible need for a ring reconfiguration, e.g., after a cable break. Following a cable break an originally short section can be replaced with a longer bypass. This can take up nearly the entire length of the ring. Using passive ring concentrators the signal would simply be weakened to such an extent as to be able to handle rerouting without intermediate amplification.

The IBM company is largely to thank for the development of the token ring network technology. IBM was also the driving force behind the standardization of this concept. An evolution of the original token ring concept was carried out further by the IEEE/ISO 802.5 working group. Three different concepts were thereby presented for the further development of the original token ring:

- **Switched Token Ring** (analog to switched Ethernet)
  The central element of switched token ring is a switch facilitating a star shape. It interconnects all linked computers via a dedicated connection with 16 Mbps

each. The switch increases the flexibility of a token ring as it can supply connected computers with 4 Mbps as well as 16 Mbps. The principle of the shared medium of the original token ring is therefore expanded to special individual connections (dedicated medium) between a switch and the computers connected to it.

- **Full Duplex Token Ring** (also Dedicated Token Ring – DTR)
  Based on the IEEE Standard 802.5r this procedure, developed at the University of New Hampshire, works analog to full duplex Ethernet. It provides every dedicated connection between switch-port and connected computer a full duplex transmission. This enable an increase in the effective data rate to 32 Mbps.

- **High-Speed Token Ring** (HSTR)
  Analog to the standard of fast Ethernet there exists in IEEE standard 802.5t a specification for a high speed token ring network that is operated with 100 Mbps instead of 16 Mbps. Furthermore, a gigabit token ring variation exists in standard IEEE 802.5v that allows operation with a bandwidth of 1 Gbps. With HSTR it became possible for the user to retain the advantages of token ring, for example the longer packet length or priorities setting. Just as Ethernet technology, HSTR offers good scalability with respect to bandwidth.

Table 4.8 IEEE 802.5: Milestones in the development of token ring.

Standard	Content
802.5	Token ring access method and physical layer specifications
802.5b	Token ring over telephone twisted pair
802.5d	Interconnected token ring LANs
802.5f	16 Mbps token ring
802.5j	Fiber optic station attachment
802.5m	Interconnection of source routed and transparent bridged networks
802.5n	Unshielded twisted pair at 4/16 Mbps
802.5r	Full duplex (dedicated token ring)
802.5t	High speed token ring with 100 Mbps (copper cable)
802.5u	High speed token ring with 100 Mbps (optical fiber cable)
802.5v	Gigabit token ring with 1 Gbps

A Comparison of Token Ring IEEE 802.5 and Ethernet IEEE 802.3

As to the question of whether token ring or Ethernet should be implemented in a company, today's choice is Ethernet technology based on its leading development and widespread use. Nevertheless, a comparison of both technologies in their respective "classical" variation is useful. This helps us to gain a better understanding of the strengths and weaknesses of the two variations and therefore a deeper understanding of both.

- **Network Adapter Structure**
  Token ring can be structured over point-to-point connections. The network adapters for individual, connected computers are made very simply and work completely digitally. Ethernet, in contrast, uses more complex components (transceivers) to ensure network access. An Ethernet transceiver contains substantially similar components that must also be capable of detecting weak signals of other computers connected to Ethernet. Furthermore, transceivers must be able to determine collisions on the transmission medium, also during the transmission process itself.

- **Cabling**
  Theoretically, token ring networks can be based on any transmission medium – from a carrier pigeon (cf. RFC 1149) to optical fibers. The normally implemented twisted pair cable (twisted copper cable) is inexpensive and easy to install. Bus cabling in Ethernet (10Base-2, 10Base-5) is expensive and subject to many restrictions regarding the fabrication and installation. However, since the introduction of twisted pair cabling for Ethernet (10Base-T) this problem no longer exists. Furthermore, token ring has the ability to localize errors in the transmission medium and, to limited extent, to circumvent them. The cable length in Ethernet is limited to 2.5 km (10 Mbps), which affects the minimum length of the data packets.

- **Overhead**
  In order to prevent incomplete data packets – caused by a collision – from reaching the transmission medium in Ethernet, the minimum data packet length is fixed at 64 bytes. If the user information to be transmitted consists of individual letters, as in the case of terminal input, a significant overhead occurs. With token ring, on the other hand, short data packets are possible. Likewise in token ring, long data packets are also possible, the length of which is restricted only by the maximum THT (token holding time).

- **Load response**
  There is virtually no waiting time with low Ethernet utilization. Access to the transmission medium is immediate. With token ring, in contrast, even with very low utilization it is necessary to wait the duration of the entire rotation of the token before a computer can gain access authorization. This behavior changes with a very high load. With Ethernet the number of potential collisions rises as the load increases, compromising efficiency. Token ring shows a very good load response, which does not decrease even as the load rises.

- **Administration**
  Ethernet has a very simple protocol without a central control instance. It is very easy to implement – as reflected in the near-monopoly position it holds in today's market. With Ethernet as well as token ring it is possible during running operation to integrate new computers into the network, or to remove them, without having to shut down network operations. In token ring, however, there must be

central control via the active monitor. Although in the event of a crash any other computer may substitute for it, the active monitor in token ring is a particularly critical component. For example, in the case of limited functionality of the active monitor, which is not yet interpreted by the other computers as a malfunction, network throughput can be significantly disturbed.

● **Fairness**
The access of a computer in the network is nondeterministic in Ethernet. This means the possibility of the theoretically highly unlikely occurrence of a computer never being able to access the network because with every attempt it is occupied. In contrast, token ring guarantees every connected computer a deterministic access, even if only following an agreed on maximum waiting interval.

4.3.4 Fiber Distributed Data Interface – FDDI

FDDI (Fiber Distributed Data Interface, fiber optic metro ring) bears a great similarity to the token ring standard. On the basis of its role as technological forerunner, the FDDI can equivocally be called the "mother" of all high speed LANs. The FDDI specification is for the two lowest layers in the network reference model: the transmission medium to be used along with bit transmission (physical layer) and access procedure (logical link layer). Unlike token ring, FDDI uses "a priori" fiber glass as the transmission medium. It is designed as a double ring with a data rate of 100 Mbps (meanwhile, also 155 Mbps and 1 Gbps) and a length of up to 100 km (meanwhile, up to 200 km) and can accommodate up to 1,000 computers.

Work on the specifications of the new high speed LAN standard had already started in 1980 with the founding of the group X3T9.5 of the American National Standards Institute (ANSI). Driven by the constantly increasing power of computers and peripheral equipment, the demand for high speed networks grew. As a group of conventional LAN standards, such as Ethernet or token ring, had already been evolved, the wish to combine these different network technologies together came to fruition. FDDI can be used in the same way as every other LAN defined in the IEEE 802 standard. In practical application it is, however, often used as **backbone** for the connection of different copper cable-based networks (cf. Fig. 4.39). FDDI is standardized in ANSI X3T9.5 (now X3T12) and ISO 9314. In 1994, the FDDI standard was expanded and transmission via shielded (STP) and unshielded (UTP type 5) twisted copper cables standardized as CDDI (Copper Distributed Data Interface). FDDI-2 as successor of the FDDI standards additionally offers the possibility of switching from a dedicated connection to a synchronous basis and thus can also be used for the transmission of real-time data, e.g., telephone or video.

The Principle of FDDI

The FDDI standard is based on a network topology split into a ring area (**trunk**) and a tree area (**tree**) (cf. Fig.. 4.40). The ring is formed by a counter-directional double

IEEE 802.4 Token Bus – A Standard Between Two Worlds

The probabilistic method used to gain access to the transmission medium in the Ethernet standard meant that an access guarantee was not possible. The manufacturing industry therefore expressed significant doubts about its application. Moreover, because the Ethernet standard did not set any priorities for data packets, its implementation in a real-time environment proved problematic. Unimportant data packets could not be allowed to hold back important ones.

If a ring topology is chosen that gives computers access authorization one after another then a maximum waiting interval and an access guarantee are ensured. However, the reliability of a ring is especially jeopardized inasmuch as the breakdown of a single connection between two participating computers means the complete breakdown of the network. Production lines also have a linear structure – which basically stands in the was of a ring topology.

As a result of such considerations, the standard **IEEE 802.4 token bus** was developed. Here, a linear physical transmission medium is used (as a rule a broadband cable). A logical ring is superimposed on the linear bus. This allowed the advantages of the robust IEEE 802.3 bus with the most reasonable worst case behavior of the IEEE 802.5 ring topology.

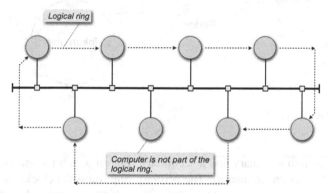

The order of the individual computers in the token bus is independent of their position, unlike in the token ring. The function of the token corresponds to the passing of the token ring. Each computer passes the token to its logical successor in the network. The final logical computer then passes it on to the logical first computer.

The logical process of the token bus is considerably more complex than in Ethernet (the protocol definitions in token bus cover more than 200 pages, with the implementation of 10 different timers and over 20 protocol variables). The cabling used by token bus is 75 Ω broadband coaxial cable. This type of cabling is also used in cable television. At the same time, the application of broadband cable demands a high degree of analog technology, for example with modems or broadband amplifiers. The token bus protocol is extremely complex and in the case of a small load not particularly efficient. It is not well-suited for implementation on a fiberglass-basis and never achieved widespread popularity, which is reflected in its lack of significance today.

Further literature:

Dirvin,R. A., Miller, A. R.: The MC68824 Token Bus Controller: VLSI for the Factory LAN, in IEEE Micro Magazine, vol.6, pp. 15-25 (1986)

IEEE 802.4: Token-Passing Bus Access Method, New York:IEEE (1985)

Fig. 4.38 IEEE 802.4 Token bus operation.

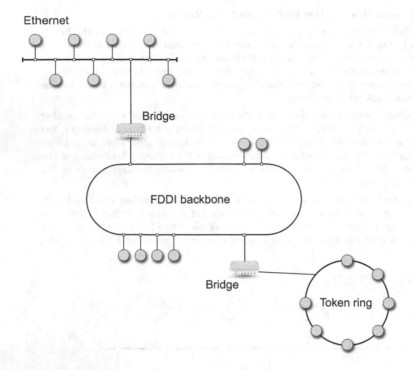

Fig. 4.39 FDDI Backbone

ring, subdivided into a primary and a secondary ring on which the individual participating computers are connected. The primary ring transmits clockwise and the secondary ring counterclockwise. Only the primary ring transmits when operation is error-free. The secondary ring functions as a backup in case of malfunction, e.g., a cable break. If only the primary ring is affected in this case then operation switches to the secondary ring. If, however, both rings are broken at the same place they connect into one ring of approximately double the size. The defect is looped. Here, the secondary ring operates as a so-called **cold reserve** . In contrast, **hot reserve** means that in the case of no errors both primary and secondary rings are operated together. If an error then occurs and operation must be carried out on just one of the remaining rings, it is possible that the transmission capacity available per computer in the ring is insufficient.

FDDI provides various options for connecting computers to the FDDI ring. Various compromises are revealed in each case in terms of effort, capacity and robustness. A connection on the primary as well as secondary ring presents the most secure, but also the most complex solution. Components interconnected this way with the FDDI ring are designated **Dual Attached Components** (Class A). Besides this, there is also the less expensive possibility of single components that connect to only one of the two rings (**Single Attached Component**, Class B). These are distinguished as:

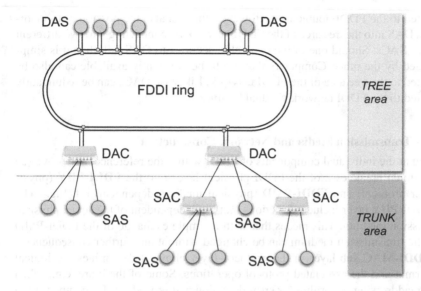

Fig. 4.40 FDDI topology

- **DAC**
  Dual Attached Concentrators are network components connected to both the primary and secondary ring. They have a bundled connection for other computers (SAS) or may be cascaded with yet other switching components (SAC).

- **DAS**
  Dual Attached Stations are end components connected to the primary as well as secondary ring. This is normally a computer that is connected directly to the double ring via its own special FDDI network adapter.

- **SAC**
  Single Attached Concentrators are single connection components that may only be connected to the primary ring. They can either be cascaded further themselves (SAC) or connected to multiple end components (SAS).

- **SAS**
  Single Attached Station are end components that have only one connection to the ring. A SAS is always shielded from the FDDI ring by at least one concentrator (SAC or DAC). In this way the operational security of the ring is increased.

Components connected to a DAC/SAC each build their own tree area in the FDDI network. The failure or shutdown of a DAC/SAC concentrator interrupts the FDDI ring and leads to a reconfiguration. In contrast, the failure or shutdown of a SAS connected to a concentrator has no influence on the primary FDDI double ring. In this case the concentrator decouples the SAS from the FDDI ring and bridges the connection. System stability in the tree area of the FDDI network can be increased by the so-called **dual homing** process. As a rule DAS computers are directly connected to the FDDI double ring. However, there is also the possibility to incorporate a DAS into the tree area of the FDDI network by connecting it to two different DACs or SACs. Should one of the two cable sections to DAS fail, then it is simply bridged by the other. Computers that are to be constantly available can also be connected to the tree area of the FDDI network. Likewise DACs can be redundantly connected to the FDDI network via dual homing.

FDDI – Transmission Media and Network Construction

Looking at the individual components of FDDI within the reference model, we see an additional breakdown of the FDDI physical layer into the **FDDI-PHY** (physical layer protocol) and **FDDI-PMD** (physical medium dependent) (cf. Fig.4.41). The FDDI-PHY layer includes functions that are independent of the actual physical transmission medium. This means that with a simple exchange of the FDDI-PMD layer, the transmission medium can be changed without any further consequences. The **FDDI-MAC** sub-layer in the protocol layer above it determines the logical data format and the associated protocol operations. Some of these are, e.g., token passing, addressing, algorithm for error detection and procedures for compensation of detected errors (error correction). The **FDDI-SMT** (system management) layer extends across all of the three sub-layers. It is used for operations affecting ring management, e.g., configuration, monitoring or error handling.

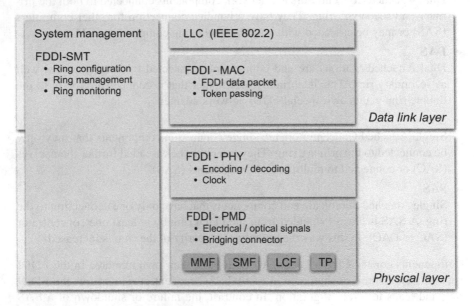

Fig. 4.41 The FDDI layer model.

The transmission medium and its dependent parameters are specified in the FDDI-PMD layer. These include the connection technology used in the transmitting and receiving components, the optical bypass and the intended damping. The FDDI standard distinguishes five alternatives:

- **MMF-PMD** (Multimode Fiber)
  Multimode fiber optics as transmission medium
- **SMF-PMD** (Single Mode Fiber)
  Single mode fiber optics as transmission medium
- **TP-PMD** (Twisted Pair)
  Twisted copper wire as transmission medium
- **LCF-PMD** (Low Cost Fiber)
  Fiber optics with reduced quality requirements as transmission medium

The most important characteristics and tasks of the FDDI-PHY layer are found in bit transmission. These include:

- **Encoding/Decoding**
  A two-step encoding procedure was chosen for FDDI that increases transmission efficiency. This is the so-called 4B/5B encoding (cf. Fig.4.42).
- **Management physical connections**
  Found between different PHY instances.
- **Transmitter clock management**
  In addition to synchronization and regeneration of the clock signal, a compensation of clock differences in various PHY instances is carried out by the so-called elasticity buffer.

The SMT component regulates on a computer within the FDDI ring the interaction and monitoring of the processes in the other FDDI components (PHY, PMD und MAC). Among these include:

- initialization of computers and network,
- addition and removal of computers in or out of the FDDI ring,
- assignment of transmission priority and administration of the available bandwidth,
- isolation of errors and the attempt to correct them or
- collection of statistical information.

FDDI uses fiber optics as the primary transmission medium. Although FDDI works via copper cable (copper distributed data interface, CDDI), the implementation of fiber optics offers a wide range of advantages. Fiber optics distinguish themselves particularly in the areas of security and reliability. They do not send out any electromagnetic signals that might be subject to potential eavesdropping. Fiber optics are also resistant to electromagnetic interference.

With the use of multimode glass optical fiber, the distance between two neighboring computers in the FDDI ring can be extended to 2 km – without signal regeneration being necessary. Using single mode glass optical fiber the distance can be even

4B/5B Encoding

In the physical layer (PHY), the Manchester encoding used in FDDI is not implemented in token ring. It would have required a transmission frequency of 200 MHz with a transmission capacity of 100 Mbps, which was seen as too expensive. Instead, a coding variation called **4B/5B** was implemented. $2^4=16$ different bit combinations are mapped onto $2^5=32$ different code words. The other half of the 32 code words are used for the coding of additional information. Three code words are used for limitation symbols (delimiters), two serve as control indicators to fix logical conditions in connection with data transmission, and a further three for hardware signals that control the conduction state. The remaining eight code words are not used. The valid codewords are chosen in such a way that never more than two zeros occur in sequence.

Group	Code	Symbol	Meaning
Line status	00000	Q	Quit
	11111	I	Idle
	00100	H	Halt
Delimiter	11000	J	1. symbol in SD field
	10001	K	2. symbol in SD field
	01101	T	end delimiter
Control	00111	R	logical zero / reset
	11001	S	logical one / set

Additionally, a so-called **NRZ-I** encoding (cf. section 3.2.1) is used. This encodes a zero with the same voltage level as the previous bit; at a one the voltage level is inverted.

Bandwidth is saved using this type of encoding, but in terms of clock generation it is not as efficient as Manchester encoding. In FDDI a long preamble is necessary as an introduction to the data packet in order to synchronize the clock between transmitter and receiver. The requirements of stability and preciseness in view of the system clocks permit a deviation of not more than 0.005%. The maximum length of a data packet is thereby limited to 4,500 bytes. This prevents the system clocks of the transmitter and receiver from deviating too far from their synchronization.

Because eight code words are not used, an overhead of 25% is produced. The result is an actual data transmission rate of 125 Mbps. The system clocks of the network adapters of the FDDI ring connected computers are accordingly clocked at 125 MHz.

Further literature:

Feit, S.: Local Area High Speed Networks. Macmillan Technical Publishing, Indianapolis, IN, USA (2000)

Fig. 4.42 4B/5B Encoding in the FDDI protocol.

greater. Single mode glass optical fiber is used particularly in long range networking from different locations. Distances of 40 km to 60 km can be bridged without additional signal amplification between two neighboring stations in the FDDI ring. Total ring lengths of 100 – 200 km can be implemented and a combination of multimode and single mode is easily possible. In today's FDDI, installations for single site networking, multi-mode fiber is typically used. Links between different sites are achieved via single-mode. At the site itself, copper cable, such as UTP-5 or STP-1, can also be implemented, made possible by the corresponding FDDI-PMD protocol layer. The maximum distance between two computers in the FDDI ring should be no more than 100 m, according to specifications. A maximum distance of up to 2 km can be bridged between two stations.

For fiber optics the FDDI specification allows for damping of less than 2,5 dB per kilometer, with a maximum bit error rate of $2,5 \cdot 10^{-12}$. As a rule, high-quality implemented components show even better results. FDDI dual attached connection are equipped with a passive mechanism **optical bypass**. This is takes the form of a relay and is first activated in the event of a cable break or computer shutdown. In the resting state, the bypass relay connects the incoming fibers of the primary ring with the outgoing fibers of the primary ring. In the event of disturbances in network operation, the incoming fiber of the primary ring is interconnected with the outbound fiber of the secondary ring or vice versa.

FDDI Data Format

FDDI data packets are very similar to IEEE 802.5 token ring data packets (cf. Fig. 4.43). Framed by the **start delimiter (SD)** and **end delimiter (ED)**, there is also an extra **preamble**, with a minimum length of 64 bits, at the beginning of the data packet. It consists of a synchronization bit sequence. There is no prescribed central system clock in FDDI – every network adapter has its own signal clock. Therefore, synchronization must be established between a received data packet and the internal clock of the receiver. Normally, the preamble is composed of 16 idle symbols (11111).

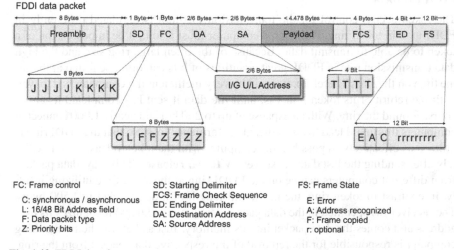

Fig. 4.43 FDDI data format.

The **frame control** field (**FC**) determines the type of data packet (control or payload packet). The **control bit** (**C**) indicates, whether the data packet originates from a synchronous (C=1) or a asynchronous transmission (C=0). The **length bit** (**L**) then follows. This indicates if the sender and receiver addresses are 16 bit (L=0) or 48 bit (L=1) . The two **frame format** bits (**FF**) in combination with the four **Z** bits indicate the exact type of the data packet. Afterwards comes the transmitter's address (**source address**, **SA**) and the receiver's address (**destination address**, **DA**. These each comply with the prescribed IEEE 802.x standard for MAC addresses and can be either 16 or 48 bits long. The first bit (**I/G**) indicates whether it is a group address (I/G=1) or an individual address (ID=0). The second bit (**U/L**) indicates if it is a local address (U/L=0), which the user can assign himself, or a universal address (U/L=1). The latter is a manufacturer-specific address assigned by the IEEE and identifiable worldwide.

The actual **payload** afterwards can consist of up to 4,478 bytes. Next, a 32 bit checksum (CRC) follows in the **frame check sequence** field (**FCS**), which secures the fields FC, DA, SA, payload and FCS. The remaining fields do not carry payload information in the strict sense.

The **frame status** field (**FS**), which follows the end delimiter (ED), has a length of 12 bits. With the help of this field the receiver can verify in which condition the data packet has been received. The bits **error detect** (**E**), **address recognized** (**A**) and **frame copied** (**C**) are required, while further fields are left up to the discretion of the manufacturer. The data in these fields is coded with the help of the symbols **R** (reset) and **S** (set). In FDDI – just as in token ring – besides the data packets that contain the actual payload there are also token-data packets to control access to the FDDI ring. The token data packet consists of the preamble fields SD, FC and ED.

FDDI Protocol

The token access control for the transmission medium is regulated differently in FDDI than in token ring. However, a computer must also first be in possession of a token to be able to transmit data . Because of the often large ring size and the high data transmission rate in FDDI, great quantities of bits can often be found at the same time on the ring. Therefore, it would be very inefficient if a computer had to wait until the return of its token – that is, until the data it sent has completed its entire course around the ring. With an expanse of up to 200 km and nearly 1,000 connected computers this could lead to a significant performance loss. So that the FDDI ring is utilized as effectively as possible, the computer with the token releases it immediately after sending the last data packet (**early token release**). This way, data packets from different computers can be on the FDDI ring at the same time utilizing it fully. In contrast to token ring, the maximum length of a data packet is 4,500 bytes. The receiver (as specified in the data packet destination address) recognizes its own address and copies the data packet into its memory. Just as before, the transmitting computer is responsible for the removal of the respective data packet from the ring. If a data packet is received that contains its own address as the source address it is no longer forwarded. Computers without transmission authorization check the passing data packets for errors. They then have the possibility to register these in the frame status field before forwarding. In Fig. 4.44 the process of the FDDI access method is shown schematically.

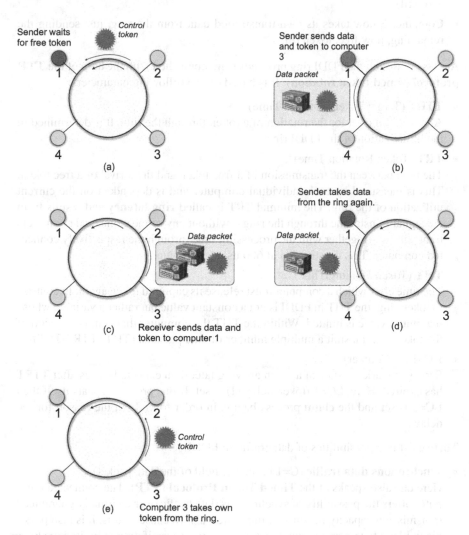

Fig. 4.44 Illustration of the FDDI access method.

- The wishing-to-transmit computer 1 waits for a free token (a).
- As soon as computer 1 is in possession of the token, it sends data to computer 3 and finishes its transmission with the received token (b).
- Computer 3 copies the data specified for it, and as it also wishes to transmit, takes the token into its possession, sends data to computer 1 and ends the transmission with the token again (c).

- Computer1 takes the data it has transmitted from the ring again, copies the data specifically intended for it as receiver and forwards this on together with the token (d).

- Computer 3 now takes its own transmitted data from the ring, just sending the remaining, now free, token (e).

Access control on the FDDI ring proceeds time-controlled with the help of the **TTR protocol** (**timed token rotation**): It is based on the following parameters:

- **TTRT** (Target Token Rotation Time)
  Allocated time set for the rotation of a token through the ring. It is determined in the initialization of the FDDI ring.

- **TRT** (Token Rotation Time)
  The time between the transmission of a free token and the arrival of a free token. This is measured by each individual computer and is dependent on the current utilization of the ring. The minimal TRT is called **ring latency** and results from the signal transit time through the ring – without any of the connected computers transmitting – together with the processing time through the respectively connected computer. This is estimated at 600 ns per computer.

- **THT** (Token Holding Time)
  The time after which a computer must release its captured token again. In contrast to token ring, the THT in FDDI is not a constant value but rather a variable whose maximum value is limited. Within the THT the computer that is in possession of the token can transmit a multiple number of data packets. (THT=TTRT-TRT)

- **LC** (Late Counter)
  Serves to mark the state of a token arriving later than expected – thus after TTRT has expired. Here, LC=1 (token delayed) is set. If this process repeats itself then LC=2 is set and the **claim** process begins in order to find out the reason for the delay.

There exist two possibilities of data traffic in FDDI:

- **synchronous data traffic** (C=1 in the FC field of the data packet)
  Here one also speaks of the **Timed Token Protocol** (**TTP**). The connected computer offers the possibility of synchronous data traffic over a virtually guaranteed transmission capacity in constant time intervals. Because of this, it is also possible in FDDI to transmit moving pictures or voice (even if only in limited quality). But it should also be noted that synchronous data traffic in FDDI specification is provided optionally and thus not supported by every manufacturer.

- **asynchronous data traffic** (C=0 in FC field of the data packet)
  Here, there are no guarantees and agreements on the rotation of the token in the FDDI ring. The connected computers must wait an undetermined period of time until they are allowed to send a data packet. This also applies when a theoretical upper limit is prescribed for the waiting time. The is the reason for the different time intervals between the individual data packets transmitted by a computer.

The **claim** process makes it possible for individual computers in the FDDI ring to reserve the token in specific, virtually constant time intervals for synchronous data traffic. All computers i that wish to transmit data synchronously, declare the key parameters for their request to send. This means, in which $\rho_t(i)$ they wish to send for a specific length of time $\delta_t(i)$. A compromise solution is then determined via the claims process. Then all computers wishing to take part in synchronous transmission must adjust their parameters according to the one whose transmission time came up the most frequently $(\min_{\forall i}(\rho_t(i)))$. Those computers whose sending frequency came up less often are now forced to send correspondingly smaller packets, i.e., to send at a higher frequency. Synchronous data traffic is granted the highest priority in FDDI, provided that such a FDDI network is designed for synchronous operation. The rest of the transmission capacity can then be used for asynchronous transmission. Eight different priority levels are distinguished. They correspond to the maximum transmission time (THT) guaranteed to a computer wishing to transmit.

FDDI – Ring administration

If during operation of a FDDI ring an error appears, such as a break in the ring or a prescribed time limit being exceeded (e.g. LC=2), then a **claim** process is initiated. If unsuccessful, a so-called **beacon** process begins. Similar to token ring, every computer connected to the ring transmits special, individual beacon data packets. Upon receiving a beacon packet from its predecessor in the ring, a computer stops sending its own beacon packet. It then proceeds to forward the data packet of its predecessor. When a connected computer does not receive a beacon packet, it reports either the proceeding computer or connection route to its predecessor as defective. Subsequently, appropriate error handling measures take effect such as a ring reconfiguration.

Further literature:

Yang, H. S, Yang, S., Spinney, B. A., Towning, S.: FDDI Data Link Development. Digital Technical Journal, (3):31–41 (1991)

Fig. 4.45 FDDI – Ring administration

FDDI-2

FDDI offers the possibility to transport both synchronous and asynchronous data streams. Synchronous data streams ensure virtually constant transmission performance in the form of regularly available guaranteed time slots. Nevertheless, this form of synchronous data transport is not sufficient for the transmission of voice or video sequences. A waiting time is necessary in FDDI for reservation of the necessary bandwidth. Even then, the bandwidth can only be guaranteed in connection with

FDDI – Error tolerance

The reliability of a FDDI network is determined by the following properties:

- a double-executed ring topology,
- the implementation of diverse connected network components and
- the possibility to simply circumvent defective computers in the network via an optical bypass .

Based on network implementation the occurring cases of error can be summed up in four basic error situations:

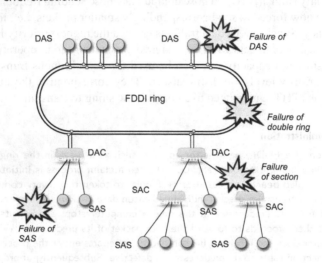

- **Failure of a DAS**
  Leads initially to an interruption in the connection of the neighboring dual attached components (DAS, DAC). The situation can be remedied through an optional optical bypass of the affected DAS, or by proceeding as in the case of interruption of the double ring.

- **Interruption of the double ring**
  The dual attached component nearest to the two effected ends creates a logical bridge between the primary and secondary ring via the BEACON process. In this way the location of the defect is treated.

- **Failure of a SAS**
  Results solely in an interruption of the primary ring. As in a SAS there is normally no optical bypass available, the SAS is bridged by the hierarchically higher DAC, i.e., logically separated.

- **Failure of a glass optical fiber section**
  Also here follows a bridging of the failed section by the superior DAC. Two functional partial networks can then result that no longer communicate with each other.

Fig. 4.46 FDDI – Error tolerance

a maximum delay (estimated with double the target token rotation time, approx. 100 ms).

Real-time media requires the possibility of **isochronous** transmission. This means that the timing conditions must be the same on both the side of the sender and the receiver. The concept of FDDI-2, for the transfer of isochronous bit streams, was developed in 1984. In 1994 it was adopted as a standard by the ANSI X3T9.5 committee. FDDI-2 provides reliable transport of real-time video data. However, the coupling of the FDDI and FDDI-2 systems proved problematic, which led to FDDI-2 eventually being pushed out of the market byB-ISDN and ATM.

FDDI-2 is capable of transmitting asynchronous, synchronous and isochronous bit streams. To ensure this process, FDDI-2 uses a hybrid ring control system. A so-called **hybrid multiplexer** (**H-MUX**) controls the fixed time sequence of the respective requests by the individual components of the MAC sublayer on the transmission medium. It contains a component for the regular, that is synchronous or asynchronous access, the so-called **packet MAC** (**P-MAC**), as well as a component that controls the isochronous access – the so-called **isochronous MAC** (**I-MAC**). If in addition to the regular data traffic (basis mode) an isochronous transmission is to take place as well, FDDI-2 works in a so-called hybrid mode. In the hybrid mode isochronous data streams always have priority. Only the unused remainder of the bandwidth is open to other data transfer.

Before data transmission can begin, the participating computers make their bandwidth request known via the station management of a special computer in the FDDI-2 network. This is called the cycle master. The cycle master then sets up isochronous channels and transmits these to the requesting computers. It continues to monitor use of the isochronous channels. The transmission of isochronous data is organized in so-called cycles with the help of 125-μs data packets. A new 125-μs data packet is generated by the cycle master 8,000 times per second. With a bandwidth of 100 Mbps this results in a length of 12,000 bits per data packet. When such a cycle makes a complete run around the FDDI ring, it is removed from the ring by the cycle master. A cycle data format is made up of the following components (cf. Fig. 4.47):

- **Preamble (PA)**
  Series of 20 bits that are exclusively used for synchronization.

- **Cycle Header (CH)**
  A 12 byte long data packet header. It contains information for H-MUX as to the the traffic types (isochronous /asynchronous) assigned to the individual channels
  .

- **Dedicated Packet Group (DPG)**
  12 bytes exclusively available for asynchronous/synchronous data transmission.

- **Cyclic Group (CG)**
  Cyclic groups of time slots, designated CH0, CH1, …, CH95. Every group is composed of 16 slots each. These can be assigned to the individual traffic types (isochronous/synchronous/asynchronous).

An **isochronous channel** (wideband channel, WBC) is implemented in FDDI-2 by reserving a certain number of slots from the 96 cyclic groups. With 96 groups, each

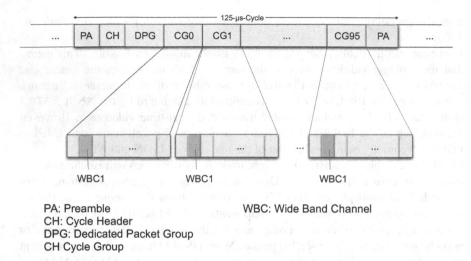

PA: Preamble   WBC: Wide Band Channel
CH: Cycle Header
DPG: Dedicated Packet Group
CH Cycle Group

Fig. 4.47 FDDI-2 Cycle data format.

with 16 time slots, it is possible to make 6 channels available, each one with a
guaranteed transmission capacity of 6, 144 Mbps. These channels can be subdivided
into 96 individual subchannels each with 64 kbps. Each of them can be reserved for
isochronous or asynchronous traffic. The assignment of cyclic slots as asynchronous
or isochronous is performed by the H-MUX. It forwards received data accordingly,
either to MAC or to IMC components.

4.3.5 Asynchronous Transfer Mode – ATM

The development of ATM (**Asynchronous Transfer Mode**) is closely connected
to the evolution of a WAN technology called **B-ISDN** (**broadband integrated services digital network**). It became clear that the existing ISDN infrastructure, with
its maximum bandwidth of 128 kbps, was not sufficient to provide transmission for
multimedia content such as video or music in CD quality. It had been implemented
with the objective of integrating conventional telecommunication services such as
telephone and telefax. From this point on, the planning of B-ISDN high-speed wide area technology began. ATM was chosen in 1986 as the basis technology upon
which the B-ISDN was built. It was standardized between 1991 and 1993 by the
ITU. Like traditional WANs, the ATM network is made up of single lines and switches (routers). The initial bandwidth of ATM was set at 155 Mbps, or 622 Mbps
with the option of expanding later into the Gbps area. The choice of 155 Mbps was
made, being the bandwidth necessary for the transmission of high-definition television. 622 Mbps was also chosen as four bundled 155 Mbps channels could be
then be transmitted. Due to the close connection with SDH multiplex technology,

the following data transmission rates can be realized today via ATM: 155,52 Mb-
48832 Gbps or 9,95328 Gbps. Originally developed as a WAN
technology, ATM is now often used in a LAN environment.

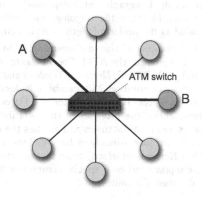

For the standardization of ATM, the so-called **ATM Forum** (today "Broadband Fo-
was founded at the initiative of the companies involved in ATM development. The development of ATM was often guided by the interests of the various
partied involved. For example, telephone companies that competed with cable television companies in the video-on-demand market. The basic element of an ATM
network is an electronic intermediary (**switch**) connecting multiple computers in a
star topology (cf. Fig. 4.48). Unlike a bus or ring topology, the data in a star network
is not forwarded to all connected computers. Only the communicating computer pair
is involved in the data transfer via the switch. The switch receives the data directly
from the transmitter and forwards it directly, or via other switches to the receiver.
Because of this when one computer or computer connection in a star topology fails
the whole network does not suffer, as would be the case in a ring topology. The
transfer is asynchronous in ATM networks. This means there is no central clock
prescribed by an entity in the network regulating how the individual data packets

The Operating Principle of ATM
ATM uses the principle of so-called **cell switching** (cf. Fig. 4.51). Here, the data to
be transmitted is packed in small packets of a constant size and transmitted over a
switch connection with a guaranteed quality. Data transmission in ATM proceeds
. While a synchronous data transmission technology assigns each
logical connection a time slot in periodic intervals, this is not necessarily the case
with ATM. ATM works according to the principle of **statistical multiplexing**. Here
as well there is a continuous stream of occupied time slots (ATM cells), but there
is no fixed relationship between the position of time slots and a certain connection

ATM Standardization

The first ideas about ATM were published in 1983 by the American research centers **CNET** and **AT&T Bell Labs**. At this time ATM was still seen exclusively in its function as WAN standard. This is why the ATM technology was also proposed for the broadband ISDN standard (B-ISDN) in 1986 by the then **CCITT** (Committe Consultatif International de Telegraphie et Telephonie) – today **ITU** (International Telecommunication Union). In 1989, the ongoing standardization dispute between America and Europe ended as the uniform length of ATM cells was finally set at 53 bytes.

With the goal of also implementing ATM in the LAN area, and in an attempt to speed up this process, the **ATM- Forum** was founded in October 1991 by the companies Cisco, NET/ADAPTIVE, Northern Telecom and Sprint. Today the renamed Broadband Forum committee numbers 750 members. It does not see itself as a standardization committee and therefore as a competitor to ITU or ISO. Instead, the committee see its role in representing the interests of the LAN industry. It expresses the legitimate concerns of users and manufacturers and enables the corresponding standard to be worked out in a timely manner. Already in July 1992 the first specification of the ATM forum (UNI 1.0 – User Network Interface) could be presented. Within the ATM Forum European interests are represented by a special committee, the **EMAC** (European Market Awareness and Education Committee).

Further literature:

Neelakanta, P. S.: A Textbook on ATM Telecommunications: Principles and Implementation. CRC Press, Inc., Boca Raton, FL, USA (2000)

Fig. 4.49 ATM standardization.

(cf. Fig. 4.50). The assignment of an ATM cell to an associated connection is purely arbitrary and depends only on the transmission activity of the respective source and network utilization. This approach is more efficient than a synchronous data transmission particularly in cases where variable and constant bandwidth requirements of different communication partners coincide. Here, only cells are occupied and – if necessary – replaced between two communication partners and therefore no bandwidth is lost.

ATM is **connection-oriented**. This means that all packets of a connection belonging to a certain data transmission are transported over the same path (virtual channel). Before the start of a data transfer, a connection must first be established to the desired communication partner. This is necessary before the actual data is sent to the receiver on the established path. While there is no guarantee for the transmission of all cells, the correct sequence of each cell is always guaranteed. Therefore, if two cells A and B are sent, and if both reach their destination, the transmission sequence A, B is guaranteed and never the reverse order B, A.

The central element within the ATM data transfer procedure is the concept of the **virtual channel** (VC) and the **virtual path** (VP). A **virtual channel** is a one-directional connection for the sequence-preserving transport of ATM cells. It is implemented over a numerical identifier (virtual channel identifier, VCI). A distinction is made between permanent virtual channels (**permanent virtual circuit, PVC**), which are permanently set up by the user manually and often exist for months or even years, and transient virtual channels (**transient virtual channel, TVC**), which

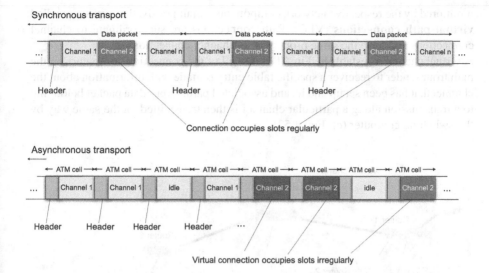

Fig. 4.50 ATM transport principle.

Cell Switching

A disadvantage of widespread packet switching in computer networks is the store-and-forward principle. This slows all technologies that build on this switching technology. All packets that arrive at an exchange must be fully stored. Checksums in the data header must be verified and addresses evaluated. To increase effectiveness of transmission capacity, concepts such as **fast packet switching** or **fast packet relaying** were developed.These follow the basic principle of forwarding packets as quickly as possible. This is only possible if the packet structure is kept simple and only a few parts of the packet have to be evaluated for forwarding. In addition, the routing algorithm, which decides about the forwarding of incoming data packets, is implemented completely in the hardware of the intermediate systems to achieve a further speed advantage.

Cell switching (cell relay) is capable of both isochronous (audio, video) and asynchronous data transmission.

Further literatur:

Garbin, A., O'Connor, R. J., Pecar, J. A.: Telecommunications FactBook. McGraw-Hill Professional,, Boston, MA, USA (2000)

Fig. 4.51 The principle of cell switching.

work only temporarily and follow the same principle as telephone calls. They are each established for a certain transmission and dismantled completly after it has ended. Virtual channels are bundled to **virtual paths** and given a numerical identifier (virtual path identifier, VPI).

A **virtual channel connection** (VCC) is a series of virtual channels between two communication partners in the ATM network. Each channel connection is assigned service parameters specifying properties such as cell loss or cell delay. Additionally, traffic parameters are negotiated for every channel connection. Adherence to them is

monitored by the respective network components. Analogously, there exist so-called
virtual path connections (VPC). Path connections are superordinate to channel
connections and have the same properties as channel connections.

A virtual channel is established in ATM. On all of the connection computers on the
path from sender to receiver a specific table entry is made with information about the
channel that has been set up and its and associated path. Every data packet belonging
to a transmission along a particular channel is then transmitted on the same way by
the switching computer (cf. Fig. 4.52).

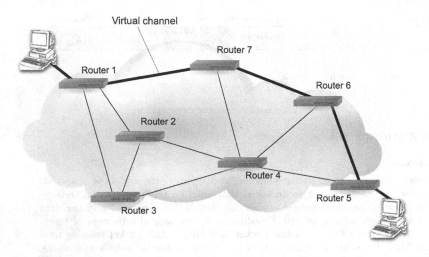

Fig. 4.52 ATM – Virtual channel

For this purpose, switching computers keep so-called **routing tables** in their memory. When a data packet reaches the switching computer, its packet header is checked
to find out which virtual channel the data packet belongs to. Afterwards, the switching computer checks the table entry for the specified virtual channel, thus determining on which line the data packet is to be forwarded. This mechanism will be
looked at more closely in chapter 6.

The ATM data format

Looking at the data format implemented in ATM, it is necessary to take two different
interfaces into account:

- **UNI (User Network Interface)**: the interface between the ATM network-connected
  computer and the ATM network.

- **NNI (network network interface)**: the interface between two switching computers in the ATM network.

The ATM cell always has a length of 53 bytes. For UNI and NNI each a 5 byte long header and 48 bytes of payload are provided. The headers for UNI and NNI are only slightly different (cf. Fig. 4.53).

UNI ATM cell

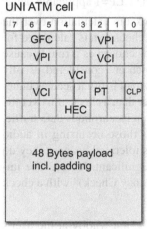

NNI ATM cell

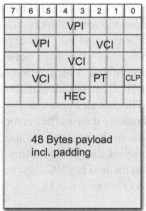

GFC: General Flow Control
VPI: Virtual Path Identifier
VCI: Virtual Channel Identifier

PTI: Payload Type
CLP: Cell Loss Priority
HEC: Header Error Check

Fig. 4.53 ATM Cell data format

- The **GFC** field (General Flow Control) only appears in data packets circulating between a connected computer and a switching computer. At the first switching computer the respective data packet reaches, the field has already been overwritten. It never reaches the actual receiver and therefore has no significance for end-to-end communication. For this reason, Tanenbaum [261] suggested we look at the GFC field as a bug in the standard definition.

- **VPI** (Virtual Path Identifier) is used for selecting a certain path on the switching computer.

- Via the **VCI** (Virtual Channel Identifier) a particular virtual channel is also chosen inside the specified virtual path. Since (in UNI data packet) VPI is defined with 8 bits and VCI with 16 bits, a computer can theoretically respond to up to 256 virtual paths with a maximum of 65,536 virtual channels each. In reality the number is somewhat less as some channels are permanently reserved for control functions.

- **PTI** (Payload Type Identifier) indicates the type of payload transported. The user provides the respective type of payload, while the network supplements information pertaining to the state of the network. For example, a cell with PTI=000 is transmitted by the sender. If congestion occurs between the computer connections of the network, then the switching computer that has identified the congestion adds PTI=010. In this way, the receiver gets advanced warning about the network problem.

- Das **CLP** (Cell Loss Priority) characterizes high and less high priority data traffic. If congestion occurs in the ATM network, time-critical data must inevitably be discarded. The switching computer first discards the CLP=1 applicable data.

- The **HEC** (Header Error Check) field serves the checksum calculation over the data header. In contrast to the usual method, only one checksum is calculated for the header and not for the payload. This approach was chosen as a rule because ATM data traffic runs over glass fiber networks, which are considered usually very reliable. Because of this, a verification of the payload is often seen as unnecessary. In addition, HEC can be calculated much faster than a checksum over the entire data packet. This has an important significance in isochronous data traffic. Moreover, in isochronous data traffic, errors such as those occurring in audio or video transmission in the payload are more or less tolerated because they do not impair the quality of data transmission to any significant degree. The implemented checksum method is CRC (Cyclic Redundancy Check)[6] with a check polynomial of 8 bits ($x^8 + x^2 + x + 1$).

After this comes 48 bytes of payload. They are, however, not wholly at the user's disposal as additional control and header information from superior protocols is transmitted with the payload as well.

ATM Interface and Network Construction

In contrast to the technologies discussed so far, such as Ethernet or token ring, the ATM standard, in view of the layer model for communication protocols, is concentrated above the physical layer. It contains no provisions at all for the technology used for bit transmission. ATM was designed with the aim of being as independent as possible from the physical transmission medium. ATM cells can therefore also be transmitted as encapsuated payload to other network technologies such as FDDI or Ethernet.

ATM defines a protocol layer model of its own, made up of three layers. Each of these layers is divided into two sublayers: a lower layer that carries out the actual tasks of the layer, as well as a so-called convergence sublayer. This provides a suitable interface for the layer above it (cf. Tab. 4.9).

- **Physical layer**
  Is not a component of the ATM specification and is divided into multiple layers (sublayers):

  - A media dependent layer (**Physical Medium Dependent, PMD**) as the current interface for the transmission medium. Here, the actual bits are trans-

---

[6] A detailed representation of the CRC method is found in chapter 3 „Fundamentals of Computer Network Communication,"in the first volume of this series: Ch. Meinel, H. Sack: Digital Communication – Networking, Multimedia, Security, Springer (2013).

Why 53 bytes?

Why do all ATM data packets have a determined length and why is this length 53 bytes? The answer to the first question is easy. Only a fixed packet length makes it possible for ATM to achieve its desired characteristics. **Switching** as a forwarding process of data packets in ATM can only be performed at a very high speed. This can only be achieved if the necessary algorithms are hard-wired into the hardware of the switching computer. In order to implement this within an acceptable degree of effort all data packet must have the same length. If this were not the case exceptional handling routines would be necessary. The **processing time** for individual data packets is reduced as well. The length is prescribed and does not have to be calculated first. Finally, the **statistic multiplexing** used in ATM can only be optimally carried out with a fixed cell length. With cells of variable length a statistical multiplexing of different data streams on a VP/VC connection would not be possible.

But why **53 bytes**? Actually the length of the payload in an ATM packet is only comprised of 48 bytes. 5 bytes are for the header of the data packets and contain relevant control information. These 48 bytes are part of the compromise that was achieved between applications of voice and data communication. The USA voted for a cell length of 64 bytes in addressing the question of ATM standardization. This length was favored for its optimal application in voice communication. The Europeans, on the other hand, who were mainly interested in data communication, preferred 32 bytes. As a compromise between the two factions a happy medium was finally decided on. Consideration factors in the choice of suitable data packet lengths were, for example, transmission speed, switching speed, delay and behavior of the queue at the switching computer (pipelining) in the network.

For the "switching" of a ATM cell that is transmitted on a 155 Mbps link are maximal

$$\frac{53 \cdot 8}{155} = 2,735 \ \mu s$$

available. Normally, an ATM switch has a large number of links at its disposal. These can work at different transmission rates. With an increase in the number of links and shorter cell lengths, the time available to forward an ATM cell decreases. Consequently, the switching speed of the ATM switch must be increased.

Further literature:

Stevenson, D.: Electropolitical Correctness and High-Speed Networking, or, Why ATM is like a Nose, in Viniotis, Y., Onvural, R. O.: Asynchronous Transfer Mode Networks, Plenum Press, New York, NY, USA (1993)

Fig. 4.54 The ATM cell

mitted and time dependencies determined and verified. Different transmission media requires different PMDs.

- Above is the **Transmission Convergence, TC** sublayer. If single ATM cells are transmitted, the TC sublayer sends these as a data string to the PMD layer. In the opposite direction, the TC layer processes the incoming bit stream, transmitted in individual ATM cells. The task of packaging in the ATM layer model is assigned to the physical layer. In other models it is usually carried out by the connecting layer above – the data link layer.

- **ATM layer**

  Is not divided further into sublayers. Its main task is found in the generation, processing and transport of ATM cells. Here, the cells are "laid out" and the header information made available. The setup and cancellation of virtual channels is carried out here as well. ATM-typical aspects of this technology occur in the ATM layer. In accordance with the TCP/IP reference model, its functionality is situated between data link layer and the Internet layer.

- **ATM Adaption Layer**

  As the majority of applications do not work directly with ATM cells, this layer was added above the ATM layer. It also allows applications of longer data packets to be passed on directly. In the ATM adaption layer these data packets are segmented in one direction into single ATM cells, and then reassembled in the other direction. Thus, this layer is subdivided into the **SAR** sublayer (**Segmentation And Reassembly**), which assumes the just described task, and the **CS** sublayer (**Convergence Sublayer**). This enables the ATM system to offer adapted services for different applications (e.g., data transfer and video-on-demand have different requirements regarding error tolerance and timing).

Table 4.9 ATM layer model and and its relation to the TCP/IP reference model

ATM layer	ATM sublayer	Task	TCP/IP layer
AAL	CS	provides standard interface	3/4
	SAR	segmentation and reassembly	
ATM		flow control,	2/3
		generation of cell header and evaluation,	
		management of virtual paths/channels,	
		cell multiplexing	
Physical	TC	generation/verification of checksums,	2
		generation of cells,	
		packing/unpacking of cells	
	PMD	bit timing,	1
		physical network access	

Additionally, the ATM layer model demonstrates a vertical subdivision into so-called **planes** (cf. Fig. 4.55). A distinction is made between a **user plane**, responsible for transport, flow control and error correction of the actual user data and a **control plane** where the connection-relevant communication takes place. Moreover, there is a **layer management plane** and a **plane management plane**, where functions for resource management and layer coordination are processed.

Principally, ATM connections are simple point-to-point connections. Unlike a bus topology such as Ethernet, more than one transmitter and one receiver never communicate over one cable. Multicast or broadcast is achieved in ATM networks in that

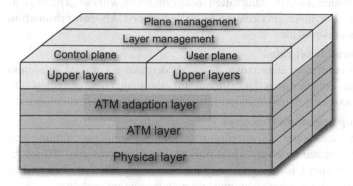

Fig. 4.55 ATM layer/plane model

a data packet arriving on one line of the ATM switching computer is directed to one multicast address (i.e., to more addresses simultaneously) and is forwarded from here to multiple lines. Every connection is unidirectional. To enable bidirectional operation two parallel connections must be switched.

ATM Service Classes

ATM is suitable for the transmission of different media formats. Video, audio or a pure data transfer place different demands on the transport medium in view of the quality of service. There are four different **service classes** (service categories) for ATM modeled on the specified service categories for B-ISDN in ITU I.362 (cf. Tab. 4.10). These can be distinguished by the following properties:

Table 4.10 ATM service classes for various applications, based on ITU-T

Service class	A	B	C	D
Time reference	yes	yes	no	no
	synchronous	synchronous	asynchronous	asynchronous
Bandwidth	constant	variable	variable	variable
Connection type	connection-oriented	connection-oriented	connection-oriented	connection-less
Service type	1/2	1/2	3/4/5	3/4/5
Example	telephone, MPEG 1	compressed voice MPEG 2	X.25, frame relay	inter-LAN

- **connection-oriented** or **connectionless**
  ATM itself functions connection-oriented. However, there are applications, such

as data transfer, which could be transmitted more efficiently without explicit switched connection. For these services the so-called **inter-LAN-communication**
the service class **D** is provided.

- **constant** and **variable bit rate**
  In addition to isochronous data streams, which demand a constant minimum
  transmission rate, asynchronous data streams are also sent. These are often subject to a high fluctuation in transmission rate regarding the transmission period.

- **time relationship transmitter/receiver**
  Above all, video, and audio transmission require close temporal coupling between the transmitter and receiver. While limited delay is acceptable, it must remain constant (variance). If the variance exceeds a threshold value, the result is
  the so-called **jitter** effect, which can interfere with correct presentation.

In addition to the ITU-T standardized service classes, the ATM Forum has its own
classification (cf. Tab. 4.11). In line with the classification of the ITU-T service
class division, the ATM Forum defined the individual classes qualitatively, without
indicating concrete numerical values. The following specified quantitative **traffic
parameters** in ITU I.356 provide a more exact description of the service requirements:

- **Peak Cell Rate (PCR)**
  Indicates the maximum transmission rate in cells per second needed for a service.

- **Sustained Cell Rate (SCR)**
  Indicates the average transmission rate that a service requires. The necessary
  mean value calculation is carried out over a longer period of time.

- **Minimum Cell Rate (MCR)**
  Indicates the minimum data transmission rate that must be available for a transmitter. This is necessary in order that the service be carried out correctly.

- **Initial Cell Rate (ICR)**
  Indicates the data transmission rate where a transmitter should start its transmission again after a break.

- **Cell Delay Variation Tolerance (CDTR)**
  Indicates the variation of time intervals between two cells following one after the
  other.

- **Burst Tolerance (BT)**
  Indicates the maximum number of cells that may be sent after a burst – i.e., back-
  to-back.

Besides the traffic parameters, the ATM Forum defines a further set of **(Quality of
Service Parameters**). These are negotiated during operation of the ATM network
before connection establishment takes place between two connected end systems:

- **Cell Transfer Delay (CTD)**
  Indicates the average as well as maximum length of time needed by a cell for
  transmission between transmitter and receiver.

Table 4.11 ATM service classes

Service class	Designation	Property
CBR	Constant Bit Rate	corresponds to circuit switching
RT-VBR	Real-time Variable Bit Rate	low delay and jitter, suitable for voice and video transmission with bursts
NRT-VBR	Non Real-time VBR	low delay variation, but high delay
ABR	Available Bit Rate	uses remaining capacity of a path, minimal data transmission for cells is guaranteed
UBR	Unspecified Bit Rate	no service guarantee of any kind
GFR	Guaranteed Frame Rate	guarantees a minimum transmission rate for TCP/IP packets

- **Cell Delay Variation (CDV)**
  Indicates the fluctuation in the transmission period between individual cells.
- **Cell Loss Ratio (CLR)**
  Indicates the number of cells that do not arrive or arrive too late at the receiver, based on the total number of transmitted cells. Cell loss can result from excessively long waiting periods at switching computers in the ATM network, or through transmission errors.
- **Cell Missinsertion Rate (CMR)**
  Indicates the number of cells that arrive at an endpoint other than the predetermined receiver during a specific time interval due to transmission error.
- **Cell Error Rate (CER)**
  Indicates the amount of cells that reach the receiver but are damaged to such an extent due to transmission error to be considered worthless.
- **Severely-Errored Cell Block Ratio (SECBR)**
  Indicates the number of blocks in which a specific amount of cells have been improperly transferred. The number of cells per block is constant and predetermined. SECBR is a measure for so-called burst error that can occur over a transmission line.

The assignment of the service class parameter to the individual service class is seen in table 4.12. If a parameter is specified in a service class, **yes** is entered in the table entry, otherwise **no**. If the application of a parameter makes no sense in a particular service class, the note **n.a.** is indicated. The column **type** shows whether it is a traffic parameter (**traffic**) or a service class parameter (**QoS**).

Mechanism for Overload Control

Just as in every network technology, in ATM precautions must be taken in order to avoid overloading the network. It is necessary that a compliance with traffic guarantees can be ensured every time. In this way different mechanisms are employed

Table 4.12 ATM service classes and service class parameters

Parameter	Service class					Type
	CBR	RT-VBR	NRT-VBR	ABR	UBR	
PCR, CDVT	yes	yes	yes	yes	yes	traffic
SCR, BT	n.a.	yes	yes	n.a.	n.a.	traffic
MCR	n.a.	n.a.	n.a.	yes	n.a.	traffic
CTD	max CDT	max CTD	avg CTD	no	no	QoS
CDV	yes	yes	no	no	no	QoS
CLR	yes	yes	yes	yes	no	QoS

that correspond to the respective properties of a specific service class. The following concepts for overload control are implemented in ATM:

- While the connection is being established the end systems involved require a certain quality of service, which is specified with the help of the traffic parameters. The ATM network verifies whether or not it is capable of fulfilling the desired traffic parameters. If this is the case then the connection is allowed, if not then connection establishment is refused. This process, also known as **Connection Admission Control, CAC**, is carried out on the interface between the end system and the network (UNI). CAC is carried out for all service classes (A-D).

- Following the requirements of the end system and subsequent acceptance by the ATM network a so-called **traffic contract** is entered into.

- After connection establishment has been completed, a **traffic control** ensures that all of the negotiated parameters are kept in the traffic contract and – if necessary – enforced. A CLP bit (CLP=1, cell tagging) is set in cells that have violated the traffic contract. In the event of congestion, cells set with a CLP bit are discarded first. This procedure is implemented in service class B. There is another possibility for discarding cells immediately that are not in compliance with the contract. Responsible for this measure (user parameter control) is, in each case, the first switching computer where the cells arrive after being sent out by the end system. Because classes C and D do not allow cell loss the CLP bit based procedure cannot be implemented here.

- To distribute the incoming data stream as evenly as possible over time, a procedure called **traffic shaping** is used. Peak data rates as well as burst lengths are limited to prevent the occurrence of local overload. Each end system, or the first switching computer, and those following thereafter are involved.

The available transmission capacity for service classes C and D is determined as follows. The end system wishing to transmit sends a specific cell (PTI field=110) for resource management to the receiver. The desired transmission rate is registered therein. Every switching computer that the cell passes by verifies whether or not it can provide the desired transmission capacity. If it is not possible, it can reduce the registered data transmission rate accordingly. The receiver verifies the entry in the resource requirements and transmits the cell correspondingly back to the original transmitter again. In this way, the transmitter can determine by evaluating the cell how high the maximum data transmission rate reserved for it is.

ATM-LAN Emulation

The transition from the existing network architecture within a company to a new LAN technology is often fraught with a great degree of technical complexity and high costs. The existing implementation often needs to be replaced completely – this means starting from scratch. ATM networks, in contrast, can be integrated efficiently into an existing LAN architecture. They offer the possibility of operating existing network applications on an emulation basis via ATM. This technique is known as **ATM-LAN emulation (ATM-LANE)**. From the side of the respective LAN application, the ATM-LAN emulation service behaves exactly like the traditional LAN software. End systems may be connected directly to the ATM network.

With the help of a bridge functionality traditional LANs can be linked directly with ATM. Because of the completely different operating principle of ATM in comparison to Ethernet or token ring networks this can, however, not happen without further provisions being made. The basic difference of both technologies are listed as follows:

1. Conventional LAN technologies such as Ethernet are **connectionless**, while ATM is **connection-oriented**. Data is sent from the traditional LAN systems in the hope that it will reach its receiver intact. As a rule there is no acknowledgment of receipt. Lost packets can first be requested again over mechanisms in protocol layers located on a higher level (e.g., TCP).

2. **Multicast** and **broadcast** can be implemented much more easily based on the shared transmission medium in conventional LAN technologies. As here all of the stations by definition are connected to the same transmission medium, every data packet reaches every station in the net segment in any case. Only a filtering, concerning the receiver address, which is carried out in the respective network adapter of the stations, gives the impression of an exclusive communication relationship. If a broadcast address is given as the receiver address the dispatch of a single packet can be communicated to the entire network segment. The connection-oriented ATM protocol provides a communication path (VP/VC) for every communication pair. Its parameters, such as bandwidth and transmission delay, are negotiated each time in the phase of connection establishment. Every data packet sent over this path reaches exactly only the designated receiver and no other computer within the network. Broadcast information to k receivers therefore demand the establishment of k different communication paths.

3. LAN MAC addresses are based on the serial number of the manufacturer of the network adapter and are independent of the network topology. The ARP protocol (cf. section 7.3.2) used for the assignment of MAC addresses to IP addresses is based on the use of broadcasts. In ATM such an assignment must be undertaken dependent on the network topology with the help of mapping tables within the switching computer in the ATM network.

The standardized LAN emulation from ATM Forum is implemented through four service modules on the ATM Adaption Layer (AAL-5).

- **LAN Emulation Client (LEC)**
  Software that carries out all necessary control operations as well as data transmission over the ATM interface. Applications in protocol layers above have at their disposal in each case a special MAC interface

- **LAN Emulation Server (LES)**
  Software for control of the emulated LANS, includes registration of LECs and assignment of MAC to ATM addresses. LEC must additionally deliver to LES the MAC addresses it represents, the corresponding ATM addresses and, if applicable, routing information. The first thing looked for in the delivery of a data packet is the ATM address of the receiver in the address table of the LES. If it is not available there, it is determined with the help of a broadcast of the BUS server.

- **LAN Emulation Configuration Server (LECS)**
  Software for the administration of simultaneous affiliation of the LEC to different emulated LANS via a special configuration data bank.

- **Broadcast and Unknown Server (BUS)**
  Software for the mediation of all broadcast and multicast data packets from LECs. Included here are data packets with native multicast and broadcast addresses, as well as data packets with MAC addresses for which the LEC has no known ATM address, or it could not be determined by the responsible LES. Also included are so-called explorer data packets of the source routing mechanism that function in determining an optimal route.

The LAN data packets themselves are encapsulated in so-called LAN emulation data packets transmitted over the AAL-5 layer. The operation between the individual software components of the LAN emulation are processed with the help of control connections (control VCC) and data connections (data VCC). Control VCCs couple the LECs with LES and LECS while the communication between BUS and LEC, as well as between individual LECS proceeds in each case over data VCCs. An LEC that registers at the LECS for participation in an emulated LAN, negotiates the corresponding connection parameters (addresses, LAN name, data packet size) over a configuration direct VCC.

Excursus 4: ATM – Cell Switching

The central element in the ATM networks are the **ATM switching computers** (ATM switching units, ATM switches, ATM cross-connects). To establish a high-speed network on the basis of the ATM switch, there must be sufficient capacity to ensure data throughput in the area of giga and tera bit. This can only be achieved if the algorithm necessary for routing is completely implemented in the hardware. Additionally, the data packet size prescribed by ATM is exploited in order to implement the necessary algorithms widely and massively parallel. The switching speed of an ATM switch exceeds the data transmission capacity of the connected stations many times over. Only in this way can all connected station use the target bandwidths to the fullest capacity.

When a data packet reaches one of the input ports of the ATM switch, the channel and path identification (VCI/VPI) of the data packet have to first be evaluated there. Afterwards, the forwarding of the packet to the designated output port of the ATM switch takes place. A differentiation is made between **ATM path mediation** (VP switches, cross-connect) and **ATM channel mediation** (VC switch, ATM switch).

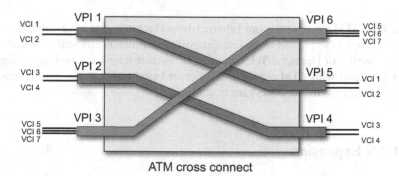

ATM cross connect

In an ATM path mediation all incoming paths are ended – including all channels found in the path – and conducted to another path. Individual ATM channels, however, remain unaffected. In contrast, ATM channel mediation ends inbound paths and channels and conducts them into new paths and channels. If a paths finishes up at an ATM switch then the channel that it belongs to it must also end.

The actual heart of the ATM switching computer is the so-called **switching fabric**, where the forwarding of the individual ATM cells is organized. In facilitating this task, the switching fabric is intended to provide dynamic transmission paths between input and output ports of the ATM switching computer in such a way that there is a minimum of internal and external conflicts. An internal conflict occurs when two ATM cells within a multi-level switching network in the ATM switching computer strive to have the same output port. If such a blockade takes place at the exit of the internal switching network of the ATM switching computer it is then considered an external conflict. The switching fabric itself is built upon single cell switching units – the switching elements. However, the number of input and output ports provided by the switching elements is normally too small to meet the demands of the ATM network. These are interconnected to become greater structures.

The switching elements are made up of so-called **interconnection networks**. These provide the actual transmission paths for the ATM data packets. There are two basic types of interconnection networks:

- **Matrix networks**
  Here all entrances of a switching element are linked with all of its exits through a network made up of transmission paths. The ATM data packets are transmitted parallel over this network matrix (**crossbar**) synchronously to the local clock of the switching element. To avoid data loss through blocking, cache memory must be set up at the input and output ports, or at the intersecting points of the transmission path, which functions as as buffer.

- **Time-division multiplexing networks**
  In this process all of the ATM data packets are either transmitted serially over a shared bus or ring structure (**shared medium switching**), or the ATM data packets are first written in a common memory via an input controller, before it is then read by an output controller (**shared memory switching**).

Further reading:

Kyas, O., Crawford, G.: ATM Networks, Prentice Hall, Uper Saddle River, NJ, USA (2002)

Sounders, S.: The McGraw-Hill High-Speed LAN Handbook, McGraw-Hill, New York NY, USA (1996)

Rather than ATM, more and more Ethernet-based technologies have come into application. They have proven to be less expensive and just as efficient. The Internet DSL connection of Deutsche Telekom as well as their telephone exchanges are no longer equipped with ATM capability. Ethernet-based technologies and IP-based-VPNs (Virtual Private Networks) are now used.

4.4 LAN Expansion

Each of the LAN technologies presented offers a particular combination in terms of minimum bandwidth, distance bridging capabilities, alloted network stations and communication costs. LANS are designed for networks within buildings and have the capability to bridge distances of several hundred meters. But it is often not possible to adhere to these length restrictions. This could be the case, for example, when communication partners are located in different buildings and the distance is too great to be bridged with the LAN technology employed. For such situations techniques must be provided to enable expansion of local networks. It should also be noted that these techniques are not only capable of expanding the wired LANs presented in this chapter, but can also be used for the expansion of wireless LAN technologies. This will be discussed in the next chapter.

4.4.1 Limitations of LAN Technology

Limitations affecting the maximum expansion of a LAN are the result of a variety of factors. All computers connected to the LAN share a single transmission medium. For this reason it must be guaranteed that all participating computers can have access to the transmission medium after a potentially limited waiting period. In order to ensure this, the maximum number of computers connected to a LAN is limited. A further important factor is the signal propagation speed. This is the length of time the shared transmission medium is occupied until news from the transmitting station reaches the receiver. The runtime of the LAN access control method – e.g. CSMA/CD algorithm in Ethernet – depends on the local expansion of each LAN. So that lengthy delays do not occur, the maximum lengths and limitations regarding the number of participants are specified. Furthermore, extent is limited by the power dissipation of the implemented LAN transmission medium. This is because computers connected to the LAN are restricted in terms of their transmission power and there is the occurrence of a signal loss proportional to signal expansion. The maximum length of the shared transmission medium must therefore be limited so that all connected computers on the LAN can still receive sufficiently strong signals.

To increase the range of a LAN, extra hardware components have been developed, for simple signal amplification or **bridges**, **hubs** and **swit-**. In addition to signal amplification by means of switching technology they can also take over tasks such as traffic loss control. There are also **routers** and **gateways**, which can achieve network expansion on higher levels of the TCP/IP protocol stack as independent network computers with their own program logic . Fig. 4.56 shows, on which layer of the TCP/IP reference model the different types of devices can be implemented in network expansion.

5	Application gateway	*Application layer*
4	Transport gateway	*Transport layer*
3	Router	*Internet layer*
2	Bridge, switch	*Link layer*
1	Repeater, hub	*Physical layer*

Devices for network expansion and their application on different layers of the TCP/IP

As previously mentioned, signal damping that occurs with the increasing spread of a signal along the transmission medium is one of the reasons for length limitation in a LAN. With increasing distance from the sender the transmitted signal becomes weaker and weaker. In order that it be received correctly it cannot drop below a certain minimum strength. In order to overcome this limitation, amplifiers (electrical or – are incorporated into the LAN. A conventional repeater is an analog electrical device that strengthens and forwards incoming electrical signals. An optical repeater (also optical modem) accepts the electrical signals of the network, transforming them into suitable optical signals. At the remote station they are transformed back into electrical exit signals. Repeaters work on the lowest layer

of the TCP/IP reference model – the physical layer. They have neither cache memory nor program logic, but rather serve only to amplify the incoming signal. Through
the implementation of repeaters, LAN expansion can be multiplied (cf. Fig. 4.57).
Time restrictions in the algorithms, which regulate access to the shared transmission medium however prevent unlimited expansion of the LAN through repeaters.
By fixing data packet length, the CSMA/CD algorithm in the classic Ethernet-LAN
limits the maximum size of the LAN and only allows the implementation of a maximum of four repeaters and an increase in network length from 500 m to a maximum
of 2,500 m (cf. section 4.3.2).

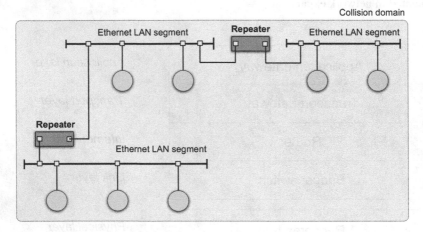

Fig. 4.57 LAN extension via repeaters.

A repeater performs signal amplification solely within the LAN and does not recognize arriving data packets as such or analyze them. This means that in forwarding
it is not capable of distinguishing between complete, valid data packets and other
electrical signal sequences. If a collision occurs on one side of the repeater, the
interference signals caused by the collision are forwarded to the other side of the
repeater in unchanged form. Similarly, interference signals, such as those created
when lightning strikes, are forwarded in amplified form.

4.4.3 Hubs

In its functionality the **hub** is a specialized, expanded form of the network repeater.
It does not only connect two network segments but can connect a multiple number
of network nodes to each other in star form. For this reason the hub is often called
a multiport repeater or repeating hub. Network nodes are connected directly to the
connectors (ports) of the hubs. They are all operated over the same bandwidth (cf.
Fig. 4.58). Arriving signals are conducted further by the hub to all of its ports. It does

not analyze them but, like a repeater, only forwards them in denoised and amplified form. If signals arrive at two or more ports simultaneously, i.e., if more data packets reach the hub at the same time, a collision results.

A star topology is implemented in the physical cabling of a hub in a network. The logical structure corresponds to a bus topology as every transmitted data packet always reaches all of the network nodes connected to the hub. Therefore all of the network nodes connected to a hub are within a shared collision domain. Through the physical star topology in a hub-based network, the degree of reliability of the entire network in view of cable break is raised significantly. This is because instead of the entire transmission medium being paralyzed, only the connection to a network node would be affected. In a hub-based network, cable breaks can be easily localized and remedied.

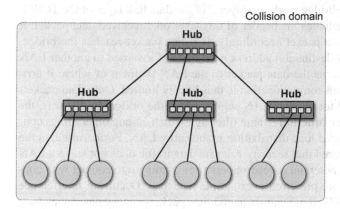

Fig. 4.58 LAN extension over hubs

In order to avoid high signal runtime, hubs should not not be cascaded arbitrarily in a network. Principally, the **5-4-3 rule** applies in the Ethernet network for the application of hubs. A maximum of 5 network segments may be connected together over 4 hubs, whereby active network nodes can only be connected on 3 network segments. With 100 Mbps Ethernet just three network segments can be connected to each other over hubs. Between any two network nodes there can be only a maximum of two hubs. Today, normally **switches** are used instead of hubs. They work on the next higher protocol layer of the TCP/IP reference model and are capable of analyzing data packets and forwarding them to specific destinations.

4.4.4 Bridges

If secondary objectives are also pursued in the expansion of a LAN, it is often preferable to employ so-called **bridges**. Among these secondary objectives are the

- connecting of LANs with different technologies, e.g., Ethernet IEEE 802.3 with token ring IEEE 802.5, or differing bandwidths.

- achieving a balanced load distribution in LAN through the encapsulation of areas that create a high load.

- bridging geographically large distances that would not be possible via repeaters or hubs.

- encapsulation of security-relevant LAN segments so that their data traffic is not forwarded to other segments in the LAN.

In contrast to repeaters and hubs, a bridge always transports complete data packets. A bridge is actually a computer connected to two different LAN segments that monitors its data traffic over two LAN interface adapters. The LAN interfaces of a bridge work in the **promiscuous mode**, which means that all data packets are received and processed. The bridge works on layer 2 of the data link layer of the TCP/IP reference model and evaluates the header of this layer on a received data packet. It can then process the data packet accordingly. If a data packet reaches the bridge, which for reasons of its destination address need not be forwarded to another LAN segment, the bridge returns the data packet to the LAN segment of where it arrived originally. Local data communication is thus locally limited. Only data packets that are really intended for another LAN segment pass the bridge. In this way, the bridge carries out a **filter function** (frame filtering), which among other things contributes to a more balanced load distribution in the entire LAN. Furthermore, in the same way it can be ensured that security-relevant data traffic does not leave a LAN segment. This prevents potential eavesdroppers from coming into contact with confidential data. Bridges are preferred over simple repeaters. Occurring interference signals (e.g., through collisions) and damaged data packets cannot pass by and are therefore not forwarded to another LAN segment. Thus every port of a bridge ends with its own collision domain. This includes the LAN segment connected to this port. Bridges that divide a LAN into different collision domains and simultaneously use the same network technology and transmission speed, only work on the MAC sublayer of the network access layer and are thus known as **MAC bridges**. Bridges appear completely transparent to all computers connected to the LAN. A sender does not know whether a data packet to be sent on its path in the LAN must pass a bridge or not.

Multiprotocol Bridges

To interconnect two LAN segments with different LAN technologies via a bridge, it is necessary that both are capable of understanding the different protocols implemented in each segment. Such a bridge is known as a **multiprotocol bridge**. While in principle every combination of prescribed IEEE 802 LAN technologies should be possible, at the same time there are several difficulties to overcome. Among them are:

- **different packet formats**
  Different LAN technologies use different packet formats. Seen technically, there

is no need for this incompatibility. The responsible manufacturers, however, see
no reason why they should change or adapt a standard they support. Because of
this, every transmission from one LAN technology to another needs additional
processing steps. Data fields have to be reformatted, checksums recalculated and,
moreover, during this process new errors can arise.

- **different bandwidths**
  Different LAN technologies do not work with the same bandwidth. In the transition from a faster LAN to a slower one it is possible that incoming data packets
  may have to be stored temporarily before they are moved on. However, such
  a temporary storage always entails capacity limitations. With a high load data
  packets could be lost. Additionally, the delay resulting from temporary storage
  can cause timing problems in higher protocol layers.

- **different packet lengths**
  The most difficult problem in coupling different LAN technologies involves different packet lengths. The IEEE 802 standard does not make provisions on this
  protocol level for the separation of data packets. Therefore, those that exceed the
  maximum length for a particular LAN technology cannot be divided into two or
  more shorter data packets (fragmentation). If these type of data packets appear
  they are inevitably lost when LAN segments are coupled via bridges.

Transparent Bridges

A primary design objective in the development of bridges was their transparency in
the LAN. Once integrated in a LAN, they work consistently without special considerations having to be taken due to their presence. Let us look at the LAN in
Fig. 4.59. Here, LAN segment A is connected to LAN segment C over bridge B1.
Both of them are connected over bridge B2 with the segments D and E. If a data
packet from segment A, which is addressed to a computer in A, reaches bridge B1,
then bridge B1 can discard this data packet immediately. A data packet originating
from A that is sent to LAN segment E is, in contrast, correctly forwarded by both
bridges.

Decisions about how data packets are to be forwarded in bridges are made with the
help of so-called **hash tables**. In such hash tables every possible destination address
is assigned to a specific port via which it can be reached. But how are the assignment of addresses and outputs established? When a bridge is switched on in a
LAN segment and begins operation its hash tables are still empty. As no destination
address is yet known a so-called **flooding algorithm** is implemented. All incoming
data packets are forwarded by the bridge to all available connections, with the exception of the port where the data packet arrived at. As every incoming data packet
has the address of its sender, the bridge learns with time to which ports the respective destination addresses are assigned. As soon as a destination address is known,
a data packet going to this address is then merely forwarded to the correct line and
not "flooded" over all the others.

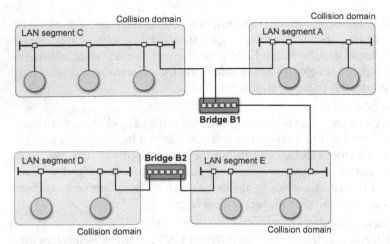

Fig. 4.59 Example of a LAN expansion via bridges.

As a LAN can change dynamically – computers and bridges can be switched off
and put into operation again at other places – appropriate measures have to be taken at the bridges in order to be able to respond to topology changes. To facilitate
this, the hash table does not only contain the address entry with the corresponding
port, but also the time when a data packet was last forwarded to the address of the
bridge in question. The bridge checks its hash tables periodically for entries that
have been inactive for an extended time, i.e., have not received a new data packet
for over a longer predetermined period of time. In this way, hash tables are adjusted
periodically and the bridge can react quickly to changes in the LAN topology.

Spanning Tree Bridges

To increase the reliability of the LAN, several redundant bridges are often connected
parallel between two LAN segments. But this can also have the adverse result of data
packets moving back and forth between the two LAN segments in infinite loops (cf.
Fig. 4.60).
To avoid this problem so-called **spanning tree bridges** are employed. These type of
bridges communicate with each other and generate from the existing LAN topology
a **spanning tree**. This loop-free graph functions as a "street map" for the LAN,
serving as a foundation for the forwarding of data packets. One of the existing LAN
bridges is deemed the root of the tree. This is done by all bridges sending their
world-wide unique manufacturer's-assigned serial number to all other lines via all
the other bridges. The bridge with the lowest serial number takes the role of the
root. Next, a tree is constructed that takes the shortest path from the root to all other
bridges. This tree establishes the LAN's street map . It specifies how the root – and
therefore all other LAN segments – can be reached from every other LAN segment.
It is updated during operation and in this way adapts to topology changes in the
LANs automatically. The installation of spanning tree bridges in the network has

The Occurrence of Infinite Loops in LAN Parallel Switched Bridges

Let us consider a data packet P from LAN 1 with a destination address that is unknown to both B1 and B2 bridges, located between the the two LAN segments LAN 1 and LAN 2. Both bridges proceed according to the flooding algorithm and transmit P in this way on to LAN 2 (a).

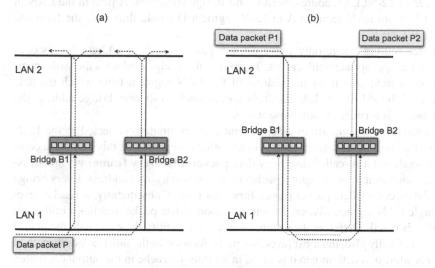

The situation is different if two data packets P1 and P2 with unknown destination addresses reach both B1 and B2 bridges (b). P1 from LAN 2 reaches bridge B1 with a destination unknown to B1. P1 is copied to LAN 1. At the same time at bridge B2 a data packet P2 arrives with an unknown destination address from LAN 2. It is also forwarded to LAN 1. Likewise P1 reaches LAN 1 via the bridge B2 and is forwarded from there to LAN 2. In this way infinite loops may occur inadvertently.

Further literature:

Tanenbaum, A.S.: Computer Networks, 4th ed., Prentice-Hall, Inc., Upper Saddle River, NJ, USA (2002)

Fig. 4.60 Infinite loops in LAN parallel switched bridges.

proven to be relatively simple. After establishing a physical cable connection, they only have to be switched on and are immediately operational. They utilize only a fraction of the available connections in the LAN topology as only the generated spanning tree is used.

Source Routing Bridges

A second alternative to LAN expansion by means of a bridge was worked out by the IEEE 802 committee. These are the so-called **source routing bridges**. Principally, this procedure is based on the presumption that a computer wishing to transmit data packets knows itself if the receiver is located in the same LAN segment or not. It makes this clear by including the destination address in the data packet of the header.

Additionally, the exact **path** to the destination computer is also given in the header of the data packet.

Every LAN segment is marked with a clear 12-bit long address and every bridge within the LAN receives a 4-bit long address. A path is therefore made up of sequences of 12-bit LAN addresses and 4-bit bridge addresses. A path in the LAN in Fig. 4.59 from LAN segment A to LAN segment D would thus have the form (A, B1, E, B2, D).

A source routing bridge only forwards data packets to those address details containing the appropriate nonlocal bit. To do this the bridge evaluates the path specification, searching in it for the address of the LAN segment from which the data packet originated. If this LAN address corresponds to its own bridge address the data packet is forwarded, otherwise it is not.

The sourcing route algorithm assumes that each computer connected to the LAN knows the most advantageous path to every other computer. For this purpose a computer sends out a so-called discovery data packet (**discovery frame**) via broadcasting, as soon as it must transmit a packet to an unknown target address. Every bridge that the discovery data packet passes forwards it on. Consequently, it reaches every single LAN segment. When the return response data packet reaches a bridge, its identity is noted. Therefore the original sender can identify the exact path the packet took, and finally also, the most favorable path. As soon as the most favorable path to a target address is determined it is saved in an internal cache in the output computer. In this way, the discovery procedure does not need to be repeated.

The disadvantage of this procedure is the possibility of frame explosions occurring due to the duplication of discovery packets. As soon as the transitions between the individual LAN segments are made redundant with multiple parallel bridges, each of these bridges sends a copy of the original discovery data packet to the neighboring segment. If a LAN consists of k segments each with b parallel bridges in between, the number of discovery data packets can multiply to b^{k-1}. A similar effect also occurs in spanning tree bridges. If a data packet with an unknown target address reaches a bridge, then it is flooded to all connected LAN segments. But as only the spanning tree and not the entire LAN is used for forwarding, the number of duplications of the output packet is only in linear proportion to the size of the network.

While spanning tree bridge topology change and errors can be determined automatically and in a simple way via their communication with each other, this procedure is more complex in the case of source routing error. If a bridge goes out, it is only registered by a sender who wishes to use this connection path that there will be a waiting time for a transmission acknowledgement. Finally, a time-out finally occurs, but the sender cannot distinguish if the target address is no longer available or if there was an error on the transmission line. A discovery data packet is sent out to investigate the problem. Especially if a bridge is affected that has a very high level of demand, a critical situation can arise as a large number of discovery data packet are then initiated. In table 4.13 the advantages and disadvantage of both procedures are shown.

Table 4.13 Spanning tree bridge and source routing bridge

	Spanning tree bridge	Source routing bridge
Orientation	Connectionless	Connection-oriented
Transparency	Fully transparent	Not transparent
Configuration	Automatic	Manual
Routing	Suboptimal	Optimal
Navigation	Backtracking	Discovery data packets
Errors	Handled by the bridges	Handled by the hosts

Remote Bridges

Bridges are commonly used to connect geographically distant LAN segments into one large LAN. In many case this is preferable to an inter-LAN connection as the resulting complete system is handled as a single LAN. Such a coupling can be achieved by installing bridges at the end points of the LAN segments to be connected to each other They are connected via a point-to-point connection, e.g., via a telephone line, a wireless route or via a satellite connection. The point-to-point connection between the two LAN segments is thereby simply seen as an additional LAN segment, which does not contain its own host. The network configuration in Fig.4.61 in this sense consists of three LAN segments that are connected to each other over two bridges. The segment with the point-to-point connection does not contain its own host.

A standardized serial protocol of the network access layer can be implemented as a point-to-point connection, such as PPP (Point-to-Point Protocol). It encapsulates the complete MAC data packets and transmits them along the wide area network nodes. Here, the LANS connected by remote bridging should use the same LAN technology. A further possibility exists in removing the headers and trailers of the MAC data packets to be sent to the bridges and encapsulating just the pure payload in a PPP data packet.

The receiving bridge can complete the accompanying control and packet information of the received MAC data packet and create a new header and trailer. If, however, checksums are also created at the receiving bridge then a new potential source of error exists. This is because the checksums calculated at the receiving bridge do not necessarily have to correspond to the original checksums calculated at the sender if in wide area data transmission, or through a memory error of the involved bridges bit errors occur. Extra attention has to be paid given the fact that the implemented point-to-point connection often works considerably slower than the LAN segments connected to it. Special filter mechanisms are employed to prevent the unnecessary transmission of data packets to the distant LAN segment. A cache is also used to buffer data packets in case of congestion.

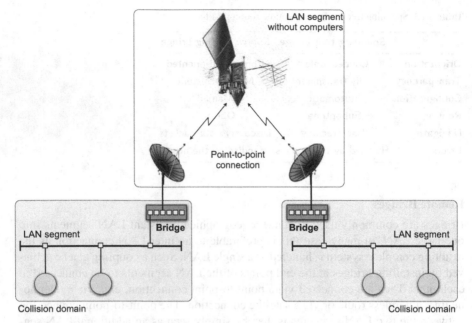

Fig. 4.61 Remote bridges for coupling geographically distant LAN segments.

4.4.5 Switches

Similar to a hub, a **switch** connects multiple computers, or multiple LAN segments
to each other. Both devices work in different ways. A hub works as a shared transmission medium for the computers connected to it. At any given time only two hosts
can communicate with each other over the hub. A switch, on the other hand, functions as a complete LAN that in and of itself connects individual segments together
each with a computer and a bridge. While the bandwidth of the hub is limited by the
maximum data transfer rate of the connected computers, the switch must provide
sufficient bandwidth for parallel data transmission of all connected computers. Just
as bridges, switches analyze the header of received data packets and are responsible
for filtering, traffic control and forwarding functions based on the evaluated MAC
addresses of the received data packets.

In contrast to a bridge, in many cases a switch administers at each of its ports not
to single LAN segments with multiple computers, but to individual computers that
are directly connected to the switch. This means that a switch must pass on every
received data packet, while a bridge discards data packets that are intended for a receiver who is from the same collision domain as the sender. For this reason switches
normally have at their disposal a considerably larger number of ports than bridges.
In addition, at every port there must be a related buffer storage provided for arriving data packages. This is to avoid data loss in high processing loads (**store-and-
forward switch**). There also exist so-called **cut-through switches**. These already

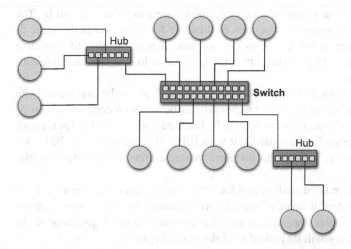

Fig. 4.62 LAN expansion with a switch as central interface between directly connected computers and two hubs, which link their own LAN segments together via the switch

begin forwarding the data packet as soon as the header has arrived at the switch and been analyzed – meaning before the entire data packet has been received. Every port on a switch has its own collision domain. Therefore, interference caused by collisions cannot take place at the switch itself. Many switches also work in the full duplex modus. Via the same port they are able to both send and receive data packets simultaneously.

A switch normally does not need to be configured manually. When a switch receives a data packet after being activated for the first time it saves the MAC address of the transmitter and the related port in the source address table (SAT). On finding the destination address in the SAT the data packet is forwarded to the port saved for it. The receiver is then in the LAN segment connected to the switch through this port. If receiving and destination segments are identical then the data packet must not be forwarded by the switch as communication can be facilitated in the segment without the switch. But if the destination address is not found in the SAT the data packet must be forwarded to all other ports (MAC flooding). Unlike hubs, virtually as many switches as desired can be connected to each other. The upper limit is not determined by the maximum cable length, but depends on the size of the SAT. If multiple switches come into application, the switch with the smallest SAT limits the maximum number of nodes in the LAN. Switches today can easily administer several thousand address entries. If the maximum number of hosts is exceeded, then data packets with destination addresses that can no longer be administered in the SAT are always forwarded to all ports. The network performance begins to sink drastically.

Similar to a hub, a star-shaped network topology is implemented with the application of a switch. This enables the connected computers a dedicated access to the transmission medium. The connected computers do not need to distribute themselves over the transmission medium, rather communication in pairs is possible. The switch activates the necessary connections exclusively for them. Yet at the same time the switch becomes a bottleneck in the network. It must be able to provide a higher bandwidth capacity than the connected computers in order to facilitate efficient communication.

If multiple switches with as high as possible bandwidth are to be connected with each other or established as high-speed connections between switches and servers, multiple connections (currently up to 8) may be bundled with each other (port bundling). This port bundling was standardized with IEEE 802.1AX-2008 in 2009. Yet the interconnection of switches from different manufacturers still remains problematic.

Switches are also offered that analyze and forward data packets on the network layer of the next higher protocol layer of the TCP/IP reference model. These switches with their so-called router capabilities are also known as level 3 switches of the multi-level switch. They will be looked at in detail in chapter 8.

4.4.6 Virtual LANs

If one looks at the development of local networks from a historical perspective, the initial focus is on the shared used of a transmission medium for reasons of cost and efficiency. With 10BASE-2 or 10BASE-5 Ethernet all computers in LAN did in fact share the same data bus, which often stretched through an entire building. While all of the connected computers were geographically close to each other, this neighborhood often had no relation whatsoever to the structure of the organization that operated the LAN. Different departments and offices used the same LAN to transport security-related information, which was not intended for all network users. Merely with a skillful configuration of the Ethernet interface (promiscuous mode) a user could read the entire network data traffic in the LAN. Technical advances led to the development of switched Ethernet technologies. Through the used of hubs and switches this allowed a separation of what had previously been a shared LAN in a simple way. The initial physical separation of individual LAN segments guaranteed the necessary security for the transport of sensitive data. At the same time it facilitated a load-based separation of the LANs. Areas with a high data transfer load could be set apart from less utilized areas so as to enable communication time with less delay. Furthermore, there was also a need to keep the LAN broadcast areas as small as possible. This was done in order not to burden the network unnecessarily with administrative broadcast information. Thus factors of security, performance and flexibility spoke in favor of a division of the company network. Because of prevalent changes affecting company organization this flexibility is necessary.

In the case of physical divisions in a company network, the system administration responsible for network technology is constantly occupied with new cable connections and the relocation of network devices. Thus the idea arose to enable a configuration and division of the existing LANs on a logical level solely through the implementation of the appropriate software. This led to the development of so-called **virtual LANs (VLANs)**. A VLAN could be put into practice with the implementation of special VLAN capable switches. There are several possibilities of how the assignment of a subnetwork to a VLAN can be carried out:

- **static assignment** via a direct port assignment on the VLAN capable switch. Port-based VLANS are the oldest form of VLANs. The VLAN capable switch with its ports is segmented here in several logical switches. Futhermore, port-based VLANs can be expanded via multiple switches. One port may only be assigned to a single VLAN.

- **packet marking** (so-called **tags**) on the data packets sent out, evaluated in the switch. VLAN specific information concerning the data packet is added through the tags. This form of the VLAN was standardized as **IEEE 802.1q** in 1998. A new field was added to the Ethernet data packet header for a VLAN tag. This addition consists of a m 2 byte long field for VLAN protocol identification that always contains the value 0x8100 and precedes the length field formerly located there. As the value is greater than 1,500 and therefore exceeds the old length limitations of the Ethernet data packet, it is automatically interpreted by the Ethernet hardware as a type field and not as a length field. After that a further 2-byte long field follows that is separated into a 12 bit long VLAN mark to serve as identification for the respective VLAN, a 3-bit long priority field to distinguish between time critical data traffic and a 1-bit long CFI field (Canonical Format Indicator). The CFI field does not have a relation to the VLAN, but indicates if a IEEE 802.5 data packet is transported encapsulated via the Ethernet LAN. The upper limit for the length of a Ethernet data packet was thereby raised to 1,522 bytes (cf. Fig. 4.63).

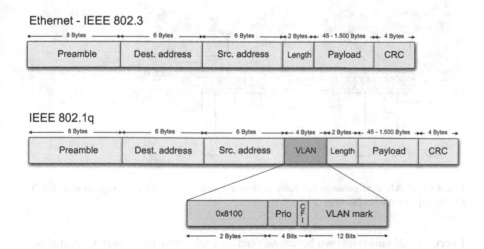

Fig. 4.63 IEEE 802.3 Ethernet data format and VLAN IEEE 802.1q data format.

- **dynamic assignment** via MAC addresses, IP addresses, or even via the use of
  TCP and UDP ports on higher abstraction layers of the TCP/IP reference model.
  This type of assignment makes it possible that a (mobile) terminal always belongs to a specific VLAN, independent of the network interface over which it is
  connected to the VLAN.

Fig. 4.64 shows two different VLAN configurations. In the upper half a separation
into two different VLANs V_1 und V_2 is shown via static port assignment. In this
case both switches A and B, respectively, must be connected over two separate lines
to each other. The lower half shows in principle the same VLAN configuration over
IEEE 802.1q. Here the switches A and B must be connected to each other only over
a single line.

(a) VLAN configuration using static port assignment

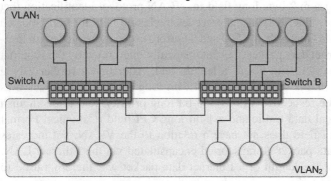

(b) VLAN configuration using IEEE 802.1q

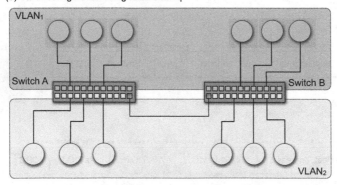

Fig. 4.64 VLAN configuration for static port assignment (a) vs. VLAN configuration for IE-
EE 802.1q data packet marking (b)

Every VLAN builds its own broadcast domain analog to a physically separated network segment. The transparent communication of the data traffic between the different VLANs requires the implementation of a **router**. This functionality is not
provided within the framework of the data link layer of the TCP/IP reference mo-

del, but rather on the next higher level of the Internet layer. Switches having this ability are therefore also called **Level-3 switches**.

4.5 Glossary

Anycast: Addressing variation similar to multicast, focuses on a group of computers that are distributed over several networks for reasons of burden sharing, thereby enabling a higher availability of the services at the anycast address.

Backbone: A backbone serves as the main connector of networks that together have an especially high capacity and bandwidth. The backbone acts as the connecting point for networks to the Internet. Normally these have a lower capacity and share the backbone resources with others connected networks.

Bit length: The length filled up on the transmission medium of a signal which carries the information content of one bit. The bit length is calculated from the quotient from transmission speed (in copper cable approximately 0,8c=240.000 km/s) and bandwidth of the signal. Therefore the result in e.g., 100-Base-T Ethernet a bit length of approx. 240,000.000 mps/100,000.000 bps = 2.4 m/bit.

Bit stuffing (byte stuffing):   The adding of one of more bites (bytes) to the data transmission information in order to transmit signal information to the receiver. The added data does not contain user information but rather fills data packets to a fixed prescribed length or synchronization.

Bridge: Network intermediate system that connects two network segments to each other on the data link layer (layer 2) of the TCP/IP reference model. Besides signal amplification, bridges can carry out filtering tasks and flow control in the network.

Broadcast: A broadcast transmission corresponds to a simultaneous transmission from one point to all network stations. Classic broadcast implementation is radio and television.

Bus topology: In a bus topology all connected computers use a **shared linear medium**, the bus, which is closed on both ends with a terminator. At any given time only one station can transmit on the bus, in contrast to **dedicated media**, which is exclusively at the disposal of only one station.

Carrier sense: Network devices wishing to send information first monitor the transmission medium (**carrier sensing**). They then receive access to the transmission medium if the transmission channel is free. If it is not free and a collision results there are several strategies that can be implemented in order to reduce the probability of a new collision.

CSMA (Carrier Sense Multiple Access):   Access procedure by which a network device monitors a transmission channel shared with other network devices (**carrier sensing**), only carrying out transmission if the channel is free.

Circuit switching: Method of information exchange via a network in which at the beginning of the exchange an exclusive, fixed connection is established between the two communicating terminals and remains for the duration of the communication. Analog telephone networks, for example, function according to this principle.

Collision domain: Term for an area of the network over which collisions of data packets can occur.

Computer network (network) :   Offers the infra-structure for data exchange to those autonomous computer systems connected to the network, each having its own memory, periphery, and computing capability . As all stations are linked to each other, the computer network offers each station the possibility to enter into a connection with every other network station.

Diffusion network (broadcast network):   In a diffusion network the signal of a transmitter is received by all computers connected in the network with consideration to the respective delay time. Each receiver must determine itself if the information is intended for it and whether or not to process it.

Flow control: A method for guaranteeing a balanced and ideally continuous data transmission between terminals that do not work synchronously. The flow control intervenes to regulate the sending sequence in the terminals and curbs transmission power to prevent potential data loss if congestion occurs along the way to the receiver.

Full-duplex: Communication variation in which every station can simultaneously send as well as receive . This form of communication is typical for a network as during transmission every station must be able to receive, in order to e.g., end the transmission in case an error occurs.

Half-duplex: Communication variation by which every station can transmit and receiver, however both never at the same time.

Hub (concentrator, multiport repeater):   A central coupling element in computer network systems that connects network-capable devices with each other via a star topology. A hub works like a repeater and conducts a signal amplification that is forwarded to all connectors on the hub in the form of a broadcast

IEEE: The Institute of Electrical and Electronical Engineering, called "I-triple-E," is an international organization of engineers based in the USA. Currently, IEEE membership numbers approximately 350,000 engineers from around 150 different countries. The main tasks of the IEEE are focused on the development, discussion, ratification and publication of standards in the area of networks. The working group **IEEE 802** is responsible for the standardization of the LAN technologies.

Logical Link Control (LLC):   Upper sublayer of the data link layer in the TCP/IP reference model. It serves the mediation between the MAC sublayer below and the Internet layer. The LLC sub layer is independent of the network technology below it and takes over tasks such as flow control, link management, error detection and correction, fragmentation, as well as data packet synchronization.

Manchester encoding: Simple encoding of binary signals based on voltage change. A binary 1 is encoded through the change low→high, a binary 0 encoded through the change high→low.

Medium Access Control (MAC):   Lower sublayer in the data link layer in the TCP/IP reference model, by which access to a shared transmission medium is regulated. It forms the interface between the physical layer and the LLC sublayer.

Message switching: A method of network communication by which the individual exchange networks save the complete message content in intermediate storage, before it is forwarded. The sender must only know the path to the next network exchange, which then forwards the message after receiving it in the same way to the next network exchange.

Multicast: A multicast transmission corresponds to a broadcast transmission made to a restricted number of stations. It involves the simultaneous transmission from one place to a specific group of network stations.

Multiple access: If the terminals connected to a network share a communication channel it is called multiple access.

Network segment: A group of network devices that uses a common physical transmission medium (not to be confused with data packet segmentation).

Packet switching   The primary communication method in digital networks. The information is divided into data packet of a fixed size. The packets are then sent individually and independently by the transmitter via potentially available exchange networks to the receiver. A distinction is made between **connection-oriented** and **connectionless** packet switching networks (**datagram network**). In connection-oriented packet switching networks, a network connection is established over fixed exchange networks before the start of the actual data transmission. In connectionless packet switching networks, in contrast, no fixed connection path is preselected.

Ports: General term for the inputs and outputs of the intermediate systems/switching systems in the network (not to be confused with TCP ports).

Remote data transmission: If the network systems where the data transmission takes place are more than a kilometer apart one speaks of remote data transmission. This boundary is, however, not fixed. There are considerable differences between procedures implemented for data remote transmission and those for systems not as far away from each other.

Repeater: Intermediate system that enables the amplification and interconnectivity of different network segments. Absolutely transparent, repeaters pass on the received data in amplified form.

Round-trip time: Round-trip time is the response time of a complete network. It is the time period required for a signal to make its way from the source across the network to a receiver, and then for the receiver to send an answer back over the network to the sender.

Simplex: Communication variation in which the information flow is only possible in one direction, namely from transmitter to receiver, e.g. in cable television.

Switch (switching hub):   Central network intermediate system via the connection between linked computers and networks that can be switched at full speed from 10 Mbps, to 100 Mbps, 1 Gbps or more. Concurrent accesses are processed sequentially.

Terminator (terminating resistor):   Used for signal damping of the respectively used transmission medium by which terminating resistors are affixed to the cable end. The resulting voltage loss prevents reflections from occurring on the cable ends.

Throughput: A measure for the performance capability of a communication system. The processed or transmitted information/ data within a specific time period is measured. The throughput is calculated from the quotient of error-free data bits and the sum of all transmitted bits in the context of a fixed time period. The measurement unit is bps (bits per second) or data packet(s).

Time slot: Division of the time axis in fixed sections of the same length. The resulting time intervals are called time slots.

Topology: The topology of a computer network is understood as the geometric form of the distribution of individual computer nodes within the network. Widely implemented computer network topologies are the: **bus topology, ring topology** und **star topology**.

Transceiver: An implemented hardware in Ethernet LAN for sending and receiving data packets as well as for monitoring network traffic. The term derives from (**Transmitter**) and (**receiver**). The transceiver is either located directly on the bus pickup or on the connected computer.

Chapter 5
Network Access Layer (2): Wireless Mobile LAN Technologies

> "Such a crowd I'd like to see,
> stand on free soil among a people free."
> – Johann Wolfgang von Goethe, Faust II

The days when access to the Internet was only possible from the home or office via a wired computer are long gone. For several years now, mobility has been the primary focus, whether it be using a laptop to access the company network while on the go or reading emails on a cell phone. This unlimited availability has been enabled by mobile communication technologies that make the normal LAN accessible anywhere. And all this without having to worry about bothersome cabling management. The wireless LAN (or WLAN) provides us today with a performance comparable to wired LAN technology. But this new found freedom is not without its darker side. With wired network technologies, unauthorized intruders and attackers had been forced to first gain physical access to the foreign network before intrusion could be carried out. The WLAN of today does away with fixed structural borders and can be received barrier-free by everyone within its transmission radius. This has made security technologies and encryption techniques inseparable with wireless network technologies. The present chapter introduces the foundations of the WLAN technology and will also look at the necessary security standards and encryption methods that allow the secure implementation of wireless technology. In addition to WLANs, close range networks, so-called Personal Area Networks (PANs), are gaining increasing importance. These have the capability of linking devices autonomously and wirelessly within the radius of only a few meters. Prominent examples of PAN technologies – Bluetooth and ZigBee – will be examined in detail.

5.1 Fundamentals of Wireless and Mobile Network Technologies

While wired LANs have dominated the network market for many years, today wireless, radio-based networks are becoming more and more widespread. Omnipresent access to the Internet is possible today from nearly everywhere, completely independent of where we are at the moment, be it on a university campus, at a public

C. Meinel and H. Sack, *Internetworking,* X.media.publishing,
DOI: 10.1007/978-3-642-35392-5_5, © Springer-Verlag Berlin Heidelberg 2013

place, in a coffee shop – or even while using public transportation. This accessibility has changed many facets of our daily life.

What had in the past been a problem due to cost factors or technical barriers presented by a wireless network, has been made possible through technological progress. This progress has gone hand in hand with the mass production of necessary components. Historically, the development of wireless computer networks goes back to 1971. That was the year Norman Abramson and his colleagues at the University of Hawaii succeeded in linking the Hawaiian Islands via a radio network called **ALOHAnet**.

The original ALOHAnet was constructed as a bidirectional star topology. Seven individual computers on four different islands were connected with the central computer on the island of Oahu, at a data transmission rate of 9,600 bps, without the employment of any cable-based transmission medium. The development of a radio-based network had been prompted by the unreliable and costly telephone connection between the Hawaiian islands. This led to Abramson's design of a wireless network based on a simple multiple access protocol (pure Aloha, slotted Aloha, see section 4.3.2). From this would later emerge the widely propagated Ethernet protocol. In 1972 the ALOHAnet was also connected to the ARPANET – first via radio and then starting in 1973 via satellite – from then on Hawaii was online.

In terms of topology, star topologies or mesh topologies are implemented in a radio-based LAN, or WLAN (wireless LAN). In a **mesh topology** (mesh network, cellular network) (cf. section 4.3.1) there exists at least two nodes with more than two, i.e., redundant, connections between these nodes. In a true mesh topology, there is a connection from each node to every other one. It is identical to a complete point-to-point network and therefore contrasts to a partial mesh topology. A true mesh topology normally remains reserved for a backbone network to guarantee a high level of reliability. Often other partial mesh topologies are linked with such backbone networks.

The **star topology** is the most widespread form of wireless LAN today. A central basis station, a so-called **access point** (AP), is responsible for the correct forwarding of data packets sent by peripheral computers. The access point is able to connect the radio network as a **bridge** with other wired networks and with the Internet. In this way it ensures an expansive network for the peripheral computers connected over radio (cf. Fig. 5.1).

In WLAN communication a distinction is made between two different working modes or types of network topologies. IEEE 802.11 defines simple cellular topologies that are independent and cover a small area as well as more complex structures that can achieve higher ranges over intermediate systems. The basic structure is always a single radio cell (**Basic Service Set**, BSS), determined by a network node radio-lit area.

- **Ad Hoc Mode**
  An ad hoc network (*ad hoc = [lat.] "made for this certain moment"*) is a WLAN in which two or more subscribers are connected to a meshed network. These types of WLANs do not need a fixed infrastructure or dedicated access points. Data is forwarded from subscriber to subscriber until it reaches the receiver. In

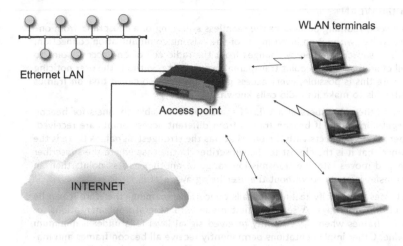

Fig. 5.1 Linking a star-shaped radio network with other networks.

this way, load is distributed more advantageously than in WLANs with a central access point. Special routing procedures ensure that the networks are continually adapted to changes in topology. Ad hoc networks are designated as **independent basic service sets** (IBSS). The range of an ad hoc network inside of buildings reaches approx. 30–50 m and in outside areas approx. 300 m.

- **Infrastructure Mode**

  An infrastructure network is a WLAN where communication of the individual terminals is enabled over a central hub (access point). In this way, it is possible to connect multiple radio cells (**basic service sets**, BSS) with each other to achieve greater range and surface coverage. The designation here is **extended service set** (ESS).

  The clients must each register with their MAC addresses at the access point. The access point can also be a mediator in another (also wireless) network (distribution system, DS).

  With the implementation of multiple access points, the entire area over which network access should be made possible is covered with single overlapping cells. In IEEE 802.11, overlapping cells can be operated without an interference problem as these work physically on different channels. Within buildings the expansion of a radio cell can reach up to approx. 100m through access points. In such a situation the access point represents the center of the BSS.

  The subscribers can move from radio cell to radio cell within an infrastructure network. Contact can be made with the respective access point without the ongoing data exchange being interrupted. This automatic radio cell exchange is known as **roaming** . It commences as soon as a subscriber moves out of the range of one radio cell into the range of another radio cell (cf. also Fig. 5.2). Free mobility through space with consistent network connectivity is ensured.

Roaming in the WLAN

In a radio network, **roaming** designates the seamless switching of a subscriber from one radio cell to another, without an interruption of the existing communication connection. In the WLAN, a subscriber thereby changes from the radio cell of one access point to the radio cell of another access point transparently – meaning, without the user noticing it. To make sure this is possible, every access point transmits so-called **beacon frames** in regular intervals to make its radio cells known.

From their side, the subscribers in a WLAN search all accessible channels for beacon frames at regular intervals. If beacon frames from different access points are received, the subscriber always connects with the one that has the strongest signal, as there is the greatest chance that it is the closest to the subscriber. In the case where the subscriber is in motion and moves into the transmission range of another access point, then an automatic transfer is prepared without the user being aware of it.

IEEE 802.11 does not specify technical details concerning roaming, therefore it can be carried out in different ways. A search is first made with several implementations for further beacon frames when the currently received signal level falls under a minimum threshold value. Other implementations permanently receive all beacon frames and manage all of the located access points parallel so that, if necessary, the implementation time for the roaming can be minimized. In addition, some access points do not communicate either the subscriber or the number of currently connected subscribers, which can serve as a decision criterion for the initiation of a roaming operation. In a similar way, potentially different data transmission rates in the respective radio cell can be taken into consideration in decisions made about roaming.

IEEE 802.11f defines the **Inter Access Point Protocol** (IAPP) for the communication between individual access points. In this way, one access point can inform a neighboring access point separately about the transfer of a subscriber between the radio cells. In the case where the old access point still has data packets ready for the switched subscriber, these can be forwarded to the adjacent access point and from there to the subscriber.

Further reading:

Roshan, P., Leary, J.: 802.11 Wireless LAN Fundamentals: A Practical Guide to Understanding, Designing, and Operating 802.11 WLANs, Cisco Press (2004)

Fig. 5.2 Roaming in the WLAN.

Besides wireless LANs, radio-based technologies come into implementation in the wide area network (MAN and WAN) as well as in the local area (Personal Area Network, PAN). For the area of the MANs and WANs, the WiMax (Wireless Interoperability for Microwave Access) was developed as standard IEEE 802.16. The first stage provides for the coverage of buildings over external antennas. In the second stage internal as well as external antennas are included. While the third stage provides for the connection of mobile terminals. Since 1999 the Bluetooth Standard (IEEE 802.15.1) has been available for the local area as well as the ZigBee Standard (IEEE 802.15.4), for the networking of intelligent sensors.

5.2 Wireless LAN (WLAN) – IEEE 802.11

For WLANs to gain widespread acceptance an industry standard was necessary, just as in the case of the network standards we have looked at so far, e.g., Ethernet or ATM. The subgroup of the Institute of Electrical and Electronics Engineers responsible for standardization in the LAN area, IEEE 802, therefore launched the associated chapter IEEE 802.11 WLAN. Already in 1997 it ratified a first standard for wireless LANs. The original standard functioned in the radio frequency range (RF) 2.4 GHz, and facilitated a data transmission rate of 1–2 Mbps. Both subsequent standards, IEEE 802.11a and IEEE 802.11b work in the range of 5.8 GHz or 2.4 GHz and reach a transmission rate of between 5 Mbps and up to 54 Mbps. 802.11b has a transmission range of approx. 50 meters. For redundancy protection and to avoid potential transmission error, the rate of data transmission in practical application is set at approx. 50–70% of what would theoretically be possible. Table 5.1 shows individual substandards and subgroups that belong to IEEE 802.11.

Table 5.1 IEEE 802.11 and its most important subgroups.

802.11 1–2 Mbps today obsolete, in 2.4 GHz frequency band (1997)

802.11a 6–54 Mbps in 5 GHz frequency band (1999)

802.11b 5.5–11 Mbps in 2.4 GHz Frequency band (1999)

802.11d Adaptation to regulatory requirements of different countries

802.11e Support of Quality of Service

802.11f Interoperability between base stations (2003)

802.11g 6–54 Mbps in 2.4 GHz frequency range (2003)

802.11h Transmission strength management and range adaptation
in 5 GHz frequency range

802.11i Improvement of security and authentication (2004)

802.11j 4.9–5 GHz expansion for operation in Japan

802.11n Up to 600 Mbps (2009)

802.11p 27 Mbps, in the 5.9 GHz frequency range
Wireless Access for the Vehicular Environment (WAVE)

802.11r Improvement of roaming between access points

802.11s Wireless mesh networks and mesh routing

802.11u Interworking with non-802 networks

802.11v Wireless network management

5.2.1 IEEE 802.11 – Physical Layer

On the lowest protocol layer of the IEEE 802.11 standard, **802.11-PHY**, it is necessary to solve the problem occurring on the frequency band allocated for data transmission. With many different computers wishing to communicate with each other, their geographic locations may overlap. But how can different communication partners in one (or more overlapping) radio network(s) be distinguished from one another? To solve this problem, different modulations procedures for data transmission are specified in the physical layer (a detailed representation is given in section 3.2.5):

- Frequency Hopping Spread Spectrum (FHSS),
- Direct Sequence Spread Spectrum (DSSS),
- Orthogonal Frequency Division Multiplexing (OFDM) and
- Packet Binary Convolutional Coding (PBCC).

FHSS divides the available frequency band into individual channels. It uses a narrow band carrier wave that permanently changes its frequency randomly, based on the so-called Gaussian Frequency Shift Keying method (GFSK). A certain protection against eavesdropping attempts is thereby made possible. An unauthorized third party is not able to predict which frequency will be switched to next and is therefore not able to receive the complete signal. This method makes it possible to simultaneously use different networks within the same physical space over FHSS, whereby the individual networks use different frequency signatures determined via GFSK.

DSSS works completely differently. It combines the data stream with a digital code of a higher speed, i.e., every data bit is mapped as a random bit sequence – the so-called chip code – that is only known to the respective transmitter and receiver. 1 and 0 are each represented by the chipping code and its inversion and in this way receive a specific bit signature by which they can be identified. With a corresponding synchronization, this type of frequency modulation even guarantees its own error correction and is therefore more robust in relation to random or intentional disturbances.

Orthogonal Frequency-Division Multiplexing (**OFMD**) is implemented to ensure the highest efficiency and the most robust data transmission possible. Here, data is transmitted parallel on different narrow band and independent channels (subcarriers), and, in relation to a comparable serial transmission, at a lower data rate. Disturbances normally only affect a narrow area of the available frequency band, so that only individual subcarriers are influenced, and therefore a higher interference resistance can be reached. OFDM additionally optimizes the use of the available bandwidth through mutually overlapping subcarriers. In order that these do not influence each other, the frequencies of the subcarrier are chosen in such a way to be orthogonally positioned to each other

In addition to FHSS, DSSS, and OFDM, in the IEEE 802.11 fixed standards for wireless LANs, another encoding procedure comes into application. This is Packet Binary Convolutional Coding (**PBBC**). Its target is to achieve higher security against disturbances, and thereby also reach a higher data transmission rate. For every 2 bits of transmitted data, a 3-bit long data sequence is transferred via a convolutional code, which with the help of 8-PSK (Phase Shift Keying) is modulated on the signal carrier. Table 5.2 summarizes the different implementations of the 802.11-PHY.

Table 5.2 Summary of the 802.11-PHY implementations.

Standard	Transmission procedure	Frequency band	Data rate
802.11	FHSS	2.4 GHz	1 Mbps, 2 Mbps
802.11	DSSS	2.4 GHz	1 Mbps, 2 Mbps
802.11a	OFDM	5 GHz	6, 9, 12, 18, 24, 36, 48 and 54 Mbps
802.11b	DSSS	2.4 GHz	5.5 Mbps, 11 Mbps
802.11b	PBCC	2.4 GHz	5.5 Mbps, 11 Mbps
802.11g	OFDM	2.4 GHz	6, 9, 12, 18, 24, 36, 48 and 54 Mbps
802.11g	PBCC	2.4 GHz	22 Mbps, 33 Mbps
802.11n	OFDM	2.4 GHz, 5 GHz	up to 600 Mbps

5.2.2 IEEE 802.11 – MAC Sublayer

The next higher protocol layer of the IEEE 802.11 standards, the Medium Access Layer (802.11-MAC), regulates multiple access on the commonly used radio data transmission medium. The method used in WLAN is very similar to the cable-based CSMA/CD algorithm in Ethernet. As in the Ethernet standard IEEE 802.3, fixed access methods provide a mutual right to access for all subscribers. In order not to trigger a collision from the beginning, in CSMA/CD a computer may only first begin a transmission procedure if no signal is discovered on the shared transmission medium.

Similarly, according to the IEEE 802.11 specification, a computer in the WLAN monitors the received energy level on an allocated radio frequency in order to determine whether or not another computer is carrying out a data transmission at that moment. If it is detected that a specific channel is free for a certain period of time – called **distributed interframe space** – then a computer can begin its transmission. The receiver confirms the reception of a complete message after the expiration of time period called **short interframe spacing**. If it is recognized that the channel is occupied at that moment, then a back-off algorithm is initiated and the computer waits a randomly selected period of time before beginning the next transmission attempt.

With the CSMA/CD algorithm, after the detection of a data packet collision the necessary measures are initiated to regulate parallel access of the different subscribers. In contrast, in IEEE 802.11 a collision avoidance procedure is implemented. It is called **multiple access with collision avoidance (MACA)**. Generally, a data transmission in which a collision has occurred must be repeated. However, as only one subscriber can access the radio data transmission medium at the same time, renewed data transmission leads to a delay for all of the subscribers. To avoid delays as much as possible and to achieve the highest possible efficiency, an attempt is made to steer clear of collisions from the beginning by using an access method that is especially tailored to the radio data transmission medium.

A further difference to wired LAN is that a subscriber cannot assume that the error-free transmitted data has in fact reached the receiver.

Electrical or optical transmission media, such as copper cable or optical fibers provide physical shielding against interference influences. The data transmission over the radio medium is, in contrast, neither protected from disturbances by neighboring systems nor WLAN-foreign interferences in the same frequency range. For this reason, appropriate error detection and correction mechanisms must already be provided on the data link layer. These are carried out in IEEE 802.11 in a similar way using an acknowledgement mechanism, as is normally first implemented in higher layers of the TCP/IP Reference Model. The receiver confirms to the sender that correct reception has taken place via an **acknowledegment**. If this confirmation is absent at the sender for a certain period of time, then the data is sent again (**retransmission**), following a short delay. The longer a data transmission takes, i.e., the longer the time needed to send a single data packet, the higher the likelihood that an error has occurred during transmission affecting the data packet. Because of this, IEEE 802.11 provides the option of **fragmentation** for large data packets, i.e., data packets are broken down by the sender into smaller units (fragments) and transmitted.

A general distinction is made in 802.11-MAC between two basic methods to accessing resources on the network:

- **Point Coordination Function (PCF)** This type of media access management is controlled centrally by a dedicated point coordinator. The network subscribers are not in direct competition with each other. A strategy of collision avoidance is followed. Normally, this task is taken over by the access point and therefore comes to application exclusively in the infrastructure mode. The time frame in which the transmission medium is centrally managed is designated as the Contention Free Period (CFP). This procedure is intended for the transmission of time-critical data, e.g., real-time audio or video, and it is necessary to guarantee media access within a certain time frame.

- **Distribution Coordination Function (DCF)**
  With this variation of the media access management there is no central management, instead the individual subscribers are in direct competition with each other. Collisions cannot be ruled out. Thereby, DCF can be implemented in the infrastructure mode as well as in the ad hoc mode. The time frame in which

the transmission medium is decentrally managed is called the Contention Period (CP).

Following in the footsteps of the wired data transmission in IEEE 802.3 Ethernet, the MACA access procedure of IEEE 802.11 works on the basis of a decentralized approach. Every subscriber bears individual responsibility for access to the transmission medium. Before a subscriber begins transmission it first verifies whether or not the transmission medium is free, only accessing it if it is not occupied. In the CSMA/CD algorithm of the Ethernets the subscribers are in a position during transmission to simultaneously monitor the transmission channel and to potentially receive any occurring collisions. In contrast, in the WLAN it is principally impossible to transmit and to receive data at the same time, except when two completely separate transmission and reception units are employed. This effort is usually not carried out and it would also not be justified. Therefore, in the WLAN a collision detection is not implemented, thus underscoring the necessity to avoid collisions from the start.

The carrier sense mechanism of the physical layer, for verification of the free status of the transmission medium, is implemented in IEEE 802.11 analogously to IEEE 802.3 Ethernet. Besides this, 802.11-MAC provides an additional **virtual carrier sense** mechanism. The foundation of the virtual carrier sense is a timer (Network Allocation Vector, NAV), which is managed mutually by all of the subscribers and indicates the estimated time that a transmission medium will be occupied for a data transmission. Only when this NAV timer has expired and the physical carrier sense function announces a free transmission medium does the subscriber access the transmission medium.

The probability of potential collisions occurring is thereby lessened. The decentral management of the NAV timer is carried out with an entry in the header of every transmitted data packet. In the duration/ID field it is indicated how long the transmission medium is occupied for the complete transmission of data. As all of the subscribers in the network range can potentially receive this data packet, they are able to continually update their own NAV timer, and the transmitting subscriber reserves the transmission medium for its exclusive use. In this way it is ensured that an ongoing data transmission is not interrupted by other subscribers.

Numerous operations in the WLAN are based on the exchange of multiple data packets. There are always short transmission breaks between these data packets. If the WLAN were only dependent on the physical carrier sense function, there would be a danger of other subscribers accessing the transmission medium during this short break and thus interrupting an ongoing operation. Despite the collision avoidance strategies implemented, the potential for collisions to occur cannot be avoided completely, nor can other disturbances that impair data transmission. For this reason, the previously mentioned confirmation of reception (**acknowledgement**) was introduced, whereby the receiver must confirm the correct reception of a data packet.

This acknowledgement is only a 14-byte long shortened header of a data packet. With broadcast and multicast data packets there is no confirmation of reception.

The time period between the correct reception of a data packet by the receiver and the transmission of an acknowledgement is called the **interframe space** (IFS). Depending on the priority of media access, different varieties of the IFS are distinguished. IEEE 802.11 follows the principle that communication with higher priority is subject to a shorter waiting time than lower prioritized communication when both are equally independent of the respective data transmission rate of the subscribers. In this way it should be assured that data with higher priority can be transmitted before data with lower priority. The following IFS types are distinguished:

- **Short Interframe Space** (**SIFS**) defines the minimum interval for acknowlededgment (ACK), clear-to-send (CTS, confirmation of receipt for a request-to-send inquiry) or subsequent data packets from a fragmented data transmission. SIFSs present the shortest time frame and in communication are reserved for the highest priority.

- **Point Coordination Function Interframe Space** (**PIFS**) is only implemented in networks that support the PCF mode and gives subscribers in the PCF mode higher priority for access to the transmission medium.

- **Distributed Coordination Function Interframe Space** (**DIFS**) is used in networks that support the DCF mode and defines the waiting time of a subscribers during which the transmission medium must be free before a transmission process is allowed to begin.

- **Extended Interframe Space** (**EIFS**) is only implemented by subscribers who work in the DCF mode and defines the shortened waiting time, in relation to the NAV-Timer, of a previously transmitting subscriber who had to interrupt transmission early due to an error.

As a rule, the received data packets in SIFS are shorter than those in DIFS, the data packets that may be sent after expiration of the SIFS automatically have a higher priority than those data packets that may be sent after expiration of the DIFS. Thereby, the receiver with higher priority gets the possibility to acknowledge the correct reception of a data packet. To prevent that after expiration of a DIFS all subscribers try at the same time to transmit on the transmission medium, each subscriber waits a certain randomly chosen time period before the actual transmission attempt. This randomly chosen time frame is determined in Ethernet with the help of a **backoff** algorithm. Only after this backoff time frame has expired do the subscribers begin to transmit with the lowest waiting time. All other subscribers receive the transmitted data packet and correct their own NAV timer, based on the Duration/Id field indicated in the header of the data packet. Here, the assumed time duration of the planned data transmission is noted, i.e., the length of transmission of the current data packet including the duration of the accompanying acknowlegements. Following the elapse of the NAV timer, the remaining subscribers will try again to gain access to the transmission medium (cf. Fig. 5.3 and 5.4).

Subscribers in a WLANs all have radio-based transmission and receiving facilities at their disposal, which for reasons of their physical conditions are subject to range limitations. Because of this, every subscriber has its own transmission and reception range and the respective geographic location of the subscriber determines who can and who cannot receive which subscriber directly. This condition can lead to **hidden**

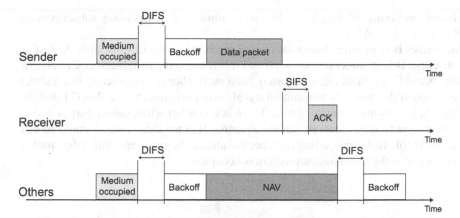

Fig. 5.3 Transmission sequence of the transmission and reception of data packets in the WLAN.

Waiting Time and Media Access in the WLAN

The actual waiting time necessary for a subscriber in the WLAN to wait for the transmission of a data packet is based on the respective rules for the choice of the interframe space waiting time, as well as on the subsequent randomly chosen backoff. This backoff time is calculated as follows:

$$\text{backoff} = \text{random(CW)} \cdot \text{slottime}$$

whereby for CW (Contention Window) applies: $CW_{min} \leq CW \leq CW_{max}$. All three dimensions, slot time, CW_{min} and CW_{max} depend respectively on 802.11-PHY and are prescribed for the different types. For example, to DSSS modulation applies: slot time=20 µs, CW_{min}=31 and CW_{max}=1031. With each failed transmission attempt the backoff time doubles itself until reaching CW_{max} (exponential backoff). If the successful transmission of a data packet was possible, the CW value reverts back to CW_{min} again.

If during the backoff waiting time a subscriber determines that the transmission medium is occupied again, then the backoff timer is interrupted. After the transmission medium is no longer occupied for the duration of a further DIFS, the backoff waiting time is resumed again.

For each repeated transmission attempt, depending on the length of the data packet to be transmitted, one of two different numbers are incremented: SSRC (Station Short Retry Count) for data packets that are shorter than the so-called RTS threshold, and SLRC (Station Long Retry Count) for data packets with greater lengths. If for one of the two counters the respective maximum value is reached (SSRC=7 or SLRC=4), then the transmission attempt is considered failed and the counters are reset.

Further reading:

Roshan, P., Leary, J.: 802.11 Wireless LAN Fundamentals: A Practical Guide to Understanding, Designing, and Operating 802.11 WLANs, Cisco Press (2004)

Fig. 5.4 Waiting time and media access in the WLAN.

station problems (cf. Fig. 5.5). This can be illustrated by assuming subscribers A, B and C are in a WLAN.

Subscriber B is in direct transmission and receiving range to subscriber A. Just as subscriber B is in direct transmission and receiving range to subscriber C. Subscribers A and C are, however, so far away from each other geographically that a direct reception of the mutually transmitted signals is not possible. Subscriber C is hidden from A, just as the opposite is true. There is a conflict when subscribers A and C both attempt to send a data packet to subscriber B at the same time. It could be that neither is able to determine that the other has already begun to transmit a data packet and therefore the transmission medium is occupied.

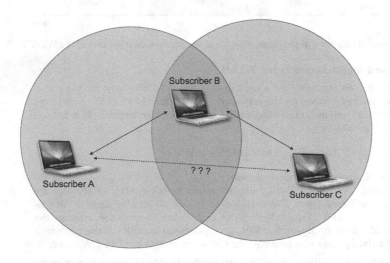

Fig. 5.5 Hidden station problem in the WLAN.

We shall now assume that in the just described scenario a further subscriber D enters the picture. Subscriber D is located in the transmission and receiving radius of subscriber C, but too far away from subscribers A and B. Just completed data communication between subscribers A and B prevents subscriber C from communicating with subscriber D, although this communication would not have been hindered by either A or B. This situation is called an **exposed station problem**.

To avoid the hidden station problem in the IEEE 802.11 standard, the **RTS/CTS procedure** (Ready To Send / Clear To Send) is used. Before a subscriber begins with transmission of the actual user data, it first sends a RTS data packet to the receiver. This contains information about the scope of the user data to be sent (via the Duration/ID field of the header). The receiver acknowledges the reception of the RTS data packet with a CTS data packet, which likewise confirms the estimated transmission duration to receive the user data. All subscribers receive the transmitted CTS data packet within its range, though not necessarily located in the range of the actual transmitter. Those subscribers not reachable by the transmitter are also informed about the upcoming data transfer and its duration and can adjust their NAV timers accordingly. This prevents the hidden station problem from occurring here. But even in this way, collisions on the radio-based medium cannot be avoided completely. This applies only to data packets that are sent out at the beginning of an intended transmission, i.e., RTS/CTS data packets with only a short length (20 or 14 bytes). Therefore, if a collision occurs the transmission medium is only paralyzed for a short time- – the transmitting subscriber cannot recognize the collision by transmission and therefore must continue the transmission procedure to its conclusion. In order to additionally avoid that the transmission medium is not unnecessarily burdened during this procedure, its implementation is first initiated at a certain data packet length. The exposed station problem is not solved by the RTS/CTS procedure. However, in comparison to the hidden station problem it is considered less serious as it leads only to a reduced throughput in the radio network.

In transmitting longer, contiguous data, the IEEE 802.11 makes the provision for the **fragmentation** of data into smaller units. On the transmitter's side, the individual fragments of a data transmission have to be consecutively numbered so that they can be reassembled correctly on receiver's side. The sequence control field is provided in the header of the IEEE 802.11 data packet to identify the fragments of a contiguous data transmission. It contains the respective fragment number and an identification of the associated data sequence (which is the same for all contiguous fragments of a data transmission). To indicate the end of a data sequence, i.e., the last accompanying fragment, a More fragment bit is set in the header. All of the fragments sent out must be acknowledged by the receiver with an ACK. In the absence of a confirmation, i.e., if there was either an error in transmission of the data fragment or of the ACK, the respective fragment is transmitted again. Only upon reception of a mutual acknowledgement does transmission carry on with the next fragment. To avoid the receiver getting the same fragment twice, and in turn processing it incorrectly, a retry bit is set in the header of the repeated transmitted fragment. Therefore, to transfer a longer, contiguous data transmission in fragmented form, every single fragment is followed by a transmission confirmation after a SIFS waiting time. After a successful acknowledgement, the following fragment is then transmitted, also after a SIFS waiting time. Thereby, the fragment sequence receives a higher priority for media access. This prioritizing also makes it possible that the NAV timer is not initialized for the length of the entire data transmission, but only for the transmission duration of a fragment and its transmission confirmation. Otherwise, unnecessarily long waiting times would arise in the case of data transmission interruption (cf. Fig. 5.8).

Not only can fragmentation be implemented in long, contiguous data transmissions, but it is also often put into practice in shorter data transmissions in environments susceptible to interference. This is because the probability of a defective transmission increases with the length of the data packet and therefore shorter fragments can transmit more reliably. However, fragmentation can only be used for unicast communication. Multicast and broadcasting transmissions do not allow themselves to be fragmented as an acknowledgement procedure with many simultaneous recipients would prove difficult to implement.

The RTS/CTS Procedure for the Avoidance of Hidden Station Problems

Let us assume that computer A would like to transmit a packet to computer B (cf. Fig.5.7).

- A first transmits a very short so-called RTS data packet (request to send) (cf. Fig.5.7 (a)). Among other things, it contains the length of the packet actually being sent, which is as a rule considerably longer.

- B answers with a CTS packet (clear to send) (cf. Fig.5.7 (b)), which likewise contains the length information already transmitted by A.

- As soon as A receives the CTS- packet it begins transmission of the actual data packet. Every computer that has received the RTS packet transmitted by A must be in close proximity to A. They are quiet for the time being, until the CTS packet has reached A again.

- Every computer receiving the CTS packet transmitted by B is in the neighborhood of B, and it must cease transmission during the time of the upcoming data transmission – the length of which can be derived from the CTS packet.

- Computer C, which is in the range of A but not of B, receives the RTS packet, but not the CTS packet. It is allowed to transmit during the upcoming data transmission between A and B as long as this does not cause a conflict with the CTS packet.

- It is a different situation with computer D, which is located in the range of computer B but not of A. D receives the CTS packet, but not the RTS packet. By receiving the CTS packet, D concludes that it is in close proximity to a computer that will momentarily receive a data packet and is quiet within the time frame given in the CTS packet.

- Despite these precautions, collisions can occur, e.g., if B and C send a RTS packet at the same time to A, this RTS packet will be lost due to a collision.

- In addition, following the successful reception of the data packet, the receiver transmits a confirmation of receipt (acknowledgement) back to the transmitter. If this confirmation fails to arrive, then following the expiration of a random time frame the transmitter begins a new transmission attempt.

Further reading:

Roshan, P., Leary, J.: 802.11 Wireless LAN Fundamentals: A Practical Guide to Understanding, Designing, and Operating 802.11 WLANs, Cisco Press (2004)

Fig. 5.6 The RTS/CTS procedure for the avoidance of hidden station problems.

5.2.3 IEEE 802.11 – Data Format

To be able to guarantee data transmission over a wireless, and therefore more disturbance-prone transmission medium, a higher coordination and control effort is necessary. This stands in contrast to technologies in wired communication. The appropriate command and control information must be exchanged, and the necessary space be provided in the header of the corresponding data format. The IEEE 802.11 data format proves more complex than the Ethernet data format, which is by comparison relatively simple. A basic distinction is made between data packets for the transport of user data (data frames), control and management information (control frames and management frames). Control frames serve to regulate access to the

Transmission and receiving radius of A

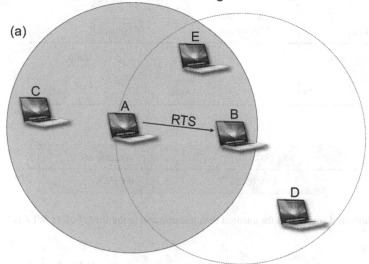

(a)

Transmission and receiving radius of B

Transmission and receiving radius of A

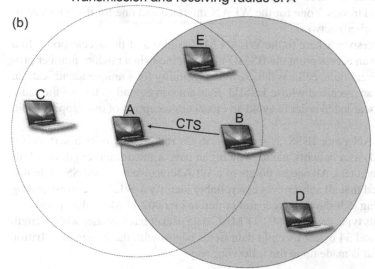

(b)

Transmission and receiving radius of B

The RTS/CTS procedure for collision avoidance. In (a) A transmits a RTS data packet to B and in (b) B answers with a CTS packet.

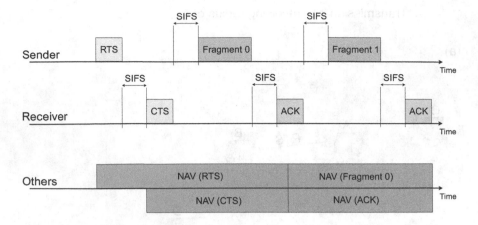

Fig. 5.8 The chronological course of a fragmented data transmission in the IEEE 802.11 WLAN.

transmission medium and to guarantee a reliable data transmission. With the help of management frames the radio network can be administered.

The identification within a WLAN proceeds over the MAC addresses of individual subscribers and access points. A radio cell (BSS) is identified over a 6-byte long Basis Service Set identification (BSSID), whose address format corresponds to the IEEE 802.3-MAC addresses. With a BSS, which contains an access point, the BS-SID corresponds to the MAC address of the access point. Normally one access point has two MAC addresses – one for the WLAN interface and one for the interface of the connected (wired) network.

The BSSID corresponds here to the WLAN MAC address of the access point. In a WLAN without an access point the BSSID is determined via a random number procedure so that every radio cell has the greatest possibility for a unique identification. If data packets are received whose BSSID does not correspond to that of the radio cell, they are discarded in order to avoid an erroneous reception of overlapping radio cells.

The entire WLAN (with IBSS as well as ESS) is represented over a service set identifier (SSID) as a network name, which can have a maximum length of 32 alphanumeric characters. All access points of a WLAN receive the same SSID. In this way it is ensured that all subscribers can reliably identify a WLANs corresponding access points. Fig. 5.9 shows the general structure of a 802.11 MAC data packet.

Depending on its type, the IEEE 802.11 MAC data packet has a header with a length of between 10 and 34 bytes. Its eight data fields begins with the 2-byte long **frame control** field that is made up in the following way:

- **Protocol Version** (2 bits) assignment per current standard set at '00'.
- **Type** (2 bits) indicating whether it is a data (10), control (01) or management frame (00).

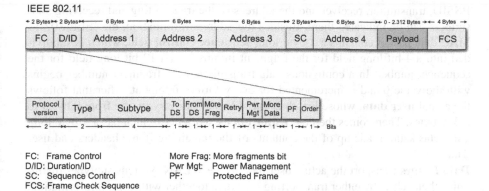

Fig. 5.9 The general IEEE 802.11-MAC data format.

- **Subtype** (4 bits) specifying for each of the three frame types (data, control, and management frames) a range of further subtypes.

- **ToDS / FromDS** (1 bit each) indicates the respective transmission path on which the transmitted data packet is located, i.e., for ToDS=1 and FromDS=1 if the data packet is intended for forwarding over the distribution system (access point), in the other case (ToDS=0 and FromDS=0) if the data packet is directly transported in an ad hoc network from subscriber to subscriber.

- **More Fragment** (1 bit) indicates whether further fragments follow that belong to the same message.

- **Retry** (1 bit) indicates whether the transmitted data packet is one that has been sent before, thereby avoiding duplicate frames on the receiver's side.

- **Power Management** (1 bit) shows whether a subscriber switches to an energy-efficient power save mode after completion of data transmission.

- **More Data** (1 bit) indicates that further data is is ready for transmission and will be sent next. Causes participants who wish to switch to the power save mode not to execute this switch in order to be able to receive further data.

- **Protected Frame** (1 bit) indicates if the data packet is transmitted in encrypted form. The encryption, however, applies only to the user data contents and not to the data packet header.

- **Order** (1 bit) indicates that received fragments of a longer data transmission should be transferred in the sequence of their reception to a higher protocol layer.

The frame control field follows the field **Duration/ID** in the header of the IE-EE 802.11 MAC data packet with a length of 2 bytes, which indicates the necessary time required for the data transmission. Depending on the frame type, the IE-EE 802.11 MAC data packet contain up to 4 or 6 byte long addresses. Which addresses are managed in the individual frame types depends on the information in the ToDS-/From-DS fields in the header. Altogether, address information about the

BSSID, transmitter, receiver and the address of the transmitting and receiving sub-scribers can be indicated there. Besides the address information, the IEEE 802.11 MAC data packet contains a 2-byte long **sequence control** field, which is subdivi-ded into a 4-bit long field for the fragment number and a 12-bit long field for the sequence number. In a contiguous data transmission the fragment number begins with the value 0 and is incremented for every further fragment. After that follows the actual **user data**, whose length is variable and can vary in size from 0 bytes to 2.312 bytes. Then comes the field **frame check sequence**, which has a 4-byte long CRC checksum made up of the contents of the remaining frame headers and user data.

Data frames transport the actual user data in the WLAN. Various subtypes distin-guish themselves by either transporting user data together with control information, or only pure user data. In an ad hoc network (IBSS) only pure user data is transmit-ted without additional control information. In infrastructure networks, in contrast, there are a variety of subtypes that allow the exchange of additional control informa-tion. The so-called **null frames** present a special case. These are data frames whose user data field remains empty, i.e., the FCS field follows the header information di-rectly. They signal an access point that one of the subscribers has switched to the power save mode. The power management bit is thereby set in the header.

Control frames serve to control access to the data transmission medium and ensure a secure data transmission. The following subtypes are distinguished:

- **Power Save Poll** (PS) for regulating the power save mode. The header of the PS data packet contains the frame control field with frame type and subtype infor-mation. It is followed by the association identity field (AID), intended to identify the subscriber to the access point, and whose transmission data should be stored by the access point for the duration of the power save mode. Two MAC addres-ses follow, the first identifying the BSSID of the radio cell and the second the transmitting subscriber. The length of the PS data packet is 20 bytes.

- **Request To Send** (RTS) announces a wish to transmit, prevents the hidden sta-tion problem. The header of the RTS data packer receives the frame control field with frame type and subtype information, Duration/ID with the intended trans-mission duration in μs (corresponds to the transmission duration of the CTS, data frames, ACK and three SIFS) and the MAC addresses of the transmitter and re-ceiver, followed by the FCS field. The length of the RTS data packet is 20 bytes.

- **Clear To Send** (CTS) to confirm a transmission request, prevents the hidden station problem. The header of the CTS data packet contains the frame control field with frame type and subtype information, Duration/ID with the remaining transmission duration in μs (corresponds to the transmission length of the data frames, ACK and two SIFS) and the MAC address of the original transmitter of the RTS data packet followed by the FCS field. The length of the CTS data packet is 4 bytes.

- **Acknowledgement** (ACK) to confirm that a data packet has been received. The header of the ACK data packet contains the Frame Control field with frame type and subtype information, Duration/ID independent of any previously sent frames (with data frames the content is set to 0, with confirmation of a fragment within a Fragment Burst the remaining transmission duration in μs) and the MAC address receiver followed by the FCS field. The length of the ACK data packet is 14 bytes.

- **Contention Free** End (CF) and Contention Free Acknowledgement show the end of a data transmission in the CFP access method procedure and are initiated from the access point of a radio cell. The length of the ACF data packet is 20 bytes and contains the field Frame Control with frame type and subtype information, followed by the Duration/ID, containing the value 0, in order that the subscribers reset their NAV value. Thereby, the media reservation for the CFP phase is removed. Two MAC addresses follow. The first contains a broadcast address, as it is necessary to notify all of the subscribers of a BSS. The BSSID of the corresponding radio cell is given via the second MAC address.

The management of a WLAN is controlled over special **management frames** . Unlike control frames they do not define their own data formats, but use data frames for the transmission of tax and management information. Normally, the three address fields each contain the address of the receiver, the transmitter and the BSSID of the respective radio cell. With the exception of beacon frames, only those management frames in a radio cell are evaluated whose address field matches the BSSID of the radio cell. The respective different management information is transferred to the data section of the management frame.

A distinction is made between the following types of Management Frames:

- **Association Request Frame** and **Association Response Frame**
  With the help of the associated request frames, a subscriber makes known that it would like to connect to an access point. Different requests are encoded in the user data, which the access point compares with its own information. It then makes a decision about a possible association with the subscriber. A positive answer proceeds over an association response frame, in which the access point transmits a valid AID (Association Identity) to the associated subscriber. In the other case, a negative response proceeds over a disassociation frame.

- **Reassociation Request Frame** and **Reassociation Response Frame**
  The reassociation of a subscriber is necessary if it has dropped out of the range of the access point and and then enters this range again, or when a subscriber enters another radio cell within the same networks (ESS). The reassociation is confirmed via a reassociation response frame.

- **Probe Request Frame** und **Probe Response Frame**
  Probe request frames can be sent to locate access points or further subscribers. Information about the SSID and the supported data transmission rate is passed along, and based on this the access point or another receiving subscriber can decide whether a requesting subscriber can be received in the radio cell. In the positive case, the request is answered with a probe response, which contains further parameter information about the host receiving the request.

- **Beacon Frame**
  An active WLAN makes its presence known by sending a beacon frame. These

are sent in the infrastructure mode by access points and in the IBSS by selected subscribers. Within a beacon frame all relevant information concerning the
connection establishment with the WLAN is made known.

- **ATIM Frame**
  Within an IBSS the ATIM frame indicates that data packets of a subscriber who
  is currently in the power save mode are buffered.

- **Disassociation Frame**
  If acceptance in a WLAN is refused, a disassociation frame is sent. It contains a
  reason for this decision (reason code).

- **Authentication Frame** und **Deauthentication Frame**
  Depending on the authentication method used, multiple authentication frames are
  exchanged between the sender and the receiver. If the connection establishment
  ends because of insufficient authentication, this takes place with the help of a
  deauthentication frame, which provides the reason for the refusal (reason code).

- **Action Frame**
  This type of frame serves to initiate various management functions in the spectrum management of the 5 GHz frequency band via TCP and DFS. Measuring
  activities and information about transmission power can be triggered and requested as well as a channel switch announced.

When compared to a wired network, other physical qualities are important for a
WLAN that directly influence access to the network and secure communication in
it. The topic of WLAN security is the focus of Section 5.2.4. Further management
processes that particularly facilitate access to the WLAN proceed in different ways
for infrastructure networks and ad hoc networks.
These management functions include:

- **Passive and active scanning**
  For a subscriber to be able to communicate in an infrastructure network, it must
  know via which access point it can connect with the network. In the same way
  an access point in a radio cell must know which subscribers belong to its radio
  cell. For these reasons, it is necessary that a subscriber first **authenticate** itself
  at a specific access point and then **associate** itself with the radio cell to take part
  in an infrastructure network. So that a subscriber knows at which access point it
  can access the radio cell, it conducts an active or passive **scanning process** (e.g.,
  after the being switched on). Thereby, the subscribers seeks beacon frames that
  are sent from the access point in regular time intervals (beacon interval, typically
  in intervals of approx.100 ms) as a broadcast to subscribers in all ranges. In this
  way, the access point and and its services are known among them. By way of the
  beacon frames, a time synchronization of the respectively assigned radio cells
  and their subscribers is also carried out through the access point.
  In **passive scanning** a subscriber listens a prescribed time period to each of the
  available channels – one after another – for the reception of a beacon frame.
  This time frame is chosen in such a way that a beacon frame can always be
  received securely and completely. If during the listening time the beacon frame

is received on a certain channel, the subscriber knows which access point on which channel is available. If multiple beacon frames from different sources and different channels are received, the subscriber generally decides on the access point with the highest signal strength and the best signal reception.

By contrast, in **active scanning** the subscriber seeks a specifically prescribed network. To do this, the subscriber sends out an initial probe request frame in which the SSID identification of the corresponding network being searched for is given, or a broadcast SSID to find an unspecified network. To prevent multiple subscribers from hindering each other by sending out probe request frames while booting up a network at the same time, a waiting time is observed before transmission begins (probe delay). Parallel to transmission of the probe request frame, an individual probe timer is started. When it expires a new attempt is started on another channel. After all of the channels are gone through, the subscriber examines the collected results. If more probe response frames are received on different channels, then the subscriber decides on the network with the highest signal and reception quality.

With an ad hoc network a subscriber must likewise find out which radio cell it can connect to. As access points do not exist here, theoretically every subscriber that is already connected in the ad hoc network can send out beacon frames. However, a subscriber only transmits a beacon frame when it has not yet received one itself. The subscriber who first sends out a beacon frame also remains responsible afterwards for the transmission of the beacon frames. A subscriber who wishes to be incorporated into an ad hoc network carries out an active scanning for this purpose and first transmits a probe request frame. The network subscriber who last sent a beacon frame answers this with a probe response frame. If multiple probe response frames are received, then the subscriber decides on the ad hoc network access with the highest signal strength and best reception quality.

- **Power Management**
A large number of mobile subscribers do not have a permanent electricity supply at their disposal, but instead are equipped with rechargeable batteries. Therefore, available energy must be used sparingly. For this reason, most mobile network-enabled terminals have the ability to throttle power consumption during longer periods of inactivity. By switching into a so-called power saving mode with reduced activity, they are able to achieve the longest running time possible. Within an infrastructure network the access point does not, however, number among the mobile devices. Because it remains active it is provided with a permanent supply of electricity. Mobile subscribers, in contrast, can change from the active state (Active Mode) to a power saving and reduced output state (Power Save, PS) during longer periods of inactivity.

If a subscriber is in the power save mode, it can neither transmit nor receive data. During this time, the corresponding access point must buffer the data for the subscriber. In specifically prescribed intervals the subscriber can cross over from the power save operation into the active mode, in order to retrieve data that has been buffered. If there is buffered data at the access point, it can append a traffic indicator message (TIM) to its transmitted beacon frames in the data section. Beacon frames are sent out in regular intervals. It is only necessary for a subscriber who is in the power save mode to switch to the active mode for a short time in order to decide, after evaluating a received beacon frame, whether it must remain active to retrieve buffered data at the access point, or if it can switch back into the power save mode. The retrieval of the buffered data is carried out via a control frame from the power save poll subtype. As long as data remains at the access point to be called up, it sets the More data bit in the transmitted data frame. To show the end of the transmission of buffered data, the More bit is reset in the last transmitted data frame. The requesting subscriber can switch back again into the power save mode.

To show that a formerly active subscriber has switched to the power save mode, the subscriber sets the bit of the power management field in the frame control field of the header to 1 in the last transmitted data packet .

If the access point receives this kind of data packet, it will buffer all of the data that accrues for the relevant subscriber. The procedure runs differently, however, for multicast and broadcast transmissions sent to all subscribers of a WLAN, as it must be ensured that all subscribers can also receive these transmissions. The access point indicates that buffered multicast or broadcast messages are ready by sending a beacon frame as before, with a TIM.

If subscribers receive the beacon frame with a multicast or broadcast message addressed TIM, these remain active for a longer period of time (DTIM interval, delivery TIM, corresponds to 3 beacon intervals) so that all stations are then able to receive the multicast or broadcast messages.

In the case of an ad hoc network, special data packets – ATIM frames (ad hoc traffic indicator message) – are sent that show whether a subscriber who receives them should remain in the active mode to be able to get data, or should switch to the power save mode. To initiate a data transmission to all subscribers in the power save mode, the transmitting subscriber waits for the so-called ATIM window. This is a special time frame in which one of the transmitted ATIM frames is received by all other subscribers, who for this short time period have switched briefly from the power save mode to the active mode to receive beacon frames or ATIM frames. With active power management in an ad hoc network all subscribers must therefore first buffer the data that a subscriber wishes to send in the power save mode.

- **Association**
  To be able to exchange data in a radio cell, a subscriber must first be associated with the corresponding access point, i.e., a unique connection must be established between the access point and the subscriber. To do this a subscriber sends an association request frame out immediately after a successful authentication (cf. Section 5.2.4). This contains the most important connection parameters for the subscriber. The access point reviews these connection parameters and, if successful, sends back an association response frame, which contains an AID. As long as there is an association, the subscriber can be uniquely identified. The subscriber confirms the reception of the association response frame with an ACK. Now the actual data exchange can begin. If the access point decides against an association

because important connection parameters do not match, it sends an dissassocia-
tion frame back to the subscriber indicating the reason for refusal. If a subscriber
changes the radio cell during an ESS, it registers at the new access point with
an reassociation request frame in which the MAC address of the last used access
point is also given. In the case where an association with the new access point is
possible, the inquiry is answered with a reassociation response frame. The subscriber receives a new AID that it uses for the duration of its time in the radio cell.
Additionally, the access point makes the reassociation known to the distribution
system, so that other access points can forward incoming or buffered data

- **Data Transmission Rate Support**
  The IEEE 802.11 standard supports a wide range of different data transmission
  rates. The components connected to a WLAN always attempt to use the highest
  data transmission rate possible. It is dependent on the distance to be bridged
  and the quality of the connection route in each case. Thereby, the IEEE 802.11
  standard leaves open how the respective data transmission rate is to be selected.
  The access point saves the data rates it supports in a "BBS basic rate set" list. All
  subscribers registering at this access point must at least support the indicated data
  transmission rates in the BBS basic rate set. Moreover, there is also a list with
  non-BBS basic set rate data transmission rates that can also be supported. If none
  of the data transmission rates listed as BBS basic set rates or Non-BBS basic set
  rates are supported, an association request is refused. Which data rates are used
  for which type of communication are not fixed in the IEEE 802.11 standard.
  Multicast and broadcast messages must always be transferred with a data rate
  from the BBS basic rate set list so that all of the subscribers in the network are
  able to receive them. ACK and CTS data packets must be transmitted with a BBS
  basic rate set data transmission rate that can only be as large as the preceding,
  to be confirmed data package. Normally, management frames are sent with the
  lowest BBS basic set rate possible. In contrast, multicast and broadcast messages,
  as well as user data, are sent with the highest.

- **Transmit Power Control**
  This function to enable adjustment of transmission power was introduced with
  the IEEE 802.11h extension of the WLAN standard to 5 GHz components. It serves to make certain that the maximum allowed transmission power per channel,
  as determined by the national regulatory board, is not exceeded. A corresponding information exchange takes place regarding the maximum and minimum
  transmission power during the association phase of a subscriber in a radio cell. If
  at this time the access point determines that the maximum or minimum prescribed transmission power of a subscriber in not in accordance with the prescribed
  frame, the association is refused. The transmit power control should additionally
  ensure that communication between the access point and an associated subscriber
  is always carried out with the minimum transmission power necessary.

- **Dynamic Frequency Selection**
  A dynamic channel change becomes necessary if during operation of a WLAN a
  user occupies a frequency range for which it is not intended, e.g., in the 5 GHz

frequency band of IEEE 802.11h radar systems or HIPERLAN/2 systems. A check is made with the use of every channel to verify whether the channel is already occupied by a user for whom it is not intended. This is also managed through the expanded information exchange during the association phase or reassociation phase. The subscriber relays to the access point the supported channels in the association request frame. It can refuse an association if channel support is lacking. The channel change itself is initiated from the access point by the transmission of a beacon frame or a probe response frame using the Quiet information element. This initiates a quiet phase on the current channel and, during this time, subscribers have the chance to receive new beacon frames on the other available channels with the associated access information. If a foreign system is detected on a channel, the delivery of data frames must be suspended from 200 ms. Before a new channel is used it should be monitored for 10 seconds to check for other systems. The verification of a channel is carried out through the transmission of a measurement request frame during the initiated quiet phase or transmission-free time. The response is sent over the measurement response frame by the associated subscribers. It is listed here on which channel a user is recognized, or which channel is free. If the check indicates that the channel is already occupied, then a new channel switch is initiated.

- **Point Coordination Function**
  In addition to the decentralized management of a WLAN, the IEEE 802.11 standard also provides for a centrally controlled access method to the transmission medium. This point coordination function (PCF) can be selected as as optional procedure in the infrastructure mode and is especially advantageous if real-time data, e.g., audio or video is to be sent over the WLAN. In the time frame of this management method, the contention-free period (CFP), the point coordinator issues a cell (mostly the access point) authorization to transmit to the subscribers via the dispatch of CF poll frames. This centrally controlled management makes sure that no collisions occur and that the necessary service guarantees can be given for the transmission of real-time data.
  IEEE 802.11 provides that CFP time frames periodically switch with CP time frames (contention period), in which the transmission medium is managed decentrally, so that also subscribers who do not support the PCF mode are assured of access to the transmission medium (cf. Fig. 5.10).
  The access point always introduces the CFP time span by sending out a beacon frame. The CFP time span corresponds to a multiple number of beacon intervals. A subscriber who does not support the PCF mode receives information about the duration of the following CFP time span over the beacon frame sent as a broadcast. This is indicated via the Duration/ID field in the header of the beacon frame that regulates the NAV timer. In this way it is ensured that a subscriber who does not support the PCF mode does not transmit during the CFP time span. The actual access of a subscriber to the transmission medium first proceeds in the PCF mode after a request by the point coordinator, who queries all of the subscribers, one after another if they have data to send (polling). The point coordinator does this by sending a CF poll frame to all subscribers wishing to transmit. They

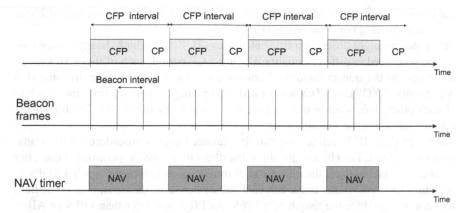

Fig. 5.10 Periodic switch between a contention-free period and a contention period in a PFC-enabled WLAN.

are first allowed to begin data transmission after receiving this management frame. As the point coordinator must have knowledge of which subscribers in its radio cell support the PCF mode, this must be made known when the association of the subscriber at the access point of the radio cell takes place. The access point enters all PCF-enabled subscribers in a polling list that is processed during every polling operation. However, a general prioritizing of audio and video data in the WLAN is also not possible in the PCF mode. Only the data traffic of a specific subscriber can be temporarily prioritized and forwarded under a service guarantee.

Which channels may be operated within the used frequency range, having specific requirements regarding maximum transmission power, is determined by the national regulatory authorities. A unified standard does not exist. To nevertheless facilitate an international exchange of WLAN equipment, e.g., for travelers, further expansions in the IEEE 802.11e standard were laid down (**international roaming**), so that a client can adjust nationally to differing requirements concerning frequency use and signal performance. An access point supporting the IEEE 802.11e standard gives information about permitted frequency use in the country field of transmitted beacon frames and probe response frames. A subscriber can then carry out configuration itself and adapt to the implemented frequency requirements based on these guidelines.

A further expansion of the original standard applies to the provision of an extended quality of service. IEEE 802.11e describes two more procedures for access to the transmission medium: **enhanced distribution coordination function** (EDCF) and **hybrid coordination function** (HCF).

With the support of IEEE 802.11e, the BSS radio cell becomes a QBSS radio cell, whose management is taken over by a hybrid coordinator (HC). While ECDF represents the standard access procedure of the QBSS radio cell, in which the subscribers

can compete with one another (contention period) while accessing media, HCF can
be implemented in both phases CP and CFP.

EDCF defines eight different traffic categories (TC), over which data transmissions
can be processed in different quality. During a CP period, each of these TCs receives access to the transmission medium for a specific length of time (transmission
opportunity, TXOP). All TCs within a station manage a backoff time independent
of each other. This waiting time depends on the priority of the TC. The higher the
priority, the lower the waiting time.

Similar to PCF, HCF defines a centrally managed access procedure on the transmission medium. The HC assigns the subscribers transmission permission one after
another. The HC initiates the allocation of transmission permission (poll TXOP) by
sending a QoS poll frame to the relevant subscriber, after the transmission medium
was not occupied for the length of a PIFS. As PIFS is shorter than DIFS or AIFS,
the HCF data traffic always receives priority over the EDCF data traffic. HC can be
initiated during the CFP as well as during the CP. Just as in the case of PCF, the CFP
phase is started over a beacon frame through the HC and ends after a specific time
period. This can be made known either over a beacon frame or by the transmission
of a CF end frame. Fig. 5.11 shows media access via HCF.

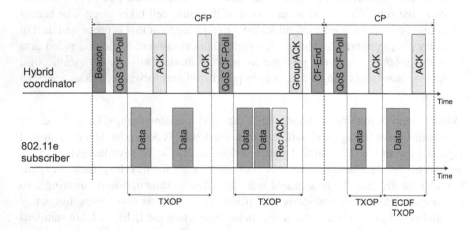

Fig. 5.11 Prioritized data transmission in the WLAN via IEEE 802.11e HCF.

5.2.4 WLAN Security

Radio-based local networks are naturally not able to offer a comparable level of security with wired networks that use spatially restricted and shielded network cables
as the transmission medium. In WLANs free space is used as the transmission medium and the dissemination of electromagenetic waves cannot be predicted exactly.

Therefore, non-authorized use or eavesdropping on data traffic is considerably easier. While a wall attenuates the transmission signal, it cannot be assumed that an outer wall can restrict a WLAN effectively. A so-called **network sniffer**, which can record the entire data traffic onto a given transmission medium and filter out security-relevant information, can be easily installed in a WLAN. Physical contact to the actual network is not necessary here, as in the case of wired networks.

This also means it is not difficult for potential attackers to discover a WLAN. Through the periodic transmission of beacon frames, with which an access point makes known its radio cells and their network access parameters, an attacker can at any time find out whether or not he is in range of a WLAN. He needs do no more than mount an antenna on the roof of his car and search for beacon frames while driving. This form of spying on radio data networks is known as a **parking lot attack**. A company can protect itself from unauthorized access in the wired Internet with a firewall, however, the access points located in a company network are still excluded from this type of protection mechanism.

Generally, the following types of attacks on WLANs are distinguished:

- Passive attacks of eavesdropping and the subsequent decryption of data traffic using methods of statistical analysis. Eavesdropping on data traffic is also known as **sniffing**.

- Active attacks, with the purpose of introducing potentially dangerous data traffic from non-authorized mobile computers into the WLAN. A attacker intervenes in the communication of authorized WLAN subscribers in order to manipulate it. Active attacks on WLANs are carried out in many ways, for example:

    - With the help of a false identity the attacker can sidestep possible access restriction to the WLAN, thereby gaining access to the data traffic and the participating network devices. This procedure is also known as **spoofing**.

    - A further attack scenario is the targeted overload of the WLAN's available resources, so that it is no longer in a position to carry out its regular service and can even be paralyzed (**denial-of-service attacks**).

    - To break into a secured WLAN so-called **brute-force attacks** might be carried out. Here, the attacker constantly tries out new password or key combination until the correct combination is found.

    - Unauthorized infiltration into foreign data traffic is also called **hijacking**. It is the starting point for the well-known **man-in-the-middle attack**.

It is a general problem that a WLAN attack can often only be determined when it is actively taking place and data is being manipulated. If data is only read passively, the unauthorized access cannot be detected. For this reason, safety-critical and private data should always be transmitted over an encrypted WLAN. Even if some of the IEEE 802.11 security mechanisms have been compromised in the meantime, it still requires quite some effort to decrypt even weakly encrypted data traffic. This effort is then expended by a hacker when he has something to gain, e.g., obtaining access data to financial transactions or other security-critical information. Moreover, publicly accessible WLANs that do not have the appropriate security measures are often abused for the delivery of SPAM.

IEEE 802.11 offers its own security measures. However, these are often deactivated on delivery and startup of the network components and must first be set up and configured by the user or network administrator. Users have the following possibilities to secure their data traffic in the WLAN:

- Activation of decryption methods, e.g., **wired equivalency protocol** (**WEP**) or **Wi-Fi protected access** (**WPA**). While today WEP is no longer considered secure, nevertheless, as an unauthorized person effort and energy are required to break down this security measure. Greater security is ensured by activation of the WPA encryption method.

- Deactivation of the shared key authentication, i.e., the implementation of a commonly used key. This method is provided by the IEEE 802.11 standard as a fundamental access limitation. As this WEP encryption is particularly easy to break during the authentication phase over the shared key procedure, its application is not advised.

- Access control via the **access control list** (**ACL**), in which the MAC addresses of all authorized subscribers of the radio cell are listed. Only subscribers that are also listed in the ACL can be connected to the radio cell.

- Hiding the SSID. If the SSID is sent openly in the beacon frames, an attacker can draw conclusions from the name used and potentially determine if it is worthwhile to break into the cell.

- Changing the initial password at the access point – as this can easily be determined by the attacker. The access point transmits the BSSID of the radio cell in every beacon frame, which at the same time corresponds to the MAC address of the access point. From the MAC address, which consists of a central IEEE registered, manufacturer-dependent section (OUI) and a section that is assigned by the manufacturer, the manufacturer of the access point can be found out easily. Operational handbooks for the access point are usually available on the Internet and these also contain information concerning the initial password via which an administrator can gain access to the access point. Once having gotten through to the access point, the attacker can change the configuration without a problem.

Wireless Equivalency Protocol – WEP

As a defense against attacks by potential hackers, who are already using the known security loopholes in the WLAN technologies, the so-called **wired equivalency protocol** (**WEP**) was defined in IEEE 802.11. This protocol had the goal of guaranteeing the private sphere in the WLAN through encryption of transmitted data packets. This had the secondary function of preventing unauthorized access and ensuring the integrity of the transmitted data. In retrospect, much too weak cryptographic procedures were used out of consideration for the usually less powerful

Spying and Break-ins in Foreign WLANs – Wardriving

Wardriving is a term used to describe spying on and infiltrating foreign networks using simple means. Many activists interpret the word "war," as an acronym for "wireless access revolution." While often viewed as a trend or entertainment past-time, wardriving can cause considerable damage. Tracking down foreign WLANs is relatively easy, as mentioned previously. The degree of danger is especially high as it can be determined that in a great number of WLANs there have been no security measures taken whatsoever. This means that they are completely open to every attacker.

The term wardriving comes from the older expression "**wardialing**," that originated in the 80's cult movie, "WarGames." In the movie, a young hacker indiscriminately dials company telephone numbers with his modem in an attempt to to break into an unprotected place.

To locate an unprotected network, wardrivers – who are normally on the road in the city – have simple mobile computers (notebooks, laptops, netbooks, PDAs, etc.) with a WLAN adapter and a suitable external directional radio antenna. It can be constructed using simple materials based on instructions gotten from the Internet. A **network sniffer**, e.g., the program NetStumbler, runs as computer software. It can also be found free on the Internet. Continuously searching all channels, it displays the located radio cells with the information it has found out about them. If the recorded transmission is coupled with a global positioning system (GPS), the position of the discovered WLANs can additionally be registered. The resulting maps are often made available on the Internet and continually updated.

Additionally, the located WLANs are often identified with special chalk-drawn markings (**warchalking**) on buildings or on the sidewalk. These symbols provide all necessary access information: SSID, available bandwidth, implemented channels and access controls. This language is taken from the signs used by hobos in 1930's America to indicate potential places to work or free meals. A closed circle means that the the described radio cell is a closed network. Two half circles placed back to back indicate an open, unprotected WLAN (cf. Fig. 5.13).

Further reading:

Ryan, P. S.: War, Peace, or Stalemate: Wargames, Wardialing, Wardriving, and the Emerging Market for Hacker Ethics. Virginia Journal of Law & Technology, Vol. 9, No. 7 (2004).

Netstumbler: http://www.netstumbler.com/

Fig. 5.12 Spying and break-ins in foreign WLANs – wardriving.

mobile devices. Reliable protection from one terminal to the next can therefore not be achieved with WEP. The WEP protocol is based on a secret, symmetrical key that is agreed on between the AP and other participating computers [1]. The payload data of the WEP data packet is encrypted with this key before it is sent. Furthermore, the intactness of the data packets is checked to make sure that they have not been manipulated while enroute. The encryption verifies that no unauthorized third party, who is not in possession of the secret key, can eavesdrop or read the transmitted data. Likewise, it ensures that no unauthorized party can itself improperly send encrypted

---

[1] A general overview of the relevant cryptographic procedures can be found in Chap. 5, "Digital Security," in the first volume in this series: Meinel, Ch., Sack, H.: Digital Communication – Networking, Multimedia, Security, Springer (2013).

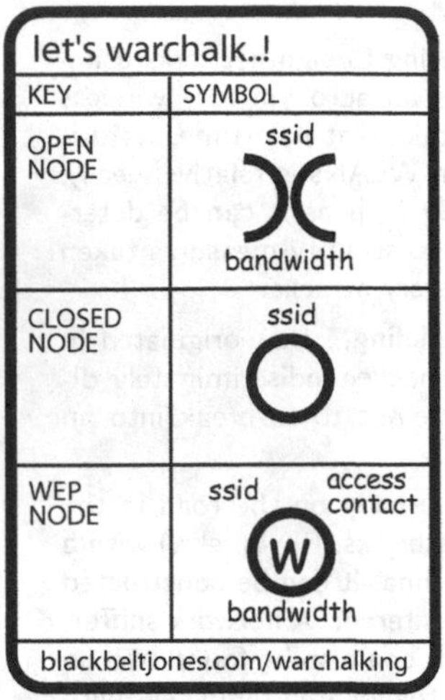

Fig. 5.13 Warchalking symbols according to `http://www.warchalking.org`

data in the WLAN that another subscriber would be able to decrypt correctly. For this reason, in most WLAN implementations a single key is determined manually and then the participating computers assigned from the access point.

A further problem of the WEP encryption can be found in the underlying encryption algorithm **RC4** itself. This is a so-called stream cypher algorithm from the company RSA Security Inc.. Stream cypher algorithms expand a short prescribed key, the so-called **initialization vector (IV)**, into an infinite stream of pseudo random keys. The IEEE 802.11 standard provides for a 40 bit long secret key (WEP40) for encryption. It was expanded to 104 bit (WEP128) by the manufactures of WLAN components for safety reasons and finally standardized with IEEE 802.11i Encrypted data is made identifiable in a 802.11 MAC data packet through the setting of the protected frame bit in the frame control field of the data packet header.

To encrypt the data to be transmitted, the key stream created with the IV is linked with the actual user data bit by bit over the Boolean operation XOR. The receiver can easily regain the original data from the encrypted data by implementing the same method. A 32-bit long checksum (integrity check value, ICV) is appended on the encrypted pay load data. It is calculated over the unencrypted user data and also transmitted in encrypted form. The ICV is used by the receiver of the user data to verify the correct decryption. The decrypted IC value is compared with a self-calculated checksum over the decrypted user data. If both result in the same value then the decryption was carried out correctly.

Additionally, the WEP data packet contains a 32-bit long IV field that is placed in front of the decrypted data and the 24-bit long initialization vector, a 6-bit long padding data field that always contains the value null, and a 2-bit long key ID, which specifies one of 4 possible keys that must be used for correct decryption by the receiver of the data packet . The structure of the WEP data packet is shown in Fig. 5.14.

WEP data packet

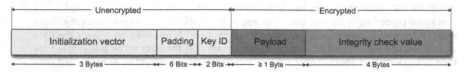

Fig. 5.14 WEP data packet.

The entire key (WEP seed) is thus made up of the IV and the secret WEP key. This is used for the initialization of the pseudo-random number generator (PRNG), which is responsible for the generation of the key stream. The process of the WEP encryption is shown again schematically in Fig. 5.15.

Manufacturers often give their WLAN products key lengths of 64 bits or 128 bits. This is in fact misleading as the 24 bit long IV is always sent as part of the WEP key in the clear text. The actual key lengths, which allow conclusions to be drawn regarding the cryptographic strength of the procedure, reach only 40 bits or 104 bits. All subscribers of a WLAN must have access to the same secret WEP key. Thereby, the IEEE 802.11 standard maintains up to 4 different keys. These must be manually entered as alphanumeric strings of 5 characters (WEP40) or 13 characters (WEP128). Before a subscriber can associate with a secure WLAN, it must first authenticate, i.e., it must believably and verifiably authenticate its identity to the radio cell. This **authentication** takes place in infrastructure networks at the access point and in ad hoc networks between two subscribers of the IBSS respectively, before they begin a data exchange. Besides general data encryption, WEP also offers a **shared key authentication** procedure for secure authentication.

In addition to the shared key authentication procedure, there is a further standard authentication procedure, the **open system authentication**, anchored in the IEEE 802.11 itself. It can, however, also not be considered completely secure. Here, two authentication management frames are exchanged between both involved parties (subscriber and access point). If the access point does not support WEP encryption, an authentication does not take place and the subscriber can connect directly with the WLAN. If the access point supports WEP encryption, the subscriber must authenticate itself with its secret WEP key to the access point. Every subscriber receives access to the WLAN with a correct WEP key. In the case where a subscriber sends an incorrect WEP key, it is still authenticated but may not carry out any data exchange with the WLAN.

The shared key procedure itself works against this with a so-called challenge-response technique. However, this procedure runs the risk of revealing the secret key on which the authentication is based (cf. Fig. 5.16). In the shared key authentication, the subscriber and access point exchange altogether 4 authentication management frames for the purpose of authentication:

WEP – Encryption and Decryption

Based on a 24-bit long initialization vector (IV), in the WEP procedure first a 40-bit or 104-bit long secret key is generated. The standard itself does not take a position on the problem of key management. It only assumes that AP and the participating computers are using the same key, which is specified over the key ID field in the WEP data packet. With the help of the **RC4** algorithm, an infinite, pseudo-random key stream is generated (WEP PRNG) from the secret key. Before the data is transmitted in encrypted form, a checksum procedure is carried out. This appends a CRC-32 checksum (integrity check value, ICV) on the clear text information to be sent and is intended to protect the data against unauthorized manipulation. The key stream and the data to be transmitted are then linked with one another over an XOR operation and transmitted as encrypted data together with the IV.

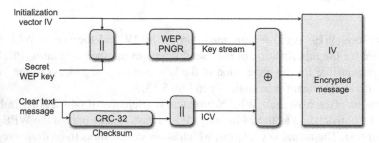

The user of a WLAN should implement different IVs so that the encrypted data packets generated are not always generated with the same key stream. First, the IV in clear text is transmitted, followed by the encrypted data stream. From the IV and the known key, the receiver can generate the key stream used for encryption, and through the simple application of the XOR link regain the original, unencrypted clear text information.

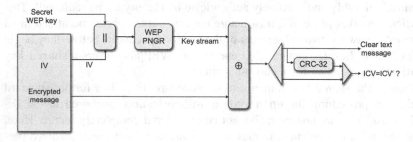

Further reading:

IEEE: IEEE Standard for Information Technology- Telecommunications and Information Exchange Between Systems-Local and Metropolitan Area Networks-Specific Requirements-Part 11: Wireless LAN Medium Access Control MAC and Physical Layer PHY Specifications. IEEE Std 802.11-1997, i -445 (1997).

Fig. 5.15 WEP protocols – encryption and decryption.

Breaking into the WEP Shared Key Procedure

The WEP shared key procedure is based on a simple challenge-response procedure. Here, a potential client identifies itself to a server aided by a secret key stream. This is generated with the help of the RC4 procedure.

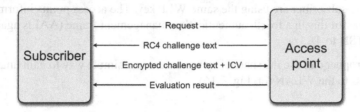

An attack on this authentication procedure can be carried out as follows:

After its request, the access point sends the subscriber the challenge text created over RC4.

The subscriber encrypts this text as indicated and sends the WEP packet (ICV + encrypted challenge text) back to the access point.

By way of unauthorized eavesdropping, the attacker can figure out the challenge text, ICV and encrypted challenge text (ciphertext). The following calculation can lead to finding out the secret key stream (cipher stream):

$$\text{challenge text} \oplus \text{ciphertext} =$$
$$\text{challenge text} \oplus (\text{challenge text} \oplus \text{cipherstream}) =$$
$$\text{cipherstream}$$

The attacker is now in possession of a valid combination made up of the key stream and ICV and can authenticate itself. If the access point sends a new challenge text, the attacker answers with a WEP packet consisting of ICV and the encrypted new challenge text, which it has calculated from the XOR combination with the known key stream. With this the attacker is successfully authenticated by the access point.

Further reading:

Housley, R., Arbaugh, W.: Security problems in 802.11-based networks. Commun. ACM 46, 5 (May. 2003), pp. 31–34.

Fig. 5.16 Breaking into the WEP shared key procedure.

In the first authentication management frame, the subscriber notifies the access point of its intention to authenticate itself (authentication algorithm identification, AAI is set to 1, likewise the authentication transaction sequence number, ATSN).

The access point answers (AAI is also set to 1, ATSN to 2) with a authentication management frame, which contains 128 bit long challenge text. This is generated with the help of the RC4 algorithm.

The subscriber calculates a checksum (ICV) for the received challenge text and a new IV, and then an encrypted challenge text and ICV with the secret WEP key.

Challenge text and ICV are sent back to the access point (AAI is again set to 1, ATSN to 3) in a further authentication management frame.

- The access point decrypts the received challenge text and ICV and compares the former with the original challenge text. A match indicates that the access point and subscriber are using the same WEP key. The access points informs the subscriber of this in a fourth authentication management frame (AAI is again set to 1, ATSN to 4).

An eavesdropper can use the shared key procedure in a simple way to gain unauthorized access to the WLAN (cf. Fig. 5.16).

Excursus 5: Critical Points Regarding the WEP Procedure

The limited key length plays a secondary role in criticism of WEP.
A more critical view is taken of the implementation of the RC4 algorithm, which for reasons of its static working method and short initialization vector offers few possibilities for variation in generating key streams. An attacker who eavesdrops on the data traffic only needs a small amount of collected information to break the secret key. A summary of weak points is that:

- the initialization vector is much too short with a length of only 24 bits.
- the 40 bit or 104 bit length of the secret key is too short.
- WEP works as a symmetrical encryption procedure with secret keys, however does not provide its own procedure for secure key management.
- the authentication implemented over WEP applies only to a WLAN adapter and not to the user itself.
- the integrity of the transmitted data is not secure, as weak WEP integrity control (CRC-32) cannot prevent the unauthorized altering of data.

The encryption algorithm that the WEP procedure is based on, RC4 , is a stream cipher, i.e., the generated cipher stream should be created in such a way that there is never a repetition of the encryption pattern. However, the length of the selected initialization vector WEP for the RC4 procedure is only 24 bits long. This means that a heavily used WLAN cannot be protected sufficiently. There is a 50% probability, that the same key will repeat after approx. 5,000 transmitted data packets[2]. Even with a moderate network load it can take just a few hours at an access point before an initialization vector repeats. The higher the number of subscribers in a WLAN, the lower the time needed for collecting the data packets to be encrypted with the same initialization vector. From these, WEP40 as well as WEP128 encrypted WLANs are affected. In spite of having different key lengths, both use equally long initialization vectors.

If two clear text messages are encrypted on the basis of the same initialization vector, the attacker can easily find out the XOR link of both clear text messages. From this, in turn, through appropriate methods of statistical analysis, the actual original data can be determined.

Encryption of Clear Text Messages with Identical Initialization Vectors
Key s is given, generated from the two identical initialization vectors, according to the RC4

---

[2] The 24 bit long initialization vector allows 2^{24}=16,777,216 different bit combinations. However, because of the "birthday paradox" there is a higher than 50% chance that the initialization vector, and therefore the entire RC4 encryption stream generated by it, will repeat already after 4,096 data packets.

procedure, v_1 and v_2 its two cipher texts that after the encryption with s incurred from the original clear text messages k_1 and k_2.

$$v_1 = k_1 \oplus s$$
$$v_2 = k_2 \oplus s$$

If the attacker knows that both cipher texts v_1 and v_2 were generated with the same key s, then it can easily gain access to the XOR link of both clear texts k_1 and k_2:

$$v_1 \oplus v_2 = k_1 \oplus s \oplus k_2 \oplus s = k_1 \oplus k_2$$

Although the attacker does not yet know the original individual messages k_1 and k_2, it can find this out easily through simple statistical procedures. If the attacker can now correlate a known pair v_1 and k_1, it can calculate k_2 simply without knowing the key s:

$$k_2 = k_1 \oplus (v_1 \oplus v_2)$$

In many cases, even the content of both clear text messages can be obtained this way. As soon as part of a clear text message can be ascertained, the corresponding part of the other clear text message is automatically present.

The more encrypted data, which is encrypted with the same key stream, collected, the higher the chance of success with this method. It is normally not a problem for an attacker to intercept multiple data packets that are encoded with the same key. As the initializati-on vector has only a length of 24 bits, the key sequence repeats on an access point that transmits 1,500-byte long data packets with a throughput of 11 Mbps after approximately 5 hours. Within this time an amount of data accrues at an access point of maximum 24 GBy-tes. An attacker can easily record and register this to find data packets with an identical initialization vector and therefore, also with an identical secret key. Because in the IEEE 802.11 standard there is nothing prescribed about the generation of the initialization vector, not all of the manufacturers of WLAN hardware utilize the available 24 bits completely so that an initialization vector often repeats much earlier.

In 2001 Fluhrer, Mantin and Shamir discovered that for the RC4 encryption algorithm, there are additionally so-called weak initialization vectors. These weak initialization vectors give an indication of the identity of a byte in the secret key created by it. If a potential attacker collects an adequate number of data packets – 6 million data packets or 8.5 GByte are already enough – these contain a sufficient number of weak initialization vectors for a complete reconstruction of the secret key.

The reconstruction of the secret key is even simpler when only digits and letters are used to generate it, rather than desired byte values, as some manufacturers allow in the user-defined determination of initialization vectors. The number of possible key sequences generated from it are therefore limited further, and potential attackers must collected even fewer data packets to be able to reconstruct the key.

Attack on Weak Initialization Vectors in the WEP Procedure (Weak Attack)

Fluhrer, Mantin und Shamir found out that in the RC4 encryption algorithm among 256 formed key streams there is a weak key stream. This is a key stream that is formed through a so-called weak initialization vector, which correlates even more strongly with the key stream than planned. Every data packet containing a weak initialization vector allows conclusions to be draw about the key byte. On the basis of this, an estimation can be made with 5% certainty about the identity of the key byte. The more data packets with a weak key vector are present, the easier it is to make a reliable estimation of the key bytes.

In this case only data packets are interesting that have a value of between i=3 and i=15 in the first byte of the initialization vector and in the second byte the value 255.

$$IV = (3 \dots 15, 255, X)$$

For every value $3 \leq i \leq 15$ approx. 60 initialization vectors and the corresponding messages encrypted with it are necessary. As the data traffic in the WLAN on higher protocol layers is processed via the TCP/IP protocol, it can be assumed that the user data section of all WEP data packets is appended with a SNAP header (cf. Chap. 8), which always has the same initial value: 0xAAAA030000. A part of the unencrypted data packet is thereby already known. This can be used to determine the following encrypted byte.

In this kind of an attack, 4 to 6 million data packets are already sufficient to break the encryption. This amount can theoretically be collected after only a few minutes. In 2007 Tews, Pychkine and Weinmann improved this attack procedure so that a data packet amount of just 40,000 − 90,000 was sufficient to successfully break the WEP128 key. Based on active attack techniques, e.g., the introduction of ARP requests, it could be shown that 40,000 data packets in a IEEE 802.11 WLAN collect in less than a minute.

Circumventing Data Integrity in the WEP Procedure

A further weak point of the WEP procedure is found in the implementation of the CRC32 checksum for calculating the integrity check values (ICV). Its actual purpose − originally carried out with the development of this checksum procedure − was the reliable detection of transmission errors. It was not implemented to ensure data integrity as it is used in the WEP procedure. On the one hand, CRC32 is a generally well-known procedure, i.e. following a manipulation of the actual user data an attacker can recalculate the checksums correspondingly to cover up its manipulation. On the other hand, RC4 as well as CRC are linear algorithms. This means that bit by bit changes of encrypted WEP data packets can be continued in the ICV even if the secret key is not known. This form of manipulation can be used to introduce manipulated data packets in the WLAN or to redirect data packets from their designated goal.

As already mentioned, the ICV in the WEP data packet is a simple CRC32 checksum whose calculation has linear calculation properties, as does the RC4 encryption procedure, and it can be used for the targeted manipulation of encrypted data.
To carry this out, the attacker intercepts a WEP message (IV,C), whereby IV designates the initialization vector and C the encrypted contents with

$$C = RC4(IV, K) \oplus (M, CRC(M)).$$

C is made up of the RC4 key stream RC4(IV,K), which is formed on the basis of the initialization vector IV and of the secret WEP key K, as well as the clear text M and its checksum CRC(M).

An attacker can construct a correctly encrypted message C', which lets itself be decrypted analog to (M',CRC(M')). For this it chooses an arbitrary message D with the same length

$$
\begin{aligned}
C' &= C \oplus (D, CRC(D)) \\
&= RC4(IV, K) \oplus (M, CRC(M) \oplus (D, CRC(D)) \\
&= RC4(IV, K) \oplus (M \oplus D, CRC(M) \oplus CRC(D)) \\
&= RC4(IV, K) \oplus (M \oplus D, CRC(M \oplus D)) \\
&= RC4(IV, K) \oplus (M', CRC(M')).
\end{aligned}
$$

To do this the attacker needs no knowledge about M or M'. But a "1" in a random position D results in a changed bit in the same position in the encrypted message M'. In this way, controlled manipulations are possible on the original clear text message M.

The attack possibilities clearly illustrate the weak spots of the WEP encryption so that today an application of this security procedure is strongly advised against. As a countermeasure, proprietary improvements of the WEP procedure are proposed or new security procedures created to guarantee increased security. **WEPplus** (also WEP+) is a proprietary improvement of the WEP procedure from the Agere Systems company. In full compatibility with the WEP standard, it suppresses the application of unreliable, weak initialization vectors through a simple filter mechanism. If there are only components in a WEPplus WLANs that support this extension, weak initialization vectors and thereby weak key streams can be avoided completely. However, because this in not official standard, the network cannot force a new subscriber to use this filter mechanism.

Further reading:

Fluhrer, S. R., Mantin, I., Shamir, A.: Weaknesses in the Key Scheduling Algorithm of RC4. SAC '01: Revised Papers from the 8th Annual International Workshop on Selected Areas in Cryptography, pp. 1–24, Springer-Verlag, London, UK (2001)

Housley, R., Arbaugh, W.: Security problems in 802.11-based networks. Commun. ACM 46:5, pp. 31–34 (2003)

Tews, E., Weinmann, R. P., Pyshkin, A.: Breaking 104 Bit WEP in Less Than 60 Seconds. In Sehun Kim and Moti Yung and Hyung-Woo Lee, editor(s), WISA, (4867):188–202, Springer (2007)

IEEE 802.11i – WPA and WPA2

Of greater significant in **IEEE 802.11i** are the security extensions that have been introduced, in which manufacturer-independent security procedures are defined. These meet today's requirements and do not demonstrate the weakness of WEP that we have seen. Maintaining compatibility requirements with older WLAN resources proved to be a problematic factor. Therefore, an optional encryption procedure was defined that is based on the original WEP algorithm but does not have the weakness it exhibited. This "short-term" solution is known as the **temporary key integrity protocol** (TKIP). It can be easily implemented by updating driver software or network firmware in existing WLAN products. In the long run, the old RC4 encryption procedure is also intended to be replaced with a modern procedure (AES CCMP, advanced encryption standard counter mode/CBC-MAC). Because of the more complex cryptographic procedure, this also makes a more powerful WLAN hardware necessary.

The IEEE 802.11i standard divides WLAN hardware into two product groups:

- Robust Network Security (RNS) WLAN components and
- Pre-RSN WLAN components (WEP)

The standard was put into application by the manufacturers with the encryption methods WPA and WPA2 (Wi-Fi protected access) .

WPA still implements the WEP procedure based on the vulnerable RC4 algorithm, while offering extra protection through dynamic keys on the basis of the temporal key integrity protocol (TKIP). In addition, authentication of the subscriber is carried out over a pre-shared key (PSK) or the extensible authentication protocol (EAP) over IEEE 802.1X.

The **temporal key integrity protocol** implements three new security mechanisms, which guarantee additional security with respect to the WEP procedure. The first

part consists of a key mixing function, combining the secret key with the initialization vector before it is used for initialization of the RC4 key stream. TKIP calculates
the specifications for the RC4 keystream generator in two key mixing stages. An attacker can therefore no longer draw any conclusions from the initialization vector
about the RC4 key. Besides this, weak initialization vectors with a high correlation to generated keystreams are prevented from occurring. Next, the length of the
initialization vector was increased from 24 bits to 48 bits. By this squaring of the
original key space it is made certain that even in a fully utilized WLAN with a
forward-looking data transmission speed of 500 Mbps, only after a good 200 years
could a repetition of the keystream be reckoned with. The 48 bit initialization vector
is thereby subdivided into a 16 bit low part, incremented from data packet to data
packet, and a 32 bit high part, correlating to the MAC address. Therefore, on the
one hand it is made certain the same initialization vector with different subscribers
leads to different RC4 keys, and on the other hand, a sequence counter is set up and
offers protection against so-called replay attacks. Finally, the original CRC 32 integrity check value is extended with a new 64 bit **Michael message integrity check**
(MIC).

Michael, the hash algorithm implemented, has a lower computational complexity to
ensure compatibility with older WLAN hardware. A further key is also involved in
calculating the MIC; a calculation that cannot be carried out without this knowledge.
To falsify a data packet the attacker must first break the MIC using brute force
methods. Fig. 5.17 shows the structure of the IEEE 802.11i data packet with TKIP.

TKIP Data Packet

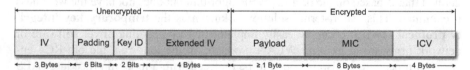

Fig. 5.17 Structure of the IEEE 802.11i data packet with TKIP.

For authentication, WPA offers two different possibilities depending on the application:

Pre-shared Keys (PSK) for the home and for smaller working groups (Small
Office and Home Office, SOHO). The PSK is generated from a password phrase
that is entered manually by all subscribers of the WLAN (subscribers and access
points). The password phrase can have a length of between 8 and 63 characters. The actual PSK has a length of 64 bytes and is made up of a pass phrase,
SSID and SSID lengths with help of a cryptographic hash function (PBKDFv2,
password based key derivation function version 2). The PSK is only used for
the generation of the secret and temporal session keys and not for the actual encryption of user data. Because of this, a markedly higher security is guaranteed
compared to WEP.

- **Extensible Authentication Protocol** (EAP) over **IEEE 802.1X** for larger WLAN networks in which a special server is employed as a authentication instance. In EAP the negotiation of the actually implemented authentication mechanism is carried out first during authentication. A subscriber who wishes to authenticate itself in a WLAN (supplicant) transmits an authentication message over the responsible access point (authenticator) to an authentication server. Thereby, different, multiple procedures are used one after the other. The authenticator is in control, and by means of transmitted requests determines the procedure. If the authentication was successful, the authentication server indicates to the authenticator to allow the supplicant access to the WLAN. The authentication server itself functions as a central interface and carries out the actual authentication of access points and subscribers. Moreover, it enables an automatic distribution of the dynamic key.

The authenticator facilitates the authentication server as proxy (EAP proxy), who provides the actual access point to the WLAN over a secured connection. It transparently forwards data packets between the supplicant and the authentication server. This is normally implemented over RADIUS (Remote Authentication Dial-In User Sevice) or transport layer security (TLS). The connection between supplicant and authenticator runs over the EAPOL protocol (EAP over LANs). Originally, EAP was developed in 1998 for authentication via dial-up connection in a network (RFC 2284, later replaced by RFC 3748). IEEE 802.1X regulates the transport of EAP messages in the LAN over EAPOL. Fig. 5.18 illustrates the format of the EAP message.

The following authentication methods are supported by EAP:

- **EAP-MD5** (EAP Message Digest Algorithm, RFC 1321)

  In a basic challenge-response procedure, the user name and password are transmitted encoded with a MD5 hashing algorithm, but without the sender and receiver authenticating themselves beforehand. This can be easily compromised by a simple man-in-the-middle-attack and therefore not recommended for WLAN communication.

- **LEAP** (Lightweight EAP)

  Proprietary solution from the Cisco company modeled after IEEE 802.1X, in which client and server initially identify each other in a challenge-response procedure. Afterwards, independent of the user, a session key is generated through its own key hashing procedure.

- **EAP-TLS** (EAP Transport Layer Security Protocol)

  Corresponds to the in RFC 2716 defined combination of EAP and Secure Socket Layer, over which a certificate-based mutual authentication takes place between client and server. This assumes that certificates will be distributed in a secure way to the subscribers of the WLAN. EAP-TLS supports user generated and dynamically generated keys and is implemented in the authentication and the subsequent negotiation of the dynamic session key.

- **EAP-TTLS** (EAP Tunneled Transport Layer Security)

Extensible Authentication Protocol – Message Format

Communication via the EAP protocol is based on four simple message types that are exchanged between the involved parties in a request-response procedure. The messages begin with a 1-byte long code field specifying one of the four message types. Then comes another a 1-byte long identifier field that identifies the respectively exchanged message during the running of an authentication process in the request-response cycle. The subsequent 2-byte long field contains the length of the message, including the header field. Finally, the transported user data follows, the length of which can be variable.

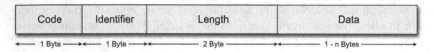

The following EAP message types are specified:

- **EAP-Request** (Code=1) for the transport of messages from the authenticator to the supplicant.
- **EAP-Response** (Code=2) for the transport of messages from the supplicant to the authenticaor.
- **EAP-Success** (Code=3) with which a successful authentication is transmitted to the supplicant.
- **EAP-Failure** (Code=4) with which a negative authentication attempt is acknowledged by the authenticator.

EAP-Request and EAP-Response additionally contain a 1-byte long type field that specifies the subtype of the exchanged message. The following subtypes were established:

- **Request-Identify** and **Response-Identify** (type=1) for querying the names of the supplicates and sending the names back to the authenticator.
- **Notify** (type=2) serves the transmission of clear text messages.
- **EAP-MD5** (type=4), **EAP-TLS** (type=13), **LEAP** (type=17), **EAP-SIM** (type=18), **EAP-TTLS** (type=21), **PEAP** (type=25) and **EAP-FAST** (type=42) serve in establishing a special EAP method for authentication.
- **NAK** (type=3) is a negative acknowledge, it is sent from the supplicant back to the authenticator when it does not support a proposed EAP method.

Further reading:

Blunk, L., Vollbrecht, J.: PPP Extensible Authentication Protocol (EAP),RFC 3748, Internet Eng. Task Force (2004)

Fig. 5.18 Extensible Authentication Protocol – message format.

Expanded form of the EAP-TLS method, in which before actual user authentication, first a secure TLS tunnel between supplicant and authentication server is established via a server certificate. A subscriber can then identify itself at the authentication server via user name and password. As no user certificate is necessary, the administrative effort is enormously reduced.

- **EAP-PEAP** (EAP Protected Extensible Authentication Protocol)
  Likewise an expanded form of the EAP-TLS, very similar to the EAP-TTLS, in which as the first step a secure tunnel is established between the supplicant and the authentication-server, without supplicant authentication. The actual authentication takes place subsequently over a clear text password authentication via MS-CHAPv2 or a token authentication over GTC (Generic Token Card).

- **EAP-SIM** (EAP Subscriber Identity Module)
  This procedure enables an authentication over the information stored on a SIM card. For example, to ensure secure communication when using a mobile telephone in a WLAN.

- **EAP-FAST** (EAP Flexible Authentication via Secure Tunneling)
  An extension of the LEAP method, developed by the Cisco company, that establishes a secure tunnel between supplicant and authentication-server with the help of a symmetrical encryption procedure. This is carried out before the actual authentication without the necessity of certificates.

In contrast to the simpler WEP procedure, different secret keys are implemented in a RSN architecture, which are derived from a **key hierarchy**. A private key is required, for example, for direct communication between the access point (authenticator) and subscriber (supplicant). Multicast and broadcast messages that must be received by multiple or all subscribers of the WLAN are encrypted with keys especially for this purpose. These must be known to all subscribers and distributed from the access point beforehand.

The key exchange handshake necessary for this procedure is divided into two phases. In the first phase, the pairwise key handshake for unicast messages is negotiated between every single subscriber and the access point. In the second phase, the group handshake takes place, in which the key for unicast and broadcast messages is exchanged. In each case the most important key is the **pairwise master key** (PMK). It is assigned to the respective communication between the supplicant and the authenticator. The PMK is either derived directly from the use of a pre-shared key or negotiated by use of an authentication server between it and the supplicant, and then subsequently sent to the authenticator. As the connection between authentication-server and authenticator is itself not a part of the WLAN, but of the wired backbone LAN, the PMK is never sent over the insecure WLAN. Via EAP, PMK is generated from the first 256 bits of the so-called AAA keys. This is the result of the handshake protocol between authentication server and supplicant.

From the PMK, with the help of a pseudo-random number generator (PRNG), the MAC addresses of the authenticator and two random numbers, originating each from the authenticator (ANonce) and the supplicant (SNonce) **pairwise transient key** is generated as temporary session key. Independent of the negotiated key procedure this has varying lengths of 384 bits (AES-CCMP) or 512 bits (TKIP). The PTK itself is subdivided into three or five keys:

- EAPOL-Key Encryption Key (KEK, 128 bit) for encryption of the data field of the EAPO- key data packets,
- EAPOL-Key Confirmation Key (KCK, 128 bit) for integrity security of the EAPOL-key data packets,
- Temporary Key (TK, 128 Bit for AES-CCMP or 256 bit for TKIP) for data encryption,
- if TKIP/MIC is used, two more 64-bit long MIC keys are added, which are taken from TK for AES-CCMP.

For the exchange of multicast and broadcast messages, a **group master key** (GMK) is generated, from which a session-related **group transient key** (GTK) is derived. The GTK has a length of 128 bits (AES-CCMP) or 256 bits (TKIP) and is formed over a pseudo-random number generator in which besides the PMK also the the MAC address of the authenticators and the ANonce value is entered. The GMK itself is generated by the authenticator and replaced by a new GMK at regular intervals.

The EAP authentication proceeds in the following way (cf. Fig. 5.19):

- Before the actual EAP authentication, the subscribers execute an open system authentication, followed by an association with the access point. Open system authentication is carried out solely for reasons of backward compatibility with older WLAN hardware, which require it in case of connection establishment with the access point without pre-shared key authentication.
- After the association, the EAP authentication is initiated with the transmission of a EAP Start message by the supplicant to the authenticator.
- The authenticator answers with a EAP identity request, in which it determines the identity of the supplicant. At the same time, the authenticator opens a 802.1X access port to the authentication-server for the secure exchange of messages between the supplicant and the authentication server.
- The supplicant sends an EAP response message back to the authenticator containing identity information.
- If an authentication method with mutual authentication was previously chosen, the authenticator answers with proof of identity via an EAP response message.
- Afterwards, further information between the supplicant and the authentication-server are exchanged. Their number and contents depend on the chosen EAP authentication method. Both authentication-server and supplicant generate the PMK, which is subsequently transmitted from the authentication- server to the authenticator via a secure connection.

- Finally, the authenticator informs the supplicant of a successful EAP authentication with an EAP success message.

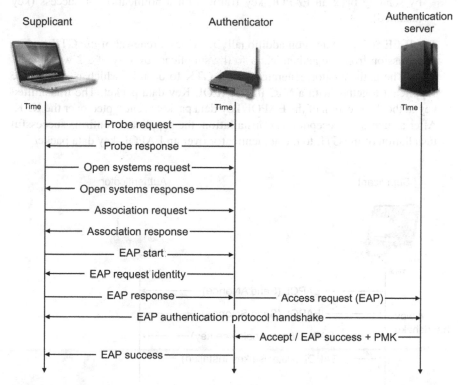

Fig. 5.19 The handshake of the EAP protocol between subscriber (supplicant), authenticator (access point) and authentication server.

After the EAP authentification has been successfully completed, in the second phase – by a 4-way handshake – a **temporary PTR** is generated for the encrypted communication between supplicant and authenticator (cf. Fig. 5.20).

- The 4-way handshake begins with the exchange of the Nonce-random value generated by both communication partners. First, the authenticator sends an EAPOL-key data packet with an ANonce to the supplicant.

- From the ANonce of the authenticator and its own SNonce, the supplicant forms a MIC and transmits an EAPOL key data packet to the authenticator. This contains the SNonce, the information element of the association request data packet and the MIC of the EAPOL-key data packet.

- The authenticator checks the MIC and transmits a EAPOL-key data packet containing the ANonce, MIC, information element of the beacon data packets or probe response data packets, and an encrypted GTK , which the supplicant should install.

- After a successful key installation, the supplicant ends the 4-way handshake by sending back an EAPOL-key frame with a notification of success (key installed).

The IEEE 802.11i extension additionally provides a renewal of the GTK and its transmission from the authenticator to the supplicant by way of a 2-way handshake. The authenticator generates a new GTK to do this, which is sent to the supplicant together with a MIC in a EAPOL-Key data packet. The transmitted key in the data section of the EAPOL key data packet is encrypted over the KEK. After a successful reception and installation, the supplicant confirms successful installation of the GTL to the authenticator over an EAPOL-key data packet.

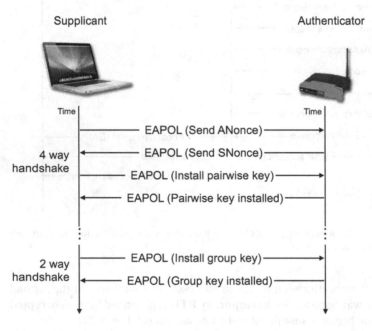

Fig. 5.20 4-way handshake following successful EAP authentication for the installation of the temporary PTK/GTK session key and the subsequent 2-way handshake to renew the GTK group key.

The most striking extension of the IEEE 802.11i standards applies to the implemented new encryption procedure **AES-CCMP**, based on the advanced encryption standard (AES). This is the successor of the well-known encryption procedure DES. For reasons of compatibility with older WLAN hardware, IEEE 802.11i still supports TKIP, however only as an optional component. AES-CCMP is designated as the standard.

WPA2 as the successor of WPA also stipulates AES-CCMP as the standard encryption procedure. In this way, on the one hand, the stream cipher RC4 implemented in WPA becomes obsolete, on the other hand, the abandonment of TKIP enables the implementation of WPA2, also in ad hoc networks. A simple conversion from WPA to WPA2 by means of a firmware update is possible in most devices, but not all. The device hardware used is often too slow to implement the AES encryption as software. The addition CCMP (Counter with Cipher Block Chaining-MAC Protocol, RFC 3610) identifies, how the AES algorithm is used on WLAN data packets [3]. An efficient implementation of the AES-CCMP protocol is only possible directly in the hardware. An implementation on a software-basis does not achieve the necessary throughout by the processors normally employed at the access points. In contrast to TKIP, the symmetrical AES in the IEEE 802.11i standard presented works with a 128 long bit key[4] and offers protection against unauthorized eavesdropping on communication as well as a guarantee of data integrity and authenticity. While RC4, on which the WEP is based, generated a continuous key stream (stream cipher), AES always works on complete data blocks, which are encrypted individually. This means that data units consisting of a length of 128 bits each are encrypted with a 128-bit long key.

AES has been considered secure up to now and has not yet been broken. For actual data encryption AES-CCMP again uses temporary session keys. Key distribution and authentication proceed over an EAP authentication or PSK. Just as AES-CCMP, AES is also operated in the so-called cipher block chaining mode. In this way, even greater security is achieved and attacks considerably more difficult to carry out (cf. Fig. 5.21).

The AES-CCMP encrypted data packet (also known as CCMP Medium Access Control Protocol Data Unit, CCMP-MPDU) contains a MAC header, a CCMP header, user data, message integrity code (MIC) and frame check sequence (FCS). From these only the user data and the corresponding MIC are each encrypted. The CCMP header is made up of 8 bytes and contains a 48-bit long packet number (PN), a 48-bit long initialization vector and a key identification (KeyID). The packet number (PN) serves in calculating a nonce value, which is instrumental in encryption and integrity control to prevent a key from being used multiple times. To this end, the PN is incremented by 1 with every message sent and initialized anew with1 as soon as the temporary key is renewed. Separate PNs are administered for PTK and GTK. To protect against replay attacks, the received PN is always compared with an internal counter. Fig. 5.22 shows schematically the process of the AES-CCMP encryption. Table 5.3 compares the various standards established in IEEE 802.11 for the safe transfer of data.

---

[3] In establishing the IEEE 802.11i standards with AES-OCB (Offset Codebook Block Mode), another encryption procedure was originally under discussion. As AES-OCB was subject to patent restrictions it was, however, rejected for AES-CCMP, in order to avoid potential license fees.

[4] AES stipulates original key lengths of 128, 192 or 256 bits.

Increased Security with the Cipher Block Chaining Mode

IEEE 802.11i with AES-CCMP implements encryption via the AES algorithm, reinforced further through the use of the cipher block chaining mode. The cipher block chaining mode is an operational method in which block ciphers can be operated like AES. Before encryption of a clear text block, it is first linked with the encrypted text block generated in the previous step with XOR. The first clear text block is thereby linked with a previously agreed upon initialization vector.

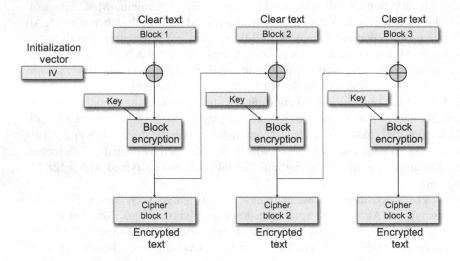

The most important advantages of the cipher block chaining mode are:

- Patterns in the clear text are made unrecognizable through the chaining.
- Originally identical clear text blocks result in different encrypted blocks.
- Attacks are made more difficult on the implemented block cipher.

As every encrypted text block only depends on the previous block, a damaged cipher block, e.g., a bit error in data transmission, does not cause great harm in decryption as only the affected clear text block and the following clear text block are incorrectly decrypted.

The cipher block chaining mode can also be used for securing data integrity. The last encrypted block as MAC (Message Authentication Code) is appended to the original clear text and this is encrypted together with the MAC.

Further reading:

Delfs, H., Knebl, H.: Introduction to Cryptography: Principles and Applications, 2nd ed., Springer, Berlin, Heidelberg, New York (2007)

Fig. 5.21 Increased security with the cipher block chaining mode.

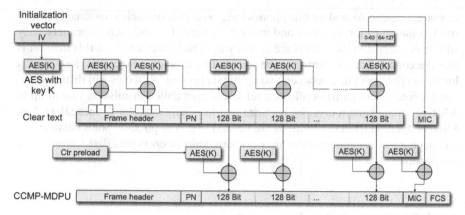

Fig. 5.22 Schematic process of AES-CCMP encryption.

Table 5.3 A comparison of the IEEE 802.11 security solutions.

	WEP	**TKIP**	**CCMP**
Algorithm	RC4	RC4	AES
Key length	40/104 bit	64/128 bit	128 bit
Key lifetime	24 bit IV	48 bit IV	–
User data integrity	CRC-32	Michael	CCM
Header integrity	–	Michael	CCM
Key administration	–	802.11i 4-way handshake	802.11i 4-way handshake

5.3 Bluetooth – IEEE 802.15

The increasingly widespread networking of data communication-enabled terminals within a few meters of each other (typically under 10 m) is based on specific technology for wireless data communication. This could apply to a printer connected to an office computer as well as to a wireless headset connected to mobile phone. This spatially limited area of data communication is known as **PAN** (**Personal Area Networks**). As a **Wireless Personal Area Network** (**WPAN**) it is subject of the IEEE 802.15 standard.

The fact is that the number of individual devices communicating with each other at this close range is on the rise. A diverse assortment of items can be found, including various peripheral devices such as the mobile phone, PDA (Personal Digital Assistant), printer or camera and even intelligent household devices. A complicated infrastructure or expensive cabling would not make sense in this case, and it would be much too costly. Among the most popular data transmission procedures in the area of the WPANs are **IrDA** (Infrared Data Association), for data transmission with the help of infrared light, and **Bluetooth**.

Countless notebooks and mobile phones today still contain such a reasonably priced IrDA communication interface and many peripheral devices, such as printers, may still be connected at the workplace in this way. The disadvantage of this method is that the communication partner must have a line-of-sight connection, i.e., objects located in the direct line between two terminals become obstacles and disturb communication. Although IrDA allows a relatively high data transmission rate of up to 4 Mbps, the maximum distance between two terminals may not be more than 2 m. Most of the time IrDA connections are limited to purely point-to-point connections and neither a connection to other network structures or encrypted data transmission are possible.

5.3.1 Bluetooth Technology

The WPAN technology made famous under the name **Bluetooth** allows terminals a wireless, radio-based communication in the 2.4 GHz ISM frequency band (i.e., between 2.402 GHz and 2.480 GHz) with data transmission rates of up to 1 Mbps (with Bluetooth 2.0+ EDR with 2.1 Mbps). The advantage of Bluetooth, in comparison to infrared based technologies, is the fact that the communication partner is no longer dependent on a direct line-of-sight connection between terminals. One of the design goals of Bluetooth was the networking of battery-operated terminals. Therefore, a primary focus in development was placed on low energy consumption. For this reason, Bluetooth devices, belonging to the group of devices designated as energy efficiency class 2, are only capable of bridging distances up to 10 m with a transmission capacity of 100 mW. The implementations of the Bluetooth interface are also kept very simple so that they can generally be implemented on a single chip. In terms of radio technology, Bluetooth works according to the frequency hopping procedure (FHSS – Frequency Hopping Spread Spectrum), changing the carrier frequency of the radio signal 1,600 times per second.

Because of the rapid frequency change, FHSS guarantees the necessary robustness against frequency selective disturbances. Additionally, a 128 bit encryption is implemented, intended to prevent eavesdropping and break-in attempts. Unauthorized eavesdropping is more difficult to carry out in Bluetooth than in the WLAN, based solely on Bluetooth's range limitation of only 10 m.

A PAN that is operated via Bluetooth is also called a **piconet**. A piconet consists of a group of Bluetooth devices all using the same FHSS hopping sequence synchronously. A special device in the piconet works as the master device. Up to seven further subscribers are connected to the master device and work as slave devices. The master determines the FHSS hopping sequence, which all slaves follow synchronously. While principally every Bluetooth node could function as master, there can only be one master acting in a piconet. All nodes in the piconet use the same FHSS procedure. Its frequency length is determined by the respective master, who also regulates access to the transmission medium through polling and reservation.

The History of Bluetooth

The namesake of the Bluetooth technology was the King of Denmark, son of Gorm, *Harald Gormsen* who lived in the 10th century (910–987) and was known for being very communicative. He united the monarchies of Norway and Denmark and laid the foundation for the Christianization of Scandinavia. As was usual at that time, Harald was also known by an epithet, which in his case was "Blåtand." This would literally be translated today as "blue tooth."

A good 1000 years later, the Swedish company Ericsson began developing a "multi-communicator link." Based on a suggestion from the developer's friends, the name of the project was renamed **Bluetooth**, after the communicative Viking King. In 1998 the five companies, Ericsson, IBM, Intel, Nokia and Toshiba founded the Bluetooth consortium with the goal of developing inexpensive, one chip solutions for close range wired network communication. In 1999 IEEE also founded its own group especially to deal with wireless networking in close range (Personal Operating Space, POS), IEEE 802.15.

Two year later in 2001, the first Bluetooth products appeared on the mass market. Today many mobile phones, notebooks, PDAs, smartphones, netbooks, tablet PCs, cameras, printers, headphones as well as numerous other peripheral devices are equipped with a Bluetooth interface.

Further reading:

Haartsen, J. C.: The Bluetooth radio system. IEEE Personal Communications Magazine, vol. 7, pp. 28–36 (2000).

Fig. 5.23 The history of Bluetooth.

Every piconet uses a different FHSS sequence, whereby a node can become a member of a piconet by synchronizing to its FHSS sequence.

In addition to masters and slaves, there are two further types of devices that are not active among the devices participating in the network communication. The first group is the so-called "parked devices," which at that moment do not have a connection to the master of the piconet but are known and can be reactiviated at any point on short notice. The second group consists of "stand-by devices." In contrast to parked devices these do not participate in the piconet. A piconet can be composed of up to 255 members, i.e., 8 active members (1 master and up to 7 slaves) and 247 inactive, parked members. A parked device can only be reactivated to participate in a piconet if there are either fewer than 7 active slaves taking part in the piconet, or if in exchange one of the active slaves switches to park status.

Theoretically, every Bluetooth-enabled device is capable of initiating a piconet as master. All subsequent devices entering the picconet then become slaves. The FHSS hopping sequence of the picconet is determined by the device identification (MAC address) of the master, who sends this out together with its current internal clock for initial synchronization of the picconet. A slave synchronizes its own internal clock with the received value and adapts its FHSS hopping sequence to that of the master's. As an active member of the piconet, every device receives a 3-bit long active member address (AMA). Parked (inactive) devices receive an 8-bit long parked member address (PMA). Devices in the stand-by mode are not assigned an address as they are not a part of the piconet.

A Bluetooth device can be registered in multiple piconets, however can only function as a master in one. Up to ten piconets form a **scatternet**. Here, the members of the individual networks can establish contact with each other. Those Bluetooth devices that do indeed exchange data with one another other must be in the same piconet. Every piconet may be identified by its own FHSS hopping sequence. This is determined by the respective master device.

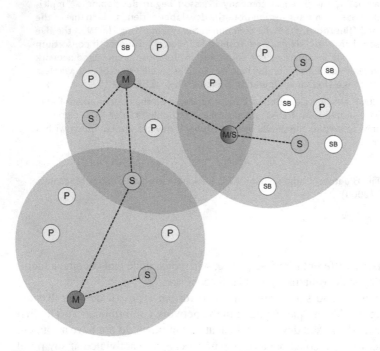

Fig. 5.24 Bluetooth scatternet with master node (M), slaves (S), parked Bluetooth devices (P) and stand-by Bluetooth devices (SB).

Piconets share the complete available bandwidth. With the rising number of members in the network the chance for the occurrence of a collision increases while, simultaneously, the performance falls. A collision happens at the moment the same frequency is accidentally used in two different FHSS hopping sequences. To participate in a scatternet a slave must only synchronize with the FHSS hopping sequence of the picconet in which it wants to become a member. In this way, it automatically drops out of its old piconet. However, before this happens the slave communicates to the former master that it will not be available for a defined period of time. If a master wishes to switch to another piconet, this is only possible if it enters the new piconet as a slave. In the new piconet it cannot become master without the old and new piconet working with the same FHSS hopping sequence, so that in this way one single big network is formed. If a master leaves a picconet, the data traffic in the old picconet is interrupted until the point when the master returns. In order to guarantee communication between two piconets, individual members can periodically switch between both networks. In this way, it is even possible for isochronous data streams to be exchanged between piconets.

Fig. 5.24 shows a scatternet made up of three piconets, with master (M), slaves (S), parked Bluetooth devices (P) and stand-by Bluetooth devices (SB). A member functions interchangeably as master in one of the piconets and as slave in another (M/S). The self-organizing scatternets have still not been standardized. One reason is that up to now no algorithm been developed that is able to address all of the requirements of a scatternet at the same time. Therefore it is not guaranteed that all Bluetooth devices actually support this scatternet operating mode.

5.3.2 Bluetooth Protocol Stack

The Bluetooth protocol stack is divided into a Bluetooth **core specification**, describing the protocols involved on the physical layer and the data link layer, and a Bluetooth **profile specification**. The latter encompasses standard solutions for differ user scenarios, thus containing a selection of protocols from all the layers of the protocol stack. While the core specification retains the classic horizontal layering of the TCP/IP Reference Model, the profile specification divides the protocol stack into vertical sections (cf. Fig. 5.25).

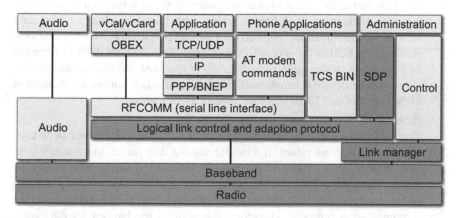

Fig. 5.25 The Bluetooth protocol stack.

Among the basic protocols of the Bluetooth core specification are:

- **Radio**
  On the physical layer, the radio layer specifies each frequency area, modulation used as well as transmission power. Bluetooth uses the worldwide license-free 2.4 GHz frequency band in the ISM frequency range. Communication takes place via FHSS, with a hopping rate of 1,600 hops per second and a time slot of 625 μs

between two hops. Every time slot uses a different, randomly chosen frequency out of a total of 79 individual channels with a distance between each of 1 Mhz. The GFSK modulation procedure is implemented as the modulation procedure. Three different power classes are specified for Bluetooth :

– *Power class 1*:
  Transmission power from 1 – 100 mW for bridging distances of up to 150 m line-of-sight with prescribed power control.
– Power class 2:
  Transmission power from 0.25 – 2.5 mW for bridging distances of up to 10 m with line-of-sight with optimal power control.
– Power class 3:
  Transmission power of maximum 1 mW.

• **Baseband**:
  Here, the basic mechanisms for connection establishment, data formats, time response and quality of service parameters are fixed. In addition to the following protocols mentioned, in contrast to other network protocol architectures, with Bluetooth it is possible to implement and subsequently use audio applications directly on the baseband. A Bluetooth data packet in the baseband is composed in the following way:

– *Access code*:
  This first part of the data packet is necessary for the synchronization and identification of the piconet. It begins with a 4-bit long preamble, a 64-bit long field for synchronization, followed by a 4-bit long trailer. The 64-bit long synchronization field is for access to the transmission medium derived from the 24 lowest bits (lower address part, LAP) of the MAC address of the master. If a particular device is summoned, this LAP is used. Additionally, there exist further pre-defined LAPs to locate specific device groups.
– *Packet header*:
  The 54-bit long packet header starts with a 3-bit address field that can be used for different address values. If a master sends a data packet to a slave, it contains the receiver's address. In the opposite situation, if a slave transmits to the master, the address field is understood as the sender's address. Only one slave can communicate with the master at one time, and the master controls the communication centrally. Therefore, this address mechanism for data communication is sufficient. The address 000 is intended as a broadcast message from the master to all of the slaves. A 4-bit long type field follows the address field. Here, different packet types for controlling or for synchronous and asynchronous data transmission can be specified.
  A simple flow control is implemented over the 1-bit long flow field. In the case where this field is set to 0, data communication is stopped and only continues when a data packet is received that has a set flow field.
  The two following fields, each 1-bit long, contain an acknowledgement and a sequence number, which is determined over the alternating bit protocol. An

8-bit long checksum (header error check) finishes off the packet header and is designed to identify transmission errors. The 18-bit long total field of the packet header is encoded with the help of a forward error correction mechanism, which expands this by a factor of 3:1, and is thereby intended to ensure additional transmission security of the important information.

– **User data**:
The Bluetooth data packet can transport up to 343 bytes of user data. The structure of the user data depends, among other things, on the individual connection type chosen.

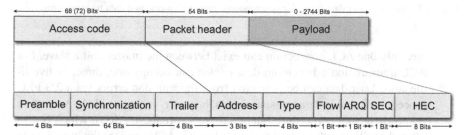

Fig. 5.26 Bluetooth data format in the base band.

On the user data layer, Bluetooth offers a synchronous, connection-oriented service for the transmission of voice signals and an asynchronous, connectionless service for the transmission of data packets.

– *Synchronous Connection-Oriented Link (SCO)*:
For the transmission of voice signals, it is implemented in the traditional telephony circuit-switched, symmetrical point-to-point connections. For this purpose Bluetooth provides a connection-oriented, synchronous service. Here, a slave receives two time slots in regular intervals, following one after another, for a connection in a forward and backward direction. A master can maintain up to three different SCO connections to slaves. A slave can support two SCO connections to different masters or three SCO connections to other slaves. A SCO connection transmits voice signals with a data rate of 64 kbps with the help of the continuous variable slope delta procedure. In order to avoid a repetition of the data transmission in error-prone and unsecure communication connections, an additional forward error correction (FEC) can be used. This increases data volume up to a factor of 3. Fig. 5.27 gives an overview of the different data schema of the user data for the SCO data transport.

– *Asynchronous Connectionless Link (ACL)*:
For data transmission Bluetooth offers an asymmetrical packet-switched transport service, which is controlled by the master of each piconet. A slave may only answer when directly addressed by the master in the previous time slot. The master proceeds by polling the slaves one after another. At the same ti-

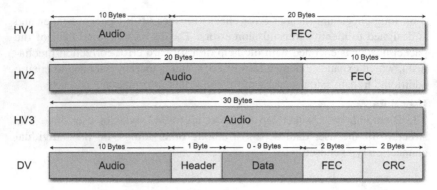

Fig. 5.27 Bluetooth user data types for synchronous, connection-oriented data transmission

me, only one ACL connection can exist between the master and a slave. For ACL transmission a Bluetooth data packet can occupy one, three, or five time slots. User data can be protected from transmission errors via a 2/3 FEC procedure. Reliable data transmission is, however, only possible by way of an acknowledgement procedure (automatic repeat request, ARQ), which is also offered by Bluetooth in the baseband. For the ACL service, the user data in the Bluetooth data packet contains a 1–2-byte long extra data header (1 byte for data packets from a time slot, 2 bytes for data packets extending over multiple time slots), contained therein is an identification for the logical channel between two L2CAP (Logical Link Control and Adaption Protocol, see Fig. 5.25) participants, a broad field for the flow control on the L2CAP level and a length field (length without data header field and checksum), followed by a CRC checksum. Fig. 5.28 gives an overview of the different data schema of user data for ACL data transport.

Link Management:
Here, the structure and the management of connections between two Bluetooth devices is specified. This includes the selection of the available security mechanisms and the negotiation of the necessary parameters that go along with it. The link management protocol (LMP) expands the service offerings provided by the baseband. Protocols on higher layers, however, can access baseband services directly. LMP offers the following services:

Authentication, pairing and encryption
LMP controls parameter exchange for the authentication carried out in the baseband. In Bluetooth, pairing is understood as the creation of a secure communication environment and mutual trust between two Bluetooth devices that have, up to that point, not communicated with each other. Finally, LMP generates and administers a connection key and is involved in Bluetooth encryption. Here, it determines the implemented key length and initial starting value (seed) for random numbers.

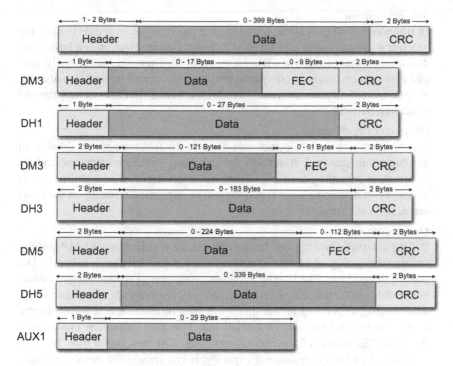

Fig. 5.28 Bluetooth user data types for asynchronous, connectionless data transmission.

Table 5.4 Different user data types in the Bluetooth baseband.

Type	Header (byte)	User data (byte)	FEC	CRC	Data rate symmetrical (kbps)	Data rate forward (kbps)	Data rate backward (kbps)
DM1	1	0 – 17	2/3	yes	108.8	108.8	108.8
DH1	1	0 – 27	no	yes	172.8	172.8	172.8
DM3	2	0 – 121	2/3	yes	258.1	387.2	54.4
DH3	2	0 – 183	no	yes	390.4	585.6	86.4
DM5	2	0 – 224	2/3	yes	286.7	477.8	36.3
DH5	2	0 – 339	no	yes	433.9	723.2	57.6
AUX1	1	0 – 29	no	no	185.6	185.6	185.6
HV1	–	10	1/3	no	64.0	–	–
HV2	–	20	2/3	no	64.0	–	–
HV3	–	30	no	no	64.0	–	–
DV	1	10 – 19	2/3	yes	121.6	–	–

– *Synchronization*

In Bluetooth communication, following the reception of each data packet, the receiver's internal clock is readjusted as exact synchronization is a basic condition for communication in a Bluetooth network. There are, additionally, special data packets that support synchronization. Bluetooth devices can also exchange synchronization information to bridge time differences in neighboring Bluetooth networks.

– *Parameter negotiation*

Not all Bluetooth devices completely support the functionality established in the Bluetooth standard. Bluetooth devices must therefore exchange and fix information regarding their own technical characteristics and abilities. It is necessary to determine what functions can be supported mutually in the communication exchange.

– *QoS negotiation*

Polling intervals, latency, and bandwidth can be regulated over LMP and determine the quality of service of communication in a Bluetooth network. The polling interval is defined as the maximum time period separating two consecutive data packet transmissions occurring between between master and slave. Additionally, depending on the quality of data transmission, a forward error correction method can also be carried out. This also influences the targeted data rate. The master of a piconet controls the communication flow. In order to increase its own transmission capacity, it can reduce the number of time slots available to a slave for transmission.

– *Power control and connection monitoring*

Devices participating in a Bluetooth network are capable of measuring the power of a received signal and can cause the transmitter to increase or reduce its transmission power in order to improve transmission quality.

– *Status and transfer mode change*

Via LMP, Bluetooth devices can switch between different operating modes. This applies to the transition from master to slave or vice versa, as well as exiting from a Bluetooth network, switching from a stand-by mode, etc.

Fig. 5.29 presents an overview of various Bluetooth operating modes in which a Bluetooth device can be located. Also identified are potential transitions occurring in operating conditions between these individual operating modes.

• **Logical Link Control and Adaption**:

The logical link control and adaption protocol (L2CAP) enable an adaption of the higher protocol layers to the possibilities offered by the Bluetooth baseband. The possibility is available to set up different types of logical transmission channels with quality of service guarantees between Bluetooth devices for ACL data traffic (SCO data transmission does not have to use the base band directly). In each case, channels are identified with a unique channel identifier (CID).

– *Connectionless, uni-directional channels*, which are mostly used for broadcast messages to slaves (CID=2). Additionally, in a connectionless L2CAP data

The Operating Modes of Bluetooth Devices

A general distinction is made in a Bluetooth network between active and inactive devices. Furthermore, there exist different levels of inactivity resulting from the power-saving possibilities.

- Principally, every Bluetooth device that is not actively participating in a piconet and is not switched off is in a **stand-by status**. Only the internal clock of the Bluetooth device is operating initially.

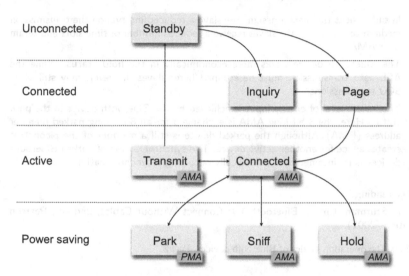

- The Bluetooth device can switch out of the stand-by status to the **inquiry status**. This means that the Bluetooth device itself can initiate a new piconet and additionally search for further Bluetooth devices within its transmission and reception range. To facilitate this, an inquiry access code (IAC) is transmitted over 32 standardized carrier sequences (wake-up carriers). Conversely, a Bluetooth device can also periodically switch to the inquiry status in order to search for IAC messages on the wake-up carrier frequencies. If one of these is recognized, the Bluetooth device transmits back a data packet with its own device address and time stamp information, thereby becoming a slave in an already existing piconet.
- Following a successful inquiry, the Bluetooth device switches to the **page status**. A master wishing to initiate a piconet waits until all of Bluetooth devices in range wishing to participate have registered. For synchronization of the piconet, the master establishes a frequency hopping order, which it communicates to the slaves. The slaves of a Bluetooth piconet synchronize their own frequency hopping order with that of the piconet.
- Following successful adaption to the frequency hopping order of the piconet, the Bluetooth device changes to the **connected status**. All devices connected to the piconet receive an active membership address (AMA).

Fig. 5.29 The operating modes of Bluetooth devices, Part 1.

The Operating Modes of Bluetooth Devices, Part 2

- Following allocation of a transmission slot by the master, active participants in a Bluetooth piconet can transmit data via the master to other participants in the network They are then in the **transmit status**.
- Via a special detach procedure, a Bluetooth device can switch back into the stand-by status. In order to work actively as well as with energy efficiency in a Bluetooth network, three energy saving modes were established:

  · In **sniff status** the master assigns the slave a reduced interval on the transmission medium to listen, i.e., the slaves receive a reduced number of time slots and retain their AMA.

  · The Bluetooth device works more economically in the **hold status**. While the ACL data transmission must be stopped here, slaves, however, may still serve SCO transmissions.

  · The lowest level of energy input is achieved by the Bluetooth device in the **park status**. Here, the 3-bit long AMA is exchanged for an 8-bit long parked member address (PMA). Although the parked device is still a member of the piconet, it creates space for another active device. The synchronization of parked Bluetooth devices is maintained by a periodically occurring resynchronization.

Further reading:

Bray, J., Sturman, Ch. F.: Bluetooth 1.1: Connect Without Cables, 2nd ed., Pearson Education (2001).

Fig. 5.30 The operating modes of Bluetooth devices, Part 2.

packet identification, to identify the receiver on higher protocol layers, and a length field must also be transmitted.

- *Connection-oriented, bi-directional channels* with double-sided support of pre-defined quality of service parameters (CID$\geq$64). Connection-oriented L2CAP data packets need no receiver identification, in contrast to connectionless L2CAP data packets. However, because of variable lengths (up to 64 kbytes), a length field is necessary.

- *Signaling channels*, which are necessary for the exchange of signal messages between L2CAP entities (CID=1).

L2CAP offers the possibility of adding slaves to existing piconets or removing them and supports multiplexing functions, QoS specifications, as well as the fragmentation and defragmentation of data packets. L2CAP data packets can be up to 64 kbytes long and must be divided into a maximum of 339-byte long fragments for transmission over the baseband.

Bluetooth supports a multi-level scalable security model. A distinction is made between three security modes with differing security mechanisms:

- *Security Mode 1 (Non Secure Mode)*: Here, an authentication is not necessary. The data is transmitted in unencrypted from. The Bluetooth devices continuously switch their frequency according to a prescribed hopping scheme in order to make eavesdropping more difficult.

- *Security Mode 2 (Service-Level Enforced Security)*: Security mechanisms are dealt with here by the host (application layer). A connection between the Bluetooth controller and the host computer is necessary.

- *Security Mode 3 (Link-Level Enforced Security)*: Security mechanisms are implemented directly through the Bluetooth module. Authentication on the link layer already takes place during connection establishment. There No additional measures are required on the host.

Among the security mechanisms implemented by Bluetooth are a challenge and response procedure for authentication (one-sided, two-sided or none), the encryption of data streams and the generation of a session key. The cryptographic procedures on the Bluetooth controller are usually of a simple type due to the power efficiency of the Bluetooth power-saving devices. Cryptographically-stronger, and therefore also more complex procedures are taken care of by higher protocol layers, just as the actual key management itself.

5.3.3 Bluetooth Security

The first step of secure Bluetooth communication already begins in the process of pair formation when two Bluetooth devices wish to communicate with each other for the first time. The input of an up to 16-byte long PIN is necessary by the users of both Bluetooth devices. Some devices with a limited user interface, e.g., headphones and headsets, already have a pre-installed PIN from the manufacturer. From the PIN of each device, address and self-generated random numbers multiple keys are generated. These are used for the authentication (connection keys) and subsequent encryption of the actual user data (user data keys). The generation of the individual keys and the authentication proceeds in Bluetooth over the symmetrical **SAFER+** block encryption method, which provides for a key lengths of 128 bits. The actual encryption of user data proceeds over the **E0** stream encryption, i.e., on the basis of the previously generated user data key as seed value, a stream of pseudo random bits is generated. This is linked with the user data over a logical XOR operation.

- **RFCOMM**:
The radio frequency communication protocol (RFCOMM) serves to emulate a wired interface, according to EIA-232 (RS-232) over Bluetooth (cable replacement protocol). In this way, all of the connections that previously could only be implemented over cable can also be used via Bluetooth, without modifications being necessary. Thus, telephony applications, for example, can use for the normal AT modem commandos, as in the case of a standard, wired modem. Calender information (vCal) and business cards (vCard) can be exchanged over the object

exchange protocol (OBEX). RFCOMM delivers a simple, reliable data stream similar to TCP.

- **Service Discovery Protocol**:
  The service discovery protocol (SDP) is located on the higher protocol layers and enables recording and searching for services. This is especially important for Bluetooth, as a Bluetooth device must be capable of networking with other devices in a foreign environment and working with them. Thereby, all devices offering a service must implement a SDP server, while an SDP client is sufficient when only searching for and using external services. SDP offers no possibility to inform Bluetooth devices about new services. Just as little does it allow access to the control of these services or to initiate forwarding. SDP uses a simple query/response model, whose individual transactions each consist of a request (request protocol data unit) and a corresponding answer (response protocol data unit). This guarantees a simple form of flow control, as a client is only allowed to send a new request when the answer of the previous request has been received. For a recording of the individual services offered by a SDP server, a service record is set up as a simple list with corresponding service attributes. The identification of a service record proceeds over a 32 bit address (service record handle). A service attribute consists of its own 16-bit long identification (ID) and corresponding value (attribute value), which can be a integer value, a Boolean variable, a string or even a uniform resource locator (URL).

- **Telephony Control Protocol Specification – Binary** (TCS BIN):
  Bit-oriented protocol for the control of telephone functions for voice and data connections via Bluetooth, including the possibility for mobility and group management.
  TCS BIN was especially introduced for wireless telephony, however did not succeed in convincing manufacturers. Today, the protocol is solely of historical interest.

- **Host Controller Interface** (HCI):
  Here, the interface between Bluetooth hardware and software is defined. Located between the baseband and L2CAP it serves to control the baseband, the connection management and access to hardware status and control register. The connection of the Bluetooth controller to the host system proceeds over the HCI interface. This makes the host available over an abstract interface for all baseband and link management functions.

- **Bluetooth Network Encapsulation Protocol** (BNEP):
  For connection to the classic TCP/IP protocol stack, the BNEP protocol, or the point-to-point protocol (PPP) as it was formerly known, is also located on the RFCOMM protocol for the emulation of a serial interface. For communication via Bluetooth, BNEP implements connection-oriented L2CAP channels.

The Development of the Bluetooth Standard

- **Bluetooth v1.0 and v1.0B** The first versions still had numerous problems regarding security and interoperability of products made by different manufacturers. The maximum achieved rate of data transmission was 732.2 kbps.

- **Bluetooth v1.1** This version was standardized as IEEE 802.15.1 in 2002. The many problems of the previous version could be remedied. In addition came the support of encrypted channels and monitoring of the input transmission strength of received signals. The maximum achieved data transmission rate was 732.2 kbps.

- **Bluetooth v1.2** This version, released in November 2003, was characterized by its decreased sensitivity to static disturbances. A new packet type for synchronous transmission (aSCO) was introduced that achieved a higher quality of audio transmission through the new transmission of corrupt data packets. The maximum data transmission rate achieved could be increased to 1 Mbps.

- **Bluetooth v2.0 + EDR** This specification is backwards compatible with v1.2 and was released in November 2004. It offers an increased data transmission rate (enhanced data rate, EDR) of maximum 2.1 Mbps.

- **Bluetooth v2.1 + EDR** This version was made available in August 2007. Besides additional improvements, it offers a higher level of security through a secure simple pairing in connection establishment as well as lower power consumption.

- **Bluetooth v3.0 + HS** This specification released in April 2009 provides for the integration of an additional high speed channel on the basis of WLAN or UWB (ultra broad band), to achieve data rates of up to over 24 Mbps. At the time of the device discovery, the Bluetooth interface, the initial connection and the profile configuration were still being used. If, however, larger amounts of data are to be transmitted an alternative basis protocol stack, that is 802.11 MAC PHY is used.

- **Bluetooth v4.0 + EDR** Version 4.0 was released in April 2010. If allowed by two Bluetooth devices, it is possible to establish a connection in less than five milliseconds and to maintain it up to a distance of 100 meters. An additional 128 bit AES encryption and a further reduction of energy consumption are planned.

Fig. 5.31 The development of the Bluetooth standard.

5.3.4 Bluetooth Profile

Data between Bluetooth devices is exchanged with the help of so-called Bluetooth profiles. These are pre-determined for specific user areas. Bluetooth profiles simplify maintaining compatibility of different hardware manufacturers and software developments. These can be used and implemented in a variety of different ways in the fixed standards of the Bluetooth specification. For this reason, additional Bluetooth profiles are fixed for the Bluetooth protocols. These are intended for implementation in certain infrastructures and for special types of application. They each fix certain sets of parameters and can be supported by Bluetooth devices to various degrees.

In this way, a profile presents a standard solution for a certain application scenario of Bluetooth technology. A specific set of protocols to be implemented and parameters to be exchanged are determined to ensure a smooth as possible and correct interoperability between Bluetooth devices. Every time a Bluetooth connection is set up between two devices, the communication partners each exchange their available profiles. In doing so, they determine which services they can make available to a potential partner and which data or commands they need for it. A simple example is a headset that requests an audio channel from a Bluetooth compatible mobile telephone and can control the volume via an additional data channel. In Table 5.5 several profiles from the constantly growing number of Bluetooth profiles are briefly introduced.

Table 5.5 Bluetooth profile.

Profile	Name	Function
A2DP	Advanced Audio Distribution Profile	Transmission of audio data
AVRCP	Audio Video Remote Control Profile	Audio and video remote control
BIP	Basic Imaging Profile	Transmission of image data
BPP	Basic Printing Profile	Printing
CIP	Common ISDN Access Profile	ISDN connections via CAPI
CTP	Cordless Telephony Profile	Wireless telephony
DUN	Dial-up Networking Profile	Internet dial-up connection
ESDP	Extended Service Discovery Profile	Expanded detection
FAXP	FAX Profile	Fax services
FTP	File Transfer Profile	Data transmission
GAP	Generic Access Profile	access control
GAVDP	Generic AV Distribution Profile	Transmission of audio and video data
GOEP	Generic Object Exchange Profile	Object exchange
HCRP	Hard Copy Cable Replacement Profile	Print application
HDP	Health Device Profile	Secure connection between medical devices
HFP	Hands Free Profile	Wireless telephony in a car
HID	Human Interface Device Profile	Input
HSP	Headset Profile	Voice output per headset
INTP	Intercom Profile	Radio
OPP	Object Push Profile	Sending objects
PAN	Personal Area Networking Profile	Network connections
PBAP	Phonebook Access Profile	Access to telephone book (read-only)
SAP	SIM Access Profile	Access to SIM card
SCO	Synchronous Connection-Oriented link	Access to headset microphone and loudspeaker
SDAP	Service Discovery Application Profile	Determining existing profile
SPP	Serial Port Profile	Serial data transmission
SYNC	Synchronisation Profile	Data matching
OBEX	Object Exchange	Generic data transmission between two devices

5.4 ZigBee – IEEE 802.15.4

The goal of the 2003 standard, ratified by the IEEE 802.15.4 working group, was the specification of a family of wireless devices networked with each other for the home environment (wireless home area networks, WHAN). They are characterized by the lowest possible power consumption and a battery life which potentially lasts for many years. In contrast to WLAN or Bluetooth, the possibility should exist for clearly less complex devices to be in the position of establishing a communication infrastructure with the least possible cost. These simple devices include: wireless light switches, household currency measuring meters, remote control and connection devices for linking entertainment electronics and household devices, as well as sensors and actuators to monitor and control industrial processes.

5.4.1 ZigBee Technology

Based on the IEEE 802.15.4 standard, **ZigBee** defines a high level protocol suite intended to offer a secure network infrastructure especially for wireless devices with low power consumption and increased battery life.

The name," ZigBee" comes from the so-called "zig-zag dance" of the honey bee. Returning to the bee hive, it shows in this way to the other bees where a source of food can be found. Similiarly, data packets in a meshed ZigBee-network should also find their way to the target. Founded in 2002, the ZigBee Alliance is a fusion of currently more than 150 companies, among them Philips, Texas Instruments, NEC, and Siemens, to name only a few. They supported and propagated the ZigBee standard ratified in 2004. The ZigBee Alliance functions for the IEEE 802.15.4 standard in a similar way as the Wi-Fi alliance for the WLAN standard IEEE 802.11. Via the certification of network products it tries to ensure their interoperability.

While IEEE 802.15.4 addresses the lower protocol layers with the definition of a physical layer and data link layer, the main focus of ZigBee is the protocol layers built on the Internet layer and application layer.

In a ZigBee network infrastructure, a distinction is made between three types of devices (ZigBee devices):

- **Simple end device** (ZigBee end device, ZED)
  Simple network-enabled devices, e.g., radio-controlled light switches, do not implement the entire ZigBee protocol suite. Because of this, they are often designated as devices with reduced functionality (reduced function device, RFD). They cannot initiate a network themselves and register at a router within their transmission range. A router forms a simple star topology with its ZigBee end devices.

- **Router** (ZigBee router, ZR)
  ZigBee-enabled devices, which implement the entire ZigBee protocol suite, are designated as completely functioning devices (full function devices, FFD). Full

function devices can serve both as end devices and as routers. If a full function device registers at a router it can support a tree topology as a further router, but also meshed topology through cross-references to other devices already connected to their own router.

- **Coordinator** (Zigbee Coordinator, ZC)
  This is a special full function-enabled ZigBee device within a ZigBee PANs. It serves as a router taking over the central role of a coordinator, i.e., it provides the fundamental parameters of the network infrastructure and manages the entire PAN. There can be only coordinator in a ZigBee network. It initiates the PAN and manages all information concerning the PAN. The coordinator is additionally able to set up connections to other networks.

In accordance with the IEEE 802.15.4 specification, the ZigBee network supports two different topologies: a star topology and a mesh topology (in the specification designated as "peer-to-peer topology"). In the star topology there are limited as well as full function ZigBee devices connected to a router (or also coordinator) as the central switching element. The peer-to-peer topology makes it possible for the participants to have direct communication with each other (cf. Fig. 5.32).

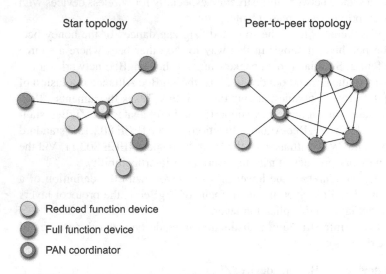

Fig. 5.32 Network technologies supported by IEEE 802.15.4.

Seen technically, the ZigBee PAN works with ranges of up to 75 m, also in the international open ISM frequency band with 16 channels of 2.4 GHz, 10 channels of 915 MHz and a channel in the European 868 MHz ISM band in the frequency spread process (direct sequence spread spectrum, DSSS). In contrast to Bluetooth, ZigBee does not use a FHSS frequency hopping procedure for channel assignment,

---

[5] The power level (decibel, dB) of a signal with respect to the size 1 mW is designated as dBm.

Table 5.6 Comparison of the IEEE 802.15.4 frequency bands.

Property	Prescribed values	
	915 MHz	**2.4 GHz**
Data transmission rate	40 kbps	250 kbps
Transmission power	1 mW	
Receiving sensitivity	-92 dBm[5]	-85 dBm
Transmission range	internal up to 30 m, external up to 100 m	
Latency period	15 ms	
Channel access	CSMA/CA	slotted CSMA-CA
Modulation method	BPSK	O-QPSK

as the effort for synchronization of the participating devices on the prescribed FHSS hopping sequence is considered too complex. Table 5.6 summarizes the differences and similarities of the ZigBee frequency bands. Various device-dependent physical layers are supported with data rates of between 20–250 kbps and a latency time of up to 15 ms. In contrast to Bluetooth networks, which only support 8 active devices, a ZigBee network can include up to 254 full function devices and 64,516 devices with reduced functionality.

5.4.2 ZigBee Protocols and Addressing

ZigBee supports two different addressing modes:

- **Direct addressing**

  In direct addressing the nodes and endpoints of the opposite side are indicated. A node, i.e., a ZigBee device, can serve multiple end points. Each node allocates up to 255 endpoints (subaddresses). Of these, many are reserved for special tasks (endpoint 0 for management tasks, endpoints 1 – 240 for application logic, 241 – 254 are reserved for future tasks, 255 for broadcast messages to all endpoints. A node is identified by a unique 64 bit long address (EUI-64, extended unique identifier).

- **Indirect addressing**

  In indirect addressing the coordinator allocates short addresses, with a length of 16 bits, to all nodes that register at the coordinator for participation in the PAN. In this way, the number of potentially participating nodes in the ZigBee network is limited to 65,536 devices. The coordinator manages the short addresses with the accompanying MAC address of the respective ZigBee device. In order to communicate with another node, a request is first sent to the coordinator, who then forwards it to the corresponding receiver node (binding).

Fig. 5.33 shows the protocol stack for IEEE 802.15.4 and ZigBee.

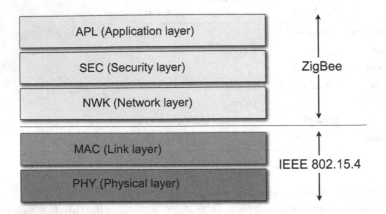

Fig. 5.33 IEEE 802.15.4 and the ZigBee protocol stack.

Media access and data formats are designed much simpler in IEEE 802.15.4 than in Bluetooth. A data packet on the physical layer (PHY) is composed of a 6-byte long header field, made up of a 4-byte long preamble, followed by a 1-byte long start of frame delimiter, which identifies the beginning of the user data. After this follows a 1-byte long length field. Up to 127 bytes of user data can follow. The data packet of the data link layer (MAC) is located in this user data. Its header starts with a 2-byte long frame control. Data packet types, address modes and further information regarding the decoding of the header are stored here. A 1-byte long sequence number follows for identification of the data packet as well as an address field of up to 20 bytes, for source and destination addresses in different formats (cf. direct and indirect addressing). After that the user data, which is up to 102 bytes long, and a 2-byte long checksum (frame check sequence) follow at the end of the MAC data packet. Fig. 5.34 shows the IEEE 802.15.4 data formats on the PHY and MAC levels.

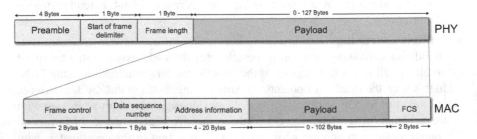

Fig. 5.34 IEEE 802.15.4 data formats on the PHY- and MAC level.

The type of data packet is determined with the help of the control field in the header of the MAC data packet. In ZigBee a distinction is made between the following types of data packets:

- Beacon data packets (code: 000) are transmitted by the coordinator in periodic time intervals to synchronize the ZigBee devices participating in the network. They are provided with information concerning the current network parameters.

- Regular data packets (code: 001) receive user data, which is processed and interpreted further by higher protocol layers.

- Acknowledgements (code: 010) for the confirmation of error-free reception.

- Control packets (code: 011) for the initiation and implementation of network management operation, e.g., network subscribers logging on and off.

ZigBee/IEEE 811.15.4 defines three different types of traffic, which are all supported by IEEE 802.15.4 MAC:

- *Periodic data traffic*, in which the application determines the frequency of data traffic. ZigBee-enabled sensors are activated and verify whether data has been sent, then subsequently go into an inactive system status again.

- *Intermittent data traffic*, i.e., data traffic that is periodically interrupted. Here again, an application or another (external) stimulus determines frequency. The ZigBee-enabled device must, however, only connect to the network when communication is really necessary. An example of such an application is a smoke detector. An alarm is only sounded in the network when it is set off by an external stimulus (here, smoke). The maximum energy can be saved with this communication variation.

- *Repetitive data traffic*, the fixed time points for connection establishment are determined in advance. With this guaranteed time slot (GTS), the ZigBee-enabled device works for a fixed period of time.

Here, the communication in ZigBee can run in one of two different modes:

- *Beacon mode* Operating mode with an optimal energy saving potential. This operating mode is employed if the coordinator is provided with power from batteries. The coordinator periodically transmits a beacon signal, which the ZigBee-enabled device is on the lookout for. Messages to the ZigBee-enabled devices are at the same time coupled with the beacon signal and are evaluated and processed by the receiving devices. When the communication is finished, the coordinator fixes a time interval. After its expiration the next beacon signal is sent. For this duration, the ZigBee-enabled devices in the network and the coordinator are in a power saving, inactive status. However, in order not to miss the next beacon signal the synchronization must either follow very accurately, or the ZigBee-enabled devices must already have switched over to the active mode at a specific time earlier. Therefore, the beacon mode for the receiving ZigBee-enabled devices is associated with an increased power consumption.

Non-beacon mode Operating mode employed if the coordinator is connected to a stationary power supply and the remaining devices are inactive most of the time, as in the case of e.g., smoke detectors or alarm systems. Here, the devices keep a vigil in randomly selected intervals, in order to confirm their affiliation with the network. If the devices are activated by an external stimulus, they immediately send a message to the permanent active coordinator. In this case there is, however, a chance (when also very small) that the communication channel is already occupied and that the receiver does not get the message.

In the beacon mode, the coordinator fixes a data packet format called a superframe. The length of the superframe is limited by the periodic transmission of beacon data packets. In between – depending on the particular application, which is fixed by the coordinator – is a sequence of single time slots of the same length. These can be used by ZigBee devices competitively for communication in the slotted CSMA/CA access method (contention access period, CAP). All transactions should be completed by the time the next beacon data packet is sent. For fast-reacting applications or applications demanding a constant service quality, a specific number of time slots can be reserved for dedicated media access (contention free period, CFP). These can be reserved as guaranteed time slots (GTS) at the end of the interval between two beacon data packets. The coordinator can establish up to seven of these GTS intervals. Their duration can also extend over more than one time slot (cf. Fig. 5.35).

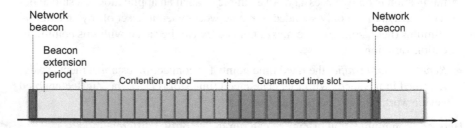

ZigBee superframe data format in the beacon operating mode.

5.4.3 ZigBee Security

On the level of the data link layer (MAC layer), ZigBee offers the possibility of the encryption of beacon data packets, acknowledgement data packets and control data packets. Thereby, in the MAC layer ZigBee supports only the encryption of data traffic between two devices in direct communication with each other (single-hop security). In order to encrypt data traffic between devices that are not directly adjacent to one another (multi-hop security), it is necessary to fall back on higher protocol layers.

In the MAC layer, the advanced encryption standard is employed to protect the confidentiality, integrity and authenticity of communicating data packets. While the MAC layer takes over the actual encryption work, the process and necessary exchange of keys beforehand is controlled via one of the higher protocols in the protocol stack.

If a device on the MAC sublayer transmits or receives an encrypted data packet, the peer address is determined beforehand and a secret associated key established. With this, the data packet must be encrypted or decrypted. For transmission of a data packet protected against integrity loss, the header of the MAC is used together with the user data to calculate a 4, 8, or 16-byte long message integrity code (MIC), which is appended to the user data (CBC-MAC, cipher block chaining). In order to ensure confidentiality in data transmission, a nonce value is generated from the user data together with a sequence and data packet counter. This is done for the initialization of encryption to prevent so-called replay attacks (AES in CTR-Mode). If upon reception of a data packet a MIC is present, it is verified and if the data present is encrypted it is then decrypted. Transmitting devices increase the data packet counter with every transmitted message. Receiving devices note the counter of the received data packet of every transmitting device. In case a data packet is discovered with an old data packet counter, a security alarm goes off.

AES is also employed on the network layer. All encryption variations (security suites) here are based on the AES-CCM* mode. Contrary to the AES CTR und CBC MAC variations implemented on the MAC layer, t AES-CCM* simplifies communication on the network layer. This is because only one variation is employed for the generation of a MICs (integrity control) as well as for encryption (or both together) with, in each case, only one secret key. If the network layer transmits or receives an encrypted data packet, it uses the services located in the operating system of the security service provider (SSP), to process the data packet.

The SSP determines from the address information which secret key is associated with the target or source address, and uses this on the data packet. At the same time, the SSP provides the network layer with suitable service primitives for the further processing of arriving or sent data packets in order that the network layer can take over their processing. However, control of the implemented security suites and the key management must be carried out by higher protocol layers.

5.5 Further radio-based network technologies

Currently, there exist numerous other proposals for the technical implementation of wireless PANs and sensor networks. It is not clear which among these could eventually become standards in the future. For the sake of completeness, they are briefly listed here.

- **Wireless USB** (also certified wireless USB, CWUSB), is a radio-based extension of the wired Universal Serial Bus Standards (USB), which seeks to combine the security and speed of wired communication with the uncomplicated quality of wireless radio transmission. Wireless USB id based on UWB technology (ultra-wide band), as specified by the WiMedia Alliance. It allows higher data transmission rates of up to 480 Mbps with a distance of up to 3 meters in the frequency range of between 3.1 GHz and 1.6 GHz.

- **DASH7** is a radio-based, energy efficient network technology. It began as a military development and follows the ISO/IEC 18000-7 standard for active RFID (Radio-frequency Identification) devices and works in the license-free frequency band at 433 MHz. DASH7 devices are designed with a battery lifetime extending over several years. They serve distances of up to 2 km between the individual network subscribers, with a transmission rate of up to 200 kbps.

- **Near Field Communication** (NFC) is a transmission standard for the wireless exchange of data over shorter distances. It was developed in 2002 by the Sony and NXP Semiconductors (formerly Philips) companies. It is intended to be used for exchanging telephone numbers, images, files or digital certificates between two devices situated close to each other. It was developed in part to facilitate payment using a cell phone at especially designed terminals. NFC works in the frequency range of 13.56 MHz and reaches data transmission rates of up to 424 kbps via a distance of only 10 cm.

- **Millimetre Wave Gigabit Wireless** (Wireless GIGE or GiFi) is a radio-based technology for data transmission in the frequency range of 60 GHz (millimeter waves). Wireless GIGE is designed to reach data transmission rates of up to 5 Gbps, the range that can be bridged is limited to approx. 10 m. Projected use of this technology is in the area of wireless networking in the home environment in media and entertainment devices such as televisions, games consoles and media centers.

- **Wireless HART** (Highway Addressable Remote Transducer) is a transmission method based on the IEEE 802.15.4 standard for the industrial production area that works in the license free 2.4 GHz ISM frequency band, according to the TDMA method (time division multiple access).

- **Z-Wave** is a standard for wireless communication that was developed by the Danish company Zensys and the Z-Wave Alliance, a group of 160 manufacturers with the goal of home automation for the wireless control of heating, ventilation, lighting and air-conditioning.
  Z-Waveworks is in the license-free 900 MHz ISM frequency range and uses GFSK modulation for data transmission. Data transmission rates of between 9,600 bps and 40 kbps are reached and distances bridged in free field of up to 200 m (in buildings approx. 30 m). Z-Wave is in direct competition with ZigBee.

- **Transfer Jet** is one of the proprietary, wireless data transmission techniques developed by the Sony company. It works in the high-frequency range of 4.48 GHz and provides data transmission rates of up to 560 Mbps. It reaches up to 3 cm over shorter distances with a very low energy consumption. Typically, two transfer jet devices should automatically assume contact as soon as they touch each other. Only simple point-to-point connections are supported.

The established specification in IEEE 802.16 for **Worldwide Interoperability for Microwave Access (WiMax)** was initially developed as a wideband radio transmission system for stationary terminals. In the meantime, it has been further developed as a a standard for third and fourth generation mobile radio systems and is considered a competitor to the UMTS successor, LTE (Long Term Evolution). This wireless technology is more often included among the WANs and is therefore discussed in greater detail in Chap. 6.

5.6 Glossary

Ad hoc network: Radio network in which two or more subscribers are connected and configured to a meshed network. There is no fixed infrastructure necessary in the form of a stationary access point.

Authentication: Serves to establish the identity of a user. In authentication, certificates of a trusted third instance are used for identification verification. For verification of message integrity, digital signatures are created and sent with it.

Beacon data packet (beacon frame):   An active WLAN makes itself known by sending out a beacon data packet. A beacon data packet contains all relevant information for connection establishment to the WLAN.

Broadcast: A broadcast transmission is a simultaneous transmission from one point to all participants. Classic broadcast applications are radio and television.

Computer network: A computer network offers autonomous computer systems, which each have their own storage, periphery and computation capacity, and are connected to the network, the infrastructure for data exchange. Because all subscribers are linked with one another, the computer network offers each participant the possibility to get into contact with all other subscribers.

Cryptography: Sub-area of computer science and mathematics that deals with the construction and evaluation of encryption methods. The goal of cryptography focuses on protecting confidential information and preventing unauthorized third parties from gaining access to it.

CSMA (Carrier Sense Multiple Access):   Access method in which a computer monitors a transmission channel (**carrier sensing**) and only proceeds with transmission when a channel is free.

CSMA/CA (Carrier Sense Multiple Access/Collision Avoidance):   A CSMA access procedure that attempts to avoid the collision of data packets, by exchanging a short packet between the sender and receiver prior to transmission of the data packet. This informs the computers within the network segments concerned about the length of the impending data transmission. It thus prompts them to discontinue their activities for this time. A simple representative of this protocol family is the MACA protocol (**Multiple Access Collision Avoid**), which is employed in the area of mobile communication.

CSMA/CD (Carrier Sense Multiple Access/Collision Detection):   In case of a collison, this CSMA access method detects it, stops transmission and waits a randomly selected period before carrying out further transmissions.

Denial-of-Service (DoS):   An attack in Internet with the intention of overloading the victim's system through targeted manipulation. The system is overloaded to such an extent that it is no longer in a position to carry out regular communication tasks, or fails completely.

Directional radio: Directional radio is understood as the transmission of electromagnetic waves with bundling, sharp-focusing antennas, simply called directional antennas.

IEEE: The Institute of Electrical and Electronical Engineering, or " I-triple-E" for short, is an international organization of engineers with its head office in the USA. There are currently approx. 350,000 engineers in the IEEE from a good 150 different countries. The main task of the IEEE is the development, discussion, adoption and ratification of standards in the network area. The working group **IEEE 802** is responsible for the standardization of LAN technologies.

ISM frequency band: Frequency area that was originally reserved for use in the fields of industrial, scientific and medical bands and is now used for radio-based, wireless data communication networks, e.g., for WLANs and Bluetooth, both of which work in the 2.45 GHz frequency band.

Man-in-the-middle attack: An attack on a secure connection between two communication partners. Here, the attacker intercepts the communication between the two communication partners (man-in-the-middle) or impersonates each endpoint without being noticed.

Medium Access Control (MAC):   A lower sublayer of the data link layer in the TCP/IP reference model. Here, access is controlled on a shared transmission medium. This forms the interface between the physical layer and the LLC sublayer.

Mesh topology: A special form of network topology in which each terminal is connected with one or several other terminals. Information is forwarded from one network node to another until it reaches its goal. In the case of failure of a network node or a line, it is possible as a rule to communicate the data further through routing.

Modulation: Modulation in telecommunication describes a procedure in which a user signal (data) to be transmitted changes (modulates) a normally higher frequency carrier signal. In this way, transmission of the user signal over the carrier signal is made possible.

Multicast: A multicast transmission corresponds to a broadcast sent to a limited group of participants. It is thus a simultaneous transmission made from one point to a specific circle of network subscribers.

Multiple access: One speaks of multiple access when network devices share a common communication channel in a network.

Packet switching: The primary method of communication in digital networks. Here, the message is separated into individual data packets of a fixed size and transmitted – independent of each other – from the transmitter to the receiver, via possibly existing exchanges. A distinction is made between **connection-oriented** and **connectionless** packet switching networks (**datagram network**). In connection-oriented packet switching networks, a connection is established in the network over fixed chosen exchanges before the start of actual data transmission. In contrast, a fixed connection path is not chosen in connectionless packet switching networks.

Parking lot attack: Attack on a wireless network in which an unauthorized party attempts to gain access to a network by intercepting the periodically sent beacon data packets by means of a mobile antenna (often from a car window in the parking lot of a company).

Pure ALOHA: Access method in which every connected computer transmits its data when needed, without listening to the transmission channel. If a collision occurs, the transmission is interrupted and starts again after expiration of a randomly selected time interval.

Roaming: Designation for the ability of a radio network subscriber to transmit and receive data in a foreign network (cells), meaning other than its home network.

Slotted ALOHA: ALOHA access method in which the time axis is divided into fixed intervals (slots). The length of a data packet transmission may not exceed the length of a slot. A computer must always wait until the beginning of a new slot until it is allowed to start its transmission.

Sniffing: Attack variation in a computer network in which a special software, or hardware (packet sniffer) is implemented. It has the ability to receive the data traffic of a network, as well as to record, represent and sometimes even analyze it.

Throughput: This is a measurement for the performance ability of a communication system. Processed or transmitted messages/data are measured within a specific time span. The throughput is calculated from the quotient of the error-free transmitted data bits and the sum of all transmitted bits within a fixed time duration. It is expressed in e.g., bit/s or data packet/s.

Topology: The topology of a computer network is understood as the geometric form of the distribution of individual computer nodes within the network. Widespread topologies for computer networks are: **bus topology**, **ring topology**, **mesh topology** and **star topology**.

Chapter 6
Network Access Layer (3): WAN Technologies

> *"He shall have dominion from sea to sea, and*
> *from the river to the ends of the earth."*
> *- Psalm 72:8.*

Crossing the close spatial borders of a LAN, WANs (Wide Area Network) offer the possibility to bridge very large distances between individual computers and local computer networks. To do this, special transmission media is needed as well as new, additional network technologies and protocols. Our access to the global Internet is carried out mostly via WAN technologies, e.g., over the telephone network, cable television network or a mobile radio network. In this chapter, the most important concepts and technologies for the establishment and operation of these types of WANs is presented. The main focus will be on addressing and routing, which are also of importance to the overlying Internet layer. On the basis of various technological examples, a closer look is taken at how WANs function.

6.1 Introduction

While Local Area Networks (LANs, local networks) are always subject to spatial and capacity limitations, the globally-wide communication network of the Internet is steadily growing. New computers and new networks are constantly being connected to this worldwide network, which, unlike a LAN, is based on a **scalable** technology. This technology of the **Wide Area Networks** (WANs) makes the interconnection of multiple local networks into one network possible. The number of end systems – individual computers or entire LANs – connected to a WAN and its spatial extension can be adapted to the individual conditions at hand. The geographic range varies from single company locations interconnected with each other in the same city, to an intercontinental networking linking entire university computer systems. The WAN technology is capable of providing sufficient power reserves to guarantee an efficient operation of the network.

C. Meinel and H. Sack, *Internetworking*, X.media.publishing,
DOI: 10.1007/978-3-642-35392-5_6, © Springer-Verlag Berlin Heidelberg 2013

The goal of coupling different end systems by means of WAN technology is to create a global network, where a seemingly uniform, homogeneous global network can be established. These so-called **internets** or **virtual networks** enable the unhindered communication between all end systems of the various interconnected network islands (internetworking). In addition to intermediate systems, e.g., switches, packet switching, routers, and a standard addressing scheme, common communication protocols are also needed (not necessary on all layers of the protocol layer model) (cf. Chap. 8). The term internet is used generically. The most famous of all internets is of course the global "Internet," which works with the IP addressing scheme and the TCP/IP protocols.

Without including all of its connected end systems, WANs principally consist of **transmision media** (e.g., cabling) and **switching elements** (e.g., routers, packet switch intermediate systems), which manage the tasks of connecting the single local subnets to each other and ensuring a seamless data transport between them. A distinction is made between connections among the individual switching elements. There are those mostly designed with a high bandwidth and those switching element connections to the respective end systems with (as the case may be) a reduced a bandwidth. Special computers are employed as switching elements. It is their task to connect two or more different types of cabling with each other. They decide over which connection, i.e. via which medium of transmission, the incoming data should be forwarded, so it reaches the desired addressee in the network. On the path that a message travels from a transmitter to a receiver there are often many different networks to bridge. It is necessary to find a suitable path (route) through the WAN so that the message can be successfully delivered. The intermediate systems involved take over this task.

After it is received by the switching element, the data is then additionally buffered, evaluated and subsequently forwarded. The forwarding of incoming data, with simultaneous buffering, is also known as **store-and-forward** switching. The process of finding a suitable pathway for the data through the WAN or in an Internet, from the transmitter to the receiver, is called **routing**.

In contrast to LANs, WANs mostly have an irregular topology. Fig. 6.1 shows an example of a WAN. The individual subnetworks A, B, C, und D are connected to each other via switching elements. The interconnections between switching elements are adapted to the respective traffic dynamics and usually designed for high bandwidths.

6.2 Packet Switching in the WAN

6.2.1 Basic Principles

Instead of a common transmission medium, as in the case of LANs, a WAN or an Internet, is a fusion of multiple, possibly different, individual networks. In order to

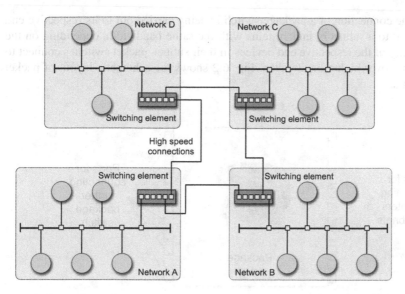

Fig. 6.1 Example of a WAN.

reach the desired **scalability**, the individual subnets are connected with each other via switching elements.

Modern computer networks function according to the principle of **packet swit-ching**[1]: The data to be transmitted is broken down into individual packets, one inde-pendent of the other, and transmitted over the network to the addressee of the data. It is certainly possible that individual data packets will reach their goal in different ways, in order to achieve the greatest possible data throughput in the network. In circuit switching, a message is always transmitted in its entirety and the transmissi-on path remains exclusively occupied for the duration of the transmission process. Packet switching, in contrast, ensures all participants fair access to the network. If all data packets are the same size, one speaks of **cell switching** .

In order to increase the size of an existing WAN, the new subnetwork to be linked to the WAN is simply connected via a switching element to the existing WAN. De-pending on need, in this way further subnets can be integrated in the WAN. The implemented switching element that handles the coupling of the networks is gene-rally known as **packet switching**. Its main task is the reception, storage, evaluation and forwarding of data packets. A packet switch is a special computer that is only used for this task. It is important to fully exploit the high bandwidth characterizing today's WANs, and at the same time achieve as short as possible switching times. To do this, the algorithm used for determining the route of a data packet to the respective receiver is implemented directly in the hardware of the computer.

---

[1] A general presentation of packet switching can be found in Chap. 3, „Foundations of Compu-ter Network Communication", in the first volume of this series: Meinel, Ch., Sack, H.: Digital Communication – Networking, Multimedia, Security, Springer (2013).

While the connection of a package switch is being carried out to its respective end system, or to a subnet of end systems with the same bandwidth, depending on the availability of the respective end devices in their subnet, packet switches connect to each other over higher bandwidths. Fig. 6.2 shows the schematic layout of packet switching.

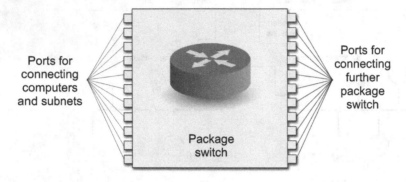

Fig. 6.2 Schematic layout of packet switching.

A wide variety of different transmission media can be used here, e.g., serial leased lined (telephone cables), fiber-optic cables, microwaves, radio, laser or satellite channels.

6.2.2 Layout of a WAN

The starting point for a the structure of a WAN are individual, geographically distant LANs, as well as the individual computers to be interconnected. For this purpose, each location has a packet switch. It is connected with the local subnet or directly with local end systems. The individual packet switches at these different locations are further interconnected via a separate - often public - network.

Fig. 6.3 shows the structure of a WAN with four packet switches. Via two packet switches entire subnets are connected with the WAN, while over the other two only single computers are connected with the packet switch. The illustration also shows that WANs need not be symmetrical. The bandwidths of the connection between the individual packet switches are chosen based on expected data traffic. Redundancy in the case of failure is also taken into account.

WANs are dependent on network operators. These operate remote networks and make them available. Network operators are public or private institutions who operate within the framework of legal regulations. Businesses can also operate their own WANs without being supported by a network operator. For this purpose, lines can be leased from network operators and then be connected to what seems to be

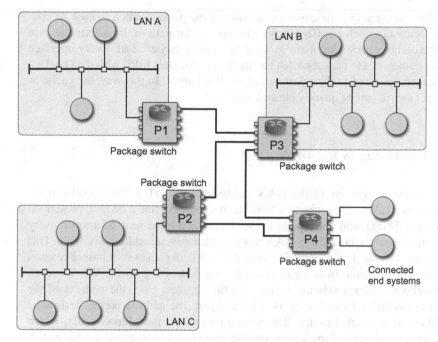

Fig. 6.3 Example of the principle layout of a WAN.

a company-owned network. The business does, in fact, bear the responsibility for operating and maintaining this corporate network .

6.2.3 Store-and-forward Switching

In a LAN, all connected end systems use a common transmission medium. Collisions or failures occurring on the shared transmission medium hinder data transmission and can also bring it to a standstill. To prevent this from happening, only one pair of end systems at a time is allowed to exchange a data packet.

In contrast to this procedure, the connected end systems in a WAN can carry out parallel communication and transmit data packets simultaneously. So that packet switches can also carry out their work correctly when there is a high volume of data packets, every received data packet is initially copied and saved in an internal cache by the packet switch. This happens before it is decided – based on the supplied address information - on which paths (i.e. via which output) the data packet leaves the packet switch again. This technique is called **store-and-forward switching**.

In store-and-forward switching, the data packets can be transmitted using the highest bandwidth supported by the hardware of the network. It is possible for the packet switch to take in many data packets at the same time in its storage. This can then act as a buffer on the slower or more loaded side of the packet switch in case of congestion. Naturally, such a buffer always has capacity limitations. In an extreme case when the cache cannot take in any more data packets they are lost. Every interface (input or output) of the packet switch has its own queue. It is filled with data packets that are then delivered to their destination via this output. In this way, the cache of the packet switch can be quickly cleared again.

6.2.4 Addressing in the WAN

Seen from the perspective of the WAN-connected terminal, a WAN works in the same way as a LAN. The respective WAN technologies define a certain, prescribed data format. This should correspond to the data packet to be sent. Moreover, every single end system connected to a WAN has its own hardware address available. This must be given as the destination address of the sent data packet so it indeed reaches its place of destination. In order to achieve clear addressing on the WLAN level, **hierarchical address scheme** is used. In the simplest case, the address of the end system is divided into two parts. The first part, the address prefix, designates the address of the packet switch. The second part shows the address suffix, which designates the address of the corresponding end system connected to this packet switch. Fig 6 4 shows an example for a hierarchical addressing in the WAN.

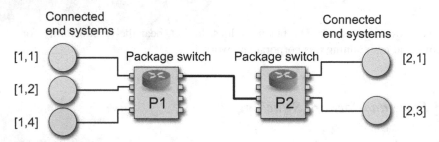

Fig. 6.4 Hierarchical addressing in the WAN.

There are three end systems at packet switch P1, each are connected to ports 1, 2 and 4. At the same time, at packet switch P2 there are two end systems connected to ports 1 and 3. The computer connected to port 4 by packet switch P1 is therefore assigned the address [1,4].

The packet switch must forward every received data packet based on its destination address. If the data packet is intended for a computer that is directly accessible over the same packet switch, the packet is simply forwarded from the corresponding output connected to the target computer. A different process is followed if the data packet is intended for a computer that can only be reached over another packet switch. Here, the packet is first transmitted via one of the high speed connections to the next packet switch on the connection route between the current packet switch and the packet switch of the target computer. Which packet switch is chosen in the the process depends on the destination address of the individual transmitted data packet and the route derived from it.

A packet switch does not have access to the complete path information of a data packet, rather, in each case, only the next partial leg to be covered (**hop**). This switching principle is also known as **next hop forwarding**. Information about the leg to be covered by a packet switch is managed there in the form of tables (routing tables) where every target is assigned to a designated output of the packet switch. Fig. 6.5 shows the next-hop table of packet switch P3. When the packet switch receives a data packet, the target address, found in the packet header, is first evaluated and a search made for the corresponding entry in the next-hop table of the packet switch. The data packet is then forwarded to the respective output indicated by the target address.

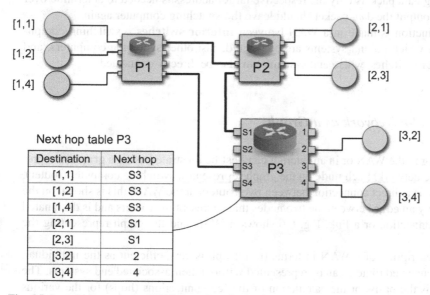

Destination	Next hop
[1,1]	S3
[1,2]	S3
[1,4]	S3
[2,1]	S1
[2,3]	S1
[3,2]	2
[3,4]	4

Fig. 6.5 Leg placement in the WAN – next-hop forwarding.

Data packet forwarding in the WAN proceeds independent of the respective source address from which the data packet was sent. This has no relevance in determining the next leg chosen and is only important if the packet switch is to be reconfigured. The forwarding of data packets follows a simple algorithm, which can be implemented in the hardware with a high degree of efficiency.

6.3 Routing

So-called **routing** determines the paths along which the data packet is transported from one transmitting end system to a receiving end system. Between the source system and target system there are (as a rule) multiple **routers**. As switching computers it is their task to transfer the data packets from one subnetwork of the WAN to the next, along the chosen path. The router determines the path (route) on the basis of the target address in the header of the transmitted data packet, which identifies the receiver end system as well as the subnetwork of this end system.

The forwarding of data packets to the respective switching computers is neither dependent on the transmitter of the data packet nor on the individual legs the data packet has already traveled but solely on its goal. This property is known as **source independence** and is one of the founding principles of network techniques. It makes the implementation of efficient **routing algorithms** possible. For the forwarding of arriving data packets only the respective target address is needed to determine over which output the data packet should leave the switching computer again.

A distinction is made in a WAN between **interior switches** – switching computers on which no end systems are connected, just other switching computers, and **exterior switches**, where end systems can also be directly connected.

6.3.1 The Network as a Graph

Routing in the WAN or in an internet can best be illustrated using a **graph** to represent the network. Each **node** on the graph represents a switching computer (router). If there is a direct connection between two routers in the WAN, this is shown on the graph by an **edge** between the two nodes that represent the routers and is designated as a connection or a link. Fig. 6.6 shows a WAN and the graph representing the WAN.

The description of a WAN in terms of a graph is very efficient as the individual, interconnected routers can be represented without their associated end systems. The graph is the basis for the calculation of the leg connections (hops) for the various routing algorithm used in WANS. In practice, an edge weight is given to the the individual edges in the graph representation of the network. The resulting costs for the use of the corresponding leg connection are indicated in such a way to give information about whether it is desirable to transmit data along this connection. In principle, the solution of the routing problem consists of finding the most cost-efficient way in the network from the transmitter of a data packet to the receiver, as represented in edge weight graphs. The costs of such a path may be simply calculated from totaling the individual costs assigned to the edges of the path.

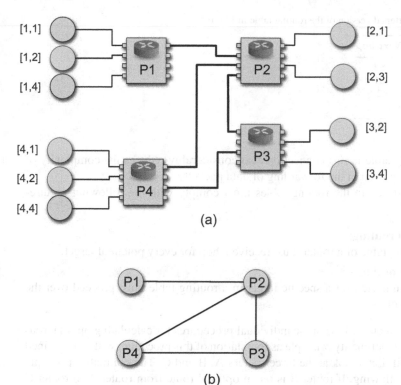

Fig. 6.6 A WAN (a) and its corresponding graph representation (b).

6.3.2 Calculation of the Routing Table in the WAN

How exactly is the route of a data packet calculated? In order to forward an incoming data packet correctly, the switching computer (**router**) has an internal table (**routing table**) at its disposal with the partial leg information necessary for forwarding. For every potential target, the router receives information indicating at which port the data packet should exit the router again. Because the routing table only contains information about the next station of a data packet on its way through the WAN, it is also known as the next hop table. If we look at the router table shown in Fig. 6.5, the advantage of hierarchical addressing in the network becomes immediately clear. All data packets transmitted to a certain subnetwork exit the router at the same port. Therefore, to facilitate a correct forwarding of data packets only the first part of the destination address must be verified. This occurs when the destination of the data packet is not located in the same subnetwork as the router in question (see Table 6.1).

The routing tables therefore contain only one entry for a specific subnetwork and one entry for every computer in the local subnetwork of the router. The possible decrease

Table 6.1 Shortened version of the routing table in Fig. 6.5.

Target	Next hop
(1,arbitrary)	S3
(2,arbitrary)	S1
(3,2)	2
(3,4)	4

of the routing table in this connection also considerably reduces the complexity of calculation involved in the forwarding of data packets.

The entries stored in the routing tables must comply with the following requirements:

- **Universal routing**:
  The routing table of a router must receive a hop for every potential target.

- **Optimal routing**:
  The hop indicated for a specific target in a routing table must proceed over the shortest path.

Before taking a closer look at the individual procedures for calculating optimal routing tables, the optimality principle as foundation of this procedure will be explained in more detail. Let's look at the three routers A, B and C. The optimality principle implies the following. If router B is on an optimal route from router A to router C then the optimal path from router B to router C is also follows this route. This assumption can be proven easily. The first partial leg of the optimal route from A to B is designated w_1 and the partial route from B to C as w_2. If a better path would lead from B to C as w_2, then we could simply connect it with w_1 and would find a route from A to C superior to our exit route, which would contradict our initial condition. From the optimality principle follows that the resulting graph if one connects the optimal routes from all possible nodes (routers) with each other $A_0,..., A_n$ to a specific target Z is a **rooted tree**, containing Z as root. This type of rooted tree is called a **sink tree**. It follows that a distance metric can be introduced to the sink tree, defined through the respective number of necessary hops. Being a tree, there are no loops contained in the sink tree. This means that every data packet following the prescribed route always reaches its target after a finite number of hops (cf. Fig. 6.7). The sink tree is not clearly defined, and there can be a multiple number of different sink trees. The goal of the various routing procedures is, in each case, to identify a sink tree and to use it for determining the optimal route.

Let us return to the subject of routing table calculation. A complete routing table for all routers from Fig. 6.6, which can be obtained from the corresponding network graphs, is shown in Table 6.2. The value pairs (u,v), which are respectively indicated in the fields "Next Hop," indicate the edge (u, v) from node u to node v, over which the data packet is to be subsequently forwarded.

It is noticeable that the indicated routing table contains many duplicate entries. Because in practical application there is only limited space available in the routing

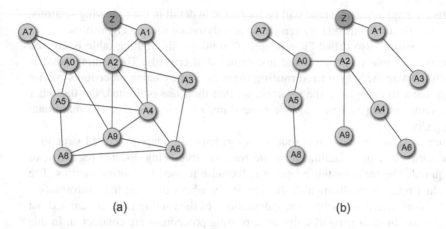

Fig. 6.7 Network (a) and associated rooted tree (b).

Table 6.2 Complete routing table for all routers in Fig. 6.6.

Node P1		Node P2		Node P3		Node P4	
Target	Next Hop	Target	Next Hop	Target	Next Hop	Target	Next Hop
1	-	1	(2,1)	1	(3,2)	1	(4,2)
2	(1,2)	2	-	2	(3,2)	2	(4,2)
3	(1,2)	3	(2,3)	3	-	3	(4,3)
4	(1,2)	4	(2,4)	4	(3,4)	4	-

table, the procedure known as **default routing** is implemented. This helps to use storage space as efficiently as possible. Thereby, a single entry in the routing table replaces a sequence of entries with identical hop values. In every table only one entry for the default routing is admissible. This has a lower priority than all other entries. If in the routing table there is no explicit entry found for a certain target, the data packet is forwarded to the route determined by default routing (see Table 6.3).

Table 6.3 Routing table based on the default routing procedures for all routers in Fig. 6.6.

Node P1		Node P2		Node P3		Node P4	
Target	Next Hop	Target	Next Hop	Target	Next Hop	Target	Next Hop
1	-	1	(2,1)	3	-	3	(4,3)
	(1,2)	2	-	4	(3,4)	4	-
		3	(2,3)	*	(3,2)	*	(4,2)
		4	(2,4)				

A manual calculation of the routing table is only feasible for very small networks, such as in our example. In practical calculation procedures, so-called **routing algo-**

rithms are implemented. These will be looked at in detail in the following sections. A basic distinction is made between **static** and **dynamic** routing algorithms.

In static routing (also called "non-adaptive" routing), the routing table is created at the start of router operation and not changed afterwards. The routing table is established with the help of fixed **routing metrics**. Static routing procedures are not able to react to changes in the network, so that their use is limited. On the other hand, static routing is a very simple procedure, requiring a lower computational complexity.

In contrast, dynamic (adaptive) routing algorithms can adapt to altered structures in the network. This is facilitated by the routing table being updated regularly, or as required. The basis for this adaption is likewise formed by routing metrics. The individual procedures distinguish themselves in where they get their information and decision criteria and when new calculations of the routing table are carried out in each case. In most networks, dynamic routing procedures are conducted. In this way, network problems are then often automatically solved. In addition, network traffic as well as the state of network hardware are permanently monitored. In the event of disturbances, network routes are correspondingly redirected.

Dynamic routing algorithms can be divided into **centralized** and **decentralized** routing algorithms. In central routing procedures, the calculation of the routing table is carried out by a central instance. Because of its comprehensive information about the network, it is able to make high quality routing decisions. However, the resulting reaction to changes in the network structure can be a relatively long time coming. Besides this, the computer on which the central routing calculations are carried out represents a critical resource. If it fails, the operation of the whole network is in danger. An additional bottleneck can also be created if the computer is not sufficiently dimensioned, significantly lowering transmission power.

With the employment of decentralized routing algorithms, the calculation of the routing tables falls to each individual router. It is based on the information presented by the respective router about the nature of the network.

In practice, most of the time decentralized routing algorithms are implemented. The most important representatives of these are presented in section 6.3.4 and 6.3.5. These are the **distance vector procedure** and the **link-state algorithm**.

In addition to this division, and independent of it, are also so-called **isolated routing procedures**. Here, every node of the network uses the locally available information exclusively to make a routing decision.

The most important representatives of the isolated routing procedures are the non-adaptive procedures of so-called **flooding** (or broadcast routing) and the load-dependent **hot potato** procedure, as well as **backward learning** and **delta routing**. They are described in detail in the following section. All of the named routing procedures and individual subprocesses may be combined with each other arbitrarily. Fig. 6.8 shows a schematic arrangement of the just-mentioned routing procedures.

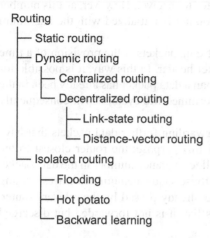

Routing
- Static routing
- Dynamic routing
  - Centralized routing
  - Decentralized routing
    - Link-state routing
    - Distance-vector routing
- Isolated routing
  - Flooding
  - Hot potato
  - Backward learning

Fig. 6.8 Schematic arrangement of the routing algorithm.

6.3.3 Isolated Routing Algorithms

We begin our presentation of different routing procedures with the isolated routing algorithm. Isolated means that each network node, i.e., a computer, only uses locally available information for routing. Its network knowledge is limited to the network node to which it is connected, with the rest of the network unknown. Locally available information can concern, e.g., the length of a queue at a router's output port or the status of a neighbor. The network node only has access to information it has collected itself. Subsequently, no exchange of routing information takes place between network nodes. An adaption to altered network topology, for example in the case of the failure of an active router or the activation of a new router, can only be carried out with limited information.

The procedure known as **flooding** (or **broadcast routing**) is a very simple and effective routing procedure. However, it usually has the consequence of a heavy burden being placed on the network. All incoming data packets are forwarded simultaneously via all the connections of the routers – with the exception of the connection over which the data packet has arrived at the router. With flooding, the data packet is multiplied in every router of the network resulting in a great number of duplicates. The number of data packets would theoretically continue to rise infinitely - resulting in a hopelessly overloaded network - if not the appropriate preventive measures were taken.

One of these measures is the provision of a counter (hop counter), which is integrated in the header of the data packet. Upon generation of the data packet, this counter is initialized with a prescribed start value and decremented with every hop. When the counter reaches the value zero the data packet is discarded and no longer forwarded. Ideally, the counter would be initialized with a value corresponding to the number of necessary hops from transmitter to receiver. However, as this number is usually unknown to the transmitter, the counter is initialized with the maximum diameter of the network.

A further measure for damming the flood of data packets is the provision of a time stamp, instead of a counter, in the data packet header. In this way, it is possible for the respective router to determine the time span a data packet has already been in the network traffic. After expiration of a predetermined life cycle, it will consequently discard the data packet.

There is the additional possibility of not forwarding further data packets that have previously been forwarded by a router. For this purpose, the router closest to the transmitter (the source router) writes a so-called sequence number in the data packet header. Every router maintains a list noting these sequence numbers corresponding to the respective source routers, which have already passed by the relevant router. If an arriving data packet is already on this list, it is not forwarded but discarded instead.

A somewhat more efficient flooding variation is so-called **selective flooding**. An arriving data packet is not forwarded by the router over all connections but only over those in the "right" direction. This is only possible if the router has information about the structure of the network so that, e.g., a data packet that should be sent west is not flooded over connections pointing east.

A further variation consists of the random choice of an output over which the data packet is forwarded (**random walk**). However, here it cannot be ensured that a data packet does in fact reach its designated receiver, or that possibly a shorter path might be taken.

For most applications, flooding procedures are not very practical, however still offer certain advantages. For example, in military applications in which a high degree of redundancy and fail-safe operation is necessary. This application is also used for the implementation of updates in distributed databases. It is important that updates are carried out as synchronously as possible in all sub-databases. Additionally, flooding is used as a metric if the quality of other routing algorithms is to be tested. Flooding in fact always finds the shortest route from the transmitter to the receiver, as all possible paths are chosen parallel. This is even the case despite the fact that duplicates arriving later are recognized as such and have to be discarded

In contrast, the procedure called **hot potato routing** avoids generating large numbers of data packet duplicates, but seldom finds the shortest path between transmitter and receiver. In this procedure, a router receiving a data packet to forward tries to get rid of it as quickly as possible. It handles the data packet like a "hot potato." The data packet is forwarded over the connection that is currently assigned the shortest queue. The advantage of this procedure is that only a low computational effort is necessary and data packets are forwarded very quickly. This means there is an optimal utilization of the network and the available capacity is filled. As a result, data packets can be subject to considerable detours as it cannot be guaranteed that in fact the most favorable path to the receiver will be chosen. It could also happen that data packets run in a cycle and pass through the same route several times. Moreover, the hot potato procedure is very susceptible to network overload. Data packets will continue to still be accepted and forwarded even when congestion has already formed on the way to the receiver. The static **cold potato routing** presents a contrasting procedure. Here, the router waits the necessary length of time it takes for a packet to be forwarded, or until a preferred output is free. Both procedures may be combined as well to a hybrid application.

- As long as the length of the queue remains below a previously determined threshold value, it is decided according to a statistical procedure which transmission line should be used for the forwarding of the data packets.
- If this threshold value is exceeded, the transmission line is chosen with the shortest queue.

The **backward learning** routing procedure is included among the isolated routing procedures as also here routers do not exchange any information among themselves. Routing information is forwarded over a simple hop counter, integrated in the header of the data packet. The data packet starts at the transmitter with the hop counter value zero, which is raised by one every time a router is passed. Every router can register via which port a data packet has arrived, which source address the data packet has, and how many routers have already traveled along the path in the network. The router puts this information in its routing table. As soon as a further data packet arrives from the same source whose hop counter shows a lower value than the corresponding entry in the routing table, this is replaced by the smaller value. The entries of the routing table are, however, also up-dated if for a certain time period no data packets with a certain hop number are received by the respective node.

It is then assumed that the originally more favorable connection no longer exists and the entry is overwritten with a new value. In this way, this isolated routing procedure has the ability to specifically adapt to changing network topologies.

However, the routing during the learning period, in which the routing tales are brought up to date, is not always optimal. If the chosen learn periods are too short, then the entries in the routing table could potentially be overwritten with worse entries. On the other hand, if the learn periods are too long an adaption to changed network topologies can only proceed very slowly. Because of this, in backward learning the choice of the corresponding learn interval is of crucial significance for the efficiency of the procedure.

So-called **delta routing** presents a special case as it can be considered a combination of centralized and isolated routing. In this procedure, each network node periodically determines costs (resulting from delay, overload etc.) accrued for the use of a network connection and transmits this information to a central management instance (routing control center, RCC). The RCC then calculates the best routes between two network nodes. Only routes are considered that can be distinguished by their initial network section. Every network node receives a list from the RCC of the shortest possible routes to all reachable network nodes. Which route is in fact chosen from this list in each case is decided locally by the individual computer.

Excursus 6: Dijkstra Algorithm

The determination of the **shortest path** from the transmitter to the receiver within a network plays a crucial role in a great number of routing procedures.
A simple technique that is often implemented in this regard is the so-called **Dijkstra algorithm**, named after its inventor *Edsger W. Dijkstra* (1930–2002).

The Dijkstra algorithm works on a graph whose nodes represent the routers of the network and whose edges the connection between the individual routers. The procedure determines the length of the shortest path from one transmitter to all the routers in the network. Through this calculation a routing table for the transmitter is compiled.

Which path is the shortest path in a network depends on the chosen metric, e.g., the number of necessary hops from the transmitter to the receiver. However, in practical application this approach is not always ideal as connections between the individual network components have different bandwidths and can extend over varying distances. Because of this, it makes sense to weight the connections based on their bandwidth and spatial extension, including among the weighting factors also dimensions such as the router switching speed or waiting times in the router. It is possible that the use of transmission lines are also subject to cost factors that additionally influence the weighting of the network edges. The Dijkstra algorithm is capable of determining the shortest path based on the chosen metric.

To illustrate how the procedure functions, let's look at the example in Fig. 6.9. The network consists of seven nodes A, B, C, D, E, F, G. The connection weights are given on the respective edges. A shorter path through the network is sought from A to G. For every node a data structure is provided, the so-called assignment, which contains each previous node along the shortest path and the currently determined distance of this node from the start node. At the beginning, for every node the previous node is unoccupied and the current distance is indicated as infinite. We distinguish permanent nodes from nodes that are still being calculated. For permanent nodes the shortest distance to the start nodes is determined. These can become the new starting point for subsequent calculations. In Fig. 6.9 permanent nodes are indicated as filled-in circles.

At the beginning of the calculation, the start node A is marked as permanent (a). Let us look at the nodes adjacent to A, nodes B and C. As these are not yet marked permanent and the distance given in their assignment is larger as the distance established to A, they receive the assignment (A,2) for B and (A,3) for C corresponding to the (weighted) distance from A. Now all of the non-permanent nodes of the graph are examined and those nodes with the smallest assignment, here B, are chosen as new output nodes and marked as permanent (b). Then D and F, the non-permanent nodes adjacent to B , are examined. Corresponding to their distance from A, which is determined from the distance from the output node B plus its assignment, the nodes receive the assignments (B,9) for F and (B,6) for D. To establish the next output node for the shortest path, again all the non-permanent nodes are examined and those with the shortest distance to A, now C, removed and marked as permanent (c). Then the non-permanent, adjacent nodes to C (D and E) are examined, and their distance established from A along the route via C (D equals 5 and E equals 5). As the route from A to D over C is shorter than the previous one, the assignment is set from D to (C,5) (d). If two or more non-permanent nodes have the same distance to the source, as in the following observation, one is randomly chosen, marked as permanent and selected as the starting point for the next round.

The procedure continues in this way until the actual target (G) is marked as permanent. The distance from starting point A simply results from the assignment of G (8). The shortest path in the opposite direction from G to A can be reconstructed following the assignment of the respective nodes (G,E,C,A).

To implement the Dijkstra algorithm, the following data structures are necessary:

- The nodes of the graph are consecutively numbered so that the node number can be used as an index for data access.

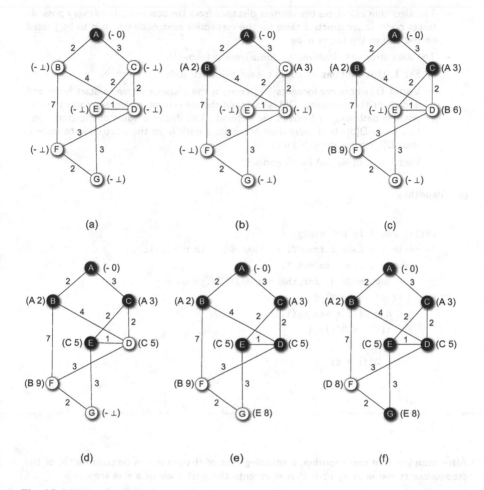

Fig. 6.9 Dijkstra shortest path algorithm.

- A distance vector **D**, whose i-te components contain the currently determined shortest distance of the i-ten node to the output node.
- A route vector **R**, whose i-te components contain the number of the node predecessor of node i, along the currently determined shortest path from A to i.
- The set **T** of the not yet permanently marked nodes, which still need to be examined. **T** can be stored as a duplicate linked list of node numbers.

The Dijkstra algorithm represented in pseudocode:

(a) Input:

– The network to be examined is represented as a graph. The source node, for which the shortest path to all other nodes should be determined is indicated. All edges of the graph bear a non-negative weighting corresponding to the chosen distance metric.

- The algorithm calculates the shortest distance from the source node to every possible target node. It constructs a table for the respective next node which is to be passed on the way to the target node.
- The data structures involved are initialized as follows:
  - Set **T**, that contains all nodes except the start node.
  - Vector **D**, whose components D[i] contain the distance between start node and node i. At the beginning, D[i] is assigned the value **max**, which is greater than all possible path lengths through the network. The distance value of the start node to itself is D[0]=0. If there does not exist a path from the start node to node i, then D[i] retains the value max.
  - Vector **R** with R[i]=0 for all nodes i.

(b) Algorithm:

```
while (set T is not empty) {
    select a node i from T, so that D[i] is minimal;
    remove i from the set T;
    for each node j, for the one edge (i,j) exists {
      if (j is in T) {
          d = D[i] + weight(i,j);
          if (d < D[j]) {
            R[j] = i;
            D[j] = d;
          }
      }
    }
}
```

After completion of the algorithm, a spanning tree of the graph can be constructed of the predecessor stored in every node that represents the start node of a **sink tree**.

Further reading:

A. S. Tannenbaum: Computer Networks, Prentice-Hall, NJ, USA, pp. 348-352 (1996)

E. W. Dijkstra: A Note on Two Problems in Connexion with Graphs, Num. Math. vol.1, pp. 269-271 (1959)

6.3.4 Distance Vector Routing

In practical application the most relevant procedures are mainly the **decentralized, distributed routing procedures**. Here, every router maintains its routing table locally and subsequently transmits messages with the routing information to neighboring routers. In this way, they are informed about the network topology from the perspective of the individual router. In this procedure, the routers send routing information periodically so that after a short lead time every router has detected the shortest path to all targets. The result of distributed routing is principally the same as in the Dijkstra algorithm. The difference is found in the fact that routers are additionally capable of adapting dynamically to changes in the topology or network utilization because they periodically exchange new routing information. If a network connection fails, a router receives no further updates from routers that were only accessible via this connection. However, if there is an alternative route, with the help of routing information from the remaining routers the router in question is still able to adapt its routing table and bypass the failed network hardware.

The procedure, known as **distance vector routing**, is a decentralized routing procedure that runs iteratively and asynchronously. Every router maintains a table (vector), which in addition to containing the best known connection to a specific target, also contains its distance. These tables are continually updated through the exchange of routing information with neighboring routers.

The distance vector routing procedure is often called the **Bellman-Ford** algorithm or the **Ford-Fulkerson** algorithm after the scientists *Richard Bellman* (1920–1984) and *Lester Ford* (*1927) and *Delbert Ray Fulkerson* (1924–1976), who developed this procedure. Distance vector routing was the first routing procedure that was developed in 1967 and was later implemented in the ARPANET. It was standardized in the Internet as **Routing Information Protocol** (**RIP**) in RFC 1058 (RIPv2 in RFC 2453). It likewise came into implementation in early versions of the DECnet and Novell IPX network software. AppleTalk and Cisco-Router use an improved version of the distance vector routing procedure.

In the distant vector routing procedure, every router in the network maintains a routing table where every other router in the network is entered. Each table entry is comprised of the connection over which each router can be reached, in addition to an estimation of the distance (temporally or spatially) to this router. The number of necessary hops, the the length of the queue or the transmission time can be used as the metric. It is assumed that every router knows the distance to its direct neighbor in the network. For example, if the number of necessary hops is chosen as a metric then the distance to the respective neighbor might be, for example, one hop. When the length of the queue is chosen as the metric, this can be determined easily by the router itself. If the transmission time is chosen, then the router transmits special data packets (ECHO packets) to its neighbors, who only need to supplement these with a time stamp. These are then immediately returned to the transmitter, enabling it to determine transmission time to its respective neighbor (see Fig. 6.10).

Just as with the Dijkstra algorithm, distance vector routing works on a graph whose nodes are the network routers and whose edges represent the connections of individual routers to each other. Every edge of the graph is assigned a weight, whereby the distance between two nodes is defined as the sum of all weights along its connection path. A router's routing table has a field containing the distance to every other router. Every node transmits the known value pairs (target nodes, distance) from its routing table to its direct neighbor. If a router receives a routing message from a neighbor, it verifies all entries in its routing table, changing any in case a neighbor

Update of the Routing Table According to the Distance Vector Procedure

In order to keep an individual routing table up to date, the incoming messages are processed with the routing tables of the neighboring router in the following way.

Let's imagine that transmission time is chosen as the metric to determine distance. Every router knows the transmission time to each of its direct neighbors. It sends these transmission times periodically in a constant time interval t. We will assume that router B and router C have just transferred their distant vectors $D_B[]$ und $D_C[]$ to router A. In both vectors each of the transmission times to router K, which is farther away in the network, is recorded. Router A therefore receives the transmission times $D_B[K]$ and $D_C[K]$ from the distance vectors of routers B and C. The transmission time from router A to routers B and C is known to A and has the value t_B bzw. t_C.

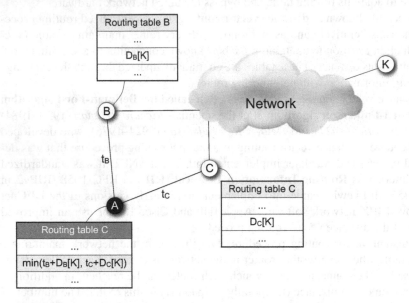

To update the entry for router K in its routing table, A compares the transmission times $t_B+D_B[K]$ and $t_C+D_C[K]$. The smaller of both is chosen and supplies the new entry in the routing table of A, $D_A[K]=min(t_B+D_B[K],t_C+D_C[K])$. The old entry of A's routing table for router K is not used in the recalculation. In this way, the best connection for all routers can always be kept up to date.

Further reading:

C. L. Hedrick. Routing Information Protocol. Request for Comments 1058, Internet Engineering Task Force (1988)

Fig. 6.10 Update of the routing table in distance vector routing.

has registered a shorter way to the target node. Fig. 6.11 show the distance vector in pseudocode.

Algorithm of the Distance Vector Routing in Pseudocode

(a) Input:

- The local routing table of the examined router, which at the beginning of the procedure need not contain an entry, a weighting function, which returns the weighting of the edges to the neighboring nodes (distance), as well as a received routing message.
- Output of the algorithm is the current routing table of the examined router.
- Entries in the routing tables (Z,N,D) always contain the corresponding target router Z, the connection over which this router can be reached (next hop) N, and a distance value D. The routing table is initialized with an entry for the local router (A,-,0).

(b) Algorithm:

```
Repeat {
    wait for routing information from a direct neighbor N;
    for each entry in the routing information {
        consider entry (Z,D):
        Dnew := D + weight (N);
        Update local routing table R:
        if (R contains no entry for Z) {
            add R with (Z,N,Dnew);
        } else
        if (R no entry for Z over N) {
            (Z,N,Dold) := (Z,N,Dnew);
        } else
        if ( R entry for Z and Dold > Dnew) {
            (Z,X,Dold) := (Z,N,Dnew);
        }
    }
}
```

Fig. 6.11 Distance vector routing in pseudocode.

Even if the distance vector routing procedure finally finds an optimal solution to the routing problem, it can take a considerably long time until it is calculated. Improvements in the network topology are registered very quickly, while if a connection fails it can take a long time until this change is detected. This can be illustrated with a simple example.

We imagine a simple network consisting of the five routers A, B, C, D and E, which are connected to each other linearly (see Fig. 6.12). The number of necessary hops between the routers is used as a distance metric. Let us assume that router A fails and this is known to all of the other routers, i.e., all of the other routers have entered the distance to A as "endless" in their routing table. If A is reactivated, the other routers learn of this through exchanged routing messages. After the first tact, B detects that router A is exactly one hop away from it and notes this in its routing table. All of the other routers still assume that A is inactive. In the next tact, A exchanges its routing information with C. Then C notes in its routing table that A is located a distance of two hops away. In this way, the routing information is propagated until this example, and after just four steps all routers have the correct routing information concerning A. If the longest path in a network consist of n hops, then every router in the network has received the activation of a new connection at the latest after n exchange operations. The table in Fig. 6.12 illustrates this example.

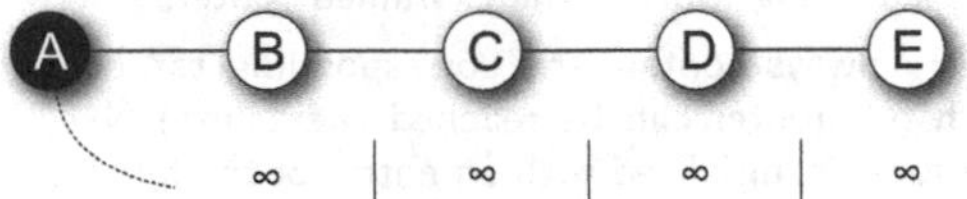

A	B	C	D	E
∞	∞	∞	∞	Start
1	∞	∞	∞	1st exchange
1	2	∞	∞	2nd exchange
1	2	3	∞	3rd exchange
1	2	3	4	4th exchange

Entries in routing table of A

Fig. 6.12 Fast update of routing information in the event of network topology improvement.

Let's now look at the opposite case. At the beginning, all of the routers are active, and then suddenly A is no longer accessible. Upon the first exchange of routing information B receives no more messages from A. However, B receives a routing table from C in which the distance from A is indicated with two hops. Incorrectly, B then determines the distance to router A as three hops and registers this information in its routing table, while the other routers leave their routing tables unchanged. In the exchange that follows, C learns that A is three hops away from each of its two neighbors B and D, and then determines the distance to A as four hops. But this is just the value that is considered to be "endless," and this behavior is therefore known as the **count-to-infinity problem**.

A possible way to solve the count-to-infinity problem is the **split horizon algorithm**, which has found widespread use. The split horizon algorithm is a simple extension of distance vector routing. Router B informs its direct neighbors about its routing table, however B does not give the distance information to another router A by way of the connection (or the connections), over which router A is accessible.

Let us look once again at the already familiar example, in which at the beginning all routers are accessible in Fig. 6.14. While here router C forwards the correct distance information for router A to router D, C passes on to router B a distance value of endless for the distance from router A. If router A now fails after the first exchange of routing information, B determines that there is no longer a direct connection to router A. In the routing information transmitted from router C to router B, the distance from C to A is indicated as "endless," as router C is not allowed to pass on

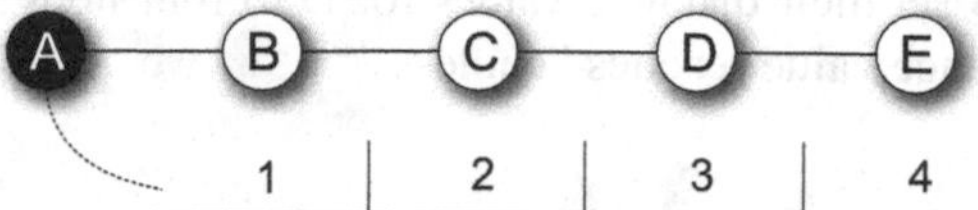

A — B — C — D — E

1	2	3	4	Start
3	2	3	4	1st exchange
3	4	3	4	2nd exchange
5	4	5	4	3rd exchange
5	6	5	6	4th exchange
...	...	...	...	...
∞	∞	∞	∞	

Entries in routing table of A

Fig. 6.13 Count-to-infinity problem in the case of router failure in distance vector routing.

any distance information in this way for A. Therefore, B likewise sets the distance value for A in its routing table at endless, while the entries of the other routers for router A remain unchanged. In the next step, C registers a distance value of "infinite" as the distance to router A, and thus continues to reproduce the information that router A is not accessible until after four steps it reaches router D.

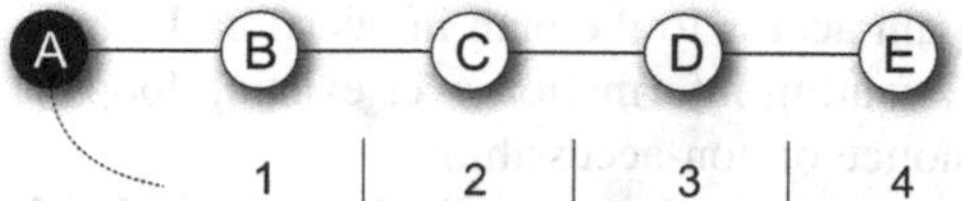

A — B — C — D — E

1	2	3	4	Start
∞	2	3	4	1st exchange
∞	∞	3	4	2nd exchange
∞	∞	∞	4	3rd exchange
∞	∞	∞	∞	4th exchange

Entries in routing table of A

Fig. 6.14 Application of the split-horizon algorithm to avoid count-to-infinity problems upon failure of a router in distance vector routing.

However, this procedure can also be unsuccessful. For example, let us look at another network consisting of the four nodes A, B, C and D (see Fig. 6.15). Initially, the distance from C to D is exactly one hop and from A and B, respectively, to D exactly two hops. If router D fails, C recognizes this immediately. As C also receives the distance value from both routers A and B: "endless", as the distance value for D, C notes this value in its routing table. However, A receives from B a distance value of two hops for the distance to D and sets the distance value in its routing table to three hops. Router B reacts likewise, as it also receives from A a distance value of two hops for D. In the next steps, A and B set their distance values for D to four hops and continue in this way, until the distance value reaches "endless."

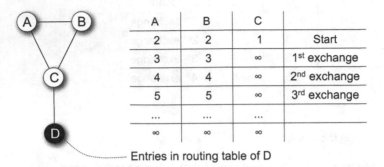

A	B	C	
2	2	1	Start
3	3	∞	1st exchange
4	4	∞	2nd exchange
5	5	∞	3rd exchange
...	...	...	
∞	∞	∞	

Entries in routing table of D

Fig. 6.15 Count-to-infinity problem despite application of the split-horizon algorithm, in the event of router failure in distance vector routing.

But also this problem of the distance vector routing procedure can be overcome in many cases. To do this, the following varieties of the split-horizon algorithm are implemented. The **split-horizon with poison reverse** procedure works analog to the simple split-horizon procedure. As soon as a router learns of a new path via arriving routing information, it makes known over the connection on which the information has arrived that it cannot gain access to the path in question. In split-horizon with poison reverse, the use of routing information over existing loops is therefore actively prevented through a notice of non-accessibility.

A further variation in avoiding defective routing information is the use of a **hold-down timer**. If a leg between two routers fails, this entry is not deleted immediately from the affected routing table, instead remains in place for a certain time (hold-down time, RFC 2091). The entries are however marked as "non-accessible." This purpose of this to allow the remaining routers to adjust to the new situation and, subsequently, that the network can return to a stable status. During the holddown time, the router does not accept any new entries for the failed connections, except for the forwarding costs (based on the chosen routing metric) if these are lower than originally or if the new entry comes from the direction of the failed router. The latter circumstance is intended to ensure that a previously failed connection can be repaired again, or an alternative path found.

The RIP routing protocol in the original ARPANET implemented split horizon with poison reverse and holddown timer to prevent the forwarding of defective routing information. Every RIP router transmits its routing table in fixed time intervals to its direct neighbor (in RIP normally every 30 seconds). In doing so, it is also possible to optionally fix that extra routing information is always transmitted exactly when a topology change is identified (**triggered update**). Similar to the split horizon algorithm, the triggered update ensures that routing loops resulting from forwarding defective (outdated) routing information are avoided. In Fig. 6.16 further routing protocols are presented based on the distance vector routing procedure.

Routing Protocols Based on the Distance Vector Routing Procedure

- **Destination Sequence Distance Vector Routing (DSDV)** This routing procedure, also based on the Bellman-Ford algorithm, was developed in 1994. In contrast to the original distance vector routing procedure, every entry in the routing table here has a sequence number. When there is a connection this sequence number is even, otherwise it is odd. The sequence number is generated by the receiver and must be used afterwards for every further update. The routing information between neighboring routers is completely exchanged in long, fixed time intervals. Incremental updates can follow in shorter time intervals. Connections that are not used for a longer time are removed again from the routing table. DSDV was developed especially for mobile ad hoc networks. Today's widely used, **Ad hoc On-demand Distance Vector routing protocol** (AODV) is based on these principles.

- **Interior Gateway Routing Protocol (IGRP)**
  IGRP was developed by the Cisco company to overcome still-existing limitations of the RIP routing protocol. These concerned the maximum distance between routers, which was limited to 15 hops, before a router was considered inaccessible. It was raised by IGRP to 255. Besides this, IGRP supports multiple distance metrics for a connection that factors in the available bandwidth, delay time, current load, transport unit limitations (maximum transfer unit, MTU), and the reliability of connections. To avoid the count-to-infinity problem in IGRP, the split-horizon procedure with route poisoning, holddown timer and triggered updates are implemented. IGRP has not been developed further and was replaced by the enhanced interior gateway routing protocol.

- **Enhanced Interior Gateway Routing Protocol (EIGRP)**
  EIGRP was developed in 1994 by the Cisco company as a successor to the IGR, with which it remains compatible. It has additional properties that are more likely to be numbered among the link-state routing protocols, so that it is often designated a hybrid routing protocol. EIGRP distinguishes itself especially through its faster convergence in the case of topology changes and through an immunity to routing loops. The **Diffusing Update Algorithm** (DUAL) is employed to calculate the optimal route. Besides the usual routing tables, EIGRP uses two extra tables (neighbor table and topology table). In these, the different neighboring relationships between routers are maintained, with only information exchanged between the routers that the respective remote station actually needs.

Further reading:

Albrightson, R., Garcia-Luna-Aceves, J. J., Boyle, J.: EIGRP – A fast routing protocol based on distance vectors, in Proceedings of Networld/Interop 94, Las Vegas, Nevada (May 1994) pp. 192–210

Garcia-Lunes-Aceves, J. J.: Loop-free routing using diffusing computations. IEEE/ACM Trans. Netw. 1, 1 (Feb. 1993) pp. 130–141

Perkins, C. E., Bhagwat, P.: Highly dynamic Destination-Sequenced Distance-Vector routing (DSDV) for mobile computers, in SIGCOMM Comput. Commun. Rev. 24, 4 (Oct. 1994) pp. 234–244

Perkins, C., Belding-Royer, E., Das, S.: Ad hoc On-demand Distance Vector (AODV) Routing, Request for Comments 3561, Internet Engineering Task Force (2003)

Fig. 6.16 Routing protocols based on the distance vector routing procedure.

6.3.5 Link-State Routing

The routing procedure known as **link-state routing** replaced the distance vector routing procedure in the ARPANET in 1979. One reason for this decision was based on the metric implemented in distance vector routing for distances in the network. Another reason was the slow convergence characteristic of distance vector routing. This proved to even be the case when improved algorithms, such as the split-horizon method, were applied. The applied metric was based solely on the queue length in the routers. Other dimensions, such as the available bandwidth, were not used in determining distance. This approach seemed justified as the ARPANET was originally operated with an exclusive bandwidth of 56 kbps. With time, however, there were additional faster connections so that the chosen approach no longer produced efficient routing results. The distance vector routing was therefore replaced by link-state routing. This procedure was first published by *John M. McQuillan* (*1949) of the American company Bold, Beranek and Newman, which played a significant role in developing the ARPANET.

In technical literature, link-state routing is often referred to as **SPF routing** (Shortest Path First), although other routing procedures also determine the shortest path. Just as distance vector routing, link-state routing is a decentralized routing procedure. The messages exchanged between each of the individual routers contain only status information about the connection to the directly neighboring routers and not complete routing tables. This status is expressed with the help of a weighting, which is calculated according to the underlying metric. In contrast to distance vector routing, the messages in link-state routing (link-state advertisements/announcements, LSA oder link-state packet, LSP) are forwarded over a broadcast to all routers in the network and not only to direct neighbors. This means every router in the network receives the complete, global status information as it affects the entire network. With the help of this information every node is able to create its own network graphs and routing tables itself. The previously discussed Dijkstra algorithm is also employed in the calculation of the shortest path in the network.

The link-state routing procedure is capable of adapting to changes in the topology and network utilization. Adaption proceeds much faster in link-state routing than in distance vector routing as here all nodes are informed at the same time about a status change. The sequence of the link-state routing procedure is summarized in Fig. 6.17.

In the worldwide Internet, the most important WAN, the link state routing is implemented in the **OSPF** protocol (**Open Shortest Path First Protocol**), the development of which began in 1988. OSPF was standardized in 1998 in RFC 2328 and is used for routing between IP routers in so-called intradomain routing, that is routing within an address domain. As such, it replaces the older RI protocol, based on the distance vector routing procedure. A current extension of the OSPF protocol is the OSPFv3, standardized in 2008. It works together with the new Internet protocol IPv6 (see also section 7.4). In contrast to alternative link state routing protocols, it was intended from the beginning that OSPF should be developed as an open standard and not as the proprietary solution of one or more manufacturers, such as in

The Link-State Routing Procedure

The link-state routing procedure is founded on the establishment of a network topology. To do this, every router carries out the following steps:

1. Search for all directly neighboring routers and learn their network addresses.
2. Measure the distance to each direct neighbor with a suitable metric.
3. Form a **LSP** (link-state packet, see Fig. 6.18) with the addresses of the neighboring router and their distance.
4. Transmit the LSP to all routers in the network (broadcast).
5. With the current LSPs of all other routers, every router gains knowledge of the complete network topology.
6. The calculation of a target route proceeds over the Dijkstra algorithm.

Fig. 6.17 The link-state routing procedure.

Exchange of Routing Information with Link State Packets

The exchange of routing information in link state routing proceeds over so-called Link-State Packets (LSP), which every router transmits. A LSP always begins with an identification of the transmitter, followed by a sequence number (Seq) that continuously increases, an entry of age (Age) and a list of the directly neighboring routers with a determination of their distance or cost.

LSPs are either created periodically in fixed time intervals, or only if there is a change in network topologies or other network properties.

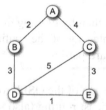

A	B	C	D	E
Seq / Age	Seq / Age	Seq / Age	Seq / Age	Seq / Age
B 2	A 2	A 4	B 3	C 3
C 4	D 3	D 5	C 5	D 1
		E 3	E 1	

A reliable distribution of LSPs is especially critical. Those routers first receiving the LSPs are also the first to change their routing tables. This can result in an temporary situation of inconsistency, with different routers having a different impression of the network topology. Distribution proceeds via a **flooding** procedure. Here, it is checked that LSPs are only forwarded the first time they reach a router. If an LSP, which is identifiable based on a sequence number, reaches the same router a second time it is then discarded. In addition, the LSP receives an age field (Age). In the first transmission it is initialized with a value that permanently counts down. When it reaches the value zero the data packet is likewise discarded.

Further reading:

Steenstrup, M. (Ed.): Routing in Communications Networks. Prentice Hall International (UK) Ltd. (1995)

Fig. 6.18 Link-state packets (LSPs) for exchange of the routing information.

the case of EIGPR from the Cisco company. As successor of the RIP protocol in the Internet, OSPF offers the following innovations:

- **Variable cost metrics and type-dependent forwarding**:
  OSPF supports a variety of various metrics for the calculation of distance between neighboring routers, focusing on a quick adaption to altered network topologies. The respective basis of metrics calculation and decisions concerning the forwarding of data traffic is additionally dependent on data traffic type. In this way, for example, real-time data can always be preferentially forwarded or specific connections with a high delay blocked for real-time data traffic.

- **Load balancing**:
  To ensure a better network utilization, OSPF has the ability of load balancing. While other routing protocols constantly forward data packets via an optimal connection path, OSPF has the possibility of splitting data traffic over multiple connection routes in order to achieve a more balanced utilization.

- **Hierarchical routing**:
  With the beginning of the development of OSPF in 1988, the Internet had grown to such a degree that the management of routing tables required by individual routers presented a considerable problem. A solution to this problem was the introduction of a hierarchical routing approach (see Excursus 7), which was likewise supported by OSPF.

- **Security**:
  OSPF supports different secure authentication procedures, thereby limiting message exchange to trustworthy routers. In this way, the integrity of the routing tables should be protected from attacks by unauthorized third parties.

- **Unicast and multicast routing**:
  With multicast OSPF (MOSPF, RFC 1584) a simple extension of the OSPF routing protocol is available, which ensures an efficient multicast routing (also see Excursus 7).

OSPF allows a hierarchical router organization within an autonomous system (routing domain). For this purpose, the autonomous system (AS) is divided into different areas. Every router in an area only forwards its link state packets to other routers within this area. Special routers, known as **area border routers**, are configured for the forwarding of data packets between different areas. In order to simultaneously manage a multiple number of areas within an autonomous system, special routers are interconnected to a **backbone**, whose primary task consists of transporting data traffic between the individual areas of the autonomous system. The backbone contains all area routers and can additionally include gateway routers (boundary routers), which take over inter-AS routing between the autonomous systems. Fig. 6.19 shows an OSPF router network in an autonomous system divided into three areas.

Link state routing protocols, such as OSPF, can become problematic as soon as the routing metric bases the calculation of routing costs on the utilization of each connection. Fig. 6.20 shows the occurrence of oscillation in a router network, preventing convergence from taking place. This is, however, only the case if the mo-

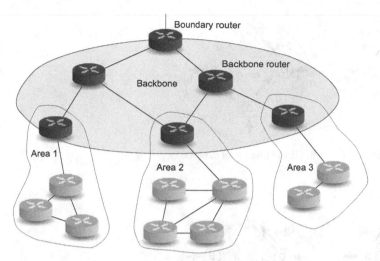

Fig. 6.19 Autonomous system with a hierarchically structured OSPF router network.

mentary burden (congestion) is used as the metric for routing decisions. In this case, routing costs are not necessarily symmetrical along a connection between two nodes, i.e., it is entirely possible that the reverse direction of a connection can involve different costs. Fig. 6.20 presents (a) the current load on the routers, which simultaneously determines the directed edge weights of the graph. On routers B and D the load is at 1, on router C the load is at x. The applied load on D is forwarded to the destination node A, and the load on C and B is likewise forwarded to A. C forwards its data in a counter-clockwise direction to node A. There is no load applied to the routers in the opposite direction.

Following an exchange of link state packets, C determines that the path to A in the clockwise direction produces only the costs 1, while in a counterclockwise direction produces the costs 1+x. For the next data transmission, node C therefore choses the alternative transmission path in a counterclockwise direction. Likewise, B determines that the best path to A is now also in the clockwise direction. This results in the edge weights shown in Fig. 6.20 (b). In the next exchange of link state packets nodes, B, C and D determine that the best connections now run again in the opposite direction, counter-clockwise, A. The result is shown in the Fig. 6.20 (c) represented edge weights. If link state packets are exchanged anew, the decision for the best routes again changes the transmission direction, resulting in the edge weights seen in Fig. 6.20 (d).

While an idea of how to solve the oscillation problem might be not to integrate the current load situation in the routing decision, this would, however, be unacceptable as overload situations should also be detected and avoided in routing. There remains the possibility of not synchronizing the operation and frequency of the link state routing operation between routers, instead running it within appropriate random intervals to avoid a synchronously clocked oscillation.

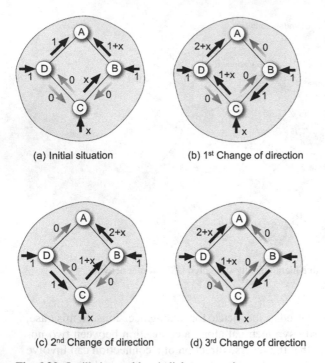

(a) Initial situation   (b) 1st Change of direction

(c) 2nd Change of direction   (d) 3rd Change of direction

Fig. 6.20 Oscillation problem in link-state routing.

The use of link state routing may additionally be found in the older **IS-IS** protocol **Intermediate System - Intermediate System**). It was originally implemented in DECnet PhaseV and standardized by the ISO in 1992 as **CNLP** (**Connectionless Network Layer Protocol**, ISO 10589). Although IS-IS is not an Internet standard, the IETF published IS-IS as RFC 1142. IS-IS was then extended with the capability of forwarding data packets of the Internet protocol (IP datagram), and designated as Integrated IS-IS. It was then published as Internet standard RFC 1195. IS-IS is implemented in many Internet backbones, e.g., in NSFNET. In contrast to OSPF, IS-IS is capable of simultaneously supporting different types of protocols in the Internet layer in the TCP/IP reference model.

Excursus 7: Special Routing Procedures

Hierarchical Routing

If networks whose routers are used for the the transmission of data packets grow rapidly, the result is a simultaneous expansion of the routing tables in each router. Besides the increase in required memory, the running time of the implemented algorithm and, consequently, the necessary computing time rises drastically. Also the bandwidth necessary for the exchange of routing information expands to such a point that the management of individual routers quickly becomes too complex. To solve this problem, the need for every router to have an entry in its routing table for every other router is eliminated. Routing is organized on a hierarchical basis, comparable with the organization of telephone numbers in a telephone book.

Here, the network is divided into individual **regions**. As previously, a router has the complete topological information for the region it belongs to. However, the topology details of the remaining regions remain unknown. If, for example, different networks are interconnected, each network is considered a separate region. This unburdens the routers in the respective network. When a large number of widely expanded networks are merged, often a hierarchy with two level is not enough. In this situation, regions are combined to even larger structures, which can in turn be aggregated into a new structure.

Let's say, a data packet is to be routed from Potsdam (Germany) to Berkeley (California, U.S.). The router in Potsdam has the complete information about the network topology in Germany. It therefore sends the data packet intended for the USA to Frankfurt, where a large number of foreign connections converge. From there, the data packet is sent directly to a router in New York, which functions as the central contact point for the entire data traffic in the USA. It is sent from the New York router to a router in Los Angeles, the contact point for the entire data traffic in the state of California. Finally, the data packet is sent on to Berkeley. Routing tables for regular and hierarchical routing are compared in Fig. 6.21,.

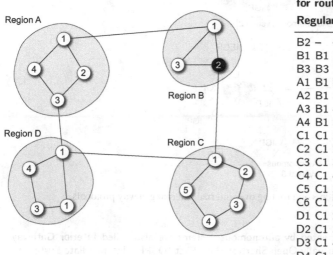

Routing table for router B2

Regular			Hierarchical		
B2	–	–	B2	–	–
B1	B1	1	B1	B1	1
B3	B3	1	B3	B3	1
A1	B1	2	A	B1	2
A2	B1	3	C	C1	1
A3	B1	4	D	C1	2
A4	B1	3			
C1	C1	1			
C2	C1	2			
C3	C1	3			
C4	C1	4			
C5	C1	3			
C6	C1	2			
D1	C1	2			
D2	C1	3			
D3	C1	4			
D4	C1	3			

Fig. 6.21 Hierarchical routing.

Instead of 17 entries, the routing table of router B2, in our example of hierarchical routing, needs only six entries. It should also be noted that hierarchical routing no longer guarantees that the optimal route will always be chosen as the connection is always switched over the designated connection router in a region. This then takes over the forwarding of the data packets within the region. At the same time, a diversion from the optimal path in a large network is minimal.

In regular routing in a network with **n** routers each router needs an entry in its routing table for every other router, i.e., every routing table contains n^2 entries. In comparison, in hierarchical routing only the routers of the own region together with the respective bordering routers of the remaining regions appear in the routing tables. This means the number of entries in the routing tables are reduced to approx. **log n** entries. According to [140], optimal routing tables in a large network are generated with **n** routers, if subdivided into **log n** levels. Whereby every router then has a routing table with **e·log n** entries (**e**=2,71828 – Euler's constant).

The routers within such a autonomous region all work with the same routing algorithm. Therefore, based on its organization form, such a region is also designated an **autonomous system (AS)**. All routers of an autonomous system have information about the other routers of the autonomous system. The respective routing algorithm employed in an autonomous system is known as the **intra-AS routing protocol**. In order to connect different autonomous systems with each other, special routers are provided to forward data packets, also beyond the borders of the system. These routers are known as **gateway routers**. In our example in Fig. 6.22, the gateway routers of the autonomous system are each shown darkened. The forwarding between different autonomous systems is also called **inter-AS routing**.

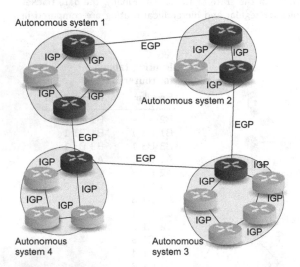

Fig. 6.22 Autonomous systems and routing over internal/external gateway protocols.

Routing protocols implemented by autonomous systems are also called **Interior Gateway Protocols** (IGP). Today, usually Open Shortest Path First (OSPF), Intermediate System to Intermediate System (IS-IS) or Extended Interior Gateway Routing Protocol (EIGRP) are used. Older routing protocols, e.g., Routing Information Protocol (RIP) or Interior Gateway Routing Protocol (IGRP) are seldom employed nowadays. In contrast to IGP, **Exterior Gateway Protocols (EGP)** regulate the connection and forwarding of data between the different autonomous systems. Fig. 6.22 illustrates the use of IGP and EGP in and between autonomous systems.

EGPs exchange accessibility information between the autonomous systems, which is then used by the routers of the autonomous systems as internal routing information for IGPs. As a rule, an EGP has the following functions:

- Two gateway routers from different autonomous systems agree on whether or not to become EGP partners.

- Within fixed, prescribed time intervals, the gateway router verifies the accessibility of the previously fixed neighbors in foreign autonomous systems.

- Upon request, the EGP partner receives a list from its neigboring routers of the respective foreign, autonomous systems accessible to them.

Before the development of the concept of the autonomous systems and the employment of hierarchical routing protocols, there was also a special routing protocol used in the Internet called **Exterior Gateway Protocol** (RFC 904). Although no longer used today, the name of this protocol is still applied to the described class of routing protocols.

The only EGP currently implemented in the Internet is the **Border Gateway Protocol (BGP)** in version 4 (BGP4, RFC 4271).. BGP works according to one of the distance vector routing methods similar to the **path vector routing method**, i.e., a BGP router does not transfer any distance information but only path information that describes the way to the goal. As no distance information is forwarded, an automatic routing decision does not follow. This means the decision about the choice of the possible path to the goal falls to the so-called routing policy of the autonomous system where the the gateway router is assigned, and thus to the responsible network administrator.

In contrast to distance vector routing, the problem of looping is avoided simply by the forwarding of all path information in a routing update to a neighbor in its network. If a neighboring router finds itself within a path of the forwarded routing information, it recognizes a loop and does not enter this path in its routing table.

In addition to the tasks of the inter-AS routing between autonomous systems (external BGP, eBGP), BGP can also be employed in the forwarding of path information within the autonomous system (internal BGP, iBGP). With the implementation of iBGP in an autonomous system, BGP connections must be set up between all routers of the autonomous system so that a complete meshed topology (full mesh) results. n routers therefore need approximately n^2 BGP connections. To avoid scalability issues in large networks, so-called **route reflectors** are used. They act as a hub of the BGP connections to all routers of the autonomous system, so that every router in the autonomous system must only keep an active connection to the route reflector. The BGP protocol has some weaknesses, such not taking into account distance information regarding hops or various connection-related bandwidths. Here, only the number of hops is used for the routing decision. Yet, such weaknesses are compensated by the prioritization of routers, which is subject to routing policies regulated manually by the responsible network administrator.

Multicast and broadcast routing

Multicasting makes it possible to transmit data simultaneously from one sender to a whole group of receivers. This form of communication is designated as **1:n communication**. If a data packet is to be forwarded to all computers in the network, it is referred to as **broadcasting**. A multicast transmission is often associated with the transmission of multimedia data, e.g., live video streaming. Among the main applications of multicast transmission are:

- Audio and video transmission,
- Software distribution,
- Updates from WWW caches,
- Conferences (audio,video and multimedia) and
- Multiplayer games.

Broadcasting can be carried out in different ways in the network:

- **Broadcasting without special routing**
  The simplest method that does not need any special measures on the part of the network and its intermediate systems. The transmitter itself sends the data to each receiver respectively. However, with this method, bandwidth and computing time are also wasted.

A significant time delay between the arrival of successive packets is the result. A list of all potential addressees is also required by the transmitter.

- **Flooding**
  Here, a router forwards the incoming data packet in replicated form over all of its lines. Although flooding would never be considered in the regular routing operation of point-to-point connections, it can make sense in broadcasting. The general problem of flooding, however, still remains: too many packets are generated and because of this bandwidth is wasted.

- **Multidestination routing**
  Every data packet is given either an address list or a bit string indicating the desired destination with this method, . The router checks the incoming data packet, replicates it and transmits it further only via those connection that are reachable over the fixed destination. In doing so, it equips the data packet with a new address list that in each case indicates only the destinations reachable over each connection line. In this way, it is assured that a data packet contains just one address after several hops and can be handled like a normal data packet.

- **Rooted tree/spanning tree methods**
  Here, a rooted trees is formed based on the router that started the broadcast. Once it is verified that every router in the network has an image of this rooted tree, it sends the replicated broadcast data packets only via those connections that list these paths in the rooted tree, with the exception of those connections over which the data packet reached the router. In this way, there are only as many data packets as necessary generated to keep the available bandwidth as efficient as possible. However, the required information is not always provided via the rooted tree (e.g., not by distance vector routing, but by link-state routing).

- **Reverse Path Forwarding (RPF)**
  An approximation of the rooted tree method can be achieved with the help of the very simple reverse path forwarding method. Here, it is not necessary for one of the routers involved to have knowledge of the rooted tree. Every router only checks if an incoming broadcast data packet has reached it over the particular connection indicated as the designated connection in the routing table for the transmitter. If this is the case, then the probability is high that the data packet in question has reached the router directly over the best possible path and is an original data packet. In the other case, if the data packet has not reached the router over the designated connection of the transmitter, then the likelihood is high that it is a copy and it is discarded.

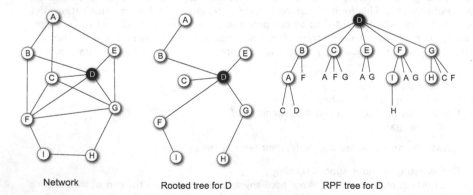

Network   Rooted tree for D   RPF tree for D

Fig. 6.23 Comparison of a rooted tree and RPF tree network.

Fig. 6.23 shows an example of the RPF method. The output network, the rooted tree for the router D, and the tree generated through the RPF method are shown here side by side. Let's look at the RPF tree with router D as the root. First, the data packet is flooded to all connections from D, so that routers B, C, E, F and G are reached following one hop. Each one of these data packets thus reaches the routers B, C, E, F and G on the designated path with respect to the transmitter. In the next step, the data packets are flooded from all of these routers over all the connections, except over the one of the original transmitter. In this way, router B sends a data packet to each router A and F. While the data packet reaches router A on the designated connection with respect to the original transmitter D, the data packet does not reach router F over the designated path with regard to transmitter D. Router A in turn forwards the data packet in the next step to all connections except to the transmitter's, while F discards the data packet. After two hops the broadcast has reached all of the participants in the network, and after three hops the data packets have already disappeared from the network.

RPF has the advantage of being easy to implement and, at the same time, works efficiently. The routers do not need to have knowledge of the spanning tree and the data packets need not carry an address list with them. Furthermore, there are special mechanisms necessary to ensure that the flooding process stops again at some point to avoid an excessive duplication of the data packets.

Multicasting is necessary if a large group of receivers is addressed at the same time. Additionally, this group is small in relation to the total number of network members so that a complex and bandwidth-devouring broadcast mechanism would not be suitable in this case. To handle such a group, special group management must be implemented, whereby it can help members to can simply join and leave groups. This management is not a part of the routing procedure. Between the end systems and the routers a special host-to-router protocol is implemented: the **Internet Group Management Protocol. (IGMP)**, standardized in RFC 1112. Over the IGMP protocol a host can register affiliation with a certain group or, as the case may be, the router queries the connected end system about this periodically. The routers pass on this information as routing information to adjacent routers. Thus, information pertaining to group affiliation is distributed over the network. An end system can join a multicasting group by registering at the nearest router for the delivery of multicast messages. The router passes on the requirements until they reach the multicasting source, or the source of the nearest router. In addition, the respective router must also be designed for a multicast operation.

Therefore, a distinction is made between **multicast routers**, meaning routers that can coordinate and forward a multicast transmission, and **unicast routers**, which are unable to do this. To still make a multicast transmission possible so-called **tunnels** are set up between multicast routers. These "tunnel under" the unicast routers located in between, i.e., the multicast data packets are encapsulated as unicast data packets and forwarded unchanged. The tunnel connection thus works in the same way as a normal point-to-point connection.

A network made up of multicast routers connected to one another that is primarily implemented to perform the task of multicasting is called **Mbone (multicast backbone)**. Mbone is a virtual network of its own, so to speak, via the WAN or Internet. Multicast routing basically proceeds as follows:

- The participating end systems register for multicasting groups over the responsible router.
- The router is in a position to identify individual group members based on group addresses.
- Data packets to be forwarded must, if necessary, be replicated by the router and passed on further over several of its connections. Additionally, procedures similar to broadcasting are implemented, only that here the connections over which a data packet is to be forwarded are chosen based on the addresses in the group.

Unlike broadcasting, in multicasting the construction of a rooted tree for the entire net-work is not required, rather the construction involves a tree just containing those routers connected to the host located in the relevant multicasting group. Multicast packets are sent along this **multicast tree** from the transmitter to all the hosts of the multicast group. The multicast tree can, however, also contain routers whose hosts are not members of the multicasting group. There are two general procedures for identifying multicast trees:

- shared multicast trees that are calculated for the entire multicast group. One of the practices implemented in identifying a commonly used multicast tree is based on the identification of a center node of the multicasting group. First, a center node is determi-ned. Routers with hosts that belong to multicasting groups send so-called join messages to the center nodes, which are forwarded over regular unicast routing until they reach either a router that already belongs to to the multicast group or the center node. The path over which the join message is sent defines a branch of the multicast tree between the router of the transmitter and the center node.

- source-based multicast trees in which a separate multicast tree is calculated for eve-ry transmitter of a multicast group. The forwarding of the multicast packets proceeds source-specifically along the multicast tree of the respective sender. Here, the already known link state routing method can be employed for every transmitter of the multi-cast group. This ensures that a source-optimal rooted tree will be designated for the respective router. The already mentioned reverse path forwarding algorithm (RPF) also delivers source-based multicast trees. While not necessarily delivering a minimal rooted tree, they have less status information to manage and exhibit a reduced computational complexity.

RPF is successfully implemented in the WAN and Internet as a multicast routing method in several routing protocols. However, there is also the possibility that unwanted conditions can arise with RPF. Fig. 6.24 shows a network with a RPF multicast tree (thick connection line). Those routers connected to their hosts and belonging to the multicast groups are indicated as darker. Following the RPF, router F forwards data packets to router G, although there are no hosts connected to it that belong to multicast groups. This situation becomes problematic when instead of router G, hundreds or even thousands of such downstream routers are reachable from router F. Every one of these routers would receive multicast data packets without being responsible for their delivery. This was just the situation that occurred in Mbone, the first multicast network of the Internet.

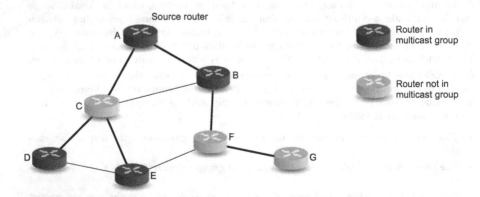

Fig. 6.24 Problems in reverse path forwarding and subsequent pruning.

To solve this problem, so-called **pruning** messages are sent in the reverse direction to the downstream routers, on which there are no hosts connected that belong to multicast groups. These convey to the upstream router that it does not wish to receive further multicast messages. If a router receives a pruning message from all of the downstream routers, it can send the pruning message to all of its upstream routers. To ensure that this procedure functions correctly, a router must always know which downstream router depends on it with respect to the multicast group. For this, RPF must additionally exchange topology information. If a subtree is pruned from the multicast tree via the pruning mechanism, possibilities must be provided that allow for a readmission of the subtree to to the multicast tree. To facilitate this, either special messages are fixed in the routing protocol, or a timing mechanism is used. Here, following expiration of a fixed time span, the pruned subtree and router will automatically be admitted again to the multicast tree (**grafting**). The readmitted router then has the possibility of avoiding unwanted multicast messages with the transmission of a new pruning message.

For the implementation of multicast routing algorithms in the Internet, the Internet router in question must first naturally be in a position to support the relevant multicasting routing protocol. In fact, only some of the implemented routers in the Internet have this capability. A virtual network can be set up between Internet routers that support multicast routing, which can also include unicast routers. The multicast messages are then sent via **multicast tunneling**. This means that a multicast data packet is packed as payload in a regular unicast data packet and forwarded along a connection line to the next multicast router, which can also contain unicast routers.

Among the methods implemented in multicast routing are:

- **Flood and prune**
  A very simple algorithm, in which the router first sends the data to all connections (flooding) and in the course of transmission, blocks those connections for multicast transmission over which there are no receivers.

- **Link State Multicast Protocols**
  In this procedure, along with their own routing information, multicast routers distribute details about the entire network for groups with direct receivers connected to them.

- **Distance Vector Multicast Routing Protocol (DVMRP)**
  A distance vector routing procedure that can forward multicast data packets. DVMRP was the first multicast routing protocol used in the Internet and standardized in 1988 as RFC 1075. DVMRP supports sourced-based multicast trees with RPF, pruning and grafting. The starting point is a routing table with the information for the shortest path to each reachable destination. This is determined with the help of distance vector routing. Additionally, DVMRP establishes a list of the dependent downstream routers for every router. Pruning messages are provided with a timeout and usually retain their validity for two hours. Previously pruned subtrees and routers can be be admitted again in the multicast tree via grafting messages..

- **Multicast Open Shortest Path First (MOSPF)**
  Multicasting-capable OPSF routing protocol. This procedure is designed for application within autonomous systems and standardized in RFC 1584. MOSPF is based on the OSPF link-state routing procedure. MOSPF routers provide for this purpose a complete representation of the network topology, including the multicast router and the multicast tunnel in between. From this, the best possible path to all multicast routers is calculated. MOSPF uses reverse path forwarding and likewise offers the possibility of pruning.

  This routing procedure was standardized as RFC 2362, and is distinguished by its two operating types, **PIM dense** and **PIM sparse**. If the bulk of end systems is in the addressed multicast group, then the PIM dense procedure is employed. Here, the network is flooded with queries to all other routers in the network. PIM dense supports RPF in combination with flooding and pruning. In principal, it works like DVMRP. However, if the groups are considerably smaller, then PIM sparse is employed. Instead of a network being flooded, specific rendezvous points are set up where all group members send their data packets. In PIM sparse, membership in a multicasting group must be announced to the central rendezvous point via periodically sent JOIN messages. As stated in its name, PIM works independently of the underlying unicast routing protocol.

- **Core Based Trees (CBT)**
  This multicast routing protocol, standardized as RFC 2201, is based on a commonly used multicast tree that is constructed around one or more "cores". The location of the cores is determined statistically. From there, the multicast tree extends to the routers of the members of each multicasting group. At the same time, a border router sends a JOIN request as an unicast in the direction of the core. The first router to receive the message, which already belongs the multicast tree, answers with a JOIN ACK message as a confirmation. The multicast tree is maintained by every downstream router, which periodically sends an ECHO REQUEST to its directly adjacent upstream router. This is then confirmed with an ECHO REPLY message. If a downstream router repeatedly fails to receive an answer to its ECHO REQUEST, it subsequently sends a FLUSH TREE message to disband the subtree in the downstream direction of the subtree.

The just-described multicast routing protocols are employed in the Internet within individual autonomous systems, i.e., all of the routers present in the autonomous system use the same multicast routing protocol. However, an **inter-AS multicast routing** also requires that multicast routing have the capability to function over different autonomous systems, where in each case different multicast routing protocols could potentially be employed. The **border gateway multicast protocol** (BGMP, RFC 3913) is a current project of the IETF for the implementation of an inter-domain multicast routing protocol. Further representatives of this class of multicast routing protocols are multicast BGP (RFC 2858) and multicast source discovery protocol (MSDP, RFC 3618).

Further reading:

Tannenbaum, A. S.: Computer Networks, Prentice-Hall, NJ, USA (1996) pp. 348-352

Deering, S. E.,Cheriton, D. R.: Multicast Routing in Datagram Internetworks and Extended LANs, in ACM Trans. on Computer Systems, vol.8 (1990) pp. 85-110

Kamoun, F., Kleinrock, L.: Stochastic Performance Evaluation of Hierarchical Routing for Large Networks, Computer Networks, vol.3 (1979) pp.337-353

Perlman, R.: Interconnections: Bridges and Routers, Addison-Wesley, Reading, MA, USA (1992)

Ramalho, M.: Intra- and Inter-Domain Multicast Routing Protocols: A Survey and Taxonomy. in IEEE Communications Surveys and Tutorials (COMSUR) 3(1) (2000)

Excursus 8: Routing Procedures for Networks with Mobile Components

In recent years mobile networks have gained unprecedented popularity. All notebook or smartphone owners also want to use their device to read an email or gain access to the WWW – independent of where they are at that moment. But it is just this wish that creates a series of routing problems in the Internet. In order to lead a data packet to a mobile end system, the network has to first be able to find the end system. The mobility of end devices involves the fact of a continually changing network topology. Not only does an end device change its position within a network, but also beyond the borders of the the network - and that (potentially) in rapid succession! To solve this problem, new routing procedures were developed and existing routing technology reworked and extended.

Mobile routing

In the case of mobile routing, a distinction is made between **stationary users**, i.e., users connected to the Internet over fixed cables and whose location does not change, and mobile users. With mobile users a distinction is made between **migratory hosts**, i.e., hosts who as a rule have a fixed Internet access that may change from time to time, and **roaming hosts**, who move around freely and at the same time still want to maintain their connection to the Internet.

It can be assumed that every user has a permanent home address in a fixed network, which does not change. If a user leaves this network, the routing cannot find it initially. While changing the address or the routing table could possibly prevent this situation, because of the multiple number of changes required this approach is not feasible in a fixed network. **Mobile IP** (RFC 3344) is an extension of the IP protocol (see section 7.3.3) and makes it possible to reach the host over its established IP address, regardless of the location. With the introduction of several additional concepts in mobile IP, the addresses and most of the routers can remain unchanged:

- **Area**
  The world is geographically divided up into individual, smaller units called areas. Typically a LAN is contained in such an area.

- **Mobile node**
  The name for a mobile end system with its original address, the validity of which is also retained in foreign areas.

- **Foreign agent**
  The designation for routers monitoring all mobile nodes that are foreign to the network. Their task is to forward the data packets they receive from the home agent of the respective mobile node. For the duration of its stay in the foreign area, the foreign agent is simultaneously the default router of the mobile node.

- **Home agent**
  Every area has a home agent that is responsible for keeping track of the mobile nodes in a respective area. The home agent knows the current location of the mobile node and sends it data packets via a tunnel over a foreign agent in the unknown area.

- **Care-of Address (COA)**
  Designates the temporary address of a mobile node in a foreign area. This is provided to the mobile node by a foreign agent in the area.

Communication with the mobile node in a foreign area typically proceeds as follows (see also Fig.6.25):

- The home agent and foreign agent make their presence known in the respective area with the help of periodic announcements. This enables the mobile node to gain knowledge of the addresses of the home agent and foreign agent.

- Upon entering a new area, a mobile node registers at the foreign agent. This, in turn, contacts the corresponding home agent of the home area and gives it the temporary new address (Care-of Address, COA) of the mobile node. As a rule this is the foreign agent's own address.

- The home agent now knows that the mobile node in question is in the area of the foreign agent. Data packets addressed to the mobile node are intercepted by the home agent, who first presents itself as this node.

- The home agent forwards the data packets by opening a tunnel to the relevant foreign agent over which the mobile node can be reached. In addition, the original data packets are encapsulated in data packets addressed to the COA.

- The foreign agent decapsulates the packet and sends it to the mobile node.

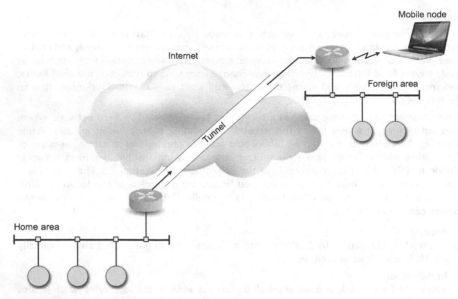

Fig. 6.25 Mobile routing.

- Communication in the opposite direction - from a mobile node to a permanent end system - is much easier. For this, the mobile node simply addresses the data packets to the relevant end address of the permanent end system, supplying its own permanent address as the sender.

Mobile ad-hoc routing in MANETs

With mobile routing it is assumed that all connecting hosts are mobile and that routers on the interconnection nodes are stationary. In cases of mobile ad-hoc networks (MANET) it must be taken into account that every (mobile) host is at the same time a (mobile) router. Mobile ad-hoc networks differ from stationary networks in that there is no predetermined topology. Furthermore, there is no fixed relation between network addresses and geographic locations or between neighbors in a network. Networks topologies, locations and neighbor relations are continually subject to change. This changing situation must also be taken into account by the routers implemented in such a mobile ad-hoc network. Two nodes in the network graph of a MANET are considered connected to each other, i.e., an edge runs between them if they communicate with their radio transmitters. Principally, the strength of a connection is dependent on the distance between the two nodes. However, obstacles such as buildings or mountains can also disrupt mutual reception, even if the geographic distance is minimal. Additionally, it needs to be considered that not all nodes have the same strength regarding transmission and receiver technology. It can happen that between two nodes A and B, a connection from A to B is possible, however not one in the opposite direction of B to A. To simplify the discussion, it will be assumed in the following illustration that all connections are symmetric.

Besides unicast, broadcast and multicast routing variations, which are based on a network topology, **position-based routing procedures** are also employed in the area of the mobile ad hoc routing. Position-based routing procedures use geodetic information concerning the exact geographic positions of the network users. This can be obtained over GPS receivers

(Global Positoning System). If all network users who are located in a specific geographical area are to be addressed, then the term **geocasting** is used. Based on the position information of the network user, the shortest path between the transmitter and the receiver can be determined.

Mobile ad hoc routing procedures may be subdivided into

- proactive,
- reactive and
- hybrid routing procedures.

Proactive routing procedures

Just as in stationary networks, the optimal path is determined here before it is actually needed. The routing information is stored in a routing table, in the form of complete paths (source routing) or next hops in the network node involved in the routing. If the data is to be sent to a known receiver, a path must not be determined first, rather the routing information can be taken from the previously calculated routing table. For the structure of the routing table, multiple messages between the network nodes must be sent that burden the entire network, although a large part of the collected information will probably not be used for later data transmission. Moreover, the often occurring topology changes in mobile networks require a renewed sending of updated information. The routing information propagated by the network is always proportional to the size of the network.

Among the proactive routing procedures for mobile ad hoc networks are:

- Ad hoc Wireless Distribution Service (AWDS)
- Babel
- Clusterhead Gateway Switch Routing Protocol (CGSR)
- Direction Forward Routing (DFR)
- Distributed Bellman-Ford Routing Protocol (DBF)
- **Destination Sequence Distance Vector Routing Protocol (DSDV)**
- Hierarchical State Routing Protocol (HSR)
- Linked Cluster Architecture (LCA)
- Mobile Mesh Routing Protocol (MMRP)
- Optimized Link State Routing Protocol (OLSR)
- Topology Dissemination based on Reverse-Path Forwarding Routing Protocol (TBRPF)
- Witness Aided Routing (WAR)
- Wireless Routing Protocol WRP)

The **destination sequence distance vector routing (DSDV)** is a typical distance vector routing procedure. Thereby, the formation of routing loops (count-to-infinity problem) is avoided by maintaining a sequence number. Every mobile host maintains a sequence number. It is increased by one every time an update message is sent to the neighboring router. Neighboring routers only overwrite entries in their own routing tables with new update messages when they have a higher sequence number than the previous entry, or if the sequence numbers are the same, however the routing distance in the update message is indicated as shorter.

Reactive routing procedures

In contrast to the proactive routing procedure, in the reactive routing procedure (also called the on-demand routing procedure) the information required for the forwarding of data packets is first determined when the actual situation presents itself. This is helpful for the often limited available bandwidth in mobile ad hoc networks, as the transmission of multiple routing messages for the construction of the routing table can be omitted. Routing information is first determined and forwarded in cases where it is in fact actually needed.

Included in the reactive routing procedures for mobile ad hoc networks are:

- **Ad hoc On-demand Distance Vector Routing (AODV)**
- Admission Control enabled On-demand Routing (ACOR)
- Ant-based Routing Algorithm (ARA)
- Ariadne
- Backup Source Routing (BSR)
- Caching And MultiPath Routing (CHAMP)
- **Dynamic Source Routing (DSR)**
- DYnamic Manet On-demand Routing (DYMO)
- Minimum Exposed Path to the Attack (MEPA)
- Mobile Ad-hoc On-Demand Data Delivery Protocol (MAODDP)
- Multirate Ad hoc On-demand Distance Vector Routing Protocol (MAODV)
- Robust Secure Routing Protocol (RSRP)
- SENCAST
- **Signal Stability Adaptive Routing (SSA)**
- **Temporally Ordered Routing Algorithm (TORA)**

Three different phases can be determined in reactive routing procedures:

- **Route Discovery**
  To convey data from a transmitter to a receiver, a way to the receiver has to first be determined. The **dynamic source routing protocol (DSR)** uses the concept of source routing to do this. Here, complete routing information is contained in every transmitted data packet. This means that all the intermediate systems along the path to the receiver are listed. First, the transmitting computer sends a so-called route request as broadcast message to all of its neighbors in the network. If a computer receives the route request and does not know the path to the described destination, it appends its own address to the route request and propagates the message to all of its neighbors in the network. In this way, paths through the network to the receiver may be determined. Looping is avoided by the router simply checking the address contained in the route request upon its reception. As soon as a route request reaches the receiver, it sends a route request reply as a unicast message on a direct path back to the transmitter together with the collected routing information. The transmitter stores the routes found in a special routing cache. In the same way, an intermediate system can send back a route reply to the transmitter if it already has an entry in its own routing cache with the sought address and a predetermined age limit has not been exceeded. Thereby, the already collected routing information in the route request is linked with the information in the routing cache and sent back to the transmitter as a route reply.

  Fig. 6.26 shows the course of the DSR routing procedure. Host A starts a route request to all of its neighbors to determine a path to host D. Host E and B receive the route request, supplement it with their own addresses and forward it to their neighbors (a). After the route request has reached host D, D transmits a route reply with the complete path information back to host A.

  The **signal stability adaptive protocol (SSA)** distinguishes between different connections based on the signal strength with which the data packet is received. To do this, regular, so-called beacon data packets are sent to the neighbors in the network to monitor

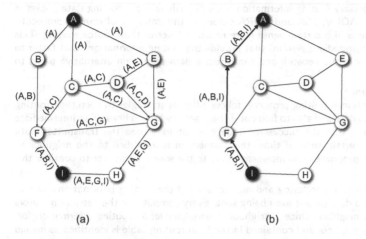

Fig. 6.26 Determining a path with the help of the DSR routing procedure in a mobile ad hoc network.

the stability of the network. Just as DSDV, SSA also sends route requests for path determination. Here, the transmitter can additionally specify a minimum signal strength that must be maintained along the individual connection in the route. Route requests are then only forwarded to those neighbors who fulfill the quality criteria established by the transmitter. Unlike DSDV, SSA uses the principle of next hop routing. As soon as the route request message reaches the receiver, a route reply message is sent back directly to the receiver in the opposite direction along the receiving route via unicast. Each intermediate system can then accordingly store the next hop to the receiver.

One of the most popular routing protocols for mobile ad hoc networks is the **Ad hoc On-demand Distance Vector routing protocol (AODV)**. It implements a topology-based reactive routing procedure, i.e., the respective router to the destination is first determined according to need. Moreover, a distance vector routing procedure is used, which is adapted to use in a mobile environment and infrastructure. This means that limited bandwidth and low battery life are calculated into the routing algorithm. AODV improves DSDV employment as only hosts along an active path in the network must exchange and provide information.

Route determination in the **temporally-ordered routing algorithm (TORA)** provides the additional possibility of multipath routing, i.e., several alternatives for a destination can be managed. For this, a height metric is performed for each host where paths from sender to receiver are always formed in the direction of decreasing height. The resulting routing network can be represented as a directed, cycle-free graph (Directed Acyclic Graph, DAG) with the receiver node as sink.

- **Data forwarding**
  Actual data forwarding in the network from the transmitter to the receiver can be carried out according to the principal of source routing or next hop routing, as the case may be. For the source routing that is implemented in DSR, for example, every data packet must have the complete path to the receiver included in its data packet header. The intermediate systems do not need to keep their own routing table, instead read the routing information from the data packet header and forward the data packet correspondingly. In contrast, in next hop routing only the address of the receiver need be listed in the data packet header. Every host in the network keeps a routing table that

provides the necessary routing information regarding where the incoming data packet is to be forwarded. AODV, SSA and TORA belong in this category of routing protocols. Next hop routing is in a certain sense more resistant to error than source routing. This is because the intermediate systems that provide the routing information react faster to topology changes in the network and can direct a data packet on alternative paths to the destination.

- **Route maintenance**
  The already mentioned routing protocols follow different strategies to keep their routing information up to date and also to find out if there are any broken links. If an intermediate system determines that the successive node has not forwarded the transmitted data packet within a certain span of time, this connection is identified to the neighbor as invalid and an appropriate announcement made to the source computer to consider this in the future.

  In next hop routing, a monitoring and maintenance of the routing information can also proceed without a data packet even being sent. Every computer in the network monitors the status of its neighbor. Once a neighbor is unreachable, all routing information forwarded over this neighbor and contained in the local routing table is identified as invalid and removed.

Hybrid routing procedures

Hybrid routing procedures combine proactive and reactive routing techniques in a joint protocol. For example, in a geographically limited area complete routing tables are generated in a proactive procedure, while the routing information for a mediation extending beyond the borders of this local area is determined with a reactive procedure.

The hybrid routing procedures for mobile ad hoc networks include:

- Hybrid Routing Protocol for Large Scale Mobile Ad Hoc Networks with Mobile Backbones (HRPLS)
- Hazy Sighted Link State Routing Protocol (HSLS)
- Hybrid Wireless Mesh Protocol (HWMP)
- Order One Routing Protocol (OORP)
- Scalable Source Routing (SSR)
- **Zone Routing Protocol (ZRP)**

Zone routing protocol (ZRP) is a hybrid routing protocol for mobile ad hoc networks. Every network node also defines a number of neighboring nodes that can be reached with a prescribed number of hops. This node set is called a zone. Within the zone the routing information is determined in a proactive way with the help of the DSDV routing protocol. Here, every node within this local area knows the paths to all accessible nodes in the zone. Whereby, the number of messages required to forward routing information is limited. For the routing between different zones (interzone routing), a reactive routing procedure is used In addition, a route request message is sent to all border nodes in the zone (border-casting), to check the routing table of their own zone to verify whether the wanted receiver is included. If the sought receiver is found within a zone, the corresponding border nodes sends back the routing information in a route reply message. Otherwise, further route request messages are initiated to the border nodes in their zones.

Mobile broadcast routing

In all mobile ad hoc networks, **broadcasting** messages occur relatively often as the route determination for reactive routing protocols is based on this technology. One simple broadcasting application is message flooding, i.e., every network node that receives a broadcasting message sends it on immediately. However, in a mobile ad hoc network the shared receiving areas of the individual network nodes often overlap each other. Because of this, flooding can lead to a high level of redundancy as well as to critical overload and collision situations, which are also known as **broadcast storm problems**. In this case, redundancy means that a network node is transmitting broadcasting messages, even though these messages have already been received by all the computers in its transmission area.

In Fig. 6.27 (a) the entire network can be covered with the transmission of two messages. The white node transmits the first message and the gray node directs the message to the other nodes. With flooding, however, four messages are sent. Fig. 6.27 (b) shows that the entire network can also be covered with the transmission of two messages, while a flood causes the transmission of seven messages. In Fig. 6.27 (c) the extent of overlapping is evident. The darker the area, the thicker the overlapping, and the more likely it is that a collision will occur.

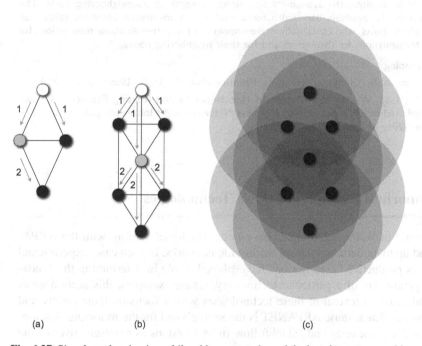

(a) (b) (c)

Fig. 6.27 Signal overlapping in mobile ad hoc networks and the broadcast storm problem.

Mobile multicast routing

For **multicast routing** in mobile ad hoc networks, a distinction is also made between procedures that a use a shared group tree for a multicasting group (core-based multicasting) and those in which different multicast trees are implemented for every transmitter of a multicast group (source-based multicasting). Because of the existing mobility of the hosts in mobile ad hoc networks and possible interference problems in radio-based communication, multicast routing procedures for MANETs are far more complicated than those for wired, stationary computer networks. The **on-demand multicast routing protocol (ODMRP)** constructs a multicast tree for every transmitter in a multicasting group. Here, the transmitters periodically send out Join data packets to the members of the multicasting group. First,

Join data packets are flooded over the network. If a host receives a Join data packet for the first time, it sends on this packet to its neighbor and establishes a reverse path direction to the previous host. Following the reception of a Join data packet, every member of the multicasting group sends a Join-table data packet to its upstream host along the reverse path. Upon reception, every host sends the Join-table data packet farther back in the direction of the original transmitter. At the same time, each host supplements its own multicast routing table for the forwarding of future to-be-received multicast messages. The previously discussed ad hoc on demand distance vector protocol (AODV)) also provides **Multicast AODV** multicast routing.

A special problem in ad hoc networks is ensuring a constant **quality of service (QoS)**, which is imperative for multimedia applications such as the transmission of audio and video data. Not only must the mobility of the host be considered, but also, for example, disruptive influences caused by neighboring transmission activities. The **ad hoc QoS multicast protocol (AQM)** is a multicast routing procedure that focuses particularly on the QoS aspect. Every network node monitors the availability and signal strength of its neighboring node. The calculation of the available bandwidth for a multicast transmission therefore takes into account all the hosts, who continually make known the respective available reservations for ongoing transmissions for themselves and for their neighboring nodes.

Further reading:

Tannenbaum, A. S.: Computer Networks, Prentice-Hall, NJ, USA (1996) pp. 348-352

Tseng, Y. , Liao, W., Wu, S.: Mobile Ad Hoc Networks and Routing Protocols. In Ivan Stojmenovic, editor(s), Handbook on Wireless Networks and Mobile Computing, pp. 371–392, John Wiley & Sons, Inc. (2002)

6.4 Important Examples of WAN Technologies

A multitude of WAN technologies have been developed starting with the ARPA-NET and up to modern high-speed worldwide networks. In each case, experimental as well as productive-based reasons have played a role in determining the fundamental parameters of a particular technology. Using examples, this section gives an introduction to several of these technologies with a focus on their variety and development. The historic ARPANET is the springboard for the following developments. X.25, frame relay and ISDN follow in its footsteps as representatives of the next generation.

At the same time, it should be noted that these WAN technologies are not well suited for the transmission of real-time multimedia data due to their limited bandwidth and quality of service. For this, broadband technologies are necessary that provide an almost constant delay in the transmitted data to enable the transmission of real-time audio and video data. In addition to the **ATM** (Asynchroneous Transfer Mode), which is based on an asynchronous time multiplex procedure, and **distributed queue dual bus (DQDB)**, primarily implemented in the MAN area, among the range of high speed extended traffic networks, the technology of the so-called **plesiochronous** and **synchronous digital hierarchy (PDH** and **SDH)** have gained a foothold. These will each be examined in a separate section, as well as the WiMax standard for radio-based high-speed networks.

6.4.1 ARPANET

Let's look first at the ARPANET, virtually the "granddaddy of all the WANs" and precursor of the worldwide Internet.[2] It was developed with the aim of creating a low-disturbance computer network. With this purpose in mind, *Paul Baran* (*1926) and *Leonard Kleinrock* (*1934) came up with the idea of a packet switching network at the beginning of the 60's – a revolutionary idea at that time. As the computers to be connected in a network (**hosts**) were operated with different computer architectures, it was decided at the end of the 60's to forgo equipping each of these computers with its own network access. The idea was to instead create a separate **subnetwork**. Its sole task should be to provide an infrastructure for communication with other nodes in this subnetwork. For the particular host computer to be connected with this subnetwork, only a simple communication interface had to be created. This would connect it with the assigned node in the subnetwork. The actual communication tasks were then carried out by the separate subnetwork.

This subnetwork consisted of special minicomputers, functioning as switching computers, connected with each other over cables. The implemented switching computers were designated **Interface Message Processors** (**IMP**). Each IMP was connected again with at least two other IMPs in order to increase reliability. This subnetwork functioned as a (packet switching) datagram network without fixed, logical connections being necessary between the communicating computers. In the case of a connection failure, the data packets were simply redirected over an alternative connection path.

At the beginning, the individual nodes of the ARPANET each consisted of an IMP and a host computer. Both stood in the same room and were connected with each other via a short cable connection. The host transmitted messages of up to 8,063 bits in length. These were fragmented by the corresponding IMP into smaller data packets of up to 1,002 bits long and then transmitted - each independent of the other - to its destination. On its way to the receiver, every data packet was stored intermediately in the switched IMP until it had arrived in its entirety, and only then was it forwarded.

To implement this network concept, the U.S. government office, ARPA (Advanced Research Agency), which later became the namesake for the network, put out a tender for the project. The company BBN (Bolt, Beranek and Newman), a consulting company with its headquarters in Cambridge, Massachusetts, won the tender. For the implementation of the IMPS, BBN chose specially modified Honeywell DDP-316 minicomputers that had a main memory of 24 kBit (12 k á 16 Bit). The IMPs did not have a hard drive, as non-stationary parts were generally considered unreliable at that time. The switching of the individual IMPs took place over leased telephone lines, operated with 56 kbps.

---

[2] A detailed presentation of the historical background of the development of the ARPANET is found in Chap. 2, „Historical Review", of the first volume in this series: Meinel, Ch., Sack, H.: Digital Communication – Networking, Multimedia, Security, Springer (2013).

The employed communication software was also divided into two separate layers: subnetwork and host. Besides the host-to-host protocols, which enabled communication between two applications on the host computer, there were, most importantly, the **host IMP** protocol regulating the connection between the host computer and the IMP, the **IMP-IMP** protocol that managed the communication between two IMPs, as well the extra **source-IMP-destination-IMP** protocol. This was responsible for increased transmission security (see Fig. 6.28).

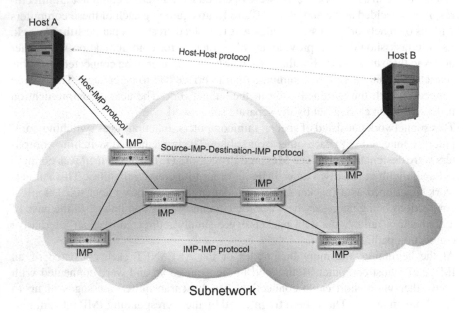

Fig. 6.28 Network design of the ARPANET.

In December 1969, the ARPANET designed this way with four nodes was put into operation at the Universities of Los Angeles, Santa Barbara, Utah and the Stanford Research Institute. The implemented communication protocol was structured in layers. The necessary communication was carried out on the application level based on programs such as **Telnet** (remotely controlled computers over text-based in and output) or **FTP** (simple data transfer between computers) via the host-to-host protocol, which in turn made use of the host-IMP protocol composing the layer below. The hosts could therefore function completely shielded from the details of IMP-IMP communication. Consequently, the fast development of new protocols upon these layers was greatly simplified (see Fig. 6.29).

In the three layers of this communication protocol model, the following tasks were carried out:

- **Host-to-IMP protocol**
  This protocol was responsible for the bidirectional communication between the host and IMP. The host first sent the message, which was to be transmitted to

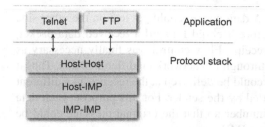

Fig. 6.29 Layering structure of the ARPANET communication protocol.

another host in the ARPANET, to its corresponding IMP. After successfully receiving the message, the IMP at the destination sent back the announcement to the transmitting host giving it this information. If the message was received successfully, the target IMP sent a RFNM announcement (ready-for-next-message). In case of error, a message was sent announcing non-completed transmission. Moreover, the receiving IMP was responsible for reassembling the individual data packets before forwarding the transmitted outgoing message to the destination host. The IMP was further capable of blocking incoming announcements so that they did not reach the destination host. The length of the announcements varied, but they could reach up to 8,096 bits. Individual addressing was not necessary on this level, as only one connection between the host and corresponding IMP was served at a time.

- **IMP-to-IMP protocol** and **Source-IMP-to-Destination-IMP protocol**
  The host computer was where the actual application programs ran and where data was exchanged. For this reason, one of the design goals of the ARPANET was to shield the host computer from the operations that were necessary for an efficient and error-free data transmission. This is why individual computers, the IMPS, were employed. In view of task assignment among the IMPs, differences can be determined between the regular IMPs of the network and the respective IMPs located at the transmitter and receiver. While the tasks of the regular IMPs consisted of receiving data packets, conducting rudimentary error detection, determining the route for the forwarding of data packets and transmitting this to the next IMP, destination IMP protocol and receiver IMP protocol had to be capable of carrying out additional tasks. While the just mentioned operations are part of the IMP-to-IMP protocol, the management of an end-to-end connection such as flow control, storage management, announcement fragmentation and reassembly are operations of the source-IMP-to-destination IMP protocol.

The division of messages into individual data packets enabled an efficient data transfer with only a limited available bandwidth. Through a pipelining of individual data packets, the waiting time of the remaining communication participants could be reduced to a minimum. The implemented error control focused on the recognition of doubled or missing data packets. Duplicated data packets could appear if an IMP ceased operation before it was in a position to send a response receipt for a received or forwarded data packet. The original transmitting IMP

then resent the same data packet. A data packet could get lost on the other side
if an IMP ceased its operation before it could forward a received data packet,
but had already sent a response receipt. Flow control was finally necessary as
data packets could be transmitted through the network on different paths. Thus it
was completely possible that they could be delivered at the receiver in a different
order than they had been transmitted by the sender. For this purpose, each data
packet was assigned a sequence number so that the original message could be
correctly reassembled at the receiver IMP.

- **Host-to-Host protocol**
  Application programs on the host computer were capable of exchanging data
  with other host computers with the help of host-to-host protocols. This protocol
  was a part of the operating system of the host computer and implemented within
  the so-called **NCP (Network Control Program)** . The NCP was responsible for
  the execution of the connection establishment and termination as well as for the
  flow control. In time, NCP was used as an alternative to host-to-host protocol and
  became the most important protocol of the transport layer of the ARPANET. It
  would later be replaced by the TCP/IP.

  In order that connections could exchange control and status information so-called
  links were set up. These were switched parallel to the actual data transfer. The
  NCP also had the task of breaking and coordinating the interprocess communication between the two communicating applications in individual messages. These
  messages were then sent to the NCP of the receiver, which reassembled and and
  processed them accordingly.

With the rapidly growing number of delivered and installed IMPs, the ARPANET
was growing constantly. To no longer have to be dependent on an extra host computer, the host-IMP protocol software was modified accordingly so that terminals
could also be directly connected to a special IMP, the so-called **terminal interface
processor** (**TIP**). Successive expansion focused on the cost-reducing connection of
multiple hosts to a single IMP, hosts that could be connected to multiple IMPs to
increase transmission security and the possibility of attaining greater distance between host and IMP. The integration of radio and satellite transmission helped the
ARPANET to bridge great distances with total coverage. The more individual networks that were taken into the ARPANET, the greater the need for an efficient,
modern Internetworking protocol software. This eventually led to the introduction
of the TCP/IP communication protocol, which was developed for this purpose.
In 1983 the original ARPANET split into two parts. One part was used exclusively
by the military (**MILNET**) and the other, which continued to operate under the
same name, was used for civilian and, therefore also, commercial purposes. With
the development of more powerful, civilian networks, such as the **NSFNET**, the
ARPANET lost its significance and ceased operation in 1990. While the military
component, MILNET, continued to operate.

6.4.2 X.25

The International Telecommunications Union (ITU), responsible for the standardization of internationally recognized norms in telephony and data transmission, defined one of the first standards for the operation of wide area networks with **X.25**. Public service operators, such as the national telephone company, had already offered this service for many years. The development of X.25 began as a research project in the late 60's at the National Physical Laboratory in England. It was carried out under the supervision of *Donald Davies* (1924–2000), who came up with the concept of packet switched computer networks independent of his U.S. colleagues *Paul Baran* (1926–2011).

X.25 was adopted as a standard in 1976 by the ITU. It described exactly the interface between the end system (data terminal equipment, **DTE**) and the network. The data transmission rates of 300 bps up to 64 kbps, and since 1992 also 2 Mbps, thereby became standard. By today's standards, X.25 defined a very slow network. X.25 was designed for the implementation of poor analog transmission routes with a high rate of error, allowing world-wide and wide area data communication coverage. The data transfer service of the German Telekom, known as **Datex-P**, is based on the X.25 standard. Although X.25 networks are still in service, their use has fallen drastically as more reliable transmission media has become available. These no longer need the complex error detection and correction mechanisms of X.25. For the most part these technologies have been replaced by newer two-layer technologies, such as frame relay, ISDN, ATM, POS, or primarily by the ever-present TCP/IP protocol family. Nonetheless, X.25 networks still remain the most inexpensive and reliable connection to the Internet in many parts of the Third World.

The X.25 protocol is also defined in a compatible ISO standard as ISO 8208 and regulates the transmission of data packets between two communicating DTEs. For this reason, it is also known as X.25-DTE-DTE. For connection to the X.25 network, equipment called **DCE** (Data Circuit-Terminating Equipment) is used. Communication between DTE and DCE is regulated by the X.25 determined DTE-DCE communication protocols. X.25 itself presents a packet switching network technology whose network internal switching nodes are known as **DSE** (data switching exchange). X.25-capable end systems, which are able to independently break down messages into data packets, can be directly connected via a DCE with the X.25 network. As many end systems are, however, not in a position to offer a dedicated X.25 conform interface, a further standard was developed with **X.28**

This could also put non-intelligent end systems, for example terminals, in a position to communicate with a X.25 network via a so-called **PAD** (Packet Assembler/Disassembler, described in the **X.3** standard). The PAD communicates with the X.25 network over a protocol standard designated as **X.29**. Together these three standards - X.25, X.28 and X.29 - are also called **Triple X**. Networks from different network operators can also be coupled with each other over their own, separate interface, standardized as **X.75** (see Fig. 6.30).

Besides packet switching network types, also other kinds of networks can be implemented for X.25. If X.25 networks are implemented as circuit switching networks,

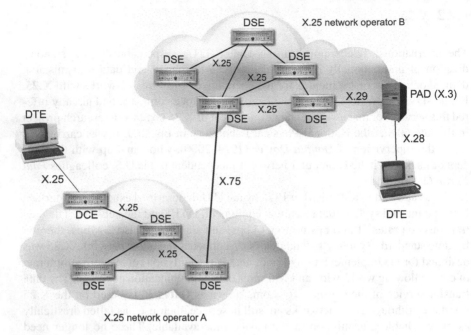

Fig. 6.30 Structure of a X.25 network.

first, a circuit connection is established via the network between the two communication partners. From the perspective of the X.25 network, this appears as a direct end-to-end connection.

X.25 works connection-oriented and is capable of supporting switched virtual networks (**switched virtual circuit, SVC**) as well as permanent, virtual networks (**permanent virtual circuits**). A SVC is generated if a computer sends sends a data packet in the network with the query to switch a connection to a distant communication partner. As soon as the connection is switched, the transmitted data packets sent along this connection always reach the receiver in the original transmission sequence. For this purpose, X.25 also provides a flow control that prevents a faster transmitter from flooding a slower or otherwise occupied receiver with its data packets.

While used in the same way as a SVC, a PVC, in contrast, is already set up in advance based on the customer preferences of the network provider. A PVC is always active so that no special command sequences are necessary for its activation. In this way, a PVC is the counterpart to a leased line.

Similar to the TCP/IP reference model, or the ISO-/OSI reference model, a layer model forms the basis for the X.25 standard. As represented in Fig. 6.31, there are three layers with the following protocols:

- **Physical layer**
  In the physical layer, the **X.21** protocols specify the physical, electrical and pro-

cedural interface between host and network. While X.21 is based exclusively on the application of digital signals, with **X.21bis** an extra standard was created that enables network access to a X.25 network over an analog line with the help of a modem.

Data link layer

The second layer of the X.25 layer model contains control procedures for the transmission of data blocks on the interface between DTE and DCE. Employed here is a variation of the so-called **HDLC** protocol (High-level Data Link Control), namely the **LAPB** protocol (Link Access Procedure Balanced).

Network layer

The protocol of the network layer is **X.25-PLP** (packet layer protocol). It is responsible for the connection establishment and termination as well as the transmission of data packets over a connection. Additional tasks such as flow control, response receipt and interrupt control are included here. The sent data packets have a maximum length of 128 Bytes, are delivered reliably and in the correct sequence. Multiple virtual connections can be operated over one physical line. In order to match a data packet to a specific connection, every data packet contains a logical channel number(LCI, Logical Channel Identifier) in its header. This is comprised of a group number (LGN, Logical Group Number) and a number within the group (LCN, Logical Channel Number). This enables the possibility of establishing multiple parallel end-to-end communications over a single physical line.

X.25 provides network nodes with a X.25 switching function via a new assignment of the LCIs. An input LCI is assigned an output LCI. In principle, this corresponds to a routing function.

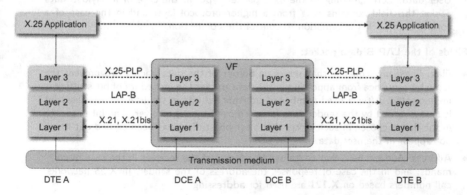

Fig. 6.31 The X.25 layer model.

Fig. 6.32 shows the data format of the X.25 protocol: a LAP-B data packet (layer 2) and an X.25-PLP data packet (layer 3). Data packets in layer1 (X.21) are identical with the LAP-B data packets and do not provide any additional header information.

X25 Data Packet Format

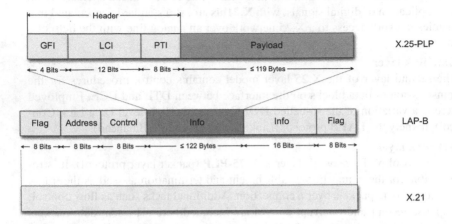

Fig. 6.32 X.25 Data packet formats.

Fields of the X.25 PLP data packets:

- **GFI** (General Format Identifier): Serves to identify the data packet as user data or control packet, for flow control and as response receipt.
- **LCI** (Logical Channel Identifier): Identifies the logical channel of the DTE/DCE interface. The network layer provides the higher level transport layer 4,096 logical channels for this purpose. These enable multiple usage of the physical line. The LCI entry has only a local significance. Every virtual connection consists of a sequence of legs, each with its own LCI value. While the connection is being established, the LCI values are determined separately for each leg and regulate the routing of the data packets.
- **PTI** (Packet Type Identifier): Identifies the data packet type. Besides user-data data packets there are 16 additional control data packet types.
- **User data**: Corresponding to the data packet type, in the case of a payload data packet this field contains data from a higher protocol layer and in the case of a control data packet information to control the connection.

Fields of the LAP-B data packet:

- **Flag**: Fixed bit sequence that serves as border of the data packet. To prevent the same bit sequence from appearing in the data packet the procedure of **bit stuffing** is implemented. After a fixed number of transmitted bits in the user data, a fixed bit is always added by the sender. It is later removed by the receiver. The bit sequence in the flag is chosen in such a way that after bit stuffing has been carried out it will not appear in the user data.
- **Address**: Address field that contains the address of the receiver in the case of commands, and in the case of responses the address of the sender. In X.25 networks, call numbers based on **X.121** are used for addressing.
- **Control**: Describes the type of data packet in more detail and additionally contains a sequence number for the flow control in layer 2.
- **Info**: Contains the X.25-PLP data packet.
- **FCS** (Frame Check Sequence): Contains a checksum.

Fig. 6.32 X.25 Data packet formats.

6.4.3 ISDN

As the primary carrier of international telecommunications, the public circuit switched telephone system has been in service for over a century. At the beginning of
the 80's a rapidly increasing need for digital end-to-end communication was predicted. This was something the old analog telephone network no longer seemed able
to provide. In 1984, under the auspices of the CCITT/ITU and the intervention of
the state and private telephone companies, the expansion of a new, completely digital, circuit switched telephone system was agreed on. It was given the name **ISDN**
(**Integrated Services Digital Network**). ISDN had the primary goal of integrating
different service offerings, such as voice and data communication, over the same
network.
In Germany ISDN began in 1987 with two pilot projects in Mannheim and Stutt-
gart, in the framework of a independent, national ISDN project. In 1994 this was
then replaced by the standardized European system **Euro-ISDN** (also **DSS1, Digital Subscriber Signalling System 1**). In the U.S. a different protocol standard
for the ISDN control and data channels was chosen. This was was standardized as
ISDN-1 (corresponding to the procedure introduced by the AT&T company, **5ESS**).
Due to a deviating coding of the transmitted data, the American ISDN standard
achieved 56 kbps per channel, while the European standard achieved 64 kbps per
data channel. The information given in this section regarding bandwidth of the different ISDN connection variations always refers to the European channel coding of
64 kbps respectively.
ISDN offers digital, circuit switched end-to-end connections. The user has a digital
interface at its disposal, and therefore is no longer required to transform digital into analog signals over a modem beforehand, as in the case of the analog telephone
network. A dial-up connection is established to the receiver. This remains available
as a transparent data transmission channel during the duration of the connection.
User data and signaling data are transmitted on separate channels. ISDN works according to a synchronous time multiplex procedure and is capable of ensuring a
constant guaranteed bandwidth and a constant transmission delay.
ISDN provides channels with 64 kbps each. This bandwidth was especially chosen
for the coding of voice information via the **PCM** procedure (Pulse Code Modulation, 8,000 samples each with 8 bits per second[3]). Depending on the connection type,
different bandwidths can be made available to the user:

* **Basis connection (BRI, Basic Rate Interface)**
  Over a physical connection with the ISDN network, the user is offered two logical connections, the so-called **B channels** (bearer channel) each with 64 kbps for
  the transmission of user data and a signal channel - the **D channel** (data channel),
  with 16 kbps for the transmission of control signals. This type of procedure, in
  which command and control information is sent on a separate channel apart from

---

[3] A detailed presentation of audio coding by means of PCM is found in Chap. 4, „Multimedia Data
and Encoding", in the first volume of this series: Meinel, Ch., Sack, H.: Digital Communication –
Networking, Multimedia, Security, Springer (2013).

the user data channel, is also called **out-of-band** signal traffic. The BRI interface is standardized in the ITU-T/CCITT ISDN standard I.430.

- **Primary multiplex connection (PRI, Primary Rate Interface)**
Over a physical connection with the ISDN network, the user is provided 30 logical connections (channels) each with 64 kbps (U.S. only 23 logical connections) for the transmission of user data and a signal channel with 64 kbps for the transmission of control signals. The PRI interface is standardized in the ITU-T/CCITT ISDN standard I.431.

User data in ISDN can be transmitted with

- circuit switched B channels,
- packet switched B channels, and also with the help of
- packet switched D channels (see Fig. 6.33).

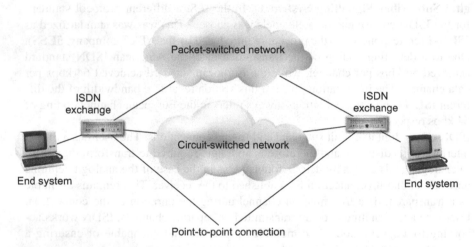

Fig. 6.33 ISDN architecture.

Additionally, the ISDN standard provides so-called **hybrid channels** (H channels). These are special channels that result from the bundling of B channels. They are used for applications requiring a higher bandwidth, e.g., the transmission of audio and video information in real time. For this, the basis channel, also called the H_0 channel, bundles 6 B channels to a hybrid channel with a bandwidth of 384 kbps. Other important hybrid channels are H_{11} (24 B channels with a bandwidth of 1,536 kbps) and H_{12} (30 B channels with a bandwidth of 1,920 kbps).

In the ISDN network, especially with regard to the **ISDN layer model**, a distinction is made between end systems, local exchanges and long distance exchanges. For the transmission of user data on the lowest layer of the ISDN protocol stack, a transparent, logical connection is established from end system to end system via local exchange and long distance exchange. The signal connections must be forwarded to

ISDN Terminology

In the ISDN standards of the ITU-T, **reference configurations** are defined. These characterize the ISDN interface. Reference configurations are made up of:

Functional groups
Included here are specific capabilities and special functions required by an ISDN user interface. These are carried out by one or more software or hardware components.

Reference points
These subdivide the functional groups, they are comparable to physical interfaces between individual ISDN components.

An ISDN user station (end system) according to the ITU-T recommendation I.411 may be structured as in the following example:

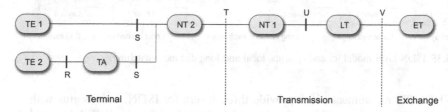

The user station consists of the network termination units **NT 1** (Network Termination 1) and **NT 2**, the terminal equipment **TE 1** (Terminal Equipment 1) and **TE 2** adaption unit **TA** (Terminal Adapter). NT 1 takes over the coupling to the connection cable, while NT 2 enables the connection of multiple TEs to a connection cable. TE 1 is a terminal equipment designed for ISDN and can be connected directly on the interface (reference point **S**), while TE 2 only has a standard (analog) interface and must be connected via an adaption module. On the exchange side, the functions of transmission technology, corresponding to the NT1 functions in the line termination (**LT**), and the functions of exchange technology are combined on the exchange termination (**ET**

Between the functional groups, reference points are defined: reference point **T** between NT 1 and NT 2, **S** between NT 2 and TE 1 or TA, **U** between NT 1 and LT, as well as **V** between LT and ET. Dependent on national and network specific regulation, the jurisdiction of the network operator ends at the reference point S, T oder U. In the case of S, the network operator for NT 1 and NT 2 is responsible, in the case of T only for NT 1 (e.g., in Germany) and in the case of U neither for NT 1 and NT 2. The reference point at which the jurisdiction of the network operator ends is simultaneously the transfer point where the network operator delivers the defined service and where it takes over maintenance responsibility.

Further reading:

Verma, P. K., Saltzberg, B. R.: ISDN Systems; Architecture, Technology, and Applications: Prentice Hall Professional Technical Reference (1990)

Helgert, H. J.: Integrated Services Digital Networks. Addison-Wesley Longman Publishing Co., Inc. (1991)

Fig. 6.34 ISDN terminology.

exchanges in the higher layers of the ISDN layer model because control and address information must be evaluated and applied there (see Fig. 6.35).

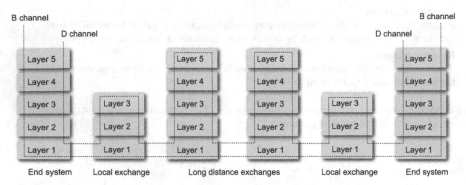

Fig. 6.35 ISDN layer model for end systems, local and long distance switching.

ITU-T/CCITT recommendations provide three layers for ISDN. This begins with the physical layer, where the specifications of the basis interface for BRI (I.430) and PRI (I.431) are fixed and continues to the data link layer, where the LAPD (Link Access Procedure for D-Channel, recommendations Q.920/I.440 and Q.921/441) protocol is implemented, on to the network layer, where the DSS1 (Digital Signaling System No.1) is implemented. DSS1 is used Europe-wide and also known by its abbreviation Euro-ISDN. Internationally, various standards are applied on the network layer, for example, the 5ESS and DMS-100 standards in the USA or NTT in Japan. In long-distant exchange, the signaling system SS7 (Signalling System No.7) is used, a set of different protocols for the communication of exchanges in the network (ITU-T/CCITT recommendation Q.7xx). The protocols brought together collectively form an individual protocol stack, just as the TCP/IP or the ISO-/OSI protocol stack. The signaling system SS7 is structured as a so-called **overlay net**. This means it is physically, and not only logically, separated from the actual user data channels. For ISDN there are five relevant layers of the SS7 protocol stack (see also Fig. 6.36), which can be divided into three groups:

- **MTP-1, MTP-2, MTP-3 (Message Transfer Part)**
  In their tasks these three layers correspond to the first three layers of the ISO-/OSI model and are responsible for the transmission of SS7 messages.

- **SCCP (Signal Connection Control Part)**
  This layer corresponds to the upper part of the ISO/OSI-layer function-wise. It allows the exchange of data without bearer channel reference and end-to-end signaling. The SCCP protocol specifies five different service classes:

  - connectionless, non-sequenced services,
  - connectionless, sequenced services,
  - connection-oriented services,

– connection-oriented services with flow control and

– connection-oriented services with flow control and error detection (or error correction).

- **ISUP (ISDN User Part)**
  This layer serves the transmission of control and reporting data It is used for the establishment and termination, as well as the monitoring, of circuit switched connections on B channels. The function of this layer is divided up into a part for ISDN telephone traffic, focusing on signal processing for voice-based communication, and a part for ISDN data traffic.

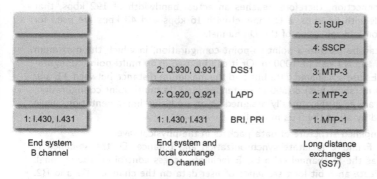

Fig. 6.36 The various ISDN protocol layers.

ISDN was a widespread attempt to replace the old analog telephone system with a digital telephone communication system capable of integrating different services, such as voice and data communication. A worldwide agreement regarding the interface for basis connections aimed to increase demand, which would result in mass production. However, the standardization process dragged out for so long that technological development had long surpassed the standard ISDN by the time it was adopted.

Excursus 9: ISDN – Data Formats

In the three layers that are provided for ISDN end systems and local exchanges for B channels and D channels, different data formats are implemented. In the **physical layer**, a **modified AMI coding** (Alternate Mark Inversion) is used to carry out the transmission of bit sequences. Unlike the conventional AMI code, a zero is transmitted as an pulse here, whereby adjacent pulses display opposite polarity and a one as a gap. In longer sequences of ones synchronization can be lost. This can be prevented with the appropriate bit stuffing.

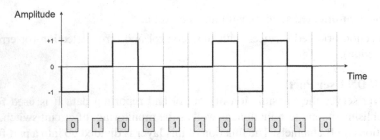

ISDN works according to a sychronous time multiplex procedure, i.e., data packets are transmitted periodically at a constant rate. The limitations of the individual data packets by means of the so-called synchronization bit proceeds with fixed bit sequences that cannot appear within the coded user data. Thereby, all 250 μs will transmit a 48 bit-long data packet. The basis connection, therefore, reaches an actual bandwidth of 192 kbps, that through 2 B channels with 64 kbps, a D channel with 16 kbps and 48 kbps are used for packet synchronization and reflection of the D channel.

A basis connection can be used in a point-to-point configuration, in which the maximum distance between TE and NT can be 1000 m. Or it can be used in an multi-point configuration where up to 8 TEs use a a shared data bus and the maximum distance between TE and NT can come to 200 m (short bus) or 500 m (extended bus). In multi-point configuration, the TEs B channels are each dynamically assigned to an exclusive implementation, while the D channel is used by all of the TEs mutually.

Fig. 6.37 shows a simplified structure of data packets in the physical layer.
The bits marked as **F** und **L** facilitate synchronization and balance. **D**, the so-called D channel bit, as well as the D channel echo bit **E** facilitate access control on the channel. **B1** and **B2** each indicate an 8 bit long sequence of user data on the channels B1 and B2. Additional synchronization bits between the user data are not included in the illustration.

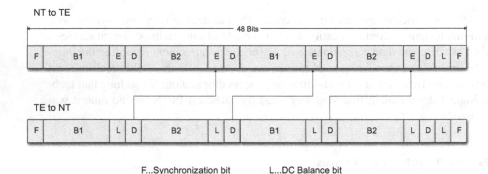

F...Synchronization bit   L...DC Balance bit
D...D Channel bit   B1...8 Bit payload B1 channel
E...D Channel channel echo bit B2...8 Bit payload B2 channel

Fig. 6.37 ISDN - Physical layer – Data format.

The data link layer and the network layer are, in contrast to the physical layer, only of significance for the D channel, or in the switching centers. The **data link layer** uses the **LAPD** protocol (Link Access Procedure D-Channel). LAPD takes over the tasks of securing all transmitted information in the D channel against transmission and sequence order errors and of allocating unique TEIs (TE identifiers). Thereby, the D channel is capable of transmitting signaling information as well as packetized user information.

The ISDN data link layer can carry out unacknowledged broadcast transmissions to all
TEs, or send acknowledged or unacknowledged messages to individual TEs. In acknowledged transmissions, the data link layer is responsible for the detection of transmission errors,
for their correction and for the control of block order. In the case of unacknowledged transmissions, the sole responsibility is for error detection. The structure of a LAPD data packet
is presented in Fig. 6.38. The structure and coding of all protocol elements corresponds
to the the standard already existing before ISDN: **HDLC**-Standard (High-level Data Link
Control). It was standardized in the norms ISO 33009 and ISO 4345. Depending on the
function of each, a distinction is made between three different data packet types:

- **I packets (Information Transfer Frames)**
  These data packets contain user data as well as information regarding flow control and
  error correction. They are implemented if acknowledged information is to be delivered
  to a certain receiver.

- **S packets (Supervisory Frames)**
  These data packets contain so-called **ARQ** information. The ARQ procedure (Automatic
  Repeat Request) facilitates error correction. Here, defective or lost data packets are
  sent again by the transmitter. Correctly received data packets are acknowledged with a
  positive response receipt (acknowledgement); defective data packets received can be re-
  requested with a negative response receipt (reject). The transmitter can conclude that
  frames have been lost in the absence of a positive acknowlegement following a timeout.

- **U packets (Unnumbered Frames)**
  U packets contain additional functions for link control. Their purpose is unacknowledged
  transmission and for this reason they also do not need sequence numbers.

Data packets of the data link layer start with an 8-bit long synchronization flag. Two
address details follow concerning the Service Access Point Identifier (**SAPI**). First, it is
indicated what type of services are provided by the data link layer (e.g., SAPI=0: signaling
information, SAPI=16: user data packet, SAPI=63: management information). After this
follows a Connect/Response bit (**C/R**). This determines whether the transmitted data
packet is an instruction (C/R=1) or an answer (C/R=0). The Address Field Extension bit
(**EA bit**) indicates if there is still an address byte to follow (EA=0) or not (EA=1). The
second address refers to the TE Identifier (Terminal Endpoint Identifier, **TEI**) and identifies
the connected end system.

In the following control fields, details are given regarding the sequence number and the
link control. The information field contains the actual user data. Its length must always be
a multiple of 8 bits. ISDN limits this field to a maximum of 280 bytes. A check sum for
error-detection follows this, and finally a synchronization flag again.

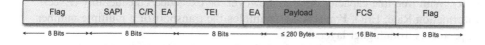

Flag	SAPI	C/R	EA	TEI	EA	Payload	FCS	Flag
← 8 Bits →	← 8 Bits →		← 8 Bits →			≤ 280 Bytes →	16 Bits →	8 Bits →

SAPI...Service Access Point   AE...Address field Extension bit
TEI...Terminal Endpoint Identifier   FCS...Frame Check Sequence
C/R...Command/Response bit

Fig. 6.38 ISDN LAPD protocol – data format.

In the **network layer** of the ISDN protocol, the connection to the chosen end system and
the selection of the desired communication service takes place. For the communication
between TE and local exchange (local ISDN exchange) the ITU-T/CCITT Q.931 protocol is

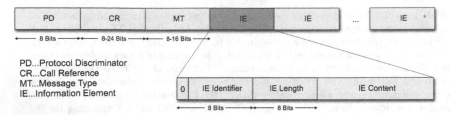

Fig. 6.39 ISDN Q.931 protocol - data format.

implemented. Fig. 6.39 shows the structure of a Q.931 data packet. Because internationally-different signaling procedures are used, the data packet starts with a **protocol discriminator**. This indicates which protocol is used (e.g., 1TR6, NTT, 5ESS or DMS-100). 16 bits follow for a **call reference**, which identifies the connection over an arbitrarily chosen number. The **message type** determines the kind of message sent (call establishment, call clearing, call information phase, among others). After that, individual **information elements** follow. These are made up of an **information element identifier** each, with accompanying length details and the actual contents itself.

Further reading:

P. Bocker: ISDN - Digitale Netze für Sprach-, Text-, Daten-, Video- und Multimediakommunikation,4. erw. Aufl., Springer Verlag, Berlin (1997)

A. Kanbach, A. Körber: ISDN - Die Technik, Hüthig, 3. Aufl., Heidelberg (1999)

6.4.4 Frame Relay

Frame Relay is a packet switched WAN technology whose development began in the 1980's. It is based on an asynchronous time multiplexing procedure, similar to ATM (which we will look at in more detail later) whereby virtual connections are set up between the communicating end systems. In contrast to ATM, in frame relay, data packets of variable lengths (up to 1,600 bytes) are transported over a line. Every connection receives a specific minimum data rate (Committed Information Rate, CIR). With free line capacity this can, however, be exceeded considerably (Excess Information Rate, EIR). Frame relay facilitates the transport of a multiple number of protocols from higher layers.

The concept of frame relay was introduced in 1984 to the ITU-T/CCITT and became widespread by the end of the 80's. In 1990 the **Frame Relay Forum** was founded as a committee for the development and the expansion of frame relay. The original frame relay specifications, standardized by the American ANSI, were adapted by the ITU-T in the middle of the 90's and further developed. Today frame relay is well established for data transport and is increasingly implemented in the area of voice transmission. In Europa, frame relay is also employed for the connection of basis stations of the GSM mobile communications network (Global System for

Mobile Communication), which receives radio signals from mobile telephones and and transfers them to the fixed-line network.

Frame relay is commonly viewed as the successor of X.25. By dispensing with several of the error correction and security measures provided in X.25, it is optimized for a fast data transfer. Modern digital transmission systems additionally show a much lower rate of error than the analog telephone network, which was developed for the X.25. For this reason, frame relay can manage with simpler to implement and less complex protocols. Higher data transmission rates can additionally be achieved for user data.

The name "frame relay" reflects fact that user data is transmitted in so-called frames, i.e., on the data link layer (layer 2) of the communication layer model. This contrasts to X.25, where the exchange takes place on the network layer (layer 3). In ANSI terminology, frame relay is also known as **fast packet switching**. The ITU-T/CCITT standards of frame relay are based to a large part on those of ISDN. These are described in the B-ISDN Protocol Reference Model (ISDN-PRM).

Differences between frame relay and X.25
The most significant differences between frame relay and X.25 are:

- Multiplexing and switching of logical connection proceed in frame relay on layer 2 (data link layer).

- Neither flow control nor error correction procedures are implemented on the frame relay transmission sections. In case it is necessary, procedures must be implemented that are used in higher protocol layers by the end systems.

- In frame relay, separated logical connections are used for signaling.

Fig. 6.40 X.25 vs. frame relay

In frame relay, the bandwidth ranges from 56 kbps or 64 kbps in integer multiples to 45 Mbps. As a rule, frame relay networks are offered for use by public service providers. It is also possible to operate these as private networks with private frame relay switches and leased dedicated lines **SDH** (Synchronous Digital Hierarchy) or **PDH** (Plesiochrone Digital Hierarchy) may be employed as the infrastructure for data transmission in frame relay.

Frame relay is not an end-to-end protocol/ technology. Rather, it describes the interface between a network node of the telecommunications provider and the end system connected to it - more precisely - the data transmission and the communication between it. Frame relay offers connection-oriented communication on the data link layer of the network. Over this, various computer worlds and platforms may be connected to each other (see Fig. 6.41). In addition, frame relay sets up a bidirectional, virtual connection between two end systems.

Regarding the use of services on layer 1 (the physical layer), frame relay is not subject to any special restrictions. Almost all of the common interfaces can be used (e.g., X.21, V.35, G.703). The key specifications are found in the data link layer, which are implemented in the protocol **LAPF** (Link Access Protocol for Frame Relay).

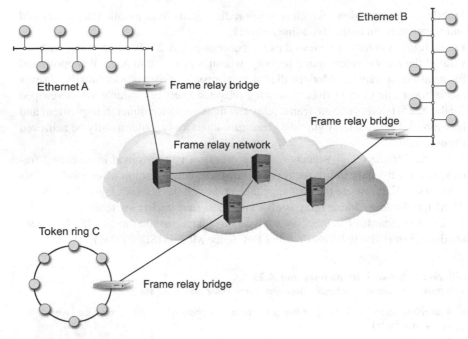

Fig. 6.41 Structure of a frame relay network.

User data is sent in containers (frames) of varying sizes in frame relay. These can take up to 600 bytes. As larger frames can be sent, in contrast to X.25, the frame relay network nodes are less preoccupied with the unpacking and packing of user data or the exchange of acknowledgement information and this results in faster transmission. Similar to the X.25 concept, frame relay also allows multiple logical connections (links) on a physical transmission route. These connections can be laid out either as a permanent connection (**PVC, Permanent Virtual Circuit**) or as a transient connection (**SVC, Switched Virtual Circuit**). In the broadest sense, PVCs correspond to a dedicated line, which is permanently in operation. They stand in the forefront of frame relay, as one of their main uses is the coupling of remote LANs. SVCs, in contrast, could be compared to a dial-up connection, which proceeds according to the scheme of connection establishment, data exchange and connection termination. Call numbers are used as addresses, based on the standards defined in X.121 or E.164.

The essential control information of the respective connection is provided via the so-called **Data Link Connection Identifier** (**DLCI**). On establishing a connection, the address is transcribed in a series of DLCIs, which are locally unique for only one link (cf. Fig. 6.42).

The provider of a frame relay network supplies the customer with a guaranteed, firmly agreed upon bandwidth (**Committed Information Rate, CIR**). If the line is not used completely, a higher bandwidth can be used temporarily. This can incre-

Data Link Connection Identifier (DLCI)

A connection in a frame relay network (PVC or SVC) consists of a series of point-to-point connections along the frame-relay switching station and is characterized by certain attributes. One of these attributes is a number assigned to the connection of each frame-relay link station, namely that of the locally valid **DLCI** (Data Link Connection Identifier). With each hop it changes for the same connection.

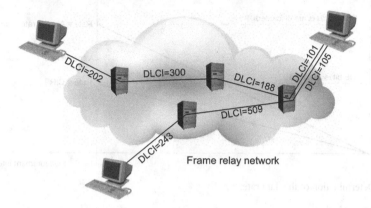

Frame relay network

Practically speaking, DLCIs are actually numbers from buffer memory for the data packets sent along a connection. The DLCI is 10 bits long, whereby the capacity limit per frame-relay switching station is restricted to 1,024 possible connections. Some of the DLCIs, however, are reserved for special tasks.

DLCI-number	Application
0	reserved for signaling purposes
1–15	reserved
16–1007	can be assigned to frame relay connections
1008–1018	reserved
1019–1022	reserved for multicast groups
1023	Local Management Interface

Further reading:

Black, U. D.: Frame Relay Networks: Specifications and Implementations. McGraw-Hill, Inc. (1998)

Fig. 6.42 Data Link Connection Identifier – DLCI.

ase the available bandwidth to the maximum of the access line (**access rate**). To determine the measurement dimensions, the number of bits B to be transmitted are measured in a prescribed measurement interval T_C (committed time). For the guaranteed bandwidth applies: CIR=B_C/T_C, whereby B_C is designated the committed burst size. If within the measurement interval T_C a bit number $B_C+B_E > B_C$ is measured, then the B_E (excessed burst) surplus bits is marked by setting a separate bit. This is the **DE** (Discard Eligible) bit. An overloaded frame relay exchange system can discard this marked data packet, if necessary, to reduce overload (see Fig. 6.43).

If during a measurement interval a number of bits even larger than B_C+B_E is measured, they are usually discarded during the first run by a provider's frame relay network node.

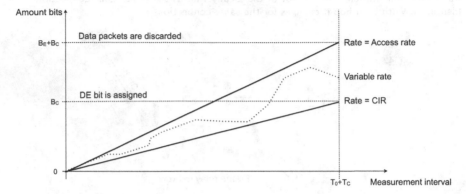

Fig. 6.43 Determination of the data rate.

A further means of controlling load is provided by the two control bits **FECN** (Forward Explicit Congestion Notification) and **BECN** (Backward Explicit Congestion Notification). An overloaded frame relay network node can set the FECN bit in passing data packets, thereby making the receiver aware of the overload. Data packets that pass the overloaded frame relay network node in the opposite direction, can convey knowledge of the overload to the upstream network nodes by setting the BECN bit (see Fig. 6.44).

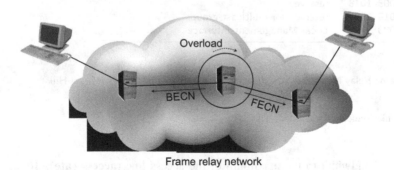

Fig. 6.44 Explicit congestion notification.

Another mechanism for overload control, the so-called **Consolidated Link Layer Management** (**CLLM**) was proposed by the ANSI and the ITU-T. In order to take overload control out of actual user data traffic, an out-of-band signaling was proposed. For this, data packets are used marked with their own reserved DLCI (DLCI=1023). These CLLM data packets are transmitted from the frame relay network nodes to the respective end systems to inform the user of the overload situation. These each receive a list of those DLCIs that are in all probability responsible for overload. If a user receives a CLLM data packet, it should temporarily interrupt data transmission so that overload can be reduced again.

On the link layer, frame relay uses the protocol **LAPF** (Link Access Protocol for Frame Mode Bearer Service, Q.922). LAPF is based on the LAPD implemented in ISDN, and therefore also on HDLC. It takes care of data transfer coordination and signaling.

A frame relay data packet begins with the synchronization flag already known from HDLC 01111110 (see Fig. 6.45). The header of the data packet contains a DLCI and several control bits and can therefore be 2, 3 or 4 bytes long. This depends on the length of the DLCI (10, 16 or 23 bits) used. The standard length of the DLCI is 10 bits. The bits BECN, FECN and DE are provided for overload control. EA1 designates the end of the header. The CR bit (Command/Response) is not used in frame relay, it can, however, be used by higher protocol layers.

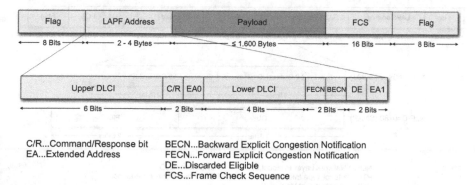

C/R...Command/Response bit   BECN...Backward Explicit Congestion Notification
EA...Extended Address   FECN...Forward Explicit Congestion Notification
                           DE...Discarded Eligible
                           FCS...Frame Check Sequence

Fig. 6.45 Frame Relay - data format.

Often data packets (Protocol Data Units, PDUs) of differing network protocols are transported in data networks . The switching systems in the network - multiprotocol routers - have to be capable of recognizing the protocol used in each case to be able to further process the PDUs correctly. As the header of the frame relay data packet does not provide a field for the identification of each protocol used, the protocol type must be established in the configuration phase. Following this, the relevant protocol is encapsulated in the frame relay data packet. The multiprotocol encapsulation for frame relay is specified in RFC 1490 and RFC 2427. It is fixed there in which way different protocols of the data link layer and the network layer are encapsulated. Additionally, the frame relay data packet is extended to four fields (see Fig. 6.46), which reduces the field previously used solely for user data. The first field (**Q.922 Control**) involves a **UI** field (Unnumbered Information), which has the value 03H and with the help of padding bits (in each case zero) is filled up to two full bytes. In the field that follows (**NLPID**) (Network Layer Protocol ID), the respective encapsulated network protocol is named. The codings for each protocol variant to be encapsulated are fixed in the norm ISO/IEC 9577 (see Table 6.4 and Fig. 6.46).

Table 6.4 Multiprotocol encapsulation for frame relay via NLPID.

NLPID	Protocol
0x08	Defining a custom format according to Q.933 (e.g. ISO8208, SNA, etc.)
0x80	SNAP format (Subnet Access Protocol)
0x81	CLNP format (Connectionless Network Protocol)
0x82	ISO ES-IS (End System to Intermediate System)
0x83	ISO IS-IS (Intermediate System to Intermediate System)

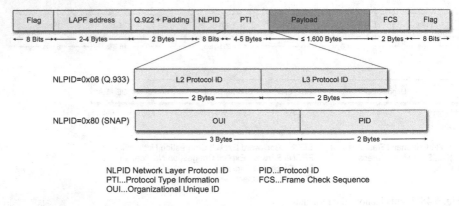

Fig. 6.46 Frame-relay multiprotocol encapsulation - data format.

Frame relay is also designed for multicast transmissions to multiple subscribers. Here, multiple permanent virtual connections (PVC) are used at the same time for the individual receivers. If a frame relay network node receives a multicast data packet, it must replicate it and forward it to all subscribers in the multicast group. These must have previously registered at the frame relay network node. For this purpose, the relevant DLCI is entered in all replicated data packets. The actual distribution over the entire frame relay network is taken over by special **multicast servers**. The Frame Relay Forum, an instance responsible for the further development of frame relay, presents a choice of three different multicast solutions:

- **One-way multicast**
  Here, there is only one dedicated sender. In a multicast transmission, all other subscribers in this multicast group can only receive data from this one transmitter. Regular data traffic between sender and individual group members remains

possible. One possible application for this technology is teleteaching, minus a feedback channel.

- **Two-way multicast**
  In this variation, there is one sender and multiple receivers. Every one of these receivers has access to the transmitter via a return channel. However, in this variation there are no more regular transmissions possible between the dedicated transmitter and the individual subscribers of the multicast group. One possible application of this technology is teleteaching with interactive student participation.

- **N-way multicast**
  In this variation, every subscriber can be the sender as well as the receiver. Every sent data packet transmitted within this group goes to every other group subscriber. One possible application of this technology is a conference call between all the members of the multicast group.

6.4.5 Broadband ISDN and ATM

In ISDN the aim is an integration of different telecommunication services, for example telephone, data communication and the transmission of image and video information. However the ISDN service offers only comparably low bandwidths. With the further development of this technology, and to integrate video and multimedia information within a standard network, **B-ISDN** (Broadband ISDN) was developed. The use of modern optical fiber technology has enabled a universal integration of multimedia information and offers, in contrast to individual, specialized networks, the advantage that, if necessary, load balancing between the different services can be carried out.

To do this, B-ISDN implements the **ATM** data transmission technology (Asynchronous Transfer Mode). The ATM basic technology is discussed in detail in section 4.3.5 "Local Area Networks." Primary interest in ATM centers on its use in the coupling of different WANs and LANs, in other words when it is employed in **internetworking**. The special role that ATM plays in this connection will be looked at in detail in Chapter 8.

Finally, the original vision of B-ISDN as a communication means of simultaneous, synchronous transmission (telephone conversations, video telephony, and video conferences) and asynchronous data transfer faded into the background with the triumphant success of the broadband internet technologies eventually replacing it. In contrast, the ATM technology still continues to exists in the form of access technologies to the WAN or Internet. An example of this is the Digital Subscriber Line (DSL), or as payload in wireless wide area technologies, e.g., WiMAX. In the backbone area, however, ATM will no longer play a major role in the future and is increasingly being replaced by more advantageous IEEE 802.3-based technologies (Ethernet) and IP-based VPNs (Virtual Private Networks).

6.4.6 Distributed Queue Dual Bus – DQDB

This technology concept designated as DQDB was originally designed for MANs (Metropolitan Area Networks), and with its potential expansion of several hundred kilometers reaches the borderline of the actual WAN. Just as ISDN or ATM, DQDB enables the integration of different services, for example voice, video, or data communication, ensuring a high speed transmission of digital data over a wide-range geographic area with data transmission rates of 34 Mbps to 155 Mbps. As a rule, a MAN is made up of multiple DQDB subnetworks, which are connected to each other via bridges, routers or gateways. Finally, through this interconnection, even larger regions can be covered.

DQDB was developed in the '80s parallel to ATM. The DQDB access procedure is based on a statistical multiplex technology, whereby all users of the network share the available transmission medium. This means that transmission proceeds via a so-called **shared medium**. This access technology is capable of transporting not only asynchronous data streams as they occur in the Inter-LAN communication, but also provides for the possibility of transmitting data streams, such as they are found in video and audio transmissions.

Together with other LAN standards, such as Ethernet or Token Ring, DQDB is specified in the IEEE 802 norm for access procedures on a shared medium as **IEEE 802.6**. IEEE 802.6 divides the DQDB procedure into two layers, which are the

- **Physical Layer**, here there are no prescribed special technologies, and the
- **DQDB Layer**, located in the lower area of the link layer of the TCP/IP protocol stack. It describes the **MAC layer** (Medium Access Control layer).

DQDB, embedded in the IEEE 802 hierarchy, implements the **LLC** (Logical Link Control) specified in IEEE 802.2 for connection management, as well as the overlying IEEE 802.1 **HLI** (Higher Layer Interface). IEEE 802.6 defines three different service qualities for DQDB (see also Fig. 6.47):

- a **connectionless service**, which regulates the transition from MAC to LLC in a similar way as in the LAN and provides the function of network access for DQDB.
- a **connection oriented service**, which enables an asynchronous transport of data via virtual channels, but does not provide any guarantee regarding the waiting time within a transmission, and because of this cannot be used for the transmission of live video and audio data.
- an **isochronous service**, which enables data transport with a constant data rate and constant waiting cycle during a transmission. It is therefore suitable for the transmission of live video and audio data.

In practical application, only the service based on the isochronous synchronization mechanism has achieved a significant role. In order to use this service, connection establishment must be made in advance between the participating DQDB network nodes to setup a virtual channel between the transmitter and the receiver.

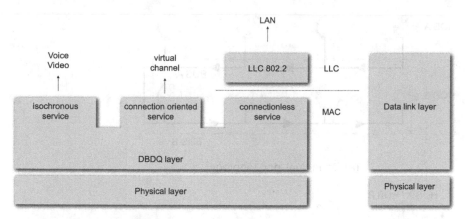

Fig. 6.47 Distributed Queued Dual Bus – DQDB.

Seen from the aspect of topology, a DQDB network consists of two opposing, uni-directional buses A and B. Every connected computer can transmit and receive on both buses. A bi-directional connection thus exists between the connected compu-ters. The computer at the head of each bus is accordingly referred to as **Head of Bus** (HOB) (see Fig. 6.48). The HOBs of both buses can either be two different computers (**Open Dual Bus Topology**) or, if the buses take a ring form, the HOBs can be installed on the same computer (**Looped Dual Bus Topology**). The HOBs serve as packet generators and generate at a clock of 125 µs free slots (cells), with a length of 53 bytes (header 5 Bytes, payload 48 bytes). These cells are compatible with ATM.

A principle distinction is made between **PA** cells (Prearbitrated Access) for isochro-nous data traffic and **QA** cells (Queued Arbitrated Access) for asynchronous data traffic. To use the possibility of isochronous data traffic, a unique channel identi-fication (VCI, Virtual Channel Identifier) must first be established by the network management for every isochronous channel. The HOB computer then periodically generates free PA cells for the channel with this number, in order to fulfill the strict time restrictions for isochronous data traffic. Those computers connected to DQDB can then use these PA cells for their respective isochronous data traffic.

For asynchronous data traffic, a distributed queuing protocol is used. The following rules must be observed when accessing QA cells:

- an end system may occupy a passing empty cell as long as no reservation exists.
- an end system located "downstream" can provide an end system located "ups-tream" a reservation on the opposite bus.
- Every end system allows as many empty cells to pass as reservations have been made.

The implementation of this distributed queuing system proceeds by way of bit-controlled access regulation. The **r-bit** (reservation, request) designates the reser-vation request for a cell in the opposite direction, and the **b-bit** (busy) designates an

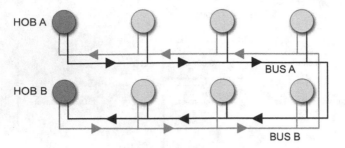

(a) Open dual loop topology

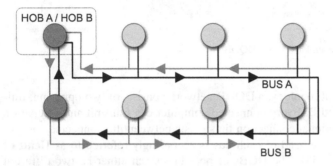

(b) Looped dual bus topology

Fig. 6.48 DQDB – Topology variations: (a) Open Dual Loop Topology, (b) Looped Dual Bus Topology.

already occupied cell. The b-bits from bus A and the r-bits from bus B are used for the control system on bus A. The opposite procedure applies for the control system on bus B In this way, a distributed queue is defined for every bus.

Every end system has two counters for each bus: **RC** (Request Counter) and **CC** (Count-Down Counter). Let's look at how the access algorithm on bus A works (access is carried out in just the opposite way for bus B). RC is incremented through r-bits that pass the end system on bus B. The respective current counter of RC indicates how many requests are waiting upstream on bus B. Free cells on bus A (b-bit=0) decrement the counter RC. If an end system wishes to transmit, RC is first copied to CC and RC=0 is set. As just described, RC is further incremented and decremented, while CC is decremented with each free cell passing on bus A. As soon as CC has reached the value zero, the end system may transmit. With the help of this distributed queuing algorithm, a **deterministic, collision-free access** is ensured for the end system on bus A (see Fig. 6.49).

The structure of the data format implemented in DQDB corresponds principally to the data format used in ATM, thus a coupling of these two technologies can be carried out with little effort. A DQDB cell has a constant length of 53 bytes, divided into a five byte long header and 48 byte payload (see Fig. 6.50). The first byte of the header is called the **ACF** (Access Control Field). It contains the b bit and a SL bit,

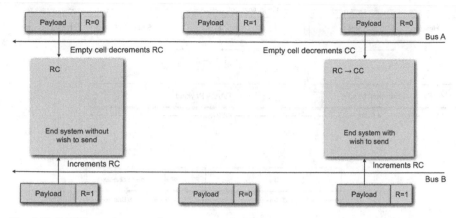

Fig. 6.49 DQDB - access procedure.

which indicates the cell type (PA or QA). After a reserved area of two bits, a field consisting of a total of four bits follows for the r-bit. This contains priority data. The next 4 bytes contain the so-called segment header. This includes the 20 bit long VCI, which identifies the respective transmission channel, the payload type (2 bits), a segment priority (2 bits) and an accompanying segment header checksum (8 bit CRC checksum, based on the polynomial x^8+x^2+x+1).

Compared to other technologies of the LAN segment with a high bandwidth, DQDB exhibits a considerably greater potential network expansion. In contrast to the classic Ethernet, which is likewise based on a bus topology, no collisions can occur on the DQDB, and when compared to token ring there is a reduction in queuing times for media access. The procedure is especially efficient in cases of high load. Here, theoretically, every free slot can be occupied and bus capacity fully exhausted. Both asynchronous and isochronous connections are simultaneously possible in DQDB, and the available bandwidth can be divided dynamically between these at any time. A ring-shaped arrangement of the double bus offers the additional possibility, in the case of line disturbance, to reface the defective area and to consider it as the new "end terminal" of the double bus.

With an increasing number of end systems connected to the double bus, the queuing time for access to the transmission medium grows as well.

To extend a DQDB network beyond its prescribed border, multiple DQDB networks have to be interconnected over the appropriate switching systems. At the same time, access procedure on the double bus is not entirely fair, i.e., not all the connected computers are treated equally. Computers located in the middle of the bus are in closer proximity to the other computers on the bus. Conversely, computers on the edge of the bus are closer to the packet generator and because of this receive more frequent access to an empty slot. In addition, a **Bandwidth Balancing Algorithm** is implemented, ensuring that the computers dependent on the successfully transmitted data packets may be allowed to pass by more free slots, with an increase in the request counter and countdown counter.

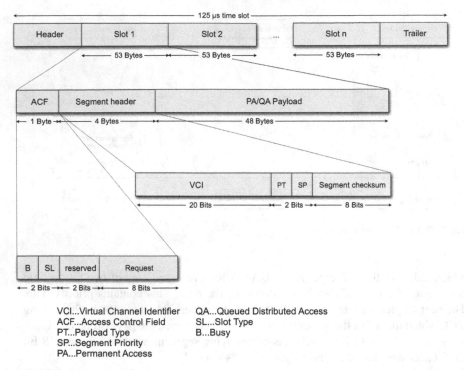

VCI...Virtual Channel Identifier   QA...Queued Distributed Access
ACF...Access Control Field   SL...Slot Type
PT...Payload Type   B...Busy
SP...Segment Priority
PA...Permanent Access

Fig. 6.50 DQDB – data format.

Based on the DQDB technology, **SMDS (Switched Multi-Megabit Data Service)**
is a broadband communication service intended exclusively for the asynchronous
transport of data. It was initially employed in the MAN area. SMDS can be used
via connections with 1,5 Mbps (T1) or 2 Mbps (DS1). Intermediate systems, for
example routers or bridges, can also be connected over a SMDS-DXI (Data Exchange Interface) with data transmission rates of between 56 kbps and 45 Mbps
(DS3). SMDS works as connectionless protocol, i.e., every data packet contains
sender and receiver addresses, so that intermediate systems can forward every data
packet individually. SMDS is currently used in the U.S. and in several European
countries. In Germany, Telekom has made a SMDS service available since 1994,
known as **Datex-M**. The Datex-M network offers its users broadband data channels from 2 Mbps to 140 Mbps, as well as synchronous transmission channels of
up to 2 Mbps. One of the primary application areas of the Datex-M service is the
temporary coupling of remote LANs.

6.4.7 Cyclic Reservation Multiple Access – CRMA

Cyclic Reservation Multiple Access (CRMA) is an access procedure within the range of high-speed networks, which has similarities to the DQDB technology. The procedure was developed in 1990 by IBM Research in the Swiss municipality of Rüschlikon. It enables a cyclic reservation for complete messages without these having to be fragmented, as is the case with DQDB. CRMA high-speed networks can extend over several hundred kilometers with a bandwidth of more than 1 Gbps. Just as with DQDB, the technology is based on a dual bus topology, but can also be carried out in the simplified version of an end-folded, single bus (see Fig. 6.51). The transmitted bit stream is divided into slots of a fixed length. Therefore, in consecutive slots longer messages can virtually be transmitted in one piece .

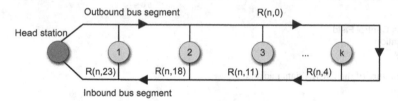

Fig. 6.51 CRMA bus topology with folded bus and reservation call R(n,x) for cycle n and the reservation request k: 4 slots, 3: 7 slots, 2: 7 slots and 1: 5 slots.

As in DQDB, a CRMA slot starts with a header (Access Control Field, ACF) providing information about reservations and their acknowledgement (allocation bit). This is followed by the busy bit, indicating whether the slot is occupied or free (see Fig. 6.52). The head station (head of bus) is responsible for packet generation. The length of data packets (cycles) can be variable. Queuing management of the registered reservations is likewise taken over by the head station. In the CRMA slot, the start of a new cycle is marked in the command field and an 8-bit long cycle number allocated at the the end of the ACF field. The bus bit, located in between, indicates which of the two buses is being used as well as providing a 2-bit long priority data. The head station transmits cyclic reservation calls that pass through the bus and the connected computers, once in a forward and once in a backward direction. The reservation call contains the respective number of the cycle for which the reservation should be made, and the information about how many slots must be occupied for a reservation.

If a computer receives the reservation call on the return path, the registered number of slots to reserve is raised corresponding to its own reservation wish. When the reservation call reaches the head station again, the number of given slots contained in it for the relevant cycle in a queue can be reserved. This will be processed according to the FIFO principle. If a critical upper limit is exceeded, the head station does not carry out the reservation and transmits a reject message. This informs the connected computers about an overload situation and that the reservation could not be carried

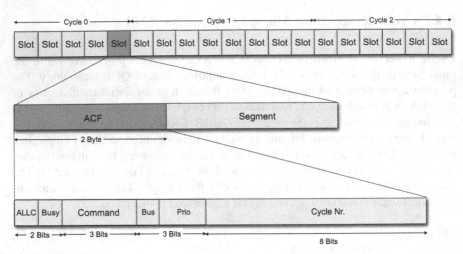

ACF...Access Control Field
ALLC...Allocation
Prio...Priority

Fig. 6.52 Structure of the CRMA data packets.

out. Based on the previous reservation, every connected computer can continuously occupy the needed number of slots in the subsequent cycle. Thus, a larger message is distributed over consecutive slots. On the other hand, before the actual transmission of data, every connected computer must first wait for the reservation and reservation acknowledgement. This waiting time can be critical, particularly in the case of real-time data. For this reason so-called gratis slots are also provided, which cannot be reserved beforehand (allocation bit=0). Non-reserved slots are generated whenever no more reservations are available. A computer that wishes to transmit data without a reservation must wait for the corresponding number of non-reserved slots to avoid fragmentation of the message.

For differently prioritized data, separate queues are managed in the head station. The queues are processed on the basis of their priority, i.e., slots with higher priority are given preference over slots with low priority and at the same time lower cycle numbers. A pending low priority transfer can be interrupted due to the preferential treatment given one with a higher priority. Because separate storage areas are provided in the connected computers, it is, however, not necessary for an explicit fragmentation to take place.

CRMA-II is an extension of CRMA, released in 1991. It supports various topologies (simple ring, dual ring, folded bus and dual bus). Any connected computer is able to carry out the function of the head station (scheduler) and take over slot reservation. If data is transmitted from a sender to the receiver, the receiver immediately frees the slots addressed to it, forwarding them so they can be used again by the following computer. Through this method the data transfer rate can be increased further in comparison to CRCM. CRMA-II distinguishes between the following slot configurations:

- FG: Slot is free and gratis,
- FR: Slot is free and reserved,
- BG: Slot is occupied and gratis,
- BR: Slot is occupied and reserved.

The scheduler allocates each computer a certain number of slots, determined by a fixed allotment strategy (see Fig. 6.53). Every computer i has two counters: the confirm counter c_i and the gratis counter g_i. The allocated slots of computer i are entered in c_i. If a computer has a wish to send and it is $c_i > 0$, then it can use both free gratis slots (FG) as well as free reserved slots (FR). If it is $c_i = 0$, the computer may only use free gratis slots (FG). In the case where a reserved slot is used, it changes its status from FR to BG and c_i is lowered by one. Using of a gratis slot, the status changes from FG to BG and g_i is raised by one. If there is no data to transmit at computer i, a free reserved slot changes its status from FR to FG and c_i is lowered by one. This is done to remove allocated, but not needed slots. In transmitting a message, the computer sends it completely – as necessary new gratis slots are generated, and arriving, already occupied slots must be delayed. To equalize this delay, at the next opportunity the appropriate number of available free slots are removed from the bus. If there is a reservation request from the scheduler, the connected computers announce the number of transmitted and still waiting slots in the last cycle back to the scheduler. Based on these values, it re-determines the next reservations i.e., c_i. At the conclusion of a transmission $g_i = 0$ is set. For still open reservations, the scheduler has its own counter m (mark counter). In the case that $m > 0$, the scheduler sets FG to FR and BG to BR and lowers m by one. If m=0 then the slots are passed on unchanged.

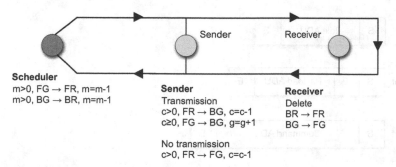

Fig. 6.53 CRMA-II access procedure.

CRMA-II slots are each made up of units consisting of 32 bits (Atomic Data Unit, ADU). Each one is framed by a start ADU and an end ADU (see Fig. 6.54). The start ADU contains an 8-bit long synchronization followed by a 24-bit long slot control field. Here, among other tasks concerning slot type, the priority, the occupation status (free or occupied), the gratis slot or reserved slot are coded. The end ADU starts with an 8-bit long end delimiter, followed by a repeat of the slot control field. The address ADU, which is only available in occupied slots, contains the network addresses of the transmitter (origin address) and receiver (destination address). The actual payload is contained in 32-bit long packages of payload ADUs. Command and control information is exchanged by way of special command slots of various lengths, which are exchanged with each other. Thus, the distribution and reservation of slots as well as the current count is administered. For the exchange of longer messages, several slots can be pooled to form multi-slots. Via CRMA-II, ATM cells can also be exchanged comprised of 17 individual ADUs (start ADU. address ADU, end ADU and 14 payload ADUs).

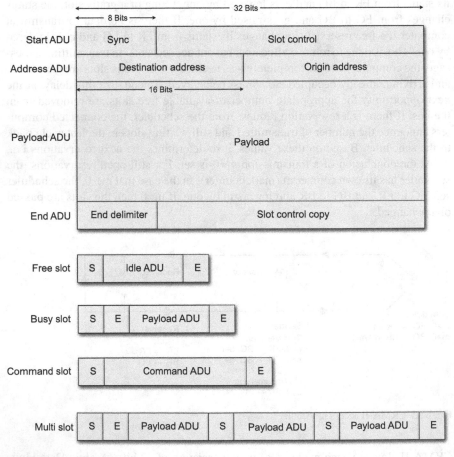

Fig. 6.54 Structure of the CRMA-II slots.

6.4.8 Plesiochronous Digital Hierarchy – PDH

The Plesiochrone Digital Hierarchy (PDH) is a synchronous time multiplex procedure that transmits signals from different sources over a shared medium. PDH is used in the transmission of language and digital data, e.g., in ISDN. Today, this network technology is used almost exclusively for data transmission rates up to 45 Mbps. Higher data transmission rates are achieved with the more powerful multiplex technique of the Synchronous Digital Hierarchy (SDH). The standardization of PDH already took place in 1972 through the ITU-T/CCITT, while different data transmission rates were determined for the regions of North America, Europe and Japan (see Table 6.5). Because of this, a repacking of multiplexed data streams is necessary with intercontinental data transmissions using PDH transmission. This is not necessary with the SDH transmission technique. The designation **Tx**, or **Ex**, and **DSx** characterize the level of hierarchy and bandwidth. Tx (Ex) applies to the transmission system and DSx to the multiplex signal.

Table 6.5 Bandwidths of the Plesiochrone Digital Hierarchy in kbps.

Hierarchy	North America		Europe		Japan	Data packet length
T1	DS1	1,544	E1	2,048	1,544	256 Bit
T2	DS2	6,312	E2	8,048	6,312	848 Bit
T3	DS3	44,736	E3	34,368	32,064	1,536 Bit
T4	DS4	274,176	E4	139,264	97,728	2,928 Bit
T5			E5	564,992	397,200	2,688 Bit

Because of the data transmission rates common in today's long-distance traffic technologies in the Gbps area, the PDH technology was often replaced by the more efficient SDH technology. However, PDH remains important today, used especially in the "last mile." The common ISDN primary multiplex connection with 30 user channels is based on the PDH bit rate E1 (2 Mbps). The bit rate E3 (34 Mbps) is frequently applied to connect individual company locations with each other over a large distance. The bit rates E4 and E5 are employed almost exclusively in large carrier networks. **Digital Signal Multiplexers**, **(DSMX)**, which combine individual channels and forward them bundled on a higher hierarchical level, serve as connective links for the structure of the PHD hierarchy (see Fig. 6.55). The name comes from the fact that in this procedure the bandwidths of the individual channels are not exactly the same, but only approximately the same (*plesios=[griech.]near, almost*). This is because every DSMX installed in different places has its own pulse-generating system clock and these cannot be perfectly synchronized. To balance out the differences between the individual channels and to make sure that with each clock adjustment none of the user data is lost, the technology of so-called **bit stuffing** is implemented.

Bandwidth and electrical, as well as physical, specifications for transmission are defined in the ITU-T/CCITT recommendations G.702 and G.703, the PDH multiplex

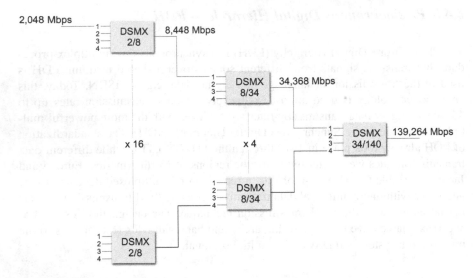

Fig. 6.55 PDH - DSMX hierarchy.

technology in the recommendations G.732, G.742 and G.751. The specified band-widths are in each case multiples of 64 kbps (T1 in North America 24 channels with 64 kbps each, Europe 32 channels with 64 kbps each).

The data packets in PDH follow a very simple structure (see Fig. 6.56). With E1 data packets 8 bits of the 32 channels are consecutively transmitted, each within a time slot of 125 µs. The result for E1 is a bandwidth of 2.048 Mbps. In transmission, channel 0 and channel 16 are not available for user data. On channel 0 a data packet password (this means a synchronization bit pattern together with error-checking information) and a data packet message word (for error management information) are transmitted interchangeably. Channel 16 serves to transport signal information. Signal coding proceeds according to the **HDB3 code**. HDB3 code (**High Density Bipolar**) involves a form of **AMI** coding in which after three logical zeros the next zero is converted into a logical one (see Fig. 6.57). Coaxial cable or twisted pair cables are used as the transmission medium, with the voltage level on the side of the transmitter at ±2,37 volts.

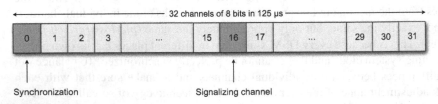

Fig. 6.56 PDH - E1 data packet.

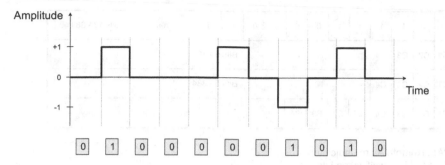

Fig. 6.57 PDH - HDB3 coding.

Data packets on higher levels of the hierarchy (E2-E5) are formed by virtue of a signal multiplexing. Here, four data packets of the underlying hierarchical level are combined. The data packets can, however, stem from different partial networks and for this reason exhibit differences with respect to bandwidth tolerances. These can be equalized through **bit stuffing**. Three distinctions are made in bit stuffing.

- In the **positive stuffing procedure**, the buffer memory of the multiplexer, which combines the data packets, is read faster than described. Depending on the level of the buffer memory, the reading is interrupted in certain areas (**stuffing places**) and a **stuff bit** added. In demultiplexing on the receiver's side, the receiver is informed via stuff information bits (see Fig. 6.58 in the example of the E3 data packet) whether stuff bits are to be added or whether user data is present in the corresponding places.

- In the **negative stuffing procedure** the buffer memory of the multiplexer is read slower than described. Depending on the level of the buffer, the stuff bits here are replaced by information bits.

- The **positive-zero-negative stuffing procedure** provides for both positive and negative stuffing places, respectively, in the data packet.

The **E3 data packet** (see Fig. 6.58), for example, begins with an identification of the beginning of the data packet (Frame Alignment) of a length of 10 bits. Bit 11 serves the transmission of alarm signals, while bit 12 is designated for national use. **C1, C2, C3** represent in each case stuff information bits. The interpretation of the stuff information bits proceeds column by column. If C1=C2=C3=0 applies in the first column, then the first stuff bit (4th line, 5th column) is in fact a stuff bit, in the other case (C1=C2=C3=1) it is a usable bit. In this way, columns 1 – 4 are assigned the stuff bits 1 – 4.

In view of the nearly synchronous working methods of PDH and the fluctuations caused by the independent DSMX clocking, PDH is only conditionally suitable for bandwidths over 100 Mbps. A further disadvantage of PDH is based on the fact that access to a E1 single signal from the top of the PDH hierarchy demands a complete demultiplexing of the aggregated data stream, which proves to be very difficult in practical application. For this reason, the development of a genuine, synchronous,

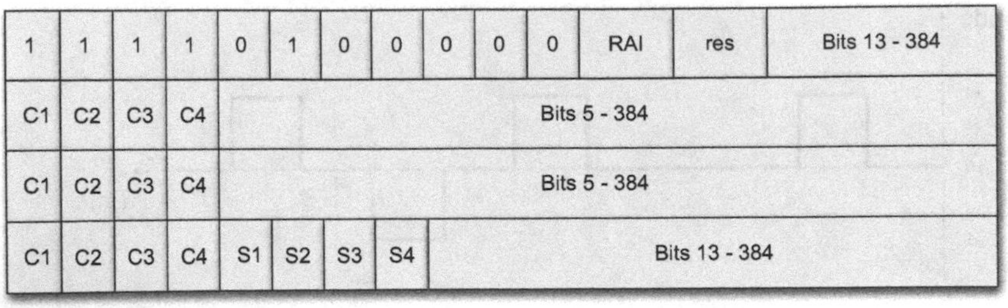

1	1	1	1	0	1	0	0	0	0	0	RAI	res	Bits 13 - 384
C1	C2	C3	C4										Bits 5 - 384
C1	C2	C3	C4										Bits 5 - 384
C1	C2	C3	C4	S1	S2	S3	S4						Bits 13 - 384

RAI...Remote Alarm Indication
Ci...Stuff bit (justification bits)
Si...Stuff bit

Fig. 6.58 PDH - E3 data packet.

signalization hierarchy (SDH, Synchronous Digital Hierarchy) made sense in order to facilitate high bandwidths in combination with a simpler access technology.

6.4.9 Synchronous Digital Hierarchy – SDH, SONET

The transmission technique known as Synchronous Digital Hierarchy (SDH) is a synchronous transmission multiplex method that was especially developed for data transmission via fiber-optic cable and radio. At the beginning of the development of the fiber-optic transmission technology, every telecommunications company that played a role in its development worked with its own proprietary multiplex procedure. Only first when the network reached a greater expansion and the necessity of a fusion of the various fiber-optic networks became increasingly imperative was a standardization established in this field. In 1985 Bellcoe and AT&T began the development of **SONET** (Synchronous Optical Network), which was taken by the U.S. ANSI as standard T1.105. A large part of the telephone traffic in the U.S. is handled on the basis of SONET. The ITU-T/CCITT recommendations for the Synchronous Digital Hierarchy (SDH, G.707, G.783, G.803) orient themselves on the SONET standard. Both systems are compatible and there are few differences between them. As PDH has a limited suitability for bandwidths over 100 Mbps, SDH was chosen as the basis for the bandwidth ISDN service (**B-ISDN**). Just as PDH, SDH encompasses an entire multiplex hierarchy with differing bandwidths from 52 Mbps (STM-1, OC-1) to currently 160 Gbps (STM-1024, OC-3072). In contrast to PDH, the clocking in the individual subnets and the implemented multiplexer proceeds strictly synchronously and always in a integer ratio to each other. Clocking takes place via a central master clock. In the case of phase shifting, which still may occur between individual subnets that are linked to each other, data is addressed via corresponding customized pointers.

The foundation for data transmission in SDH is based on a simple principle. Individual data streams from n different sources with bandwidth B are combined using a synchronous multiplex procedures into a single data stream with a bandwidth of n·B. In this way, it is possible to generate a multiplex signal of the n+1th hierarchy level directly from the signals of all the hierarchy 1,2,3,...,n below it. Similarly, in contrast to PDH, a signal from one of the lower hierarchy levels can be separated directly and simply from the data packet of one of the higher hierarchy levels. SDH likewise allows the transport of ATM cells or PDH multiplex signals (**cross connect**). Table 6.6 presents the SDH hierarchy levels (STM-x, Synchronous Transfer Module) in relationship to the respective SONET hierarchy levels (STS-x, Synchronous Transfer Signal (copper cable) or OC-x, Optical Carrier). In practical application, primarily STM-1 (Synchronous Transfer Module 1), STM-4, STM-16 and STM-64, are used. In each case, 5% of the bandwidth is reserved for command and control information.

Table 6.6 SDH and SONET hierarchy levels in Mbps.

SDH	Bandwidth	SONET
STM-0	51.84	STS-1 / OC-1
STM-1	155.52	STS-3 / OC-3
STM-2	207.36	– / –
STM-3	466.56	STS-9 / OC-9
STM-4	622.08	STS-12 / OC-12
STM-6	933.12	STS-18 / OC-18
STM-8	1,244.16	STS-24 / OC-24
STM-12	1,866.24	STS-36 / OC-36
STM-16	2,488.32	STS-48 / OC-48
STM-32	4,976.64	STS-96 / OC-96
STM-64	9,953.28	STS-192 / OC-192
STM-128	19,906.56	STS-384 / OC-384
STM-256	39,813.12	STS-768 / OC-768
STM-512	79,626.24	STS-1536 / OC-1536
STM-1024	159,252.48	STS-3072 / OC-3072

SDH offers the functionality of the physical layer found in the communication protocol layer mode, here divided here into four separate sublayers (cf. Fig. 6.59):

- **Physical Interface**: Specifies the technical parameters of the respective transmission technology (fiber optics, radio, satellite link).

- **Regenerator Section**: Serves to refresh the damped or distorted signal or clock and amplitude.

- **Multiplex Section**: Specifies the bundling of plesiochronous and/or synchronous signals for the very high bandwidth SDH data streams, as well as the reverse process of the decoupling of individual signals from an aggregated SDH data stream.

Virtual Container (VC): User data is transported in containers, VC-4 regulates the coupling/decoupling (mapping) of 140 Mbps signal, and VC-12 the mapping of 2 Mbps signals.

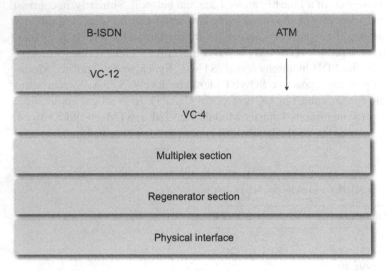

Fig. 6.59 SDH in the layer model (simplified example).

In accordance with the SDH definition, the following elements (network elements) are distinguished in the SDH network:

Terminal Multiplexers (TM) facilitate the access of an end system to the SDH network, combining multiple incoming signals to a SDH hierarchy level. For this the TMs have mutliple plesiochronous or synchronous interfaces for end systems, as well as one or two synchronous interfaces for the actual SDH network.

Regenerators (REG) facilitate amplification of the optical signal along the SDH transmission line. In contrast to a purely optical amplification, a received optical signal is first transformed into an electrical signal. The electrical signal is amplified, synchronized and corrected in its form. Afterwards, it is transformed back into an optical signal.

Add-Drop-Multiplexers (ADM) work in a similar way to TMs. Individual channels from the multiplex data stream can be filtered out with these in a simple way or be added again.

Digital Cross Connect systems (DXC) serve as further switching elements on the basis of ADMs. These can switch individual channels between different incoming and outgoing lines and regulate flow control in the SDH network.

Fig. 6.60 provides an example for the combination of SDH network elements.
As a rule, SDH primary networks are based on a double ring. In normal operation, only one ring is operated with the second available as a cold reserve in case of a

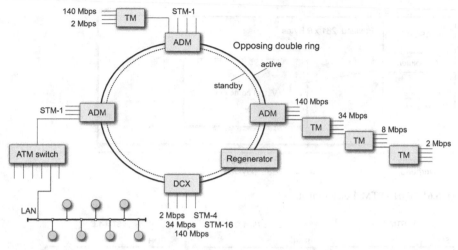

Fig. 6.60 SDH - Design of a SDH ring structure.

disturbance. Through the application of a DCS, individual rings can be coupled. A simplified version of the double ring failover is the 2 fiber MS-SPRing (Multi-Section Shared Protection ring), in which half of the available bandwidth is kept free or is only operated with low-priority data traffic. In cases of error, this lower prioritized data traffic is discarded and the failed ring route shut down.

The structure of the SDH **data packets** is divided into a user data section and into a section for command and control information (RSOH, Repeater Section Overhead and MSOH, Multiplexer Section Overhead). Data packets from a higher order of hierarchy are always formed through the multiplexing corresponding to the number of data packets of the respective hierarchical level underneath. Fig. 6.61 shows the schematic structure of a STM-1 data packet. The STM-1 data packet consists of 9 lines and is sent line-by-line from left above to right below. The AU pointer in the overhead section (the first 9×9 octets) indicates the position of the respective virtual container inside the user information.

Virtual containers in the STM-x data packet are not bound to prescribed packet limitations. In contrast to PDH, the AU pointer allows the direct addressing of a user data signal in a high bit signal without demultiplexing the entire signal. Because of the possible occurrence of phase shifts, an adaption of the AU pointer can be necessary if a virtual container has been displaced.

6.4.10 Worldwide Interoperability for Microwave Access (WiMAX) – IEEE 802.16

The Worldwide Interoperability for Microwave Access implementation (WiMAX) is a broadband radio transmission technology for mobile and stationary end sys-

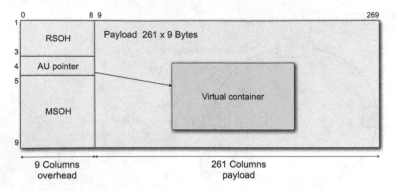

Fig. 6.61 SDH - STM-1 data format.

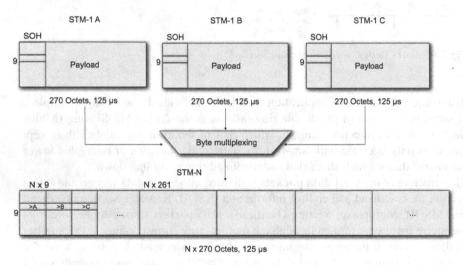

Fig. 6.62 SDH - multiplexing.

tems. Beginning in 1999, it was standardized as IEEE 802.16 by the working group Broadband Wireless Access (BWA). In contrast to the rest of the IEEE 802 standards for LANs and PANs, BWA encompasses a much larger geographic area (up to 50 km, wireless MAN) offering data transfer rates from 70 Mbps and higher. IEEE 802.16 (2004) is designed for implementation in stationary end devices in wireless, wide-area networks (Fixed WiMAX). IEEE 802.16e (2005) expands the standard to mobile end systems and the required fast cell exchange necessary during operation (mobile WiMAX).

Interest groups from industry and commerce joined together in 2001 to form the WiMAX Forum[4] in support of the WiMAX standards. The new technology offers a standard for bridging the last mile in broadband radio networks; an efficient alternative to wired access technologies, such as DSL (see also section 6.5.3), or mobile

---

[4] http://www.wimaxforum.org/

technologies of the 3rd and 4th generations (UMTS and LTE). Therefore, WiMAX is also designated wireless DSL (WDSL). The network expansion of a single Wi-MAX basis station is comparable to that of a cell in a mobile network. In this way, WiMAX offers country locations a genuine alternative where the laying of a cabled network infrastructure for high-speed networks would be too costly. The WiMAX Forum has taken over the task of certifying the compatibility and interoperability of products that are made according to the IEEE 802.16 standard.

The IEEE 802.16 wireless MAN standard specifies transmission technologies for systems in frequency ranges between 10 and 66 GHz. In this frequency range the stationary operation with visual connections (fixed WiMAX) is of special interest, as high transmission rates of more than 70 Mbps, with distances of more than 50 km, are possible. A visual connection cannot always be ensured in mobile application. If this situation occurs, a lower frequency area with a likewise lower data transmission rates is used. The expansion IEEE 802.16a broadens the IEEE 802.16 standard regarding the use of regulated and license-free frequency areas from 2–11 GHz. Here, data transmission rates with limited mobility are possible of up to 20 Mbps, over distances to 5 km (indoor antenna) or 15 km (outdoor antenna). The expansion of IEEE 802.16c creates additional system profiles for 10–66 GHz and defines necessary interoperability specifications. IEEE 802.16e combines the expansions for mobile WiMAX in the frequency range between 0.7 – 6 GHz. Thereby data transmission rates of up to 15 Mbps, with distances between 1.5–5 km are assumed (see Table 6.7). The frequency range reserved in Germany for the area of WiMAX is 3.4–3.6 GHz and 2.5–2.69 GHz.

Table 6.7 IEEE 802.16 and WiMAX.

	IEEE 802.16	IEEE 802.16a/d	IEEE 802.16e
Standardiization	2002	2004	2005
Frequency	10–66 GHz	2–11 GHz	0.7–6 GHz
Max. bit rate	≤134 Mbps	≤75 MHz	≤15 MHz
Max. range	≤ 75 km	≤5 km (indoor antenna) ≤15 km (indoor antenna)	≤5 km 1.5 km (typical)
Receiver	fixed	fixed (outdoor antenna), limited mobility (indoor applications)	mobile (nomadic use)

WiMAX enables a bundling of various services, e.g., telephony, video-on-demand and the Internet via a single standard. This ability is marketed as „Triple-Play" and allows a service provider of formerly heterogeneous technologies, such as DSL, cable television and the classical analog or digital telephone network, to offer a full range of services. Currently, WiMAX supports the following operational variations (see also Fig. 6.63):

- **Point-to-MultiPoint mode** (PMP):
  Stationary or mobile network nodes that can be operated as end nodes in the

WiMAX network. These are designated as Subscriber Stations (SS) or Mobile Stations (MS). In the PMP mode all SS or MS are directly connected to a Basis Station (BS), meaning all devices must be located within the radius of the signal range of the WiMAX network. Subscriber Stations are coordinated from the basis station. The connection proceeds encrypted.

- **Point-to-Point mode** (P2P):
  To enable the mutual communication between two basis stations these can be networked with each other over a radio-based point-to-point connection along a Line of Sight (LOS).

- **Mesh mode**:
  It is not necessary for all of the devices to be located within the signal radius of the basis station in the mesh mode. Individual SS or MS can forward the signal of the BS as a relay station to devices that are located outside the range of the BS. In this way, a multi hop relay network is created that provides larger geographical coverage.

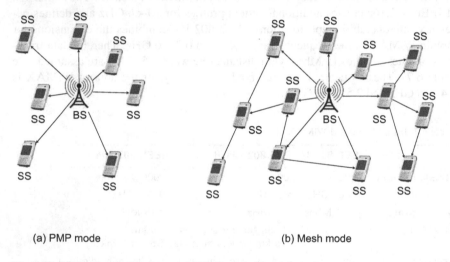

(a) PMP mode   (b) Mesh mode

Fig. 6.63 WiMAX operational types, (a) PMP mode (b) mesh mode.

The WiMAX architecture defines three fundamental units (see Fig. 6.64): Mobile stations or subscriber stations, the Access Service Network (ASN) and the Connectivity Service Network (CSN). As end devices, MS and SS are connected over a connection-oriented radio interface (R1) with the Basis Station (BS) in the access service network. While from the perspective of MS/SS this is a point-to-point connection, the basis station operates a point-to-multipoint connection here. The ASN is comprised of one or more basis stations, which are connected to one or multiple ASN gateways (Access Router), and establish the connection to the Wi-MAX network of the Network Access Providers (NAP). Over a further interface (R3), the ASN gateway connects the access service network with the connectivity service network, over which the network service provider supplies the connection to the Internet. Between MS/SS and the connectivity service network, an interface (R2) is defined over which the general management tasks and mobility-related tasks are processed. The interface to the visited CSN first then becomes active if the home CSN is left (roaming). Further interfaces are defined between individual ASNs (R4) for micro mobility management as well as between Home CSN and Visited CSN (R5) for macro mobility management.

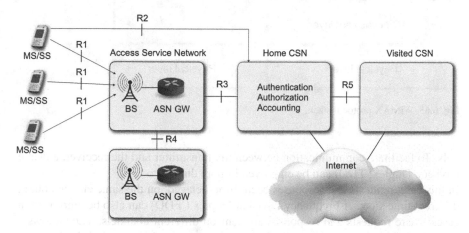

Fig. 6.64 WiMAX architecture.

The protocol architecture of WiMAX is comprised of the physical layer (PHY) and the data link layer, whereby the data link layer is subdivided into the MAC Privacy Sublayer, MAC Common Part Sublayer and Service Specific Convergence Sublayer (see Fig. 6.65). In the service specific convergence sublayer, two services are provided for adaptation to the services of the MAC sublayers: the cell-oriented ATM Convergence Sublayer (ATM CS) and the packet-oriented Packet Convergence Sublayer (Packet CS). Additional tasks make up the quality of service and broadband management. In the MAC Common Part Sublayer (MAC CPS), fragmentation functions for the unpacking of the data packets in MAC frames, Automatic Repeat Request (ARQ) and quality of service functionalities are supported. In a separate sublayer, the MAC Privacy Sublayer (MAC PS) supports the security relevant functions for authentication, secure key exchange and encrypting by means of the Data Encryption Standard (DES), Cipher Block Chaining (CBC) or Advanced Encryption Standard (AES). In each case, only the payload of the MAC frames is encrypted with the transmitted header information remaining unencrypted.

WiMax employs OFDM (Orthogonal Frequency-Division Multiplexing) as a multi-carrier modulation method on the physical layer. It uses multiple orthogonal carrier signals for data transmission (see section 3.2.3). This procedure is also used for the mobile WiMAX as Orthogonal Frequency-Division Multiple Access (OFDMA). It facilitates multiple access to the same signals for different users on different chan-

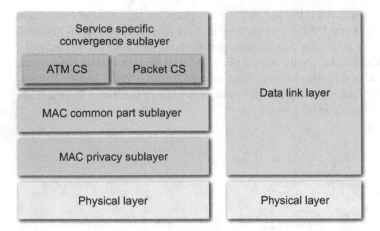

Fig. 6.65 WiMAX protocol stack.

nels. To facilitate communication between the transmitter and the receiver, a (Time Division Duplex, TDD) can be employed in both directions

In this case, requests and responses occur time-delayed, on the same channel but in different time slots. A Frequency Division Duplex (FDD) can also be employed in cases where requests and responses are sent on different channels. TDD allows a more flexible division of the available frequency range, whereas here it is necessary that users be exactly time synchronized to prevent transmissions in neighboring cells from disturbing each other.

TDD prevailed in the area of mobile WiMAX. A data packet (frames) was split into two subframes, in each case for downlink (BS after SS/MS) and uplink (SS/MS after BS), was separated by a short security interval ((Transmit/Receive Transition Gap, TTG). These are sent one after the other in the same frequency band (see Fig. 6.66). Every downlink subframe starts with a preamble that is used for synchronization and for cable determination. To equalize disturbances caused by mobility problems and channel impairment, WiMAX allows a multiple repetition of the preamble during transmission. Therefore, in uplink subframe short versions of the preamble (midambles) can occur after 8, 16 or 32 OFDM symbols, while in downlink subframe every bundled data transmission (data burst) is preceded by a preamble.

6.5 Access to the WAN

The normal end user generally does not have a dedicated WAN access. Direct access to a fast, wide-area network is usually only the privilege of companies and scientific and military facilities, who gain access over special leased lines. Whoever does not have direct WAN, gains access via a special **access network** in conjunction with the actual network (core network). This access network must be laid out in such

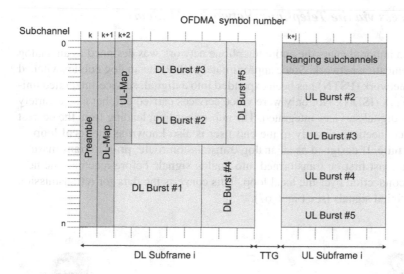

Fig. 6.66 WiMAX data packet, consisting of a Downlink subframe (DL) and Uplink subframe (UL).

a way that it allows for the integration of various services and applications, just as the actual network. It should also be open for further extensions in regard to the offered functionality. This type of extensible network is also known as a **Full Service Network** (FSN).

Basic distinctions are made in an access network between:

- **Wired access networks**
  This category includes, e.g., the regular analog **telephone network** or **ISDN** whereby these only allow for narrow bandwidths of up to 128 kbps (2 channel ISDN). Higher bandwidths are first possible with transmission technology such as **DSL**. At present, access is usually gained via symmetrical copper cable, while direct access over fiber-optic cable is already widely used and will certainly dominate in the future. Cable television networks can also be implemented as access networks if a suitable return channel is available. Other providers use the electricity network at their disposal.

- **Wireless access networks**
  This includes access over **mobile communications** or **radio relay systems**. Satellite communication connections have also established themselves here, while a purely optical transmission can only be used with the aid of lasers for short distances.

Most of the time, the access network does not offer symmetrical access to the end user, i.e., communication in the direction of the network to the end user (**downstream**) is generally possible with a higher bandwidth, as is the reverse direction from end user to the network (**upstream**).

6.5.1 Access via the Telephone Network – Modems

For nearly a hundred years the entire telephone network was designed as an analog telecommunication network. Since approximately 1989, the analog public switched telephone network (PSTN) has been expanded into a digital, service-integrated universal network (ISDN). Not only were voice services carried out, but also a variety of different digital services integrated. The subscriber line, leading from the nearest wide area connections directly to the end user is also known as the **local loop** or as the **last mile**. If designed as an analog transmission route, prior to transmission, digital data must first be transformed into analog signals before reaching the next wide area connection over the local loop. This converts the data for retransmission back into digital signals (see Fig. 6.67).

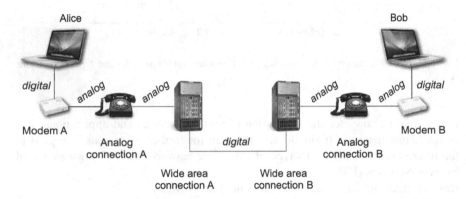

Fig. 6.67 Analog and digital data transmission with modems.

Every end user who wishes to carry out a data transmission over the analog telephone network needs to have a **modem**. The name "modem" is made up of the two tasks which must be performed: **modulation** and **demodulation**. In modulation, the frequency spectrum of the original digital signal (basis band signal) is shifted into the frequency range transmitted by the analog telephone network of 300 Hz to 4000 Hz, while demodulation reverses this shift. Analog signal transmission is based on the transfer of time variable voltage level via an electrical line. Analog transmission cannot be carried out without the occurrence of measurable losses. For this reason, analog signal transmission is subject to strict restrictions, which are determined by three different types of disturbance.

- **Signal attenuation**
  As signal spread increases, the output signal becomes increasingly weaker (in electrical cables, for example, logarithmic to the distance covered). Signal attenuation is indicated in decibels per kilometer and dependent on the implemented signal frequency.

- **Signal disortion**
  If a signal is observed over time as a function made up of individual fixed-frequency components of a Fournier analysis, it is evident that individual signal components spread out at different speeds. When transmitting originally digital data it can happen that fast components of a transmitted bit "pass up" the slow components, resulting in transmission error.

- **Signal noise**
  Noise along the transmission medium can also be caused by sources of energy other than those of the transmitter's. So-called **crosstalk** designates a type of disturbance occurring if induction effects are generated between two cables in close vicinity to each other.

To avoid the problems mentioned, an attempt is made to keep the transmitted frequency within as narrow frequency band as possible. However, digital signals produce square waves, which are especially prone to the effects of distortion and attenuation due to their wide frequency spectrum. Therefore sinus carrier waves are used in the area between 1000 Hz – 2000 Hz, whose **amplitude**, **frequency** or **phase** can be modulated in order to transmit information. The method of amplitude modulation (Amplitude Shift Keying, ASK), frequency modulation (Frequency Shift Keying, FSK), phase modulation (Phase Shift Keying, PSK), which are used for the analog transmission of digital signals, and the implementation of combined bandwidths for an optimal use of the available bandwidth, such as Quadrature Amplitude Modulation (QAM), are examined in detail in section 3.2.3.

In modem operation, a distinction is made between **full duplex** and **half duplex transmission**. The transmission channel is used at the same time in both directions in full duplex communication. This can be implemented most simply by means of a 2-wire line. If communication takes place in both directions with different bit rates, one speaks of **asynchronous** communication, otherwise of **synchronous** communication. Transmission and reception take place in a prescribed clock that is generated by the modem itself. In half duplex communication the modem is capable of using the available bandwidth to the sender in only one direction.

The ITU-T/CCITT defines different types of operation for modems that allow transmission with different performance data and characteristics. Table 6.8 shows a selection of the ITU-T/CCITT specified modem standards.

For V.90, even up to 56 kbps can be reached, however only in the direction of the access mediation to the user (downstream). This is enabled by omitting the digital/analog transformation between the access mediation and the server selected by the user. V.90 is only able to work with 56 kbps if a digital network (e.g., ISDN) is used between the server and access mediation. That means, on the connection route between user and server only one digital-analog transformation is allowed to take place, whereby the existing analog transmission route between access mediation and user is subject to strict, qualitative guidelines and length restrictions.

Communication between the server and access mediation transpires on a digital 64 kbps channel. Per each transmitted byte, one bit is used for error detection (thus only 56 kbps). In the reverse direction (upstream), the digital data transmission bet-

Table 6.8 ITU-T/CCITT modem standards.

Type	Operating mode	Max. bandwidth	Modulation	Transmission
V.21	async., sync.	up to 300 bps	2-FSK	full duplex
V.22	async., sync.	300, 600, 1,200 bps	4-FSK	
V.22bis	async.	2,400 bps	16-QAM	full duplex
V.23	async., sync.	600, 1,200 bps	FSK	half duplex
V.26	sync.	2,400 bps	4-DPSK	full duplex
V.26bis	sync.	1,200, 2,400 bps	2-PSK, 4-PSK	half duplex
V.27	sync.	4,800 bps	8-PSK	full duplex
V.29	sync.	9,600 bps	4-QAM, 16-QAM	full duplex
V.32	async., sync.	9,600 bps	16-QAM	full duplex
V.34	async., sync.	33,600 bps	256-QAM/TCM	full duplex
V.90	async.	56,000 bps	PCM, digital	downstream
		33,600 bps	256-QAM/TCM	upstream
V.92	async.	56,000 bps	PCM, digital	downstream
		48,000 bps	PCM, digital	upstream

ween subscriber switching system and server cannot be used. Upstream V.90 uses the encoding procedure known from V.34 with a maximum of 33.6 kbps. The 2002 ratified V.92 standard also allows a higher transmission data rate upstream of up to 48 kbps, through exploitation of a digital communication channel between access mediation and server. Still higher data transmission rates via the analog telephone network are only possible with extra data compression procedures.

Many modems are equipped with compression and error detection procedures. Modems transmit bit streams transparently and, for this reason, need not be concerned with protocol functions on the higher communication protocol layers. The correction mechanisms available over the modem facilitate improve data transmission, without making changes in existing protocol software necessary. The family of the **MNP protocol** (Microcom Networking Protocol) includes ten different variations (levels) of error detection, error correction and data compression. **MNP3** offers error correction through a checksum procedure, just as **MNP4**, which offers error correction according to the **ARQ** procedure (Automatic Repeat Request). In the ARQ procedure, the bit sequences recognized as defective are corrected by renewed request and transmission from the sender.

MNP5, the most widespread procedure of the MN protocol family, includes the methods of MNP3 and MNP4. Further, it implements a run length method for data compression. The **V.42l** protocol was specified by the ITU-T/CCITT and also has the error correction procedure used in MNP3 and MNP4. While the **V.42bis** protocol uses a LZW procedure for data compression and contains the functionality of MNP5[5].

In connection with the compression procedure V.42bis, V.90 achieves data transmission rates of between 56 kbps and 220 kbps. Combined with the V.44 compression

---

[5] see also section 4.3 "Text Data Formats and Compression," in the first volume in this series: Meinel, Ch., Sack, H.: Digital Communication – Networking, Multimedia, Security, Springer (2013).

procedure it reaches rates of between 56 kbps and 320 kbps in downstream data transmission. The data transmission rate reached in each case depends on the type of data that is to be transmitted. Simple text data achieves a high degree of compression and therefore high transmission speeds. Already compressed data, such as JPEG graphic data, does not allow itself to be further compressed and therefore reaches only the lower limit of 56 kbps.

6.5.2 Access over ISDN

If the ISDN network is used as an access network to Internet, the transmitted data is delivered directly in digital form. An analog-digital transformation, as in the analog telephone network, is not necessary.

ISDN enables the end user utilization of various services, such as language or data communication, via a single network access. As a rule, a so-called **basis connection** is available to the end user, providing two user data channels (B channels). In each case, there is a band width of 64 kbps and a control channel (D channel) with 16 kbps. Transmission reaches a maximum of 144 kbps in full duplex operation. In order that the existing end user connection can continue to be used for language communication, two data transmission methods are available:

- **Time division method**
  This method is also known as the ping pong or time fork procedure. It uses time multiplexing to separate the signals for both transmission directions. The necessary interface for this was standardized by the ITU-T/CCITT as U_{P0}. Data packets, 20 bits in length each, are transmitted alternatively in both directions (see Fig. 6.68 (a)). To avoid a possible overlapping of opposing data packets, a suitably long waiting interval between the data packets must be maintained. On the basis of this waiting time, the actual bandwidth must be more than twice as large as the targeted data transmission rate of 144 kbps. A bandwidth of 384 kbps is in fact used. Due to the signal run time and high data transmission rate, relatively high attenuation effects occur, this means that the maximum distance to be bridged shrinks to 2-3 km.

- **Same position with echo compensation procedure**
  The interface implemented here was specified by the ITU-T/CCITT under the name U_{K0}. An increase of bandwidth, as is the case in the time division method, can be avoided here. In fact, analog to the telephone network, a full duplex operation is implemented with the help of a hybrid circuit As in the standard analog telephone network, the hybrid circuit enables an implementation between two-wire and four-wire lines with simultaneous separation of direction (see Fig. 6.68 (b)). To facilitate this, it must be ensured that the receiver, on one side, only receives the signal of the opposite side each time. However, this solution can also causes the occurrence of **echo streams**,. This means that the receiver also gets unwanted parts of signals from the transmitter that can still be present on the line

as an echo of an earlier sent data packet. Therefore, an **echo compensation** is carried out on the receiver's side, whereby an anticipated echo signal is always subtracted from the received signal. The precalculation of the echo signal is done with adaptive procedures and ensures that the distance to be bridged increases to 4-8 km.

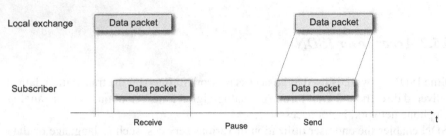

(a) Time division method (expiration)

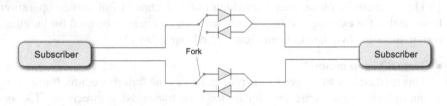

(b) Same position procedure (schematic)

Fig. 6.68 ISDN data transmission to the end user: (a) time fork procedure, (b) same position procedure for full duplex operation.

Line coding and data formats implemented are described in section 6.4.3 "ISDN."

6.5.3 Access via Digital Subscriber Line – DSL

Besides the now-dated analog telephone network, or the digital ISDN network, there are currently far more powerful access possibilities offered to the end user. Belonging to this category is the **DSL** family (**Digital Subscriber Line**), as well as access via alternative networks, such as the **cable television network** - insofar as it has a suitable reverse channel - or the **electricity network**. Furthermore, there also exist wireless varieties, for example the **mobile radio network** (GSM, UMTS und LTE) or **radio**.

The digital subscriber line, in contrast to ISDN, offers a considerably higher bandwidth for the end user, although as a rule the same cable technology (twisted pair

copper cabling) is used. The expression commonly used by the telecommunication companies describing DSL as " broadband communication" is purely a marketing term. Although this indicates that compared to older technologies a higher data
transmission rate can be achieved with DSL, the old cabling is still employed - getting in the way of further development. The proposed next step envisions fiber-optic
cabling directly to the home, and has thus been named **Fiber-to-the-Home** (**FTTH**).
Principally, there are two modulation procedures for digital data transmission available for DSL. Both are based on the frequency splitting technology:

- **Carrierless Amplitude Phase (CAP)**
  The available frequency spectrum is hereby divided into three parts:

  – The frequency range from 0–4 kHz is used for voice-based communication
    and the services of the standard telephone system (Plain Old Telephone Sys-
    tem, POTS).

  – A narrow frequency range from 25–160 kHz (ADSL) is designed for upstream
    data transmission.

  – The frequency range of 240 kHz–1.6 MHz is designed for downstream data
    transmission.

  Because of this clear separation of individual frequency ranges, interference between the different frequency bands is avoided.

- **Discrete Multitone (DMT)**
  With this variation, the available frequency range is subdivided into 256 subchannels of 4.3126 kHz each. Just as with CAP, the first subchannel is designed for
  voice-based communication. Upstream and downstream are not handled in different frequency ranges here but partly overlap each other. As the implemented
  copper cables have better transmission properties in lower frequency ranges, these are used for a bi-directional transmission, while the higher frequency ranges
  remain reserved for downstream transmission. DMT is more flexible and better
  adaptable than CAP to the transmission medium employed. At the same time,
  DMT requires a more complex implementation.

The family of the DSL variations, also known as **xDSL**, provides the following
versions:

- **ADSL (Asymmetric DSL)**
  With the asymmetrical variation of the DSL procedure, different bandwidths are
  used in each direction: from the user to the network access point (upstream)
  and from the network access point to the user (downstream). The downstream
  bandwidths are normally considerably higher. For a ADSL connection, a modem,
  a so-called splitter and a two-wire line are necessary. The splitter takes over the
  task of separating the frequency range according to the modulation procedure
  used.
  The modem on the user side is designated as **ATU-R** (ADSL Terminal Unit – Remote), and on the side of the switching center as **ATU-C** (ADSL Terminal Unit –

Central Office). On the switching side, multiple users can be combined to a **DS-LAM** (Digital Subscriber Loop Access Multiplexer) and from there transmitted via a fiber-optic cable to a concentrator (DSL-AC). They are then transferred to the backbone of the provider (see Fig. 6.69).

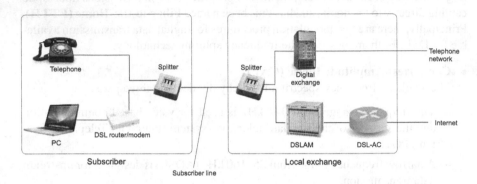

Fig. 6.69 ADSL - Asymmetrical DSL procedure.

In ADSL, the implemented bandwidth can be adapted to the respective distance to be bridged in steps of 32 kbps. Subsequently, with ADSL2 or ADSL2+, downstream from the switching center to the end user data transmission rates of theoretically up to 25 Mbps can be reached. In the transmission direction (upstream), conversely, up to only 3.5 Mbps are possible.

- **UADSL (Universal ADSL, ADSL light)**
  Designation for a simplified ADSL variation that works without complex splitters and is exclusively offered in the U.S. In contrast to ADSL, instead of 256 carrier frequencies only 128 are used. UADSL is specified in ITU-T/CCITT G.992.2 and allows, under ideal conditions, a bandwidth of up to 1.5 Mbps downstream or 512 kbps upstream with a range limitation analog to ADSL. Based on the high rate of error due to interference caused by simultaneous telephoning and data transmission, the providers guarantee a transmission rate of only 64 kbps.

- **HDSL (High Bit Rate DSL)**
  This is the oldest variation in the DSL family. It is exclusively designed for data transmission and, consequently, needs no splitter. The bandwidth is symmetrical in this procedure, i.e., in upstream as well as downstream, the end user has 1.544 Mbps (with two two-wire lines, T-1 lines in the U.S.) available or even 2.048 Mbps (with three three-wire lines, E-1 lines in Europe). The line length in HDSL can extend up to 4 km.

- **SDLS (Symmetric DSL, Single-Pair DSL)**
  In contrast to the HDSL method, SDLS requires a two-wire line with corresponding decrease of the transmission capacity. The transmission is carried out symmetrically, just as in HDSL, with a bandwidth to 2.36 Mbps and a range of up

to 8 km. Transmission takes place symmetrically, just as in HDSL, with a bandwidth of up to 2.36 Mbps with a range of up to 8 km.

* **VDSL (Very High Bit Rate DSL)**
  VDSL provides a higher bandwidth than ADSL, yet just as ADSL is based on copper wire cable in the last mile to the end user.
  VDSL enables data transmission of up to 52 Mbps downstream and 11 Mbps upstream, whereby the usable transmission bandwidth sinks considerably with increasing distance. Already starting at 900 m, the data transmission rate sinks to approximately half of the original level, and at 2 km reaches approximately ADSL standard values. VDSL2 is based on the DMT modulation procedure and designed to facilitate data transmission rates over 100 Mbps. The theoretical maximum is 250 Mbps directly at the source, yet as this distance increases a drastic reduction sets in. After 500 m a maximum of 100 Mbps is possible, while after 1.6 km the ADSL2+ level is reached.

All DSL procedures are based on a full duplex transmission. Additionally, modulation and echo compensation procedures are implemented, as in standard modems. The transmission proceeds either from upstream or downstream signals with the help of a frequency multiplexing procedure. Here, both components occupy different frequency bands, or share a frequency band, in which case it is necessary to work with echo compensation. Voice communication or the standard telephone service are transmitted in a separate frequency range. This is separated from the remaining frequency spectrum via a splitter.

6.5.4 Wireless Access to the WAN – GSM, UMTS and LTE

One of the advantages of wireless network access is the implementation of a widespread network with less costs in relation to infrastructure. There are neither expenditures for laying work or for a cable network. In a wired network, the end user is additionally bound to the access point of the geographic location of his stationary network, whereas in a wireless network the user can move about freely - is mobile. In wireless network access types a distinction is made between radio application, where the end user is required to be stationary, and mobile radio applications, in which the end user is in fact location-independent and (to a certain extent) mobile.

Global System for Mobile Communication – GSM

The most widely used mobile network at this time, worldwide, is the digital **GSM** mobile radio network (**Global System for Mobile Communication**), a second generation mobile radio system. At its high point in 2008, it was made up of more than 3 billion subscribers. The GSM network is built on individual, overlapping **cells** (see Fig. 6.70). The size of each of these varies depending on the density in the number of subscribers. The cell structure was already implemented in the analog predecessor system (the C network in Germany). It is conceived in such a way that a connection is not broken with the switching of a subscriber from one cell to an adjacent one, rather it is simply passed on further (handover). To ensure that the accessibility of a subscriber is fully covered, a switched on mobile telephone notifies the local basis station about the switch into a neighboring cell. It then passes on this information to a central database (**roaming**). In this way the current location of a network subscriber with a switched on mobile telephone is always known and the connection can then be guaranteed. Every subscriber in the GSM network receives a so-called "SIM card" (Subscriber Identity Module). This clearly identifies the subscriber and can be used in different end devices.

In Germany, two different variations of GSM are used. The first, GSM 900, works within a frequency range of 880 MHz to 960 MHz. This is where the German mobile radio network, called **D-Netz**, is located. The second, GSM 1800, whose frequency range is between 1710 MHz to 1880 MHz, is where the German **E network** and the **D network** are found. For data transmission in GSM 900, the frequency range between 935 and 960 MHz is used by the respective base station (downlink). The frequency range between 890 and 915 MHz serves as the transmission range for the mobile stations (uplink). The frequency range for uplinks and downlinks are in each case divided into 124 channels (frequency bands), with a bandwidth from 200 kHz, which via a time multiplex procedure (TDMA, Time Division Multiple Access) is used by different subscribers at the same time. The implemented time multiplex procedure uses 8 time slots with a length of 0.577 ms each. Every TDMA time slot carries 148 bits of data. For reasons of synchronization they each start and end with 3 null bits. After that, following at the front and in the back of each data frame, are 57 bits of user information, including a control bit that indicates whether the following data file contains voice information or data. In between is a synchronization field of approximately 26 bits in length . Based on the high rate of error involved in a radio transmission, mechanisms for automatic error detection and error correction are absolutely necessary. The net data rate of a single GSM data channel is 13 kbps, which because of the complex error detection and correction measures is in real terms reduced to 9.6 kbps. **GMSK** (Gaussian Minimum Shift Keying) is used as a modulation procedure.

In addition to voice services, GSM also offers data transmission. There is a principle difference made between basic services (**bearer services**), divided into data transmission and teleservices, among which are services of the telephone and telefax, as well as additional services such as the Short Message Service (SMS). Fig. 6.72 shows the embedding of the GSM network in the structure of the public network.

The GSM carrier services offer synchronous, asynchronous, line or packet-oriented services in the bandwidth range of 300 bps up to 9,600 bps. The individual services are addressed by means of corresponding numbers. The service platform designated as **MExE** (Mobile Station Execution Environment) in GSM forms the basis for the **WAP** protocol (**Wireless Application Protocol**). With this, mobile access to the WWW is made possible.

A GSM network consists of three subsystems (see Fig. 6.73):

Organization of the GSM mobile radio network - Cells

Wireless data communication is based on the spreading of electromagnetic waves through space. It functions according to the principle of an antenna transmitting nearly **isotropically**, meaning evenly in all directions, at a relatively low frequency. Therefore, all receivers located within a certain radius of the transmitter can pick up the transmission. In the case of bidirectional communication, the transmitter and the receiver must share the available frequency and transmit in turns using a time multiplex procedure. In this way, only a half-duplex connection can be implemented. For a full-duplex communication a second frequency must also be available.

The frequency range available for radio transmissions is always limited - a receiver must share the frequency range with many others. Because of this, so-called **cells** are set up. This ensures wide-range coverage as well as the reuse of frequencies. Every cell has its own base station, which isotropically covers a specific area, mostly within the radius of 10-20 km, with a low transmission capacity. Attention is paid that neighboring cells are always operated with different frequencies to prevent interference.

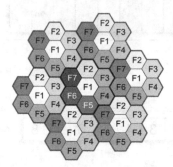

Although cells are actually circular, to be able to easier determine area coverage a model made up of hexagonal cells is used. With careful planning, an area can be covered with cells using only seven different frequencies.

At any given time, a mobile network subscriber is logically assigned to one cell. If a subscriber is ready to leave a cell, the base station responsible for the cell recognizes that the signal of the subscriber is getting progressively weaker. It proceeds to inquire at the neighboring cells as to how strongly they are able to receive the subscriber in question. Finally, the base station hands over "care" of the network subscriber to that cell who shows the strongest reception of the corresponding network subscriber's signal. Informed about the switch, the network subscriber changes the frequency to that of the cell now supervising it. This procedure of switching to another cells is known as a **handover**, and is carried out in a few hundred milliseconds.

Further reading:

Schiller, J.: Mobile Communications, 2nd ed., Addison-Wesley (2003)

Fig. 6.70 Organization of the radio network in individual cells.

- **Radio Subsystem (RSS)**

  Here are all radio-specific components, such as mobile stations (MS) and base stations (Base Station Subsystem, BSS, also called site areas). The BSS provides all of the functions that are necessary for a permanent radio connection to a MS.

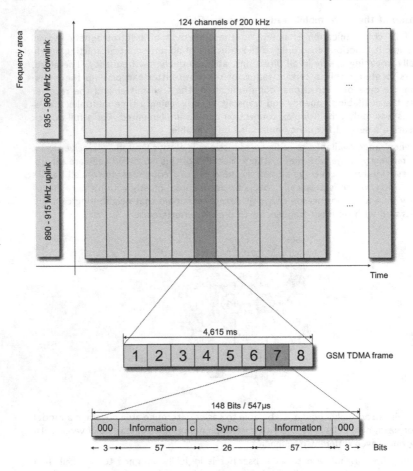

Fig. 6.71 FDMA and TDMA components in the GSM network.

Among these are the encoding and decoding of language information for the telephone service and, possibly, the necessary adaption between the data transmission rates of different network components. Included in the BSS is a transmission and receiver station (Base Transceiver Station, BTS), which encompasses
all radio-engineering facilities, e.g., antennas, signal-processing components for
transmission and amplification. Each GSM cell has a BTS, which together with
one of the BSS assigned Base Station Controllers (BSC) is connected with an
A interface. This is capable of transmitting up to 30 simultaneous connections
of 64 kbps each. Each GSM cell can cover an area of approx.100 m to 35 km,
depending on the degree of development and amount of radio traffic to be dealt
with.

• **Network and Switching Subsystem (NSS)**
The mobile connections network represents the core of the GSM network and

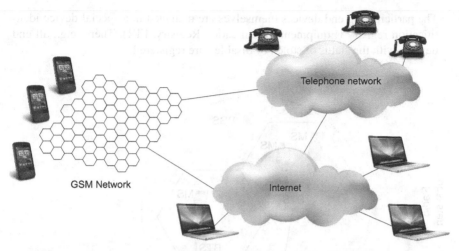

Fig. 6.72 Embedding of the GSM network in the public network structure.

connects the wireless network with the public wired networks. In addition, it is responsible for the connection of the base station systems (BBS) with each other and provides functions for worldwide tracking, roaming and accounting from subscribers between different network operators in different countries. The Mobile Service Switching Centers (MSC) are the central hubs forming the backbone of the GSM networks. They connect all of the base station systems (BSC) with each other. Special gateway MSCs connect the GSM netwwork with other public networks. Every service exchange is connected to a Home Location Register (HLR) that keeps the local subscriber data in readiness, e.g., telephone numbers, available services, an international subscriber identity together with dynamic data such as the current location of the subscriber. Foreign subscribers are managed in the Visitor Location Register (VLR). Every time a subscriber enters a new cell, all subscriber-relevant data is copied from the responsible Home Location Register into the current Visitor Location Register.

- **Operation Subsystem (OSS)**
  The operation and maintenance system provides the necessary services for all participants in the GSM network. The entire operation is controlled by the Operation and Maintenance Center (OMC), which controls all of the other network components. Numbering among the duties of the OMC are the monitoring of data traffic, collecting of statistical data, security management, subscriber management and billing and invoicing. Additionally, an Authentication Center (AuC) is responsible for paying attention to the special security situation of radio-based networks and guaranteeing the identity of the subscriber as well as handling the security of the data transmission. Accordingly, the AuC is usually located in an particularly protected area. The AuC maintains all of the necessary keys and generates all of the parameters required for subscriber authentication in the home location register.

The participating end devices themselves are managed in a special device identification registry (Equipment Identification Registry, EIR). There, e.g., all end
devices with the status of stolen or disabled are registered.

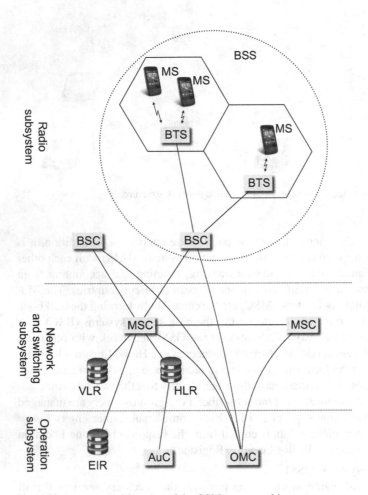

Fig. 6.73 Structure and function of the GSM system architecture.

Because the bandwidth of data transmission in GSM is limited to 9,600 bps (with
some network operators to 14.4 kbps), different procedures were developed to get
around this restriction. Nevertheless, these data transmission services are no longer
sufficient for today's Internet use. When the GSM network was developed, no one
ever imagined the huge popularity and related growth of Internet services versus
voice services. There are only two possibilities to still be able to raise the power
capabilities of GSM without directly intervening in the implemented modulation
procedure and data structures:

- **Channel bundling**
  As GSM offers a connection-oriented service and a bandwidth of 9.6 kbps per connection, more channels can simply be bundled into one connection
- **Packet-oriented data transmission**
  Through a departure from a connection-oriented service paradigm to a wireless packet-oriented service, higher data transmission rates can generally be achieved.

The following data transmission procedures were developed for GSM in order to raise its performance capability:

- **HSCSD (High Speed Circuit Switched Data)**:
  Here, through the bundling of multiple GSM channels, bandwidths of up to 57.6 kbps can be reached.
  Theoretically, through the merging of all 8 time slots of the TDMA frames, it should be possible to reach 115.2 kbps, however, this would assume that the end device (MS) is also in a position to transmit and to receive. While in comparison to the GSM, the HSCSD shows a considerably higher data transmission rate, it is ineffective for the typical communication situation in the Internet. By retaining connection-oriented channels, the entire available bandwidth is permanently occupied. Data transmissions in the Internet, however, mostly takes place in batches, i.e., phases of high communication activity are interrupted by longer phases of less activity. Most of the time, the reservation of a channel is connected with direct costs, as reserved channels cannot be used by other subscribers. For this reason, HSCSD is not well-suited for data transmission in the Internet.
- **GPRS (General Packet Radio Service)**:
  A combination of packet data transmission and channel bundling allows a maximum bandwidth of up to 170 kbps. The design as a packet-switched network allows a more efficient use of the channels, which in each case are only occupied when there is in fact something to send. For the end user the advantage of this wireless network access is seen in the fact that costs are only incurred for the actually transmitted data volume. In principle, a permanent "network connection" always exists (always-on), and does not first have to be established before a data transmission in the form of a dedicated connection. For a GPR channel, between one and eight time slots can be occupied in the TDMA framework. The difference to HSCSD is found in the fact that the occupation of time slots does not proceed based on a prescribed pattern, but takes place based on need.
- **EDGE (Enhanced Data Rates for GSM Evolution)**:
  This is a further development of GPRS and HSCSD, which currently supports up to 220 kbps, retaining the entire mobility broadband. With EDGE, the GSM data services, GPRS and HSCSD, are expanded to E-GPRS (Enhanced GPRS) and ECSD. EDGE achieves the increase in data transmission rate by switching to a more efficient modulation procedure. Instead of implementing the GSMK procedure, standard in GSM, an 8-PSK procedure (Phase Shift Keying) is used. A data transmission rate of up to 59.2 kbps per TDMA time slot is possible. The upper limit in bundling all eight time slots is theoretically 473 kbps. To enable compatibility with GSM/GPRS as well, the change to the 8-PSK modulation procedure

proceeds only on the frequency channels occupied by EDGE-capable end devices. This allows a simultaneous, disturbance-free use of GSM/GPRS and EDGE in the same cell. EDGE supports different quality levels and assigns the data to be transmitted to different service classes. The lower end of this class is the **conversational data** service class. While this enables an isochronous transmission of data, because of its isochronous properties it does not allow for repeated transmission in the case of error. This means that only one data transmission, potentially containing errors, can be performed. Small errors and disturbances can, however, be tolerated in the transmission of voice and video data. At the upper end of the scale is the service class known as **background data**, which guarantees a reliable data transmission. If a transmission error is detected, then via an **ARQ** procedure (Automatic Repeat Request error correction procedure), the faulty data packet is requested again from the transmitter. This service class is suitable for the transmission of files or emails.

Universal Mobile Telecommunication System – UMTS:

The Universal Mobile Telecommunication System is a third generation mobile radio standard. It reaches considerably higher data transmission rates than those of the preceding standard, based on the second generation of GSM (GPRS, HSCSD, EDGE). Already in 1992, the World Radiocommunication Conference (WRC) decided to reserve a 230 MHz wide frequency band, in the range of 2 GHz, for the coming mobile radio standard of the third generation (3G, see Fig. 6.74). The intention was to establish an entire family of high-speed radio technologies under the designation IMT-2000 (International Mobile Telecommunications 2000). These should be designed for both mobile telephone services as well as for data communication and multimedia services. In 1998, the regional telecommunications organizations of Europe, the U.S., Korea and Japan merged to form the consortium of the 3rd Generation Partnership Project (3GPP). Their goal was to achieve a worldwide coherence of the 3G development. The first mobile radio system of the third generation worldwide was established in Japan by the telephone company NTT DoCoMo, and given the name FOMA (Freedom of Multimedia Access). In Europe, experimental operation began in 2002 with the introduction of full-coverage in 2005. The auctioning of license rights, in July and August 2000, for 20-year limited use of the UMTS frequency range generated even more interest in the new mobile standard. The Federal Republic of Germany raised 98.8 billion German marks (approx. 50 billion euros) with this effort.

In the course of its development, UMTS went through different established versions (Releases), determined by the standardization committee 3GPP . While the first Release could be identified by its corresponding year, the following steps in the evolutionary process were consecutively numbered beginning with Release 4. Every new Release contains new possibilities, and as such improves on the preceding version.

What are the capabilities of 3rd generation mobile radio systems?

Based on a definition of IMT-2000, mobile radio systems of the third generation offer general language and broadband multimedia communication for the mass market. The following special features are among the specifications of the 3G mobile radio networks.

- Broadband communication, i.e., optimal data transmission rates of 2 Mbps and more
- Circuit-switched language services with the additional services known from ISDN, based on GSM
- Further development of the packet-switched data transmission services (3G GPRS) from GPRS
- Permanent access to the Internet (always-connected and anywhere/anytime)
- Provision of different Quality-of Service classifications (QoS) for various services
- IP-based multimedia platform with separated interfaces for user data and signalizing based on H.323 (ITU-T), as well as Session Initiation Protocol (SIP, IETF).
- Convergence of fixed and mobile radio networks, public and private networks, as well as a fusion of voice and data communication.

Further reading:

Cox, Ch.: Essentials of UMTS, Cambridge University Press (2008).

Fig. 6.74 What are the capabilities of mobile radio systems of the 3rd generation?

- **Release 99**

  Contains the specifications for the first level from UMTS. As a clear progressive development, when compared with GSM, the new access network (UMTS Terrestrial Radio Network, UTRAN) is specified. The time and frequency multiplex procedure already existing in GSM have been replaced by a code multiplex procedure (wideband CDMA). A data transmission rate of max. 384 kbps in the downlink and 64 kbps in the uplink were implemented for Release 99. The subsequent core network on the UTRAN interface involves the already existing GSM/GPRS core network, where additional interface definitions enable a shared operation.

- **Release 4**

  The most important innovation in Release 4 is the Bearer Independent Core Network (BICN). Here, formerly circuit-switched voice and data services in the core network are forwarded via ATM or IP data packets, instead of via time slots. This step was removed from the goals of the network operators in merging circuit-switched and packet-switched core networks. Because there was a continual decrease in the number of circuit-switched voice connections, as compared to packet-switched data connections, cost advantages could be realized.

- **Release 5**

  While in Release 4 only the core network was converted to packet-oriented switching, with Release 5 a further step in the direction of All-IP networks was undertaken. Now, voice connections could also be transported from end subscriber to end subscriber in the area of the UTRAN over the Internet Protocol (IP). Up to now the data connection had been briefly interrupted when switching cells.

Yet, in Release 5, the network management for the handover in the air interface is no longer taken over by the end device but by the network itself. This enables a interruption-free handover. Additionally, a new data transmission procedure, (**High Speed Download Packet Access**, **HSPDA**) was introduced. With the help of new modulation procedures and code bundling, data transmission rates of up to 14 Mbps in the downlink are possible.

- **Release 6**
  In addition to the available HSPDA in Release 5, in Release 6 the **High Speed Upload Packet Access** (**HSPUA**) data transmission procedure was introduced for the uplink. It enables data transmission rates of up to 5.76 Mbps.

- **Releases 7, 8, and 9**
  There were further improvements in the Releases that followed, and the targeted transmission rates made possible by the introduction of new modulation procedures. In Release 7, **Evolved High Speed Packet Access** (**HSPA+**) with 64-QAM modulation and the bundling of multiple antennas (**Multiple-Input Multiple-Output**, **MIMO**), theoretically offers data transmission rates of up to 56 Mbps in the downlink and 22 Mbps in the uplink. In Release 9, a doubling of the frequency area, **Dual-Cell HSPA** (**DC-HSPA**), makes it possible to theoretically raise this capacity again to approximately 84.4 Mbps.

UMTS was introduced parallel to the existing GSM mobile radio networks, and retains existing elements and infrastructures. The prescribed frequency range for UMTS joins the area used by GSM-1800 of 2 GHz . Fig. 6.75 shows an overview of the frequency bands in this area employed in Europe. For the uplink, the area between 1.710 and 1.785 Mhz is used by GSM-1800, and for the downlink the range between 1.805 and 1.880 Mhz (also designated DCS-1800, Digital Cellular System). Located between 1.880 – 1.900 MHz is the DECT area (Digital Enhanced Cordless Telecommunications), used by wireless landline telephones The frequency range from 1.900 – 1.980 Mhz is provided for the UMTS uplink (UTRA-TDD and UTRA-FDD), and 2.010 – 2.025 MHz and 2.110 – 2.170 MHz for the UMTS downlink. The range in between 1.980 – 2.010 MHz, and the subsequent 2.170 – 2.200 MHz, is reserved for future satellite-supported communication (MSS, Mobile Satellite Service).

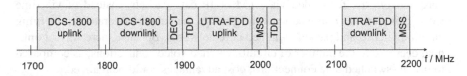

Fig. 6.75 Frequency distribution for the UMTS telecommunication in Europe.

In a greatly simplified form, the UMTS reference architecture may be divided into three areas (see Fig. 6.76):

- **User Equipment** (UE):
  The user equipment is a part of the user equipment domain, which is specifically assigned to a single end user and provides all necessary functions for access to the UMTS service. In addition, are the USIM domain (UMTS Subscriber Identity Module), which keeps the SIM ready for UMTS and provides functions for encryption and authentication of end users, and the mobile equipment domain (end device domain). This provides the end user all functions necessary for radio transmission and all interfaces.

- **Universal Terrestrial Radio Access Network** (UTRAN)
  The UTRA network provides functions for the management of mobility and contains multiple radio subsystems (Radio Network Subsystems, RNS). Among the tasks of a RNS are the encryption of a radio channel, the control of connection transfer (handover) when changing cells, and the management of radio resources. A Radio Network Controller (RNC) is located in a RNS. This controls multiple base stations (B node) and parallels a GSM base station controller. All radio resources of the connected bases station are managed via the RNC. To enable that the handover between two cells is as error-free and comfortable as possible (soft handover), the RNC communicates with its neighbor. This makes it possible to merge radio signals from multiple base stations. Conversely, multiple base stations can also transmit the same radio signal, thus raising the quality of reception and avoiding connection loss (macro diversity). The UTRAN is connected over the air interface U_u with the mobile end device (user equipment). Communication of UTRAN with the core network is carried out over the I_u interface. All terrestrial interfaces are based on the ATM technology. The UTRAN defines the access network domain, which is part of the infrastructure domain.

- **Core Network** (CN)
  The core network provides all mechanisms for a connection handover to other systems or interfaces to other fixed networks or mobile systems. Furthermore, the core network has functions ready for the management of the current location, in case there is no explicit connection between the user equipment and the UTRAN. The core network domain is the second component of the infrastructure domain. Functionally, it is made up of the service network domain, containing all of the functions needed by an end user to gain access to the UMTS services, the home network domain, where all functions and information regarding the home network of an end user are managed. And finally, the transit network domain, which manages the potentially required connection from the service network to the home network.

For the so-called air interface (U_u) exists two different procedures to facilitate "direction separation." This is necessary to enable a full-duplex operation, meaning simultaneous transmission and reception, between the sender and the receiver. The **Frequency Division Duplex** (FDD) implements two different frequency ranges for uplink and downlink. Accordingly, data traffic between mobile end device and base station uses two different frequency bands for transmission and reception. This was a technique already used in GSM. Besides this, UMTS offers the possibility of **Time**

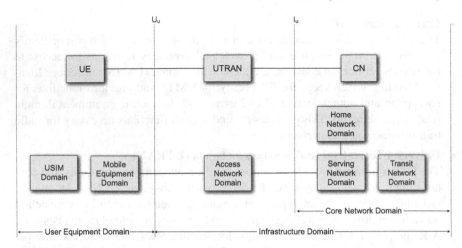

Fig. 6.76 Main components of the UMTS reference architecture.

Division Duplex (TDD). Here, transmission and reception channels use the same frequency, but work separately from each other time-wise. Information is transmitted with the help of a pre-determined timer, time-delayed and in short sequences. The switching between the transmission and receiving mode transpires so quickly that it goes unnoticed by the user. This technology is implemented as well in wireless landline telephones (Digital Enhanced Cordless Telecommunication, DECT). In contrast to GSM or DECT, for simultaneous multiple use (Multiplexing) of the UTMS frequency range a code multiplexing procedure is implemented (Code Division Multiple Access, CDMA, see also section 3.30). In order to distinguish between individual users of the frequency band, their data streams are coded with the help of special, if possible orthogonal, spreading codes. This makes it possible to gather the original data again on the receiver's side through a separation process, by means of a new correlation with the respective spreading code sequence.

Currently, the UMTS network operators are constructing their networks in the UMTS-FDD mode, until a geographically extensive coverage is possible. Using this procedure in the downlink, a data transmission rate of max. 384 kbps can be achieved. In the UMTS-TDD mode, a frequency band is divided into 15 time slots, whose entire transmission length is 10 ms. Every time slot is split into multiple radio channels over CDMA. Aided by the W-CDMA spreading code procedure (Wideband CDMA), data transmission rates of up to 1.920 kbps may be reached in the downlink.

As with other radio networks, the UMTS network is structured geographically hierarchical. Every cell is supplied with radio signals from its own base station and holds a limited number of communication channels. The smaller the size of the cell, the higher the infrastructure costs for the network operator, but also the greater network capacity can be reached. The GSM network uses smaller cells in metropolitan areas than it does in the country. Every one of these cells is assigned multiple frequency bands in GSM. Although due to the FDMA multiplex procedure used in GSM, it is necessary to make a clear distinction between the frequency channels of neighboring cells to avoid reciprocal interference. As signal attenuation for higher frequencies is stronger, cells for the GSM-900 (900 MHz) system can be larger than cells for the GSM 1800 (1.800 MHz) system. UMTS uses frequencies in the range of 2 GHz and also requires the installation of small cells, based on higher signal attenuation. In UMTS all of the neighboring cells can be operated in the same frequency range. This is because user separation is implemented via a code multiplexing procedure (W-CDMA). To make wide-coverage UMTS possible, cells of different sizes with stepped-performance data are structured hierarchically, overlapping (see Abb. 6.77).

Picocells

Microcells

Macrocells

World Cell

Fig. 6.77 Radio cell structure in the UMTS network.

The following cell types can be distinguished:

- **World cells**
  This cell type, with a planned future implementation, enables world-wide coverage through UMTS via satellite connection, particularly in thinly populated regions, such as desert or ocean areas. World cells are implemented by way of an independent radio network.

- **Macrocells**
  Macrocells likewise form an independent radio network, with a spatial extension of up to 2 km. They provide extensive UMTS coverage for a larger area with reduced data transmission capacity. The data transmission rate in macrocells is limited to 144 kbps, whereby the mobile subscriber within the cell is able to move with a maximum speed of up to 500 km/h.

- **Microcells**
  In contrast to macrocells, microcells have a maximum radius of about only 1 km and can therefore ensure higher maximum data transmission rates of up to 384 kbps. However, because the construction of microcells is associated with high infrastructure costs, they mainly find implementation in heavily popula-

ted metropolitan areas. For the end user, mobility with a maximum speed of 120 km/h is possible.

- **Picocells**

  Picocells form the smallest cell unit, with a maximum diameter of 60 m. Within this narrow geographic area they enable data rates of up to 2 Mbps and more. Picocells are used in so-called "hotspots," where there is a need for higher transmission speed and where the necessary investment for the network operator pays off. Picocells are normally utilized in the indoor areas, e.g., at airports, in train stations, conference centers, business parks or stock exchanges. In comparison to macrocells and microcells, the maximum speed at which an end user can move without losing the connection is limited to walking speed, i.e., max. 10 km/h.

Long Term Evolution – LTE

The succeeding generation built on UMTS is designated Long Term Evolution (LTE). It is also known as **4G** or **Next Generation Mobile Networks** (**NGMN**). Special performance characteristics of this technology can be identified such as its higher data transmission rate, of up to more then 100 Mbps (326.4 Mbps as the theoretical maximum data rate in the downlink and 86.4 Mbps in the uplink), in connection with a very low latency time of less than 10 µs. This makes the technology especially attractive for isochronous multimedia data streams and Internet telephony. The implemented end devices remain permanently connected with the mobile radio network or with the Internet, without a disconnection necessary for reasons of infrastructure and in cases of mobility ("always on"). 4G technologies, just as IEEE 802.16 WiMAX, are based on an OFDM (Orthogonal Frequency-Division Multiplexing) modulation procedure. However, unlike WiMAX, the specification of the respectively used frequency bands is carried out adaptively in relation to the respective requirement (frequency bandwidths of 1.25 MHz to 20 MHz). In this way, an efficient use of infrastructure is made possible. The standardization of 4G technologies has been in operation since 2005 by the standardization committee, Third Generation Partnership Project (3GPP). Its commercial implementation started in 2010. A successor of LTE technology is already in the planning phase. It goes by the name LTE-advanced, and supports data transmission rates in the range of 1,000 Mbps. The extensive geographical structuring of the mobile radio network is also carried out hierarchically, with small, powerful cells up 5 km in diameter and cells up to 100 k with reduced performance. The operation of LTE is carried out in coexistence with the already existing 3GPP standards: GSM, GPRS, EDGE, UMTS, HSPA and eHSPA.

LTE works in the frequency areas of 800 MHz and 2.6 GHz. In the area of 2.6 GHz, frequency ranges are reserved for the operation via frequency multiplexing (FDD) as well as for operation via time multiplexing (TDD). For the FDD operation, a paired frequency space (operation band 7) is reserved with 70 MHz each, which several network operators are required to share. For the uplink, the frequency range from 2500 MHz to 2570 MHz is available, for the downlink the frequency range from 2620 MHz to 2690MHz. Based on the low bandwidth (for this operational mode, UTMS has likewise just over 60 MHz available), it is unlikely that a network operator could operate in this mode with the full bandwidth of 20 MHz. For the TDD operation, a total spectrum (operation band 38) of 50 MHz is reserved in the 2.6 GHz-area. This extends from 2570 MHz up to 2620 MHz.

To be able to achieve high data transmission rates, the modulations 4PSK, 16QAM– 64QAM are supported by LTE in the downlink with the simultaneous application of MIMO (currently 2 transmission and reception antennas each, in a later expansion level up to 4). Moreover, in the development of LTE importance was placed on a simple as possible (flat) network architecture. There should be few components to define in order to keep the complexity of the network infrastructure as low as possible. In doing so, costs could be cut and time delays in the switching process minimized (latency time). This applies to the radio network (Evolved UTRAN, E-UTRAN), as well as to the actual core network (Evolved Packet Core, EPC), which is completely designed as a packet switched network. This meant that line-switched components for voice and real-time data were deliberately omitted. These are now handled entirely over the Internet protocol (Voice over IP, VoiP).

6.5.5 Alternative Access Procedures

The **broadcast cable network** has been used for many years to transmit radio and television signals. It is therefore possible to build on an existing infrastructure that has nearly complete coverage. The cable networks were originally broadcast systems in which the implemented amplification elements were solely designed for forwarding the signal from the direction of provider to end user. To enable interactive applications or bidirectional data exchange, the cable network must have a so-called **return channel**. In a few cases this was realized over the existing telephone network, however, the end user did not have any real access advantage over a telephone-based network. The necessary modernization - ongoing for many years - is complex and expensive. A return channel in a cable network can be implemented as either a separate data line or a specific frequency splitting procedure. In the latter case, a certain frequency is used for each one of the two directions.

Most of the time, copper coaxial cable (**broadband**) served as the cabling basis. However, when it is necessary to replace legs in the line, fiber-optic cable is now used. The transmittable frequency range of analog, copper coaxial cables is separated into many channels, each serving its own range. This means that many different television and radio stations can also be operated as one or more data channels.

Because only analog signals can be forwarded, digital transmissions must be transformed beforehand by means of a modem. To gain access to the cable network, the end user needs a special **cable modem**. This was specified in the ITU-T/CCITT recommendation **DOCSIS** (Data-over-Cable System Interface Specification) and adopted by the American ANSI. Cable modems can be connected to the computer at the end user via an Ethernet network adapter. In contrast to analog modems, cable modems partially take over even such tasks as network management and diagnostics, which are actually in the router's sphere of responsibility.

Furthermore, they are frequency-agile, i.e., they look for the most suitable channel in an assigned frequency range and adjust themselves to it. Thus, frequency ranges can be used in such a way that a greater transmission capacity is available in the direction of the subscriber (downlink) than in the opposite direction (uplink). An optimal frequency assignment is important as the cable network takes the form of a tree topology. All subscribers connected to such a tree must share the frequency range reserved for data transmission.

Due to the tree structure of the cable television network, different modulation procedures are used in both the transmission and receiving directions. In the downlink, frequency ranges above 450 MHz are implemented with a quadrature amplitude modulation (QAM), in the uplink, according to Euro-Docsis 2.0, a frequency range from 10–65 MHz is used with quadrature phase-shift keying (QPSK). The bandwidth reached in a cable network can in the downlink (i.e., from the provider to the end user) be as much as 120 Mbps. While on the return channel (uplink) it can reach about 5 Mbps.

Paralleling the prevalence of cable television networks, most of the countries in the world have an extensive **electricity supply network** at their disposal. The condition for using this type of infrastructure for data transmission is the ability to guarantee a sufficiently high bandwidth, as well as the guarantee of electromagnetic compatibility with the other types of use employed in this frequency spectrum. Energy-producing companies themselves have used high voltage (110–380 kV) and medium voltage networks (10–30 kV) for a long time to transmit data and for control and monitoring purposes. Also low voltage networks (to 0,4 kV), which ensure extensive access to the electricity network, can be used as an access network for the general data networks. The low voltage network supplying individual households has a tree structure with a low voltage transformer station as root node. It can therefore function as a shared transmission medium. In the low voltage transformer station, specially setup head stations serve as interfaces to the backbone of a data network. Data transmission via the electricity network proceeds over modulation of the user signal, with the help of the carrier frequency system. (PowerLAN, Powerline Communication). However, the use of electrical supply lines also entails overcoming technical difficulties. Among these are:

- Energy intensive processes can result in disturbances in data transmission. The tolerance threshold must be set accordingly, so that occurring transmission errors never surpass a prescribed maximum limit. It is otherwise impossible to ensure a reliable data transmission.

- Very high attenuation effects occurring within the low voltage network need to be compensated.

- The very low, and generally strongly fluctuating, performance impedance of the low voltage network must also be compensated.

- Broadband data signals have to be shielded from strong radiation effects. This could otherwise lead to a disturbance of other services sharing that particular frequency.

The frequency range used for Powerline Communication is between 1–30 MHz. Data transmission rates between subscriber and the forwarding transformer station is possible between 1.5 Mbps and 205 Mbps. This range must be shared by the network subscribers who are actively connected to the transformer station. Wide-coverage broadband Internet access is usually possible only in (densely) populated areas. Without the suitable terrestrial infrastructure, a setup is complex and expensive in cases where the cable television or electricity network is not available already. **Satellite networks** remain a possible alternative worldwide in thinly populated and remote areas. Satellite-supported broadband Internet accesses are implemented over geostationary satellites, whose entire radiation area (footprint) can be used. There are two basic variations of network connections via satellite:

- **2-way satellite connection (pure satellite connection)**
  The transmission route via satellite is used in both directions, from transmitter to receiver and back, in this variation. In the downlink, the user has at its disposal data transmission rates between 64 kbps und 5.12 Mbps, depending on the provider. In the opposite direction (uplink), the data transmission rates are currently between 64 kbps and 1 Mbps. Higher data transmission rates could also be technically possible here. Satellite connections of this type are often used as an additional redundant data connection because of their independence from a terrestrial network infrastructure and consequently failsafe performance.

- **1-way satellite connection (satellite connection with terrestrial return channel)**
  Here, only the connection route from the Internet to the end user (downlink) is implemented as a satellite connection. The reverse direction (uplink) is carried out over the telephone, cable television or mobile radio network. The end user currently has at its disposal - depending on the provider - data transmission rates of between 256 kbps and 2,048 kbps. The advantage of a 1-way satellite connection with terrestrial return channel, as compared to the 2-way satellite connection, is the lower costs for satellite transmission and the simpler and therefore less expensive end devices.

Because of the great distance to be bridged (in each case 36,000 km with a satellite in the geostationary orbit), using satellite transmission as an access network results in high latency times. Signal propagation time from the terrestrial station to the satellite and back is at least 239 ms. The route must be run through twice with bidirectional communication. Combined with further delay factors, the result is a latency time of between 500 and 700 ms. In comparison, the normal latency time for DSL is only around 20 ms.

With data transmission in the Internet via the Transmission Control Protocol (TCP, see Chp. 8), further technical problems occur as a result of high signal run times in satellite transmission. The Round Trip Time (RTT) results from the signal runtime for the duration of the signal's route from transmitter to receiver and back again. It leads to the consequence of the TCP connection, following the so-called slow start, does not raise the data transmission rate anymore to any significant degree. In unfavorable conditions this means that only a fraction of the theoretically available data transmission rate can be used. This problem can be avoided by employing special proxy servers.

6.6 Glossary

Access network: Generally-speaking, most end users do not have their own direct access to a wide area network (WAN). Due to the high costs involved, this remains the privilege of large companies or scientific and military organizations. The end user gains access to the WAN, most of the time, via an already existing access network, e.g., the analog telephone network.

ARPANET: The packet-switched WAN was developed at the end of the 1960's by the Advanced Research Projects Agency (ARPA) for the U.S. Department of Defense (DoD). The ARPANET is considered a precursor of the global Internet.

Asynchronous Transfer Mode (ATM):   From the ITU for wide area networks with high bandwidth standardized data transmission procedure. It is based on the principle of asynchronous time multiplexing. ATM is also used in LANs. The basic unit for information transmission is the so-called ATM cell. This is a data packet with a fixed length (53 bytes). Virtual connections can be permanently switched between two linked ATM end systems or established, as needed. ATM defines multiple service classes for the support of multimedia applications, among these are an isochronous mode, which emulates the behavior of a line-switched connection and is used for the transmission of video and audio data streams.

Bandwidth: The bandwidth of a connection route in a network is a physical size, which is given in hertz (1 Hz = 1/s). In the analog range the bandwidth indicates the frequency area in which electrical signals are transmitted with an amplitude decrease of up to 3 dB. The bigger the bandwidth, the more information can theoretically be transmitted in a time unit. The term "bandwidth" is also often used in connection with the transmission of digital signals, while here the **transmission rate** is meant. Nevertheless, there is a direct connection between bandwidth and transmission rate. The transmission speed reached in data transmission depends directly on the bandwidth of the network. The maximum bandwidth utilization for binary signals is 2 bits per hertz bandwidth.

Bit stuffing: In synchronous data transmission, the clock of transmitter and receiver are synchronized with each other by means of a so-called synchronization bit at the beginning of a data packet. The synchronization bits form a fixed bit pattern and are ordered in such a way that they can never be mixed up with the payload information. If e.g., 01111110 is used as a synchronization bit sequence, care must be taken that no sequence of six consecutive ones could be identified in the payload as a synchronization bit sequence. Because of this, after five ones a zero is added by the transmitter and then removed again by the receiver.

Broadcasting: In the broadcasting procedure a transmission is made to all connected systems in the network at the same time. If, in contrast, there is a 1:1 connection with only one communication partner, the procedure is designated as an unicast transmission.

Default routing: To avoid the occurrence of multiple identical entries in routing tables, every router is assigned a **default route**, and the corresponding entry in the routing table is simply deleted. If an entry does not exist in the routing table for a particular destination, the default route is used to forward the relevant data packet.

Digital Subscriber Line (DSL):   Digital access procedure for wide area networks. DSL enables high bandwidths for the connected end users; it is at the same time subject to strict length and quality restrictions regarding the implemented network infrastructure.

Dijkstra algorithm: Named after Edsger W. Dijkstra, it is the algorithm for determining the shortest path from a specific start node in a graph. The Dijkstra algorithm is employed in the routing procedure. It is used to attempt to find the shortest connection path from transmitter to receiver.

Distance vector routing: A routing procedure in which all routers in the network receive their routing information through exchange with the directly adjacent router. It is assumed that every router is capable of establishing the distance to its direct neighbor itself. The routing information determined this way is subsequently exchanged with all direct neighbors. This continues, in a multi-stage process, until, finally, all routers in the network have received the complete routing information. Distance vector routing, a very simple method, was employed as the first routing procedure in the ARPANET.

Distributed Queue Dual Bus (DQDB):   DQDB is a technology concept for MANs, based on two parallel, counter-powered buses. End systems wishing to transmit receive the access right to a bus by means of a divided queue. They can then transmit cells of a constant length. DQDB is efficient when there is a heavy load, but not always fair. DQDB networks can extend over several hundred kilometers.

Flooding: Isolated routing method in which an incoming data packet is forwarded by a router to all connections, except for the one by which the data packet reached the router. To avoid overburdening the network, the life-time of a forwarded data packet can be limited through different procedures. Although this method produces a high redundancy, it always finds the shortest route.

Frame Relay: A connection-oriented, packet-switched network technology that was developed in the mid-80s especially for wide area networks. Frame Relay is based on an asynchronous time multiplex procedure. While transmission errors are detected, they are not automatically corrected. In a Frame Relay network, virtual connections are set up between the communicating end systems.

Global System for Mobile Communication (GSM):   Mobile radio system of the second generation with currently the largest distribution worldwide. GSM is based on individual cells, whose expansion varies depending on subscriber density. GSM operates in the time multiplex procedure and offers different carrier services for data communication. These are based on line or packet-switched methods on individual channels, each with data rates of up to 9,6 kbps.

Hop: Designates the leg of a route from one end system to the nearest switching computer, or between two adjacent switching computers, or from the switching computer to a connected end system. A route through a network, from a transmitter-end system to a receiver-end system, is made up of multiple hops.

Hot-Potato routing: An isolated routing method in which the router tries to forward an incoming data packet on the quickest path. The forwarding connection of the router is chosen with the shortest queue. The forwarded data packet is often forced to accept detours with this procedure.

ISDN: The WAN technology, Integrated Services Digital Network, integrates various services such as voice, data, and image communication in a single network. ISDN provides channels with 64 kbps each. ISDN was developed as a substitute for the old analog telephone network and has been expanded with worldwide, extensive coverage since the mid-80's.

Isolated routing: Designation for a routing procedure that works locally and does not exchange information over the network topology with its neighbors. Examples for this method are flooding or hot potato routing.

Link-state routing: Decentralized, adaptive routing procedure (also known as SPF routing/ Shortest Path First routing), in which every router sends the routing information it can determine about its direct neighbors to all other routers in the network via a broadcast With the help of this information they are constantly kept informed about the current topology of the network. Routes are calculated in link-state routing with the help of the Dijkstra algorithm.

Local Area Network (LAN):   Spatially restricted computer network that can accept only a limited number of end devices (computers). A LAN enables an efficient and fair communication of all end systems connected to it. As a rule, all of the connected computers additionally share a transmission medium.

Local Loop: This is a designation for the subscriber line from the local exchange to the end user. The distance to be bridged to the local exchange is also known as the **last mile**, and it is implemented over a dedicated transmission medium (in contrast to a shared LAN medium).

Mobile routing: A special form of routing for networks with mobile network nodes. A mobile network node can move freely through different wireless networks (WLANs). This is why a router must first determine its location before a data packet can be forwarded to the mobile network node. The procedure used (**mobile IP**) assumes that the mobile node always has a fixed assigned address in a home network. If the mobile network node enters a foreign network, it must register there at a special router (**foreign agent**). This establishes contact to a router in the home network of the mobile node (**home agent**). Together they organize the correct forwarding of the data packet addressed to the mobile node

Modem: For data transmission over an analog telephone network, a modem must be employed to facilitate the transformation of digital signals into analog signals and back again. The frequency spectrum of the original digital signal is moved into the frequency range of the analog telephone network with so-called **modulation**. This shift is undone with **demodulation**.

Multicasting: In a multicast transmission, a source transmits to a group of receivers simultaneously. This involves a 1:n communication. Multicasting is often used for the transmission of multimedia data.

Next hop forwarding: This is a switching variation in a packet-switched network. Here, the relevant switching computers do not store information about the entire route that a data packet must cover to another end system, but only for the next leg to be undertaken (hop).

Packet switching   The dominant communication method in digital networks. The message is segmented into single data packets of a fixed size and the packets are sent individually and independent of each other by the transmitter to the receiver. A distinction is made between **connection-oriented** and **connectionless** (datagram network) packet switching networks. In connection-oriented, packet-switching networks, before the beginning of actual data transmission a connection is set up over fixed, selected local exchanges. In connectionless packet switching, in contrast, no fixed connection path is established

Plesiochronous Digital Hierarchy (PDH)   The PDH defines a synchronous time multiplexing-based WAN technology, whereby signals from different sources are sent over a shared channel. The bandwidth of the different channels are, however, in reality not exactly the same, rather nearly the same (thus the designation from plesios=[Greek] close). Because this is a synchronous procedure, differences must be balanced out. This is possible with so-called **bit stuffing**.

Router: A switching computer that is able to connect two or more subnetworks with each other. Routers operate on the transport layer (IP layer) of the network and are able to forward incoming data packets on the shortest route through the network based on their destination address.

Routing: In a WAN there are often multiple switching elements along the path from the transmitter to the receiver. These take over the transfer of the data sent to the corresponding receiver. The determination of the correct path from transmitter to receiver is called routing. Dedicated local exchanges (**routers**) receive the transmitted data packet, evaluate its address information and forward it accordingly to the designated receiver.

Routing table: Principle data structure of routers containing partial route information regarding forwarding of the incoming data packets. Every target address in the routing table is assigned to a specific exit (port) of the router, via which the next leg - **hop** - of the data packet proceeds to its receiver.

Routing algorithm: A calculation procedure for the creation of the routing table. There is a distinction made between static and dynamic routing algorithms, which can adjust dynamically (adaptive) to changes in the network topology. Routing algorithms can be executed from a central place in the network or from every router itself.

Signaling: Designation for the exchange of all information that is required for the establishment, monitoring and dismantling of connections in the telecommunications network. This task is also known as **call control** or **connection control** . A fundamental distinction is made between **in-band signaling**, in which the signal information is transmitted in the same logical channel as the payload, and **out-of-band signaling**, whereby a separate logical channel is used for the transmission of command and control information.

Spanning tree: The spanning tree of a graph designates a subgraph containing all nodes and edges of the original graph necessary to maintain original accessibility relations without looping. Those parts of the network connections are thereby designated that link all routers and prevent an occurrence of looping.

Store-and-forward switching: Switching procedure in a network, whereby a fixed connection between the transmitter and receiver is not required to be switched. The message received from the transmitter is sent to exchanges in the network where, in each case, it is buffered before being forwarded. A difference is made between **message switching**, in which the transmitter's complete message is buffered before being forwarding to the exchange, and **packet switching**, in which the message from the transmitter is segmented into individual data packets. These are sent over the network independent of each other.

Synchronous Digital Hierarchy: The data transmission technology known as synchronous digital hierarchy presents a synchronous multiplexing based transmission technique. It was especially conceived for data transmission via fiber-optic cable and radio. SDH currently represents the primary standard for networks in the WAN field, and will continue to gain importance in the future.

Universal Mobile Telecommunication System (UMTS): Mobile radio system of the third generation, whose application from 2003 allows data rates of up to 2 Mbps (in virtually steady-state operation) in data communication.

Wide Area Network (WAN): Freely scalable computer network that is not subject to spatial or capacity-related limitations. Individual partial networks are connected by routers, which coordinate data transfer in the WAN. The WAN technology supplies the foundation for **internetworking**.

WiMAX: Under the name Worldwide Interoperability for Microwave Access, and standardized as IEEE 802.16 wireless broadband wide area network technology, it presents a broadband radio transmission system technology for mobile and stationary end devices. In contrast to the other IEEE 802 standards for LANs and PANs, WiMAX encompasses a much larger range of up to 50 km (wireless MAN), with data transmission rates of 70 Mbps and more.

X.25:   A packet switching technology for wide area networks that was standardized at the beginning of the 70s. It enables the construction of multiple, virtual connections between two end systems (within this norm, designated as DTE - Data Terminal Equipment). Considered the first generation of the public data transmission technology, X.25 is still supported today by many network operators, particularly in Europe.

Chapter 7
Internet Layer

„On the Internet, nobody knows you're a dog"
– George Steiner, in „The New Yorker"
(Vol.69 (LXIX)), 1993

Born out of everyday practical need, and far from being considered the ultimate solution, TCP/IP with the Internet Protocol Version 4 (IPv4) represents the core of quintessential Internet technology today. And this even after 30 years. The ISO/OSI reference model for communication protocols, developed parallel by ISO and intended as the universal and final standard in the area of network communication, could not prevail against TCP/IP. While the latter has not always been adequate, it has proven exceptional in its robust practical application and reliable working method. It has enabled the rapid growth of the Internet as the global communication infrastructure of modern society. The next generation of the IP protocol is well equipped to meet future demands with IPv6, which has been waiting in the wings for some time to succeed IPv4. Yet the transition from the old version to the new version has proven more difficult and lengthier than imagined, especially because the actual problem is not to be found in the technical changes alone.

Up to now we have perhaps thought ourselves to be in a homogeneous network infrastructure, imagining that identical protocol can be found on the respective protocol layer of every computer in the network. In reality, a completely different picture presents itself. An unclear number of providers and protocols compete on individual network islands, which in keeping with a common goal are combined together on a higher level to a network infrastructure, or **internet**, presenting an illusion of homogeneity to the user.

Until just a few years ago, the **Internet Protocol (IP)** used here was just one in a group of manufacture-specific protocols, e.g., DECNet, Internetwork Packet eXchange (IPX) or Systems Network Architecture (SNA). However, with the triumphal success of the global Internet, the Internet Protocol evolved into the standard protocol for network comprehensive computer communication. In order to offer reliable communication services via the unreliably working IP, the **Transmission Control Protocol (TCP)** was developed. Today, it forms a cornerstone of the global Internet and will be further discussed in section 8.3. The present chapter examines those

C. Meinel and H. Sack, *Internetworking*, X.media.publishing, 455
DOI: 10.1007/978-3-642-35392-5_7, © Springer-Verlag Berlin Heidelberg 2013

problems that must be solved if two or more, heterogeneous networks are indeed working together on the protocol level. We will also look at how the Internet layer of the TCP/IP reference model is managed by the Internet Protocol, and additional protocols building on it.

7.1 Virtual Networks

From the user's perspective, the Internet seems to be a unified, large and inter-connected network whose components mesh seamlessly with one another. But appearances are deceiving. A vast number of different network technologies and network protocols are involved. Moreover, only an extremely complex protocol software is capable of concealing the details of the physical and logical connections and individual components from the user. The Internet creates such a **virtual network**. With a combination of hard and software, the illusion of a unified network is given, which in fact does not exist in this form.

If multiple networks of different types are to be operated together a problem immediately becomes evident. Principally, only those computers can communicate with each other initially that are connected to the same network. Communication with computers connected to other networks, with different technical and logical parameters, requires the development and provision of appropriate translation and conversion mechanisms. This problem first became evident at the beginning of the 70's, when various large companies began to simultaneously operate multiple and different networks. An urgent demand arose for computer communication across the respective borders of the individual network islands.

In addition, there were other needs, e.g., if different tasks were to be carried out in individual networks requiring a company employee or manager to have access to results in various networks and in geographically distant computers. Driven by these demands, as well as by the awareness of the great productivity and innovation potential, possible solutions were investigated to enable any two computers to communicate with each other when they belonged to different networks.

To solve this problem, a **universal service** should be developed, with which a user from any computer of the company could transmit messages or data to another user at any other computer in the company. In this way, all required tasks could be taken care of by a single computer and a need for separate computers to handle different tasks would be eliminated. With such a universal service all available information could be obtained from every computer in the company.

At the same time, great difficulties must be overcome to implement a universal service as the different LAN network technologies are typically completely incompatible with each other. As already discussed in Chapter 4, LANs are subject to restrictions in regard to their geographic extension as well as their capacity, meaning the number of connected computer systems. Besides that, in each case different LAN technologies have special, individual technical-physical and logical parameters. This makes a "simple" fusion of different technologies completely impossible.

A universal service must therefore be in a position to accommodate these incompatibilities. To connect multiple heterogeneous networks with each other, extra hardware and software is required. The additionally implemented hardware allows the physical coupling of the incompatible network types. At the same time, the extra software has task of installing the universal service on all of the components of the now connected networks. This enables the possibility of transparent communication. The structure created by the fusion of different networks with the help of such a universal service is called an **internet**. Naturally, the global Internet is also such an internet, and to be sure the most famous one.

As discussed in the previous chapter, the individual networks of the worldwide Internet are linked to each other, just as in a WAN, by special packet switches (routers). So that the network system coupled together by the routers also shows its intended uniformity outwardly, the following design directives are observed in the communication protocols acting on it:

- The services offered by the employed communication protocols should be independent of the implemented technologies in the respective routers.

- The subsequent transport layer in the TCP/IP reference model and its communication protocols should be shielded from the technical details of the router communication.

- The addressing scheme provided for the protocols of the transport layer should follow a uniform concept and be implemented across network borders in LANs and WANs.

Building on these few basic principles, there remains the greatest possible freedom for an implementation of services on the Internet layer, based on the available communication protocols.

If multiple heterogeneous networks are connected to each other to form an internet, different combinations of network types can occur. As shown in Fig. 7.1, the following combinations are theoretically possible:

- **LAN–LAN**
  Within the same company, data can be exchanged between different LANs. Thus, for example, an employee in the sales department can request the vacation schedule from someone in the human resource department.

- **LAN–WAN**
  A company employee communicates with a customer who is connected to a distant WAN. From the company LAN a connection to the addressed WAN is established to exchange data.

- **WAN-WAN**
  Two private people, each connected to a WAN over a commercial Internet provider, exchange data.

- **LAN-WAN-LAN**
  Employees of two companies, which each have their own LAN, exchange messages with each other, whereby the LANs involved are connected to each other via a WAN.

The various connection possibilities are each shown in Fig. 7.1 by a separate, dotted line. In order to establish these connections a special connectivity hardware is required in each case. In the illustration, this is indicated by the small boxes between the schematically represented networks. This special connection hardware is necessary to carry out the required physical and logical conversion between the network types involved.

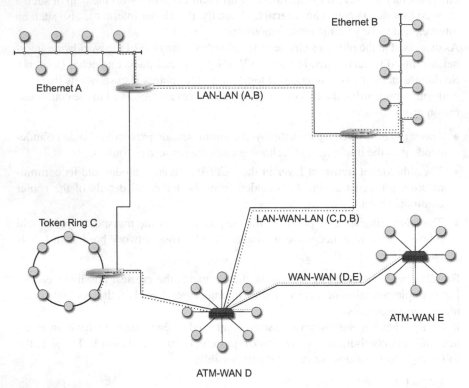

Fig. 7.1 Internetworking.

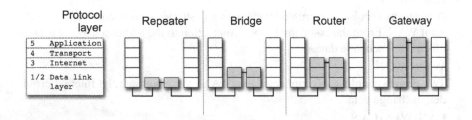

Fig. 7.2 Intermediary and switching systems in the Internet.

The connectivity hardware necessary for connecting two networks receives its designation based on the layer of the Internet protocol stack on which it works (cf. Fig. 7.2). The following terminology is used to designate the various layers:

- **Layer 1**
  So-called **repeaters** are implemented in the lowest layer. Their task consists purely of signal amplification in order to increase the range of LANs. Repeaters do not have any type of temporary cache storage and simply pass on the incoming bits.

- **Layer 2**
  On the link layer, individual network segments can be connected with each other via **bridges**. In contrast to repeaters, bridges have their own cache storage where the incoming data packets are temporarily stored before being forwarded. After arrival of the data packets, a simple error detection is carried out via the checksums sent in the data packets. Bridges can perform changes in the header of the data packet. However, this only applies to the headers in layer 2. Headers in the above layers are treated solely as user data, without it being possible that information from them is understood and implemented.

- **Layer 3**
  Routers mediate between different networks on the Internet layer. One of their tasks is to determine the next leg of the path of incoming data packets as they each make their way to a specific destination. Incoming data packets are temporarily stored, the packet header evaluated, and their forwarding organized, based on the indicated address information. If the data packet is forwarded to a network in which another network protocol, with another data packet format is implemented, the router must then undertake the appropriate reformatting. Routers capable of performing this are known as **multiprotocol routers**.

- **Layer 4 and Layer 5**
  To enable a connection above the link layer, so-called **gateways** are necessary. On the transport layer of the network, transport gateways connect different byte streams with each other. Application gateways are used on the subsequent application layer. The task of such a gateway involves connecting two components of a distributed application with each other that are working on computers in different networks. For example, it is possible to use a special email gateway to forward emails, which is also capable of mediating between different email applications.

However, such strictly defined distinctions are seldom maintained in practical application. The intermediate systems available on the market often combine properties of bridges, routers and gateways. If a gateway is used in an application between the networks of two large organizations, or even two countries, a configuration of the connectivity hardware from both organizations must be carried out. The bigger the parties involved, the more complicated the regulations in managing and controlling access to the interface. Because of this, the connectivity hardware is usually distributed. In each case, it is controlled by one of the connection partners. In contrast to a single **full- gateway**, two **half-gateways** are employed. The partners connected in this way must agree on a shared connection protocol implemented between both half-gateways. The configuration of each half-gateway is subject only to the control of one party, and therefore has a much less restrictive form than when two parties must be coordinated (see Fig. 7.3).

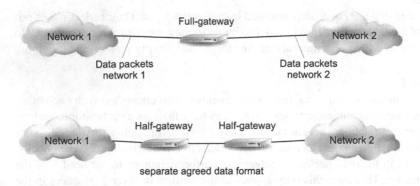

Fig. 7.3 Connections via a full-gateway and a half-gateway.

7.2 Internetworking

In practical application, connecting networks with different technologies is quite a challenging undertaking. This is mostly due to the great variety of physical-technical and logical parameters that distinguish network technologies from each other. If between the transmitter and the receiver there are multiple, different network types to bridge, there is likewise an array of conversion problems to be solved at the interfaces between the individual networks. These stem from different operating parameters and data formats. Numbering among them are:

- **Connection-oriented and connectionless service**
  In the conversion of data packets from networks offering a connection-oriented service to those only providing a connectionless service, the forwarded data packets quickly lose their original order. The receiver is no longer able to correctly interpret the original data if the receiving order for the correct sequence is not kept.

- **Communication protocols**
  In different networks, different communication protocols can be employed, such as IP, IPX, AppleTalk, DECNet or SNA. For this reason, protocol conversions must take place at network interfaces. However, this is not always possible to a full extent as not all of the communication protocols have an identical functionality and therefore information can get lost.

- **Addressing**
  Differing network technologies lack a uniform network-wide addressing. For this

reason, an appropriate address conversion must take place at the interface of the network. To accomplish this, directory services are necessary that allow a classification of the various address types.

- **Multicasting**
  If on its path through the Internet a multicast message encounters a network that does not support multicasting, the message must then be replicated accordingly and forwarded individually to all the members of the relevant multicasting group so that it reaches all the addressees.

- **Maximum packet size**
  Different network technologies have different maximum packet sizes, i.e., data packets can only be transported by the network technology up to a specific maximum size. If a data packet should pass through a network whose packet size is longer than is allowed in that network, then it must (possibly once more) be fragmented. It cannot always be assumed that there is a guaranteed quality of service. Therefore, mechanisms responsible for the newly disassembled data packet reaching the receiver correctly and in its entirety must be in place, as well as those to handle its reassembly in the correct sequence.

- **Quality of service**
  Many networks are not capable of handling real-time data, which is delivered virtually isochronously. If this type of real-time data passes through a network not able to offer any guarantee, then the correct delivery of the relevant data to the receiver remains uncertain.

- **Error handling**
  In the event of transmission errors, mechanisms for error handling react in different network technologies in different ways. These include the simple discarding of a defective data packet, to a new request in the case of a defective packet, to automatic error correction. If a transmission error in fact occurs, this difference in behavior can lead to problems in the transmission of the networks involved.

- **Flow control**
  Here as well, different network technologies can be applied in various procedures that help to avoid an overload of the end systems. Above all, problems can arise in connection with various mechanisms for error handling, overload control and with different packet sizes of the networks involved. (see additionally section 8.3.1).

- **Congestion control**
  To avoid overload in the intermediary systems in individual networks, different algorithms are used in part. In connection with various mechanisms for error handling, data packet loss can occur as a result of router overload. (see additionally section 8.3.1).

- **Security**
  Besides a variety of mechanisms for data encryption, different regulations can be applied in networks involving the management of confidential, private data. A coordination and conversion of individual regulations is complex and not always possible to realize fully.

- **Accounting**
  In different networks various accounting systems can be employed. These generate invoices for the use of the network based on different parameters, e.g., connection time or the transmitted amount of data. Even the transmitted amount of data can be calculated differently, in bit, byte or the number of data packets calculated.

7.2.1 Connection-Oriented Internetworking

Just as with single networks, in linked multiple networks a distinction can be made between a connection-oriented and a connectionless data transmission service[1]. If, for example, individual, connection-oriented networks are switched together and a connection should thereby be established, one speaks of **connected virtual circuits (concatenated virtual circuits)**.

A connection to an end system in a remote network is generally established in the same way as in a connection-oriented single network (see Fig. 7.4). At the initiative of the transmitter, the subnetwork of the transmitter recognizes that a connection is to be established to an end system outside of the subnetwork. First, a connection is switched to the router that is closest to the network of the addressed end system. From there, a connection to the external gateway, usually a multi-protocol router is switched. This leads over a path to the network of the addressed end system. The gateway registers this virtual connection in its routing table and proceeds to establish a connection to the appropriate router in the next network along the connection path. This process continues until a connection to the originally wished addressee is established.

As soon as the actual data traffic along the switched connection is admitted, every router forwards the incoming data packets implementing all the necessary translations of data formats and communication parameters required for each network involved. Additionally, a fixed identification number is assigned to every single switched connection within a network, which also must be implemented in the switched intermediary system. For this purpose, the routers manage the identification numbers of the connections to be switched in their routing tables. The transmitted data packets pass the router along one of these types of connections, always in reverse sequence. In this way, the data packets arrive at the receiver in the original order.

This type of connection works most reliably when the connected networks bear a strong similarity to each other. Only when each of the involved networks offers a reliable service, can the network system offer a reliable service as well.

Theoretically, the same advantages and disadvantages apply to connection-oriented internetworking as to a simple, connection-oriented network: buffer storage can already be allocated before the data transfer, an adherence to the original sequence

---

[1] An overview of connectionless and connection-oriented services can be found in Chapter 3, „Foundation of the Communication in Computer Networks", in the first volume of this series: Meinel, Ch., Sack, H.: Digital Communication – Networking, Multimedia, Security, Springer (2013).

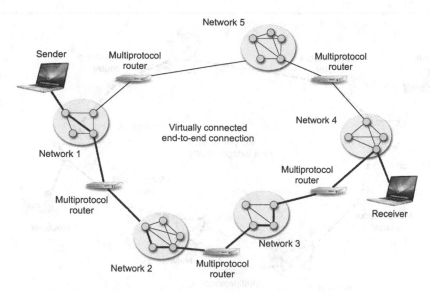

Fig. 7.4 Virtual circuits (concatenated virtual circuits).

is guaranteed, the necessary overhead of additional control information in the data header is small and problems with delay or duplicated data packets do not occur as a rule. On the other hand, additional storage space is needed in the routing tables for every existing connection in the participating routers. If an overload or the loss of an intermediary system occurs, it is not possible to take an alternative route.

7.2.2 Connectionless Internetworking

In contrast to connection-oriented implementation, datagram networks are interconnected in this variation. These are networks that each offer only a connectionless service. To carry out a data transfer to a remote end system, an explicit connection does not have to be established here, rather all data packets of the transmitter depart on their journey to the designated receiver independent of one another. This means that each data packet can reach its destination on a different path over different (intermediate) networks. As the participating routers do not assign individual connection identification, as in the case of connection-oriented internetworking, no dedicated connections from the routers can be manged. For every single data packet this means that a separate routing decision must be made in each case as to what path the data packet takes to reaches its goal.

As in the case of an individual, connectionless network, here also no service guarantee can be given as to whether the data packet arrives in the original order when it reaches its destination or even if it reaches its goal at all.

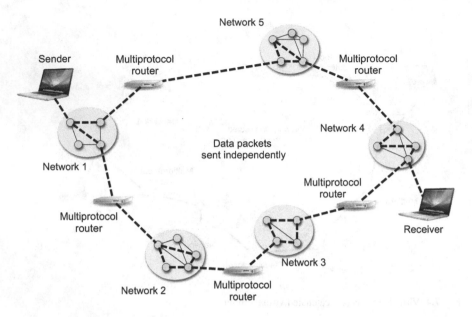

Fig. 7.5 Connectionless internetworking.

The advantages and disadvantages of connectionless internetworking correspond to those of a standard connectionless service. While an overload is more likely to occur here than in a connection-oriented service, the connectionless service provides special detour and avoidance mechanisms for overload situations. These are also implemented in the case of router loss. As individual data packets make their way to the respective receiver independent of each other, the control information in the data header of the packet must be more detailed than in the case of virtual connections. This can thus limit the efficiency of network throughput. The big advantage of this implementation is, however, that this procedure can also be employed when the involved networks are of different types in view of service guarantees.

7.2.3 Tunneling

As previously described, the explicit conversion of data packets from one network to another proves to be a complex and not always solvable problem. However, a simpler case presents itself if a transmitter and receiver network are each of the same type and networks of other types must only be bridged. Transmitters and receivers can process data packets with identical data formats, only the participating intermediate systems (multiprotocol routers, gateways) have to carry out the necessary conversion and reformatting necessary for forwarding over different network technologies. The original data packet sent on the way from the transmitter remains unchanged. It is encapsulated in its entirety or in individual pieces (fragments) in the payload field of the different network data format and at the end unpacked and reassembled (cf. Fig. 7.6).

An Example for the Tunneling of IP Data Packets

In the example presented, both the transmitter and the receiver networks are of the Ethernet type, while a WAN with different network technology must be bridged. Both of the Ethernets to be connected are TCP/IP-based, i.e., on the Internet layer the Internet Protocol (IP) is implemented. For communication in the network that must be bridged, a so-called **tunnel** is set up. This makes data transfer across the borders of different networks considerably easier. The Ethernet data packet of the transmitter is given the IP address of the receiver and sent to the multi-protocol router, which connects the Ethernet-LAN with the WAN. The multi-protocol router takes the IP data packet out of the received Ethernet data packet and sends it in payload field in a data packet of the next WAN. In a case where a WAN data packet cannot take the forwarded IP data packet as payload, it is disassembled in order to fit (**fragmented**). Having reached the multi-protocol router of the opposite side, which connects the WAN with the destination-LAN, the received WAN data packet from the original IP data packet is unpacked, transformed back into an Ethernet data packet and sent via the LAN to the designated destination system. If the original data packet is transmitted in single, disassembled fragments, it is necessary that the fragments are first put back together before being forwarded to the destination-LAN.

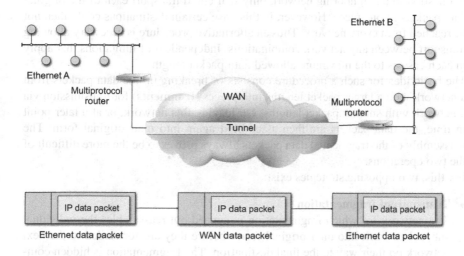

Fig. 7.6 An example for the tunneling of IP data packets between different network technologies.

In this organization of communication, first the original data packet is appropriately packed. The path traveled through the intermediary networks proceeds from one multi-protocol router to another as if through a **tunnel**. This is as if the transmitter and the receiver were connected to each other by a simple serial cable. In this way, no information from the IP data packet is lost through possible translation. Only the multi-protocol routers at the input or output of their networks, or tunnel, have to be capable of processing the original data packet as well as the data packets of the neighboring networks. It is not necessary for the transmitter and receiver in either of the remote LANs to have any information about the network located between them

7.2.4 Fragmentation

Every network limits the maximum length of a single data packet that is to be transmitted to a specific value. This is due to limitations on the hardware side of the operating system implemented, protocols and standard conventions or restrictions caused by competing goals. Among these are e.g., the reduction of transmission fields in the individual data packets (the longer the packet, the more possible fields), or the shortest possible occupation time of the transmission medium by an individual data packet.

The maximum length of the transported payload ranges from 48 bytes in a ATM data packet to 65.515 bytes in IP data packets (IPv4). On higher protocol layers even bigger packet lengths are possible. If different types of networks are connected with each other, the data packets of one network could be too large for the next network. This problem could simply be solved by allowing the forwarding of the data packet in question to a neighboring network only if it could transport each of the original data packets in one piece. However, in this case certain destinations could then not be reached from every network. Thus an alternative procedure is necessary allowing transport between any network combinations, independent of limitations that apply in each case as to the maximum allowed data packet length.

The basic idea for such a procedure consists of breaking up the data packets from a network with a larger packet lengths into pieces (**fragments**) for transmission via a network with smaller packet lengths. On leaving this network, or at a later point in time, the data packets are then assembled again into their original form. The reassembly of the fragmented data packets always proves to be the more difficult of the two operations.

For this, two opposing strategies exist:

- **Transparent fragmentation**

  The network for which fragmentation is carried out reassembles the individual subpackets back into their original form, before they are forwarded to the next network on their way to the final destination. The fragmentation is hidden completely from the next network along the path to the final destination network (cf. Fig. 7.7).

  If a too large data packet reaches the switching system of a network functioning according to the principle of transparent fragmenting, it is first broken up into single subpackets. Each one of these subpackets then receives the destination address of the multiprotocol router. This router connects the current network with the subsequent network along the prescribed path to the final end system. After having arrived there, the subpackets are assembled again into the original packet.

To do this, it is necessary that every subpacket be given an appropriate identification. The receiving multiprotocol router must also be capable of determining
whether in fact all of the subpackets of the original data packet have arrived. A
special counter is provided for this purpose, which also facilitates the reassembly of the arriving data packets into their correct sequence and a special end bit
identifying the last subpacket of the data packet.

It must further be ensured that all of the subpackets are indeed traveling on the
path over the designated multiprotocol router and that no subpackets take another route as they proceed to their final destination. Although this step can lead
to a drop in transmission efficiency. A further disadvantage is the potential high
processing overhead resulting when data packets are fragmented in multiple networks along the path to the destination system and must be subsequently put
back together.

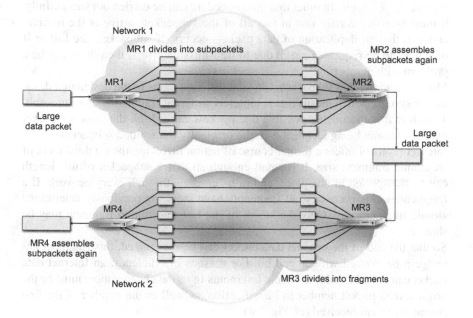

Fig. 7.7 Transparent fragmentation.

- **Nontransparent fragmentation**

  Unlike transparent fragmentation, a reassembly of the fragmented subpackets is
  deliberately avoided in the switching systems between the individual networks.
  If an over-sized data packet is disassembled, then each subpacket is treated as a
  separate data packet and forwarded. Reassembly back to the original data packet
  is first carried out at the destination system. For this to be possible, every end
  system must be in a position to assemble a fragmented data packet again. In contrast to transparent fragmentation, no additional overhead occurs in processing

at the intermediary systems, however the transported data overhead is bigger. This is because once a fragmentation is performed it is maintained until arrival at the destination system. Therefore, each one of the shorter subpacket's corresponding data headers must be added. This procedure has the advantage that all subpackets can reach their destination on different paths through the individual networks, which can mean an increase in throughput. However, this procedure does not have any special advantage when virtual connections are used. In this case, the same route is used for every subpacket.

The difficulty of this procedure lies in taking precautions so that also multiple fragmented data packets can be restored back to their original state. Numbering is used for this which takes the forms of a tree, as soon as a new fragmentation is conducted. For example, if a data packet with the identification number 0 is segmented, the subpackets are numbered 0.1, 0.2, 0.3, etc. If one of the subpackets is split up again, e.g., 0.2, then the fragments are numbered 0.2.1, 0.2.2, 0.2.3, etc. In order that this procedure can be carried out successfully, it must be made certain that in fact all of the subpackets arrive at the receiver and also that no duplication of data packets occurs. If subpackets are lost or if they have to be requested again, due to transmission errors, then there can be a problem with numbering.

Moreover, individual subpackets can take different paths to the designated destination system, which means that completely different fragmentations can result. Therefore it is thoroughly possible that packets can exhibit the same subpacket number without being identical with the originally transmitted subpacket.

Internet protocol offers a further course of action involving the establishment of an atomic fragment size. It is small enough so that a subpacket of this length can be transported in one piece, and in its entirety, through every network. If a fragmentation is necessary, all fragments have the same previously determined atomic fragment size. Only the length of the last transmitted fragment may be shorter.

So that the end of the original data packet can be recognized, this fragment has to again be marked with an end bit. For reasons of efficiency, an Internet data packet can contain multiple atomic fragments In its data header there must be the original data packet number as identification, as well as the number of the first atomic fragment received (cf. Fig. 7.8).

7.2.5 Congestion Avoidance

One speaks of **congestion** to describe the situation that occurs when too many data packets are transmitted in a communication network and because of this the data transmission output in the network drops. As long as only as many data packets are transmitted as the switching routers in the network can actually forward, no data packet gets lost in a datagram network.

An Example for the Fragmentation of IP Datapackets

So that an IP data packet can be correctly reassembled after a (multiple) fragmentation has been carried out, the header of an IP data packet always contains two sequence numbers:

- the packet number of the IP data packet and
- the number of the first atomic fragment contained in the IP data packet.

An additional control bit (end bit) signals the end fragment of a data packet:

- end bit=0: beginning or intermediary fragment of an IP data packet.
- end bit=1: last fragment of an IP data packet.

It is ensured in the IP protocol that the fixed length of the individual atomic fragment is always smaller than the minimum data packet size of all the connected networks. Therefore, successive fragmentations - right down to atomic fragmentation - can be carried out and subsequently reversed at the receiver without a problem. The following example shows the process of the multiple fragmentation of an IP data packet. The limit value of a bite was chosen for the atomic fragment size. Based on the packet and fragment number in the data header, the original data packet can always be reconstructed.

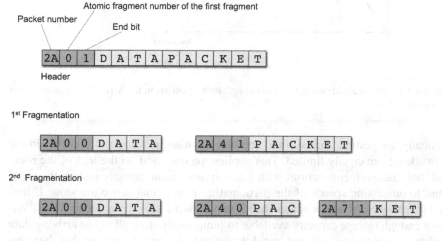

Further reading:

Postel, J. (eds.).: Internet Protocol – DARPA Internet Program, Protocol Specification, Request for Comments 791 (1981)

Fig. 7.8 An example for the fragmentation of IP data packets.

The number of sent and delivered data packets rises proportionately until the transmission capacity of the network is reached. Or, in other words, it rises until the maximum number of data packets is reached that can be forwarded by the router in the network.

Yet if the load continues to increase, the data packets arriving at the routers can no longer be temporarily stored and more and more data packets get lost.

This can reach a point where the transmission output of the network falls to such a degree that nearly no more data packets can be forwarded. (cf. Fig. 7.9).

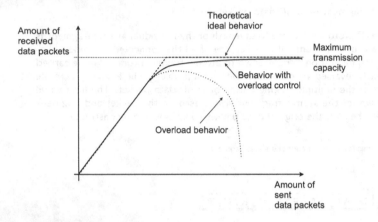

Fig. 7.9 Congestion avoidance helps avoid congestion situations in the network to a great extent or remedies them,

Basically, congestion occurs in the network because the resources available in the network are principally limited. This applies, on one hand, to the legs of the route and those network connections with a low transmission capacity and on the other hand to processing speeds of the participating routers that prove too slow. If more data packets arrive at a router than it can forward directly, a queue forms. Without enough storage capacity available to temporarily store all of the arriving data packets, some of them will get lost. Interestingly, it could be proven that there are only limited advantages to an increase in router storage. When a data packet remains in a queue for a longer stay this results in a timeout. Data packets that have not yet been forwarded are then sent again as duplicates.

While both describe different situations, there is a direct connection between congestion avoidance and flow control. Congestion avoidance ensures that every section of a network is capable of handling the applied data traffic. In this way, avoidance is the task of all the involved network components. In contrast, flow control applies only to the control of data traffic between a dedicated transmitter and receiver pair. Here it should be made certain that a fast transmitter does not overload a slow receiver with a transmission. To enable this, there needs to generally be direct communication between the transmitter and the receiver.

The goal of connection avoidance generally consists of transmitting as many data packets as possible within a network without an overload occurring. From the perspective of control engineering there are two basic solutions paths for congestion avoidance. Solutions known as **open loop** describe methods that avoid a subsequent regulation of the original system through preventive planning and design. Among these are included decisions made in advance about when data is to be forwarded in the network and when it is to be discarded. In contrast, solutions with **closed loop** are based on the approach of a feedback loop, i.e., the system always takes into account the current network conditions.

Overload control based on the principle of the closed loop is composed of three parts:

- **Network monitoring**
  At the beginning of closed loop stands the monitoring of system parameters. Here, different metrics are implemented which determine, for example, the amount of discarded data packets, the average queue lengths, the number of the data packets sent again due to a timeout, the average delay in data transmissions or their variations or standard deviation. Thereby, it is intended to determine when and where an overload situation occurs.

- **Announcement of overload information**
  If an overload situation occurs in the network, this information has to be forwarded from the point at which it happens to an instance that is in a position to enact the necessary countermeasures. It would be easiest if the router on which an overload was determined were to send special data packets to announce this situation. Yet, this would cause a further increase in the load of the relevant network. Reserving a special field in the header of every data packet to indicate an overload is a better solution. In an overload situation, the affected router can then transcribe a predefined value (e.g., its own network address) in this field, thereby bringing the overload situation to the attention of the adjacent router

- **Solving the overload problem through adaption of the network operation**
  Following the distribution of a message announcing an overload situation, the notified network nodes take the appropriate countermeasures to remedy the problem. Principally, two different algorithms types fall into this category.

  - Algorithms with **explicit feedback** initiate a router, where an overload has taken place, to send the affected data packet back to the transmitter to warn it of the overload.

  - Algorithms with **implicit feedback** determine the occurrence of an overload situation from indirect, local indicators, e.g., from the time that elapses until confirmation of a successful data packet delivery.

In an overload situation where the current network load is greater than what the infrastructure can manage, the available resources of the infrastructure can either be raised or the applied load decreased. An increase in the transmission capacity can be achieved, e.g., by additionally directing the data traffic over alternative routes. If there is no possibility to increase transmission capacity, the applied system load must be reduced. In addition, data transmission can be delayed or even refused, depending on its priority.

Overload avoidance can take place in the network on different layers of the TCP/IP reference model. Measures can thereby be distinguished that access the link layer, the Internet layer or the transport layer.

- **Link layer**

  - **Renewed transmission of data packets**
    The transmission of a data packet starts a timer. After the timer has run out, a new transmission of data takes place in the case of non-delivery. This applies either only to data whose timer has run out (selective new transmission) or to all (or a prescribed amount of) data sent since transmission of the outstanding data packets. The latter, however, means a higher load for the network so that overload avoidance of a selective new transmission represents the better strategy.

  - **Temporary storage of data packets**
    This applies to the handling of data packets that are sent out of turn. If a receiver or intermediate system discards all data packets transmitted out of turn, these must be transmitted anew and the load on the network rises.

  - **Transmission of response receipts**
    If every single data packet is immediately invoiced following its reception with its own response receipt, a significant load is generated on the network On the other hand, the collection and bundled delivery of response receipts cause new timeouts and related the data transmissions. These also burden the network. For this reason the choice of the bundle size (window) is a critical factor in the control of overload or the entire network flow.

  - **Flow control**
    The tighter the flow control, i.e., the shorter the time interval chosen for controlling network data flow, the greater the reduction in data transmission rate. For this reason, flow control has a direct influence on control avoidance.

  - **Timeouts**
    Data packets whose timer has expired are discarded by the routers and must be sent anew by the transmitter. While the current load on the individual router is reduced, as a whole data transmission is merely delayed so that an overload situation can occur again with a new transmission.

- **Internet layer**

  - **Use of alternative subnetworks**
    On the Internet layer choices can be made in regard to the forwarding of data in subnetworks between virtual connections and pure datagram subnets. With both, different algorithms are implemented for the avoidance, or remedying, of overload situations.

- **Queuing management**

A router can have its own queue for each one of its inputs, its outputs, or for both. The distribution and the amount of available storage set aside influence the behavior of overload just as the sequence of the processing of the data packets in the queue. These can be processed according to the first-in-first-out principle or on a priority basis.

- **Discarding of data packets**

If the queues at the input, or output, of the routers are occupied, new incoming data packets can no longer be temporarily stored and are discarded. Different rules and strategies also exist as to selecting the data packets to be discarded in order to reduce a present overload situation.

- **Routing algorithm**

Special routing algorithms help to quickly reduce a overload situation that occurs. Among these procedures are hot potato routing (cf. section. 6.3.3), in which the choice of the router output is determined by the respective length of the output queues.

- **Lifespan of data packets**

The selection of the timeout for transmitted data packets influences overload behavior and network flow. If the time period chosen is too long, data packets that have been lost remain in the the network too long and burden the data traffic in the network, On the other hand, if the time is too short, data packets are discarded too quickly and must be transmitted anew, thereby also influencing network throughput.

• **Transport layer**

- **Renewed transmission of data packets**

As in the link layer, the selection of the timeout for controlling new transmission of lost data packets is critical for the control of overload and network flow. In contrast to the link layer, on the transport layer parameters such as transmission delay and data transmission time cannot be predicted exactly because of the routing located in the layer below. For this reason, adaptive algorithms are used that adapt to the current situation in the network.

- **Temporary storage of data packets** (cf. link layer)
- **Transmission of response receipt and flow control** (cf. link layer).

While the stated measures for congestion avoidance principally target avoiding an overload situation, in the feedback-based procedure (closed loop) a distinction must be made on the Internet layer between procedures that work on subnetworks with virtual connections and procedures on datagram subnetworks.

In **subnetworks with virtual connections** congestion avoidance takes the form of either prohibiting new virtual connections (admission control), in the event of a congestion situation, or of establishing new virtual connections only along routes that circumvent the overloaded areas of the network until the data burden has normalized again. Beyond that, connection parameters can be negotiated when establishing

a virtual connection between a router and a subnet. These can orient themselves on
the current network utilization to avoid this type of overload situation in advance.
In **datagram subnets** every router continually monitors the utilization of its connection lines. Congestion avoidance keeps an overview of current utilization as well as
its history to make decisions about the proper action to be taken. Included among
the congestion avoidance procedures in datagram subnets are:

- **Overload flag**
  If a router determines an overload situation, it sets a special bit at one in the
  packet header of the outgoing data packet as an overload flag. The receiver of
  the data packet copies the overload flag in its response receipt which it then
  returns to the data transmitter. In this way, the transmitter learns of the overload
  situation and reduces the rate of traffic. A rise in the rate of transmitted data
  traffic occurs first when the acknowledgement response reaches the transmitter
  without the set overload flag. This is however only the case if no router along the
  path from transmitter to receiver indicates an overload situation. This procedure
  is implemented in the Internet as the Explicit Congestion Notification (ECN) for
  the avoidance of overload situations. Its disadvantage is the necessity for both
  transmitter and receiver to support such a data packet.

- **Choke data packet**
  Routers can also provide information about a current overload situation directly
  to the affected transmitter. Here, a special **choke data packet** is sent back to
  the transmitter telling it to reduce its transmission power (data transmission rate)
  by a specific, prescribed percentage. Additionally, in the regular, forwarded data
  packets an overflow flag is also set. In this way, not every router along a congested
  route is forced to send its own choke data packet to the transmitter. If a transmitter
  receives a choke data packet it reduces the flow of traffic. As it is probable that
  numerous data packets are already traveling in the network on the overloaded
  section of the route, further choke data packets from the same router will arrive
  at the transmitter. These are ignored for a certain span of time until a further
  reduction of the transmission power take place. An increase in the transmission
  power can follow at a lower percentage after throttling in order to avoid a renewed
  overload in advance.

- **Hop-by-hop choke data packet**
  If the number of routers to be bridged between the transmitter and the receiver is
  too large, an adaption of the transmission power can only take place slowly with
  a choke data packet. This can be remedied in that the choke data packet already
  becomes effective at each of the hops on the route, reducing traffic there, rather
  than first at the transmitter. In this way, considerably faster relief for the overload
  situation can be achieved.

- **Load shedding**
  If the overload situation cannot be remedied using other methods, then it is inevi-
  table that data packets will be lost. The targeted discarding of data packets for the
  purpose of overload control is known as **load shedding**. Older data packets are
  more important than newer data packets and are protected accordingly when the

choice of data packets to be discarded is made. This is due to the probability that an older data packet may be a part of a data transmission that has already begun following this data packet. If an older data packet is discarded on which previously sent data packets depend, then the entire data transmission based on the older data packets becomes incomplete. Also the newer, already transmitted data packets must be resent in the course of a new transmission. If, on the other hand, a new data packet is deleted then only those data packets have to be resent that are dependent on this data packet. Seen statistically, the number of data packets resent as a result of deleting an older data packet is greater than the number of those in connection with a newer data packet. A further variation of targeted load shedding consists of giving the data packets priority based on the effect of their absence. The router then always begins load shedding with the lowest priority data packets.

- **Random Early Detection (RED)**
  All of the aforementioned methods fail if the wait to implement them is too long. Therefore, if overload control is already active when the first signs of a bottleneck are threatening, the likelihood is great that a total failure can be avoided. For this reason, the discarding of data packets is already begun before the queue of a router is completely filled up. The **Random Early Detection (RED)** algorithm works according to this principle. The length of a router's middle queue is continually ascertained, and when a previously determined value limit is exceeded the random discarding of data packets begins. This random deletion is based on the reason that it is not possible to determine with exact reliability which transmitter bears responsibility for a particular overload situation. The Weighted Random Early Detection (WRED) protocol, as an extension of the RED procedure, allows the management of multiple, different value limits per queue. In this way, it is possible to carry out a differentiated handling of differently prioritized data packets. The transmitter of the deleted data packet can also be informed now about a choke data packet or an overload situation. Yet, exactly this kind of transmission creates additional data traffic in the overload situation. As the deleted data packet does not arrive at the receiver, the response receipt is absent as well. This means a timeout is determined for the sent data packet at the transmitter and it must be sent anew. Timeouts are usually caused by an overload of discarded data packets. Because of this, the transmitter can indirectly identify an overload situation and automatically reduce its data transmission after a timeout has been determined. However, this only makes sense in the case of wired networks as data packets are often lost as a consequence of disturbances in the radio link in wireless networks.

7.2.6 *Quality of Service*

A primary expression for measuring performance in the network is the **Quality of Service**, with which the requirements of a data flow in a network are expressed[2]. Service quality is dependent on the following parameters:

- throughput and transmission delay as quantitative ascertainable measurements,
- fluctuation in performance (jitter) in relation to the occurring transmission delay and rate of error,
- reliability as a qualitative measurement with the criteria of the completeness and uniqueness of a data transmission as well as the granting of security standards in terms of confidentiality, data integrity, authenticity, liability and availability.

Guaranteeing certain levels of quality is closely related to addressing overload in a network. Three levels of service quality are distinguished in a packet switched network (see also Fig. 7.10):

- **Guaranteed Service (Integrated/Predicted Service)**
  On this quality level, the transmitter requires the reservation of a part of the available transmission capacity for a data packet from the router along the determined route. In this way, a logical (virtual) channel is defined along which a specific data transmission rate is guaranteed, also in cases of overload.

- **Differentiated Service (Imperfect Service)**
  Data traffic on this service level is given preferential treatment by every intermediary system in the network regarding available transmission capacity. In case of overload, as few data packets will be lost as possible.

- **Best Effort Service**
  The remaining data transmission capacity can be used by data traffic on this lowest possible service quality level. In the case of an overload, data packets with this quality of service will be the first to be lost and must be transmitted anew.

Different measures and procedures exist to facilitate fixing a quality of service in advance:

- **Provision of a sufficient infrastructure**
  To guarantee a defined quality of service in advance, a suitable infrastructure must be provided beforehand, i.e., a router with high processing and switching speed, buffer storage, and sufficient bandwidth. However, the demands of the infrastructure rise with the number of users and increased distribution of multimedia data services.

---

[2] A general representation of the service quality in computer networks is found in section 3.3.3 „Quality of Service", in the first volume of this series: Meinel, Ch., Sack, H.: Digital Communication – Networking, Multimedia, Security, Springer (2013).

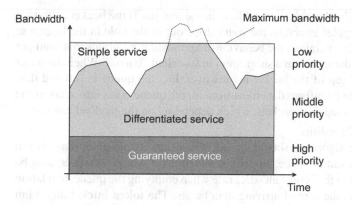

Fig. 7.10 Quality of Service levels in a packet switching network (from [236]).

- **Temporary storage of data packets (buffering)**
  If a buffer storage is set up on the receiver's side, this helps to smooth out fluctuations in the transmission delay of individual data packets. Incoming data packets are initially collected in a buffer storage before being forwarded to the processing application. This initially has no effect on the the bandwidth or the reliability of data transmission. Multimedia applications, e.g., video or audio streaming, however, rely on an even transmission delay. In cases where deviations from the medium delay in transmission (jitter) are too great, interruptions and gaps in the video or audio-streams appear. The respective buffer size stands in direct relation to maximum compensable transmission fluctuations. The larger the buffer chosen, the greater the fluctuations that can be compensated. At same time, with upstream buffer storage there are also delays in the playback of video and audio-streaming. Before the start of playback, the buffer must first be filled to a certain level. The total delay makes the use of buffer storage in real-time data, for example in video telephony, problematic.

- **Traffic Shaping**
  While buffer storage at the receiver provides data for maintenance of quality of service, at the transmitter there can also be a regulation of the data traffic to avoid overload situations and guarantee an agreed upon quality of service. In **traffic shaping** the average data transmission rate is regulated by a previously reached agreement between the transmitter and the implemented subnet (Service Level Agreement). The transmitter informs the necessary infrastructure resources while the subnet operator guarantees the provision of the required resources to the transmitter, as long as these are not exceeded from the transmitter's side.

- **Leaky Bucket Algorithm**
  If on the transmitter's side a buffer storage is also set up, whose output sends data packets with a constant transmission rate into the network, a nearly constant transmission delay can be achieved from the transmitter's side. This procedure is known as a **leaky bucket**, as it can best be described by this metaphor. Here,

we can imagine a bucket with a small hole in the bottom. If the bucket is flooded with water in irregular intervals, the water drips out of the hole in the bottom at a constant rate. This parallels the behavior of application programs that send out large amounts of data, each in a short time, in so-called "bursts." When the water level exceeds the top of the bucket it runs over, i.e., the queue is full and data packets are lost. In a uniform data transmission, the queue takes into account the size of the transmitted data packets, which depending on the payload can vary.

- **Token Bucket Algorithm**
  The leaky bucket algorithm should enable as constant as possible transmission delay on the transmitter's side. In certain situations it would, however, also be desirable to adapt to the "constant" data rate when emptying the queue, in relation to its fill level and the size of arriving data bursts. The **token bucket** algorithm enables such a flexible adaptation. In advance, certain contingents – also called tokens - are distributed for the forwarding of data packets. These are utilized or can be collected up to a specific predefined limit until the "bucket" is full. Every token stands for a certain amount of data that may be transmitted. When a data packet is transmitted a token is removed from the bucket. When all of the tokens are used up, a data packet to be transmitted is either placed in a queue until a token is distributed or it is discarded. As an alternative, the data packet can be forwarded in spite of the missing token. But in this case, it is marked as a data packet that can be discarded if a bottleneck situation should arise in another place. If during a longer period no data packet is sent, a stock of tokens is collected making it also possible on short notice to transmit larger amounts of data (data bursts) at one time. However, in the long run the fixed upper limit determined by the regular distribution of tokens with respect to the maximum data transmission rate remains in place.

- **Resource reserving**
  Regulating the data stream with the mechanisms discussed here raises the chances of ensuring the necessary quality of service. Implicitly this regulation also means that the resulting data flow must always take the same path to ensure the service quality required. If a certain path is determined for a data flow beforehand, then resources can be reserved along it to guarantee this quality. The following resources may be reserved beforehand:

  - **Bandwidth**
    Corresponding to the available maximum bandwidth on a transmission line, partial capacities for a data stream may be reserved, whereby attention must be given so that a line is not overlapped.

  - **Buffer cache**
    Delivered data packets are placed in the buffer cache of the router if they cannot be forwarded immediately. In the event that the fill level of the buffer storage is exceeded at a certain point, newly arriving data packets are discarded. Different data streams compete for the available storage space in the buffer cache. If a fixed storage space in the buffer cache is exclusively reserved for a data stream in advance, the data packets in this data stream are no longer in

competition with the remaining data streams. For this reason, there is always free storage space for this data stream in the buffer cache, up to a prescribed limit.

- **Computing capacity (CPU cycles)**
  Every arriving data packet is analyzed at a router, i.e., address information in the data packet header is read and processed The address information determines over which output the data packet is forwarded by the router. Computing capacity for processing the data packets is limited. Because of this, before being forwarded, arriving data packets compete for available computer time so that they can be forwarded in a timely manner.

A data stream is marked by a series of descriptive parameters, known as a **flow specification**. In negotiating a data flow through a network by routers many parties may be involved, whose data streams to be forwarded are described with the help of flow specification. A flow specification is generated by the transmitter, corresponding to the requirements of the planned data transmission, and sent to the receiver. On its way to the receiver, the flow specification is examined by every router and the parameters changed in correspondence with the free capacity of the router. Thereby, the individual, indicated parameters can only be reduced in each case but not increased. Having arrived at the receiver, the finally available, and therefore reservable resources of the network can be determined.

- **Differentiated routing**
  In the processing of multiple data streams at a router there is a danger that a single, large data stream might cause the others remaining to "dry up," as data packets are usually processed in the order they reach the router. One possibility of getting around this problem is providing queues for every single data source in a router (**fair queueing**). The router runs through all queues assigned to a single output line in a round-robin fashion. In this way, a data packet is forwarded interchangeably from every transmitter. However, the original algorithm did not take the size of the data packet into account, and so transmitters with larger data packets were clearly given preference. This algorithm was therefore expanded so that for every data packet to be transmitted, the end time of the transmission is calculated according to its size. The data packets are then sorted in the order of their end time and transmitted in this sequence. Additionally, a weighting of the individual data streams can follow in which the number of data streams coming from a transmitter are also taken into account. Every transmission process has the same bandwidth available to it (**weighted fair queuing**).

To guarantee a sufficient service quality for special services with high resource demands (e.g., multimedia streaming), a series of data-flow based algorithms (flow-based algorithms) or integrated services have been defined. These are discussed in detail separately in Chapter 9 in connection with real-time and multimedia applications. A simpler variation, the **class-based quality of service**, scales better in contrast to integrated services and can be implemented in every router locally, i.e., without inclusion of the entire data transmission route. It was standardized over so-called "**differentiated services**" (DS, RFC 2474, RFC 2475). For the implementation of differentiated services, groups of routers are combined within an administrative domain. The instance responsible for administration defines a number of different service classes and establishes for them differentiated forwarding rules. Users of such administrative domains can show the chosen service class via a special service class attribute in the data packet header. Certain service classes have preference over other - possibly "more cost effective" - service classes. Every router in the administrative domain can decide the forwarding of a data packet based alone on the service class, without prior resource reservations and other agreements.

However, on the Internet, forwarding over multiple, different administrative domains is also necessary. The chosen service class should, if possible, be supported by all administrative domains it passes. To enable this, network-independent service classes are defined, such as **expedited forwarding** (RFC 3246) or **assured forwarding** (RFC 2597). In the simplest variation, expedited forwarding allows the definition of two different service classes: normal (regular) and express (expedited). The bulk of transmitted data belongs to the service class "normal." Only a small part is given preferential treatment and transported "express." For the implementation of this concept, there are always two instead of one queue at every router, for both of the two service classes. The incoming data traffic is assigned to the appropriate queue, based on its service class, and the "express" data traffic is forwarded preferentially. By comparison, assured forwarding allows a differentiated handling of data traffic by providing four different service classes with (in each case three) different possibilities for discarding data packets in the case of a overload situation.

7.2.7 Internetwork Routing

Routing in an internet environment is carried out similarly to that in an individual network (cf. section 6.3). In an internet, an attempt is first made to analyze in a graph the existing topology of the multiprotocol router, which connects the individual networks to each other. Every multi-protocol router has, so to speak, a direct connection to every other multiprotocol router with which it is connected through a maximum of one network (see Fig. 7.11). The example shows that router MR1 can reach routers MR2 and MR3 over the network N1, as well router MR5 over network N4.

On the topology of the multiprotocol router represented in the graph, the routing algorithms discussed in the WAN chapter (Chap. 6), such as link-state routing, can already be applied. Internet routing is carried out as a two-step routing process. For routing in the involved network, a so-called **Interior Gateway Protocol (IGP)** is implemented and between the respective multiprotocol routers a so-called **Exterior Gateway Protocol (EGP)**. Every one of the networks involved acts independently of the others. This means, it is also possible that different routing procedures are employed in individual networks. The individual networks are therefore also referred to as **autonomous systems** (**AS**, also see Excursus 6 "Special Routing Procedures").

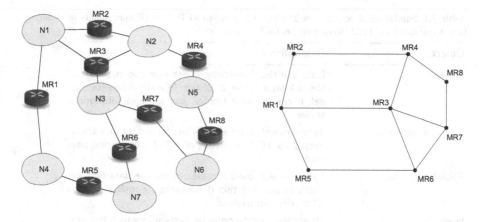

Fig. 7.11 Internetwork routing.

If a data packet should be sent from a transmitter in one network to a remote network, it is first sent within the transmitter's network to the respective multiprotocol router. This connects the transmitter's network to the next network on the path to the target network. The address and routing information from the MAC layer of the link layer (layer 2 in the TCP/IP reference model) is used for this. Having arrived at the multiprotocol router of the transmitter network, the routing information of the next higher network layer (layer 3) is then used to determine the subsequent multiprotocol router. The data packet is then forwarded there. The forwarding is based on the conditions and parameters of the network to be crossed. Or, in other words, based on the communication protocol implemented, the data packet must be fragmented, or as encapsulated payload, tunneled through the network. This process repeats itself until the target network is finally reached.

A further fact that makes internetwork routing more complicated than routing in a network is that operators must often be changed between networks. Here, attention must be paid to changing sovereign and national rules and regulations. In the individual networks different calculation procedures are also implemented. This means the additional problem exists of determining reasonably priced ways of data transmission.

7.3 Internet Protocol – IP

One of the two main components of the TCP/IP protocol family is the **Internet Protocol (IP)**, which lays the foundation for the worldwide Internet. In the following section, we will first look at the net structure and the manner of computer addressing in the worldwide Internet. Then the mapping of Internet addresses onto local network addresses will be examined. Over so-called "address binding" a coupling

Table 7.1 Fundamental concepts of TCP/IP, according to RCF 1009 (Requirements for Internet Gateways) and RFC 1122 (Requirements for Internet Hosts).

Object	Description
Frame	Data unit that is exchanged between two instances of the link layer (layer 2) of the TCP/IP reference model. It consists of a header, payload and subsequent trailer.
Message, segment	Synonymously used expressions for a data unit transmitted via TCP, consisting of a TCP header and payload.
Packet, datagram	Synonymously used expressions for the data unit exchanged between two IP instances, consisting of an IP header and payload.
Host	Designation for a computer system in which the user of an internet service (client) works.
Router, gateway	A mediating intermediary system between different network segments. Mediation taking place on the network layer (layer 3) of the TCP/IP reference model uses a so-called multiprotocol router if the router is capable of mediating between networks of different types. In the TCP/IP reference model, the older designation, gateway, is often used synonymously with multiprotocol router, although the meaning more closely describes communication on a higher protocol layer.
Interior Gateway Protocol (IGP)	Routing procedure that is implemented in a single network.
Autonomous System (AS)	Subnets and systems controlled by an individual operator and using a shared routing procedure.
Exterior Gateway Protocol (EGP)	Routing procedure employed between individual autonomous systems.

of the TCP/IP reference model Internet layer with the link layer and its local technology is carried out. Message formats and protocol functions complete the function description of the Internet protocol. At the end of the section, the focus will be on the Internet protocol's successor: the Internet Protocol Version 6 (IPv6).

From the perspective of the internet layer of the TCP/IP reference model, the worldwide Internet may be viewed as a huge collection of individual, autonomous systems (subnets). Core of the coupling of these autonomous system are so-called **backbones**. These are high-speed connections with an extensive transfer capacity, connected with each other over fast routers on so-called **peering points**. **Regional nets (midlevel networks)** are linked to backbones, on which LANs greatest operators, such as universities, companies or Internet Service Providers, are already connected (see Fig. 7.12). In this way, the Internet's organization structure is virtually hierarchical.

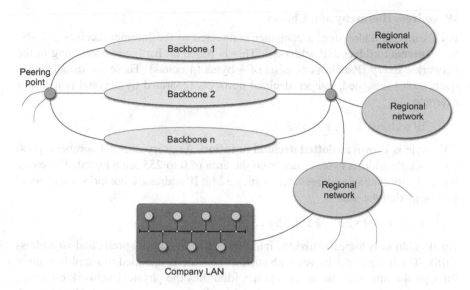

Fig. 7.12 Hierarchical structure of the Internet.

All networks connected to each other use the Internet Protocol. This ensures that data packets always find their way from the transmitter to the designated receiver, independent of which subnetwork transmitters or receivers are connected.

7.3.1 IP Addressing

As has already been shown, the goal of every internetworking centers on giving the user the illusion of a large, uniform network. This includes hiding the actual physical details of the network underneath. The global Internet is thereby like every other internet: nothing more then a pure software product essentially consisting of protocol software, in particular the IP protocol. The user is thus provided with mandatory directives, independent of the particular hardware implemented, concerning all of the necessary communication parameters, such as data packet size and format. In addition, it supplies a universal addressing scheme enabling direct communication with every single computer connected to the Internet. Thanks to the IP addressing scheme, the global Internet does in fact appear to the user as a single, homogeneous and universal network. The details of the many individual nets, comprising the Internet, remain fully concealed from the user's eyes. The actual IP addresses remain hidden from most users as well. In their place, easy to remember, hierarchically-structured logical names may be used. Before taking a look at the uniqueness of this name allocation and the rules for implementing names in IP addresses (cf. section 9.2.1), the IP addressing scheme itself will be presented.

IP Address Hierarchy and Classes

For its unique addressing, a computer connected to the Internet receives a 32-bit long **Internet address(IP address)** . This is based on hardware addressing in the networks. Every IP address consists of 4 bytes (4 octets). These are usually a sequence of 4 unsigned, integer, decimal numbers separated by decimal points. For example:

IP address: 155.136.32.17

This style is known as **dotted decimal notation**. As every decimal number represents a single address byte, values from the area of 0 to 255 are adopted. The entire theoretically available address area of such 32 bit IP addresses, comprises in decimal representation the area

0.0.0.0 to 255.255.255.255

An IP address is logically divided into two parts: an address prefix and an address suffix. The hierarchy achieved with such a structure is intended to simplify routing through the Internet. The address prefix identifies the physical network on which the corresponding router is connected and is also known as **network ID** (Network Identification, NetID). In contrast, the address suffix identifies a specific computer in a network designated by the network ID. The address suffix is therefore also called **host ID** (host identifier).

While network IDs as well as a host ID are always uniquely allocated in one specific network, it is thoroughly possible that in different networks identical host IDs can be used. This scheme of address hierarchy supports a global management of network IDs, while the management of host IDs can proceed locally through each network administrator. A problem stemming from this allocation of IP addresses in network IDs and host IDS was that while on one hand enough space must be provided to uniquely address all physically existing networks, on the other hand, for every network there must also be a sufficient number of addresses available, even with a large number of connected computers. With the fixed address length of 32 bits it was necessary to decided what part of the available address space should be used for network IDs and which for host IDs. As there are often a completely different numbers of computers on the Internet-connected networks, and indeed only a few very large networks but a multiple number of smaller ones, a compromise was reached. The IP address space was divided into five different address classes, whereby in three classes the address prefix and address suffix have a different share of the available 32 bit IP addresses (see Fig. 7.13). The address class can be determined from the first bits of an IP address, and therefore the division of the address into the address prefix and suffix. Fig. 7.13 shows the permissible IP address classes. Based on the valid conventions for TCP/IP protocols, the bits are read from left to right, respectively, whereby the first left bit is designated as the 0. bit.

Classes A, B and C are designated as **primary IP address classes**. Besides these, there also exist the special classes D and E. Class D is reserved for multicast addresses and class E for experimental purposes. The number of available host IDs for

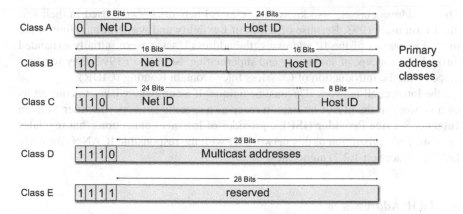

Fig. 7.13 Address classes of IP addresses and the division into network ID and host ID.

every network class and accompanying address space in decimal representation is shown in Table 7.2.

- **IP Addresses of Class A** were intended for very large networks in which more than 2^{16} computers were connected. Included among the holders of class A networks are huge firms such as IBM or the American government. In class A IP address 7 bits are provided for the network ID and 24 bits for the host ID. Theoretically, up to 16,888,214 computers can be connected in a single class A network.

- **IP Addresses of Class B** were intended for medium-size networks. There is a much larger number of these, and therefore 14 bits are used for their network ID. 16 bits are left for host IDs. A class B network allows the connection of up to 65,534 individual computers.

- **IP Addresses of Class C** have the greatest dissemination. The 21 bits for the net ID of an IP address of class C allows a maximum number of 2,097,152 unique addressed networks. At the same time, the available space for host IDs is limited to 8 bits. As two of each of the possible host IDs are reserved (local host and broadcast), there can be a maximum of 254 computers connected in a class C network.

Table 7.2 IP address classes and properties.

Class	Number of nets	Number of hosts	Address space
A	126	16,777,214	1.0.0.0 – 126.0.0.0
B	16,834	65,534	128.1.0.0 – 191.255.0.0
C	2,097.152	254	192.0.1.0 – 223.255.255.0
D			224.0.0.0 – 239.255.255.255
E			240.0.0.0 – 255.255.255.254

The IP address classes, also known as "classful networks," were used in their original form until 1993. Because of a lack of flexibility and wasteful handling of the limited resources of the IP addresses, the address classes were initially expanded through the concept of subnetting and supernetting. Starting in 1992, they were replaced with the introduction of Classless Inter-Domain Routing (CIDR).

On the Internet all network IDs must be unique. If a company wishes to connect its own network to the global Internet, it receives a specific network identifier from an **Internet Service Provider (ISP)**, a provider of Internet connections that regulates the issue and registration of this network ID with the responsible ICANN (formerly IANA, see section 1.2.2) authorities.

Special IP Addresses

In addition to the possibility of addressing every single host, it is useful to enable the simultaneous addressing of all computers in a network (**broadcast**). For this reason, the IP address standard provides several special IP addresses that cannot be given to individual computers, but are reserved for special use.

- **Network addresses**
  If the value 0 is used as host ID, this is the address of the network specified by the network ID and not that of a special computer connected to this network. In the local network itself, the host ID 0 always indicates the individual host computer (**local host**). Similarly, an address with the network ID 0 always refers to its own, local network.

- **Broadcast addresses**
  Via a special broadcast address, all of the connected computers can be addressed at the same time in their networks by their network ID. To facilitate this, all bits of the relevant host IDs are set to 1. An IP data packet containing the broadcast address is then forwarded through the local network until it has been received by all of the computers in the network. If the broadcast address of a class C network is, e.g., 136.77.21.255, then 136.77.21 designates the network ID and 255 the host ID provided especially for the broadcast. A special case is the IP address in which all bits are set to 1 (255.255.255.255). This is the broadcast address of the local network (network ID equals 0), where the transmitting computer is connected. The router forwards a data packet with the local broadcast address only within its own network and does not let it enter the Internet. In the new Internet protocol IPv6 the broadcast addresses were done away with and replaced by multicast addresses.

- **Loopback addresses**
  For test purposes it is often useful for a transmitting computer to receive back a data packet it has sent out. In this way, it is possible to draw conclusions about response time behavior and network errors. The special addresses reserved for this purpose are known as loopback addresses. The IP standard reserves for it the network ID 127 in class A. Because all host IDs are handled equally, the host

ID used is irrelevant. Conventionally, the host ID used is 1, so that the common loopaddress is 127.0.0.1.

- **Private addresses**

  Special IP addresses are not forwarded in the global Internet but reserved for use in private networks that are not connected to the global Internet. As specified in RFC 1597, the address fields available for this purpose are 10.x.x.x, 172.16.0.0 to 172.31.254.254, as well as 192.168.0.0 to 192.168.254.254.

- **Multicast addresses**

  Instead of addressing individual hosts (unicast) or entire networks (broadcast) at the same time, special IP addresses are available to enable addressing of groups of related computers (multicasting). Every computer group receives an **IP address of class D** for this purpose, which uniquely addresses this group. Fixed multicast addresses exist that are established on a permanent basis for special groups, as well as temporary ones, which are available for private use. A maximum of 2^{28} different multicast groups are possible. To assign a computer to a specific multicast group, a corresponding request has to be made at the nearest multicast-router (cf. section 6.3.5).

Routers, or **multiprotocol routers** serve to link individual networks connected with each other on the Internet. As a link between two or more networks, these are logical components of the connected networks and must therefore have their own host. An IP address does not identify a specific computer but only the connection of a computer with a certain network. A computer connected to multiple networks simultaneously must have its own IP address for every connection. Fig. 7.14 shows two routers each connecting different networks to each other, with the associated IP addresses.

It is also possible for computers to be connected simultaneously to multiple networks This case is known as a **multihomed host**. (**multiple residence host**) With the simultaneous connection to multiple networks, the availability and failsafe performance of the computer can be increased. If one of networks fails, the computer can still be reached over a further network. The same applies to a situation of router overload, where the affected router can simply be bypassed through the use of an alternative net connection.

Subnet Addressing

In the case where a network grows beyond the reliable number of available host IDs, the operator is faced with several serious problems. One possibility of managing this overload can be the reservation of an additional network ID at the responsible Internet Service Provider. However, it is then necessary to operate two physically independent networks whose configuration and maintenance pose additional effort. On the other hand, a company can also operate a priori multiple, independent LANs and configure these under different network IDs.

If a computer should then be passed from one network to another, in each case an individual network configuration to that host is required. Furthermore, this change

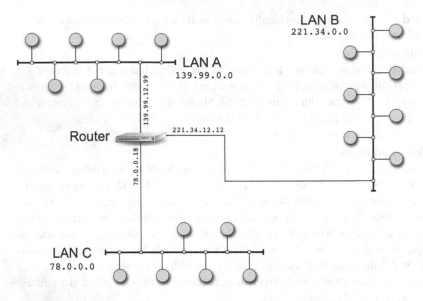

Fig. 7.14 IP address assignment to routers that connect different networks to each other.

of network ID must first achieve world-wide validity before all messages that are intended for this computer also reach its new address. **Subnet addressing (Subnetting)** offers an efficient approach for the flexible division of address space in an address class. A refinement of IP address classes can be achieved through subdividing networks of different classes into physically-separate from each other independent subnets. They continue to be addressed over a uniform network ID. Subnet addressing was introduced in 1985 and in 1992 standardized for the first time in the RFC 950.

In subnet addressing, a specific number of bits from the host ID are reserved as the **subnet ID**. Therefore, e.g., a network of class B whose host ID is made up of 16 bits, could be subdivided into a 6-bit long subnet ID and a new 10-bit long host ID, so that an addressing of 62 subnets is possible (0 and 1 are reserved) with 1,022 available host IDs each (0 and 255 are reserved). From the outside, this division into subnetworks is not visible and therefore also not traceable. Address assignment (ICANN) can thus take place without the involvement of the regulatory authorities, and is solely the responsibility of the owner of the network ID. To distinguish the host address from the network address (network ID and subnet ID) a so-called **subnet mask** is used. The subnet mask is like an IP address, also 32 bits long, whereby the network address area is replaced with 1 bit and the area of the host ID with 0 bits. The evaluation of the subnet address and forwarding to the relevant subnet proceeds via the router. In its routing table, there are initially two types of addresses (not including subnet addressing) stored:

- **Network addresses**: (network ID, host ID=0) for the forwarding of data packets in remote networks and

- **Host addresses**: (Network-ID=0, Host-ID) for the forwarding of data packets in the local network.

Every entry in the routing table is assigned a certain router's network interface for the forwarding of data packets. When a new data packet reaches the router, its target address is searched for in the routing table. If a data packet is involved that is to be forwarded to a remote network, on the path to the target network it is sent to the router in the next network over the interface indicated in the routing table. If it is intended for a local computer it is delivered directly. When the data packet is meant for a network that is not yet noted in the routing table, forwarding proceeds over a so-called **default router**. This has more detailed routing tables and takes over the forwarding of the data packet. As a rule, routers therefore have only the information concerning the individual computers of the local network, as well as the network addresses for forwarding to remote, non-local computers. If the router works with **subnet addressing**, the entries are supplemented in the routing table with a new format:

- **Subnet addresses**: (network ID=0, subnet ID, host ID=0) and
- **Host addresses**: (network ID=0, subnet ID=0, host ID).

In this way, it is ensured that a router located in a certain subnet, e.g., the subnet with the subnet ID=34, knows how data packets are to be forwarded in all other subnets, or how data packets should be delivered in the local subnet (subnet ID=34). Details about individual computers located in other subnets do not need to be administered to enable a more efficient utilization of the routing table. With the help of **subnet masks** provisions for determining the necessary interfaces of the routers for forwarding from the routing table can be made considerably easier (see Fig. 7.15 and 7.16). For this purpose, the network software calculates the bit-wise logical AND between the subnet mask and the target address of the data packet to be forwarded. If the network address calculated in this way corresponds to an entry in the routing table, the interface of the router corresponding to the entry is used for forwarding.

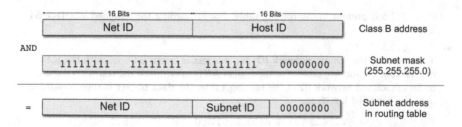

Fig. 7.15 Network addresses and subnet masks.

A so-called **subnet directed broadcast** is performed when all computers in a subnet are to be addressed at the same time. Here, all bits of the associated host ID, just as in a regular broadcast, are set to 1, while the bits of the subnet ID remain unchanged. If, however, all of the computers from all local subnets are to be addressed, a

Example for Routing Calculation in Subnet Addressing:

As shown in the diagram, router R1 has four interfaces, which are each assigned two local computers, a further local subnet and three remote networks in the routing table.

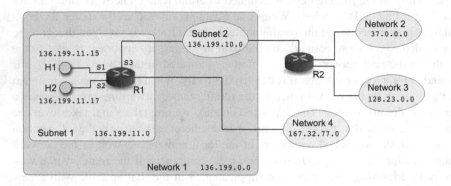

The routing table of R1 looks like this:

Network ID	Subnet ID	Host ID	Subnet mask	Interface
0.0	10	0	255.255.255.0	S3
0.0	0	15	255.255.255.0	S1
0.0	0	17	255.255.255.0	S2
37		0.0.0	255.0.0.0	S2
128.33		0.0	255.255.0.0	S2
167.32.77		0	255.255.255.0	S4

Data packets with the following address details reach R1:

- `136.199.11.15`
  As a local computer is involved (network ID: 136.199), only the **host ID** needs to be evaluated. The host ID is derived from the address by means of linking the **bit-wise inverted subnet mask** and the address by means of a binary **AND** operator. This is carried out for every entry in the routing table until the resulting address is found.

$$136.199.11.15 \textbf{ AND } 0.0.0.255 = 0.0.0.15$$

  0.0.0.15 is present in the routing table. For forwarding the assigned interface **S1** is used.

- `37.16.23.222`
  To determine the remote **network ID**, the address is linked bit-wise with the subnet masks from the routing table over the logical **AND** operator. If the result corresponds to an associated network ID in the routing table, the data packet is then forwarded over the assigned interface.

$$37.16.23.222 \text{ AND } 255.0.0.0 = 37.0.0.0$$

  The data packet is forwarded over the interface **S2**.

Further reading:

Mogul, J.C., Postel, J.: Internet Standard Subnetting Procedure, Request for Comments 950, Internet Engineering Task Force (1985)

Fig. 7.16 Example for subnet addressing.

so-called **all-subnets directed broadcast** is carried out. Here, all bits of the host ID as well as the subnet ID are set to 1. To designate a certain subnetwork, the first address in this subnet is given together with the subnet mask or prefix length. Because of it being more compact, **prefix notation** has prevailed over mask notation. Only the prefix length is given, i.e., the number of the leading 1 bits in the subnet mask. For example, instead of writing `172.16.4.0/255.255.252.0` (mask notation), one writes `172.16.4.0/22` (prefix notation).

Classless Addressing and Classless InterDomain Routing

Subnet addressing was developed at the beginning of the 80's as a way to use the scarce Internet address space more efficiently. Divided as it were into different network classes, it was realized early on that available address space would be used up quickly. This problem was discussed under the name **Running Out of Address Space** (ROADS). While the first work was also taking place on the development of a successor for the established IP protocol standard, **IPv4**, it was clear that much time would pass before it was introduced more widely. In order to still be able to cope with the fast growth of Internet address space, a temporary decision was made for so-called **supernetting** or also **classless addressing**.

Classless addressing follows an complimentary approach to subnet addressing. While in subnet addressing different physical networks can be addressed with a single address prefix (network ID), classless addressing allows the parallel application of different address prefixes for a single network. A network of class C allows the connection of up to 254 computers, while a network of class B, in which over 65,000 computers can be connected, is difficult to come by due to high demand as only 17,000 of these networks can be granted. In addition, it rarely makes sense that an organization wishing to connect e.g., 4,000 computers on the Internet, for example, occupies address space for a total of 65,000 computers. The concept of classless addressing allows the allocation of contiguous address blocks for individual organizations and companies, thus supporting more efficient use of the available address space.

However, this procedure brings other new problems with it. Let us assume that a company reserves 128 class C networks instead of one class B network. The address space would, indeed, be better used than in the assignment of a class B network where the half of the address space remains unused. However, now instead of a single entry, up to 128 entries must be allocated in the routing tables of remote routers in order to forward data packets to a computer of this firm. A technology designated as **Classless InterDomain Routing** (CDIR), which was standardized in 1993 in RFC 1519, was intended to solve this problem. CDIR forms a block of interconnected class C addresses as a single entry in the routing table that consists of

(network address, counter)

whereby the smallest address of the allocated network address block and number of the allocated network addresses is given. For example, the entry

(136.199.32.0, 3)

specifies the following three network addresses:

136.199.32.0, 136.199.33.0 and 136.199.34.0.

In fact, the employment of CDIR is not limited to class C networks, rather it makes it possible to combine blocks of addresses in powers of two. Let us assume a company needs 1,024 contiguous addresses that should begin with the address 136.199.10.0. The necessary address space therefore ranges from 136.199.10.0 to 136.199.13.255.

Table 7.3 Example CDIR – address space limits.

Limits	Address	Binary address
lower	136.199.10.0	10001000 11000111 00001010 00000000
upper	136.199.13.255	10001000 11000111 00001101 11111111

As shown in Table 7.3, the limits of the upper and lower borders of the address space distinguish themselves in their binary presentation first starting from bit 22. For this reason, CDIR requires a 32 bit address mask, in which for example given, the first 22 bits are set to 1 and the remaining bits to 0.

11111111111111111111111100000000

This address mask is used in the same way as address masks in subnet addressing. A division of addresses can be achieved from bit 22, and the basic address for the entry in the routing table calculated through an AND operation of the current address value with the mask.

To be able to identify a CDIR address block, the basic address of the block and the mask are always necessary to determined the number of addresses in the block. The information is generally given in an abbreviated form, the so-called **CDIR-Notation** (slash notation). As in the prefix notation, the mask is given as a decimal number separated with a slash from the basic address. It gives the number of bits after which the 1en of the mask changes to the value 0.

In the example given in Fig. 7.16, just

136.199.13.0 /22

is written. The routing protocol, Border Gateway Protocol (BGP), Routing Information Protocol v2 (RIP v2) and Open Shortest Path First (OSPF) use CIDR. With the introduction of CIDR, the commonly used IP address classes and their designations were eliminated for the most part. Only various-sized sections of formerly class A or class B networks are allocated today by Internet Registries. If, as in the case of CIDR, an address class is not used, the forwarding of the data packet in the router is made more difficult. When until now address prefix and suffix could be separated in a simple way based on the address class, in CIDR every entry must be extended by a 32 bit net mask. All address details in incoming data packets must be masked and checked for conformity with the table entry. It can thereby happen that several entries (with different net masks) match each other. In this case, the longest net mask is always used.

7.3.2 Binding of Protocol Addresses

IP addressing is a purely virtual addressing scheme administered by the network software. The hardware involved in the LAN or WAN does not have any knowledge of the relationship between an IP address prefix with a certain network, or an IP address suffix with a certain computer in the network. To reach a target computer at all, a data packet in the target network must be marked with the the hardware address of the corresponding target computer To do this, the network software in the target network must convert the given IP address into a corresponding hardware address.

With this in mind, let us look once more at the processes necessary to send a data packet by means of IP from the transmitter to the receiver in a remote network. The IP-capable transmitter packs the data to be sent into IP data packets each one identified with the IP address of the receiver. This IP address is evaluated by the routers and hosts which the data packet passes by on its way through the network. From this IP address information the next **hop** is determined in each case, i.e., the address of the subsequent intermediate system along the path to the receiver. While the implementation of IP software creates the illusion of a large, homogeneous net, i.e., target address as well as the address of the next hop are IP addresses, yet the network hardware itself cannot use these addresses, as it does not understand them. The network hardware uses in each case its own data packet format and its own form of addressing. For this purpose, the IP data packets must be repacked and the IP addresses used must be translated into hardware addresses. This address translation is known as **address resolution** . A computer can only resolve IP addresses for the network on which it is physically connected. Address resolution is therefore limited to a specific local network.

For example, an IP data packet is to be forwarded over an Ethernet LAN. To do this, the 32 bit IP address for the next hop in the Ethernet LAN must be converted into a 48 bit Ethernet hardware address. This is because the Ethernet hardware only works on the data link layer (layer 2) of the TCP/IP reference model and therefore does not understand IP addresses. In address resolution, different resolution techniques are possible. These can be divided into the following categories:

- **Table lookup**
  The binding between the different address formats is stored in a table. This is managed in the main memory of the computer. Every IP address is assigned a hardware address in this table. In order to save space, the network IP of the IP address is left off in each case if a separate table is created for every net. For address resolution, the network software table searches the table in the memory of the computer. The necessary algorithm for determining the hardware address

is universal, uncomplicated and simple to program. For small tables a sequential search algorithm is sufficient, however, if the number of table entries grows,
techniques such as **indexing** or **hashing** are employed.

- **Direct calculation**
  Table lookup is generally implemented in small networks with a statistical assignment of hardware addresses to IP addresses. Particularly in networks that
  allow a free configuration of hardware addresses, procedures are implemented
  which enable a direct calculation of the hardware addresses from the given networks. The hardware address of a computer is then chosen in such a way that can
  be determined with a simple arithmetical procedures from the IP address of the
  computer. If the hardware address can be freely chosen, it can be determined in
  a way to be, e.g., identical with the host ID of the IP address and can be calculated very quickly using a simple, binary logical operation. For networks with free
  configurable hardware addresses, this procedure is always given preference over
  table calculation as it can be carried out more efficiently (cf. Fig. 7.17).

Example for Address Resolution with Direct Calculation

For example, let us look at the class C network with the network address 136.199.10.0.
There are two computers R1 and R2 connected to the net. They have the IP address
136.199.10.15 and 136.199.10.17, respectively. As it involves a network of class C,
the corresponding host ID can be calculated easily by means of a bit-wise, logical AND
connection of the 32 bit IP address with the value 255 (11111111).

$$\text{Host ID} = \text{IP Address AND } 11111111$$

The resulting host IDs (15 for R1 and 17 for R2) are in this case directly the hardware
address to be calculated .

Fig. 7.17 Example for address resolution with direct calculation.

- **Message exchange**
  The address resolution methods presented so far assume that every computer in a
  network is responsible itself for the management, or calculation, of the hardware
  addresses of the other computers in the network. If an IP address or a hardware
  address changes, extensive reconfiguration is necessary in order to bring the table for address calculation into a consistent state again. Another way is achieved
  by procedures, in which address resolution is carried out with the help of messages exchange between the computers in the network. The computer wishing
  to translate an IP address into a hardware address, sends out an enquiry. This
  contains its translation wish together with the IP address to be translated. If the
  address information is centrally managed by a special server computer, then the
  computer sends its enquiry directly to the responsible server. As an alternative,
  the management of address information can also be carried out fully decentralized and every computer itself manage the allocation of its own hardware address
  to the corresponding IP address. A request for address resolution must then be
  directed via a broadcast to all computers in the network. In the first case, the ser-

ver computer answers the requests for address resolution In the second case, the relevant computer itself, recognizes its own IP address in the broadcast request and responds by sending back its own hardware address as answer. The so-called **Address Resolution Protocol** (**ARP**) is provided by the TCP/IP protocol family especially for this purpose.

Address Resolution Protocol - ARP

Address Resolution Protocol (ARP, RFC 826) is a procedure implementing mutual message exchange for address resolution. The messages exchanged in this procedure are put into a uniform format. For address resolution, the ARP protocol is implemented in combination with IPv4. In the succeeding standard, IPv6, this task is taken over by the Neighborhood Discovery Protocol (see section 7.4). The ARP standard defines two different types of messages:

- **ARP request**: Contains the IP address to be translated and the request for translation into a hardware address.
- **ARP reply**: Contains the requested IP address together with the corresponding hardware address.

Fig. 7.18 shows an example for the process of address resolution through ARP.

The ARP standard does not provide a fixed format for the transmission of ARP messages. This is because the transport of an ARP message is the responsibility of the respective network hardware. The ARP message is only encapsulated in the payload section of a network hardware data packet. At the same time, the hardware of the receiver must be in a position to recognize an ARP message as such. This is ensured with the help of a special field in the data packet header of the network data packet. Every network technology provides a field in the data packet header where the type of transported payload is described in greater detail. Specifically, there is a special field type value available in every technological variation for the identification of an ARP message. For Ethernet networks this field type value is, e.g., 0x0806, other network technologies use other identifiers. The receiver of the ARP message must then only evaluate the payload section of the network data packet to be able to determine whether it is an ARP request or an ARP reply.

The ARP standard can be used for every kind of address translation. It is not mandatory that only IP addresses are converted into hardware addresses via the ARP protocol. For this reason, the ARP data format is also laid out quite flexibly so as to facilitate the conversion of address in very different formats.

Fig. 7.19 shows the structure of an ARP message. Every ARP message begins with two 16 bit type identifiers. First, the hardware address type is specified and then the protocol address type to be translated. For example, if an IP address is translated into an Ethernet address, the hardware address type receives the value 1 and the protocol address type the value 0x0800. After that, two fields follow giving the length of the hardware address and the protocol address, respectively. Then comes the so-called operation field that shows whether it is an ARP request (Op=1) or an

Address Resolution with the ARP Protocol.

In the given network, computer R1 attempts to determine from the IP address of computer R3 (I_{R3}) its hardware address P_{R3}. To do this, R1 sends a data packet, that contains I_{R1} based on the format requirements of the network, via broadcast, to all computers in the network. Every computer in the network then checks whether the ARP request in the received data packet also contains its own address. If this is not the case, the request is ignored.

Only computer R3, which recognizes the IP address I_{R3} specified in the ARP request as its own, answers with a data packet directly to transmitter R1. It contains both the requested IP address I_{R3} as well as its hardware address P_{R3}.

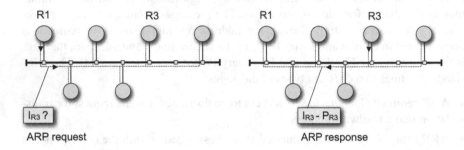

ARP request   ARP response

Further reading:

Plummer, D.: Ethernet Address Resolution Protocol: Or Converting Network Protocol Addresses to 48-bit Ethernet Address for Transmission on Ethernet Hardware, Request for Comments, Request for Comments 826, Internet Engineering Task Force (1982)

Fig. 7.18 Address resolution with ARP.

answer ($Op=2$). The protocol and hardware addresses of the transmitter and target computer come next. With an ARP request, the hardware address field of the target computer is filled with zeros as this information is are not yet known.

0	8	16	24	31
Hardware address type		Protocol address type		
HW addr length	Pr addr length	Operation		
Sender HW address (octets 0-3)				
Sender HW addr (octets 4-5)		Sender Pr addr (octets 0-1)		
Sender Pr addr (octets 2-3)		Target HW addr (octets 0-1)		
Target HW address (octets 2-5)				
Target Pr address (octets 0-3)				

Fig. 7.19 ARP - data format.

If, however, for every data transmission an address resolution must be carried out over the ARP protocol, this can influence the throughput of the network negatively. For every data packet transmission, an ARP request must first be sent via broadcast, followed by a waiting period for the ARP answer. This proves extremely inefficient in the long-run. Computers in a network using ARP keep small tables for intermediate storage for this reason. These contain the last resolved protocol hardware address pairs. This procedure is called **ARP caching**. If an IP address is to be resolved in a hardware address, its own cache table is first inspected to find the needed hardware address. When this is sufficient, address resolution can proceed immediately, without sending out an ARP message. If the hardware address is not in the cache table, then the regular ARP request and reply cycle takes place. The hardware address determined this way is added as protocol hardware address pair to the cache table. When filled, the oldest entry in the cache table is erased and replaced with a new entry, or after a set time period an expired cache entry loses its validity and is also erased. This regular cleansing of the cache table is especially important as hardware addresses lose their validity when errors occur, e.g., if network hardware is exchanged in the case of a replacement unit and hardware addresses change.

The ARP cache has an additional implementation. Following a system start, over ARP a computer distributes the locally stored protocol hardware address pairs via broadcast to all other computers in the network. This procedure is called **gratuitous ARP**. In this way, all connected computers can update their cache tables. If, contrary to expectation, an ARP answer should come back, it can then be concluded that the protocol address in question has been assigned twice and that there has been an error. The attempt via ARP to determine the hardware address of a computer located beyond a router in another LAN is doomed to failure. The broadcast (broadcast in layer 2) instigated by this router is not forwarded and therefore does not reach the receiver at the foreign network. There are two possibilities in this situation. In the first, it is recognized immediately that the requested IP address belongs to a computer in a remote network, and the data packet to be transmitted is sent to the standard router of this local network.

Network errors caused by ARP are not always easy for the user to trace as the communication via ARP is carried out transparently, i.e., is not registered by the user. The validation of an ARP entry in the ARP cache needs several minutes most of the time. If it is defective, there can be no communication with the affected computer for the length of time that the false entry is present. Incorrect entries can result from an e.g., overloaded computer that has an old, no longer valid IP address. Because of the overload, this computer is the last to answer an ARP request and can then overwrite a possibly correct entry in the ARP cache. The targeted manipulation of the ARP cache with false IP addresses is called **ARP spoofing** and presents a major security problem.

Reverse ARP – RARP

While an address resolution from protocol addresses into hardware addresses may be performed over the ARP protocol, Reverse Address Resolution Protocol (RARP,

RFC 903) provides address resolution in the opposite direction. Via RARP, the corresponding IP address can be determined from a given hardware address. The necessity for address resolution in this direction happens, e.g., at the system startup of a computer that does not have its own permanent memory (e.g., hard disk). Such a computer can receive the operating system to be loaded from a remote file server. As in a network there are often many computers without their own memory, it is an advantage for reasons of efficiency if all are offered an identical, uniform download of the operating system. This precludes the possibility that the operating system download contains each of the different IP address of the computer to be supplied. To determine one's own IP address in a system startup, the computer concerned sends a RARP message with its own hardware address via broadcast to all connected computers. A specially set up RARP server, which manages a table with all assignments between hardware and IP addresses, sends the requesting computer back a RARP message with its own IP address.

Because a RARP broadcast is not forwarded by the network routers beyond the network's borders, every network that has computers without their own hard disk must provide a RARP server that manges the requested IP Addresses and delivers them per RARP.

The **Bootstrap Protocol (BOOTP**, RFC 951, RFC 1542, RFC 1532) was developed as an alternative to RARP. BOOTP is based on the protocols UDP and TFTP, which can also be transmitted beyond the borders of routers. Computers without hard disks can therefore be managed via BOOTP, without the necessity for every single network to have its own server for downloading the operating system and for determining the IP address. BOOTP is included among the protocols of the application layer, and thus belongs to the highest layer of the TCP/IP reference model.

The **Dynamic Host Configuration Protocol (DHCP)** is an extended version of the BOOTP protocol. In addition to the range of functions in BOOTP, it offers further configuration possibilities for end systems. DHCP is specified in the RFCs 2131, 1531 and 1541. The most important innovation is the capability of automatic and dynamic assignment of an end system with IP addresses that are not statically allocated and can be used again. This dynamic allocation of IP addresses is of great importance particularly for radio LANs. Thereby, end systems are provided with an IP address and subnet mask only for a specified time (leasing). Just as with BOOTP, DHCP is an application layer protocol. It is meant to replace BOOTP in the long run (cf. section 9.6.1).

7.3.3 IP Datagram

The Internet Protocol (IP) offers the user, or a user program, a **connectionless service** (datagram service). This service manages the delivery of IP data packets (**datagrams**) beyond network borders to remote receivers. At the same time, however, the service offered by IP does not offer any service guarantee for the user. If and when a sent data packet in fact reaches the designated receiver is not guaranteed. This is even the case assuming that as a rule delivery proceeds as reliably and quickly as possible (**best effort**). Yet errors or overload in the individual networks can lead to a packet that is defective, in transit for an unreasonably long period of time, duplicated or even lost. IP is therefore termed an **unreliable** service. However - and this is the great advantage of the IP protocol - IP conceals the various network hardware and details of the network software installed on it from the eyes of the Internet user, thereby presenting the illusion of a uniform, homogeneous large network.

How does an IP data packet reach its predefined destination? The transmitter indicates the IP address of the receiver in the data header of the IP data packet and sends the IP data packet on its way through the local network. The IP data packet is then additionally packed in a network data packet (layer 2) ("encapsulated"). Corresponding to the requirements of the local network, it is sent to the nearest router (default router). This unpacks the IP data packet, evaluates the given IP destination address and sends the newly encapsulated IP data packet through the next network to the subsequent router along the path to its predefined goal. This procedure repeats itself until the IP data packet reaches a router that can transport it to its final destination in one of the connected networks. As the IP data packet must be transported through various types of networks with different characteristics and parameters, a format was chosen for the IP data format that is completely independent of these parameters in its underlying hardware.

The most widespread IP protocol to date is **IP Version 4 (IPv4)**. Although the shortcomings of this no longer up-to-date, protocol have been known for a long while, it will surely retain its importance for a time to come. **IPv6**, the designated successor, has been waiting in the wings for some time, however, a worldwide changeover to the new Internet protocol has been slow getting off the ground (see section 7.4).

Just as many other protocol data formats, the IPv4 datagram follows the structure of the already known scheme of a separate **IP data header**. Here, command and control information are contained that are necessary for the correct forwarding of the datagram as well as the **IP payload** with the actual data to be transmitted. The size of a IPv4 datagram can be determined by the user, or an application itself, i.e., the data volume to be transported is not fixed. An IPv4 datagram may only not exceed the maximum length of 64 kBytes. The longer the datagram to be transported, the better the relation between transported useful information and overhead. This is due to the command and control information in the header. The IP data header consists of a 20-byte long fixed section and an optional section of variable length. The length of the transported IP user data may vary within the given length limits. The transmission of the datagram proceeds in so-called **Big Endian Order**, i.e., from left to right. Transmission begins with the high-order bit of the first field.

Fig. 7.20 shows the structure of the IP data header:

- **Version**
  The 4-bit long version field contains the version of the IP protocol used and is filled with the value 4 in IPv4. In a specific framework during a transition time the coexistence of different IP versions should thereby be made possible.

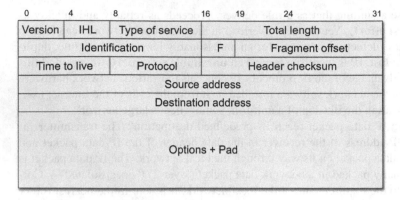

IHL   Internet Header Length
F   Fragmentation

Fig. 7.20 IPv4 datagram header - data format

- **Internet Header Length**
  As the length of the IPv4 data header is not constant, in the next IHL field (Internet Header Length) follows a length specification. This is given in units of 32 bits. The minimum length is 20 bytes. This can be extended in 4-byte large steps through the use of optional header fields. Because the length of the IHL field is 4 bits, the maximum length of the IPv4 data header is limited to 60 bytes.

- **Type of Service**
  The TOS (Type of Service) field provides a description of the required quality of service, however because of the properties of IP (best effort transmission) these are not relevant. At this time, this field is mostly ignored by the routers responsible for forwarding datagrams. The 8-bit long TOS field is divided as follows:

  - **Precedence**: 3-bit long field that establishes the priority of the datagram (0=normal, 7=high).
  - **Delay**: 1 bit, prioritized short as possible waiting times,
  - **Throughput**: 1 bit, prioritized as high throughput as possible,
  - **Reliability**: 1 bit, prioritized, as high reliability as possible.

  The remaining two bits of the TOS field are not used.

- **Total Length**
  The TL (Total Length) field indicates the entire length of the IPv4 datagram in bytes - meaning the length of the data header and payload. The maximum length is 65,535 bytes.

- **Identification**
  Next comes the ID (identification) field. It serves to identify a connected data unit that can be present in multiple fragments distributed in the payload section of the

IPv4 datagram. All fragments have the same identification. A fragmentation of the IPv4 datagram can be necessary if the datagram must pass a network whose maximum allowed data packet length is smaller than the existing IPv4 datagram. In this case, the datagram is segmented into individual fragments. Each fragment has its own data header. The connected fragments of the header distinguish themselves only in the fields connected with the fragmentation. The receiver of the fragmented IPv4 datagram puts back together the individual fragments in the reassembly process (defragmentation).

- **Fragmentation**
  The 3-bit long F-field (fragmentation) is comprised of three individual bits:

  - **M**: 1 bit, announces further fragments (More Fragments).
  - **D**: 1 bit, instructions to a switching system not to carry out fragmentation (Don't Fragment).

  The last bit is not occupied.

- **Fragment Offset**
  The following 13 bits are in the FO (Fragment Offset) field This gives the running number of the first byte of a fragment relative to the first byte of the entire datagram (cf. Fig. 7.8). If fragmentation is not carried out, the FO field receives the value 0. Every fragment in a datagram, except the last one, must have the length of a multiple of 8 bytes. A maximum datagram length of 64 kBytes and the 13-bit long FO field result in a maximum 8,192 fragments of 8 bytes in length.

- **Time to Live**
  This 8-bit long counter is set to a start value when an IPv4 datagram is transmitted. It is decremented by every intermediate system along the path to the destination system. When the value zero is reached, the IPv4 datagram is discarded, whereby the transmitter is informed of the deletion by means of a special ICMP data packet. The danger of an Internet overload caused by undeliverable or wandering data packets is reduced by this procedure.

- **Protocol**
  The 8-bit long PR (Protocol) field identifies to which protocol the datagram should be passed to in the next higher protocol layer of the TCP/IP reference model. The coding is fixed in RFC 1700 "*Assigned Numbers*," e.g., TCP: 7, UDP: 17, ICMP: 1.

- **Header Checksum**
  The 16-bit long HC (Header Checksum) field serves to detect errors, specifically in the IPv4 datagram header. The algorithm used here adds 16-bit long blocks in the order of their arrival in one's complement arithmetic. The checksum result is one's complement of the calculated sum. For this reason, after successful transmission the calculated checksum is assumed to be zero. Attention must be paid that at every hop the HC field is recalculated upon passing a router. This is because in the IPv4 datagram header at least one field changes - namely the TTL field - with every hop.

- **Source Address**
  32 bit IP address of the transmitter.

- **Destination Address**
  32 bit IP address of the receiver.

- **Options and Pad**
  Via the Options Field extra possibilities are offered for the control and monitoring of IP data transmissions. The maximum length of the options is 44 bytes. Bits that are not used are filled up with stuffing bits (pads). Currently, the following five options exist:

  - **Security**:
    Serves to identify the degree of secrecy of the transported contents, however, is not directly connected with a possible encryption of the datagram contents.

  - **Strict Source Routing**:
    Here, a complete list of the intermediate systems (routers) that the IPv4 datagram must pass on its way to the the destination system is given (also known as Poor Man's Routing).

  - **Loose Source Routing**:
    In contrast to strict source routing, a list of intermediate systems are given here that should be run through on the way to the destination system. This list is, however, not necessarily a complete list of all intermediary systems that will be passed.

  - **Record Routing**:
    Here, all intermediate systems passed by are instructed to append their own IP address in a option field. In this way, the path that the IPv4 datagram takes to its destination can be traced.

  - **Timestamp**:
    All intermediate systems passed are recorded here in such a way that in addition to the IP address, time stamp information is also added. It is not only possible to document the path through the Internet but also the time point when an intermediate system is passed.

A special option (End of Option List) in the IPv4 datagram header indicates the end of the option list. At the time of the ARPANET the maximum possible 44 bytes were sufficient for options to completely document the route of a IPv4 datagram (no data packet at that time had to pass through more than 9 routers on its way from transmitter to receiver). This amount of space is much too small to meet current demands.

Excursus 10: IP Encapsulation and IP Fragmenting

This excursus addresses the fragmentation of the IPv4 datagram. While IPv4 fragmentation is taken over by the routers in the network, IPv6 shifts the problem of fragmentation to the participating end systems (see section 7.4). Should an IPv4 datagram be transmitted from one computer to another via a series of heterogeneous networks, the IPv4 datagram is repacked in the valid hardware during the transition between networks. In the transmission of an IPv4 datagram the respective network software bears the responsibility to make sure that it is sent "encapsulated" in a data packet of its network format. The technology described as **IP encapsulation** provides a complete IPv4 datagram to be packed in the payload area of the network data format. The respective network hardware handles this data packet with the encapsulated IPv4 datagram like every other data packet. It forwards it on the path to the destination system via the next receiver, as prescribed by the router. The contents of the payload area is thereby neither checked nor changed in any way (see Fig. 7.21).

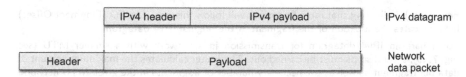

Fig. 7.21 IP encapsulation.

So that the receiver can correctly unpack and interpret the IP datagram from the payload area of the network data packet, there must be a type field in the header of the network data packet occupied with a specific value. Based on this, the transmitter and receiver in the local network can communicate concerning the type of transported data.

The IP encapsulation is only valid for one leg of the path. The local receiver who gets the encapsulated IPv4 datagram in the network data packet, removes the IPv4 datagram from the payload section of the network data packet, which is then discarded. For possible forwarding, the local receiver reads the IPv4 destination address given in the IPv4 datagram. If it matches the IPv4 address of the local receiver, the IPv4 datagram is considered to have been received. Whereupon the the content of the IPv4 payload section is forwarded from the network software to an instance of a protocol software located higher in the protocol hierarchy. If the local receiver is not the final receiver, the IPv4 datagram must then be encapsulated again for forwarding in a network data packet of the subsequent network along the path to the final receiver. In this way, an IPv4 datagram can travel through different network types on the way to its final destination.

IP Fragmentation
Every network type involved in data transmission in the Internet has it own specified data format. It is allocated a maximum length, which is always to be observed, known as the **Maximum Transmission Unit (MTU)**. The encapsulation of an IPv4 datagram in a network data packet can only proceed if it shows a smaller length than the payload of the prescribed MTU of the network of transmission. There are no exceptions allowed as the MTU is fixed. Fig. 7.22 shows two connected networks that each have a different MTU. Network N1 has an MTU of 1,000 bytes, while network N2 only allows an MTU of 800 bytes. To still be able to transmit IPv4 datagrams, which can extend up to 65,535 bytes, a **fragmentation** is carried out. If a router has to forward an IPv4 datagram whose length extends beyond the MTU of the subsequent network, it will be disassembled into individual fragments, which are then transmitted separately.

Every fragment of an IPv4 datagram retains the same format as the original IPv4 datagram, i.e., the complete data header - with the exception of details for fragmentation - is adopted. It can be communicated in the IPv4 F field whether a complete datagram or a fragment is concerned. If the MF bit in the F field is set then it is a fragment and is signaled (MF=More

Fig. 7.22 Example for the necessity of fragmentation by data transmission in networks with different MTUs.

Fragments), indicating that more fragments will follow this one. The FO (Fragment Offset) field indicates the location of the fragment in the original IPv4 datagram.

To prepare an IPv4 datagram for transmission in a network with a smaller MTU (see Fig. 7.23), several steps must be carried out. The router calculates the maximum amount of data per fragment as well as the necessary number of fragments in the given MTU network. The size of the original IPv4 datagram header is also calculated. Then the fragments are generated. In this procedure, the router removes the amount of data intended for a fragment from the payload, adding to it the original IPv4 datagram header. This is supplemented with the necessary details for fragmentation (Fragment Offset and More Fragments). The router continues this process until the entire payload section of the original IPv4 datagram is segmented into individual fragments and transmitted in an encapsulated network data packet in the order it was generated in.

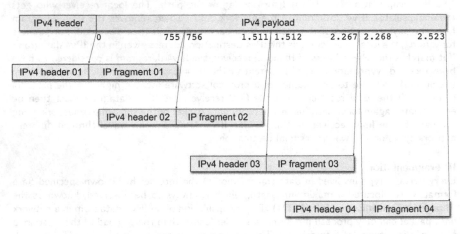

Fig. 7.23 Fragmentation of an IPv4 datagram.

Example for IP fragmentation

If a IPv4 datagram with 2,534 bytes payload content is to be transmitted from router R2, in Fig. 7.22, over the network N2 with a MTU of 800 bytes, fragmentation takes place as follows (see Fig. 7.23):

Assuming the header of a network data packet from network N2 has a length of 24 bytes, then a maximum of 776 bytes for the payload data contents are available. If the IPv4 datagram header has a length of 20 bytes then in every fragment of the IPv4 datagram a maximum of 756 bytes of the original IPv4 datagram payload contents can be transported.

The first fragment therefore takes the user data from byte 0 to byte 755 of the original IPv4 datagram.

The MF bit in the data header of the first fragment is set to 1, the fragment offset to 0. The second fragment comprises the user data from byte 756 to 1,511, the MF bit of the data header is also set to 1, the fragment offset to 755. The third fragment comprises the user data from byte 1,512 to 2,267, the MF bit is set to 1, the fragment offset to 1,511. In the last fragment the remainder of the user data of the IPv4 datagram is transported, from byte 2,268 to 2,533. The MF bit is set to 0, as this is the last fragment of the relevant IPv4 datagram, and the fragment offset is 2,267.

After transmission, the individual IP fragments must be put back together again. The IP standard prescribes that the original IPv4 datagram is reassembled first at the destination address of the given IPv4 datagram target computer. This can arrange and reassemble the individual fragments based on their identification and of then given offset, which in each case corresponds to the original IPv4 datagram. The end of a fragmentation is recognized by a fragment whose MF bit is set to 0.

Advantages of IP Fragmentation

The key advantage of this procedure is that the individual fragments do not have to take the same route through the Internet but can reach their goal in different ways. This means that there is an increase in throughput. Furthermore, the processing effort at the individual routers is reduced as the single fragments are first reassembled at the destination system. Again, this means an increase in processing time.

The individual fragments of an IPv4 datagram do not all reach the receiver on the same path. Because of this, the receiver must temporarily store all arriving fragments until the last one comes in before it can reassemble them again back into the originally sent IPv4 datagram. If only one fragment is lost or if it arrives at the receiver with a significant delay, then all of the fragments that have already been received and stored are erased after a prescribed waiting time to avoid using up the available storage space. For this reason, a timer is started at the arrival of the first fragment of an IPv4 datagram. In the case that all fragments reach the receiver before expiration of the time, the IPv4 datagram is reassembled at the receiver, otherwise all of the already received the fragments are discarded. This " all or nothing" strategy makes sense as the IP protocol does not make any provisions for the renewed transmission of individual fragments. It is not be possible to assure that an IPv4 datagram will take the same path in a new transmission, which means it could arrive at the receiver in a completely different fragmented form.

On the way from transmitter to receiver it is entirely possible that successive fragmentations can be carried out. As fragments from fragments can also have the same identification as the original IPv4 datagram, and a correspondingly calculated fragment offset is added, the receiver can reassemble multiple-fragmented parts without a problem. Generally, it has proven advantageous to establish one valid atomic fragment size for all network types that corresponds to the smallest appearing MTU. In this way, fragments of the same size are generated and an unlimited cascading fragmentation is avoided, therefore saving processing time in routers and receiving computers.

Further reading:

Clark, D. D.: IP datagram reassembly algorithms. Request for Comments 815, Internet Engineering Task Force (1982)

Hall, E. A.: Internet Core Protocols, O'Reilly, Sebastopol CA, USA (2000)

Mogul, J. C., Kent, C.A. Partridge, C., McCloghrie, K.: IP MTU discovery options, Request for Comments 1063, Internet Engineering Task Force (1988)

Mogul, J. C., Deering, S. E.: Path MTU discovery. Request for Comments 1191, Internet Engineering Task Force (1990)

Postel, J. (eds.).: Internet Protocol - DARPA Internet Program, Protocol Specification, Request for Comments 791, Internet Engineering Task Force (1981)

7.4 Internet Protocol Version 6 – IPv6

In 1983, when the still current protocol standard IPv4 for Internet data communication was introduced, it was only given a short life expectancy. Yet despite any current problems, this version of IP has proven itself to be long-lived and extremely successful. Credit for the huge growth of the global Internet can be attributed to IPv4. Its own IPv4 datagram format and the mechanisms of IPv4 to connect different network types, allow the global Internet to appear as a uniform, homogeneous network, hiding from the user the details necessary for network hardware and hardware communication software. Thanks to its ingenious design, the IPv4 protocol has survived a series of hardware generations. To a large degree this can be credited to its high degree of scalability and flexibility.

So, why should a protocol be replaced that has proven so robust? The main reason can be found in the limited IP address space. When IP was developed in the mid-70s, no one could foresee the explosive growth of the data network of the Internet. The 32-bit long IP address was considered completely sufficient, and has made possible the connection of millions of different networks. The growth of the global Internet, however, proceeded exponentially, the number of the connected computers doubling continually in less than a year. While the personal computer market seems to be nearly saturated, the next wave has already appeared on the horizon. Mobile telephones, sensors, RFID devices and the not yet widespread "end devices" of the future from the "Internet of Things," such as chip cards, household electronics, or automobile license plates. These are all equipped with processors that communicate with each other over the global Internet. A necessary unique address is the precondition for global (inter-) connectivity. Extending the limitations of the IPv4 address space is therefore the first, and presumably most important, goal of an IP renewal.

IPv4 works as a connectionless, reliable service. While functioning according to the best effort principle, it cannot guarantee quality of service. Today's Internet applications, however, work increasingly with multimedia data contents that practically demand real-time forwarding. The second important reason for a renewal of IPv4 was found in its inability to ensure service guarantees.

The common allocation practice of IPv4 addresses to end users proceeds as a dynamic address assignment via an Internet Service Provider (ISP). The end user connects his Internet-capable device over a suitable access network with the ISP, and is assigned an IPv4 address from the available address allotment for a certain period of time. After its validity expires (usually after 24 hours), a new IPv4 address must be assigned. During this procedure, existing communication connections are interrupted. This also applies to the area of mobile end devices if a new IPv4 address must be allocated when a connection is delivered during radio cell exchange (handover). However, modern applications often demand so-called "always on" connectivity, i.e., the connection between the end device and the network must be permanent and therefore may not be interrupted. This requirement could only be fulfilled with the help of IPv6.

Already in 1990, IETF began working on a project to develop a successor for the IPv4 protocol. The initial name of the project was **IP – The Next Generation (IPnG)**. Following the end of the definition phase, a final name was sought as "next generation" was in common use among other projects. The decision was made to chose a new version number for this IP protocol. Version 5 had already been given away to designate the experimental **Stream Protocol Version 2** (ST2), so Version 6 was selected. The ST2, in the meantime abandoned, was not planned as an IPv4 successor, but rather as an optimized communication protocol to be used simultaneously for streaming. Thus it came to pass that **IPv6** became the succeeding version of the current IPv4.

7.4.1 Properties and Features of IPv6

Many of the features that characterized the success of IPv4, were retained in the new version. Just as before, IPv6 centered on a connectionless datagram service. Every IPv6 datagram contains the addresses of sender and receiver. The time-to-live mechanism, determining the number of possible hops of an IP datagram, was kept, just as was the possibility of additional header detail options. Although the basic concepts of the successful IPv4 standard have remained, in certain respects considerable changes have been made in view of the respective details:

- **Address size and address management**
  The previous address size was extended from 32 bits to 128 bits. Analogous to the decimal point representation of the IPv4 address, the address is given with a maximum of 8 four-digit hexadecimal numbers, separated from each other by a colon (e.g., 231B:1A:FF:02:0:3DEF:11). Attention is paid, in particular, to a flexible as possible division of the address space and use of the address bit.

- **Header format**
  IPv6 has an individualized header format that was developed independent of IPv4.

- **Multiple header**
  The extended address in IPv6 has led to a considerable extension of the header. In order to keep the unavoidable overhead as low as possible, the concept of the optional header was introduced. It is used only if there is a need. In contrast to IPv4, an IPv6 datagram can have multiple headers. After the first, obligatory basic header, optional multiple extension headers can follow before the subsequent actual user data.

- **Video and audio support**
  For IPv6 the possibility was planned to transmit multimedia data , e.g., audio and video data, in real time. This mechanism can additionally be used to transmit individual datagrams through the Internet via more cost-effective paths.

- **Auto-configuration**
  The configuration of an IPv4 network is very work-intensive and complex most of the time. Even if address configuration tools and protocols, such as, e.g. DH-CP, make this work easier, the problems of a simple TCP/IP administration are only solved to a partial extent. Therefore, in the design of IPv6 special import-ance was placed on simplifying and automatizing the current, complex network configuration as much as possible.

- **Multicasting**
  While IPv4 allows the possibility of comfortable multicasting addressing and management in connection with additional protocols, IPv6 already natively sup-ports different multicasting variations.

- **Security**
  At the time that the design of IPv4 was determined, Internet security aspects did not as yet play a crucial role. No one had reckoned with the huge spread and popularity of the Internet medium and only a relatively manageable number of individual networks had to be connected. This situation changed dramatically. While encryption and authentication were only implemented optionally over ad-ditional protocols (IPsec, see section 7.5) in IPv4, IPv6 itself offers the support of sophisticated security technologies.

- **Mobility**
  When the design of IPv4 was developed, Internet mobility did not play a role. First, with the arrival of mobile end devices and the attempt to connect these with the worldwide Intent came the development of special, mobile Internet protocols (MobileIP, see section 7.7). IPv6 is based on this development, already natively offering support for mobile networks.

- **Extensibility**
  While the possibilities of IPv4 were defined completely, from the beginning IPv6 provided the possibility for an extension of the protocol standard, whereby ad-ditional developments can be integrated with the IPv6 datagram. It is intended that a high flexibility and adaptability be achieved with respect to future impro-vements.

In order to develop a protocol that could do justice to all of these demands, the IETF organized with the RFC 1550 (IP: Next Generation IPng White Paper Solicitation) a public invitation to tender. Twenty-one proposals were submitted to the IETF. These ranged from smaller changes to the existing IPv4 protocol to completely new deve-lopments. From the proposals three were shortlisted - among them the later IPv6. A decision was finally made for IPv6, which retained all of the advantageous pro-perties of IPv4, while eliminating, or at least mitigating the disadvantages to large extent. It also introduced new benefits where they were needed the most. Particu-larly important for the choice was, however, the fact that other TCP/IP protocols, such as TCP, UDP, ICMP, OSPF, IGMP, BGP or DNS, could be retained without significantly large changes being made.

7.4.2 The IPv6 Datagram

IPv6, standardized as RFC 2460, has a datagram format that distinguishes itself completely from IPv4 (see Fig. 7.24). An IPv6 datagram always starts with a **basis header** of a fixed, predetermined length. Following this can be one or more optional **extension headers**. First then comes the actual payload of the IPv6 datagram.

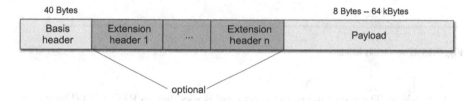

Fig. 7.24 IPv6 - datagram format

IPv6 Basis header

Although the basis header of an IPv6 datagram is only has twice the length of an IPv4 datagram header, it is possible to accommodate both the the transmitter's and receiver's address there, a four-fold length of 128 bits each as compared to IPv4. This paradox can be explained by the fact that the other information contained in the IPv4 header is left off and stored in the optional extension header, which is only used if needed. In comparison to the older IPv4 datagram header, the IPv6 header also reveals fundamental simplifications (see Fig. 7.25). In addition to both the 16 byte (128 bits) each long addresses from the transmitter to receiver - the most significant change with respect to the IPv4 – is contained in the six fields of the IPv6:

- **Version**
  Just as in the case of the IPv4 datagram header, the base header begins with a 4 bit long version field indicating what type of IP protocol is implemented. In IPv6 this field always contains the value 6.

- **Traffic class**
  The 8-bit long traffic class field is used to distinguish the priority of the present datagram relative to other datagrams. It also allows routers in the Internet to assign the datagram to a specific data traffic class, for example data, audio or video. The association with a data traffic class determines the behavior of the router in relation to prioritizing, queue management and the discarding of datagrams due

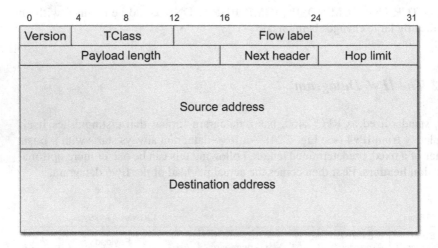

Fig. 7.25 IPv6 - basis header.

to overload. The use of the traffic class field is specified in RFC 2474 (Definition of the Differentiated Services Field in the IPv4 and IPv6 Headers).

- **Flow label**
  The 20-bit long FL (Flow Label) field provides space for a randomly chosen identification number (in the area between #00001 and #FFFFF) for a virtual end to end connection, with which specific datagrams can be identified for a special transmission on the global Internet. Datagrams with identical flow labels must also contain the same source and destination addresses. For this reason, on the basis of flow labels, routers are in a position to directly forward the concerned datagram without having to evaluate the rest of the header. Moreover, routers can make special transport decisions based on flow labels about how to handle these datagrams.

- **Payload length**
  A 16-bit long field giving the length of the payload following the header fields (basis header and extension header). The indication of length follows in bytes, with a resulting maximum length of 64 KBytes (65,536 Bytes) for the payload in a IPv6 datagram. If an extension header of the type "fragment" is implemented, then even larger amounts of payload are possible with the fragmentation of IPv6 datagrams. The payload length 0 usually indicates a so-called **jumbo packet (jumbogram)**.

- **Next header**
  The 8-bit long NH field (Next Header) tells what type of header the next one will be. This can either be an extended header of the IPv6 datagram, or a header of a data packet from a higher protocol layer (e.g., TCP or UDP) to which the payload will subsequently be transported. The next header value 59 has a special function, indicating that neither headers nor payload follows. These special datagrams can

be used as control or test data packets. Table 7.4 gives an overview of the values the next header field can take and to which subsequent header/protocol type they refer.

- **Hop-limit**

  The former time-to-live field of the IPv4 datagram header was replaced by the 8-bit long HL field(Hop Limit). Its functionality completely matches that of the IPv4 header field. The starting field value is decremented from the corresponding router by value 1. The particular router that enters the value zero in the field discards the IPv6 datagram and sends a ICMPv6 message („*Hop Limit exceeded in transit*") back to the transmitter.

- **Transmitter and destination address**

  128 bit addresses of transmitter and receiver. If an extension header of the type "routing" is used, the address of a subsequent router can also be given as the receiver address.

Table 7.4 Next header values and associated header types and protocols.

Value	Description
0	Hop-b-Hop Option Header
1	Internet Control Message Protocol (ICMPv4)
2	Internet Group Management Protocol (IGMPv4)
4	IPv4
6	TCP
8	Exterior Gateway Protocol (EGP)
17	User Datagram Protocol
41	IPv6
42	Routing Header
44	Fragment Header
45	Interdomain Routing Protocol (IDRP)
46	Resource Reservation Protocol (RSVP)
51	Authentication Header
58	ICMPv6
59	No sequence header or protocol
60	Destination Options Header
89	Open Shortest Path First (OSPF)
115	Layer 2 Tunneling Protocol (L2TO)
132	Stream Control Transmission Protocol (STCP)
135	Mobility Header
136–254	Unused
255	Reserved

IPv6 Extension header

So-called **extension headers** are provided for additional command and control information in the header of the IPv6 datagram. For reasons of efficiency, incorporating fields in the header not needed by all datagrams for specifying additional options in the specification of the IPv6 basic header was avoided. If needed, these can be included in extension headers. Up to a total of 6 extension headers can be used in a IPv6 datagram. These provide either an indication of the type of subsequent extension header or as conclusion header the type of protocol whose header is transported first in the payload area (see Fig. 7.26).

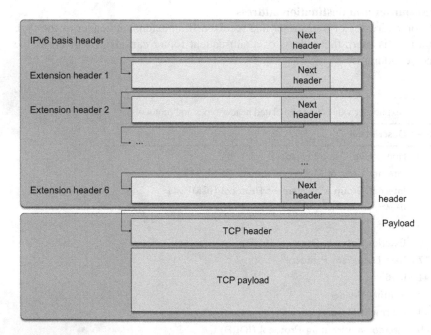

Fig. 7.26 IPv6 - extension header principle

With exception of the hop-by-hop header, extension headers are exclusively read and processed by the target computer to which the IPv6 datagram is addressed. The following six extension header types are available in IPv6. Each header type, with the exception of the destination option header, may only be used a single time in an IPv6 datagram.

- **Hop-by-hop options header**
  This header type contains optional details in the form of so-called **Type-Length-Value information**. TLV information is interpreted by every router along the path from transmitter to sender. The hop-by-hop header always follows first after the basic header to save processing time at the routers. Just as the destination option header, the hop-by-hop options header has a variable length and can thus

be adapted to specific applications. Two special applications of the hop-by-hop options header apply to the transmission of longer than normal datagrams (jumbograms) as well as to command and control datagrams. These can be used for resource reservation at the routers.

- **Destination options header**
  The destination options header can appear a maximum of two times in an IPv6 datagram. It contains information that is important for forwarding routers as well as for the receiving computer. It must be interpreted by both. If router information is involved, the destination option header follows the hop-by-hop options header directly even before the routing header. If, however, information for the receiver is involved then the destination options header is always in the last place directly before the transported IP payload data. Fig. 7.27 shows the structure of a destination options header, or hop-by-hop options header.
  The header always begins with a 1-byte long next header field, followed by length details (header extension length), also 1 byte long. The options that follow are each composed of 1-byte long type information (option type), followed by 1-byte long length information (option data length) and the option for more specified content-related data (option data) of variable length. The type information itself is made up of three parts (option type subfields): 2-bit long control instructions are provided in case that the receiving device does not recognize the specified option (unrecognized option action). This is followed by a flag bit that gives information about whether or not the relevant options along a route can change (option change allowed flag). The remaining 5 bits are provided for the specification of maximum 32 different instructions regarding the transferred options.

- **Routing header**
  This header provides a list of routers (intermediate systems), which should be visited on the path from transmitter to receiver. The established standard in RFC 2460 offers the possibility of different routing variations. Until now only the variation of the **loose source routing** has been used. While the routers indicated in the header should be visited, it is also possible that the data packets can pass other routers.

- **Fragment header**
  With the help of the fragment header, a IPv6 datagram, whose length exceeds the MTU (Maximum Transmission Unit) of the underlying network, can be segmented into individual fragments. The fragment header contains the information necessary to reassemble the individual fragments of a IPv6 datagram at the receiver.

- **Authentication header**
  Header containing a checksum, which enables verification of the sender's identity (authentication) (see section 7.5).

- **Encapsulation Security Payload**
  Header that receives a specific key number in the encryption of the payload or header data. (see section 7.5).

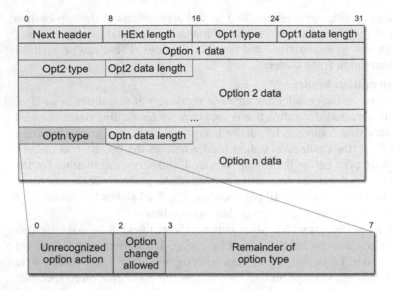

Fig. 7.27 Structure of the IPv6 extension header in a hop-by-hop option header/destination option header example.

The reasons why the concept of different, optional extension headers was chosen for IPv6 are, as already indicated, the following:

- **Economy**
  IPv6 has great flexibility as many different options can be fixed with the help of different headers. In reality, not all of these options are normally used at the same time. This means that the header can simply be left off when the corresponding option is not used. In IPv4, e.g., a large part of the datagram is not fragmented. Yet the header of the IPv4 datagram still always contains fragment bits and fragment offset. IPv6, in contrast, uses a fragment header only when in fact a fragmentation is to be carried out. This improves the relation between the transmitted payload information and the transmitted control information.

- **Extensibility**
  If a protocol such as IPv4 should be expanded in its functionality, a new design of the complete data format is required. The datagram header must be extended a new field and the entire network hardware and software adapted accordingly. This is different in IPv6. In order to take on a new functionality, only a new extension header type must be defined. If the new header type is not known to an Internet intermediate system, it is simply ignored. In this way, new functions can first be tested in a small subsection of the Internet before the hardware and software of the global Internet are subject to the necessary adaptations.

7.4.3 IPv6 Fragmentation, Jumbogram and IPv6 Routing

In contrast to IPv4, the IPv6 basic header does not contain a field for controlling the fragmentation of an IPv6 datagram. A separate extension header is available for this. If fragmentation is necessary, i.e., the MTU of the network to be bridged is smaller than the IPv6 datagram to the transmitted, the entire IPv6 datagram (header and payload data) is segmented into individual pieces, which are set in the payload data area of the new fragments. Basic headers, i.e., those extension headers that must be present in all fragments (hop-by-hop options header or destination options header), are excluded and not disassembled (non- fragmentable portion). Every fragment contains its own basic header and an additional extension header of the type of fragment header that identifies the IPv6 datagram as a fragment. It also contains a value for the fragment offset analogous to the IPv4 header so that the original IPv6 datagram can be reassembled again at the receiver.

For reasons of efficiency, the fragmentation of IPv6 takes another form than in IPv4. While in IPv4, every router is capable of fragmenting a too-large datagram into suitable fragments for the network to be bridged, in IPv6 the transmitting computer itself is responsible for this kind of segmentation. This means the router in an IPv6 Internet always expects to receive correctly-sized IPv6 datagrams, or fragments. If a too-large datagram reaches an IPv6 router, it will simply not be forwarded. The transmitting computer has two possibilities for "correctly" adjusting the size of the IPv6 datagram it is transmitting. The first is to use the guaranteed **minimum MTU** of 1,280 bytes, which every IPv6-capable network must be able to forward in unfragmented form. The second is to determine which fragment size must be chosen for the intended path to the receiver with the help of the **Path MTU Discovery** procedure. To establish the so-called **path** MTU, meaning the smallest MTU along the path from the transmitter to the receiver, the transmitter uses an iterative learning process. It continuously transmits IPv6 datagrams of various sizes. As soon as they are too large, this is acknowledged with an error message.

The purpose of this end-to-end fragmentation is to relieve the router and intermediate systems of the Internet as much as possible in order that they can reach a higher throughput. However, in this process sacrifices have to be made in terms of flexibility. In IPv4, the individual fragments can simply reach their goal on different paths, and in the case of overload or error be redirected by other routers. With IPv6 this is no longer so easy. A detour from a leg of the route could mean a smaller MTU. The transmitter would then have to determine a new path MTU and transmit the IPv6 datagram with a new fragment size. Fig. 7.28 shows IPv6 fragmentation in terms of a simple example.

In contrast to a IPv4 datagram, which is allowed a maximum length of 65,535 bytes, a size limitation can be avoided in IPv6 with the help of a option designated as "**jumbo payload**." RFC 2675 enables the transmission of overly long data packets via the hop-by-hop options, which allows a length of up to 4,294,967,335 bytes ($=2^{32}$-1 bytes). Adaptations to higher protocol layers are, however, necessary for such data packet sizes adaptations, as these are often only prescribed a 16-bit long size specification (see Fig. 7.29).

Example of an IPv6 Fragmentation

An IPv6 datagram to be transmitted has a length of exactly 390 bytes. Following a 40-byte long basis header are four 30-byte long extension headers and 230 bytes of payload data. Two of the extension headers may not be fragmented. The IPv6 datagram should be transmitted over a network connection that has a MTU of 230 bytes. In order to transmit the IPv6 datagram, a total of three fragmentations are necessary:

- **Fragmentation 1**
  The first fragment consists of the 100-byte long unfragmentable part, followed by an 8-byte long fragment header and the first 120 bytes of the original datagram, i.e., the first two fragmentable extension headers and the first 60 bytes of payload data. The remaining 170 bytes of the original datagram must be divided further.

- **Fragmentation 2**
  The second fragment consists of the 100-byte long unfragmentable part, followed by the fragment header and 120 bytes of payload.

- **Fragmentation 3**
  The third fragment consists of the 100-byte long unfragmentable part, followed by the fragment header and the last 50 bytes of payload.

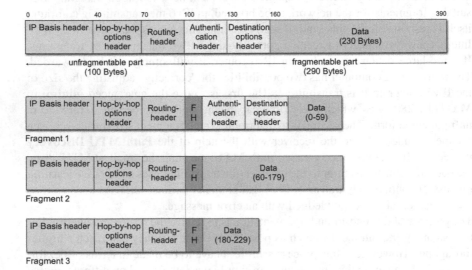

Further reading:

Deering, S. Hinden, R.: Internet Protocol, Version 6 (IPv6) Specification, Request for Comments, Request for Comments 2460, Internet Engineering Task Force (1998)

Fig. 7.28 Example of an IPv6 fragmentation.

IPv6 Jumbo Payload

A regular IPv6 datagram can, just as a IPv4 datagram, transport a a maximum payload data length of 65,535 bytes. However, the IPv6 standard also provides the possibility to transport longer data packets, especially important for applications in the area of supercomputing where many gigabytes of data have to be transported. The identification of an overly long IPv6 datagram, a so-called **jumbogram**, proceeds with the help of hop-by-hop options extension headers. If a payload length of zero is given in the basic header, it marks the transported data as a jumbogram, whose length is contained in the hop-by-hop options extension header in a 32-bit long field.

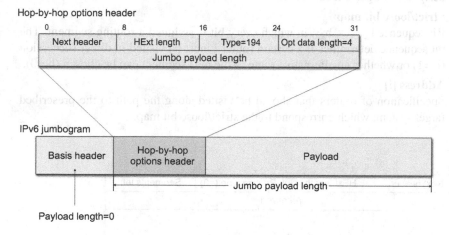

Further reading:

Borman, D., Deering, S., Hinden, R.: Pv6 Jumbograms, Request for Comments, Request for Comments 2675, Internet Engineering Task Force (August 1999)

Fig. 7.29 IPv6 - jumbogram.

IPv6 supports different possibilities of **source routing**, just as IPv4. The transmitter of a datagram specifies the path the datagram should take through the Internet. While one of the option fields of the header is used for the relevant information in IPv4, IPv6 provides its own extension header, the **IPv6 routing extension header**, for this. Fig. 7.30 shows the structure of the IPv6 routing extension header. At this time, only one routing type (type=0) is defined, which roughly corresponds to loose source routing in IPv4. The individual fields in the IPv6 routing extension header have the following meaning:

- **Next header**
  Type of the sequence header.

- **Header extension length**
  Indicates the length of the IPv6 routing extension header in units of 8 bytes each (whereby the first 8 bytes of the header are not included). These can be different lengths, depending on the number of given intermediate systems along the route

to follow. In the case of the loose source routing (type=0), this value is twice as long as the number of addresses embedded in the routing extension header.

- **Routing type**:
  Information regarding what variation of the source routing should be used. At this time only routing type 0 is specified.

- **Segments left**
  The number of routing segments (hops) left to be covered along the path to the designated target is located here.

- **Strict/loose bit map**
  Bit sequence $b_0, \ldots, b_{23}$, in which every bit is assigned a routing segment. The bit sequence determines if a certain router i must be visited in direct succession ($b_i=1$) or whether an alternative route to the next segment can be chosen ($b_i=0$).

- **Address [i]**
  Specification of routers that should be visited along the path to the prescribed target system, which correspond to the strict/loose bit map.

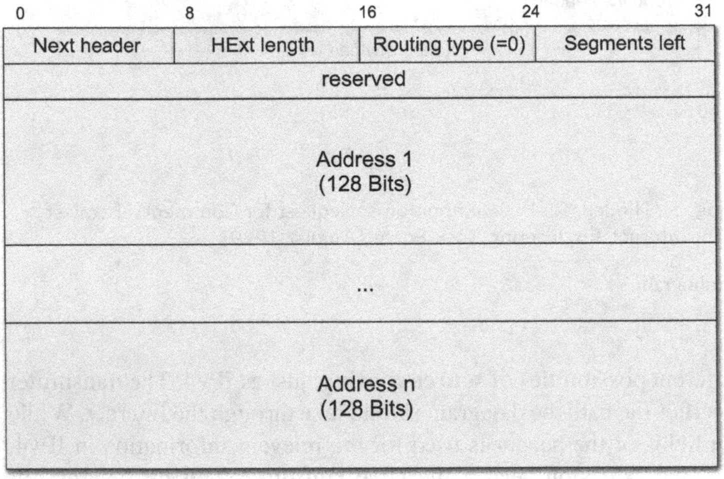

0	8	16	24	31
Next header	HExt length	Routing type (=0)	Segments left	

reserved

Address 1 (128 Bits)

...

Address n (128 Bits)

Fig. 7.30 IPv6 - routing header.

Based on the assignment of the bit vector in the field **strict/loose bit map**, the path to be taken is specified. Thereby, all of the routers indicated in the proceeding address vector must be visited. This is independent of the value given for the router i in b_i. If $b_i=1$, this simply means that router i is the direct subsequent router of router i-1 and that no alternative path from i-1 to i may be chosen.

It should also be noted that in RFC 5095 the replacement of type 0 from the described IPv6 routing extension header is recommended. It can be misused to excessively increase data traffic along a remote network section and consequently paralyze it (denial-of-service attack).

7.4.4 IPv6 Addressing

It was already determined at the beginning of the 1990s that the Internet address space available with IPv4 would not suffice for long, given the explosive growth of the Internet. Therefore, an extension of the available address space was one of the main reasons to develop a successor to IPv4. The address length was extended from 32 bits to 128 bits in IPv6. This meant that the available address space was multiplied by the factor 2^{96} (see Fig. 7.31).

The Enormous IPv6 Address Space

The extension of the address space, resulting from a quadruplication of the address length from 32 bits to 128 bits, is difficult to visualize. Every single bit added to the original 32-bit long address, means a doubling of the number of available addresses. Thus, the original address space encompassing $2^{32} = 4,294.967,296$ addresses doubled a total of 96 times. The resulting number of 2^{128} addresses can also be approximated in scientific notation as $3.4 \cdot 10^{38}$ – a number with 38 zeros or 340 sextillion.

The following example offers an analogy. The earth is estimated to be approximately 4.5 billion years old. If at the same time as the earth's creation one had begun distributing IPv6 addresses from the existing address space, at a rate of 1 billion addresses per second, until today less than a trillionth of the address space would be given away.

Further reading:

Hinden, R., Deering, S.: IPv6 Addressing Architecture, Request for Comments, Request for Comments 4291, Internet Engineering Task Force (2006)

Fig. 7.31 IPv6 - address space.

Just as in IPv4 , an IPv6 address does not identify an individual computer but only a network interface that is assigned to a certain computer. Systems can have multiple interfaces at their disposal and, e.g., work as routers. The division into an address prefix, which identifies the network, and an address suffix, for host identification, was also taken over by IPv4. Only the introduction of the address classes was done away with from the beginning. In order to read the part of network address from the IPv6 address, the notation (prefix notation) already established with CIDR is used. While IPv4 only supports the address types unicast, broadcast and multicast, for IPv6 a further address type, "anycast" is introduced. A general distinction is made in IPv6 between the following address types:

- **Unicast addresses**
  A unicast address identifies an individual network interface, thus an individual computer. Unicast addresses are used for the support of point-to-point connections.

- **Multicast addresses**
  A multicast address identifies a group of multiple network interfaces that belong together. An IPv6 datagram with a multicast address in the receiver field is forwarded to all of the members of the relevant groups. Although mutlicasting is also supported by IPv4, its dissemination and use have been delayed due to the

lack of support by many hardware components. In contrast to IPv4, multicasting in IPv6 is no longer an optional component but must be supported.

- **Anycast addresses**
  An anycast address (cluster address) identifies a group of computers (network interfaces), which are locally in a functional relationship with each other. An IPv6 datagram, with an anycast address, is initially sent to a certain router in the target network (usually to the nearest). This takes over the forwarding to specific computers in this network. The idea behind the definition of anycast addresses was to be able to distribute the network services offered to several computers in a target network, thereby increasing processing performance. The idea of strict address uniqueness was done away with. Multiple computers share the same anycast address and the transmitter of a data packet does not know from the beginning to which computer in the anycast group the data packet will be forwarded first.

Broadcasting is not supported directly by IPv6. It can, however, be realized through the implementation of multicasting.

While the new IPv6 addresses expand the available address space enormously, their presentation is rather confusing if the usual decimal point notation is used, as can be seen in the following example of a 128-bit long address:

`103.230.140.100.255.255.255.255.0.0.17.128.150.10.255.255`

In order to make this kind of address more readable, the developers of IPv6 decided on a hexadecimal colon notation. Here, units of up to 16 bits each are combined into a hexadecimal number and separated with colons. For example:

`67E6:8C64:FFFF:FFFF:0000:1180:96A:FFFF`

To make things even clearer, leading zeros can be left off and a series of zero values replaced by " ::", (**zero compression**). Therefore, for example, the address

`000E:0C64:0000:0000:0000:1342:0E3E:00FE`

is written in abbreviated form as

`E:C64::1342:E3E:FE`

The replacement of successive chains of zeros through " ::", may, however, only be done in one place in an IPv6 address. Otherwise, the address cannot be unambiguously reconstructed. Particularly with special addresses, e.g., general loopback addresses, the replacement of the series of zeros has great advantages. Therefore, by replacing the zeros in

`0:0:0:0:0:0:0:1`

the result is simply

`::1 .`

A statistical distribution of the address space into separate address classes, as in IPv4, based on the distinction between network ID and host ID, is no longer found in IPv6. The leading bits of the IPv6 address are designated the **format prefix**. With their help, certain address types can be identified (see RFC 2373 and RFC 3513). A similar schema was followed as in IPv4 address classes. However, from the beginning a variable number of bits were planned for use as a format prefix and space was left for possible expansion. Table 7.5 shows the currently used format prefixes together with information for their use.

Table 7.5 Use of IPv6 address format prefixes.

Format prefix	Share of address space	Use
0000 0000	1/256	unallocated
0000 0001	1/256	unallocated
0000 001	1/128	Network service access point addresses
0000 01	1/64	unallocated
0000 1	1/32	unallocated
0001	1/16	unallocated
001	1/8	Global unicast addresses
010	1/8	unallocated
011	1/8	unallocated
100	1/8	unallocated
101	1/8	unallocated
110	1/8	unallocated
1110	1/16	unallocated
1111 0	1/32	unallocated
1111 10	1/64	unallocated
1111 110	1/128	unallocated
1111 1110 0	1/512	unallocated
1111 1110 10	1/1024	Link-local unicast addresses
1111 1110 11	1/1024	Site-local unicast addresses
1111 1111	1/256	Multicast addresses

Besides the distinctions made between multicast and unicast addresses, the following unicast address classes are distinguished:

- **Provider-based global unicast addressing**
  These address were first specified in RFC 1884 and in the meantime have been replaced bye aggregatable global unicast addresses, which will be examined in more detail later. Provider-based global unicast addresses, intended for regular point-to-point addressing, are examined here briefly for the sake of completeness. IPv4 addresses are divided into static address classes. In contrast, the attempt was made to introduce hierarchical structuring according to a geographical assignment in provider-based global unicast addresses. As shown in Fig. 7.32, a provider-based unicast address is made up as follows:

  - **Registry ID**:
    Designates the respective international organization where the address is re-

gistered (e.g., `RegID=100000` designates the ICANN (Internet Corporation for Assigned Names and Numbers) or `RegID=01000` die RIPE (Réseau IP Européan, Regional Internet Registration Authority for Europe).

- **National Registry ID**:
  An international organization for the registration of IPv6 addresses, it can coordinate multiple, national sub-organizations in allocating addresses. In this case, an identification of the national registration authorities can follow the international registry ID. In Germany, e.g., registration is taken over by DENIC (`www.intra.de`).

- **Provider ID**:
  The following specification contained the identification of the Internet provider, who as a provider of Internet services passes on this address to its customers. The length of the field is variable. As such, different address classes can be established for the provider by the respective registration authorities. This is associated with the length of the subsequent subscriber ID in such a way that a short provider ID can respectively manage a large number of subscriber IDs and vice versa.

- **Subscriber ID**:
  The subscriber ID represents the identification of the operator of a private network and in this way is equivalent to the IPv4 network ID. It is connected with the provider ID so that together both fields have an available length of 56 bits.

- **Intra-subscriber ID**:
  The following 64 bits defined the internal network structure of the network operator. Thereby, 16 bits are provided for a **subnet ID**, serving to identify the subnet. The last 48 bits (**interface ID**), in contrast, are provided for the identification of the addressed end system. The interface ID of an IPv6 address corresponds to the host ID of an IPv4 address. It is noteworthy to mention that in a LAN with a shared transmission medium (shared medium LAN) every network address (MAC address) is also 48 bits long. Because of this, the interface ID can be directly embedded in the LAN address. A calculation and address translation step may therefore be saved at the router of the target network if the IP address is implemented in the physical network address. The application of ARP protocol, necessary in IPv4 for address resolution, can be dispensed with in IPv6.

- **Aggregatable global unicast addresses**
  Just as provider-based global unicast addressing, this form of global unicast addressing offers regular point-to-point addressing. The prefix reserved for this address type, `001`, reserves 1/8 of the total address space. Aggregatable global unicast addresses (AG addresses) replace the concept of provider-based global unicast addresses and are specified in RFC 2374. The design of this address type was based on a strict hierarchical structure of the Internet world. At the highest point of this hierarchy (top level) international and national organizations are

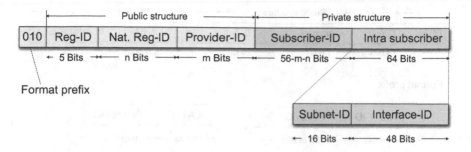

Fig. 7.32 IPv6 - provider based global unicast address (outdated).

distributed over the available address space. The next hierarchy level takes in organizations that likewise function as administrators of IPv6 addresses, but also as direct providers of Internet services. These can also be in a hierarchical relationship to each other. Individual organizations as end users of Internet services are on the the last hierarchical level. The structure of the aggregatable global unicast addresses is presented in Fig. 7.33 and contains the following structural elements:

- **Top Level Aggregation Identifier (TLA ID)**:
  13-bit long identification of the top level organization.

- **Next Level Aggregation Identifier (NLA ID)**:
  On 8 reserved and so far unused bits follows the identification of the organization on the next lower hierarchy level. It has a length of 24 bits and can be structured further itself to also image the hierarchical structure of the Internet in addresses.

- **Site Level Aggregation Identifier (SLA ID)**:
  The following 16 bits serve to identify an organization on the lowest hierarchy level, that is, virtually the end user. The SLA ID can likewise be further hierarchically structured to image a subnet structure and to subdivide a large physical network further.

- **Interface identifier**:
  A 64-bit long field that identifies the network interface of an end system. Here, the physical network addresses of a LAN end system can be specified directly.

- **Local use unicast addresses**
  Here, a distinction is made between two address types: link local use unicast addresses and site local use unicast addresses. As shown in Fig. 7.34, addresses of these two types are only intended for local use, e.g., exploration of the local network environment (neighbor discovery) or for automatic address configuration. The Link Local Use Unicast (LLU) address contains no identification of subnetworks and therefore can only be used inside isolated subnetworks. LLU

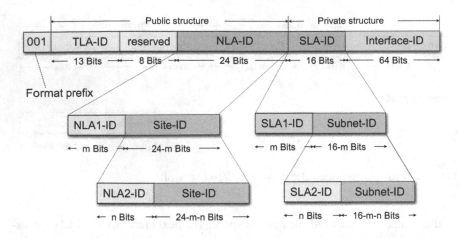

Fig. 7.33 IPv6 – aggregatable global unicast addresses.

addresses have the format prefix FE80::/10 and are not allowed to be forwarded by routers externally, but only in their own, local subnetwork.

In contrast to a LLU address, a Site Local Use Unicast address (SLU address) contains a subnetwork identification, but otherwise no structure information about higher-located levels within the Internet hierarchy. A SLU address can be sent beyond subnetwork borders, but only within an isolated site. SLU addresses contain the format prefix FEC0::/10 and are not forwarded by routers in the global Internet. They allow the assignment of unique addresses within an organization that is not connected to the global Internet, without having to use globally unique addresses. In order to enable unrestricted communication, an end system must be assigned a global unicast address.

LLU Address

1111 1110 10	00000 ... 000	Interface-ID

Format prefix ◄────────── 54 Bits ──────────►◄────────── 64 Bits ──────────►

SLU Address

1111 1110 11	000 ... 00	Subnet-ID	Interface-ID

Format prefix ◄──── 38 Bits ────►◄─ 16 Bits ─►◄────────── 64 Bits ──────────►

Fig. 7.34 IPv6 - Link Local Use/Site Local Use Unicast addresses.

- **Special unicast adresses**
  The special unicast addresses, described as follows, are defined in IPv6.

- **Unspecified address**:
The unspecified address 0:0:0:0:0:0:0:0, or "::" for short, can be used as the sender address if an end system has not yet received its own IP address. In order to ask about its own address, an appropriate IP datagram with the un-specified sender address is used.
- **Loopback address**:
For test purposes there are, just as in IPv4, loopback addresses 0:0:0:0:0:0:0:1, or "::1" for short. A datagram with a loopback address does not leave the com-puter via its interface, but only appears to be sent by it and received again. With the help of loopback addresses network applications can be tested befo-re they receive access to the actual network itself.
- **IPv6 addresses with encapsulated IPv4 addresses**:
These addresses are especially intended for the migration of IPv4 to IPv6, and thus enable the coexistence of both IP versions. Here, 32-bit long IPv4 addresses are expanded to 128-bit long IPv6 addresses. In the first option, a total of 96 0 bits are set in front of the 32-bit long IPv4 address (format prefix ::/96), to define a "IPv4 compatible IPv6 address," for application in a dual stack environment, meaning the systems in question must be both IPv4 and IPv6-capable. If in communication from an IPv6 network to a IPv4 network, an IPv4-compatible IPv6 address is used, there automatically follows an en-capsulation of the IPv6 datagram into a IPv4 datagram. In the header of the IPv4 datagram, the removed IPv4 address section from the IPv4-compatible IPv6 address is then automatically used .

As an alternative, so-called "IPv4 mapped IPv6 addresses" can be formed. Here, the 80 leading 0 bits follow a 16-bit long prefix made up of 1 bits, which appears in front of the IPv4 address (format prefix ::FFFF/96). These addresses are used for native IPv4 systems that are not IPv6-capable. In the case of two networks, when one is only IPv6-capable and the other one only IPv4-capable, a protocol translation is used between IPv6 and IPv4.

- **Multicast addresses**
IPv6 does not make provisions for broadcast addresses. This function is taken over by IPv6 multicast. Here it is possible to communicate directly with an entire group of different end systems simultaneously via a single address. The broadcast address therefore corresponds to a multicast address, assigned to all the compu-ters in the network. Fig. 7.35 shows the structure of a multicast address that al-ways begins with the format prefix 11111111, or FF::/8, and thus takes over1/256 of the available address space. The following fields are contained in the multicast address:

- **Flags**:
A 4-bit long field whose first three bits are currently reserved and must always be set to 0. The last bit (transient bit, T bit) serves to distinguish between a permanently assigned (T=0) or a temporarily assigned multicast address (T=1).

- **Scope**:
  This 4-bit long field specifies the scope of a muliticast address, beginning with a local scope (scope=1 – node local scope / scope=2 – link local scope) up to a global scope (scope=14 – global scope). This difference is important as multicast addresses with global scope must be allocated unambiguously throughout the entire Internet (see Fig 7.36). Multicast addresses with only limited, local validity, in contrast, must only be unambiguous within the scope for which they were established. Therefore, an especially high flexibility is possible with multicast addressing. The specifications of scope allows the router to decide how far multicast datagrams are to be forwarded, i.e., the data traffic is thereby limited to that part of the network fixed by the scope.

- **Group identifier**:
  This 112 -bit long field identifies the addressed multicast group. The multicast groups with the identifiers 0, 1 and 2 are reserved as so-called "well-known multicast addresses." 0 is not used, while 1 stands for all nodes in the relevant scope and 2 for all of the routers in the respective scope. In addition, every unicast address in a network is assigned a special multicast address (solicited-node multicast address). In this way, in comparison to the ARP protocol, a more efficient procedure for address resolution within the local network is implemented. The solicited-node multicast address is formed according to a fixed procedure from the last 24 bits of each unicast address of the relevant network node.

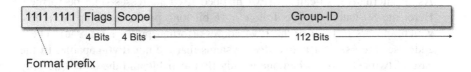

Format prefix

Fig. 7.35 IPv6 – multicast addresses.

- **Anycast addresses**
  Anycast addresses are among the new features of IPv6 over IPv4. Their functionality differs fundamentally from unicast and multicast. While a unicast address specifies one unique computer (more precisely, a specific network interface of that computer), a multicast address stands for all computers in a particular, established group to which an addressed datagram is sent. In the case of an anycast address, however, exactly one computer out of a specific, predetermined group of computers in the network is addressed. Normally, the computer of the anycast group is chosen closest to the transmitter in terms of the underlying routing metric. As a result, a datatgram is sent via anycast to the nearest computer of a group determined by the anycast address. All computers in an anycast group are capable of offering the same service. Services subject to a high traffic load can be scaled in this way without the requesting computer being required to find

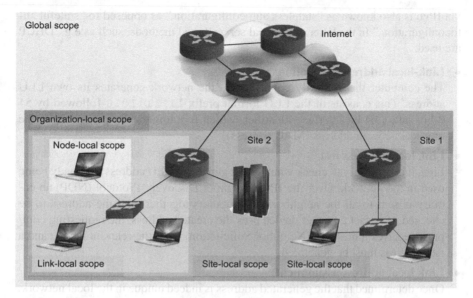

Fig. 7.36 IPv6 – multicast scope defines the validity area of the multicast.

out itself which computer from the group can deliver the requested service the fastest with the least effort (load sharing). This same functionality and flexibility is difficult to implement in IPv4.

There is no special address schema for anycast addresses. A unicast address automatically becomes an anycast address, as soon as it is assigned more than once. However, the farther the respective computers are from each other, the more complicated the administration of the anycast addresses becomes. The actual intention behind the definition of anycast was to house the computer in question if at all possible in the same network segment. In this way, management expenditures should be kept to a minimum while at the same time offering the required flexibility.

7.4.5 IPv6 Autoconfiguration

One of the most striking features of IPv6 addressing is the possibility of **autoconfiguration**. IPv6 was designed in such a way that an autoconfiguration, i.e., the assignment of an IP address to a computer newly connected to the network is possible without use of a dedicated server. Previously in IPv4 this service was either carried out manually or over a configuration protocol, e.g., the Dynamic Host Configuration Protocol (DHCP), which provides a special server computer in the network to give the required address information. Via the previously discussed LLU addresses, one's own IP address can be requested from a router. The autoconfiguration realized via IPv6 is also known as "stateless autoconfiguration," as opposed to "stateful autoconfiguration." In the latter, additional server-based methods, such as e.g., DHCP are used.

- **Link-local address generation**:
  The computer that is newly connected to the network generates its own LLU address. This consists of the LLU format prefix `1111111010` followed by 54 0-bits and a 64 bit interface identifier, which is automatically generated by the physical network address (MAC address).

- **Link local address test**:
  First, it is necessary to check whether the generated LLU address is already being used in the network. Over the IPv6 Neighbor Discovery Protocol (NDP) an enquiry is sent to all the neighboring computers together with the address to be checked. If in fact a further device uses the created address, the requesting computer receives a message (Neighbor Solicitation) from the relevant device and a new address must be generated.

- **Link local address assignment**:
  Once determined that the generated address is indeed unique in the local network, the address is assigned its own network interface. With the thus-generated LLU address, the computer is capable of unlimited communication in its own network. Nevertheless, access and reachability from the global Internet is still not possible.

- **Router discovery**:
  In order to be able to communicate with computers outside of one's own local network, autoconfiguration is continued in search of the nearest router. To do this, the configuring computer attempts to receive the **router advertisement** datagrams periodically sent by the router. Or, it sends itself a **router solicitation** in the local network, so that the nearest router transmits a router advertisement. As soon as the computer receives the router advertisement as an answer, it declares the answering router as its **default router**.

- **Router instructions**:
  The router advertisement sent by the router contains information concerning how autoconfiguration is to be carried out. Either the router informs the computer that an automatic address configuration is not possible and must proceed manually or via DHCP protocol (stateful autoconfiguration), or the router informs the computer how it should generate its global valid unicast address.

- **Global address configuration**:
  In the case where the local network allows an automatic autoconfiguration, the originally requesting computer generates its now globally valid unicast address, based on the information that the router has transmitted. Normally at this time the router communicates the network prefix that is to be used, which is combined with the interface identifier of the device (see step 1).

In the realm of autoconfiguration also falls the renumbering of entire address areas (address renumbering). In IPv4, address numbering is problematic for networks

IPv6 Address Generation from Hardware Addresses

The generation of new IPv6 addresses for a network-capable device proceeds according to a simple address imaging procedure. This images a device-specific 48 bit IEEE 802 MAC address onto the last 64 bits of an IPv6 address. The 48 bit IEEE 802 MAC address is normally comprised of two parts, each 24 bits long. The first is the Organizationally Unique Identifier (OUI), which identifies a specific hardware manufacturer. The second, a further 24 bits, identifies an individual device of this manufacturer. Additionally, the IEEE further defined a 64-bit long address format for device identification: the 64 bit Extended Unique Identifier (EUI-64). Besides a 24 bit OUI, it provides a 40 bit device identification field. A special variation of this format, the modified **EUI-64** address, is intended for use in the IPv6. Thereby, only the 7th bit of the EUI-64, the so-called universal/local bit, is set from value 0 to value 1.

Today, mostly 48 bit MAC addresses are used. These are transformed into an EUI-64, and subsequently into a modified EUI-64, as follows:

1. The 48 bit MAC address is divided into the 24 bit OUI and the remaining 24 bit address suffix. The original OUI occupies the first 24 bits of the EUI-64, the address suffix the last 24 bits of the EUI-24.

2. The remaining 16 bits, in the middle of the EUI-64, receives the following bit pattern: 11111111 11111110 (FFFE as hexadecimal number).

3. The address is now in the EUI-64 format and only the 7th bit must be set to value 1. In this way it is transformed into a modified EUI-64, which now makes up the last 64 bits of the new IPv6 address. The front 64 bits of the IPv6 address comprise each network prefix respectively.

Example:

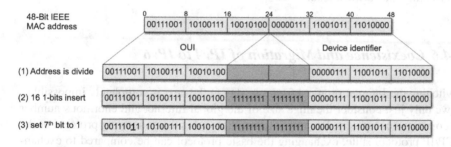

Further reading:

Hinden, R., Deering, S.: IPv6 Addressing Architecture, Request for Comments, Request for Comments 4291, Internet Engineering Task Force (2006)

Fig. 7.37 IPv6 – LLU address generation from hardware addresses.

starting from a certain size, even when they are supported by protocols and procedures, such as DHCP. In this procedure, the network devices involved must be
reachable for a period of time at both the old and new IP addresses. This is especially difficult if a change is to be made from one Internet Service Provider (ISP)
to another. in this case, it is only possible when the network is operated as "multihomed," meaning that during a certain time span it is served by more than just
one ISP, with IP connectivity and IP address areas. Address renumbering with the
help of the Border Gateway Routing protocol (BGP) is a way to enable this in IPv4.
But, at the same time, this leads to a fragmentation of the address space, as a multitude of small networks reach as far as the the routing tables of the core area of
the Internet in the process and the routers involved must be laid out accordingly. In
contrast, the problem of address numbering was taken into account in advance when
the IPv6 protocol was designed and standardized in RFC 4076. Entire networks can
be renumbered in IPv6. To do this, during autoconfiguration routers can provide the
network prefixes sent to the hosts with an expiration interval. It is sufficient when a
new network prefix is sent via a router. Hosts in IPv6 are capable of generating a new
unicast address by simply appending their host identifier on the network prefix. The
parallel operation of multiple IP address areas also proves uncomplicated in IPv6.
In RFC 3484 the procedure of how the source and target addresses are selected is
fixed when more are available. The network operator is responsible for enabling an
uncomplicated exchange between ISPs as well as a lasting parallel operation with
multiple ISPs. The focus is on encouraging competition, increasing reliability and
distributing the network load.

7.4.6 Coexistence and Migration of IPv4 to IPv6

Switching the Internet protocol from one day to the next is completely impossible -
if we only just consider the huge size of the global Internet and enormous number
of connected computers. Because Internet Protocol is the underlying protocol of the
TCP/IP protocol suite, exchanging the basic protocol can be compared to exchanging the foundation of a building. Even if this image might first seem a bit incon-
gruous, it is clear that exchanging such a foundation can only proceed with great
care. If we take this metaphor one step further, it is possible to compare the global
introduction of the new IPv6 protocol with the replacement of all building foundations in the entire world. While the conceptual specification of the IPv6 protocol in
fact already took place over a decade ago, in practical application the changeover to
the new Internet protocol hides multiple difficulties that are not only technical but
above all economic in nature.
The installed IPv4 hardware basics are enormous and a changeover to IPv6 must
be carefully planned and implemented. Almost unnoticed by end users, this has
already proceeded step-by-step for years through the realization of IPv4 to IPv6
coexistence and migration scenarios. Nearly all operating systems deployed today
support IPv6. Likewise, almost all IPv6 hardware components today support the

IPv6 and Privacy

The possibility of stateless auto configuration offered in IPv6 enables the generation of an individualized IPv6 address solely from the respective network prefix and the IEEE 802 MAC address of the network end device. The MAC address consists of the manufacture-specified Organizationally Unique Identifier (OUI) as well as a device specified part (Interface Identifier), which serves to identify the network end device. The calculation of the IPv6 address is carried out in accordance with a specified schema (modified Extended Unique Identifier, modified EUI-64, see Fig. 7.37) and thus, vice versa, enables the identification of the network device via the assigned IPv6.

Most of the network end devices are used by just one person (personal computer, smartphone, etc.). Therefore, knowledge of the IPv6 address - left behind by an end device with Internet service use - allows unambiguous conclusions to be drawn about the identity of the end device, and therefore about its user. In this way, it is thus possible for outsiders to draw conclusions about user behavior and the activity profile already on the IP level of the network.

In IPv4 such possibilities to make inferences about the identity of a user were not available. IP address allocation within a network was usually regulated by the Dynamic Host Configuration Protocol (DHCP), which simply assigned the IP address temporarily on the basis of availability. This process was made more difficult by the Network Address Translation (NAT) Technology, which enabled the setup and management of entire networks under a single IPv4 address.

For this reason, with RFC 4941 so-called "**IPv6 privacy extensions**" were introduced to protect the privacy sphere. These create random interface identifiers for the last 64 bits of the IPv6 address, which in each case are generated with the help of a MD5 hash function (Message Digest Algorithm 5) from the original interface identifier, or from the last calculated MD5 value and retain their validity for a limited time. In this way, not only is an identification of the user prevented, but also the establishment of a permanent IPv6 address.

In addition to the IPv6 privacy extensions, further possibilities exist to conceal user identity. For example, instead of the Neighbor Discovery Protocol (NDP), the Secure Neighbor Discovery Protocol (SEND) is implemented. With the help of cryptographic methods, it calculates new interface identifiers (Cryptographically Generated Addresses, CGA) from the fixed network prefix, a public key and a nonce value. IPv6 privacy extensions, as well as CGA, use a situation-dependent nonce value for calculation of the interface identifier, meaning that with every new network connection a new interface identifier is also generated.

Further reading:

Narten, T., Draves, R., Krishnan, S.: Privacy Extensions for Stateless Address Auto-configuration in IPv6, Request for Comments, Request for Comments 4941, Internet Engineering Task Force (2007)

Arkko, J., Kempf, J., Zill, B., Nikander, P.: SEcure Neighbor Discovery (SEND), Request for Comments, Request for Comments 3971, Internet Engineering Task Force (2005)

Fig. 7.38 IPv6 privacy extensions.

handling of IPv4. Only older, purely IPv4-based hardware components cannot communicate with components connected exclusively over IPv6 without an appropriate translation.

The key to a successful changeover to IPv6 is found in a migration that is long term and as cost-efficient as possible. In principle, this migration can happen in three different ways.

- **Parallel Implementation (Dual Stack Implementation)**

  In this variation, systems such as routers are employed that are capable of understanding both IP versions at the same time and of communicating in both versions. A dual-stack-capable network end device is also known as an IPv4/IPv6 node. Any communication it has with an an IPv6-capable network end device proceeds via IPv6. Communication with an older, exclusively IPv4-capable network end device is carried out via IPv4. An IPv4/IPv6 computer thus has at least two IP addresses each: an IPv4 address and an IPv6 address. Address configuration that is either a static configuration or a configuration via DHCP is implemented on the IPv4 side. In contrast, the IPv6 side can carry out its own IPv6 auto configuration. In IPv6 special attention must be paid to the assignment of domain names. This point will be looked at more closely in the section 9.2.1 Domain Name Service.

  To be able to take up the dual stack operation in a network, the complete network software that in the network router must be renewed accordingly with the help of an appropriate upgrade. Existing routing protocols must be available in duplicate form for IPv4 and for IPv6 each. Fig. 7.39 shows the communication in a dual stack network:

  Computer A is configured for dual stack operation (dual stacked host) and has an IPv4 and an IPv6 address available. If computer A wants to communicate with computer B, it inquires at the responsible DNS server (Domain Name System) about an address translation of the domain name from computer B into an IP address. In the case of computer B, the DNS Server gives back an IPv4 address via which an IPv4-based communication can then take place between A and B. If computer A wants to communicate with computer C, the DNS server returns an IPv6 address and communication between A and C proceeds over IPv6.

- **IPv4/IPv6 Translation (IPv4/IPv6 Translation)**

  A special variation of the dual stack operation distinguishes itself in that several systems are capable of accepting inquires from IPv6 hosts, of converting them into IPv4 datagrams and of forwarding them to an IPv4 addressee. The answer of the IPv4 device is processed in the reverse way. This variation should, however, only then be employed if an IPv6-only network end device wishes to communicate with an IPv4-only network end device. As both IPv4 and IPv6 protocols are not completely compatible, information loss takes place in the translation process. In addition, the Internet layer in the network protocol stack receives an added level of complexity Currently, the **Stateless IP/ICMP Translation (SIIT)** has established itself as a standard solution in this area. To facilitate communication between an IPv4-capable and an IPv6-capable device, RFC 2765 defines a

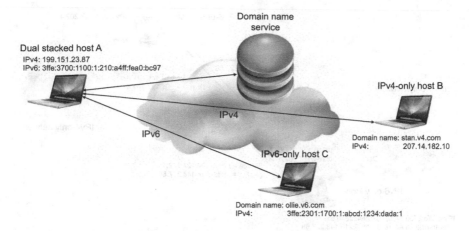

Fig. 7.39 IPv6 - dual stack implementation.

procedure for translating the IPv6 data packet header (and ICMPv6 data packet header) in the corresponding IPv4 data packet headers (or ICMP data packet header) and vice versa.

Structuring a new network segment that works purely in IPv6 operation is one possible application of the SIIT procedure. Its data traffic is connected to regular IPv4-capable network segments with the help of the SIIT protocol translation, as well as to the standard IPv4-capable network. For this purpose, a special IPv6 address type was introduced, the "IPv4-translatable address." These addresses have the address prefix $0::ffff:0:0:0/96$. As host identifier an IPv4 address, which comes from a special address pool, is appended to this address prefix. The address prefix was chosen to always produce the value zero in the checksum calculation occurring on higher protocol layers and therefore does not influence the checksum.

IPv6 expansion headers, e.g., routing headers, hop-by-hop options headers and destination options headers allow just as little reciprocal translation as the IPv4 header options does. It is also not possible to translate IPv6 multicast data traffic because IPv4 multicast addresses cannot be converted into IPv6 multicast addresses. If an IPv4 data packet reaches a router which is intended for protocol translation via SIIT, based on the IPv4 address it can decide whether it is a data packet intended for an IPv6-only host and for this reason has to translated. To carry out translation, the IPv4 data packet header is removed and the details contained therein translated into a new IPv6 data packet header. This is then placed in front of the original payload. Fig. 7.40 shows address conversion via SIIT. Computer A is located in an IPv6-only network and computer B in an IPv4-only network. Correspondingly, computer A has an IPv6 address, which for reasons of reachability via IPv4, was chosen as an IPv4-translatable address. Computer B has only an IPv4 address. From the IPv6-only network this can also be addressed as an IPv4-translatable address.

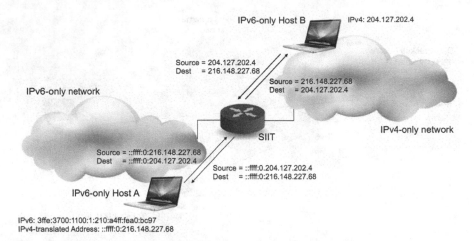

IPv6-only Host B   IPv4: 204.127.202.4

Source = 204.127.202.4
Dest   = 216.148.227.68

IPv6-only network

Source = 216.148.227.68
Dest   = 204.127.202.4

SIIT

IPv4-only network

Source = ::ffff:0:216.148.227.68
Dest   = ::ffff:0:204.127.202.4

Source = ::ffff:0.204.127.202.4
Dest   = ::ffff:0:216.148.227.68

IPv6-only Host A

IPv6: 3ffe:3700:1100:1:210:a4ff:fea0:bc97
IPv4-translated Address: ::ffff:0:216.148.227.68

Fig. 7.40 IPv6 - protocol translation via Stateless IP/ICMP Translation (SIIT).

There are additional IPv4/IPv6 translation variations as well.

– **Network Address Translation – Protocol Translation (NAT-PT)**
   Based on the network address translation protocol (see section 8.4 for a detailed presentation of the NAT technology). This procedure implements a protocol translation that was intended for IP address translation as well as for the
   handling of checksums for the protocols IP, TCP, UDP und ICMP. However,
   due to various problems, this protocol proposed as RFC 2766 was taken out
   of circulation and relegated to the status "deprecated."

– **Network Address Port Translation – Protocol Translation (NAPT-PT)**
   This protocol translation procedure is almost identical to NAT-PT. Besides the
   translation of IP addresses, it was intended for the translation of TCP and UDP
   port numbers (see Chap.. 8.3) and ICMP messages. Just as NAT-PT (and for
   the same reasons) NAPT-PT was abandoned.

– **Transport Relay Translation (TRT)**
   This protocol translation procedure implements the specified methods established in NAT-PT and NAPT-PT and is based on the implementation of DNS
   entries.

– **NAT64**
   NAT64 allows the communication of IPv6-only end devices with IPv4-only
   servers. An NAT64 server has at least one IPv4 address of its own and a IPv6
   network segment the size of 32 bits (64:FF9B::/96). The IPv6 client em-
   beds the IPv4 address, with which it wishes to communicate, with the help
   of these 32 bits. It then sends its data packets to the resulting address. The
   NAT64 server maps this IPv6 address on the IPv6 address contained therein.

• **Tunneling**
   In this variation of IPv4/IPv6 protocol implementation, IPv6 datagrams in a IPv4

network are forwarded unchanged. An interpretation of the IPv6 datagram only takes place in systems that are IPv6-capable. IPv6 datgrams are completely encapsulated as payload in IPv4 datagrams and sent through the IPv4 network. In this case, IPv6 is operated like a protocol attached to IPv4 - as if located in a protocol stack on a higher protocol layer. The specification used in the IPv4 datagram regarding the protocol type of the transported payload is 41 for IPv6 . The tunneling of IPv6 datagrams is necessary if all network nodes between both IPv6 communication end points do not have dual stack capability. The reverse situation can also occur where IPv4 datagrams must be transported through an IPv6 network. In this case, IPv4 data packets are transported as payload in IPv6 datagrams. Fig. 7.41 shows the forwarding of IPv6 datagrams via an IPv6 tunnel inside an IPv4-only network.

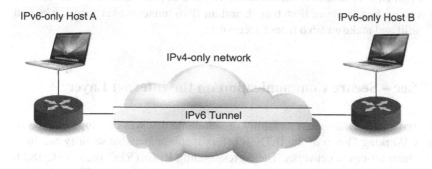

Fig. 7.41 IPv6 – a tunnel through an IPv4-only network.

Different technologies are implemented here:

– **6to4**

The 6to4 procedure for IPv6 protocol implementation is standardized in RFC 3056 and implements an automatic IPv6 tunnel procedure through IPv4-only networks. This means the routing infrastructure is capable of determining the tunnel end points without manual configuration. The end points of the thus- realized IPv6 tunnels are implemented with the help of known IPv4 anycast addresses on the remote receiver's side and via embedding of the IPv4 address information on the local transmitter's side in IPv6 addresses. The use of 6to4 is widespread today, e.g., in most of the UNIX variations or in Microsoft Windows Vista and Windows 7.

– **Teredo**

The Teredo method is standardized as RFC 4380 and also implements an automatic IPv6 tunnel procedure on the basis of encapsulated UDP data streams. Multiple NAT bridges can likewise be tunneled through. Due to native support from the side of Microsoft Windows Vista and Windows 7 the use of Teredo is very widespread.

- **Intra-Site Automatic Tunnel Addressing Protocol (ISATAP)**
  The ISATAP procedure for the IPv6 protocol implementation is specified in
  RFC 5214 and deals with an IPv4-only network as a virtual local IPv6 connection, i.e., every IPv4 address is mapped via a special mapping procedure onto
  a IPv6 link-local address. In contrast to 6to4 and Teredo, ISATAP is only deployed as a intra-site tunnel procedure within a corporate network.

- **6in4**
  The 6in4 procedure for IPv6 protocol implementation is specified in RFC
  4213 and, in contrast to 6to4, implements a statistical IPv6 tunnel procedure
  in which the tunnel end point must be configured manually.

- **Tunnel Setup Protocol (TSP)**
  The tunnel setup protocols, specified in RFC 5572 is deployed during the setup of an IPv6 tunnel. It automatically negotiates parameters between a client,
  who wishes to use an IPv6 tunnel, and an IPv6 tunnel broker server, who can
  build and make an IPv6 tunnel accessible.

7.5 IPsec – Secure Communication on the Internet Layer

IP Security (IPsec) designates a family of standards that was developed by the IP
Security Working Group of the IETF. It offers a comprehensive security architecture for Internet-based networks. The corresponding RFCs (RFC 1825, RFC 1829,
RFC 2401–2409, RFC 4301–4309, RFC 4430) were published in 1998 and all refer to the Internet layer of the TCP/IP reference model. They describe data formats
for different methods of encryption and authentication of IP datagrams. IPsec was
developed as an integral component of IPv6. It can also be used optionally for applications with IPv4. Table 7.6 gives an overview of the most important IPsec Internet
standards.

Table 7.6 The most important IPsec Internet standards.

RFC	Title
2401/4301	Security Architecture for the Internet Protocol
2402/4302	IP Authentication Header
2403	The Use of HMAC-MD5-96 within ESP and AH
2404	The Use of HMAC-SHA1-96 within ESP and AH
2406/4303	IP Encapsulation Security Protocol (ESP)
2408	Internet Security Association and Key Management Protocol (ISAKMP)
2409/4306	The Internet Key Exchange (IKE/IKEv2)
2412	The OAKLEY Key Determination Protocol
3602	The AES-CBC Cipher Algorithm and Its Use with IPsec
4430	Kerberized Internet Negotiation of Keys (KINK)

7.5.1 IPsec Security Architecture

The summary of protocols and services under IPsec serve different security goals. As security architecture is already implemented on the Internet layer, all of these services can be used by all protocols and applications on higher protocol layers without them having to implement individual security mechanisms. This distinguishes IPsec from security architectures on higher protocol layers of the TCP/IP reference model, such as the Secure Socket Layer (SSL) or the Transport Layer Security (TLS). Special measures must be taken to use these based on their Internet application. Included among the security goals addressed by IPsec are:

- Encryption to maintain secrecy of messages transmitted by the user.
- Securing the integrity of transmitted messages so that these cannot be altered by an unauthorized party during transmission.
- Protection against certain variations of denial of service attacks, e.g., so-called replay attacks, in which access to a network connection is disturbed by the repeated, mass transmission of data packets.
- The possibility of negotiating the respective security algorithm used between communicating end devices based on their security requirements.
- Supporting two different security modes (transport mode and tunnel mode) to thereby support differentiated security demands.

To enable secure IP-based communication on the Internet layer of the TCP/IP reference model, the involved network devices (either the network end devices or the Internet intermediate systems, e.g., routers or firewalls) must set up a secure path, which potentially numerous unsecure intermediate systems may need to pass. The network devices involved need to fulfill at least the following requirements:

- The involved network devices must agree to the basic number of respective security protocols to be used prior to communicating with each other.
- The involved network devices must agree on the encryption procedure used.
- The involved network devices must be capable of exchanging cryptographic keys, to subsequently be capable of decrypting encrypted data.
- After the conclusion of this approval process, the protocols and procedures which the network devices have agreed on are then used for safe communication.

Two core protocols were introduced for the implementation of the IPsec basic functionality, **IPsec Authentication Header (AH)** und **Encapsulating Secure Payload (ESP)**. AH serves the purpose of safeguarding the authenticity of each transmitter and guarantees the integrity of the transmitted information. ESP is responsible for the encryption of the transmitted payload. Although AH and ESP are designated as protocols, they are only each responsible for special data packet headers added to the IP datagram. They are supported by the following IPsec support components (cf. Fig. 7.42):

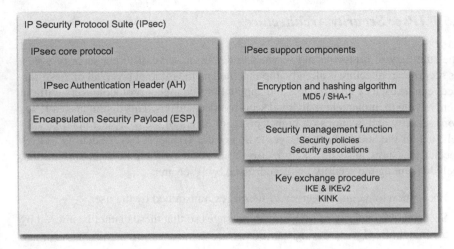

Fig. 7.42 Overview of the IPsec security architecture.

- **Encryption and hashing algorithm**
  AH and ESP are generic in nature and therefore do not specify the actual encryption algorithm. IPsec thus retains a flexible architecture as well as allowing the integration of different encryption procedures and methods for their negotiation. Two simple cryptographic hashing methods implemented in IPsec are **Message Digest 5** (**MD5**) and **Secure Hash Algorithm 1** (**SHA-1**).[3]

- **Security management function**
  IPsec grants permission to the network devices involved concerning negotiations of security mechanism and procedures. Therefore, rules and ways of proceeding (**security policies** and **security associations**) must be established as to how this is to be carried out.

- **Key exchange procedure**
  This is a procedure for the communication of two network devices via a cryptographic key exchange. It is necessary that an advance exchange of cryptographic keys takes place to enable a decryption of the encrypted information. This is done in IPsec over the **Internet Key Exchange protocol** (IKE and IKEv2) as well as the **Kerberized Internet Negotiation of Keys** (KINK, RFC 4430).

RCF 2401 and RFC 4301 propose different implementation forms of the IPsec security architecture. These mainly distinguish themselves according to the network component where the security mechanisms are implemented:

- **End-Host implementation**
  If IPsec is installed on all devices connected to a network the greatest flexibility

---

[3] A detailed presentation of cryptographic hashing procedures is found in excursus 17 „Cryptographic Hash Functions" in the first volume of this series: Meinel, Ch., Sack, H.: Digital Communication – Networking, Multimedia, Security, Springer (2013).

and security possible with the deployment of IPsec is gained. Virtually any one network device can communicate with another.

- **Router implementation**

  In this variation, the IPsec components are only installed on the routers of one network. This means much less cost and effort. However, at the same time, secure data transport can only be guaranteed between the routers of the network. This procedure is sufficient to set up a Virtual Private Networks (VPN) within a network. However, the connection between an network end device and the next router remains unprotected. As this unprotected connection is usually located within a company's borders it is, however, viewed as a less dangerous risk.

- **Integrated architecture**

  Ideally, IPsec should be directly integrated in the IP protocol. This would mean that all security functions would be directly integrated in the basic protocol of the Internet layer already and no extra software components or layers of architecture would be necessary. This was taken into consideration with the conception of IPv6 and IPsec is an integral component of the new Internet protocol. A belated integration in the IPv4 protocol would require modifications of all currently existing IPv4 implementations on all network devices, making this impractical.

- **Bump in the Stack**

  In this alternative, IPsec can be viewed as an additional protocol layer in the TCP/IP reference model. This variation, called "Bump in the Stack" (BITS) has IPsec intercept regular IP datagrams, modify them based on the available security functions and pass them on to the data link layer for forwarding. This means the protocol layer defined by IPsec is located between the Internet layer and the data link layer. IPsec can be easily installed at a later time in every network device as it does not interfere with the IP protocol. Admittedly, this procedure involves an additional communication overhead in contrast to integrated architectures. IPv4-capable network devices are usually equipped with IPsec via BITS.

- **Bump in the Wire**

  This variation known as "Bump in the Wire" (BITW) introduces additional hardware components in order to provide a dedicated connection in the network with IPsec functionality. This can take place without intervening in the network software of network devices. Between two routers, two new IPsec-capable intermediate systems are brought into the network in each case. These secure the originally unencrypted data traffic between the original routers explicitly via IPsec. Because in BITW additional hardware must be brought into the network for every router connection, this solution is particularly complex and costly.

Depending on the type of communication that takes place via IPsec, a distinction is made between two **IPsec operation modes**. IPsec either connects two communication end points directly with each other (transport mode) or is used to connect two subnets with each other over two routers (tunnel mode). Both modes are closely associated with the respective core protocols AH and ESP. In this way, not only their functionality is influenced, but also the chosen mode determines which parts of the IP datagram are protected and how the datagram header arranges the core protocol.

- **Transport mode**

  In the transport mode, the IPsec datagram header is added between the IP data-gram header and the payload, which is directly passed on from the above trans-port layer of the TCP/IP-reference model. This means the IP header follows the IPsec header and afterwards the TCP/UDP header of the forwarded data from the transport layer. In this case, the IP header remains unchanged (see Fig. 7.43). Only the transported payload is protected by IPsec and not the original IP hea-der. After receiving the IPsec datagram, the original payload (TCP/UDP packet) is unpacked and forwarded to the transport layer located above The transport mo-de is particularly suited for host-to-host or host-to-router, applications, e.g., for network management.

- **Tunnel mode**

  In contrast to the transport mode, the complete original IP datagram is encapsu-lated by IPsec in the tunnel mode , i.e., the IP payload as well as the IP datagram header. The encapsulated IP datagram is thereby prefixed with an IPsec header and this in turn given a new IP header addressing the tunnel end points (see Fig. 7.43). The actual start and destination addresses of the IP datagram are lo-cated in the inner IP header. This is encapsulated by IPsec and first evaluated after it passes the router at the end of the tunnel (security gateway) Normally, in the tunnel mode router-to-router as well as host-to-router connections are imple-mented. If the communication end points are at the same host as the tunnel ends, host-to-host connections are also possible in this mode. One of the advantages of the tunnel mode is that IPsec has to only be installed and configured at the tunnel ends.

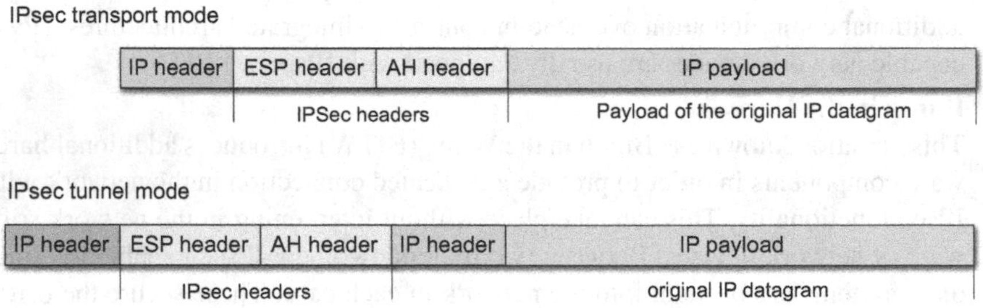

IPsec transport mode

IP header	ESP header	AH header	IP payload
	IPSec headers		Payload of the original IP datagram

IPsec tunnel mode

IP header	ESP header	AH header	IP header	IP payload
	IPsec headers			original IP datagram

Fig. 7.43 IPsec-datagrams in the transport and tunnel modes.

7.5.2 IPsec Authentication Header (AH)

The IPsec authentication header, one of the two IPsec core protocols, serves to pro-vide security for the authentication and integrity of the transmitted IP datagram. For this purpose, the original IP datagram is given a new header, which is calculated from the contents of the transmitted IP datagram. Which parts of the original datagram are included in the calculation of the AH depends on the respective transport mode and version of the IP protocol (IPv4 or IPv6) used. The processing of the AH header proceeds in the same way as the checksum calculation in the protocols of data link layer or Internet layer. Based on a predetermined cryptographic procedure, the transmitter calculates a hash value from the payload to be protected. This is prepended to the IP datagram in the AH header. The receiver carries out the same calculation and compares its result with the hash value of the received AH header. If both match, the datagram is considered correctly authenticated. If both do not match, values of the original data packets have been modified somewhere along the line. In this way, AH is responsible for maintaining the integrity of the transferred data, however, not for its secrecy. In contrast to checksum calculation, the authentication in AH can only be carried out with the help of a cryptographic hash function together with a cryptographic key. Both may only be known to the transmitter and to the receiver. The hash function as well as the key are agreed upon between the communication partners prior to the the the IPsec secure data transfer with help of the security association and key exchange protocols. The test value calculated in this way is known as the **Integrity Check Value (ICV)**.

To indicate that an IP datagram is an IPsec datagram in IPv4, the protocol field of the IP datagram header is set to the value 51 to refer to the inserted authentication header. This in turn refers to the header of the transported payload (TCP/UDP header) in the transport mode and in the tunnel mode to the header of the encapsulated IPv4 datagram (see Fig. 7.44).

In IPv6, the authentication header is treated as a normal IPv6 extension header, meaning the prepended IPv6 datagram header in the next header field refers to the authentication header. This, in turn refers in its next header field either to a further IPv6 extension header or directly to the header of the transported payload (TCP/UDP header) (see Fig. 7.45).

The length of the entire authentication header must be a multiple of 32 bits (IPv4) or 64 bits (IPv6) and is always structured as described in Fig. 7.46:

- **Next header**
  1 byte, indicates the type of transported payload, analogous to a IPv4 header.

- **Payload length**
  1 byte, in contrast to its designation does not indicate the length of the transported payload but the length of the AH. The length is given in units of 32 bits each, whereby the value 2 is subtracted in order to ensure consistency in calculating the header length.

- **Security Parameter Index (SPI)**
  4 bytes, specifies the type of security procedure used. Different parameters are necessary here (e.g., procedure used, implemented cryptographic key, validity duration of the key etc.) and are summarized in a so-called **Security Association (SA)**. The SPI only gives an indication of the actual SA entry of the security association database. This is a database containing all available parameter com-

Original IPv4 datagram

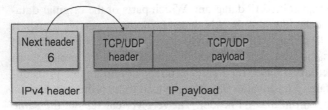

IPsec IPv4 authentication header datagram - IPsec transport mode

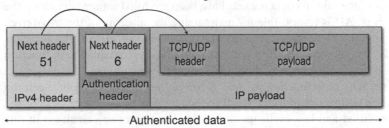

IPsec IPv4 authentication header datagram - IPsec tunnel mode

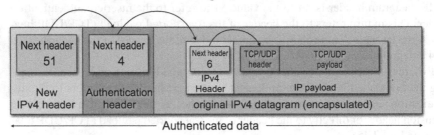

Fig. 7.44 IPv4 datagram with IPsec authentication header.

binations and which must be available to both the sender as well as the receiver (see also section 7.5.4). To specify the SA entry unambiguously, the SPI as well as the IP destination address are both necessary.

- **Sequence number field**
  4 bytes, which number the exchanged datagrams between the sender and receiver. This procedure is intended to prevent replay attacks, which are detected by the security system based on previously sent and successfully received IP datagrams. Nevertheless, following the exchange of 2^{31} IP datagrams at the latest, a new SA must be established to completely eliminate the possibility of replay attacks. In each case, sequence numbers are added by the sender.

- **Authentication data**
  Details about authentication in variable lengths (Integrity Check Value). The length is dependent on the respective authentication procedure described in the

Original IPv6 datagram

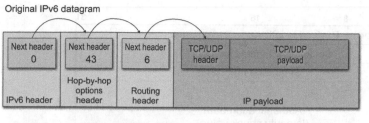

IPsec IPv6 authentication header datagram - IPsec transport mode

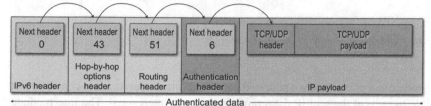

IPsec IPv6 authentication header datagram - IPsec tunnel mode

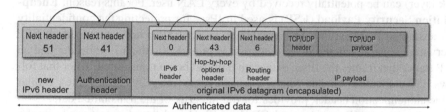

Fig. 7.45 IPv6 datagram with IPsec authentication header.

RFCs 2403 (MD5) and 2404 (SHA-1), and must be a multiple of 32 bits in each case. It is important in the calculation of a message digest for authentication that the calculated authentication value is set to 0, since this authentication value itself may not be included in the calculation. The other values that change during transport (e.g., TT field, fragment offset, header checksum), must also be set to " 0" in the calculation of the message digest. In implementing the options "loose source routing" and "strict source routing," in which a series of routers to be passed are predetermined, attention must be paid that always only the last address from the provided address list is used for calculating the message digest (in other words, the actual target).

7.5.3 IPsec Encapsulation Security Payload (ESP)

Besides authentication, secrecy maintenance of the transferred message by means of an appropriate encryption procedure is certainly the most important safety requirement in the global Internet. Because every single IP datagram of a transmitted Inter-

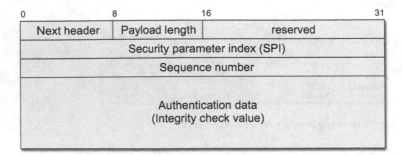

Fig. 7.46 Authentication header of the IPsec datagram.

net message can pursue its own path to the predetermined receiver, it is generally not possible to prevent the sent IP datagrams from being intercepted. The intermediate systems involved are accessible, and a transmitted datagram on the LAN level (data link layer) can be potentially received by every LAN user. For this reason, **Encapsulation Security Payload (ESP)** is used in IPsec for protecting the confidentiality of transmitted IP datagrams. Just as AH, the use of ESP is not compulsory. If ESP encryption is deployed, the IPsec datagram is provided with an ESP header and an ESP trailer that frames the IPsec datagram payload being transmitted. After that follows ESP authentication data whereby a hash value (integrity check value), similar to the authentication header, provides for the integrity of the transmitted data.

Depending on the IP version used (IPv4 or IPv6) and the transport mode implemented (tunnel mode or transport mode), the placement of the ESP header in the IP datagram is carried out similarly to the authentication header. Thus, the ESP header in the transport mode in IPv6 is handled the same way as an IPv6 extension header. It is placed after all other IPv6 datagram headers, but before an IPv6 destination options header, which contains control information for the target computer. In the tunnel mode, in contrast, the original, completely encapsulated IPv6 datagram is placed in front of a new IPv6 basis header followed by the ESP header (see Fig. 7.47). In IPv4, the ESP header is followed in each case by the IPv4 header in the transport mode the transported payload, and in the tunnel model the completely encapsulated IPv4 datagram (see Fig. 7.48). In both IPv4 and IPv6, the ESP trailer follows up the transported payload data of the IP datagram. Normally, the ESP trailer contains filler data (padding). This is necessary as different encryption procedures require the data to be encrypted to have a fixed block size. In ESP, the transported payload as well as the appended ESP trailer are encrypted. Unlike the authentication header, the ESP header does not contain a next header field that shows the type of protocol header to follow. The next header field in ESP is located in the ESP trailer. The subsequent authentication data refers to the entire ESP datagram, i.e., the ESP header, the transported and encrypted payload as well as the ESP trailer.

ESP header and trailer consist of the following fields (see Fig. 7.49):

- **Security Parameter Index (SPI)** and **Sequence Number**
  These two-bit long fields of 4 bytes each are identical with the corresponding

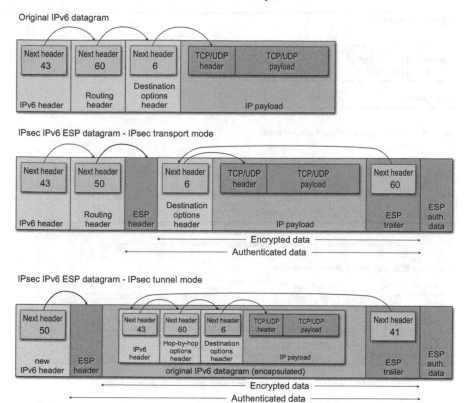

Fig. 7.47 IPv6 datagram with IPsec ESP header - the specified values in the next header field indicate the type of the sequence header (cf. Tab. 7.4)

fields in the AH. Here, the SPI with the target address indicate a security association containing an encryption procedure and its necessary cryptographic parameters (e.g., key or initialization vector), and not an authentication procedure (hash procedure) as in the case of the AH.

- **Padding**

  As the transported payload can be of different lengths, padding must be added so that the subsequent ESP authentication data can always begin at a 32 bit word boundary, or so that the data to be encrypted always maintains the predetermined block size. If as an encryption procedure a block cipher is chosen such as Advanced Encryption Standard (AES) or Data Encryption Standard (DES), it is important to note that these always process 8 or 16 bytes. The number of padding bits must therefore be calculated so that the length of the payload - including padding, padded bit length and subsequent next header field - is a multiple of this block length. The padding length (1 byte) must always be indicated in order to remove the padding after reception of the IPsec datagram.

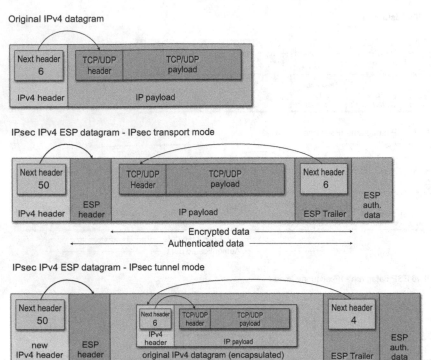

Fig. 7.48 IPv4 datagram IPsec ESP Header - the specified values in the next header field indicate the type of the sequence header (cf. Tab. 7.4)

- **Next header**
  This field corresponds to the next header field of the regular IPv4 datagram that describes the type of transported data.
- **ESP authentication data**
  As in implementation of the AH, a message digest for authentication of the transported data can be transported over ESP. Here, however, the calculation is easier as only static fields are incorporated

7.5.4 IPsec Support Components

Before a secure data exchange can take place via IPsec, security protocols and cryptographic keys must be negotiated between the communication partners involved. While the data transmitted via IPsec can be intercepted and read by uninvolved third parties,when implementing ESP the data is in encrypted form, which means that manipulation cannot be carried out in AH without it being noticed. In order to

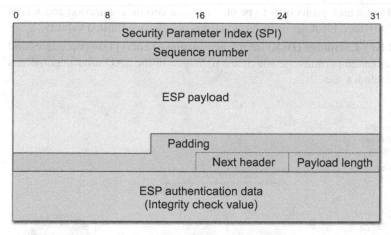

Fig. 7.49 Encapsulation security payload in the IPsec datagram.

be able to read and change data in clear text, the authorized communication partners must have access to the confidential cryptographic keys. These are exchanged over a special protocol, the **Internet Key Exchange Protocol** (**IKE**), which will be discussed in greater detail in the excursus "IPsec – Key Management." In the simplest case, the keys to be used are already at the respective end system (pre-shared keys). In the other case, the keys are determined by way of a random value to be used one time (shared session key). This is exchanged between the communication partners with the help of an asymmetrical encryption procedure.

Looking at an IPsec implementation shows that it consists of a series of individual software modules (see Fig. 7.50):

- The **AH/ESP module** must interconnect the generation and processing of the **Authentication Header** (**AH**) and **Encapsulation Security Payload** (**ESP**) of the IP datagram in the network software on the network level. Incoming encrypted IPsec datagrams are decrypted, checked and passed back again to the network software as regular IP datagrams. The IP datagrams to be sent are received and based on the respective specifications forwarded unchanged, encrypted, authenticated or discarded.

- **Security Policy Database** (**SPD**) has static regulations concerning the handling of IP datagrams that are to be sent between two communication partners along a certain connection. The SPD module must not necessarily be implemented in the form of a database, but can be laid out as a routing table. As such, IP datagrams to be encrypted are simply routed to the network adapter of the AH/ESP module.

- **Security Association Database** (**SAD**) with each available variation of the security mechanism and the accompanying parameters (security associations). The AH/ESP module uses the SAD module by referencing the routers used via - SPI and IP target addresses. Security associations each have a specific status which, in comparison to the entries in SPD, can change relatively often. They contain

details about which protocol is to be chosen for a certain connection and which
keys are provided for it Security Associations only have a limited validity.

- **Internet Key Exchange (IKE)** for the negotiation of encryption procedures and
  the implemented parameters, which are managed in the SAD after negotiations
  and available for use.

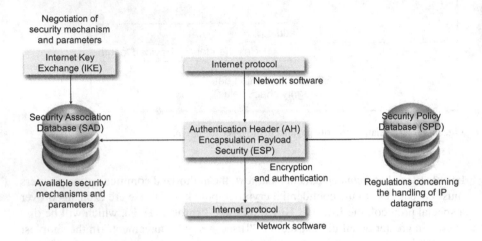

Fig. 7.50 IPsec – architecture.

To illustrate how an IPsec datagram is sent from computer A to computer B, the
example of an ESP encrypted IPsec datagram is chosen. The negotiation of the encryption procedure and parameters is already finished so that SAD and SPD receive
the necessary entries on how to proceed with the transmission of the IPsec datagram.
The transmission process then takes place as follows (see Fig. 7.51):

1. The IP software of the ready-to-transmit computer delivers the complete IP datagram which is to be sent, handing it over to the AH/ESP module.
2. The AH/ESP module contacts the SPD module to know how to proceed with
   IP datagram to be sent. In the example, it can be assumed that all of the IP datagrams to be sent to B should be encrypted. Along with the SA applicable to
   B, describing the encryption method to be implemented and the accompanying
   parameters, the AH/ESP module receives the directive for encryption of the IP
   datagram.
3. The AH/ESP module requests from the SAD the referenced information about
   the SA about the encryption procedure to be used, as well as the cryptographic
   parameters to be used. It then encrypts the IP datagram accordingly.
4. The AH/ESP module hands back the IPsec-datagram, consisting of the encrypted
   payload, the ESP header and trailer as well as the original IP datagram header, to
   the network software which now sends the IPsec datagram.

5. Computer B receives the IPsec datagram. In an evaluation of the datagram by the network software it is determined that an IPsec datagram (protocol field) is involved. Hence, the datagram is forwarded to the AH/ESP module.

6. The AH/ESP module extracts the SPI (Security Parameter Index) from the ESP header. It references the necessary information for decryption of the IP datagram regarding the encryption method used as well as the crypto parameters from the SAD.

7. The AH/ESP module encrypts the IPsec datagram and hands the encoded datagram back to the IP software, which takes over appropriate further processing.

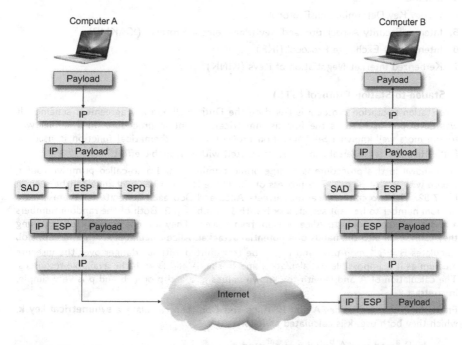

Fig. 7.51 IPsec - the procedure in sending an IPsec datagram

Among various possibilities for the use of IPsec, deployment in **Virtual Private Networks (VPN)** is very popular. In the past, connecting geographically distant LANS to a logical, connected network (e.g., the LANs of different branches of a company) usually meant leasing private, dedicated lines. In this way, a closed network could be built up that ensured security in the case of unauthorized attacks by outsiders. A more cost-efficient solution is naturally offered by the networking of individual LANs via public Internet. Yet at the same time this involves the increased risks of unauthorized attacks. If IPsec is deployed in the forwarding and encryption of data traffic, security is guaranteed as no outsider is in a position to read or falsify the data traffic. Virtual data channels (so-called virtual private networks) are thereby formed.

Excursus 11: IPsec – Key Management

The necessary key management for IPsec to facilitate the exchange of cryptographic parameters between communication partners is an independent application. It is first activated when a requested Security Association (SA) is not found in the local Security Association Database (SAD). Unsecured UDP or TCP connections are used for the key management itself.

The following basic protocols for key management in the Internet are presented in this excursus:

1. Station-to-Station Protocol (STS)
2. Photuris
3. SKEME
4. Oakley Key Determination Protocol
5. Internet Security Association and Key Management Protocol (ISAKMP)
6. Internet Key Exchange Protocol (IKE)
7. Kerberized Internet Negotiation of Keys (KINK)

1. Station-to-Station Protocol (STS)

The station-to-station protocol is based on the **Diffie-Hellman** key agreement scheme . It was developed in 1976 as the first asymmetrical encryption procedure. In a similar way to the more well-known RSA encryption procedure, a mathematical function is used by Diffie-Hellman. Its reversal cannot be calculated with reasonable effort.[4]

Well-known in this procedure is a large prime number p and a so-called primitive root g *mod* p with $2 \leq g \leq$ p-2. The process of the Diffie-Hellman key exchange can be seen in Fig 7.52. The two communication partners Alice and Bob each generate for themselves a random number to be kept secret, a or b with $1 \leq$ a,b $\leq$ p-2. Both of the random numbers a and b are known only to Alice or Bob, respectively. They are not transmitted, meaning they cannot fall into the hands of a potential attacker. Alice calculates $A = g^a$ *mod* p, Bob calculates $B = g^b$ *mod* p. A and B can be transmitted without danger over the insecure medium as it is impossible to calculate a and b from A and B with the available resources. The calculation of A and B with the known values g, a and p or g, b and p is very simple, in contrast.

From the exchanged values A and B, Alice and Bob then calculate a **symmetrical key k**, which they both use. k is calculated as follows:

$$k=B^a \text{mod } p = A^b \text{mod } p = g^{ab} \text{mod } p.$$

However, as the Diffie-Hellman procedure can be successfully challenged by a man-in-the-middle attack, an additional authentication mechanism to prove Alice and Bob's identity may be added (see Fig. 7.53). The values A and B exchanged between Alice and Bob are authenticated. This is done by forming two different message digests from these and signing each with Alice or Bob's secret key, as the case may be. The digital signatures thus received are themselves encrypted with the symmetrical session key k, generated in the key exchange, before being sent. In order to verify the signatures, Alice and Bob each need a certified public key from their communication partner.

---

[4] The discrete logarithm is used for this purpose in the Diffie-Hellman procedure. Also called the modulo logarithm, it involves the inverse operation of modulo exponentiation. If $a, b, n \in \mathbb{N}$ are given, then the discrete logarithm is the number x, for which applies: $a^x = b \bmod n$. Basic principles of digital security, e.g., encryption procedures, digital signatures or public key infrastructures are described in detail in Chap. 5 „Digital Security" in the first volume of this series: Meinel, Ch. Sack, H.: Digital Communications – Networking, Multimedia, Security, Springer, (2013).

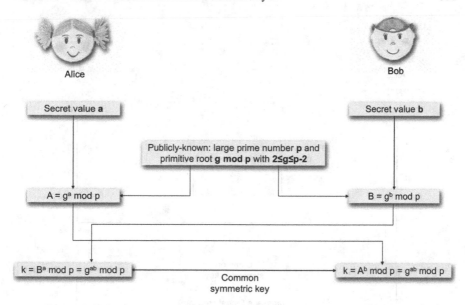

Fig. 7.52 The Diffie-Hellman procedure for the generation of a mutual symmetrical key.

The STS protocol guarantees **perfect forward security**. This means that even if the long-life, public key used for authentication is broken, only future data traffic can be read or influenced with this knowledge. It is not possible to make inferences about any past communication. To do this it would be necessary to break all of the implemented session keys.

2. Photuris

In the STS protocol, an attacker can initiate a denial-of-service attack by transmitting an excessively large number of randomly chosen A or B values. The public-key operations used in each case link a high computing capacity of the executing computer. The protocol scheme **Photuris** (RFC 2522), which did not evolve into an Internet standard, was the first to encounter an attack based on a concept involving exchanged **cookies**. This concept was later taken up in the OAKLEY protocol (RFC 2412) and IKE.

Unlike the HTTP protocol, in which cookies are also used, in Photuris cookies are not implemented for the storage of status information. Rather, a 64 bit long random number is concealed behind a cookie. The communication between Alice and Bob begins with the sending of a cookie request from Alice to Bob. This contains an initiator cookie. Bob answers with a cookie response containing the responder cookie. This is either determined randomly or can be calculated from the initiator cookie. Also contained are suggestions regarding which subsequent procedure should be used for key exchange. All data packets transmitted afterwards must contain both cookies in the header, otherwise the opposite side will not react.

3. SKEME

The Secure Key Exchange Mechanism Protocol (SKEME) facilitates an authentication in three phases (see Fig. 7.54):

- **SHARE**
  At the beginning of the SKEME protocol Alice and Bob each exchange two "half-keys,"

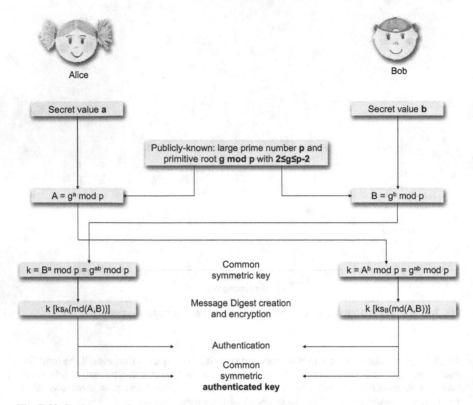

Fig. 7.53 Station-to-station protocol.

which are encrypted with a certified public key of the communication partner. Alice sends the half- key k_A together with her own address id_A, to identify herself to Bob. Both are encrypted with Bob's public key kp_B. Bob answers this with his own half-key k_B, which is in turn encrypted with Alice's public key kp_A.

Both decrypt the half-keys received with their own private key and are now in possession of two half-keys. These are combined via a hash function into a symmetrical session key $k1=hash(k_A,k_B)$. The exchange is carried out with the help of the certified public key from Alice and Bob, meaning that Alice and Bob can be sure that only the correct communication partner is in possession of both half-keys.

- **EXCH**
  In this phase, the normal exchange of key values, based on Diffie-Hellman, takes place to generate the actual session key k2.

- **AUTH**
  For authentication of the session key k2, a digital signature is calculated and exchanged in this phase with help of the symmetrical key k1, from the Alice and Bob's addresses A and id_B, as well as the exchanged Diffie-Hellman values A and B. Because both communication partners have access to all parameters involved and no public keys are used in this phase, Alice and Bob can verify the received digital signature themselves.

4. OAKLEY Key Determination Protocol

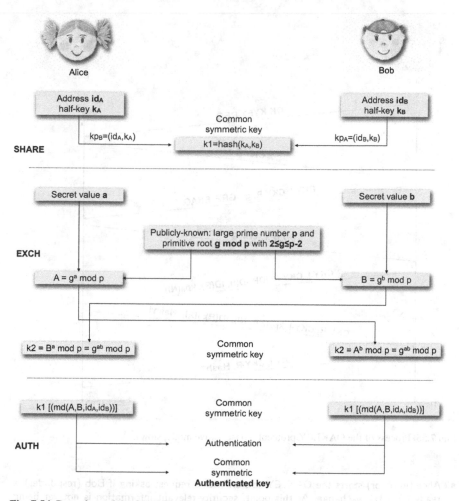

Fig. 7.54 Procedure of the SKEME protocol.

The procedure for key exchange called the OAKLEY protocol is based on concepts from STS, SKEME and Photuris and is specified in RFC 2412. OAKLEY takes over the general principle of encryption agreement according to Diffie-Hellman from STS, as well as its authentication. SKEME adds further authentication variations, while the cookie concept for prevention of denial-of-service attacks is taken from Photuris.

The special characteristic of OAKLEY is that, for one thing, the implemented security mechanism can be negotiated first. No fixed order of the messages to be exchanged is dictated.

The information is exchanged step-by-step and in each step both communication partners decide how much information to pass on. Whereby, the less information revealed in each step, the safer it is.

In a slow mode (conservative mode), an OAKLEY protocol session can proceed as follows (see Fig. 7.55 and table 7.7):

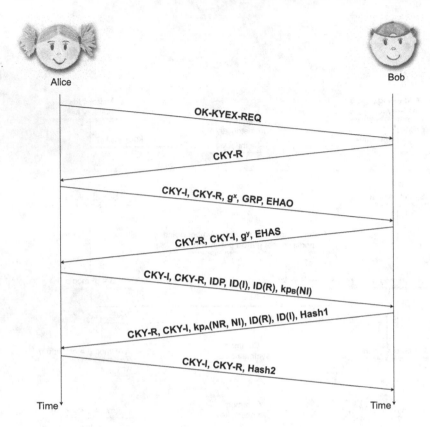

Fig. 7.55 Process of the OAKLEY protocol (conservative mode, slow).

- Alice (initiator) starts the OAKLEY session with a request asking if Bob (responder) is ready for a key exchange. At this point, security-relevant information is not yet being transmitted.
- Bob answers with a cookie generated according to the Photuris procedure.
- Alice sends the cookie received with her own back to Bob. Alice additionally transmits her part of the Diffie-Hellman value for the key exchange along with a list of the cryptographic procedures (encryption, message digest, authentication) she supports.
- Bob selects one of the suggested cryptographic algorithms and sends it back to Alice with his part of the Diffie-Hellman value for the key exchange.
- From this point on, Alice and Bob are in a position to encrypt the messages they exchange.
- For the subsequent authentication, Alice sends her identity and the presumed identity of the communication partner together with a random number, which she encrypts with the public key of her presumed communication partner. This message is encrypted with the symmetrical key obtained via the Diffie-Hellman procedure.
- Bob decrypts the received random number with his private key, thereby proving his identity. He now sends it, encrypted, with Alice's public key back to Alice. In addition, Bob also generates a random number and encrypts it with the public key from Alice.

In doing so, he verifies the identity of his communication partner. Then all transmitted information is authenticated via a hash value.

Table 7.7 Parameters for the OAKLEY key exchange protocol.

Parameter	Description
CKY-I	Cookie of the initiator (Alice)
CKY-R	Cookie of the responder (Bob)
MSGTYPE	Type of proceeding message
OK-KYEX-REQ	OAKLEY Key exchange request
ISA-KE&AUTH-REQ	Key exchange and authentication request
ISA-KE&AUTH-REP	Key exchange and authentication reply
ISA-NEW-GRP-REQ	Negotiation of a new Diffie-Hellman group (request)
ISA-NEW-GRP-REP	Negotiation of a new Diffie-Hellman group (reply)
GRP	Designation of a implemented Diffie-Hellman group
g^x/g^y	Diffie-Hellman message
EHAO	List of offered cryptographic algorithms
EHAS	List of selected cryptographic algorithms
IDP	Flag indicating whether data is encrypted
ID(I)/ID(R)	Identity of initiator/responder
NI/NR	Random number (nonce) of initiator/responder

The disadvantage of this extremely safe procedure is the high time effort it requires. In the faster mode, designated as the more aggressive variety, defense from denial-of-service attacks is dispensed with. The exchange of cookies thus only serves to identify the transmitted messages. Likewise a more secure protection of the communication partner's identity is also omitted. Through omission of both of these safety features, an exchange of only three messages is necessary in the aggressive mode versus seven in the conservative mode.

5. Internet Security Association and Key Management Protckoll (ISAKMP)

Parallel to the development of protocols for the secure exchange of keys to protect data traffic on the Internet, work was also being done on a message format for security association. This was specified in RFC 2408, **Internet Security Association and Key Management Protocol (ISAKMP)**. Two phases are distinguished in ISAKMP:

- **Phase 1**: Negotiation of the ISAKMP-SA (bidirectional).

- **Phase 2**: Implementation of the negotiated ISAKMP-SA for the encryption of the transported payload. Here, the SAs for IPsec, AH and ESP are negotiated. As each of the involved parties can choose a different SA, each format has only a unidirectional validity. For this purpose, ISAKMP requires a key exchange protocol in this work phase such as OAKLEY or IKE. To first be able to determine which cryptographic parameters have to be negotiated for the security protocol used, ISAKMP consults a so-called **Domain of Interpretation (DOI)**.

The ISAKMP data format can be used for different protocols pertaining to negotiation agreements. These protocols for key exchange communicate via TCP or UDP, i.e., they behave in the same way as protocols on the application layer of the TCP/IP reference model. ISAKMP was assigned the standard UDP port 500 by the IANA. The SPD of every IPsec implementation thus follows a rule that UDP datagrams addressed to port 500 must be sent in unaltered form.

After the successful conclusion of negotiations of the SA to be used for encryption of the actual data, this is then forwarded to the security protocol IPsec over an Application Programming Interface (API) . The interplay of the individual components of the communication architecture of ISAKMP is shown in Fig. 7.56.

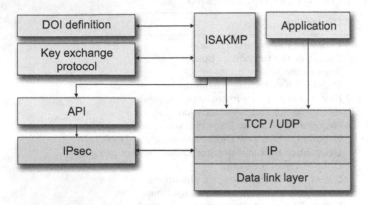

Fig. 7.56 ISAKMP – communication architecture.

An ISAKMP message consists of an ISAKMP header and multiple payload fields. At this time, 13 of them are fixed (see Fig. 7.57). The coding used for each, indicating the payload type, is to be taken from a DOI. For the implementation of IPsec, the corresponding DOI model is described in RFC 2407. The ISAKMP header begins with both 64 bit long cookies (initiator and responder, CKY-I und CKY-R). This is followed by an 8-bit long field (next payload), which describes the content of the first payload field, followed by the 8-bit long version number of the ISAKMP protocol.

Via the next payload field follows a link to the payload field contained in the ISAKMP message. Next comes an 8-bit long field (exchange type) describing the implemented mode of key exchange, similar to the fixed, different modes for OAKLEY. After that come flags for communication control, an identification of the message (message ID) and the extent of the message (message length).

6. Internet Key Exchange Protocol (IKE)

The **Internet Key Exchange** protocol (**IKE**, RFC 2409, RFC 4306) uses the prescribed message format from ISAKMP for the IPsec key management. IKE uses constructs that are already present in OAKLEY and SKEME, in a limited form, to make the implementation of the procedure easier. For example, two modes (main mode and aggressive mode) are available for the ISAKMP-SA in ISAKMP Phase 1. Similar to the OAKLEY supported modes these distinguish themselves based on the length of the necessary message exchange and its resulting security. In the less secure aggressive mode it is also possible to ensure an authentication of the communication partner over the implemented public key authentication in SKEME.

As the messages sent in the second phase of the ISAKMP message exchange are already protected by ISAKMP-SA, deployment of the main mode or aggressive mode for negotiation of the IPsec AH and ESP is dispensed with in this phase. Only the quick mode is applied in the negotiation of the implemented key procedure.

While the IKE described procedure is relatively secure, simultaneously it also requires a great deal of time. When in the main mode six messages must be exchanged between

ISAKMP message

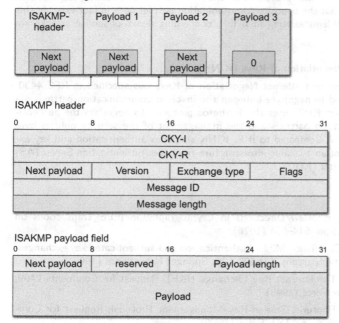

Fig. 7.57 ISAKMP - message format.

communication partners for the negotiation of ISAKMP-SA, in phase two the exchange of three further messages is necessary for the negotiation of the IPsec encryption. During the complete runtime these nine exchanged messages have to be taken into account prior to every encrypted communication. Under certain circumstances this can lead to considerable delays. A further disadvantage is that IKE does not offer effective protection against denial-of-service attacks. Cookies are indeed used but because IKE must store them it is possible for this resource binding to become the target of an attack.

The specification of IKE is distributed among a total of four different RFCs and contains a great many different options. This has resulted in the evolution of incompatible implementations. For this reason, and because of the disadvantages mentioned, a successor version, **IKEv2**, was proposed in RFC 5996. It includes the following extensions:

- Simpler and shorter message exchange with only four messages in contrast to the eight messages in IKEv1.

- Support of mobile application and users (MOBIKE).

- Support of NAT (Network Address Translation) networks, which are protected by firewalls.

- Support of Internet telephony and voice-over-IP (SCTP).

- Easier implementation, because fewer cryptographic procedures are used and a harmonization with the procedures implemented in ESP takes place.

- IKEv2 uses sequence numbers and a confirmation mechanism for a more reliable processing of communication.

- IKEv2 only begins the complex processing of the involved communication partners when it is first determined that the opposite communication partner does in fact exist. In this way, some of the problems existing in IKEv1 concerning denial-of-service attacks are counteracted.

7. Kerberized Internet Negotiation of Keys (KINK)

The protocol called Kerberized Internet Negotiation of Keys was specified in RFC 4430. Just as IKE, it is intended to negotiate between the involved communication partners via IPsec security associations. KINK uses the Kerberos protocol, to negotiate the authentication of the communication partners and the management of the security policies over a central trusted authority. Compared to IKE, KINK simplifies authentication and key exchange through the use of an available infrastructure from the Authentication Server (AS) and Key Distribution Center (KDC).

Further reading:

Diffie, W., Hellman, M. E.: New Directions in Cryptography, in IEEE Transactions on Information Theory, no.6, pp. 644–654 (1976)

Diffie, W., Oorschot, P. C., Wiener, M. J.: Authentication and authenticated key exchanges, in Designs, Codes and Cryptography, (2):107-125, Springer Netherlands (1992)

Harkins, D., Carrel, D.: The Internet Key Exchange (IKE), Request for Comments 2409, Internet Engineering Task Force (1998)

Karn, P., Simpson, W.: Photuris: Session-Key Management Protocol, Request for Comments 2522, Internet Engineering Task Force (1999)

Kaufman, C.: Internet Key Exchange (IKEv2) Protocol, Request for Comments 4306, Internet Engineering Task Force (2005)

Krawczyk, H.: SKEME: a versatile secure key exchange mechanism for Internet, in Proc. of Symposium on Network and Distributed System Security, IEEE Computer Society, Los Alamitos, CA, USA (1996)

Maughan, D., Schertler, M., Schneider, M., Turner, J.: Internet Security Association and Key Management Protocol (ISAKMP), Request for Comments 2408, Internet Engineering Task Force (1998)

Orman, H.: The OAKLEY Key Determination Protocol, Request for Comments 2412, Internet Engineering Task Force (1998)

Sakane, S., Kamada, K., Thomas, M., Vilhuber, J.: Kerberized Internet Negotiation of Keys (KINK), Request for Comments 4430, Internet Engineering Task Force (2006)

7.6 Internet Control Message Protocol – ICMP

As previously described, the IP protocol provides an unreliable datagram service, thereby fulfilling its function according to the "best effort" principle. It nevertheless offers modest possibilities of error recognition, e.g., through the use of checksums. If a receiver recognizes that during transmission an error has occurred based on the transmitted checksum, then the data packet concerned is discarded. For the transmission of corresponding error messages and further administrative tasks, a special protocol was developed that is based directly on the IP protocol, using IP datagrams for the encapsulation of its message. This important protocol from the TCP/IP protocol family is the **Internet Control Message Protocol (ICMP)**. Although ICMP presents an independent protocol, it is considered an integral component of the Internet protocol IP, i.e., every Internet end device must be in a position to receive and understand ICM messages. The transport of the ICMP message proceeds encapsulated as payload of a regular datagram. In addition to the transmission of error messages via ICMP, messages and status information from the Internet can also be transported. ICMP was specified already in 1981 with the RFC 792 and then later extended again in compliance with RFC 1256. For the new Internet protocol IPv6 there is a special version of the ICMP protocol with ICMPv6.

7.6.1 Function of ICMP

Among the most important functions of the ICMP protocol are:

- Support in error diagnostics (e.g., support in the application of **ping**, for testing the reachability of an Internet system),
- Support in recording timestamps and output of error messages when elapsed timestamps in IP datagrams are recognized.
- Management of the routing table,
- Notification of flow control to avoid overload situation at routers and
- Help in locating the maximum allowable datagram size in IP net segments (Maximum Transmission Unit, MTU).

ICMP actually numbers among the protocols on the network layer (layer 3) of the TCP/IP reference model. At the same time, ICMP data is encapsulated in IP datagrams, exactly as if they belonged to a higher protocol layer. In order to identify ICMP messages in an IP datagram, the PR field (protocol field) in the IP datagram header is given the value PR=1 (see Fig. 7.58).

ICMP messages should inform about error situations on the Internet. Therefore, it is possible that computer A detects an error and announces it to computer B. Computer B could then identify an error in the message from computer A and send a further error message in turn to computer A. In this way, a mass transmission of ICMP messages could result between both computers. This could consequently paralyze data traffic in the affected network segment. To avoid unnecessary network traffic and in particular to prevent overload situations caused by ICMP, several rules were laid out in RFC 792 that can help prevent the occurrence of circular and excessively sent ICMP messages from developing. Accordingly, ICMP messages should never be sent as a reaction to the following situations:

- As an answer to an ICMP error message, to avoid overload situations as described (conversely, it is thoroughly possible that an ICMP message may be sent to answer another ICMP message).

- As an response to a multicast or broadcast message. If a multicast message containing an error is potentially sent a thousand times and consequently every receiver sends back an ICMP error message, the original transmitter of the multicast message would receive several thousand error messages.

- As an answer to a defective IP datagram fragment, except the first fragment. If a defective IP datagram is fragmented, the errors usually multiply and enter the individual fragments as well. To avoid unnecessary communication, an ICMP error message is only send to the first defective fragment of an IP datagram.

- To IP datagrams that have not given a unicast address as a sender's address. This prevents ICMP error messages being sent multicast or broadcast.

7.6.2 ICMP Message Format

An ICMP message has a simple format, which is shown in Fig. 7.58. Embedded in an IP data packet, the ICMP message always begins with a 32-bit long header. This is structured the same in all ICMP message types. This is followed by the ICMP payload, which depending on the ICMP message type can have a different structure and content. The shared ICMP header contains the following fields:

- **Type**
  The 8-bit long type field contains information to distinguish it among the individual ICMP message types. The different ICMP message types are compiled in table 7.8.

- **Code**
  A further means of distinguishing between certain types of messages is the code field that follows with its length of 8 bits.

- **Checksum**
  A 16-bit long checksum concludes the ICMP header. This is, however, only calculated from the ICMP payload and does not include the header field transmitted in the ICMP message.

All ICMP messages containing an error message retain a part of the original IP datagram that caused the error in the first place. This is transported as a part of the ICMP message body. Normally, this contains the complete IP datagram header and a part of the datagram payload, which has the protocol header of higher located protocol layers.

7.6.3 ICMP Error Messages

The primary application of the ICMP protocol serves the reporting of various error situations. An end system or a router sends an ICMP error message back to the

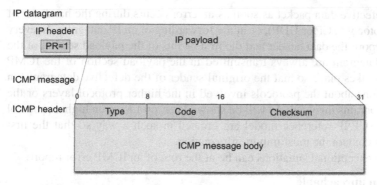

Fig. 7.58 ICMP – message format.

Table 7.8 ICMP – different message types.

Type	ICMP-Message
0	Echo reply
3	Destination unreachable
4	Source quench
5	Redirect
8	Echo request
9	Router advertisement
10	Router Solicitation
11	Time exceeded
12	Parameter problem
13	Timestamp request
14	Timestamp reply
15	Information request
16	Information reply
17	Address mask request
18	Address mask reply
30	Traceroute
31	Datagram conversion error
32	Mobile host redirect
35	Mobile registration request
36	Mobile registration reply
37	Domain name request
38	Domain name reply

sender of a defective data packet as soon as an error occurs during the handover of a transport protocol (TCP or UDP), or in the forwarding of an IP datagram. In every ICMP error report, the data header and the first 64 bits of the payload section of the defective IP datagram are always transmitted in the payload section of the ICMP message. This takes place so that the original sender of the defective datagram can draw conclusions about the protocols involved in the higher protocol layers or the application programs involved. As will be shown later, protocols on higher protocol layers of the TCP/IP reference model are created in such a way so that the first 64 bits always contain the most import information.

The following exceptional situations can be at the root of an ICMP error report:

- **Destination unreachable**
  A transmitted IP datagram cannot be delivered to the addressed destination computer. As a reaction, the message **destination unreachable** is sent by the last responsible router, which should have taken over sending the delivery to the addressed end system, to the original sender. This indicates that the designated receiver is unreachable. There can be various reasons for this situation, e.g., the destination computer no longer exists or there is no fitting protocol available that would be able to take over the payload section of the datagram.

- **Time exceeded**
  Every datagram contains a time-to-live field in its header. This gives the maximum number of hops that an IP datagram is allowed to carry out. The time-to-live field of the IP datagram is lowered by one every time an intermediate system is passed by. When the field reaches the value zero, the message **time exceeded** is sent by the router - which removes the relevant IP datagram from circulation - back to the original sender.

- **Parameter problem**
  The header of the IP datagram contains invalid parameters that are not understood by the receiving computer. This makes the correct processing of the datagram impossible. In this case, the message **parameter problem** is sent back by the receiver, or an intermediate router, to the original sender of the IP datagram. This error message is intentionally kept general. This means that in the header of the ICMP message there is no information about which IP datagram header field caused the error. Instead of a special error code, the ICMP message body, in which the original IP datagram header of the defective datagram is returned in accordance with RFC 792, contains a pointer in the field containing the error. In some cases the specifications of a pointer has no relevance, for example, in the case where specifically required options in the IP datagram are not sent at the same time. In these cases, the code field of the ICMP header does not contain the value 0, as in the normal case, but 1 (missing option) or 2 (faulty length specification).

- **Source quench**
  If an end system or one of the intermediate systems involved in forwarding IP datagrams is not able to process the number of transmitted datagrams in a timely manner, the message **source quench** is sent to the original sender so it can

interrupt the transmission of IP datagrams for a certain time span. This can be the case, for example, if the receiver of the datagram is an especially popular website that is flooded with HTTP requests or if a fast transmitter communicates with a slow receiver that has only a low processing capacity. A router can also be affected if it receives IP datagram via a high-speed connection and these must be forwarded over a comparatively slow interface. In this case, the internal queue where the datagrams are cached overflows. As soon as a network device can no longer process a datagram and it is lost, a corresponding ICMP message is triggered communicating this information to the sender. However, once the congestion at the affected network device clears up again, no further ICMP messages are sent. A receiver can only determine this by the absence of ICMP source quench messages, at which time he can gradually increase his transmission rate again. A targeted variation of network flood control is first implemented by the TCP protocol on the next higher transport layer of the TCP/IP reference model.

- **Redirect**
  In case an intermediate system, forwarding IP datagrams, determines a better path to the target system leading over another router than the present one, it can recommend it via a **redirect** message. The address of the recommended router is then sent in the payload of the ICMP message, followed by the header of the original IP datagram. The original sender of the IP datagram, who receives the ICMP redirect message, is informed this way about which address the redirect message concerns. And this can be taken into account in the future. The specification of a corresponding code value in the ICMP header can be further differentiated. It can be indicated whether a redirect applies only to a certain end device (code=1) or to the entire subnetwork (code=0) where an end device is located. The ICMP redirect message is, in fact, not an error message at all but rather a message about improving an inefficient situation.

7.6.4 ICMP Requests and Informal Messages

Not only error messages can be forwarded over the ICMP protocol, but status information can also be requested and the results reported back. This applies to test and diagnostic functions as well as the distribution of critical information necessary for the error-free operation of the network. The following status information can be made available with ICMP:

- **Echo function**
  The application program **ping** serves to determine the reachability of a computer in the Internet and uses the ICMP protocol to facilitate this. For this purpose, an **echo request** is sent to all computers to be tested. Every IP-capable computer must answer an echo request message with an **echo reply**. It is possible that a computer receives echo requests not just from a single network device but from multiple ones as well. This makes it important to be able to assign an echo reply

to a previously sent echo request. At the beginning of the ICMP message body, the ICMP message provides two 16-bit long fields (identifier and sequence
number). These uniquely identify echo request/reply messages.

- **Timestamp function**
  An end system or an intermediate system can send out an inquiry with a **timestamp request**, asking about the current date and time . A computer that receives
  such an inquiry answers in turn with a **timestamp reply**, containing the requested
  information. The individual end devices each maintain their own system clocks
  on the Internet. Even when these work with great accuracy, it can often happen
  that time lags occur between the individual devices. Some operations, however,
  require that the communication partners work synchronously and with as little time lag as possible. For this purpose, synchronization messages can be exchanged
  over the ICMP timestamp function. In order to facilitate this, the message body
  of the ICMP message contains two 16-bit long fields (identifier and sequence
  number). This is to ensure a unique assignment of the sent messages, such as in
  the echo function (see Fig. 7.59). After that follow three 32 bit-long timestamps.
  The first timestamp (originate timestamp) gives the time of the transmission of
  the original timestamp request message. This is followed by the time of the reception of the timestamp request message (receive timestamp) in the timestamp
  reply message. Finally comes the time point of the timestamp reply message
  transmission (transmit timestamp). From these three timestamps the requesting
  computer can determine the runtime of the datagram between the network sender and receiver in the network as well as the time needed for processing the
  timesstamp ICMP message on the receiver side. IP only works according to the
  best effort principle. The transmission of the timestamp message does not always
  proceed over the same connection path, and it is also possible for packages to get
  lost enroute. Therefore, today's modern network end devices use **Network Time
  Protocol** (**NTP**), especially created for synchronization. This protocol allows a
  reliable time-synchronization to be carried out.

0	8	16	31

Type	Code	Checksum	
Identifier		Sequence number	
Originate timestamp			
Receive timestamp			
Transmit timestamp			

Fig. 7.59 ICMP timestamp and timestamp reply message format.

- **Information request**
  ICMP offers a computer the possibility to inquire about its IP address by way of
  an ICMP **information request** message. This is sent by an address server with
  an **information reply**. For this purpose, however, the DHCP protocol is now
  deployed, meaning this function of the ICMP protocol is rarely used today.

- **Address mask request**
  Via the **address mask request** message a computer can inquire about its subnet address mask. In a subnetwork that supports this function, one or multiple subnet mask servers (normally a router) are active. They answer this type of inquiry with an **address mask response**. This contains the desired information about the local network. This ICMP function was specified in RFC 950, which also describes the IPv4 subnetworking. The requesting of submasks via ICMP is not required for IP. This means that the subnet mask can be manually configured in the computer concerned or alternatively determined automatically via the DHCP protocol.

- **Router discovery**
  Every computer in a network has to know the IP address of a router connecting that network with the Internet world. This IP address is normally fixed in the network configuration as a so-called **default gateway**. In addition to manual configuration, ICNP offers the possibility of automatically determining the address of the default gateway via the exchange of messages (see Fig. 7.60). The procedure for this (router discovery) was first specified in RFC 1256. The computer seeking its assigned router sends a so-called **router solicitation** ICMP message ("router sought . . . ") as a multicast or broadcast into the local network. In the destination address of the encapsulated IP datagram either the multicast address 244.0.0.1 or the broadcast address 255.255.255.255 is given. As part of the local network, the router receives the router solicitation and answers with a **router advertisement**. Router advertisements are also sent by routers into the network periodically without prior requests. The space between each time interval can be fixed when the router is configured. If a network end device receives a router advertisement message, this is processed and synchronized with its own routing information. Besides the normal fields in the ICMP message header, the router advertisement contains the following fields:

  - **Number of addresses**
    8-bit long field giving the number of router addresses transmitted in the answer. It is possible for a physical network interface in the router to be assigned multiple IP addresses. These are all transmitted in the router advertisement answer.

  - **Address entry size**
    8-bit long field indicating the number of 32-bit words provided for every routing address.

  - **Lifetime**
    16-bit long field indicating the number of maximum seconds for which the indicated addresses can be considered valid.

  - **Router Address[i] / Preference Level[i]**
    IP address of the i-th router and its preference. The higher the two's preference value indicated for a router, the more it should be preferred.

- **Route tracing and Path MTU Discovery**
  The last important task of the ICMP protocol is determining the **Maximum**

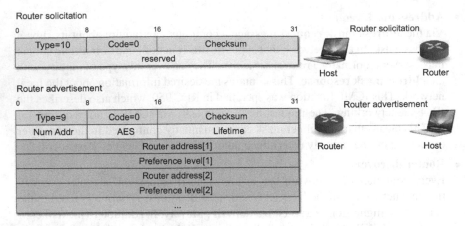

Fig. 7.60 ICMP – determining a router via router solicitation/router advertisement.

Transmission Unit (**MTU**). Defined here is the maximum fragment length for transmission via multiple networks with different data format lengths in the global network. The path taken by a datagram from the transmitter to the receiver is traced. The procedure for determining the MTU is designated **Path MTU** (**PMTU**) **Discovery** and specified in RFC 1191. The PMTU Discovery procedure is shown in Fig. 7.61.

Special ICMP messages are deployed in transmitting the answer of the involved routers, located along the connection path. Besides the usual ICMP message, header fields contain a further 16 bits of information about the MTU of the respective next hop (Next Hop MTU). This is followed by the IP header and the first 64 bits of the originally sent IP datagram (see Fig. 7.62).

In order to be able to trace a path taken in the network, **traceroute** application programs are used. These are especially employed in the RFC 1393 specified CMP traceroute messages. In addition to containing the general ICMP message fields type (type=30), code and checksum, they include the following details (see Fig. 7.63):

- **ID number**
  A 16-bit long identification number to connect the traceroute message sent back with the traceroute message originally transmitted by the sender.

- **Outbound hop count**
  A 16-bit long counter giving the number of routers that the originally sent traceroute message has already passed on its way to the receiver.

- **Return hop count**
  A 16-bit long counter that gives the number of routers the response to the original traceroute message has already passed.

Procedure for determining the Path MTU:

1. The sender transmits a regular IP datagram set with a DF bit (Don't Fragment) to the designated receiver. The set DF bit prevents a fragmenting of the datagram as soon as it encounters a network whose MTU is smaller than the length of the datagram. The length of the sent datagram corresponds exactly to the maximum MTU of the local network and thus to that of the first hop.

2. If the datagram reaches a transfer network whose MTU is smaller than the datagram length, an error is detected at the transmitting router. The IP datagram is discarded and the original sender receives an ICMP message **destination unreachable** with the status code **fragmentation needed and DF set**. The router also adds the currently allowed MTU to the next hop field of the ICMP message and transmits it back to the original sender.

3. The sender registers the error message upon receiving the ICMP message. It then matches the length of the datagrams to be sent with those of the MTU contained in the ICMP message.

4. The process is carried out as long as it takes for a transmitted IP datagram to actually reach the receiver. To be able to adapt to any possible alternations along the Internet connection route, such as changed routers, the procedure is periodically repeated.

Further reading:

Mogul, J. C., Deering, S. E.: Path MTU discovery. Request for Comments 1191, Internet Engineering Task Force (1990).

Fig. 7.61 ICMP - Path MTU discovery.

0	8	16	31
Type=3	Code=4	Checksum	
unused		Next hop MTU	
IP- Datagram header + 64 Bits of the original payload			

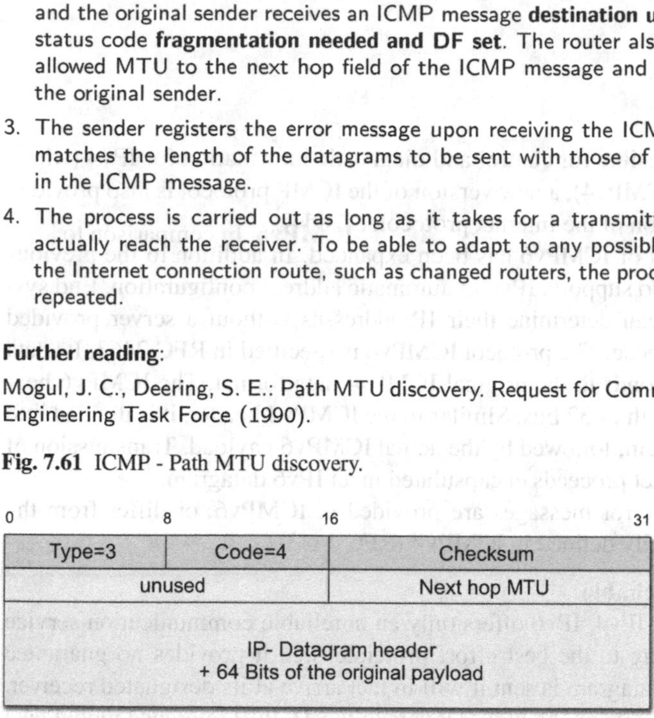

Fig. 7.62 ICMP - Path MTU discovery data format.

- **Output link speed**

    32-bit long information regarding the connection speed at which the traceroute message is sent, measured in bytes per second.

- **Output Link MTU**

    32-bit long size specification of the connection MTU via which the traceroute message is sent, measured in bytes.

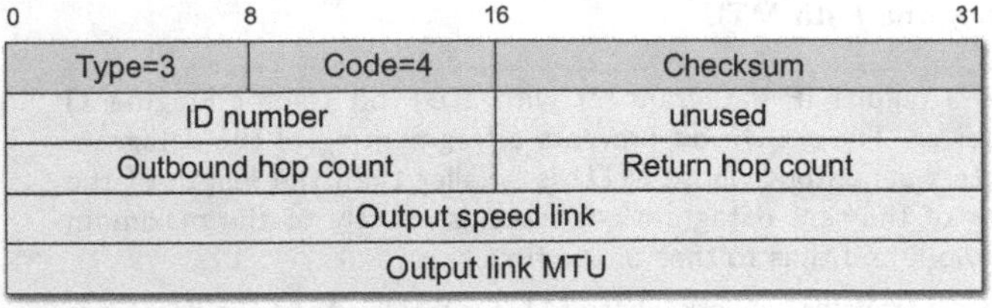

0	8	16	31
Type=3	Code=4	Checksum	
ID number		unused	
Outbound hop count		Return hop count	
Output speed link			
Output link MTU			

Fig. 7.63 ICMP - traceroute message format.

7.6.5 ICMPv6

Just as ICMP is responsible for control and diagnostic information for IPv4 (which is also referred to as ICMPv4), a new version of the ICMP protocol is also provided for the successor version of the Internet protocol, called ICMPv6. In comparison to ICMPv4, the task area of ICMPv6 has been expanded. In addition to the previous functions, ICMPv6 also supports IPv6 in automatic address configuration. End systems in the network can determine their IP addresses without a server provided especially for this purpose. The protocol ICMPv6 is specified in RFC 2463. Its data format largely corresponds to the original ICMPv4 data format. The ICMPv6 header extends over a length of 32 bits. Similar to the ICMPv4 header, it is divided into type, code and checksum, followed by the actual ICMPv6 payload. Transmission of the ICMPv6 data packet proceeds encapsulated in an IPv6 datagram.

The following ICMP error messages are provided in ICMPv6, or differ from the error messages originally defined in ICMPv4 (RFC 4443):

- **Destination unreachable**
  In the same way as IPv4, IPv6 offers only an unreliable communication service that works according to the best effort principle, i.e., it provides no guarantee that once an IPv6 datagram is sent it will in fact arrive at its designated receiver. Normally, the protocols on the higher layers of the TCP/IP reference model take over the task of guaranteeing reliable data transport. They make the appropriate mechanisms to carry this out available, for example the Transmission Control Protocol (TCP) on the transport layer. If a datagram is discarded for reasons of router overload, the receiver is unable to confirm an expected reception of the transmitted data on the transport layer. This results in the corresponding instance of the TCP protocol on the sender's side triggering a new transmission of the affected data. Nevertheless, there are also situations where this mechanism is not successful. For example, if a receiver is not reachable because the address information in the IP datagram is defective, a renewed transmission of the IP datagram triggered by TCP will also not be of any help. For such situations, there is a special ICMP message type available in ICMPv6, just as in ICMPv4. In contrast to ICMPv4, the number of communicated causes of error via ICMPv6 is limited. Some of these causes simply no longer exist in IPv6. Values permitted for the code field in ICMPv6 are:

- **No Route to Destination**, Code=0, there does not exist a route to the given destination.
- **Communication with Destination Prohibited**, Code=1, the datagram could not be delivered to the indicated receiver, as its contents are restrained by filters or blocking mechanisms (e.g., firewalls).
- **Address Unreachable**, Code=3, the datagram could not be delivered to the indicated address in the target network. This means that the address itself is defective or that an error occurred during address translation on the network layer (level LAN/WAN).
- **Port Unreachable**, Code=4, the indicated address port (see section 8.1.3) is defective or not available to the receiver.

- **Packet too big**
  One of the basic differences in data transport under IPv6 when compared to IPv4, is the handling of larger IP datagrams. These are datagrams exceeding the maximum allowed transport size of the network underlying the protocol stack. In IPv4 every end device (up to a certain predetermined maximum size) could send virtually any size datagram. These were fragmented by the access routers of the networks to be passed through and in the form of smaller datagram fragments further transported (see section 7.2.4). In contrast, in IPv6 fragmentation is solely up to the end devices themselves (see section 7.4.3). If a router in IPv6 receives a datagram that is too large for further transportation over the network it is connected to, there is nothing left to do but discard the datagram. In this case, the router sends an ICMP **Packet Too Big** message back to the sender of the over-sized datagram along with the current details of the allowed MTU for the affected network.

- **Time exceeded**
  Both IPv4 and IPv6 provide a maximum lifetime for sent datagrams in the network. This is to prevent datagrams from circulating endlessly in the network and causing overload due to routing errors. For this situation, the *Time to Live* (TTL) field was provided under IPv4 in the IP datagram header. When a router was passed it always counted down with the value of 1 until the value 0 was reached. If a router received an IP datagram with the TL value 0, this datagram was discarded. This concept was retained with the so-called *hop limit* field under IPv6 . Here, if for the same reason a router discards an incoming IP datagram, an ICMP **Time Exceeded** message is transmitted to the sender. Another example of when this error message must be sent might occur in fragmentation. When a datagram is segmented into multiple, single fragments in the Internet these are transmitted independent of each other and not necessarily on the same network route. Waiting times can arise at the receiver between the arrival of the first fragment and the last, before the datagram can be completely reassembled. If it happens that a specific timespan is exceeded, the not yet fully assembled datagram is discarded at the receiver and an ICMP time exceeded message sent back to the transmitter.

- **Parameter problem**
  Similar to IPv4, faulty information in the IP datagram header in IPv6 can cause incorrect delivery to the predetermined receiver. If the error that occurs is serious enough that the datagram can no longer be delivered unreliably, then the ICMP parameter problem message is triggered. The ICMPv6 parameter problem message is kept general in the same way as the corresponding ICMPv4 message. Therefore, it can refer to any field in the IP datagram header. For identification of the relevant defective field, a pointer is provided in the payload section of the ICMPv6 message. It indicates the IP datagram header of the following original IPv6 datagram header. In contrast to the 8-bit long pointer field in ICMPv4, the size of the pointer for ICMPv6 was increased to 32 bits.

The following general ICMP status messages are provided in ICMPv6, or deviate from the original error notifications in ICMPv4:

- **Echo request**
  As under ICMPv4, ICMPv6 provides a simple procedure for testing the reachability of a communication partner in IPv6. This is carried out via a simple question-answer schema. The requesting computer sends an ICMPv6 echo request to the computer being tested, which in return answers with an ICMPv6 echo reply giving information regarding the status of its reachability. Usually, ICMPv6 echo messages are implemented in the framework of the previously mentioned IPv6 **ping** application.

- **Router advertisement and solicitation**
  Router advertisement proceeds in IPv6 with the help of ICMPv6 messages in a similar way as in IPv4 and ICMPv4. In IPv6 the functionality of router determination is integrated in the Neighbor Discovery Protocol (NDP) and specified in RFC 4861. For this purpose, router advertisement messages are provided to make known the available routers, as are router solicitation messages for inquiring about the reachable router in the local network, just as in ICMPv4. The message format of the **ICMPv6 router advertisements** differs slightly from the format of the corresponding ICMPv4 message and contains the following fields (see Fig. 7.64):

  - **Type, code and checksum**
    These fields correspond to the matching fields in the ICMPv4 message format and indicate the type of message (type=134). The code field is unused and the checkcum calculated from the header of the ICMP message.

  - **Current Hop Limit**
    8-bit long field indicating the router-recommended hop limit for end devices of the local network to facilitate further communication. If the value is 0, no recommendation is given and the end devices can set this value independently.

  - **Autoconfig Flags**
    8-bit long field containing 2 command and control bits that give various options for IPv6 autoconfiguration (see section 7.4.5).

– **Router lifetime**

16-bit long field that gives the time span in seconds in which a specific router is to be used by the network end devices. If the value is 0 then that particular router should always be used as the default router.

– **Reachable time**

32-bit long field providing the time span in milliseconds in which an end device can be considered reachable by a neighbor. This occurs after a confirmation regarding reachability of the neighbor in question has been received.

– **Retransmission time**

32-bit long field giving the length of time in milliseconds an end device should wait before resending a neighbor solicitation message to determine its neighbor in the local network.

– **Options**

Router advertisement messages may contain three different options:

· The network address of the sender (source link-layer address) if the router, sending the router advertisement, knows its network address (data link layer, MAC address).

· The MTU of the local network, in case the end devices of the network concerned does not have this information available.

· The IPv6 network prefix that must be used by end devices in the local network for addressing.

The message format of the ICMPv6 router solicitation message does not distinguish itself from the ICMPv4 version. ICMPv6 router solicitation messages are usually sent to the IPv6 "all routers" multicast addresses, as routers in IPv6 must always log in to this multicast group.

If an ICMPv6 router advertisement message is being sent to a router solicitation, it is sent back directly to the requesting computer as a unicast.

In the case of routinely sent general ICMPv6 router advertisements, the message is sent to "all computing nodes" multicast addresses in the local network.

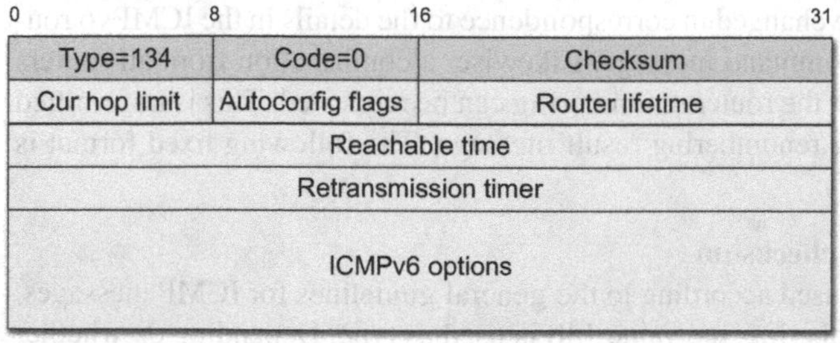

0	8	16	31
Type=134	Code=0	Checksum	
Cur hop limit	Autoconfig flags	Router lifetime	
Reachable time			
Retransmission timer			
ICMPv6 options			

Fig. 7.64 ICMPv6 - router advertisement message format.

- **Neighbor advertisement and solicitation**
  In a similar way to router advertisement, ICMPv6 provides a message exchange in the framework of the neighbor discovery protocol to establish address information in the neighboring end devices in the local network. Via the ICMPv6 neighbor solicitation message it can be verified whether neighboring computers exist in the local network and are reachable. This means, if it is possible to initiate an address resolution. The existence of a neighboring computer can be confirmed over the ICMPv6 neighbor advertisement message and, if requested, an appropriate network address (MAC address) sent back.

- **Redirect**
  Normally, a network end device in an IPv6 network does not need information about predetermined routers, i.e., computers in the local network are usually addressed directly and data traffic that should leave the local network is sent to the default router of the network. However, networks often contain multiple routers and an end device does not always have information about which router in a network is the most efficient for the upcoming data transfer. If a router in the local network determines that the data transfer it is processing could be handled more efficiently by another router in the same network, it sends an appropriate ICMPv6 redirect message to the original sender. The ICMPv6 redirect message contains two IPv6 addresses in the payload section. The first is the address of the new router over which future data traffic should be directed (target address). The second is the address of the relevant end device, whose data traffic should be rerouted (destination address). Both address details are 28 bits long each.

- **Router renumbering**
  With its 128-bit long IPv6 address format, IPv6 also offers the possibility to migrate larger network segments, or entire networks, into another address area, i.e., to renumber it internally. This procedure is specified in RFC 2894 and includes the use of ICMPv6 router renumbering command and routing renumbering result messages. The ICMPv6 router renumbering command message contains a list of network prefixes to be renumbered. Every router to receive such a message checks whether its routing table has address prefixes matching those network prefixes transmitted in the routing renumbering command. If this is the case, the relevant prefixes are changed in correspondence to the details in the ICMPv6 routing renumbering command message. Likewise, a confirmation from all routers that have carried out the router renumbering can be requested. This is transmitted as a ICMPv6 router renumbering result message. The following fixed format is given (see Fig. 7.65):

  - **Type, code and checksum**
    These fields are used according to the general guidelines for ICMP messages. For router renumbering, the value 138 is used as type. Depending on whether a request (router renumbering command, code=0) or a confirmation (router renumbering result, code=1) is involved, different values for the code field are used.

- **Sequence number**
  This 32-bit long field prevents so-called replay attacks as the receiver can easily identify duplicates or transmitted messages that are irregular.
- **Segment number**
  8-bit long field intended to distinguish between different, but valid, router renumbering messages with the same sequence numbers.
- **Flags**
  8-bit long field that contains 5 individual flags. These control the execution of the renumbering commands.
- **Maximum delay**
  16-bit long field that contains a time specification in milliseconds. It corresponds to the maximum time span a router is allowed to wait before sending an answer to the router renumbering command.
- **Message body**
  This field of variable lengths, made up of two parts, exists for the router renumbering command. The first part (match-prefix part) has the network prefixes that should be replaced for renumbering. The second part (use-prefix part) contains the new network prefixes to be used. In a router renumbering result message this field contains all to-be-renumbered network prefixes, which the transmitting computer has processed. There is additional status information as to whether the change could be carried out successfully.

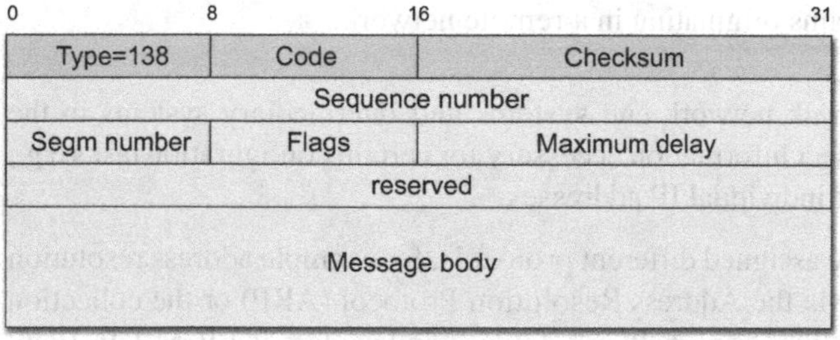

Fig. 7.65 ICMPv6 - router renumbering command.

7.6.6 Neighbor Discovery Protocol – NDP

When the new Internet protocol IPv6 was developed, one of the design principles was to keep as many of the operational properties of IPv4 as possible. The basic IP protocol was retained as well as the most important functions supporting the ICMP protocol. While functionality was extended, it was not changed essentially. Furthermore, a new, similarly functional supporting protocol was developed that did not have a direct predecessor in the old IP version. It bundled and expanded some already existing functionalities, and was given the name **Neighbor Discovery Protocol (NDP)**. The internetworking layer of the TCP/IP reference model provides for the combining of heterogeneous individual networks into one common internet. It is intended that the protocols located in the higher protocol layers should appear as one homogeneous network. This means that a network node in the local network - seen from the perspective of a higher protocol layer - cannot be distinguished from a network node in a remote network. Yet the difference is fundamental to the protocol located lower in the protocol stack. Among the tasks a network node must carry out in its local network, together with other network nodes are:

- **Direct delivery of datagrams**
  In the local network datagrams are exchanged between individual network devices directly, i.e., without the involvement of a router or another transmitting intermediary system.

- **Network addressing** To facilitate a direct delivery of datagrams in the local network, the network end devices must know the respective network addresses (MAC addresses) of the other network devices in the local network.

- **Router identification**
  To transport a datagram to a computer outside of one's own local network, a network device must know the necessary router for forwarding to send it the relevant datagram.

- **Router communication**
  Vice versa, a router must know all local network devices in the local network to deliver those datagrams originating in a remote network.

- **Configuration tasks**
  Network devices check network end systems and intermediary systems in the local network to collect information necessary for certain configuration tasks e.g., the determination of individual IP addresses.

In IPv4 these tasks were assigned different protocols, for example address resolution for the data link layer via the Address Resolution Protocol (ARP) or the collection of information about neighboring network devices and routers via ICMP. In IPv6, in contrast, the task are bundled and assigned to the Neighbor Discovery Protocol. In 1996 the Neighbor Discovery Protocol (NDP) was specified in RFC 1970 and has since been revised several times. The current version of the standard is found in RFC 4861. It is based on the ICMPv6 protocol and its informal message types. The standard describes altogether nine different functions. These are divided into host-host functions, host-router functions and a redirect function:

- **Host-host communication functions**
  This class of functions applies in general to the information exchange between network end systems and intermediary systems, i.e., host as well as router:

  - **Address resolution**
    This applies to the translation of an IP address (Internet layer, layer 3 address)

in a MAC address (data link layer, network address, layer 2 address). NDP takes over this function by sending an ICMPv6 neighbor solicitation containing the IPv6 address of the receiver in the local network. The intended receiver answers with an ICMPv6 neighbor advertisement. This contains the desired layer 2 address for the IPv6 address.

– **Next-hop determination**

Before a datagram can be sent to its receiver, it must be determined where the datagram will be directly forwarded to (next hop). This is based on the receiver's address. If a receiver in a local network is involved, the next-hop address corresponds to the receiver's address. In the other case, the datagram is forwarded to one of the local routers. The decision can be made based on the network prefix of the IPv6 address.

– **Neighbor update**

Unlike router advertisements, status notifications are not sent via the neighbors in the local network (neighbor advertisement) on a regular basis. This is because, the situation in the local network does not change permanently and a regularly-sent neighbor advertisement there would mean a waste of resources. However, a neighbor advertisement can be sent for instance if a computer determines a hardware error on the network interface.

– **Neighbor unreachability and caching**

Except for the purpose of address resolution, neighbor advertisements and solicitation messages also serve for checking the reachability of end devices in the local network. If a local network is no longer reachable, the neighboring nodes change their transmission behavior. They do not do this by permanently resending the undeliverable datagrams and thereby overloading the network, but by waiting a previously determined period of time before attempting to transmit anew. For this purpose, network nodes also keep a neighbor cache. Here, the reachability of every neighbor is temporarily stored as soon as a datagram is received from this neighbor. This cache entry is provided with a timer. After the timer's expiration, the neighbor is (initially) considered unreachable until it has received a new datagram or via neighbor solicitation its reachability can be verified.

– **Detection of duplicate addresses**

In the course of autoconfiguration a network end device is assigned a new IPv6 address. To verify that it has not already been assigned in the local network, the concerned network end device sends a neighbor solicitation message to the determined address. If it receives a neighbor advertisement as an answer, the address is already being used by another device.

• **Host-router communication functions**

NDP host-router communication functions correspond to router determination in the local network as well as to the general communication between routers and local network end devices. This applies to the following functions:

– **Router determination**
Router identification is the core functionality of this communication function. Routers make their presence known to computers in the local network via
router advertisement messages, sent in regular intervals. Vice versa, computers can inquire in the local network as to the router responsible for them,
which - with the help of the router solicitation message - connects the local
network with the worldwide Internet

– **Prefix determination**
Via router solicitation and advertisement messages the valid network prefix
for the local network can be determined in addition to determining the responsible router for a network end device. Based on this, the network end
device can create its own IPv6 address in the course of autoconfiguration. It is
also decided whether a datagram to be sent as next (next hop) is first sent to a
router or directly delivered to a computer in the local network.

– **Parameter determination**
Similarly, additional parameters in the local network between router and host
can be exchanged over router solicitation and router advertisement. For example, the maximum size of the datagram payload to be transmitted (Maximum
Transmission Unit, MTU) or the hop limit to be used.

– **Address autoconfiguration**
With help of the network prefix, determined via router solicitation and advertisement message, a network device can create a valid IPv6 address from its
own hardware address (MAC-address, layer 2 address).

• **Redirect**
Another important task of the NDP protocol is the redirection of what had previously been mostly inefficient traffic. If a router determines that the sender of a
datagram has made an error concerning the next receiver (next hop), or if another
next hop receiver is more suitable to forward the datagram to its final receiver,
this is communicated to the sender via an ICMPv6 redirect message by the router
so it changes its next hop table.

7.7 Mobile IP

The term **mobile computing** describes systems and communication networks whose mobile network end devices can be moved from one place to another. One immediately thinks of the wireless, radio-operated networks and systems that allow
location changes on a large scale with high speed. Seen from the perspective of the
Internet Protocol, the difficulty is found more in the change of location than the
fact that the move there is carried out quickly. There is little that changes from the
IP point of view if a network end device, moving fast within a network, remains
assigned to the same transmission and receiving stations. But as soon as it leaves
the local network - and this is even the case when a stationary computer is dis-

mantled, rebuilt in another place and reconnected to the network - a new network configuration must follow. The number of mobile end devices has grown rapidly in recent years. Besides laptops, the area of mobile telephones and smartphones has particularly expanded. All of these devices require full access to the Internet today, regardless of their location. When Internet protocol was developed computers were still usually quite large and today's sense of mobility was beyond anyone's imagination. Because of this, in designing the IP protocol no provisions were made for the possibility of any special measures that might be taken. These became necessary in the course of limitless mobility. Only in the '90s was a first specification for an IP protocol with mobility support developed. This was then standardized in 1996 in RFC 2002.

7.7.1 Fundamental Problems and Requirements

One of the fundamental problems for a "mobile Internet" was motivated initially by the predefined IPv4 address scheme. This has a hierarchical structure and in each case is comprised of a network class, network ID and host ID. If a data packet is sent to a specific computer in the Internet via IP, virtually any Internet router can determine where the data packet should be forwarded based simply on the network ID. Even when the addressed computer is brought from its home LAN to a remote network, the data packets sent to it still continue to be transmitted to the home network.

To solve this problem either

- the complete IP address of the computer must be changed so that it receives the network ID of the new network, or

- the routers must be informed that it is necessary to forward data packets addressed to the relocated computer.

At first glance, both solutions appear feasible, yet when more than just a few computers are involved such procedures carried out on a wide scale quickly prove inefficient. If the IP address of the relocated computer is altered, an avalanche of changes in applications and routers is triggered. Should forwarding be carried out by these computers, this leads to network overload as the number of computers continuously increases. The already extensive routing tables then quickly become overloaded. Neither solution is therefore easily scalable.

Thus, the Internet Engineering Task Force (IETF) started its own working group (IP Mobility Support Group) to deal with the problem of integrating mobile computers in the Internet as well as the changing requirements of the basic IP protocol involved. This working group first complied a technology-independent list of requirements necessary for a mobile IP solution:

- **Transparency**
  The achieved mobility should not be visible for protocols on a higher protocol

layer - meaning transport and application protocols. Nor should it be visible for routers. Even all opened TCP connections should be capable of withstanding the location change while remaining ready for operation, as long as no data transmission is taking place.

- **Compatibility**
  The mobility that has been achieved does not require any changes be made to routers and stationary computers. This means, particularly, that a mobile end device can readily communicate with a stationary computer operated with IP and that no new addressing schema must be implemented. The IP addressed mobile end devices should have the same format as the addresses of stationary computers.

- **Security**
  The message traffic needed to integrate a mobile end device in a network must always be authenticated for security reasons. This counters attempts to pass off a false identity, which could otherwise allow an unauthorized party access to the network concerned.

- **Efficiency**
  Mobile end devices often only have narrow band access to the stationary part of the Internet. Therefore, the amount of data to be transmitted should always be small.

- **Scalability**
  The new procedure should be capable of managing a very large number of mobile end devices without complex changes being necessary.

- **Macro-mobility**
  Rather than concentrating on managing fast network transitions of mobile end devices, such as is usually the case with mobile telephones, IP instead focuses on the limited transfer of an end device in a remote network. For example, a notebook taken along on a business trip and operated for a limited time at the destination.

7.7.2 Principal Operating Sequence

The biggest problem in developing a procedure for a mobile IP protocol that can meet the above requirements involves the change of location. Here, an end system keeps its address without the Internet router having to define new paths for this address in its routing table. In this way it should be possible for normal network end devices, without mobile IP, to also communicate with mobile devices as if they were still in their home network. The solution found for Mobile IP provides two IP addresses simultaneously in each case for every mobile end device:

- **Primary IP address**
  The first, stationary IP address of the end system (**home address**), which never changes, is received by the mobile end system as IP address at its local home network. Applications on the mobile end device always use the primary address.

- **Secondary IP address**
  A temporary IP address (**care-of address**) changes whenever the mobile end devices has a change of location. It retains its validity only as long as the relevant end system remains in the new network. The mobile end device receives this address upon registering at a foreign network and logs in at a special **agent** at the home network (**home agent**, which is normally a router). The agent at the home network then forwards data packets that are directed to the primary address of the mobile end devices, via **IP-to-IP** encapsulation to the secondary address. For this, the complete IP datagram directed to the primary address is simply forwarded to the secondary address in a new IP data packet as payload section. If the mobile end device changes its location again and receives a new secondary address, this must be made known in the same way to the agent at the home network again. Likewise, a logging off must take place at the agent if the mobile end device again returns to the home network again. This is necessary in order that the forwarding of IP data packets prompted by the agent is suspended. Care-of addresses are not used by applications on the mobile end devices and remain unknown to them. Only the IP software of the mobile end device and the agent at the home network use these addresses. There are two forms of secondary addresses:

  - **Co-located care-of address**:
    In this form, the mobile end device must take over all forwarding management itself. The mobile end device uses both addresses - care-of address and primary address - simultaneously. While lower protocol layers implement the care-of address for receiving IP datagrams, applications always work with the stationary primary address. The care-of address is assigned to the mobile end devices using the same mechanisms for assignment of the network address at stationary computers (e.g., through DHCP). For this purpose, special, temporarily valid IP addresses must be set aside. In IPv4, the supply of non-assigned local IPv4 addresses can be used up easily.
    Routers at foreign networks cannot distinguish whether the relevant computer is a mobile or stationary computer. The received care-of address is made known to the agent at the home network that takes over the forwarding of IP data packets to the mobile computer. The advantage of this variation is that the available infrastructure at the foreign network can be used without the necessity of a special agent (**foreign agent**) being provided there for mobile computers.

  - **Foreign agent care-of address**:
    In this variation a special agent - the **foreign agent** (usually a router) - is responsible for the assignment of a care-of address to the new mobile end device. However, it must take up contact with the foreign agent beforehand and apply for participation in the network operation. The foreign agent does not need to allocate a new IP address to the mobile end device. It takes over the forwarding of datagrams itself, intended for the mobile end system. To do this it normally uses direct network hardware addresses (MAC addresses, layer 2 addresses). This has the advantage that only one foreign agent care-

of address is necessary for all mobile (foreign) computers located in the local network. They are all reachable at the same care-of address. The foreign agent then takes over further distribution to the mobile end devices.

The procedure for forwarding IP datagrams from a home network to a foreign network is presented in detailed form in excursus 8 "Routing Procedures for Networks with Mobile Components."

7.7.3 Mobile IP Message Format

At a foreign network the foreign agent is responsible for the assignment of a care-of address to the mobile end device wishing to engage in "its" network. For this purpose the mobile end device must first register at the foreign agent. The finding of the foreign agent, (**agent discovery**) takes place with the help of the ICMP **router discovery** procedure. To discover the responsible router, the end system sends an ICMP router solicitation message in a local network, which is confirmed by the responsible router with a ICMP router advertisement message, containing the desired information. In the course of the mobile IP agent discovery process, ICMP agent advertisement messages are sent in regular intervals by the foreign agent. The purpose being to inform the new mobile network devices of all information needed to establish an IP network connection. In turn, a mobile end device can send out an ICMP agent solicitation message to a foreign network after being switched on to initiate a foreign agent to send an ICMP agent advertisement message.The ICMP agent solicitation message corresponds exactly to the ICMP router solicitation message (see Fig. 7.60).

Fig. 7.66 shows the data format of the ICMP **Mobility Agent Advertisement Extension**.This is sent by an agent as an answer to a mobile end device inquiry. It distinguishes itself from the ICMP router advertisement message and consists of the following fields:

- **Extension type**
  8-bit long type specification of the ICMP message (here =16).

- **Length**
  8-bit long specification of message length given in bytes, without a type and length field.

- **Sequence number**
  16-bit long sequence number that allows the receiver to determine if a message is lost.

- **Registration lifetime**
  16-bit long time specification given in seconds, showing how long the foreign agent is prepared to accept a registration.

- **Flags**
  8-bit long information field, providing information about the condition and function of the foreign agent. The individual bits have the following meaning:

  - **Registration required**:
    If set, a registration at the foreign agent is absolutely necessary, even if a co-located care-of address is known.

  - **Busy**:
    If set, the foreign agent is momentarily "busy," i.e., not reachable for registration inquires.

  - **Home agent**:
    The relevant agent is prepared to also act as a home agent for the forwarding of datagrams (i.e., it is thoroughly possible for a computer to function as a foreign agent and a home agent at the same time).

  - **Foreign agent**:
    The relevant agent is prepared to act as a foreign agent.

  - **Minimal encapsulation**:
    The agent is capable of receiving encapsulated IP datagrams with minimal encapsulation (RFC 2004), in addition to the usual IP-to-IP encapsulation.

  - **GRE encapsulation**:
    The agent is capable of receiving encapsulated IP datagrams with GRE encapsulation (Generic Routing Encapsulation, RFC 1701), besides the normal IP-to-IP encapsulation.

  - **Reserved**:
    This bit is not used and is always set to the value 0.

  - **Reverse Tunneling**:
    The agent supports the reverse tunneling procedure.

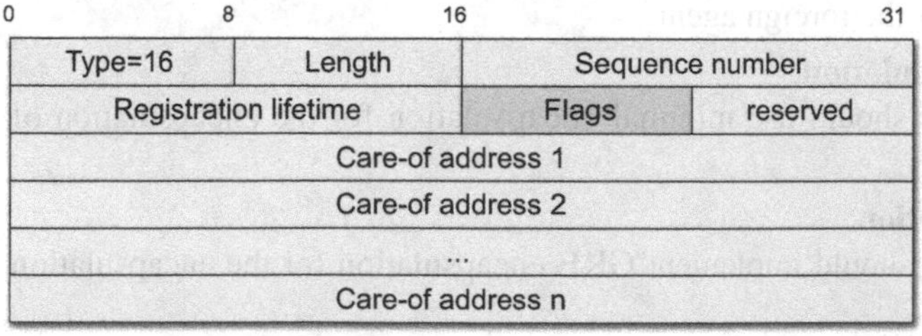

0	8	16	31
Type=16	Length	Sequence number	
Registration lifetime		Flags	reserved
Care-of address 1			
Care-of address 2			
...			
Care-of address n			

Fig. 7.66 ICMP - mobile agent advertisement extension.

Before a mobile end device can transmit and receive data in a foreign network, it must first login by means of a registration process. When a mobile end device has ended the process of agent discovery, it knows whether it is in its home network or in a foreign network. In a foreign network it must communicate over mobile IP. For this purpose, it is required to register at its home agent in the home network in order that the data can be forwarded accordingly (**home agent registration**). In doing so, it must inform the home agent of its current care-of address so that it knows where the data directed to it is to be forwarded. After a successful registration at the home agent, the care-of address is bound to the primary address (**mobility binding**). Registration is only valid for a limited time, therefore a re-registration of the mobile end device must take place in regular intervals. The data traffic transmitted by registration consists of two message types, a registration request and a registration reply. In contrast to the agent discovery, these messages are not exchanged via ICMP, but rather with the help of the user datagram protocol (see section 8.2). Registration takes place at the agent via UDP over the port number 434. In addition to the mandatory fields of fixed lengths there are possible extensions that can be of differing lengths. For example, every registration inquiry must contain an authentication of the mobile client. With this, the agent at the home network can verify the identity given. A mobile IP registration notification contains the fields (see Fig. 7.67):

- **Type**:
  8-bit long field indicating the type of registration message (`Type=1`: Registration Request, `Type=3`: Registration Reply).

- **Flags**:
  8-bit long information field with command and control bits, where details about the forwarding of IP datagrams are fixed. In the case of an inquiry (`Typ=1`) the flags are defined as follows:

  - **Simultaneous binding**:
    The mobile node requires that the previous mobility bindings remain in place besides the current request.

  - **Broadcast datagrams**:
    Broadcasts at the home network that should be forwarded to the mobile node.

  - **Decapsulation by mobile node**:
    The encapsulated datagrams are directly unpacked by the mobile network node rather than by the foreign agent.

  - **Minimal encapsulation**:
    The home agent should use minimal encapsulation for the encapsulation of the datagrams.

  - **GRE encapsulation**:
    The home agent should implement GRE encapsulation for the encapsulation of datagrams.

  - **Reserved**:
    This bit is not used and is always set to the value 0.

  - **Reverse tunneling**:
    The home agent should use reverse tunneling.

  - **Reserved 2**:
    This bit is not used and is always set to the value 0.

With the answer (Type=3) this corresponding field shows the result of registration. If the registration is accepted, it contains the value 0. In the other case,
this field contains an error code that indicates the reason for the refusal of the
registration request.

- **Lifetime** :
  16-bit long time specification in seconds that gives the validity of registration (0:
  immediate logoff, 65.535: unlimited).

- **Home address**:
  32-bit long primary address of the mobile end device.

- **Home agent**:
  32-bit long IP address of the agent at the home network.

- **Care-of address**:
  32-bit long secondary address of the mobile end device.

- **Identification**:
  64-bit long identification number generated by the mobile end device that provides the corresponding answers to inquires as well as maintaining the sequence of
  registration messages.

- **Extensions**:
  Field of variable length, that e.g., can contain the authentication of the mobile
  end device.

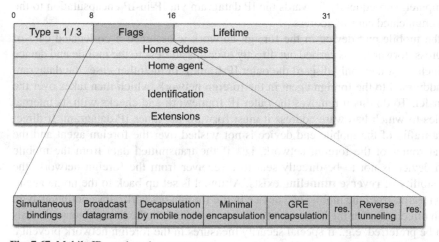

Fig. 7.67 Mobile IP - registration message.

The registration procedure distinguishes itself according to the type of care-of address used. If a co-located care-of address is used, a registration can be directly
carried out at the home network, i.e., the mobile device sends a registration request
straight to the home agent and receives an answer from it. In the case of a a foreign
agent care-of address, registration proves to be somewhat more complicated. Here

the mobile device must first send a registration request to the foreign agent of the foreign network. In attempting to make a first contact with the foreign agent at the foreign network, the mobile end device does not yet have its own IP address in this network. And in fact the rules concerning the use of valid addresses from local to LAN are loosened.

The mobile end device can initially use its primary address (home address), which is also used by the foreign agent, to send its answer to the requesting mobile computers. However, with the use of this home address in the foreign network it is not possible for an ARP request for address resolution to take place at a network hardware address. For this reason,the foreign agent must note the hardware address of the mobile end device and, with the help of its internal table, carry out an address resolution. The foreign agent processes the registration request from the mobile end device and forwards it on to the home address at the home network. The registration answer of the home agent is likewise first sent back to the foreign agent, before this gives the answer to the mobile device.

After a successful registration at the foreign agent, the mobile computer can begin its Internet data exchange. To communicate with any computer the primary address is used. This is entered in the sender field of IP datagram. The datagram is routed from the foreign network directly to the the receiver address. If the reverse situation takes place and an answer is to be sent to the mobile computer in the foreign network, this cannot take place directly. The answer is sent to the primary address of the mobile end device and therefore first routed to its home network. The home agent at the home network, which is informed about the new location of the mobile computer, subsequently forwards the IP datagram via IP-in-IP encapsulation to the communicated care-of address.

If the mobile end device in the foreign network is assigned a co-located care-of address, forwarding is carried out directly over this address to the mobile end device (which must then only discard the outer IP frame). In the other case, the datagram is addressed to the foreign agent in the foreign network, which then takes over the transfer. To do this it removes the outer IP framework and checks with its internal tables to which hardware address it must forward the inner IP datagram. If direct data traffic of the mobile end device is not wished over the foreign agent and the local router of the foreign network, i.e., if the transmitted data from the mobile end device is not to be directly sent to a receiver from the foreign network, the possibility of **reverse tunneling** exists. A tunnel is set up back to the home agent. Data traffic from the mobile end device in the foreign network runs continuously in both directions over the home agent and not via a router in the foreign network. This can be preferred, e.g., if special security measures in the foreign network prevent a network end device located there from sending data with another network prefix than the local one.

7.7.4 Mobile IP and Routing Efficiency

As previously described, a mobile end device in a foreign network always uses its primary address as the sender address. This is the IP address at its home network. Yet according to the locality principle, (**spatial locality of reference**), there is a high probability that the mobile computer in the foreign network also communicates often with its local computers. An IP datagram sent by a mobile end device is locally routed for this, i.e., it does not leave the local network. However, the reply is transmitted to the primary address of the mobile end device at its home network. The IP datagram is received there by the home agent and sent back again to the foreign network. The reply of a local computer in the foreign network must therefore proceed twice on its path through the Internet in order to reach the mobile end device. This problem is known as the **two-crossing problem** (2X-Problem). This case becomes even more complex if reverse tunneling is additionally employed for the data traffic originating from mobile devices. Principally, the problem can be solved by setting up a host-specific route especially for the mobile end device at the local router. This must then naturally be deleted upon log-off of the mobile end device. Yet the problem remains for all other networks, even if, e.g., communication is only desired with one network that is directly adjacent to the foreign network. Equally problematic is the communication of a mobile end device in the foreign network with a computer in the home network. All IP datagrams sent from outside of the home network to the mobile end device are transmitted directly to the home agent, which is then responsible for forwarding these to the mobile end device. However, if a local computer sends an IP datagram from the home network to the mobile end device, it is not forwarded to a router because it is directed locally to a computer within the home network. In this case, the network software of the sender carries out an ARP request (Address Resolution Protocol). This determines the hardware address of the mobile computer to which the data packet in the local network is to be forwarded.

So that the home agent can specifically intercept this data packet it uses a procedure known as **proxy ARP**. Here, the home agent first intercepts all ARP requests transmitted by the sender from the home network for determining the hardware address of the mobile end device. The home agent answers these by transmitting its own hardware address. Proxy ARP is carried out completely transparently. Instead of the requesting computer receiving the hardware address of the mobile end device it simply receives the hardware address of the home agent, who sends a data packet addressed to the mobile end device in the foreign network.

7.7.5 Mobile IP Version 6 – MIPv6

Mobile IPv6 (MIPv6) was developed as a part of the IPv6 standards to facilitate mobile communication connections, i.e., IP-based connections with and between mobile network devices, while offering them a continuous and cross-network mobility. Thus, MIPv6 represents an expansion of the Mobile IP standard based on IPv4. It was specified in 2004 in RFC 3775. Numerous mechanisms and functions in Mobile IP that had to be developed and integrated especially for guaranteeing mobile communication were already an integral component of IPv6. For example, IPv6 guaranteed the secure transmission of registration messages without the need for additional measures to be taken. IPv6 auto configuration supports the automatic procurement of care-of addresses. The discovery of the topological neighbor in the network (neighbor discovery) is likewise a component of the IPv6 protocol, so that special foreign agents are no longer needed for this purpose. In addition, every computer connected via IPv6 to the network is capable of sending update messages to other computers so that a mobile node can send its care-of address directly to a corresponding node or to the home agent.

In mobile IP a foreign agent is necessary for operation of a mobile device in a foreign network. In comparison, in MIPv6 the mobile device retains its primary address and uses a co-located care-of address. This is linked to the primary address. In this way, the operation of a foreign agent becomes unnecessary. In the case that the mobile end device is located in a foreign network, follows a forwarding of the data traffic intended for the end device via a home agent in the home network. This home agent intercepts the datagrams intended for the mobile device, transferring them to the current location of the mobile device.

The focus was on implementing the following design goals when MIPv6 was developed:

- **Static IP addresses**
  A mobile end device is assigned a static IPv6 address in the home network. It also retains it in mobile operation in different networks and network technologies.

- **Always-on connectivity**
  A mobile device remains reachable, even across network borders. Existing connection are not interrupted in transitioning from one network to another- The mobile end device always has the same IPv6 address assigned to it.

- **Roaming between different layer 2 technologies**
  A mobile end device transitioning between networks should also keep various technologies, such as WLAN, WiMAX, UMTS or LTE, connected to the Internet. Existing connections continue to remain intact and the assigned IPv6 address retained.

- **Roaming between different (sub)networks**
  Most of the time large WLAN networks use multi-layer 3 subnetworks. If a mobile device moves from one subnetwork to another, existing connections remain intact and the assigned IPv6 address is retained.

- **Session persistence**
  Also when there is a transition between different networks and network technologies, communication connections are maintained without interruption and can be used permanently.

- **Server-ability mobile devices**

  The above properties make it possible to operate mobile devices also as servers, in the sense of a reliable client-server communication.

In contrast to mobile IP, mobile IPv6 needs fewer additional mechanisms and functions to ensure mobile communication. A special foreign agent is no longer necessary in mobile IPv6. A mobile node requires only the ability to decapsulate datagrams, which are encapsulated by the home agent for forwarding. It must also be able to determine when it needs a new care-of address or if actualization messages must be sent to the home agent or to further nodes.

7.8 Glossary

Address resolution: Translation of IP addresses into the respective corresponding hardware address of the network technology. A host or router uses address resolution if data is to be sent to another computer within the same physical network. Address resolution is always limited to a specific network, i.e., a computer never resolves the address of a computer from another network. Vice versa, the fixed assignment of a hardware address to an IP address is designated **address binding** .

Address Resolution Protocol (ARP): Protocol of the TCP/IP protocol family used for address resolution from the protocol address to the hardware addresses. To determine a hardware address, the requesting computer sends the corresponding protocol address as an ARP message via broadcast. The requesting computer recognizes its own protocol address and is the sole one to transmit its hardware address as ARP message directly back to the requesting computer.

Autonomous System (AS): Designation for a number of subnets and computer systems that are under the control of a single operator and use a common routing method. The routing procedure within an AS is known as **Interior Gateway Protocol** (IGP). In contrast, the routing procedure between different AS is called **Exterior Gateway Protocol** (EGP).

Big Endian Order: Different computer architectures interpret bit sequences, and information coded as bit sequences, in various ways. One interpretation is the Big Endian Order. Here, a bit sequence is read from left to right starting with the highest-value bit. The interpretation in the opposite direction is called **Little Endian Order**. IP datagrams are always transmitted in Big Endian Order. This means that computers whose architecture is based on the Little Endian Order standard (e.g., Pentium), must always carry out a datagram conversion when sending and receiving IP datagrams.

Bridge: Intermediate systems that serve to connect individual segments of a LAN with each other. Bridges work on the transport layer (layer 2) of the network and are capable of carrying out filter functions to limit local data traffic on the respective LAN segment.

Broadcasting: Designation for the simultaneous addressing of all computers in a network. In the Internet a distinction is made between a broadcast in the local network and a broadcast over all networks in the Internet. If a message is to be forwarded to only one computer - corresponding to the usual addressing - then one speaks of a **unicast**.

Computer network: A computer network (**network**) is a data transmission system between the network-connected autonomous computer systems, which have their own storage, periphery and computing ability. A computer network offers every participant the possibility to become connected with every other network participant.

Connection-oriented/connectionless service: A fundamental difference is made between **connection-oriented** and **connectionless** services in the Internet. Connection-oriented services must first set up a connection over firmly agreed on switching stations in the network before the actual data transmission begins. The established connection path is used for the duration of the entire communication. Connectionless services do not chose a fixed connection path from the beginning. The transmitted data packets are each sent independent of each other on potentially different paths across the Internet.

Dotted decimal notation: For reasons of better legibility, 32-bit long IP addresses are often written as a sequence of 4 unsigned, whole decimal numbers (ocetets), separated by decimal points. To do this, the IP addresses are simply subdivided into four 8-bit long sections that are each interpreted as a decimal number (e.g., 136.199.77.1).

Flow control: In a communication network, flow control prevents a fast sender from flooding a slow receiver with transmitted data and thereby causing **overload** (**congestion**). Normally, the receiver has buffer storage available to temporarily store incoming data packets until their subsequent processing. To avoid an overflow of this temporary storage it is necessary to provide protocol mechanisms which enable the receiver to hold off receiving further data packets until buffer storage has completed its current processing.

Fragmentation/defragmentation: Due to technical restrictions, the length of data packets that a communication protocol sends in a **packet-switched network** is always limited below the application layer. If the length of the message to be sent is longer than the prescribed data packet length, the message is disassembled into individual (**fragments**) that meet length restrictions. In order that the individual fragments can be correctly put back together at the receiver following transmission (**defragmented**), they must be given a sequence number. This is because an adherence to the sequence order cannot be guaranteed when an Internet transmission is carried out.

Gateway: Intermediate system in the network that is capable of linking individual networks together to create one new system. Gateways enable communication between user programs at different end systems and are located on the application layer of the communication protocol stack. Because of this, they can translate different application protocols into one another.

Internet: Merging of multiple networks, which are incompatible among each other, into one. Appearing to the user as a a homogeneous universal network, it enables the transparent communication of each computer connected to a single net in this union with every other computer in the network. An internet is not subject to any limitations in view of its expansion. The concept of **internetworking** is very flexible and an extension of the internet is possible any time without limitations. The global Internet offers amazing proof of the internet's functional ability.

Internet Protocol (IP or, more precisely, IPv4):   Protocol on the data link layer of the TCP/IP reference model. As one of the cornerstones of the Internet, IP enables that a unified, homogeneous network appears from an internet made up of many heterogeneous single networks. A uniform addressing schema (**IP addresses**) is responsible for independent, unique computer identification for the respective network technology. IP's **connectionless, packet-switching datagram service** is not able to fulfill service guarantees, but always works according to the **best effort** principle. The **ICMP** protocol serves as an integral component of IP to communicate control information and error messages

IP address: 32-bit long binary address that designates a computer uniquely in the global Internet. For better readability, this address is subdivided into four octets that are interpreted as unsigned decimal integer and each separated by a decimal point (e.g., 232.23.3.5). The IP address is separated into two parts, the **address prefix** (**network ID**), which uniquely identifies worldwide the network the addressed computer is in, and the **address suffix** (**host ID**), which identifies the computer uniquely in its local network.

IP address classes: IP addresses are divided into different address classes. These are determined by the length of the respective network ID and host ID. The primary address classes A, B and C are first distinguished. There are 126 networks in address class A, each with up to 16,777,214 connected computers, in address class B there are 16,384 networks with up to 65,534 connected computers and in class C there are 2,097,152 networks with a maximum of up to 254 connected computers. Addresses in class D are designated as multicast addresses and addresses in class E are reserved for experimental purposes.

IP datagrams: The data packets transmitted over the IP protocol are called datagrams. This is due to the fact that the IP protocol only provides a connectionless and unreliable service (**datagram service**).

IPv6:   Protocol standard that succeeded the Internet protocol **IPv4**. It provides a considerably expanded functionality. The limited address space in IPv4 - one of the main problems of the normal IP standard - is drastically expanded in IPv6 with an extension of the IP address from 32 bits to 128 bits.

Loopback address: Special IP addresses (127.0.0.1 in IPv4, ::1 in IPv6), that can be set in a datagram as a receiving address and serve only for test purposes. A datagram with a loopback address does not leave the computer via its network interface, but only appears to be sent and received again. With the help of loopback addresses, network applications can be tested, before they receive access to the actual network.

Mobile IP: Expansion of the IP protocol that was especially developed for mobile end users. The mobile end user can leave its home network and register in a foreign network, thereby retaining its original IP address. The mobile computer logs off at an agent in its local network (home agent) and contacts a special computer in the foreign network (foreign agent). This assigns it a temporary address and is responsible for forwarding communication intended for it in collaboration with the home agent.

Maximum Transmission Unit (MTU):   Maximum length of a data packet that can be transmitted over a physical network. The MTU is determined by the type of network infrastructure implemented. In the transition of a data packet from a network with a large MTU in a network with a smaller MTU, the data packet must be fragmented if it is longer than the MTU of the network concerned.

Multicasting: In a multicasting transmission, a source transmits simultaneously to a group of receivers The transmission type is a 1:n communication. Multicast is often implemented for the transmission of real-time multimedia data.

Port number: 16-bit long identification for a TCP connection that is always associated with a specific application program. The port numbers 0–255 are reserved especially for TCP/IP applications (**well known ports**), the port numbers 256–1,023 for special UNIX applications. The port numbers 1,024–56,535 can be used for one's own applications and are not subject to fixed assignment.

Repeater: Intermediate system that serves in the amplification of the signals in the interconnection of different network segments. Repeaters are absolutely transparent and transmit each of the received data packets in amplified form.

Reverse Address Resolution Protocol (RARP):   Protocol for address resolution from the hardware address to the protocol address - thus the reverse of ARP protocol. RARP is deployed as a rule so that computers without a hard drive can determine their own IP address at system startup. To do this, the starting computer sends a RARP message via broadcast containing its hardware address. The RARP server of the network then answers the requesting computer with the corresponding IP address.

Router: Switching computer that is capable of connecting two or more subnets with each other. Routers work on the transport layer (IP layer) of the network. They are capable of forwarding incoming data packets on the shortest path through the network based on their destination address.

Routing: On the path between the sender and the receiver in an internet are often multiple intermediate systems (**routers**). These take care of the forwarding of transmitted data to the respective next router (next hop) along the way to the receiver. The determination of the path from the transmitter to the receiver is called routing. Routers receive a transmitted data packet, evaluate its address information and forward it accordingly to the next router or to the final receiver, as the case may be.

Subnetworking: Subnet addressing was developed at the beginning of the '80s with the intention of using the limited address space of the Internet more efficiently. A network is divided logically as well as physically into individual subnetworks. This is done using so-called subnet masks. The regular division of the IP address into the network ID and the host ID is expanded. The field of the host ID is split up into a subnet ID and the remaining host ID. A local router must now no longer manage all individual computers in the network, but only those in the local subnetwork. Computers assigned to another subnetwork are only entered grouped into a single entry in the routing table. This is identified by the specific subnet ID.

Subnet mask: The subnet mask is used for subnet addressing. With its help, it is possible to determine the network ID, subnet ID, or host ID, from one address representation, which are necessary for forwarding at the routers.

Supernetting: The concept of supernetting allows a more efficient use of the available IP address space . Here, an individual organization operating a network assigns entire blocks of network addresses a lower address class, in case a higher address class of the organization has been assigned too large of an address space.

Transmission Control Protocol (TCP):   Protocol standard on the transport layer of the TCP/IP reference model. TCP provides a reliable, connection-oriented transport service, which many Internet applications are based on.

User Datagram Protocol (UDP):   A simple protocol standard on the transport layer of the TCP/IP reference model. UDP provides a connectionless transport service that sends IP datagrams over the IP protocol. The principle difference between IP and UDP is that UDP is capable of managing port numbers by way of which different computers communicate with each other via the Internet .

Universal Service: A universal service has the task of making available all accessible information on all computers, which can also be located in different networks.

Virtual connection: A connection between two end systems that are only generated by the software installed on each of the end systems. The actual connection network does not need to provide resources for this purpose, but only to guarantee a transport of the data. As the connection is not fixed across the network in switched form - only the illusion of an existing connection is presented by the installed software at the end system - it is designated as virtual.

Chapter 8
Transport Layer

„Much might be done – did we stand fast together.
– Friedrich Schiller (1759-1805), "Wilhelm Tell"

The Internet protocol provides the unifying connection for a variety of network tech-
nologies. It creates a uniform, homogeneous and global network from different ar-
chitectures that can be likened to a colorful patchwork. But this is by no mean the
end of the story. On its own, IP provides an unreliable and insecure service. Data
is transported into the receiver's network, each individual piece independent of the
other, and without a guarantee of secure, reliable delivery. Reliable communication
is, however, the prerequisite for efficient and safe data traffic in the network. An
application must be able to rely on once-sent data actually reaching its designated
receiver. At the very least, it is necessary to determine whether or not the transmitted
data has in fact reached its goal, and if so, whether this happened in a timely and
reliable manner. The establishment and management of a secure, reliable connection
is one of the tasks of the next higher protocol layer in transport layer of the TCP/IP
reference model. This is where the complex and reliable Transmission Control Proto-
col (TCP) works, and the speed-optimized User Datagram Protocol. These provide
the user comfortable tools for connection management. The transport-layer protocols
place a further abstraction layer on the Internet and implement a direct end-to-end
connection, without attention having to be paid to the details of the connection,
e.g., routing of the data traffic.

Seen from the perspective of the network layer of the TCP/IP reference model,
the worldwide Internet is a huge collection of single, autonomous systems (subnetworks) connected with each other via multiprotocol routers. Networking is implemented by the Internet Protocol (IP) on the Internet layer of the TCP/IP reference
model. IP makes it possible for programs and applications working on one of the
higher layers in the protocol stack, to communicate with each other over the Internet The transmission and reception of IP datagrams, however, proceeds based on
a "best effort" service. This means that there is no guarantee for a reliable, error-
free and secure transmission of data. The receiver cannot depend on the correct and
complete transmission of data received on the Internet layer. While simple error
detection technologies - such as a checksum in the IP datagram header - are alrea-

C. Meinel and H. Sack, *Internetworking*, X.media.publishing,
DOI: 10.1007/978-3-642-35392-5_8, © Springer-Verlag Berlin Heidelberg 2013

dy employed on the Internet layer, nevertheless the Internet Protocol does not have any explicit mechanisms for error detection, or error handling. This procedure focuses on efficiency, as an error-free transmission is not always necessary, nor is the effort needed to achieve it worthwhile. In contrast to the Internet layer, protocols supplement the subsequent transport layer of the connectionless datagram service with extra functions. These allow a reliable and secure data transmission between two end systems.

Numerous programs and applications require a reliable path of data transmission without themselves providing the special mechanisms to carry this out. These mechanisms have already been adopted by the transport layer of the TCP/IP reference model and are discussed in detail in the sections that follow. Afterwards, a look will be taken at the two most important protocols of the transport layer: Transmission Control Protocol (TCP) and User Datagram Protocol (UDP). Network address translation is an important technology that can circumvent some of the problems in IPv4 stemming from the limited IPv4 address resources. This technology, which is widespread today, is located on the transport layer of the TCP/IP reference model and will be introduced after our look at TCP and UDP. At the end of the chapter we will examine fundamental security procedures employed on the transport layer.

8.1 Tasks and Protocols of the Transport Layer

The protocols of the transport layer are located on the connectionless datagram service IP, on top of the network layer of the TCP/IP reference model. The TCP/IP protocol family provides two very different transport protocols for data transfer, both based on the unreliable datagram service IP.

- **User Datagram Protocol (UDP)** A simple transport protocol that handles an insecure, connectionless communication between two end systems.

- **Transmission Control Protocol (TCP)** Implements a secure, connection-oriented full duplex data stream between two end systems.

Additionally, the TCP/IP protocol family has further transport protocols based on TCP or UDP, which use TCP or UDP data structures as a transport container. Among these are:

- ISO Transport Protocol ISO DP 8073 (RFC 905)
- OSI Connectionless Transport Services based on UDP (RFC 1240)
- ISO Transport Service based on TCP (IITOT, RFC 2126)
- NetBIOS based on TCP/UDP (RFC 1001/1002)
- Real-time Transport Protocol (RTP, RFC 1889)
- System Network Architecture (SNA) via UDP (RFC 1538)
- Datagram Congestion Control Protocol (DCCP, RFC 4340)

- Explicit Congestion Notification (ECN, RFC 3168)
- Licklider Transmission Protocol (LTP, RFC 5326)
- Stream Control Transmission Protocol (SCTP, RFC 4960)
- Lightweight User Datagram Protocol (UDP-Lite, RFC 3828)

While UDP was hardly subject to change based on its simple range of functions, TCP has been constantly developed and its control mechanism continually refined. This has resulted, in the meantime, in the development of a very complex protocol. The transport layer provides the application layer above it with a series of so-called **service primitives**. Via the application program these can process and manage their data connection in a simple way without having to know the properties of the Internet layer or technologies below it. This simplifies the administrative burden and makes possible application development independent of the network technology. Furthermore, on the transport layer it is ensured that all transmitted data arrives at the receiver in a form that is error-free, complete and sequentially correct. This **reliable data transmission** is made possible by the use of sequence numbers, timers, flow control and an acknowledgment mechanism for the received data. Correctly received data is confirmed and faulty data that is incompletely transmitted requested again. In contrast to the Internet layer, in the transport layer a message of almost any length, so-called "streams" can be sent. These are subdivided into segments for transport. Additionally, in the transport layer an overload control is provided. This takes care of a load-independent regulation of the transmission volume so that connections within a network are not burdened excessively.

8.1.1 Services of the Transport Layer – an Overview

A fundamental difference between protocols of the Internet layer and the transport layer of the TCP/IP reference model is that communication on the transport layer is always carried out between the two end systems in the form of an **end to end transmission**. The protocols of the transport layer must therefore only be installed at the end systems. Intermediate systems, e.g., routers, work only on the Internet layer and are not integrated in the direct communication on the transport layer. In this way, the burden on routers in the Internet is reduced. Numerous tasks to guarantee the reliability of communication are carried out by the end systems themselves regarding the connections they maintain. If data communication breaks off due to error, the affected end systems communicate to each other how much of the data to be exchanged has already reached the receiver. Also communicated is where the interrupted transmission is to be taken up again. The intermediate systems along the communication path are not burdened as a result. A further advantage of this end-to-end communication is that the data to be transported can be interpreted as a data stream. In this way, additional effort, above all in the programming of network applications, is avoided that would otherwise be necessary if the packet-oriented transmission method of the Internet layer below would have to be taken into consideration. A distinction is made on the transport layer between **connectionless and connection-oriented data transport**. How the connectionless service on the transport layer is implemented with the user datagram protocol is similar in many aspects to the connectionless service on the Internet layer. Conversely, the Transmission Control Protocol implements a connection-oriented service between two end systems. Data transmission proceeds in three phases:

- **Connection establishment**
  A virtual connection must initially be established between the two end systems with the help of protocols from the lower protocol layers of the TCP/IP reference model.

- **Data transmission**
  Once the connection between the two end systems has been established, the actual data transmission can proceed. Messages in the IP datagrams are transmitted encapsulated along the virtual connection to the receiver. IP itself handles the transported messages of the transport layer as pure payload, i.e., the Internet protocol does not carry out any interpretation of the command and control information contained in the messages of the transport layer. Intermediary systems can therefore forward the encapsulated IP datagrams without a problem.

- **Connection release / connection sharing**
  Upon completion of a data transmission, the virtual connection that was previously set up between both end systems must be released again. It is necessary to make sure that all data transmitted beforehand by the sender has already arrived at the receiver.

Fig. 8.1 shows the virtual connection between two end systems on the transport layer in interplay with the other layers of the TCP/IP reference model.

Data transmission in the transport layer proceeds on the basis of a **byte-oriented data stream**, i.e., applications must not exchange complete messages as is necessary in the Internet layer below when data transfer takes place. In fact the data to be exchanged is viewed as a continual stream of bytes that is controlled by the application in the communication. Data transmission on the transport layer guarantees the correct order of transferred data in transport and delivery (**same order delivery**). This is implemented with the help of segment numbers, which ensure that data is received in the same order in which it was sent. The transport layer protocol receives the individual data packets from the protocol instance of the Internet layer below and passes them on in the correct oder to the calling application. However, it is also possible for delays to occur during data delivery. These are caused by a transposition of the order sequence on the Internet layer during transmission (head-of-line blocking).

Transport layer protocols can enable a **reliable data transmission**. In the case that data packets in the Internet transmission are lost, e.g., through router overload (network congestion) or network error, or if their contents are damaged, this can be determined by the receiver side and suitable counter-measures be taken. It can easily be determined if there is damage to the contents of a transmitted data packet

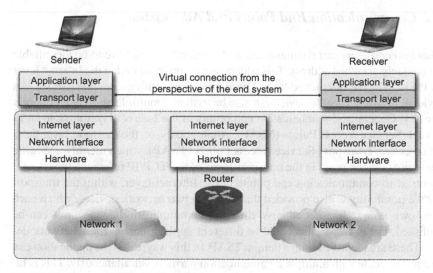

Fig. 8.1 Virtual connections on the transport layer of the TCP/IP reference model.

with a checksum comparison. If a data packet is received in a correct and undamaged form, the receiver sends a confirmation of receipt (acknowledgment). In the other case an acknowledgment is not sent. At the start of data transmission a timer is started on the sender side. If the sender gets an acknowledgment from the receiver before the timer runs out, correct reception is thereby confirmed. However, if the timer runs out before an acknowledgment has been received, it is then necessary to carry out a new transmission of the relevant data. Data may also be requested again from a sender through explicit repeat-requests.

An additional **flow control** is implemented by the protocols on the transport layer. Possibilities have to be provided to control the rate at which the sender and receiver send their data. A faster sender must be throttled if the receiver cannot process the incoming data fast enough. Otherwise, the result will be an overflow of the receiver's available buffer storage. The sender's transmission rate can be increased in the opposite situation - if the receiver is able to process the arriving data faster then it can be delivered by the sender. In an extreme situation an **overload** can occur in the network. Here, transmitted data may be lost or entire network segments can fail. Automatic repeat-requests - that is, repeated requests for data - can quickly overload a network. Because of this, the protocols of the transport layer provide additional measures in addition to flow control in an attempt to avoid overload situations. The transport layer offers the possibility for multiple (virtual) communication connections to be established and managed between two computers in the network. This can take place at the same time and independently. So-called **port multiplexing** makes it possible to serve multiple receivers simultaneously over the same message connection. For this purpose, so-called ports are set up at every host. Via these ports, different network services can communicate with each other at the same time.

8.1.2 Communication End Points and Addressing

So that two end users can communicate with each other they have to be identifiable
via a uniquely assigned address. This addressing is implemented in the Internet layer
with the help of the Internet address. These uniquely identify a network device. An
individual network-capable device can also be assigned multiple, individual Internet
addresses. Generally, communication end points in the Internet layer are known as
Network Service Access Points (NSAP). Analogously, on the transport layer there
exist so-called **Transport Service Access Points (TSAP)**, which identify a specific
communication end point in the transport layer of the TCP/IP reference model.

In contrast to communication end points in the Internet layer, within the transport
layer the possibility is also provided that between two network devices, which each
has its own unique Internet address, different communication connections can be
established. For example, this could be different applications on these network devices. These are identified with a unique TSAP. In this way, the same application can
also communicate with multiple communication partners simultaneously. This is facilitated by establishing individual instances on the application layer of the TCP/IP
reference model, i.e., copies of an application as a so-called **process**. Application
processes can connect with a TSAP on the local computer to establish a connection
to a remote TSAP. These connections must then run via an assigned NSAP. For this
purpose a **multiplexing** of the different transport connections must take place over
the NSAP of a computer. It must be possible to address the TSAPs according to a
unique schema.

Vice versa, on the foreign computer one of possibly multiple TSAPs is chosen for
the transport connection that is to be set up by the local NSAP. Based on the connection requirements, it is assigned to a specific application on the remote computer by
way of an application process. On the input side a TSAP is therefore assigned a specific application so that already through the connection establishment the selected
connection partner process can be addressed (cf. Fig. 8.2). In this way, a **demultiplexing** is carried out over a mutually implemented NSAP transport connection.

Normally, both communicating processes work with each other in a complementary
manner on the side of the sender and the receiver in the TCP/IP reference model.
Thus, the process on the sender side, the so-called **client**, requests a service on
the receiver side. On the receiver side the so-called **server** provides the requested
service, unless this is prevented by error conditions or access restrictions.

8.1.3 TCP and UDP Ports and Sockets

To maintain a TCP connection a sender and a receiver must each establish the necessary end points of communication – so-called **sockets**. The assignment of the
sockets in a TCP communication is limited to the involved communication partners
and duration of the connection. Every socket has a reserved memory storage available in the computer as a communication buffer. Here, the data is deposited which is

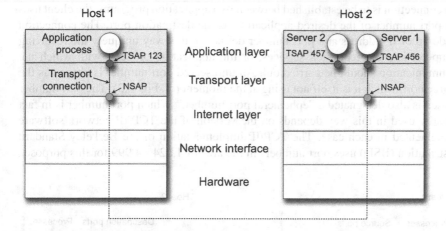

Fig. 8.2 Transport connections via TSAPs and NSAPs.

to be transmitted or which is received. For identification purposes, every socket has its own socket number. This is composed of the IP address of the computer and a locally assigned 16-bit long number, which gives the so-called **port**:

< IP address >:< port number >

A service access point on the transport layer (TSAP) is designated as a port. Ports with the numbers 0 to 1,023, the so-called **well-known ports**, provide a reservation worldwide that is unique for standard services, such as HTTP, FTP or telnet (cf. table 8.1, RFC 1700). The area from 1,024 to 65,535 can be assigned on the computer to any desired application program. A general distinction is made between:

- **Reserved ports** (privileged ports): Applies to the port numbers 0 - 255 for TCP/IP applications and 256 -1,023 for specific UNIX applications. These ports are assigned TCP/IP basic applications by the Internet Assigned Numbers Authority (IANA), which is also responsible for the general coordination of IP address assignment. They are all fixed in the RFC standardization process, or will be standardized at some point in the future. Therefore, these port numbers are also known as system port numbers.

- **Registered ports**: Applies to port numbers 1,024 - 49,151 registered by (IANA), but not assigned as TCP/IP basic applications, which are standardized via an RFC. The purpose of registration is to prevent applications in this area from hindering each other through the unintentional use of identical port numbers. These port numbers are also known as user port numbers.

- **Private, dynamic ports** (ephemeral, temporary ports): Applies to the remaining possible port numbers between 49,152 and 65,535. These are neither reserved nor registered by the IANA and are freely available to all users without restrictions.

If a connection is to be established between two application programs, the client uses the port number of the desired application as the destination port. The connection made up of IP address and port number defines in this way uniquely the receiving computer and the requested services, or the application process, with which the communication should be carried out. An unreserved port number is chosen as the source port, which it is itself not using at the moment (cf. Fig. 8.3). The port number chosen is also designated a "ephemeral port number." Which port number is in fact actually used in this way depends on the version of the TCP/IP network software implemented in each case. The TCP/IP implementation of the Berkeley Standard Distribution (BSD) uses port numbers in the area of 1,024 - 4,999 for this purpose.

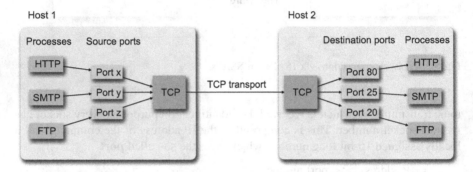

Fig. 8.3 A TCP connection between different application processes via TCP ports.

To facilitate communication between two end users both must know the respective socket number being used. The sender choses a socket number that is already assigned a specific communication application. On setting up the transport connection, the receiver must, however, be informed of the socket number of the sender so that it actually gets the answer the receiver sends it. A **TCP connection** can be uniquely identified by means of both socket numbers on the sender and receiver side via the tupel (socket sender, socket receiver). It is possible for a socket to serve multiple connections at the same time. So-called **service primitives** are implemented such as request, response, confirm, read, write, etc. These allow the user, or an application program, use of the transport service provided to it.

8.1.4 Service Primitives on the Transport Layer

Abstract, implementation-independent processes for the use of services on a certain level of the TCP/IP reference model are called service primitives (also service elements). They serve to describe communication processes and can be used as abstract guidelines in the definition of communication interfaces. Via service primitives only

Table 8.1 Several important reserved TCP port numbers (well-known ports) according to RFC 1700.

Port number	Application
7	Echo Protocol
20	FTP (File Transfer – Data)
21	FTP (File Transfer – Control)
23	Telnet
25	SMTP (Simple Mail Transfer Protocol)
37	Time Protocol
53	DNS (Domain Name Service)
69	TFTP (Trivial File Transfer Protocol)
80	HTTP (Hyper Text Transfer Protocol)
110	POP3 (Post Office Protocol 3)
119	NNTP (Network News Transfer Protocol)
123	NTP (Network Time Protocol)
143	IMAP (Internet Message Access Protocol)
194	IRC (Internet Relay Chat)

the data to be exchanged between communication partners is fixed and not how this procedure takes place (cf. Fig. 8.4).

Table 8.2 Service primitives to establish a simple connection-oriented service in the transport layer.

Primitive	Send	Explanation
LISTEN	–	Waiting until a connection is set up
CONNECT	Connection request	Active establishment of a connection
SEND	Data	Sending of data
RECEIVE	–	Waiting until data is received
DISCONNECT	Disconnection request	Process wishing to terminate a connection

The necessary service primitives for definition of a simple connection-oriented service on the transport layer are shown in table 8.2. In order to establish a connection, the client sends the server a connection request. When the server receives it, a check is first made as to whether the server is in waiting status (LISTEN). The waiting status is over and the server confirms the connection request. The client is blocked while waiting for the server's answer. As soon as the server's positive confirmation has arrived, the connection is established. Data can now be exchanged via the connection that has been set up with the help of the service primitives SEND and RECEIVE. One of the two communication partners can additionally execute a RECEIVE blocking, until the other communication partner frees the blocking again

Service Primitives

Service Primitives define abstract processes in the communication processes between the service user and the service bringer (Service Access Points, SAP), within the framework of a communication model. Principally, four groups of service primitives are distinguished:

- **Request (REQ)**
  A request describes the service queried by the service user at a SAP.
- **Confirmation (CONF)**
  Confirmation of the provision of service at the SAP to the requesting service user.
- **Indication (IND)**
  An indication shows the service request at the SAP to the service user.
- **Response (RESP)**
  An answer to an indication from the service user at the SAP.

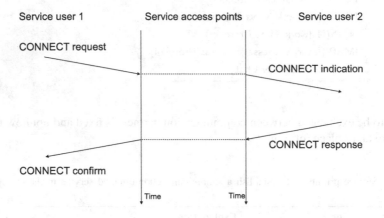

Further reading :

Tanenbaum, A.S., Wetherall, D.J.: Computer Networks, 5th edition, Prentice Hall (2010).

Fig. 8.4 Service primitives.

with a SEND and is able to receive data. In this way, data can be exchanged reciprocally. When the data transmission is complete the connection is then terminated. A distinction is made between a symmetrical and an asymmetrical connection termination. In an **asymmetrical connection termination** every direction of the connection must be ended separately via a DISCONNECT.

8.2 User Datagram Protocol – UDP

The Transmission Control Protocol (TCP) is one of the name-sake protocols of the TCP/IP reference model. It provides a reliable connection-oriented transport service. A second, considerably simpler, transport protocol also exists in the transport layer: the User Datagram Protocol (UDP). For data transmission via TCP, a connection to the desired communication partner must first be established. Following the successful data transmission it must be explicitly terminated again. On the other hand, UDP provides a simpler, connectionless transport service. It works more efficiently than TCP and can be implemented in a simpler way, with far less effort.

8.2.1 Tasks and Functions of UDP

The alternative to a complex, connection-oriented transportation is provided by a connectionless interface. This makes it possible to send messages at any time, to any destination without having to worry about the establishment of a connection beforehand. In contrast to TCP, the **User Datagram Protocol** (**UDP**) is a connectionless, unreliable transport service. It is specified in RFC 768. Unlike many other Internet protocols, its only 3 page long specification has not been revised since its publication in 1980. For over 30 years now it has remained unchanged. The development of this simple transport layer protocol began in 1977. The need for a easier and faster protocol than the connection-oriented and reliable TCP became pressing with the transmission of voice over the Internet. This protocol was only intended to ensure the safety of addressing, without being responsible for a reliable data transmission, as this led to transmission delays that caused disturbances in real-time voice transmission. UDP involves the simplest, connectionless service of the TCP/IP family, which is located directly on the IP. Similar to IP, UDP offers neither a secure transmission nor a flow control. There is accordingly no guarantee that the data packets sent via UDP actually arrive at their designated receiver. The Internet protocol only implements a "best effort" service, meaning that IP datagrams can get lost during Internet transmission, e.g., in overload situation at routers. UDP does not provide a possibility to register the loss of data packet nor to react to it. Above and beyond the functionality of IP, UDP offers two communication partners a transport service - in this case connectionless and unreliable - that connects the respectively communicating applications over assigned communication ports (**UDP ports**) with each other. Additionally, an optional checksum is included with which transmission errors can be discovered. UDP can be used for different application layer protocols, which are each identified by a special port number. Included among these are client/server- based applications whose communication is limited to a single, simple question/answer exchange. These applications favor UDP as a transport protocol because an extra connection does not have to be established and managed for one single transaction. UDP takes data from different application programs on a computer over the ports that are assigned to each application. It then passes it on as

UDP datagrams in encapsulated form to IP for forwarding over the Internet (**port multiplexing**). Vice versa, UDP accepts encapsulated UDP datagrams from IP and make them available to applications via the appropriate ports (cf. Fig. 8.5).

Tasks of the UDP Protocol

The simple UDP protocol encompasses the following three basic tasks:

- **Higher Layer Data Transfer**
  An application program or service of the application layer that passes a message to the UDP protocol for transmission.

- **UDP Message Encapsulation**
  The message to be sent is encapsulated in the payload section of the UDP data packet. Source and destination port are given in the header of the UDP data packet. If necessary, another checksum is calculated for the to-be-transmitted data.

- **Data Transfer to IP**
  The UDP data packet is passed to the Internet protocol for transmission.

On the receiver side, these operations proceed in the reverse order and direction.

Further reading:

Postel, J.: User Datagram Protocol. Request for Comments 768, Internet Engineering Task Force (1980)

Fig. 8.5 Tasks of the UDP protocol.

8.2.2 UDP Message Format

The header of a UDP data packet has a very simple structure (cf. Fig. 8.6):

- **Source port**
  16-bit long port number for the sender.

- **Destination port**
  16-bit long port number for the receiver.

- **Length**
  16-bit long field giving the length of the entire UDP data packet in bytes, including the header. The minimal value for the length of the UDP data packet is 8 bytes, which corresponds to the length of the header.

- **Checksum**
  The 16-bit long checksum used for error detection is optional. If it is specified with zero, this means that no checksum is implemented. The algorithm for calculation of the checksum corresponds to the algorithm used in IP. However, the IP checksum calculation extends only to the IP header, so that it is not advisable to omit calculating the UDP checksum. UDP checksum calculation includes the UDP header and the so-called **pseudo header**. This pseudo header is made up

of the IP addresses of the sender and receiver and the protocol information from the IP datagram header. It also includes length information from the UDP data packet header. The pseudo header totals 12 bytes in length (cf. Fig. 8.7). It is only used for checksum calculation and because of this is placed before the actual UDP data packet header. The receiver can therefore always check whether the data packet is in fact actually delivered to the correct addressee.

0	8	16	31
Source port		Destination port	
Length		Checksum	
Payload			

Fig. 8.6 UDP datagram - data format.

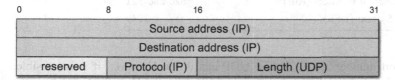

0	8	16	31
Source address (IP)			
Destination address (IP)			
reserved	Protocol (IP)	Length (UDP)	

Fig. 8.7 UDP - pseudo header data format.

8.2.3 UDP Applications

In contrast to the reliable, connection-oriented TCP protocol, UDP provides an unsecure transport service that offers almost no other functionality. This is also the reason that classic Internet applications dependent on reliable data transmission, such as email or the Hypertext Transfer Protocol (HTTP), do not implement the simple UDP protocol. On the other hand, there are numerous applications in which transmission capacity, i.e. quantitative throughput, is the main focus rather than reliable and error-free transport. These include the real-time transmission of multimedia data, e.g., video-streaming or Internet telephony. If an error-free transmission was given preference in these applications, a new sending of data damaged in transport would result in unwanted transmission delay. This would have a noticeable effect on the multimedia data transmitted in real-time, taking the form of audio breaks or "jerks" in the video image. Data loss in a multimedia real-time data stream can be tolerated to a certain extent without sensory perception being dramatically disturbed.

A further category of application that uses the UDP protocol focuses on the transmission of relatively short messages. For reasons of efficiency, the UDP protocol is preferred. A reliable TCP connection would present a significant overhead here in view of transmission length and the amount of date to be transmitted. Numbering among the services on the application layer that use UDP as a transport protocol are those applications shown in the table 8.3.

Table 8.3 Services that use UDP as a transport protocol.

Service	UDP port number
Domain Name Service (DNS)	53
Bootstrap Protocol (BOOTP)	67 and 68
Dynamic Host Configuration Protocol (DHCP)	67 and 68
Trivial File Transfer Protocol (TFTP)	69
Network Time Protocol (NTP)	123
Simple Network Management Protocol (SNMP)	161 and 162
Remote Procedure Call (RPC)	111
Lightweight Directory Access Protocol (LDAP)	389
Routing Information Protocol (RIP)	520 and 521
Network File System (NFS)	2049

The fixed port numbers for TCP and UDP can differ. However, if it is possible to use a service with UDP as well as with TCP (e.g., the Domain Name Service (DNS), then standardized port numbers are used. The **Lightweight User Datagram Protocol (UDP-Lite)** is a special variation of the UDP protocol. It is specified in RFC 3828. UDP-Lite is especially designed for the transmission of data which requires the least amount of delay possible and where smaller transmission errors can be dealt with easily. For example, in the case of live transmissions of audio or video data streams. UDP-Lite has the same data format as UDP, just with the length limitation of 65,535 bytes (16-bit length field) lifted. The length field indicates over which length the checkum should be calculated (checksum coverage). The actual length of a UDP-Lite data packet must be figured from the length specifications in the IP datagram header. If the length specifications are different, the UDP-Lite data packet receives additional unverified data.

8.3 Transmission Control Protocol – TCP

The just discussed user datagram protocol is designed particularly for the fastest and most efficient transport of data possible and places less emphasis on an error-free and reliable transmission. However, these properties are a necessary fundamental condition for many Internet applications. The sender must have the assurance that the data it has transmitted arrives at the receiver in in the exact form in which it has been sent. Transmission errors and data loss must be registered by the communication partners on both sides and remedied by both parties so that the prerequisites for a reliable transport service are guaranteed. The **Transmission Control Protocol** constructs out of an unreliable and connectionless Internet protocol a reliable and connection-oriented transport service. To make this possible, TCP bundles multiple procedures for the connection management, the correction of transmission errors, flood control and overload control.

Seen historically, the development of today's TCP had already begun in the early 1970s. The ARPANET, precursor of the modern-day Internet, was expanding rapidly, with more and more new networks seeking to connect to it. In 1973 Robert E. Kahn and Vinton G. Cerf started development of the first "Transmission Control Program." This became the successor of the first used network management software, the Network Control Program (NCP), which had become more and more burdened with the management and operation of the growing Internet. This first version differed from that currently implemented as it was responsible for both the connectionless transmission of datagrams and routing (Internet layer functionality) as well as for connection management and reliable data transport (transport layer functionality). Because of this, conflicts were created with both modularity principles as well as the strict layer structure of protocol functionality. It was no longer possible to provide Internet layer functionality and transport layer functionality separate from one another. Thus, TCP evolved into a very inflexible and inefficient protocol focused on applications and services that only needed a simple, connectionless and unsecure communication. With the creation of the fourth version - specified in RFC 793 - it was decided in 1981 to finally divide the Transmission Control Program into two separate protocols: the Internet protocol IPv4 and the Transmission Control Protocol. Each of these offered the necessary functionality and services for the Internet or transport layer. In addition to the fundamental RFC 793, there also exist numerous other RFCs in which the procedures and algorithms of the TCP protocols are further defined and expanded (cf. Fig. 8.4).

Client/server applications use TCP as a transport protocol for two-way communication. In the interaction with the TCP protocol software, an application must give details that specify the desired communication partner and initiate the communication. The interface using an application in the interaction with the TCP protocol software is generally known as an **Application Programming Interface** (**API**). An API defines a series of operations that can carry out an application in interacting with the TCP protocol software. The TCP protocol has just a small number of interfaces that are defined in detail in RFC 793. These refer to the establishment and termination of connections as well as the control procedures for data transmission:

- **Open**

  Opening of a connection. Here, parameters must be established indicating whether an active or a passive connection opening is involved, the port number and IP address of the communication partner, the local port number and a value for the timeout. As return value, the routine returns a local connection name over which the connection can be referenced.

Table 8.4 Important TCP Internet standards.

RFC	Title
813	Window and Acknowledgment Strategy in TCP
879	TCP Maximum Segment Size and Related Topics
896	Congestion Control in IP/TCP Networks
1122	Requirements for Internet Hosts – Communications Layer
1146	TCP Alternate Checksum Option
1323	TCP Extensions for High Performance
2018	TCP Selective Acknowledgement Options
2581	TCP Congestion Control
2988	Computing TCP's Retransmission Timer
3168	Explicit Congestion Notification

- **Send**
  Transmission of data to the TCP send buffer, which is subsequently sent over the TCP connection.
- **Receive**
  Passing on the received data from the TCP receive buffer to the application.
- **Close**
  Ending of the connection. Before this is done all data from the TCP receive buffer is passed on to the corresponding application and a FIN segment sent.
- **Status**
  Output of status information concerning the existing connection.
- **Abort**
  Immediate interruption of the send and receive processes.

In the following section, the functions and tasks of TCP are described. A look will be taken at the fundamental procedures and algorithms for reliable data transport. Finally, command functions for flow control and overload are presented.

8.3.1 Functions and Tasks of TCP

TCP offers a truly reliable transport service for sending messages over the Internet. For this purpose, TCP provides its own addressing scheme for the identification of communication end points in a connection in the form of TCP ports. As soon as a connection is established between two communication end points, a continuous (bidirectional) exchange of messages can begin via the connection. Data is divided into so-called **segments** and transmitted as a byte stream. It is then forwarded to the Internet protocol below it and in the form of datagrams further fragmented and transmitted. Finally having arrived at the receiver in reverse order it is then reassembled. To enable this, TCP defines a number of different procedures and algorithms that guarantee a reliable, consistent and timely data transmission procedure. TCP has the following basic properties to accomplish this:

- **Connection-oriented data transmission**
  TCP offers a connection-oriented service. This is the reason a connection must first be set up between both communication end points to the designated receiver before the actual data transmission can take place. After the completion of the data transmission the connection is terminated. The connection provided by TCP is known as a **virtual connection**, because it is achieved purely as a software technology. The actual Internet offers neither the necessary hardware nor software to support dedicated connections for this purpose. Thus, the two TCP processes on the sender and receiver side only give the illusion of this type of connection through the manner of their message exchange (cf. Fig. 8.1). TCP uses the IP datagram service for data transport. Every TCP message is encapsulated in an IP datagram transmitted via the Internet and then unpacked. IP itself handles the TCP messages purely as user data, i.e. IP does not undertake an interpretation of the TCP data or the TCP data header. As can be seen in Fig. 8.1, only the respective end system has a TCP implementation, while for the intermediate systems a TCP implementation is not necessary.

- **End-to-end transmission**
  TCP only allows a data transfer between exactly two dedicated end points. The connection therefore runs directly from an application at the sender computer to an application at the receiver computer. The establishment of a virtual connection is the responsibility of the involved end systems alone. The network located between both end systems with its intermediate systems serves solely in forwarding. Multicasting or broadcasting is not possible with TCP. Between two end devices, multiple connections can exist parallel. These can each be used by different applications, services or users. Every connection is therefore managed independent of the other connections, conflict-free.

- **Reliable transmission**
  TCP should always guarantee an error-free data transmission without data loss or the sequence exchange of received data. In each case it is therefore checked whether all data transmitted by the sender is in fact received. Also, a check is made of whether the integrity of the sent data has been ensured and, if it is necessary to resend the data. In this regard, TCP uses the following technologies:

  - **Positive Acknowledgement with Retransmission, PAR**
    A technology initiated by TCP that contributes significantly to reaching a high degrees of reliability is data **retransmission**. The correct receipt of a sent TCP message is always confirmed by the receiver with an acknowledgement. This is transmitted back to the sender of the data. Before every data transmission the sender starts a timer. If this timer runs out at the sender before the confirmation from the receiver has arrived, the message is considered lost and the sender starts a new transmission (cf. Fig. 8.8).

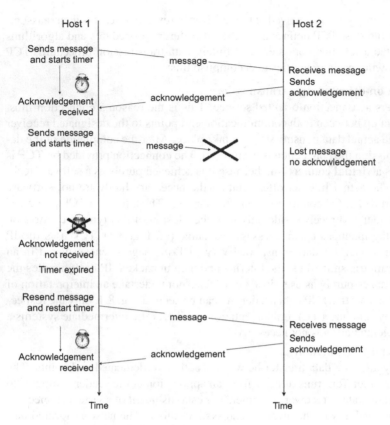

Fig. 8.8 TCP – retransmission in the case of a timeout.

The described procedure is inefficient because a new message can first only be sent upon arrival of a confirmation. If the sent message is marked by an indicator (message ID), then multiple messages can be sent at the same time. However, to prevent creating an overload, the receiver can limit the maximum number of parallel transmissions allowed. At the same time, an individual timer is necessary for every parallel transmission taking place. This improved PAR procedure is shown in Fig. 8.9. Host A transmits message 1 to host 2. Even before the complete message has reached host 2, host 2 sends the information about the maximum accepted number of parallel message connections, i.e., the maximum number of outstanding messages at any one time. In the given example, the maximum number of open message transmissions equals two. Host 1 can therefore continue directly after transmission of message 1 and send message 2. Before it can continue with message 3, the maximum limit of 2 outstanding (unconfirmed) connections from host 2 reaches it. Because of this, message 3 can only first be sent when the confirmation for message 1 has arrived.

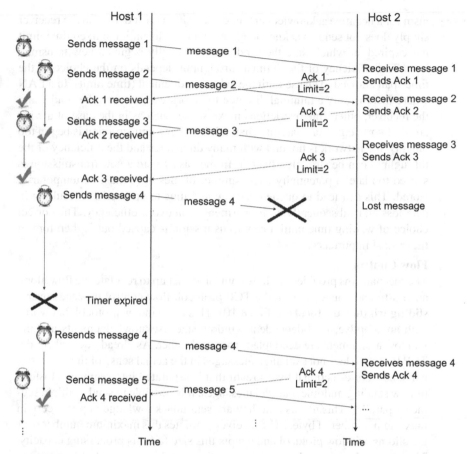

Fig. 8.9 TCP – improved PAR procedure.

– **Adaptive retransmission**

The developers of TCP quickly recognized that in choosing a fixed timeout it would never be possible to treat all situations fairly. For this reason, TCP was equipped with an adaptive mechanism to determine an appropriate time limit for a new transmission. TCP therefore monitors the payload for every connection. This procedure measures the **round-trip time** - the time from the sending of a message to the complete receipt of the reply. This measuring process is carried out for every sent message and any deviation appearing in a particular monitored connection is recorded and flows into the new calculation of the valid timeout. This results in a **smoothed round-trip time**. Therefore, in high load situations TCP can raise the timeout and then subsequently lower it again when the net load returns to a normal level.

In contrast to other troubleshooting mechanisms, TCP offers the receiver no possibility to force the new transmission of a faulty segment, as no mechanism for a negative acknowledgement exists. Upon error detection the receiver simply does not send an acknowledgement and waits until the fixed time limit has expired, at which time the sender automatically begins a new transmission. The efficiency of the optimal throughput depends on the choice of the time span allowed to elapse until the acknowledgement (time limit). In a LAN where there is only a minimal distance to be bridged between the sender and the receiver, other time restrictions make sense, such as in the case of a wide area network, e.g., via a satellite channel. If the new transmission is begun too quickly, the network is flooded with many duplicates and the efficiency of the throughput can be strongly affected. In the case where a new transmission is started too late, a potentially large number of messages must be temporarily stored. This can lead to queue overflow and thus to an overload resulting in data loss and subsequently an impairment of network efficiency. The correct choice of waiting time until a new transmission is carried out is therefore of the utmost importance.

– **Flow Control**

The mechanisms provided for flow control are set up to regulate the flow along an end-to-end connection. In the TCP protocol, flow control proceeds over a **sliding window protocol** (cf. Fig. 8.10). This is a window protocol that works with an adaptive, load-dependent window size. Acknowledgements and the window assignment are decoupled from each other. As already described, the TCP Protocol does not exchange messages in the actual sense of the word, rather sends a bi-directional byte stream that is separated into segments. Instead of now sending multiple, parallel messages, it is possible to send multiple segments parallel. This means that they are sent unacknowledged up to a certain maximum number of bytes. The receiver regulates this maximum number over the sliding window protocol and adapts this size to fit its processing capacity. Therefore, when more data is delivered than can be processed, the amount is reduced and this information passed on to the sender.

The procedure of the sliding window protocol in TCP may be described by way of an example (cf. Fig. 8.11). Sender and receiver must agree to the use of the same segment numbers for the transmitted data. Normally, this synchronization takes place when the TCP connection is established. The point of departure here are the two computers - host 1 and host 2 - which wish to carry out a data transmission via TCP. The window size of the connection of the sender, computer host 1, to the receiver, computer host 2, is 1,500 bytes and a message of 2,500 bytes is to be transmitted.

·   After connection establishment, host 1 sends the first segment with a length of 1,000 bytes. Host 2 receives the segment, writes the data in its input buffer, reducing the window size to 1,500–1,000 = 500 bytes (W=500) and acknowledges receipt of the first 1,000 bytes (ACK=1,000).

·   Similarly, in the second segment based on the allowed window size, 500 bytes are transmitted, the arrival is acknowledged (ACK=1,500) and the window size reduced to 0 bytes (W=0).

Sliding Window Protocol Procedure:

An increased data transmission capacity can be reached with acknowledged transmission (i.e., every sent message requires an acknowledgement) when multiple unacknowledged messages can be sent before a corresponding collective acknowlegement arrives at the sender. Procedures that work according to this principle are called **window protocols** The size of each window indicates the maximum number of messages (or their length in bytes) that can still be sent unacknowledged. The individual messages receive a sequence number for unique identification in this procedure.

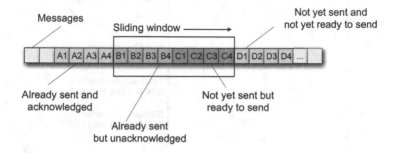

Messages to the left of the sending window were already sent and acknowledged (A_i). Messages in the sending window can be sent unacknowledged. It is necessary to distinguish between the messages in the sending window on the left (B_i), which have already been sent and whose acknowledgement is outstanding, and the messages on the right in the sending window (C_i), that are waiting to be sent unacknowledged. The respective window size corresponds to the number of bytes in categories B and C. Messages to the right of the sending window first move into the sending window following an acknowledgement receipt.

Further reading:

Comer, D. E.: Internetworking with TCP/IP, Volume 1: Principles, Protocols, and Architecture, Prentice Hall (1995)

Fig. 8.10 Sliding window protocol.

- The application program at host 2 cannot accept the arriving data as fast as it is being delivered. Because of this, host 1 must wait as long as necessary until the application at host 2 has read the 1,000 bytes from the input buffer, raised the window size again to 1,000 bytes and transmitted this (W=1,000) to host 2 together with an acknowledgement via the sequence number of the last received byte (ACK=1,500) to the sender.

- Corresponding to the new window size, host 1 can send the remaining data to be transmitted to host 2. The latter acknowledges each receipt and reduces the window size accordingly.

- The application program at host 2 again removes 1,000 bytes from the input buffer. This (W=1.000) is passed on to host 1 together with the sequence number (ACK=2,500) of the last received byte.

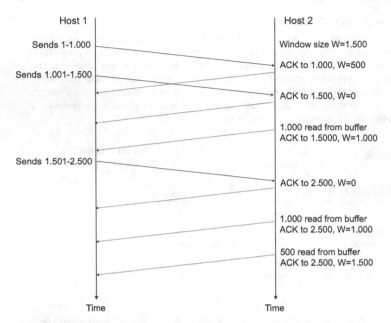

Fig. 8.11 TCP – flow control with the sliding window protocol.

· As the last sequence of the message has now been sent, host 1 does not need to undertake any further steps.

· The application program at host 2 again removes the last 500 bytes from the input buffer (W=1,500). This is communicated to host 1 along with the sequence number (ACK=2,500) of the last received byte. The message transfer is now concluded.

– **Congestion control**
Congestion control presents one of the most difficult problems faced by TCP. The load situation can be monitored at each end devices involved in the communication and communication flow correspondingly controlled by changes in the window size. Nevertheless, the load situation at the end systems along the data transmission path cannot be directly detected and influenced. Another problematic area is found in the properties of the IP protocols that TCP is based on and upon which it relies to acquire information. IP is a connectionless and stateless protocol, i.e., transmitted data packets can be transported in different ways through the Internet and there is no information about any incidents that may have occurred along the way.

For this reason, it is rarely possible to obtain information regarding an overload at the involved intermediate systems. If this can be done at all, it is only possible indirectly. Additionally, different TCP instances in the Internet do not have a possibility to cooperate with each other to recognize and control overload situations. This is why e.g., the number of lost segments - that is, segments whose transmission process is not acknowledged by the receiver within the fixed time span and are thus considered lost and must be resent - are taken as a measure for overload in the Internet. This overload is controlled by a window protocol to regulate congestion situations. Congestion control at the transport layer is necessary. Otherwise, congestion situations occurring in the intermediate systems along the communication path would result in an ever-increasing loss of segments. These would then have to be resent in a new transmission, raising the load at the intermediate systems more and more.

Numbering among the procedures implemented for congestion control are:

· **Slow-start algorithm**
  This algorithm is used to optimally adapt window size in the sliding window protocol.

· **Congestion avoidance algorithm**
  Based on the number of segments to be transmitted new, this algorithm indicates overload situations in the network.

· **Nagle algorithm**
  Algorithm for the avoidance of unnecessary overhead caused by acknowledgements.

· **Karn/Patridge algorithm** and **Karel/Jacobsen algorithm**
  These algorithms provide the optimal determination of the respective network round trip time for adaptation of the retransmit interval in the new transmission of lost segments.

A detailed description of this algorithm is given in section 8.3.3.

• **Full-duplex transmission**
  TCP allows a bi-directional end-to-end data transmission. The sender and receiver can even transmit at the same time. So that the sender and the receiver can implement a parallel processing, a buffering of the relevant data must be carried out upon both sending and receiving.

• **Stream interface**
  Applications using a TCP transport send a continuous **byte stream** over the established data connection rather than consecutive individual messages. Message borders are therefore also not maintained with an end-to-end connection. The stream interface of the receiver computer passes to the receiver application the data of the transferred byte stream in exactly the same order in which the sender application has transmitted it.

• **Connection management (TCP Connection Management)**
  Before a data transmission can begin in TCP, a communication connection must first be established between the two partners involved. Critical for reliable connection management is, first, connection establishment and, then, following a successful data transmission, the connection release. Care must be taken that a data transmission is only first begun after the connection has been reliably established and the procedure necessary for connection establishment has been successfully concluded. Connection establishment can first proceed as soon as all

data of the previously carried out data transmission has been sent and obtained by the receiver. TCP provides separate mechanisms for both processes that will be described in the following.

Different parameters have to be reserved, exchanged and stored in managing connection status. This proceeds via the **Transmission Control Block** (**TCB**), which must be reserved for both end devices involved in the connection management. In addition to other information, TCP contains the socket information of the communication partners involved, pointers to the buffer memory areas for incoming and outgoing data, as well as numerous counters that monitor the amount of exchanged data for the sliding window protocol. Before actual connection establishment takes place, both communication partners set up a TCB especially for this connection.

- **Reliable connection establishment**
  TCP requires from both communication partners - the sender and the receiver - an agreement concerning the connection to be established. Possible duplicate packets from previous connection are thereby to be ignored. The following functions are associated with connection establishment:

  - **Establishing contact and communication**
    Sender and receiver take up contact with each other to establish a communication connection and to facilitate mutual message exchange. The receiver does not know in advance who will initiate contact with it, only learning this upon connection establishment.

  - **Synchronization of the sequence numbers**
    Each of the involved partners communicates the initial sequence numbers for the beginning of the communication to its counterpart.

  - **Parameter exchange**
    Further parameters for control of the data exchange are swapped between the sender and the receiver during connection establishment.

  The procedure used to establish a TCP connection is known as a **3-way handshake** as only three messages must be exchanged (cf. Fig.. 8.12). Two different message types are thereby exchanged between both communication partners.

  A connection is initiated by the so-called **synchronization segment** (**SYN segment**), a segment in whose header the synchronization bit is set to one.

  A connection wish is confirmed by an **acknowledgement segment** (**ACK segment**) sent via a SYN segment. The acknowledgement bit in the header is set to one in this segment.

  The swapping of four messages would actually be necessary for a connection exchange. However, for reasons of efficiency, the acknowledgement of an initially received connection registration and the login of a return connection in the opposite direction are pooled together. This results in the 3-way handshake.

  1. To signal the establishment of a connection via the 3-way hanshake, a SYN segment is sent from the sender to the receiver. This contains the initial segment number of the sender x (SYN).

2. The receiver of the SYN segment acknowledges the sequence number x, by sending back an ACK sequence number x+1. The receiver leaves the synchronization bit set, additionally setting the acknowledgement bit, thus returning its own sequence number y (SYN + ACK).

3. Following receipt of this message, the sender acknowledges receipt of the synchronization answer with the sequence number y by sending an ACK segment with the sequence number y+1 (ACK). Upon receipt of this sender acknowledgement, the 3-way handshake is completed.

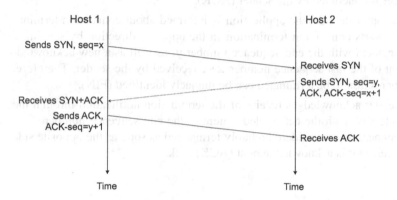

Fig. 8.12 TCP - connection establishment with the 3-way handshake.

The 3-way handshake protocol fulfills two important tasks. The first is ensuring that both sides - sender and receiver - are ready for a data exchange. The second is providing the opportunity for both sides to exchange important sequence numbers at the start of the data transfer. TCP is likewise capable of handling two parallel connections between two communication partners who both wish to communicate with each other actively. Here, the 3-way handshake becomes a simple, simultaneous exchange of the SYN segments, with which both communication partners express their wish to communicate and exchange the initial segment numbers. The SYN segments are in each case answered reciprocally with an ACK segment.

- **Careful connection termination**
  As soon as one of the two parties wishes to end the established connection, TCP ensures that all of the transmitted data that has not yet arrived at its destination is in fact delivered before the connection is dismantled. For this purpose, a slightly modified 3-way handshake is used. However, the reliable termination of a connection requires the exchange of four TCP segments (cf. Fig. 8.13). Two application programs that simultaneously communicate via TCP bidirectionally are seen as two separate, unidirectional connections.

1. If an application program ends the connection it sends an **end segment** segment). In its header the FIN bit is set to one and there is an end sequence number x.

2. The opposite side acknowledges receipt of the FIN segment and takes no further segments for this connection, meaning segments with a segment number bigger than x. The TCP instance informs the local application program about the termination of the connnection (FIN).

3. With the receipt of an acknowledgement - in which the received end sequence number x+1 is again confirmed - the connection in the direction of sender to receiver is concluded by the sender (ACK).

4. On the opposite side, the application is informed about connection termination. It starts connection termination in the opposite direction by sending a FIN segment with the end-sequence number y, as well as a new acknowledgement of the last sequence number x+1 received by the sender. Therefore, the connection to be terminated can be uniquely identified (FIN).

5. The sender acknowledges receipt of the termination notification through the opposite side with the acknowledgement of the end sequence number The connection is considered reliably terminated as soon as the opposite side has received this acknowledgement (ACK) back.

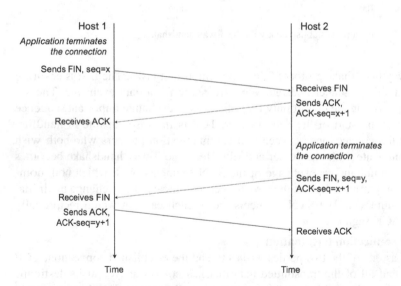

Fig. 8.13 TCP – reliable connection termination with the modified three-way protocol.

Connection termination can also proceed between both communication partners in a way that is virtually simultaneous. In this case, the wish of one partner to terminate the connection does not trigger connection termination in the opposite direction.

Instead, both partners seek to end the connection and both send a FIN segment - before receiving one from the counterpart (simultaneous connection termination).

Excursus 12: TCP – Connection Management

Protocols are often formally defined in computer science with the help of so-called **Finite State Machines** (FSM), which can be clearly represented with the help of transition diagrams.

A finite state machine is in a perpetual **state** (depicted by a circle), whereby certain **events** (represented by arrows) can cause a **transition**. The **start state** and **end state** are two definitions of specific states. The arrows leading from one state to the next, are labeled with an event/action pair. The event determines the corresponding transition state, which is triggered by a specific **action**. In a protocol defined by a finite machine, the event usually corresponds to the receipt of a specific message and the action triggered to the sending of a specific message.

A finite state machine representing connection establishment and connection release in the TCP protocol must access the following states:

- **CLOSED**
  There is no active or pending connection.

- **LISTEN**
  A server waits for a request.

- **SYNC RECEIVED**
  A connection request is accepted; waiting for ACK.

- **SYNC SENT**
  An application starts a connection opening.

- **ESTABLISHED**
  Normal state; data transmission takes place.

- **FIN WAIT1**
  An application is completed.

- **FIN WAIT2**
  The counterpart agrees to release.

- **TIMED WAIT**
  Waiting until all outstanding segments have arrived or for a timeout.

- **CLOSING**
  Both sides are trying to end the connection simultaneously.

- **CLOSE WAIT**
  The counterpart begins connection release.

- **LAST ACK**
  Waiting until all outstanding segments have arrived or until a timeout has occurred.

The complete transition state diagram for the TCP connection management is shown in Fig. 8.14. The thicker lines indicate an active connection, e.g., from a client, while the dotted lines show the corresponding process on the other side - a passive connection on the server's side. The thinly drawn lines indicate unusual events and operations, which for the sake of completeness must also be taken into consideration.

- The starting state for connection management is always the state CLOSED. This state is first exited when a request for connection release is either received (**passive open**) or sent (**active open**).

- An event can thereby either be initiated by the user (CONNECT, LISTEN, SEND, CLOSE) or triggered by an arriving segment (SYNC, FIN, RST). A dash "−" signals that no event or action has taken place.
- The procedure can be understood most simply if first the active client side (thick lines) is followed and subsequently the occurrences on the server side (dotted lines).
- **Client side**
  - An application on the client side wishes to open a connection by transmitting a CONNECT request. For this purpose, the local TCP instance generates a Transmission Control Block (TCB) for management of the TCP connection. It enters the SYNC SENT state and sends a SYNC segment, to open the 3-way handshake procedure. As a multiple number of connections must be managed permanently, every state defines only one connection.
  - As soon as the client receives the SYN+ACK, the TCP instance sends back the last ACK of the 3-way handshake and enters the state ESTABLISHED. In this state, the regular data transmission can now begin.
  - Upon completion of the application, the local TCP instance receives a command to close the connection (CLOSE). It then sends a FIN segment and waits for an corresponding acknowledgment (ACK). As soon as the acknowledgement has been received, the TCP instance enters into the state FIN WAIT 2. One direction of the bidirectional connection has therefore been concluded.
  - As soon as the server side has also ended its connection it acknowledges this with a FIN segment, which must also be acknowledged by the TCP instance on the client side (ACK).
  - The connections in both directions are now finished However, the TCP instance on the client side waits as long as necessary until a timeout occurs. This is to ensure that all transmitted segments have indeed arrived. Finally, all management data structures provided for this connection are deleted

- **Server side**
  - On the server side, the local TCP instance waits (LISTEN) until a connection opening is requested by a client (SYN). When the server receives a SYN segment, the receipt is acknowledged (SYN+ACK) and the TCP instance enters the state SYNC RECEIVED.
  - As soon as the SYN segment sent back by the server is acknowledged (ACK), the 3-way handshake is concluded and the TCP instance enters into the ESTABLISHED state, in which normal data transmission is conducted
  - If the client wishes to end the connection, it sends a FIN segment, which reaches the server. The server acknowledges receipt (ACK) and the local TCP instances enters into the CLOSE WAIT state. The connection has already been ended at this time by the client side; the passive connection release on the server side starts.
  - If the application on the server's side is ready to end the connection (CLOSE), the server sends a FIN segment to the client. The TCP instance enters into the LAST ACK state and waits for client's confirmation (ACK) before the final connection release. After a subsequent acknowledgement, the connection is terminated.

Further reading:

Tannenbaum, A. S.: Computer Networks, Prentice-Hall, NJ, USA, pp. 529-539 (1996)

Postel, J.: Transmission Control Protocol (TCP), RFC 793, Internet Engineering Task Force (1981)

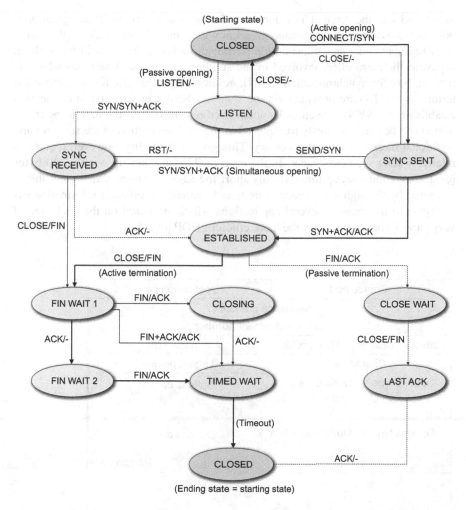

Fig. 8.14 TCP – transition state diagram of a finite machine for definition of the TCP connection management.

8.3.2 TCP Message Format

TCP transmits the messages to be exchanged in individual data blocks called **segments** via an existing connection. In the division of a message into individual pieces - the **segmentation** - each TCP segment is given a TCP header. This contains the command and control information for guaranteeing the reliable transport of the segment. Additionally, before the beginning of actual data transfer, i.e., during connection establishment, a maximum segment size must be agreed upon between both parties. For the actual transmission, the TCP segments are fragmented into IP datagrams and encapsulated. After arriving at the destination computer they are reassembled into the original TCP message. In the case where an IP datagram does not reach the destination computer, no acknowledgement is sent back for that particular segment and a new transmission (retransmission) of the segment takes place. To avoid the extra effort involved in the sending of command and control information, bits for synchronization (SYN), acknowledgement (ACK), or connection termination (FIN) are provided via the segment header bits so that in connection establishment (SYN) as well as its acknowledgement (ACK) the data to be transmitted can be simultaneously transported in the payload section of the segment and additional messages are not necessary. This greater flexibility, however, requires a relatively long segment header. In the case of TCP, it takes up a whole 20 bytes (excluding additional options). If only short messages are to be transferred, this is impacted by the additional header length and prevents an efficient information exchange. For this reason, several applications which are based on the exchange of very short information prefer the more efficient UDP protocol.

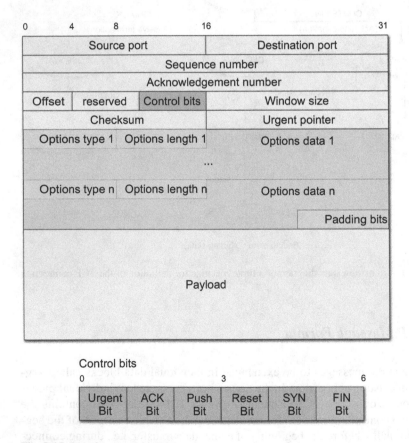

Fig. 8.15 Data format of the TCP segment.

Fig. 8.15 shows the structure of a TCP segment consisting of segment header and segment user data. The TCP header is made up of the following fields:

- **Source port**
  The 16-bit long field identifies the source port, i.e., the starting point of the TCP connection. This is associated with the specific application process that started the connection. In an active connection this can be a freely chosen port number (ephemeral port number) or in an answer segment a known or registered port number assigned to a specific (server) application.

- **Destination port**
  16-bit long field for the destination port. This is the endpoint of the TCP connection where the data is addressed. Again an application process is associated with the destination port that receives the data transmitted over the connection. In the case of an active connection this can be a known or registered port number that designates the addressed (server) application. In an answer segment this can be a randomly chosen port number that was communicated in the previously received segment.

- **Sequence number**
  Based on the 32-bit long sequence number, the transmitted data segments are numbered consecutively in the transmission direction. On establishing a TCP connection, both communication partners generate a start sequence number (Initial Segment Number, ISN). When the connection is established in the 3-way handshake these sequence numbers are mutually exchanged. So that sequence numbers always remain unique for a particular connection, the repetition of sequence number within the fixed lifetime of a TCP segment must be ruled out. Additionally, the sender raises the current sequence number to the number of already sent bytes.

- **Acknowledgement number**
  The 32-bit long acknowledgement number follows the sequence number. It serves as a confirmation of the received data segments on the receiver side. The acknowledgement number is set by each receiver to communicate to the sender up to which sequence number of sent data has been correctly received.

- **Data offset**
  A 4-bit long field (data offset or also Header Length, HLEN) that gives the TCP header in 32-bit words and as an offset value at the beginning of the actual user data in the transmitted segment. Following the data offset field there is a 6-bit long field, which is not used.

- **Control bits**
  The following six control bits determine which respective fields in the header are valid and in this way control the connection. The control bits have the following meaning when set to the value one:

  - **Urgent Bit (URG)** Activates the urgent pointer in the TCP header (cf. below).

- **Acknowledgement Bit** (ACK)

  Activates the field for the acknowledgement number.

- **Push Bit** (PSH)

  Activates the so-called **push function**. This allows the transmitted data in the user data section to be passed on immediately to the next higher protocol layer. The transmitted data is therefore not written in an input buffer that must first be emptied when it is completely full. This function serves to speed up the transmission of data, which should be carried out as quickly and efficiently as possible.

- **Reset Bit** (RST)

  Resets a connection that has gotten mixed up as the result of an unforeseen event (e.g., computer crash). RST is also used to dismiss an invalid sequence number or reject a connection establishment.

- **Synchronization Bit** (SYN)

  Signals the establishment of a connection. In a connection inquiry SYN=1 and ACK=0 are always set to show that an acknowledgement field is still inactive. First the answer to a connection inquiry contains an acknowledgement, and thus SYN=1 and ACK=0 are set there. In a connection establishment with a set SYN bit, the ACK bit distinguishes whether an inquiry or an answer is involved.

- **Finish Bit** (FIN)

  Signals a one-sided connection release; the sender has no more data to transmit.

- **Window size**

  The 16-bit long window size serves flow control with the help of the sliding window protocol. This allows the receiver to control the data stream it receives. The transmission window indicates how many bytes, starting with the acknowledgement number, the receiver can still accept in its input buffer. If the sending computer receives a TCP segment with the sending window size at zero, it must hold off sending until it receives a positive sending window.

- **Checksum**

  In order to ensure a high reliability of the TCP segments, a16-bit long checksum is included in the TCP segment header. The checksum is calculated from the TCP header - whereby the checksum itself and possible padding bits are equal to zero - the TCP user data and the contents of the **pseudo header**. The pseudo header is actually not a part of the information contained in the TCP segment, rather is only temporarily calculated from the IP and TCP data to be transmitted. Included in the pseudo header are the IP addresses of sender and receiver, the protocol number and the TCP identified, as well as the length of the complete TCP segment. The pseudo header is used to identify potentially misdirected TCP segments, however actually contradicts the fixed protocol hierarchy as it includes IP addresses from the network protocol layer. The data format of the pseudo header differs depending on whether IPv4 or IPv6 is used as the network protocol. Fig. 8.16 shows the described structure of the TCP pseudo header in IPv4. According to RFC 2640, in IPv6 after the respective 128-bit long IPv6 addresses for sender and receiver follows the length of the TCP segment. Instead of the IPv4 protocol specifications comes an IPv6 next header field that gives the type of transported data (here TCP) (cf. Fig. 8.17).

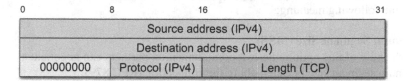

Fig. 8.16 TCP – pseudo header data format for IPv4.

To calculate the checksum, all incoming 16-bit words are summed up in the one's complement, and the one's complement of the results is saved. If the receiver calculates the result of the checksum with the correctly transmitted checksum value contained in it, the value zero must result in the case when the transmission is error-free.

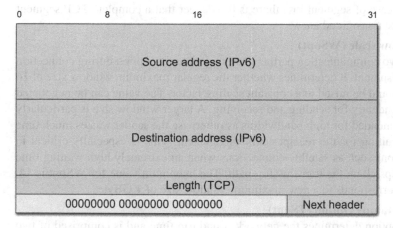

Fig. 8.17 TCP - pseudo header data format for IPv6.

- **Urgent pointer**
  The TCP protocol provides a possibility to transmit urgent data (out-of-band data), e.g., through interrupts, along with the data that is to be sent regularly to the communication partner. This urgent data should be passed on as quickly as possible in the receiving process without, e.g., first being stored in the input buffer. The 16-bit long field of the urgent pointer thereby serves as an offset specification and is only active if the URG bit is set to one. As a rule, urgent data follows

the TCP header directly and is therefore at the beginning of the TCP payload. The regular payload begins later, i.e., after the byte length indicated in the urgent pointer.

- **Options**
The options indicated here allow the integration of additional functions under TCP that are not covered in the remaining header fields. The first byte in the options field determines the option type (cf. RFC 1323 and RFC 2018), which can have the following meaning:

  – **Maximum Segment Size (MSS)**
  With the help of this option, the two communication partners can negotiate a maximum segment size in establishing the connection (RFC 879).
  The greater the implemented segment length, the less the overhead of 20 bytes of the TCP header reduces the efficiency of data transmission. If this option is not used in connection establishment, a standard payload length of 536 bytes is implemented. Every computer in the Internet must therefore be capable of accepting a TCP segment with a length of $536 + 20 = 556$ bytes. Transmission efficiency depends on choosing the best MSS. If it is set too low, the overhead in the TCP header grows proportionately. If, on the other hand, it is too high, this leads to increased IP fragmentation on the Internet layer, whereby new overhead is created by the extra IP datagram header required. Furthermore, in the case of segment loss there is the danger that a complete TCP segment must be transmitted anew.

  – **Window scale (WSopt)**
  The two communication partners can negotiate this option during connection establishment. It determines whether the regular maximum window size of 16 bits should be raised to a constant scaling factor. The value can be negotiated independently for sending and receiving. A larger window size is particularly recommended for high bandwidths as otherwise the sender wastes much time idly waiting for the receipt acknowledgement. This is especially critical in situations such as satellite connections when an extremely high waiting time develops along the transmission path. The maximum value for WSopt is 14. This corresponds to a new maximum window size of 1 GByte.

  – **Timestamps Option (TSopt)**
  This option determines the network round trip time and is comprised of two parts: the Timestamp Value (TSval) und Timestamp Echo Reply (TSecr). An exact measurement value for the round trip time of a TCP segment can be determined from the difference between the timestamp set by the sender and the receiver, which enters into calculating the mean round trip time.

  – **Selective acknowledgement**:
  The procedure known as selective acknowledgement is implemented in the TCP overload control (cf. RFC 2018). It ensures that after cumulative timeouts only a single data packet is transmitted. In spite of their lateness, the remaining data packets do in fact arrive at the receiver.

- **Connection Count (CC)**
  This options field supports a transaction-oriented expansion of TCP and is accordingly called **Transaction TCP (T/TCP)** (cf. Fig. 8.18).

- **Padding bits**
  Padding bits fill the length of the TCP segment header to the word limit of 32 bits.

Transaction TCP (T/TCP), (RFC 1379, RFC 1644)

TCP is built on a symmetrical transport connection between two communication partners. Yet in reality, Internet communication seldom transpires symmetrically. In response to a short inquiry (**request**) a detailed answer follows (**response**). This mode of operation, known as a **transaction principle**, occurs mainly in client/server communication, e.g., in HTTP (Hypertext Transfer Protocol) in the World Wide Web. At the same time, the efficiency of data transfer in this transaction-oriented processing is considerably reduced due to the protocol overhead of the TCP operational mode (connection establishment with 3-way handshake, data transfer, connection establishment with 4-way handshake).

For the efficient execution of transaction processes T/TCP offers two TCP extensions (cf. RFC 1644):

- **Connection count**
  Each TCP segment includes a transaction counter for every request/response as an option in the TCP header. This is used to identify the transaction.
- **TCP Accelerated Open (TAO)** This procedure circumvents the 3-way handshake in opening a new connection and uses the connection count instead.

Further reading:

Braden, R.: Extending TCP for Transactions – Concepts, Request for Comments RFC 1379, Internet Engineering Task Force (1992)

Braden, R.: T/TCP – TCP Extensions for Transactions, Functional Specification, Request for Comments RFC 1644, Internet Engineering Task Force (1994)

Fig. 8.18 Transaction TCP (T/TCP).

8.3.3 TCP – Reliability, Flow Control and Overload Control

The main task of the TCP protocol is centered on the reliable sending and receiving of data. In this way, it is not much different than many other communication protocols. However, with the help of the integrated sliding window procedure, TCP also facilitates an adaption of data transmission based on the performance of the respective communication partner and its current load, i.e., it is capable of adaptively managing and controlling the communication flow. This form of flow control can only have an active influence on the communication partners directly involved. The intermediate systems along the communication path are located beyond the control of the TCP protocol. Seen from the perspective of the TCP/IP reference model, the flow control and overload prevention at the intermediate systems are handled by the Internet layer protocols and theoretically play no role in the protocols of the transport layer. Yet in practical application, the intermediate systems are also involved in data transmission between the communication end subscribers. At the same time they manage the data transfer of numerous other communicating end systems. Due to this situation, a high traffic load can occur at the intermediate systems influencing communication on the transport layer. If the load at the intermediate systems becomes too great, not all of the arriving IP datagrams and IP datagrams can be processed and are then discarded in order to regulate the load. This data loss at the intermediate systems also means that the transmitted segments on the transport layer cannot be delivered correctly to the receiver. Their reception is not acknowledged and consequently a a new transmission is triggered. Because a TCP segment can often be quite extensive, it may be broken up into multiple IP datagrams to facilitate transmission. This requires the new transmission of a complete segment at the same time as the new transmission of the numerous connected IP datagrams. Through the new transmission of TCP segments, the additional load increases which is already overburdening the intermediate systems. In order to ensure efficient data transmission, TCP procedures must also be provided that may be put to use in the case of overload or try to avoid this situation before it even happens. Among the most important procedures and algorithms implemented to guarantee high reliability are:

- **Retransmission timer, retransmission queue and selective acknowledgement**
  In order to ensure reliable TCP data transmission every transmitted segment must be acknowledged by the receiver. If this acknowledgement is outstanding then the segment must be sent anew. As previously mentioned, TCP manages a clock (retransmit timer) primarily for every transmitted segment. This clock waits a certain, prescribed time span for the reception of an acknowledgement before a new transmission of the relevant segment is initiated.
  For reasons of efficiency, the administration of the retransmit trainer proceeds as follows:

  1. As soon as a segment is to be sent, a copy of the segment is written in a data structure of the retransmission queue, which is especially intended for this purpose. Simultaneously, a separate retransmit timer is started for the new entry in the retransmission queue. Every segment to be sent is also in the retransmission queue, sorted based on the still remaining time span of the retransmit timer corresponding to the individual entries. Those entries with the shortest remaining time are found at the beginning of the retransmission queue.

  2. If an acknowledgement is received for a segment found in the retransmission queue before its retransmit timer has expired, the entry, together with its timer is deleted from the retransmission queue.

  3. If the retransmit timer of a segment runs out before a transmission acknowledgement arrives for it, then a so-called retransmission timeout goes into effect. The relevant segment is automatically sent anew. The segment remains in the

retransmission queue and the retransmit timer is started anew with the prescribed time span.

To avoid a segment from being continually resent - e.g., if the receiver is no longer reachable - the connection is broken off following a prescribed maximum number of transmission attempts. It is crucial that attention be paid that segments via TCP are acknowledged individually with a **sequence number** sent from the receiver. A segment is considered completely received if all transmitted sequence numbers of the segment are smaller than the last acknowledged sequence number. Fig. 8.19 shows the process of a complete segment reception from host 1 to host 2, using an example. The following four segments are to be sent continuously: segment 1 starts with sequence number 1 and a segment length of 100. Segment 2 runs from sequence numbers 101–200, segment 3 from 201 to 350 and segment 4 from 351 to 500.

Host 2 receives the first two segments and confirms them with the acknowledgement sequence number 201. Thus, host 1 learns that the first two segments have been successfully transmitted. They are removed from the retransmission queue. Let us assume that segment 3 is lost after transmission. Segment 4 has already been sent shortly after segment 3 and received by host 2. Segment 4, however, has not been initially confirmed, but rather copied in the receive buffer. The transmission is not yet allowed to proceed, otherwise TCP would conclude from the segment 4 acknowledgement number that segment 3 has also been successfully transmitted. In the retransmission queue from host 1, the retransmit timer from segment 3 finally runs out and a new transmission is triggered. As soon as segment 3 has been received by host 2, then the already received segment 4 is considered correctly delivered. The acknowledgement sequence number 501 confirms the successful receipt of segment 3 and segment 4.

As previously seen, in data transmission via TCP it is not longer necessary for each part of a segment to be acknowledged individually. With the confirmation of sequence numbers (acknowledgement number) in the acknowledgement process, all transmitted segments with smaller sequence numbers are confirmed automatically. While with this method TCP seamlessly handles consecutive segments very efficiently, it cannot be implemented if the sequence is interrupted. If a segment is lost, as in Fig. 8.19, and its retransmission is also unsuccessful, this can lead to problems in executing the sliding window protocol. The acknowledgement for the lost segment remains outstanding and further segments can no longer be transmitted beyond the current window. Moreover, neither does the receiver have the possibility to communicate to the sender that segments have been correctly sent after those that still remain outstanding. TCP has two different strategies to deal with this problem:

– **New transmission of lost segments only**
This strategy is in line with the path followed up to now of only transmitting segments whose retransmission timer has run out. It is assumed that the segments to come - if indeed some have been sent - were transmitted correct-

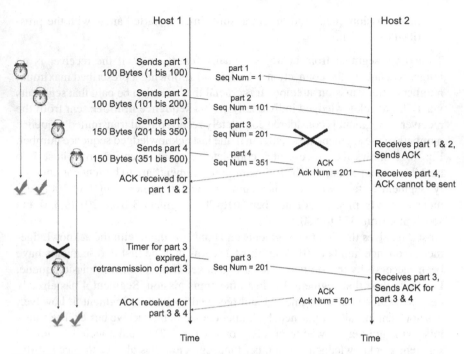

Fig. 8.19 TCP – transaction example with renewed transmission (retransmission).

ly. If the segments to follow also get lost, then a separate timeout must be determined for each one and a new transmission initiated.

– **New transmission of all outstanding segments**
This strategy assumes the pessimistic case scenario. In the timeout of a segment's retransmission timer, the segments to come - as far as some have already been sent - have gotten lost. Instead of the new transmission of a lost segment, all of the previously sent segments are retransmitted.

A more efficient method was implemented with a later extension of the original sliding window protocol specified in RFC 1072 or RFC 2018. This is the so-called "**Selective Acknowledgement**" (**SACK**).

If this method is used, it must be implemented by the sender and receiver during TCP connection establishment. This is done by the setting and acknowlededement of the SACK permitted option in the TCP segment header. This option contains a list of sequence numbers representing individual areas of already received, but not yet acknowledged unrelated segments.

The retransmission queue of sender and receiver must additionally manage a flag (SACK bit). It is set as soon as the corresponding segment is selectively acknowledged through receipt of a sequence number from the returned list (SACK options). In this way, only those segments that have not yet been selectively acknowledged are actually transmitted again.

As stated previously, in new transmissions the time span of the retransmission timer and that on which a new transmission is based is crucial for the efficiency of data transmission. If the time period selected is too short, unnecessary new transmissions are triggered despite that fact that the relevant segment has been received and its acknowledgement is outstanding. Ideally, the selected timeout should be just slightly bigger than the mean **Round Trip Time** (**RTT**) between sender and the receiver. In other words, the time a message needs to be transmitted from the sender to the receiver and back again. Nevertheless, determining the exact time span is problematic. It must be applicable to both a connection in a high-speed network as well as to a multi-forwarded connection in the worldwide Internet, which could include e.g., slow and error-prone radio networks. This time span is furthermore dependent on the currently applicable load of the respective network connection and varies over time. For this reason, there is no static size chosen in TCP for this time period, but there exists various procedures for dynamic determination (adaptive retransmission). In this way, TCP can adapt flexibly to different connection paths and to different volumes to be processed. Various procedures for the calculation of the retransmissions timers are discussed in detail in RFC 2988.

1. **Original implementation of the RTT calculation**
   Th RTT can vary dramatically with every new data transmission between sender and receiver. Because of this, the RTT is measured for every transmitted segment and from it a weighted mean calculated:

$$RTT_n = a \cdot RTT_{n-1} + (1-a) \cdot RTT_{akt}.$$

RTT_{akt} stands for the current measurement and RTT_n for the specific mean according to n measurements.
a is a smoothing factor that determines the influence of the current measurement on the calculated mean value. $RTT_0 = 2s$ is taken as the start condition. An estimate of the timeout results from:

$$Timeout = b \cdot RTT_n$$

Normally, the parameters $a = 0{,}9$ and $b = 2$ are chosen.

2. **Karn/Partridge Algorithm**
   If the original implementation is followed, a lost segment already falsifies the determined value for the round trip time. This results in an underestimation based on incorrectly measuring the new transmission time of the segment up to the receipt of a delayed acknowledgement based on the first transmission (acknowledgement ambiguity). This time period can be significantly smaller than the actual round trip time. The algorithm from *Phil Karn* and *Craig Partridge* resolves this problem by only entering those segments of the round trip time into calculation that are actually acknowledged by the receiver. Additionally, the timeout is raised (timer backoff) after every successful delivery to give a new transmitted segment enough time for an acknowledgement:

$$Timeout = 2 \cdot Timeout.$$

The timer increases with every new transmission until a maximum value is reached. The value of the timer remains increased until acknowledgements of regularly sent segments arrive and with them correctly measured roundtrip times. Thus, the timer can be adapted to circumstances responsible for data loss over a longer time period or those that create a temporary delay within a network, to then return to the exit value afterwards.

3. **Jacobson/Karels Algorithm**
   The Karn and Partridge implementation was perfected further by *Van Jacobson* and *Michael J. Karels* in the formula named after them: the Jacobson / Karels Algorithm. Here, the fluctuations of the measured round trip time

$$RTT_n = RTT_{n-1} + g_0 \cdot (RTT_{akt} - RTT_{n-1})$$

are weighted with $g_0 = 0,125$. Additionally, the calculation of the timeout is adapted by means of an auxiliary variable delta:

$$Delta = Delta + g_1 \cdot (RTT_{akt} - RTT_n)$$

with $g_1 = 0,25$. The new resulting timeout is

$$Timeout = p \cdot RTT_n + q \cdot Delta,$$

and for both parameters p and q, the empirical values $p = 1$ and $q = 4$ are chosen.

- **Congestion window size**
  The congestion window in the sliding window protocol makes it possible to control the amount of unacknowledged and received data and therefore also the network flow. Here, a distinction must be made between the speed at which the sent data is delivered to the receiver, and the time that it needs to be passed on from the congestion window to a further processing application. If the delivery speed of the data is so great that the data is not removed from the congestion window and processed in a timely manner, the threat of overflow and data loss results. It is therefore necessary that the size of the congestion window be adapted to the respective load conditions in the network. This is carried out so that with an acknowledgement of successfully received segments the space remaining in the current congestion window is also conveyed to the sender. In this way, from the beginning, there is not an excess amount of data sent that cannot potentially be processed by the receiver in a timely manner. This remaining space is reduced as long as the sending rate is greater than the amount of data that can be processed. When the value zero is reached one speaks of a "closed congestion window." At this point, the sender stops transmitting further data until the window size is raised again. In the case where a window size has already been announced by the receiver, it would prove problematic if due to operational measures on the recei-

ver's side this size were to be compromised by unexpectedly having to reduce the congestion window. Here, the sender of the data would assume a too large window size and the situation could arise of the delivered data not being written completely in the available window. Data loss would result and a new transmission would subsequently be triggered. TCP solves this problem by not allowing an operating system-related reduction of the congestion window.

A further problem can occur when the congestion window is closed on the receiver side. This means that the still remaining window size is set to zero based on a high data volume or a high processing load. The sender is not able to send any further data until the window size is reset to a positive value by the receiver via a further acknowledgement. But an acknowledgement can get lost during data transmission and the sender never receive the sending release. To avoid this situation, in the case of a closed congestion window TCP provides the periodical transmission of probe segments from the sender side. These initiate the receiver to send back a receipt confirmation together with the currently free window size.

- **Silly Window Syndrome (SWS)**

  The silly window syndrome describes a situation that often occurred in early TCP implementations. The receiver repeatedly transmits only a very small available window size. In response, the sender only transmits very small segments (cf. Fig. 8.20). This occurred mostly when the available window size was exhausted, i.e., the value zero reached and a first update of the window size determined too quickly and sent back to the transmitter. In a fast connection this can result in the window size value continually wavering around the zero and the available transmission path being used insufficiently. This phenomenon was dubbed the "silly window syndrome," based on the sliding window protocol whose implementation resulted in this behavior.

  Normally, the segment size in TCP is controlled via the Maximum Segment Size Parameter (MSS). This, however, only gives a maximum value for the segment size so that no unnecessary fragmentation occurs in the Internet layer underneath, thereby causing additional overhead. A minimum segment size to maintain an efficient relation between control information in the header and the transported payload can normally not be specified. Because of this, heuristics on the receiver side ensure that a response concerning available window size first occurs after reaching the zero value when at least 50% of the maximum window size allowed is reached. Vice versa, in the case of an existing connection and still outstanding acknowledgement on the sender side it is ensured that the data to be sent is held back in the output buffer until either a new acknowlegement with a correspondingly large window size arrives or at least a segment can be sent that does not completely exhaust the remaining maximum window size.

- **Nagle Algorithm**

  Like the silly window syndrome, the Nagle algorithm, developed by *John Nagle* and described in detail in RFC 896, also focuses on data transmission overhead in the relationship between transmitted header information and payload. Often the TCP segments to be transmitted are only of a short length so that the overhead of the command and control information in the header is a dominant part

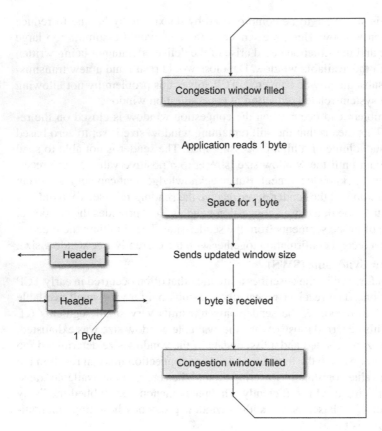

Fig. 8.20 TCP – silly window syndrome.

of the entire communication. In order to avoid or at least limit this, the acknowledgement of a segment is delayed as long as possible so that it can then be sent
together with its own payload data or together with other outstanding acknowledgements. The Nagle algorithm, used to this end, collects all user data until the
next acknowledgement is due and then transfers everything together in a single
segment.

- **Congestion Avoidance Algorithm**
  TCP determines the overload in the network based on the number of segments
  that have gotten lost. A segment is considered lost if the acknowledgement for it
  does not arrive within the fixed time span. The overload, i.e., the actual data loss,
  already occurs on the Internet layer of the TCP/IP reference model if routers serving as intermediate systems in the Internet can no longer handle the traffic load.
  Then the input buffers overflow and the incoming datagrams have to be discarded. If a TCP segment has to be transmitted anew because of data loss caused by
  overload, the load in the network is not lessened because of it. On the contrary,
  potentially necessary IP fragmentation arising from the loss of an individual da-

tagram can trigger the new transmission of a whole series of IP datagrams and the load can raise even more (congestion collapse) . For this reason, care must also be taken on the transport layer to recognize overload situations and, if necessary, by means of appropriate mechanisms to either stop these situations or at least to avoid them. These mechanisms are specified in RFC 2001. If segment loss increases then the situation of an existing network overload is identified and the data transmission rate decreased. This is done to prevent the existing overload situation from being further aggravated by the continuously repeated transmission of segments believed to be lost. Later it is then slowly increased again in individual steps (cf. slow start algorithm).

- **Slow Start Algorithm**
  For a better adaptation to window size, at the start of a connection TCP uses a small window (as a rule a segment, i.e., 536 bytes). This is then slowly raised sequentially until the rate of the sent segments and its acknowledgements match each other. The increase progresses exponentially and only ceases when segment loss is triggered by one of the acknowledgements signaling that a fixed waiting period has been exceeded. When this happens the congestion avoidance algorithm previously described goes into effect.

8.4 Network Address Translation – NAT

Appropriate transition technologies have been sought to keep the IPv4 protocol alive until the new IPv6 protocol is introduced in a form with sufficient technological sophistication and comprehensibility. In spite of the prevailing shortage of IPv4 address space, these technologies seek to allow for the operation of more end devices and larger networks than would be possible based on the level of IPv4 address distribution. The technology known as **Network Address Translation** (**NAT**) enables the operation and dynamic management of a large number of computers in a shared network via a small number of public IPv4 addresses through the use of private IPv4 address space. The devices operated in a NAT network remain reachable publicly via the Internet, although they do not have their own public IP addresses and can thus only be addressed over an appropriate NAT gateway connected to technology on top of the transport layer of the TCP/IP reference model. This procedure also offers protection in the face of Internet attacks, as direct addressability to the network end devices brought together by NAT is not possible from the Internet. In this section, the functions and tasks of NAT technology will first be introduced before a closer look is taken at possible applications as well as their pros and cons.

8.4.1 NAT – Functions and Tasks

NAT was primarily developed to deal with the lack of address space in IPv4. While it would be theoretically possible to operate NAT in conjunction with IPv6, there is simply no need for it. Under IPv6 every end device can have one or more of its own permanently assigned addresses.

As the lack of available address space in IPv4 became noticeable in the '90s, the IETF (Internet Engineering Task Force) took on the problem and proposed the Network Address Translation (informal) standard[1] In addition to addressing IPv4, NAT addressed a series of other problems:

- **Costs for reservation of IPv4 address space**
  Even if the shortage of IPv4 addresses were not an issue, the reservation of a larger, interconnected address space involves higher costs for a company than the reservation of only a few addresses does. The arrival of NAT technology was timely in focusing on the issue of conserving valuable address space while, at the same time, lowering costs for address reservation.

- **Problematic Internet security** As the popularity of the Internet rose at the beginning of the '90s, the number of criminal activities connected with it also grew. With more and more company computers being directly connected to the Internet there was a perpetually greater risk that one of these computers would become the target of a criminal attack.

For this reason, the solution was decided that as few company computers as possible should have a direct connection to the Internet. The remaining computers would only be allowed indirect access, hidden behind a security gateway or firewall. This was easy to implement since most of the computers in a company did not provide any server-service capabilities to the outside. As clients, they only used the Internet-provided server services. Communication was usually initiated from a company computer in the direction of the Internet, and not the other way around. Direct addressability of a computer was secondary, with a much greater importance placed on the outgoing direction. Besides, the chance of all computers in the company network accessing the Internet at the same time remained small. Access to the Internet is normally transaction-oriented. This means information is retrieved that must first be processed or read by the local computer before the next access can take place. In this case, the potential bottleneck of an additional central Internet gateway proves to be only a small risk. At least in direct Internet connections, data traffic flows mostly over its own (or sometimes multiple) special company routers, which control and regulate company-wide communication with the Internet being the central communication point.

NAT is based on the use of the private IPv4 address area for the company network. This address area is not routed in the Internet, i.e., forwarding to these addresses

---

[1] NAT was specified as an informal standard in RFC 1631, i.e., seen from a technological perspective, NAT is not an official Internet standard.

does not proceed over the Internet router and it can be used by all Internet subscribers for private purposes. In addition, the company network receives one or more so-called NAT routers that are directly connected with the Internet and take over the administration of the company network and the private IPv4 address area used there. The private IPv4 address space is shown in Tab. 8.5. For NAT mostly the area starting with 10.0.0.0 is used.

Table 8.5 Private IPv4 address space

Start	End	Number
10.0.0.0	10.255.255.255/8	16.777.216
172.16.0.0	172.31.255.255/12	1.048.576
192.168.0.0	192.168.255.255/16	65.536

The NAT router now takes over central mediation between the public Internet and the private enterprise network. IP datagrams are not only forwarded, but access to their internal structure also follows, i.e., they are translated. IPv4 addresses are rewritten into internal private addresses. Fig. 8.21 shows the schematic structure of a NAT-based enterprise network and its connection to the global Internet.

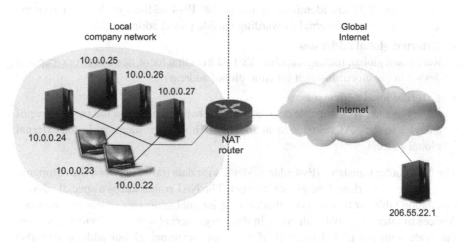

Fig. 8.21 Network Address Translation (NAT) and an overview of its function.

The following types of addresses are generally distinguished in the NAT:

- **Internal and external addresses**
  Every network-capable device inside the private enterprise network that NAT uses is known as an internal address (**inside address**). Every network-capable device outside of the local enterprise network, in other words, the global Internet, is reachable via an external address (**outside address**).

- **Local and global addresses**
  Depending on which area (local enterprise network or global Internet) a network-capable device is addressed from, a distinction is made between local and global addresses. Addresses known as **local addresses** are those used in an IPv4 datagram in the private company network. This applies to both internal as well as external addresses. In contrast, all addresses used in IPv4 datagrams in the global Internet - meaning outside of the private enterprise network - are known as **global addresses**. This is also independent of whether they identify internal or external addresses.

Within the private enterprise network only local addresses are always used, independent of whether the transmitted datagram is directed to a receiver in the local enterprise network or to a receiver in the global Internet. Outside of the private enterprise, in contrast, local address cannot be used. Four different address types can be distinguished (cf. also Fig. 8.22):

- **Internal local addresses**
  Refers to a network-capable device in the private enterprise network addressed within the local network (inside local address).

- **Internal global address**
  Refers to a global routing-capable IPv4 address that represents a network-capable device inside the private enterprise network. Normally, the network-capable devices in the NAT are identified by the global IPv4 address of the NAT routers. This takes over the internal forwarding (inside global address).

- **External global addresses**
  Refers to a global, routing-capable IPv4 address that identifies a network-capable device in the global Internet (outside global address).

- **External local addresses**
  Refers to a network-capable device in the global Internet from the perspective of a device in the NAT (outside local address). This can be identical to the external global address.

The NAT router translates IPv4 address details for data traffic between the company-own private network and the global Internet. The NAT router keeps a special address translation table for this purpose that assigns internal addresses of internal network devices to internal global addresses. In this way, internal network devices can communicate with the global Internet. If necessary, external global address can also be mapped on external local addresses. Address mapping can proceed statically or dynamically. In static mapping the allocation between global and local address representation is carried out permanently without further changes taking place. In dynamic mapping the NAT router automatically carries out an address allocation for as long as it is necessary. Thereby, the NAT router has a predetermined quantity of internal global addresses from which one is selected, based on need, and assigned for as long as an internal device needs it for a communication operation. Afterwards, the internal global address is free again and can be used by other network devices.

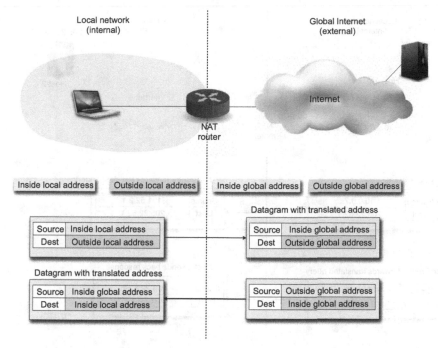

Fig. 8.22 Address types in the NAT.

8.4.2 NAT – Application Possibilities

The original idea behind the definition of the NAT technology in RFC 1631 was to enable network devices inside a private network access to the global network by sharing and mutually using one or more IPv4 addresses. It was assumed that the internal network devices request primary services from computers in the global Internet, i.e., that the connection had its source in network devices in the private network. Because of this, the traditional variation of NAT is also referred to as **unidirectional NAT** (also traditional NAT or outbound NAT).

Fig. 8.23 shows the function of the unidirectional NAT based on a simple example. Let us assume that the internal network devices are addressed over the internal local addresses 10.0.0.0 and those following. An internal computer (host 1) with the internal local address 10.0.0.7 wishes to communicate with a computer (host 2) in the global Internet with the external global IPv4 address 205.55.62.1 . The responsible NAT router in the private network has a dynamically managed address pool with global IPv4 addresses.

1. The internal computer sends an inquiry to the NAT router. The transmitted IPv4 datagram contains as sender the internal local address 10.0.0.7 and as receiver the external local address 205.22.62.1.

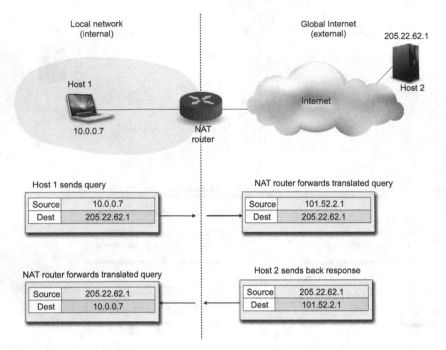

Fig. 8.23 Traditional NAT.

2. The NAT router translates the sender address 10.0.0.7 - which it recognizes as an internal local address - into a free address from its global address pool, e.g., 101.52.1.2. It then forwards the IPv4 datagram further to the global Internet. The receiver address is not translated in the unidirectional NAT, i.e., the external local address is simultaneously the external global address 205.22.62.1. At the same time, the NAT router writes the address mapping in its address translation table.

3. The receiver processes the IPv4 datagram without being able to tell that it comes from a NAT network. The sender address 101.52.1.2 is treated as a normal IPv4 address and a return answer is sent back in a new IPv4 datagram. The returned IPv4 datagram contains the sender address 205.22.62.1 (external global address) and the receiver address 101.52.1.2 (internal global address).

4. The NAT router of the private network receives the IPv4 datagram intended for the network and replaces the receiver address (internal global address) based on its entry in the address translation table in internal local address 10.0.0.7. The datagram is then sent back to its original sender.

The NAT router translates not only the addresses involved but must also carry out further changes in the IPv4 datagram. Among these are the new calculation of the checksum in the IPv4 datagram as well as in the header of the transported TCP or the

UDP header, as the addresses involved in checksum calculation change in address translation.

If a private network has multiple NAT routers, they must all use the same address translation tables. In the case that, e.g., an IPv4 datagram is sent via the NAT router 1 and the answer reaches the private network via NAT router 2, this must be capable of translating back the internal global address there into the correct internal local address.

The term **bidirectional NAT** (also called two-way NAT or inbound NAT) is used if an internal network device in the private network is to be addressed from the global Internet. Communication then follows from out of the global Internet into the private network. It is problematic that while from out of the private network the external global addresses of the network devices in the Internet are known however, vice versa, a computer in the private network is not directly visible from the outside and also its internal global address, which is managed by the NAT route, is not known.

The internal local/global address of a network device may also be known globally in the case of NAT static address allocation. In dynamic address allocation, address resolution of Internet NAT addresses is usually carried out with the help of the Domain Name Service (cf. section 9.2.1). The Domain Name Service (DNS) translates so-called domain names into IP addresses.

With the help of domain names, an additional abstraction layer is placed over the Internet addressing. Here, the IP addresses, which can only be noted with difficulty, are masked with the help of hierarchical, mostly self-explanatory names. These are translated by way of special DNS servers into IP addresses. The interplay of NAT and DNS is specified in RFC 2694. Address resolution in the interaction between DNS and NAT proceeds as follows:

1. Host 2 (205.22.62.1) sends a request from the global Internet to host 1, who is located in a local, private network. In doing so, host 2 uses the domain name of host 1, e.g, host1.private.net. Additionally, an initial DNS request is sent to the responsible DNS server in order to translate the domain name into a valid IP address.

2. The responsible DNS server receives the DNS request and transforms it into an internal local address of the private network (10.0.0.7). This corresponds to the transmitted domain name.

3. The internal local address 10.0.0.7 is forwarded directly to the NAT router who transforms it into an internal global address. In doing so, the NAT router helps itself from the local pool of global IPv4 addresses and carries out the assignment (e.g., to 101.52.1.2), which it enters in its internal address translation table. The internal global address is sent back to the DNS server.

4. The DNS server forwards the the internal global address, 101.52.1.2, received from the NAT router to the requesting host 2, who can use it to communicate with host 1 directly.

Therefore, the internal global address of the computer located in the private network has to be known by the computer requesting from the outside before communication can be carried out. Either a statically assigned internal global NAT address can be used for this or a dynamically assigned internal global address determined over a DNS request. If the internal global address is known, communication proceeds over NAT as follows (cf. Fig. 8.24):

- Host 2 sends a request to the NAT router. It uses the internal local address from host 1 (101.52.1.2) as the receiver address, and as sender address its own external global address (205.22.62.1). The sent IPv4 datagram is forwarded to the responsible local router of the NAT network, i.e., usually the NAT router.

- The NAT router translates the receiver address into an internal local address (10.0.0.7) according to the already existing entry in the address translation table. It forwards the IPv4 datagram to the receiver.

- Host 1 receives the IPv4 datagram, generates an answer and sends it back to 205.22.62.1 (external local address). The IPv4 datagram is initially forwarded to the NAT router.

- The NAT router replaces receiver address 10.0.0.7 (internal local address) with the corresponding external local address 101.52.1.2 and forwards the IPv4 datagram to the designated receiver 205.22.62.1 (external global address).

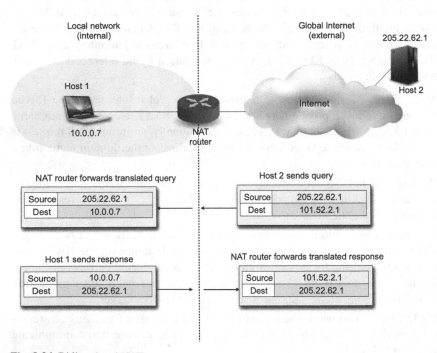

Fig. 8.24 Bidirectional NAT.

NAT technology is based on mapping one or more available, global IPv4 addresses onto a large number of private IPv4 addresses. Yet this described procedure has a serious disadvantage. If all available internal global IPv4 addresses are used up, this means that virtually no other network device can communicate from the private network before the communication of another is finished. To get around this problem, NAT uses the address port provided by protocols of the transport layer. This makes it possible to have multiple communication connections over the same network device - i.e., over the same IP address - from the protocols of the transport layer.

Together with the IP address, the port number forms a so-called socket. This identifies uniquely an Internet communication connection. The socket-based communication is understood as an overloaded internal global address (**overloaded NAT** or **Network Address Port Translation, NAPT**). Thus, NAT is made more flexible and at the same time it becomes possible to manage thousands of port numbers over a single internal global IPv4 address. The price to pay is increased complexity, with NAT raised from the Internet layer of the TCP/IP reference model to the transport layer. The communication process itself progresses in the same way as described, only that in addition to IP address translation the respective port numbers on the NAT routers are translated as well (cf. Fig. 8.25).

- Host 1 sends a request to host 2. Host 1 uses its internal local address 10.0.0.7 and the transmission port 6000. The request to host 2 should be sent to a WWW server (i.e., to port number 80). The request is forwarded first to the Internet NAT router.

- The NAT router of the private network translates the internal local address and the sender port 10.0.0.7:6000 into an internal global address and a new sender port 101.52.1.2:6112. The address allocation is noted in the address translation table of the NAT router. Then the IPv4 datagram is forwarded to the receiver 205.22.62.1:80.

- Host 2 receives the request, processes it and sends back an answer. The receiver address of the IPv4 datagram is 205.22.62.1:80 and the receiver address 101.52.1.2:6112 (internal global address).

- The NAT router of the private network receives the IPv4 datagram and translates the receiver address and receiver port number in the corresponding internal local address 10.0.0.7:6000, as dictated by the entry in its address translation table.

NAT, or NAPT, makes it possible to bypass the problem of limited IPv4 address resources by using the private IPv4 address space in connection with the 16-bit long port number. At the same time, this course of action is not always unproblematic. Let us assume that two networks use NAT with the same or an overlapping private address space. If these networks are brought together, as happens e.g., in the fusion of two companies, problems with address translation may result. There are general problems with the overlapping of private and global address areas. This is because the NAT router is not able to know if a receiver address is located in its own network or in a foreign one. This situation is solved by the NAT router not solely translating the sender address or the receiver address, as the case may be, but both addresses.

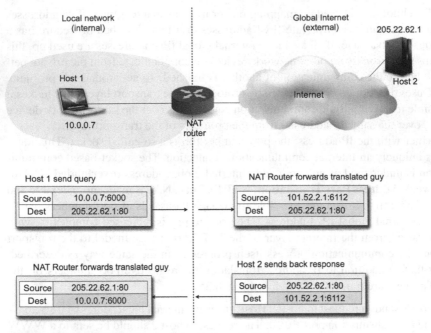

Fig. 8.25 Network Address Protocol Translation – NAPT.

Unlike the previous situation, here both the external global address and the external local address are different. For this reason, this procedure is also known as **overlapping NAT** or **twice NAT**. In the local address translation table of the NAT router, not only the address translation for network devices from the individual private network is noted, but also that of other networks whose address area overlaps one's own. As in the case of bidirectional NAT, an interaction with DNS is necessary.

Once employed, NAT does not work completely transparently with all network devices it comes into contact with. In addition to address translation, which actually represents a problem of the Internet layer of the TCP/IP reference model and therefore actually only applies to the Internet protocol, there are a whole series of other protocols that are affected by NAT address translation. A NAT router must therefore not only intervene on the Internet layer, but also consider the other protocols and protocol layers affected. NAT has an influence on the following areas:

- **TCP and UDP**
  As soon as the IP address is changed, the header of the IP datagram changes, i.e., a new checksum must be calculated for the IP datagram. On the transport layer, TCP and UPD also calculate a checksum over a so-called pseudo header. This includes the IP addresses of the sender and the receiver. Because of this, the checksum on the transport layer must also be newly calculated for TCP and UDP.

- **ICMP messages**
  The Internet Control and Messaging Protocol (ICMP) is an auxiliary protocol of the Internet protocol. It handles numerous command and control tasks. IP addresses are often contained in ICMP messages as parameters, for example, when the transmission of non-reachable IP addresses is involved (destination unreachable). If applicable, these IP addresses must also be taken into consideration by NAT routers.

- **Internet applications that embed IP addresses**
  Just as with ICMP, several Internet applications implement IP addresses as parameters. These are transported in the messages exchanged. For example, the File Transfer Protocol (FTP) sends address and port information as clear text in exchanged FTP messages. In order to also operate this application in NAT, the NAT router must identify and factor in this address information. If the number of digits are changed in the IP address - the address is transmitted as text and not in binary form - the size of the transported payload changes as well. This results in additional effects on the exchanged TCP segments and their sequence numbers.

- **IPsec problems**
  If the IPsec protocol is used for more secure data transfer, both the IPsec authentication header as well as the encapsulating security payload use an integrity control that is calculated from the transported contents. In the case where NAT changes checksum values for TCP or UDP, the transported contents also change and an integrity check by IPsec fails. It cannot be determined anymore whether the transported contents have been manipulated or not by unauthorized third parties. Therefore, NAT cannot be operated securely together with IPsec.

8.4.3 NAT – Advantages and Disadvantages

NAT has been further developed since its introduction and represents for many companies today an indispensable link between the corporate network and the global Internet. Individual IPv4 addresses can be simultaneously used by numerous network-capable end devices to gain Internet access with the help of the port-based NAT (NAPT). At the same time, twice NAT supports the establishment of virtual private networks in which address translation is carried out in the private NAT network for both outgoing as well as for incoming datagrams. The continuing growth of the Internet with its new mobile end devices, network-capable entertainment and household electronics as well as sensor networks, is responsible for what seems a never ending demand for new IP addresses. This can no longer be satisfied by the current IPv4 address space even if NAT is implemented as a multiplier technology. While NAT offers a wide array of advantages, these are often paid for at the cost of serious disadvantages. The main advantages of the NAT technology are:

- A large number of network-capable end devices can share with each other one or more public IPv4 addresses. This saves on costs and valuable IPv4 address space.
- NAT gives company network administrators more efficient and more comprehensive control possibilities over the private network. These possibilities are greater than those to be had with a public network that is simultaneously connected with the global Internet.
- NAT makes it easier for a company to change an Internet Service Provider (ISP). Normally, when there is a change of provider the complete network must be supplied with new IPv4 addresses. With NAT, in contrast, the organization structure and the private address space remain in place in the private company network, while only the public IPv4 addresses change. These addresses are used by NAT to communicate with the global Internet.
- NAT enables simple growth for company networks. This is because new network devices can easily be addressed over the available private address space without the necessity of providing new public IPv4 addresses or complete IPv4 networks.
- An important advantage is the increased security accompanying the establishment of NAT. The NAT router functions simultaneously as a company firewall over which the private company network can be centrally protected from outside attacks. Through the implementation of NAT it becomes considerably more difficult for an unauthorized third party to gain access to an end device located in a private network.
- Usually, the effort involved in changing the (public) IPv4 network configuration of a NAT is only connected with the configuration of the responsible corporatewide NAT router. The remaining computers in the private company network remain unaffected.

Besides these advantages, the use of NAT in a company also poses disadvantages:

- Being an extra network technology, the establishment of a NAT also means extra configuration and administration effort.
- As the network devices in the private company network in the NAT do not have their own real IPv4- addresses, communication and network applications which require this are not possible in a NAT. For example, this applies to a secure end-to-end communication, such as can be carried out via IPsec.
- Further compatibility problems arise with Internet applications and protocols in connection with address translation in the NAT, which use the original (real) IP addresses to make extra functions available. This is the case, for example, with the new calculation of the checksum necessary for IP, TCP and UDP, as well as in Internet applications such as the File Transfer Protocol (FTP). Here, the Internet address used is transmitted as text and in the process of address translation a problematic change occurs in the amount of data transferred.
- Security functions, such as a guarantee of the transmitted data remaining intact (integrity) as taken from IPsec, can no longer be ensured in NAT. This is because in the address translation process the contents of the transmitted datagram

is changed. This change cannot be distinguished from what could otherwise be malicious tampering by unauthorized third parties.

- NAT makes it difficult for an outside attacker to penetrate a computer in the private network. However, it is exactly this difficulty that complicates so-called peer-2-peer applications, making them virtually impossible to be used by a network device inside a NAT.

- Besides the added complexity of NAT, an address translation between the private network and the global Internet also costs more time. In addition to address translation, the new calculation of checksums in different parts of the transmitted datagram must be carried out.

An extensive introduction of the IPv6 network technology makes the long-range implementation of NAT unnecessary. The main advantage of multiplying the sparsely available address space loses its importance in light of the address space available in IPv6 . Nevertheless, given the further advantages of NAT, the voices in favor of its retention under IPv6 have grown louder.

8.5 Security on the Transport Layer

While the Internet layer has its own secure communication protocol with IPsec, at the transport layer of the TCP/IP reference model we do not find a special, secure transport protocol. In most cases this is not necessary because a secure communication protocol already exists with IPsec. However, it should be remembered that IPsec involves an independent protocol on the Internet layer. Applications wishing to communicate securely over the Internet layer must be capable of using this communication protocol. Another way was chosen for the transport layer. Instead of the transport layer using an independent protocol, Internet applications can communicate over a secure transport infrastructure. This is found as an intermediate layer between the transport layer and the application layer in the TCP/IP reference model. To enable secure data transport over the World Wide Web (WWW), the first browser manufacturer, Netscape, used its own secure transport infrastructure: **HTTP over SSL** (HTTP over Secure Socket Layer, HTTPS).

This secure transport infrastructure was created by taking on an additional protocol layer in the TCP/IP reference model that was solely responsible for secure transport. It was implemented between the standard WWW protocol HTTP (Hypertext Transfer Protocol) on the application layer and TCP, the transport protocol below it. The Secure Socket Layer (SSL) and Transport Layer Security (TLS), which take over these tasks are described in the following sections.

8.5.1 Transport Layer Security and Secure Socket Layer – TLS/SSL

An advantages of the chosen protocol architectures for the provision of a secure transport infrastructure over the Internet is that it can be used by communication protocols and applications specified on a higher protocol layer in the TCP/IP reference mode - without requiring further changes be made to them.

Secure Socket Layer (SSL) was originally designed by Netscape in 1994 and further developed though 1996 to the version SSLv3. When SSL was designed, neither the Internet protocol IPv4 nor TCP had their own mechanisms for securing communication. Therefore, developers decided to expand the available TCP socket interface to include the appropriate cryptographic components. The development of the Transport Layer Security (TLS), as the Internet standard RFC 2246, is based on SSLv3 and was published in 1999. With TLS 1.1 (RFC 4346) and TLS 1.2 (RFC 5246), the standard was expanded in 2006 and 2008 to include additional functionality. TLS/SSL works transparently and therefore enables protocols without their own security mechanisms the possibility of communicating over a secure connection. As shown in Fig. 8.26, this architecture is also suitable for the securing of further services, such as FTP, IMAP or TELNET. These can also use the secure connection provided by TLS/SSL.

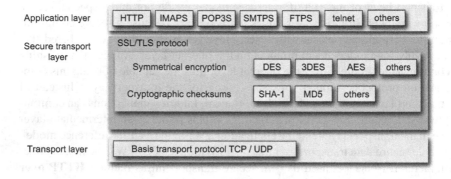

Fig. 8.26 Communication protocols in the layer model – TLS/SSL.

SSL gets its name from the programming interface of the TCP/IP protocol - the so-called socket. However, SSL does not define a fixed programming interface, rather it defines only the actual, secure protocol. In this way, different implementations with different programming interfaces and starting points are available for the application program located in the next higher protocol layer.

From the standpoint of its historical development, TLS/SSL is primarily employed in connection with the reliable TCP protocol. However, there are also implementations for datagram-oriented transport protocols, for example, UDP or the Datagram Congestion Control Protocol (DCCP). These variations of the TLS/SSL were standardized separately in RFC 4347 as **Datagram Transport Layer Security (DTLS)**. DTLS is employed particularly in connection with Voice over IP (VoIP) in Internet telephoning. VoIP data is transmitted via UDP and not via TCP. The focus is on communication in real-time conditions and transmission errors and data loss can be tolerated to a certain degree. TLS is not designed for this application scenario as the cryptographic procedure used for authentication cannot deal with data loss.

Secure communication via TLS/SSL offers two communication partners protection against eavesdropping and unauthorized tampering with transmitted messages. Altogether, TLS/SSL includes the following cryptographic functions:

- **Guaranteeing private connections**
  After an initial handshake procedure to establish a secret key, the data to be transmitted is encrypted with a symmetrical cryptographic method.

- **Authentication**
  The reliable identification of the author of an encrypted message is carried out by means of a public encryption procedure (public key encryption, asymmetrical cryptography).

- **Guaranteeing reliable connections** The message transport checks the integrity of the transported data via a so-called **Message Authentication Code (MAC**, cryptographic checksum).

Interoperability stood primarily in the forefront when TLS/SS was developed. It was meant that programmers be in a position to develop applications independent of one another. Set on TSL/SSL, these were to interact with each other without mutual knowledge of the source code of the respective application being necessary. Additionally, TLS/SSL offers the possibility to easily integrate additional, new cryptographic methods and has a high degree of flexibility and **extensibility**.

As cryptographic methods often demand an extremely high computational effort, special attention was paid to an efficient caching procedure in the development of TLS/SSL. The focus was on keeping the number of new connections to be established in a session as low as possible. Irrespective of the degree of authentication selected, TLS/SSL distinguishes between different types of communication connections:

- **Anonymous connections**
  Neither the identity of the client nor that of the WWW server is verified While in this operational mode TLS/SSL offers protection against simple eavesdropping, a man–in–the–middle attack is still possible. Therefore, this operational mode is not advisable.

- **Server authentication**
  The WWW server presents the client with a certificate, which proves its identity with the help of a public key infrastructure and must be accepted by the client. While the identity of the client is not verified, it can rely on the identity of the WWW server being correct.

- **Client/server authentication**
  Here, both communication partners verify their identities by means of certificates.

The intermediate layer defined independently by TLS in the TCP/IP reference model is divided into two sublayers: the record layer and the handshake layer (cf. Fig. 8.27). Segment processing above TCP takes place in the record layer together with the encryption and authentication by TLS.

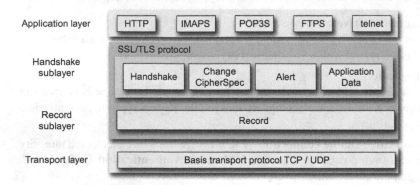

Fig. 8.27 TLS/SSL layer model.

Based on this, the following sub-protocols are found in the handshake layer:

- **Handshake**
  During connection establishment the handshake protocol takes over the task of negotiating the cryptographic parameters between the communication partners involved.

- **Change cipher spec**
  The cryptographic procedure negotiated by the handshake protocol is activated by the change cipher spec protocol.

- **Alert**
  If problems and exceptional situations arise during TLS operations and secure data transmission, the alert protocol generates and sends status and error messages.

- **Application data**
  The application data protocol comprises the interface for applications that use the TLS/SSL for secure data transport. The application transmits the data to be sent to the application data protocol. Vice versa, the application data protocol forwards the data transmitted from the TCP protocol to the application.

The cryptographic procedure negotiated in the handshake between communication partners is defined in TLS/SSL with the help of so-called **cipher suites**. A cipher

Cryptographic procedure in TLS/SSL

The TLS/SSL protocol supports a multiple number of various cryptographic algorithms that serve in the processes of authentication, certificate transmission or session key selection. Thereby, the TLS/SSL handshake protocol takes over the job of negotiating the method to be used for solving this task between client and server.

- **DES** (Data Encryption Standard), symmetrical encryption procedure,
- **DHE** (Diffie-Hellman Key Exchange), secure key exchange procedure in which a symmetrical key can be exchanged safely between two communication partners.
- **DSA** (Digital Signature Algorithm), part of a digital authentication standard,
- **KEA** (Key Exchange Algorithm), algorithm for secure key exchange,
- **MD5** (Message Digest Algorithm 5), algorithm to generate a digital fingerprint (Message Digest),
- **RC2** and **RC4**, encryption algorithm for so-called stream ciphers to enable encryption of bit streams.
- **RSA** (Rivest-Shamir-Adleman Data Security), asymmetrical encryption procedure,
- **SHA-1** (Secure Hash Algorithm 1), algorithm for the generation of a digital fingerprint (message digest),
- **SKIPJACK**, symmetrical key procedure developed by the US government and
- **3DES**, triplicate applied DES algorithm. Via KEA or RSA symmetrical keys are generated for the client and server. These are then implemented in the TLS/SSL session.

Further reading:

Dierks, T., Rescorla, E.: The Transport Layer Security TLS Protocol Version 1.1. Request for Comments, 4346, Internet Engineering Task Force (2006)

Meinel, Ch., Sack, H.: Digital Communication – Networking, Multimedia, Security, Springer, (2013)

Fig. 8.28 A selection of several cryptographic procedures supported by TLS/SSL.

suite always contains a key exchange procedure, an encryption procedure for the transmitted data and an authentication procedure (cf. Fig. 8.28).

It generally applies that a cipher suite is only as safe as the weakest of these three implemented procedures. Determining which procedures is offered by the cipher suite on the client or server side is in the hands of each administrator, who can activate or deactivate the relevant algorithm.

During the handshake, the communication partners - the requesting client and the answering server - negotiate which TLS/SSL cipher suite should be employed. The client proposes various cipher suites that can be implemented for secure data transmission at the time of the handshake. From these proposals the server chooses exactly one cipher suite. This is to be used for the remaining communication time. Fig. 8.29 presents examples of several of the TLS/SSL cipher suites.

Data communication via SSL proceeds in two steps. First, a handshake protocol is worked out in which the cryptographic abilities of both communication partners involved is verified. Included among these abilities can be an authentication of cli-

TLS/SSL Cipher Suites

The names of each cipher suite always begin with the identifier TLS followed by the designation of the key procedure. After that comes the term WITH, followed by the designation of the encryption and authentication procedure. At the beginning of a connection, the cipher suite TLS_NULL_WITH_NULL_NULL is always activated.

- TLS_RSA_WITH_3DES_EBE_CBC_SHA
  The RSA public key encryption procedure is used here for the key exchange, i.e. for authentication of the communication partner and for encryption of the key generated by the client (sender). The data transmission itself is guaranteed through the symmetrical 3DES encryption procedure. It is operated in the EBE-CBC mode. Data integrity is ensured with the help of the SHA-1 procedure.

- TLS_DHE_RSA_WITH_3DES_EBE_CBC_SHA
  The key is generated with the help of the Diffie-Hellman key exchange procedure. Both the sender and the receiver have a hand in generating the key. The data exchanged during the key transmission is authenticated with the help of RSA. Data transmission and the securing of data integrity is carried out as previously.

- TLS_DHE_DSA_WITH_3DES_EBE_CBC_SHA
  Authentication of the data transmitted during the Diffie-Hellman key exchange is implemented with the help of the DSS procedure. The further functionality is identical to the previous procedures.

- TLS_DHE_anon_WITH_3DES_EBE_CBC_SHA
  The data transmitted in the Diffie-Hellman key exchange is not authenticated by the communication partners. This, however, does not offer any protection against a man-in-the-middle attack.

- TLS_DHE_RSA_WITH_NULL_SHA
  Here, data transmission is not encrypted. Only an authentication of the communication partners and an integrity protection of the transmitted data is carried out.

Further reading:

Medvinsky, A., Hur, M.: Addition of Kerberos Cipher Suites to Transport Layer Security (TLS), Request for Comments 2712, Internet Engineering Task Force (1999)

Fig. 8.29 Selected examples of several TLS/SSL cipher suites.

ent and server, as well as a determination of the suitable cryptography procedure mastered by both. The strongest cryptographic procedure available to both is chosen. In the first phase of session establishment strong cryptographic methods are used to exchange the session key. The actual encryption of the data exchanged during a session then follows with less complex cryptographic methods. These offer a compromise between attainable security and the computational effort necessary for coding/decoding.

Excursus 13: TLS/SSL Handshake Procedure

The TLS/SSL protocol uses a combination of methods implementing public and symmetrical keys. Symmetrical encryption procedures work faster, but procedures with public keys offer safer authentication methods. A TLS/SSL session always begins with a message exchange. This is called a **handshake**. In the handshake procedure, the server first identifies itself to the requesting client with the help of a public key. Afterwards, the client and server cooperate to generate a symmetrical key that is implemented during the session for fast encryption, decryption and to discover attempts to manipulate the transmitted data. The authentication of the client is optional and is only then required by the WWW server when it is considered necessary. Fig. 8.30 shows the principle process of the TLS/SSL handshake protocol and Fig. 8.31 the messages exchanged between the client and server during the handshake.

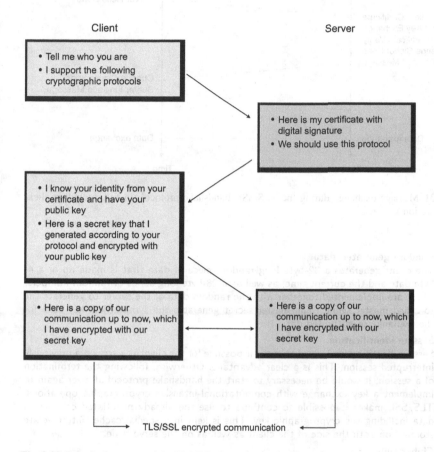

Fig. 8.30 Principal process of the TLS/SSL handshake protocol in its simplest form (without client authentication).

1. **Initial message from client to server**

   Client Hello

   The client begins a new session with the transmission of a client hello message to the server. The client hello message contains the following information:

   - **Versions number**
     The client sends the version number of the highest TLS/SSL version it supports. The value 2 corresponds to version SSLv2, 3 to version SSLv3 and 3.1, 3.2, 3.3 to versions TLS 1.0, 1.1 and 1.2.

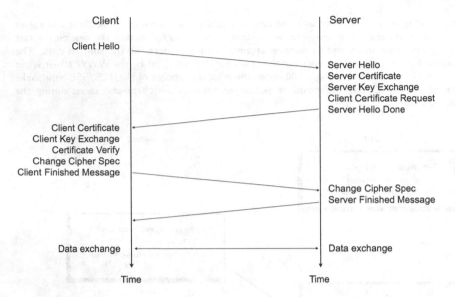

Fig. 8.31 Message exchange during the TLS/SSL handshake protocol (in full, including client authentication.)

- **Random generated data**
  The client generates a 32-byte long random piece of data that is made up of a 4-byte date and the current time, as well as a 28-byte long series of random numbers. These are implemented together with the random data of the server to generate the so-called master secret. The master secret generates the symmetrical key used for the exchange of encrypted data.

- **Session identification**
  Session identification is used to make it possible for the client to carry on a previously interrupted session. This is a clear advantage. Otherwise, following the termination of a session, it would be necessary to start the handshake protocol all over again to implement a key exchange with computational-intensive cryptographic operations. TLS/SSL makes it possible to continue to use the already negotiated connection data including the cryptographic key. This is kept in a special cache (intermediate storage) on both the side of the client as well as on the server side.

- **Cipher suite**
  The client sends a list of the cipher suites it supports. These identify the cryptographic procedure for key exchange, authentication, encryption and ensure data integrity.

- **Compression algorithm**
  The client can additionally provide a data compression algorithm to be used for client/server data transmission.

2. **Initial answer from server to client**
   Server Hello
   The server answers the client's initial request with a server hello message containing the following information:

   - **Version number**
     The server sends the highest version number, which is supported by the client as well

as the server, back to the client. This is the smaller of the two maximally supported version numbers from the client hello message and from the server-supported version.

- **Random generated data**
  Like the client, the server generates a 32-byte long series of random data. This is made up of a 4-byte long date and the current time as well as a 28-byte long series of random numbers. This random data is used together with the random data sent by the client to generate the so-called master secret. The master secret generates the symmetrical key used during the encrypted data exchange.

- **Session Identification**
  Three possibilities exist:
  - If no session identification was sent by the client in the initial client hello message, this indicates that the client has no wish to resume a possibly terminated session. The server then generates a **new session identification**. This also happens in a case where the client wants to continue a terminated session, however the server is unable to fulfill this wish.

  - If a session identification was handed over by the client in the initial client hello message, the client refers to the previously terminated session now to be resumed. Able to fulfill this wish, the server sends a **resumed session identification**, i.e., it confirms the session identification sent by the client.

  - The client signals the start of a new session. The server complies with the request, however sends no **acknowledging session identification**, as it does not wish to resume the session after a possible termination.

- **Cipher suite**
  The server chooses from the list of possible cryptographic procedures sent by the client in the initial client hello message. It selects the cryptographically strongest cipher suite supported by both client and server. If the communication partners have no mutual access to cipher suites both can support, the handshake protocol ends here with a handshake failure alarm.

- **Compression algorithm**
  The client can additionally provide a data compression algorithm. This is to be used for data transmission between the client and the server.

Server certificate
The server sends its certificate to the client. TLS/SSL uses certificates based on the ITU-T X.509 standard. This is the certificate standard most often used today[2]. The server certificate contains the public key of the server. The client uses this public key for server authentication and in the key exchange procedure for encryption of the so-called premaster secret.

Server key exchange
The server generates a temporary key that it sends to the client in this optional step. This key can be used by the client to encrypt messages sent by the client in the following key exchange This step is only then necessary when the implemented public key algorithm provides no key for encrypting the message in the key exchange. For example, if there is no public key indicated in the server certificate.

Client certificate request
This step is likewise optional and is only carried out when the server requires the authentication of the client. An authentication of the client is always necessary when particularly

---

[2] Detailed information about cryptographic certificates and certificate management may be found in the first volume of this trilogy, Meinel, Ch., Sack, H.: Digital Communication – Networking, Multimedia, Security, Springer (2013), in section 5.4 „Public Key Infrastructure and Certificates".

sensitive information is being transmitted, e.g., account and transaction information in online banking.

Server Hello Done
As the last message in this transaction phase of the handshake protocol, the server signals its completion of the current phase and is now awaiting the client's answer.

3. **Client's answer to the initial answer of the server**
Client certificate
If the server has requested a client certificate in its previously transmitted answer, the client sends back the requested certificate. The certificate of the client contains the client's public key. This is to be used for authentication of the message sent in the key exchange procedure.

Client key exchange
The client sends a client key exchange message to the server after it has calculated the premaster secret from the random data of the client and server transmitted in the initial phase of the handshake. With the help of the public key from the server, the premaster secret is transmitted encrypted to the server. Server as well as client subsequently calculate the mutual master secret from the premaster secret originating on the opposite side. The session key used for data transmission is derived from this. If the server is in a position to read the data transmitted by the client, which is encrypted with the public key of the server, the client can be sure that the server has the corresponding private key for encryption. Server authentication is proven in this way, as only the server with the private key corresponding to the public key delivered in the server certificate can be the right communication partner. This message additionally contains the version number of TLS/SSL protocol that has been used. Upon receipt, the server compares it with the version number originally sent in the initial handshake phase.
In this way so-called rollback attacks can be prevented. Here, an unauthorized third party manipulates the exchanged message with the intention of switching back to an older, i.e., cryptographically weaker protocol version.

Certificate verify
This message is only transmitted if the client has sent back a requested client certi-ficate message beforehand. An authentication of the client is carried out at the same time. With the help of its private key it uses a hash value for the digital signature. This is calculated from all the messages that have been transmitted up to now. The server can decrypt this signed hash value with the help of the public key contained in the client certificate and through its own calculation verify the hash value from the transmitted messages. The authentication of the client is thus ensured. Only the owner of the mat-ching counterpart to the public key in the client certificate can in fact have encrypted the hash value transmitted.

Change cipher spec
With this message the client communicates to the server that from this point on all of the transmitted information is encrypted with the help of the negotiated cryptographic procedures and parameters.

Client finished
This message contains a hash value calculated from all of the messages transmitted up to now between the client and server. Encryption proceeds with the help of the negotiated encryption procedure and the session key generated mutually in the key exchange. This message is no longer transmitted via the handshake protocol, but is the first message transmitted encrypted by the record layer protocol.

4. **Final answer of the server to the client**
Change Cipher Spec The server confirms with this message that all data from now on is to be transmitted encrypted with the exchanged cryptographic procedures and parameters from the first phase of the handshake.

Server finished message
This message contains a hash value calculated from all the the transmitted messages between the client and the server up to now. This is encrypted together with a Message Authentication Code (MAC). If the client is subsequently in a position to decrypt the message and to correctly validate the hash value, it can be assumed with a high degree of certainty that the TLS/SSL handshake has been carried out successfully and that the key on both the sides of the client and server correspond to each other.

Encryption procedure in the TLS/SSL Protocol

- **Asymmetrical procedure**
  Before a symmetrical session key can be generated, data has to be exchanged with the help of an asymmetrical encryption procedure. Public and private keys can be exchanged with this procedure. For authentication of the server, the client encrypts a premaster secret with the help of the public key of the corresponding server. The public key of the server originates in the server certificate The transmitted data can only be decrypted again with the private key held by the relevant server. Subsequently, the symmetrical session key is generated from the premaster secret. The authenticity of the server is ensured through the implementation of the key pair made up of public and private keys. For client authentication, the client encrypts data with the help of its own private key, i.e. it generates a **digital signature**. The public key issued in the client's certificate is only able to validate the data if it has indeed been encrypted with the private key of the client. Otherwise the server is not able to validate the client's digital signature and the session is ended.

- **Symmetrical procedure**:
  After the premaster secret between client and server has been successfully exchanged, it is used together with other data available to the server and to the client in calculating a mutual master secret. The session key is derived from this. This can then be used throughout the entire session for fast encryption and decryption.

Server authentication
As already described in step 2 of the TLS/SSL handshake protocol, the server sends a certificate that it uses to authenticates itself to the client. The client uses this certificate in step 3 to verify the identity of the server.
It is necessary to check the association between the public key and the server key, as identified in the certificate, and in this way to authenticate the server. The TLS/SSL-enabled client is therefore obligated to answer the following questions (cf. Fig. 8.32):

1. **Has the request been made within the validity period of the certificate?**
   The client verifies the validity period of the certificate. It stops the authentication procedure if the current request has been made outside of this time period. Otherwise, the client continues with authentication.

2. **Is the authority who issued the certificate trustworthy?**
   Every TLS/SSL-enabled client manages a list of trustworthy issuers of certificates (**Trusted Certificate Authorities, CA**). Based on this list, it is determined which server certificate the client accepts. If the name of the authority CA, (**Distinguished Name, DN**) matches a name from the list of trustworthy client CAs, the client carries on with server authentication. However, if the authority CA is not on the client's list, the server is not authenticated. An exception is made if the server is able to send a certificate chain containing a CA that is on the client's list.

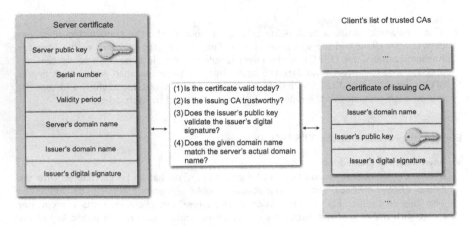

Fig. 8.32 SSL handshake – server authentication.

3. **Can the digital signature of the issuer be validated with the public key of the issuer?**
   Here, the client uses the public key of the CA from its own list of trustworthy CAs to
   validate the digital signature on the certificate of the server. The server is not authenticated if the information in the server certificate is changed after the CA has signed it.
   This is also the case if the public key from the list of clients does not correspond to the
   private key used to create the signature of the server certificate. If on the other hand
   the digital signature can be validated, then the certificate is considered genuine.

4. **Does the domain name of the server in the certificate match the current domain
   name of the server?** This test is intended to ensure that the server is indeed among
   the network addresses of servers provided. While from a technical point of view this test
   is actually not a component of the TLS/SSL handshake protocol, its implementation
   is intended to prevent a man-in-the-middle attack. With the matching of both domain
   names, the client carries on with the handshake.

5. The server is now successfully authenticated and the client carries on with the handshake
   protocol. However, if for whatever reason the authentication procedure does not get to
   this step the server cannot be authenticated. The user is then informed that an encrypted
   and secure connection cannot be established.

6. **Client authentication** is now carried out if required by the server.

7. The server then uses its own private key for decryption of the premaster secret. This is
   sent by the client in step 4 of the TLS/SSL handshake. Otherwise, the TLS/SSL session
   is terminated.

Client Authentication
TLS/SSL-enabled servers can be configured in such a way as to require an authentication
from the requesting client. This is especially true when sensitive data is being transmitted
from the server to the client and must be prevented from falling into the wrong hands. The
client then sends a client certificate (TLS/SSL handshake step 6) together with a digitally
signed message to authenticate itself. The server uses this signed message to validate the
public key of the client found in its certificate. It thus confirms the identity of the client as
verified by the certificate. The contents of the signed message are known only to the client
and the server. The digital signature is generated with the private key of the client, which in
a positive case scenario corresponds to the public key from the client's certificate. In order

for the authenticity of the requesting client to be successfully confirmed, the following questions must be answered by the TLS/SSL-enabled server (cf. Fig. 8.33):

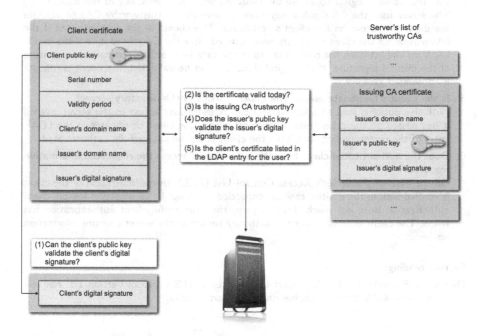

Fig. 8.33 TLS/SSL handshake – client authentication.

1. **Can the client's public key validate the digital signature of the client?**
   The server checks whether the digital signature of the client can be calculated from the public key in the client certificate. If successful, the server can then be sure that the public key purportedly belonging to the client does in fact correspond to its private key, from which the digital signature is generated. This confirms that the transmitted data has not been changed since the time of the signing. However, the connection between the public key and the **Distinguished Name** (**DN**) provided in the certificate of the authority CA cannot yet be determined. The possibility still remains that the certificate could have been generated by an unauthorized third party posing as the client. In order to correctly establish the connection between the public key and the DN the following additional steps need to be taken:

2. **Is the certificate still valid?**
   The server verifies the valid time span of the certificate. If the current request is outside of this time span, the authentication procedure is stopped. Otherwise, the server continues authentication.

3. **Is the issuer of the certificate trustworthy?**
   Every TLS/SSL-enabled server manages a list of trustworthy certificate issuers (Trusted Certificate Authorities, CA). On the basis of this list, it is decided which certificates are accepted by the server. Matching the unique name (**Distinguished Name, DN**) of the issuer CA, with a name from server's list of trustworthy CAs, the server continues authentication of the client. However, if the authority CA is not on the server's list, the client is not authenticated. An exception is the client who is capable of sending

a certificate chain containing a CA found on the server's list. The respective server administrator manages the list of trustworthy CAs.

4. **Can the issuer's digital signature be validated with the public key of the issuer?**
The server uses the CA's public key from its own list of trustworthy CAs to validate the digital signature on the client's certificate. The clients is not authenticated if the information on the client certificate was changed after the CA signed it or if the public key from the list does not correspond to the private key used to generate the signature of the client's signature. If the digital signature can be validated, the client certificate is considered genuine.

5. **Is the client's certificate among the LDAP entries in the directory server?**
With this optional step an administrator can refuse a user access to the server, even if it has passed all the other tests. If identical with those client certificates in the LDAP entry of the corresponding user, then the client could be successfully authenticated.

6. **Does the authenticated client have the authority to access the requested information resources?**
On the basis of the server's **Access Control List** (ACL), the authorization of the client is verified and, in the positive case, a connection is set up. If, for whatever reason, client authorization does not reach this last step, then successful client authentication has failed. The client does not have the authority to access the server's secure information resources.

Further reading:

Dierks, T., Rescorla, E.: The Transport Layer Security TLS Protocol Version 1.1. Request for Comments, 4346, Internet Engineering Task Force (2006)

Schematically, the protocols of the handshake sublayers in the TLS/SSL protocol are on a higher layer in the TCP/IP reference model, whereby they are virtually encapsulated by the **TLS/SSL record protocol**. Depending on whether the sending or receiving of data is concerned, the TLS/SSL record protocol makes various services available:

- On the **sender side** the TLS/SSL record protocol offers the following services:

  - Receipt of the data to be transmitted from the application,

  - Fragmentation of the message into easy to handle data blocks (up to 2^{14} bytes long. Various sets of data of the same type can be combined in such a block and transferred).

  - Optional data compression (the compression agreed upon by the communication partners must work loss-free and data length may never exceed 1,024 bytes),

  - Authentication of the sender and the transmitted data with the help of algorithms (hash functions). These generate a fingerprint (message digest) of the transmitted data, as defined in TLS/SSL using the procedures **MD5** and **SHA-1** as the **HMAC** algorithm. Its use guarantees the authenticity of the communication partners and the integrity of the transmitted data, and

  - Encryption of data and its delivery.

- On the **receiver side** the TLS/SSL record protocol makes the follow services available:
    - Receipt of messages from the transport layer protocol,
    - Encryption of the received data,
    - Verification of authenticated data,
    - Decompression of compressed data,
    - Reassembly of fragmented data into complete messages,
    - Authentication and
    - Forwarding of the data to the receiver.

8.6 Glossary

Address resolution: The translation of IP addresses into the corresponding hardware addresses of the network hardware. A host or a router uses address resolution if data within the same physical network is sent to another computer. Address resolution is always restricted to a particular network, i.e., a computer never resolves the addresses of a computer in another network. Vice versa, the fixed assignment of a hardware address into an IP address is known as **address binding**.

Address Resolution Protocol (ARP): An implemented protocol from the TCP/IP protocol family for address resolution of protocol addresses into hardware addresses. To determine a hardware address, the requesting computer sends the corresponding protocol address as an ARP message via broadcast. The requesting computer recognizes its own protocol address and is the only one to send its hardware address as an ARP message directly back to the requesting computer.

Application Programming Interface (API): An interface between the application and protocol software; the so-called application programming interface makes routines and data structures available for use and control of the communication between applications and protocol software.

Authentication: Proves the identity of a user or the integrity of a message. In authentication, certificates of a trustworthy instance are used for identity verification and digital signals are generated and sent with a message for checking its integrity.

Broadcasting: Term used for addressing all computers in a network simultaneously. A distinction is made in the Internet between a broadcast in the local network and a broadcast over all networks. If a message is to be forwarded to just one computer - which corresponds to standard addressing - the term **unicast** is used.

Certificate: Digital certificates are the electronic equivalent to a personal ID card. They assign their owner a unique **public key** and therefore a **digital signature**, which can only be generated with the corresponding private key. Certificates must be issued and signed by a trustworthy third party known as a certification authority.

Certificate Authority (CA): A **certificate authority** authenticates public keys of registered users with the help of certificates, according to the Internet standard RFC 1422. It is implemented in user identification. The public key of the user is digitally signed with the name of the user and the control details of the CA and issued in this form as a **certificate**.

Congestion control: If a sender is transmitting faster along a communication connection in a network than the receiver can process the received data, the data is then stored temporarily in a special buffer expressly for this purpose. Once the buffer is full the incoming data packets can no longer be secured and must be discarded - with resulting congestion. Because the transport layer only establishes and controls end-to-end connections there is no possibility of directly influencing the intermediate network systems along a connection. This is the reason that possible delays due to congestion have to indirectly be taken into account when determining the round-trip time between sender and receiver. Data flow between sender and receiver must likewise be throttled via flow control to counteract any arising congestion situations.

Connection-oriented/connectionless service: A distinction is generally made between **connection-oriented** and **connectionless** Internet services. Before the actual start of data transmission, connection-oriented services must establish a connection via fixed, negotiated exchanges in the network. The thus-established connection path is used for the duration of the entire communication. Connectionless services do not choose a fixed connection path in advance. The transmitted data packets are sent - independent of each other - on possibly different paths through the Internet.

Cryptography: The branch of computer science and mathematics concerned with the construction and evaluation of the key procedure. The aim of cryptology centers on protecting confidential information from being accessed by unauthorized third parties.

Data Encryption Standard (DES):   Symmetrical block encryption standard that was published in 1977 and updated in 1993 for commercial use. DES encodes blocks of 64 bits each with an equally long key (effectively 56 bits). The DES procedure is made up of a total of 19 rounds. The 16 inner rounds are controlled by the key. The DES procedure represents a 64 bit substitution encryption procedure and can be easily broken today by relatively simple means. A multiple implementation of DES is carried out using different keys for heightened security, e.g., **triple DES** (3DES).

Data integrity: While in cryptography there is no way to prevent data or messages from being changed by an unauthorized third party during transport, the changes can be made identifiable through the use of so-called hash functions that send a digital fingerprint along with the transmitted data.

Diffie-Hellman Procedure: The first, officially known, asymmetrical encryption procedure developed in 1976 by W. Diffie, M. Hellman and R. Merkle. Similar to the RSA procedure, Diffie-Hellman is based on a mathematical function whose reversal - particularly the problem of the discrete logarithm — is virtually impossible to calculate with a reasonable effort.

Digital signature: Is used to authenticate a document and consists of the digital fingerprint of the document encrypted with the private key of the originator.

Flow control: In a communication network, flow control prevents a faster sender from flooding a slower receiver with transmitted data and causing **congestion**. As a rule, the receiver has buffer storage where incoming data packets can be stored until subsequent further processing. Protocol mechanisms must be provided to prevent this intermediate storage from overflowing. This means the possibility for the receiver to inform the sender to wait until the buffer storage has been processed before sending more data packets.

Fragmentation/defragmentation: The length of data packets sent by a communication protocol in a **packet-switched network** is always limited below the application layer due to technical restrictions. If the length of the message to be sent is larger than the respective prescribed data packet length, the message is broken up into message pieces (**fragments**). These correspond to the prescribed length restrictions. Individual fragments are given a sequence number so that after transmission they may be reassembled correctly (**defragmented**) at the receiver. This is necessary as transmission order cannot be guaranteed in the Internet.

Gateway: Intermediate system in the network that is capable of connecting individual networks into a new system. Gateways allow communication between application programs on different end systems and are located in the application layer of the communication protocol model. They are thus able to translate different application protocols into each other.

Internet: The merging of multiple, mutually incompatible network types with each other. Appearing to the user as a homogeneous universal network, it allows all computers connected to a single network in this union the possibility to communicate transparently with every other host on the the the internet. An internet is not subject to limitations in terms of its expansion. The concept of **internetworking** is very flexible, with an unlimited extension of the internet possible at any time.

Internet Protocol (IP): Protocol on the network layer of the TCP/IP reference model. As one of the cornerstones of the Internet, IP is responsible for the global Internet, made up of many heterogeneous individual networks, appearing as a unified, homogeneous network. A standard addressing schema (**IP addresses**) offers worldwide unique computer identification. For this, IP provides a **connectionless packet-switching datagram service**, which works according to the **best effort** principle rather than fulfilling a service guarantee.

IP datagram: The data packets transmitted via the IP protocol are referred to as datagrams. This is because the IP protocol only provides a connectionless and unreliable service (**datagram service**).

IPv4 address: 32-bit binary address uniquely identifying a computer in the global Internet. This address is subdivided into four octets for better readability. These are interpreted as unsigned decimal integers in a binary code, each separated by a decimal point (e.g., 232.23.3.5). The IPv4 address is subdivided into two parts: the **address prefix** (**network ID**), which uniquely establishes worldwide the network where the addressed computer is located, and the **address suffix** (**host ID**), uniquely identifying the computer within its local network.

IPv6: The successor protocol standard of the **IPv4** Internet protocol, offering a considerably expanded functionality. The limited address space in IPv4, one of the major problems of the popular IP standard, was drastically increased in IPv6 from 32 bits to 128 bits.

Key: A message can be transmitted safely via an insecure medium if its contents remain hidden from unauthorized third parties. This is done with the help of an encryption procedure (**cipher**). The original message, the so-called **plain text**, is used for encryption with a **transformation function** contained in the encrypted message (**cipher text**). The transformation function for encryption can be parameterized via a key. The size of the key space is a measure for the difficulty of an unauthorized reversal of the transformation function.

Man-in-the-middle attack: An attack on a secure connection between two communication partners. The attacker intervenes between both ("man in the middle"), intercepts the communication and manipulates it - unnoticed by the communication partners.

Multicasting: A source transmits to a group of receivers simultaneously in a multicasting transmission. A 1:n-communication is involved. Multicasting is often used for the transmission of real-time multimedia data.

Network Address Translation (NAT): With a small number of public IPv4 addresses, NAT technology makes it possible to operate a much larger number of computers in a shared network and to manage them dynamically using the private IPv4 address space. The NAT-operated devices remain publicly reachable via the Internet, although they do not have their own public IP address and can only be addressed over an appropriate NAT gateway.

Port number: 16-bit long identification for a TCP connection that is always associated with a specific application program. The port numbers 0–255 are reserved for special TCP/IP applications (**well known ports**) and the port numbers 256–1,023 for special UNIX applications. The port numbers 1,024–56,535 can be used for the individual's own applications and are not subject to fixed assignment.

Privacy: Contents of a private message may only be known to the sender and receiver of the message. If an unauthorized third party "listens in" on a communication (**eaves-dropping**), confidentiality can no longer be guaranteed (**loss of privacy**).

Public key encryption: In the cryptographic procedure known as "public key," every communication partner has a pair of keys consisting of a so-called **public key** and a **secret**, **private key**. The public key is available to all participants with whom communication is desired. The participants wanting to communicate with the holder of the public key encrypt their message with its public key. A message encrypted this way can only be decoded by the public key holder with the help of the corresponding secret key held securely by the owner.

Public Key Infrastructure (PKI):   In implementing the public key encryption procedure, every participant is required to have a **key pair** consisting of a key accessible to everyone (**public key**) and a secret key to which only it has access (**private key**). To eliminate abuse, the assignment of the participant to its public key is confirmed by a trustworthy third party, the **Certificate Authority (CA)**, by means of a **certificate**. In order to be able to evaluate the security of a certificate, the rules as to how this certificate is created (**security policy**) must be made public. A PKI contains all of the organizational and technical measures required for a secure implementation of an asymmetrical key procedure for the encryption or that are necessary for the digital signature.

Router: A switching computer capable of connecting two or more subnets with each other. Routers work in the transport layer (IP layer) of the network and are capable of forwarding incoming data packets on the shortest route through the network based on their destination address.

Routing: There are often multiple intermediate systems (**routers**) along the path between the sender and the receiver in an internet. These handle the forwarding of transmitted data to the respective receiver. The determination of the correct path from sender to receiver is known as routing. Routers receive a transmitted data packet, evaluate its address information and forward it accordingly to the designated receiver.

RSA procedure: This is the most well-known asymmetrical encryption procedure and is named after its developers: Rivest, Shamir and Adleman. Just as Diffie-Hellman encryption, the RSA procedure works with two keys. One is a public key, available to everyone, and the other is a secret, private key. RSA is based on facts from number theory - the problem of prime factorization. A decryption with reasonable effort is not possible without knowledge of the secret, private key.

Secret key encryption: Oldest family of encryption procedures with the sender and receiver both using an identical secret key for the encryption and decryption of a message. A distinction is made between **block cipher**, where the message to be encrypted is segmented into blocks of a fixed length before its encryption, and **stream cipher**, where the encrypted message is viewed as a text stream. A one-time key of identical length is generated and encryption of the message is carried out character by character. Symmetrical encryption involves the problem of keeping the key exchange secret from third parties.

Service primitives: Abstract, implementation-independent processes for the use of a service on a certain level of the TCP/IP reference model, also called service elements. They define communication processes and can be used as abstract guidelines in defining communication interfaces. Only the data to be exchanged between communication partners is defined via service primitives and not how the process is carried out.

Socket: The TCP protocol provides a reliable connection between two end systems. For this purpose, sockets are defined at the endpoints of participating computers. They are made up of the IP address of the computer and a 16-bit long port number. These uniquely define this connection together with the corresponding equivalent of the communication partner. Via sockets so-called **service primitives** are available, which allow a command and control of the data transmission. Sockets associate incoming and outgoing buffer storage with the connections they have started.

Transmission Control Protocol (TCP): Protocol standard on the transport layer of the TCP/IP reference model. TCP provides a reliable, connectionless transport service upon which many Internet applications are based.

User Datagram Protocol (UDP): Protocol standard on the transport layer of the TCP/IP reference model. UDP provides a simple, non-guaranteed, connectionless transport service over which IP datagrams are sent via the IP protocol. The principle difference between IP and UDP is actually only that UDP is capable of managing port numbers, which allow applications on different computers to communicate with each other via the Internet.

Virtual connection: A connection between two end systems solely created by the installed software at the end systems. The actual connection network must therefore not provide any resources but only guarantee data transport. Because the connection is not present in real form across the network, rather its illusion created by the software installed at the end systems, the term virtual connection is used.

Chapter 9
Application Layer and Internet Applications

> *"All theory, dear friend, is grey,*
> *and green is the golden tree of life."*
> *– from Faust I, J.W. von Goethe (1749 – 1832)*

We have so far seen how data is transported from one computer to another over the global Internet, yet an interface is still missing where this basic technology can be used for special services and applications. Tasks that have now become an integral part of life, such as sending electronic mail or using interactive information resources on the World Wide Web, are all founded on a client/server-based interaction model that takes advantage of the possibilities offered by the Internet and its protocols. This chapter focuses on the application layer of the TCP/IP reference model and deals with a variety of services and applications located at this layer. Following a description of the client/server interaction model, directory and name services will be presented before turning to the most important Internet application besides the World Wide Web - electronic mail and the protocols that make it possible. Particular attention will be paid to multimedia applications, such as the popular media streaming, as well as a wide range of other applications.

9.1 Principles, Function and Overview

After having looked at all the necessary fundamentals of Internet communication, we will now turn our attention to the application layer of the TCP/IP reference model. It is in the layers below the application layer that all details of the data transfer are regulated and a reliable transportation service provided. Yet this goes unnoticed by the user. The interfaces he or she directly interacts with already offer completely functional applications, whether they be transferring data, sending and receiving email or accessing HTML pages. The connections and communication protocols of a network are necessary for survival on the Internet, yet the directly usable functionality is first provided by the highly sophisticated application software on top of these services and used to implement them.

C. Meinel and H. Sack, *Internetworking*, X.media.publishing,
DOI: 10.1007/978-3-642-35392-5_9, © Springer-Verlag Berlin Heidelberg 2013

The Internet can be compared to a telephone network. While the protocols responsible for delivering the infrastructures to make communication possible are necessary, the participants are equally important. Those who want to communicate via the telephone and make use of the wide variety of services offered, for example, fax, are essential for making the telephone a truly useful service.

Just as in telephone service, the participant expects interaction of the Internet application programs wishing to communicate. An application on one computer contacts an application on another remote computer to exchange the data it needs. Also applications not among those in the TCP/IP reference model, as well as those directly operated by the user, implement the transport services of the TCP/IP reference model with their own protocols. They need suitable, abstract interfaces, which the application layer of the TCP/IP reference model also provides. Therefore, the application layer provides an application-to application communication, while the transport layer, situated underneath, enables a host-to-host connection. Seen from the perspective of the operating system of the respective end system, one speaks of **processes** rather than applications. The application layer thereby installs a network-wide process-to-process communication. A sending process generates and sends messages over the network to a receiving process. The receiving process gets the transmitted message and potentially relays an answer back to the sender.

9.1.1 Internet Services and Application Layer Protocols

The term **Internet services** classifies those applications situated in the reference model above the TCP. These services use a transport protocol for communication over the Internet that is either UDP or the more complex TCP. Some services may even have the choice of using both protocols. Fig. 9.1 provides a summary of different Internet services. These are also specified in RFC 1123 explicitly as protocols of the application layer, with the transport protocols they use.

The following Internet services use the **Transmission Control Protocol** (**TCP**) in different ways:

- Simple Mail Transfer Protocol (SMTP),

- Hypertext Transfer Protocol (HTTP),

- Remote Procedure Call (RPC),

- Multipurpose Internet Mail Extension (MIME),

- File Transfer Protocol (FTP),

- Telecommunication Network Protocol (TELNET),

- Domain Name System (DNS),

- Border Gateway Protocol (BGP).

The **User Datagram Protocol** (**UDP**), in contrast, is used to carry out the following Internet services :

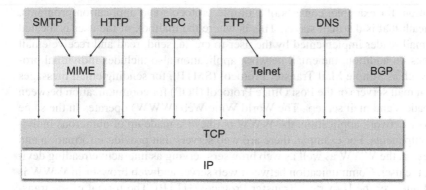

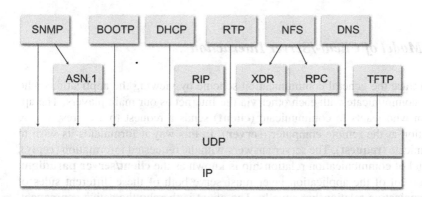

Fig. 9.1 An overview of various Internet services.

- Trivial File Transfer Protocol (TFTP),
- Domain Name System (DNS),
- Simple Network Management Protocol (SNMP) in connection with Abstract Syntax Notation 1 (ASN.1),
- Bootstrap Protocol (BOOTP),
- Dynamic Host Configuration Protocol (DHCP),
- Routing Information Protocol (RIP),
- Realtime Transfer Protocol (RTP),
- Remote Procedure Call (RPC),
- Network File System (NFS) in connection with External Data Representation (XDR).

It is necessary to distinguish between network applications and protocols of the application layer. Most of the time the protocols of the application layer each have a network application behind them. Therefore, the email system as a network application, for example, consists of multiple protocols and application programs, underneath that is the mail server. This is where the mailbox of the user is located or the mail reader implemented by the user to create, send, read and receive email messages. In addition, the email network application also includes individual protocols such as Simple Mail Transfer Protocol (SMTP), for sending email messages between mail servers or the Post Office Protocol (POP) for communication between mail readers and mail servers. The World Wide Web (WWW) operates in the same way. As a network application, the WWW is likewise made up of numerous individual components. For example, there are web servers that provide information and services via the WWW as well as web browsers serving as interactive/reading devices for the user. Communication between web server and web browser in WWW is carried out over the Hypertext Transfer Protocol (HTTP). The format of the transmitted document is determined via the Hypertext Markup Language (HTML).

9.1.2 Model of Client-/Server Interaction

We can trace the general communication scheme by viewing the applications who wish to communicate with each other via the Internet as our main players. The application who wants to communicate (**client**) sends a request to a corresponding application at the remote computer (**server**). In this way it formulates its wish to communicate (**request**). The server answers with the requested information (**reply**). This kind of communication relationship is known as the **client/server paradigm**. The protocol of the application layer must serve both of these different sides of a communication relationship equally. The client is thereby the active component, who initiates the communication relationship, and the server is the passive component. The server waits for a client to make a request (cf. Fig. 9.2). The information can flow in both directions: from the client to the server or from the server to the client. Looking at the WWW, we see that the user's web browser on the client side establishes the communication relationship. The web browser sends a query based on a specific information resource to a web server who establishes the communication relationship on the server side. The HTTP protocol must therefore be provided with client functionality on the web browser side and server functionality on the web server side.

It is completely possible that different services (servers) are offered at each computer. In each case, a special server program is necessary for their operation.

This makes sense and is very economical - a server program is not using up any resources while it waits for a client query. Yet at the same time, the clients need to have the possibility to be able to address specific server services at a remote computer. The tools provided by the protocols of the transport layer to enable this are the so-called **ports**. Every application program that can communicate over the Internet is assigned to a certain port, identified by a unique **port number** (cf. Chap. 8.1.3). This port number is assigned to the application program by the network transport software. Client queries are then forwarded by the network software via the speci-

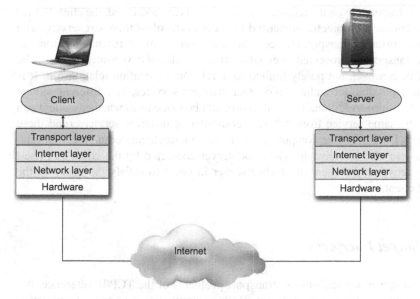

Fig. 9.2 Client and server communication over the Internet.

fied port number to the server. The client adds the specified port number from the server service to its query. It is then possible for the network software on the server side to forward the query to the appropriate server.

It rarely happens that a server processes only one client query at a specified time. Normally, there are multiple open queries from different clients simultaneously. These are all processed parallel without a client being forced to wait until the interaction with another client has been completed. Generally, every single client query is operated with a copy of the original server program. This makes it possible to take care of short queries very quickly, without having to wait for the conclusion of an extended query.

The copies of the server application for the parallel operation are generated dynamically, i.e., when required. The server generates a special, so-called **thread** (process, task) for every client query. Every server program is generally made up of two parts: a main thread, which is always active and waits for the incoming client inquires, and a set of dynamic threads. These generate a copy for each new incoming client query, these copies are processed parallel and independent of the main thread. At the same time the main thread continues to work regularly and receive client queries. For example, if communication is being carried out simultaneously with n clients, then altogether n+1 server threads are active: a main thread and n dynamic threads. So that every server thread can work correctly with its clients without confusion, the transport protocols identify every single connection based on a unique identifier. This uniquely specifies the respective client (e.g., client IP address) and the respective service (e.g., port number). This combination of client and server identifier supports the selection of the "correct" server in the parallel server operation. Depending on the requirements of the service addressed, the client/server services can use a **connection-oriented** or a **connectionless** transport service. The connection-oriented transport service is normally served by TCP, while the connectionless transport is processed over UDP. Internet client/server interaction in the Internet is, however, not purely limited to a 1:1 communication relationship. It is completely possible for a client to contact different services, i.e., different servers (1:n relationship). Additionally, the decision can be made as to whether a client can request the same service from different computer or different services (and these sometimes from the same computer). Transitive, hierarchical communication relationships are also conceivable. Thus, the server contacted by the client can itself become active as the client of another server in order to satisfy the query of the original client.

9.1.3 Socket Interfaces

Client and server applications use transport protocols of the TCP/IP reference model to communicate over the Internet. Both communication partners (application programs) must indicate what is to be done in each case with the received data - where and in what form the data is to be forwarded to the application program. The interface that uses an application program for communicating with the transport software is known as the **Application Programming Interface** (**API**). The API specifies all operations that an application can carry out in communication with the transport protocol and provides the necessary data structures. The **socket API** is the standard interface for the transport protocols of the TCP/IP reference model (cf. Chap. 8.1.3). It was developed in the '70s along with the operating system BSD-UNIX at the University of California, Berkeley. The widespread distribution and popularity of the UNIX operating system, which contained the communication protocols of the TCP/IP protocol family, played an important role in the dissemination of the free TCP/IP communication software. With its increasing use and porting on other operating systems, socket API became the standard interface for the communication of application programs over the Internet. In accordance with the client/-server paradigm, an Internet application is always made up of two processes on two different network participants (hosts). The two processes communicate by sending and receiving messages via their sockets. In this way, sockets form the central interface between the computer-bound processes and the communication infrastructure. Fig. 9.3 shows two processes communicating with each other over the Internet via their sockets.

For the developers of an application this means they have control of all activities on the application side of the socket, yet hardly any control of the socket transport layer, except for the choice of transport protocol implemented and control of several communication-related parameters.

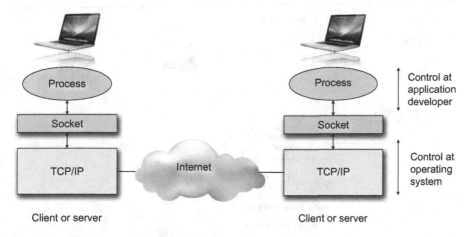

Fig. 9.3 Processes communicating with each other using sockets.

9.2 Name and Directory Services

To be able to access a specific service on a remote computer over the Internet, the application, i.e., the user, who is operating the application program must first identify the relevant computer by means of its worldwide unique IP address. Rarely does the user remember this 32-bit long IPv4 address or 128-bit long IPv6 address. The IPv4 address is comprised of four octets - four decimal numbers in the range between 0 and 255. The IPv6 address, in contrast, consists of 8 groups, each 16 bits - thus 16 octets. These are difficult to memorize, particularly when they occur in clustered form. We can be sure that every user who has accessed a remote computer over the Internet did not have to input the said IP address but could identify the computer via an easy to memorize **symbolic name**. Symbolic names providing address details in the World Wide Web can be found in, e.g., email addresses or Uniform Resource Locators (URL). Yet while symbolic names are less complicated for people to use in practical application - being much easier to remember by associative connection - the use of symbolic identifiers in computer processing remains a comparatively more challenging task. The simple reason being that 32 or 128 bit IP binary addresses can be represented more compactly and are easier to process.

The solution for this problem was found with the introduction of so-called **name services**. These carry out a standard translation of symbolic computer names into unique IPv4 / IPv6 addresses. Most of the time, the translation is carried out automatically and not recognized by the application or user. The actual IP address determined this way is then only used inside the computer by the protocol software for the exchange of IP datagrams (cf. Fig. 9.4).

In order to be able to access not only computers but also their resources, these must naturally also be uniquely addressable. A service that carries this out and specifies

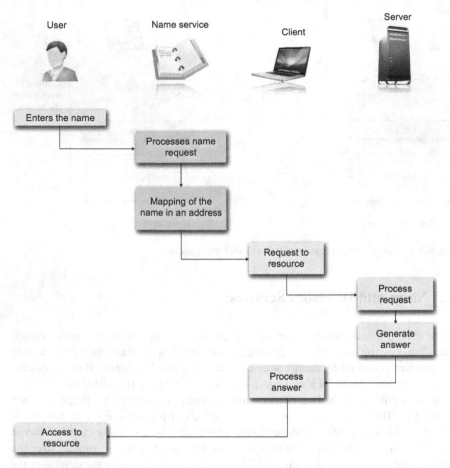

Fig. 9.4 Use of an Internet service via a name service.

resources uniquely, making them available over the Internet, is generally known as a **directory service**.

9.2.1 Domain Name System – DNS

While most people and most application programs use symbolic names for addressing remote computers in the Internet, for reasons of efficiency and uniqueness the network software only uses 32-bit or 128-bit long binary address.

In the early days of the Internet, at the time of the ARPANET, the allocation between symbolic computer names and their binary addresses was organized by means of a centrally stored file host.txt. This was to be continuously updated. Each night all computers in the ARPANET requested this file to update their own address assignment lists. Such a course of action was, however, only feasible as long as the network in question consisted of only a few hundred computers. With the explosive growth of the Internet it quickly became clear that central management of a name-address allocation no longer made sense to a large degree simply for the reason that the assignment tables had rapidly become too big to manage (cf. Fig. 9.5).

Because of the constantly increasing size of the central host file and the permanent changes and modifications taking place, it was necessary for every computer in the network to repeatedly copy this file anew from the central directory via the network into its local file system. The need of necessary bandwidth to keep the name allocation table consistent grew constantly and beyond all measure. Moreover, this flatly organized name space had yet another problem: any symbolic names assigned twice would have to be recognized and remedied at the cost of considerable administrative effort.

To solve this problem, the so-called **Domain Name System (DNS)** was developed by *Paul Mockapetris* (* 1948). It was first defined in 1983 as RFC 882, and RFC 883, and published as an Internet standard. In 1987, this specification was succeeded by RFCs 1034 and 1035. DNS works as a distributed database application, assigning symbolic names to binary IP addresses. To organize the distributed application as efficiently as possible, the provided address space is hierarchically structured and different sections are each given their own DNS server. These take over mapping in the corresponding address space section.

ADNS at a Glance

Thus, the use of symbolic names was introduced to replace the hard to remember binary computer addresses. At the same time, the weaknesses of the initially implemented central system of name resolution, via a centrally maintained name table, soon became apparent. The higher administrative effort as well as the necessity of frequent and complex synchronization with the central table led to the the idea of expanding and managing the original "flat" organized name space hierarchically. Additionally, in 1981 the idea of giving this name space a structured topology by consolidating the related sections into so-called **domains** was conceived. This idea was first defined in RFC 799, "Internet Name Domains." The developments and extensions that followed led to the RFCs 1034 and 1035 in 1987. These remain the basic standards of the domain name system up to today. Indeed, it was the hierarchically-structured organization of the name space that lead to the explosive growth of the Internet as we know it. Organization, administration and extension could be controlled decentrally and with run with little effort. Table 9.1 summarizes the most important standard of the domain name system.

One of the biggest challenges facing the development of the Domain Name System was that by the 1980's the Internet had already grown to include a significant number of participating computers. This meant that transition to a new name resolution would involve considerable effort. Furthermore, the developers knew that the new system would also need to stand the test of time. A further system changeover in the

TCP/IP Host Table Name System

With the beginning of the ARPANET, the precursor of today's Internet, it soon became clear that one could not only communicate with computers linked to each other via their (usually binary) addresses, but that symbolic, and therefore easy to memorize computer names, would have to be assigned to the addresses. At first the address consisted of a simple combination of computer number and port number over which the computer was connected to the communication network. However, as soon as the ARPANET expanded to include several dozen individual computers a solution had to be found to enable an allocation of the symbolic name with the computer address.

This name and address assignment was initially carried out in a small table called a *host table*. From 1971 this existed locally on every computer in the ARPANET. Naturally, host tables on all the network computers have to be identical and therefore need certain synchronization mechanisms. These were first defined in RFC 226. This synchronization was first conducted manually. The respective assignments were published in subsequent RFCs and the necessary changes of the host table files were done by hand.

Clearly, this solution was not ideal. Changes in the network topology could only be first carried out on other computers following publication of the corresponding RFCs. It was also possible that file errors by the network administrators could occur during the manual maintenance of the host table. Therefore, RFC 606 and RFC 608 proposed a centrally issued host table file. This should be automatically copied by the individual computers in the network. This principle was retained until the beginning of the 1980s. It was also adapted in the new TCP/IP network software and accordingly in the IP addresses used from then on in RFC 810 .

The **host table name system** is extremely simple, comprised for the most part of the maintained host table (text-) file on every participating computer. It is filed in the UNIX systems as /etc/hosts (under MS Windows as HOSTS, normally in the Windows main directory). In the hosts file, after several commentary lines, follow pairs made up of IP addresses and their symbolic accompanying names (cf. Fig. 9.6).

The translation of the symbolic names into addresses is very simple as well. Every computer reads the file in the main memory during system startup. Users may now use symbolic names instead of computer addresses. If a name is discovered in an expected address entry, which is located in the hosts file, then it is simply replaced by the associated IP address. **Name resolution** proceeds wholly locally.

This method is problematic only when changes occur in the network topology that are first carried out locally on a computer. To maintain global consistency it is necessary that all network applications are forwarded to a central instance that provides a global hosts file for all other network participants.

Further reading:

Karp, P. M. .: Standardization of host mnemonics. Request for Comments, 226, Internet Engineering Task Force (1971)

Feinler, E. J., Harrenstien, K., Su, Z., White, V.: DoD Internet host table specification. Request for Comments, 810, Internet Engineering Task Force (1982)

Fig. 9.5 TCP/IP host table name system.

```
# Host Database
# This file should contain the addresses and aliases
# for local hosts that share this file.
#
# Each line should take the form:
# <address>            <hostname>
#
127.0.0.1              localhost
141.89.225.120         www.hpi.uni-potsdam.de
62.50.45.35            www.springer.de
```

Fig. 9.6 Example of a TCP/IP host file.

Table 9.1 Important Internet standards for DNS.

RFC	Title
799	Internet Name Domains
1034	Domain Names: Concepts and Facilities
1035	Domain Names: Implementation Specification
1183	New DNS RR Definitions
1794	DNS Support for Load Balancing
1886	IPv6 DNS Extensions
1995	Incremental Zone Transfer in DNS
2136	Dynamic Updates in the Domain Name System
2181	Clarification in DNS Specification
2308	Negative Caching of DNS Queries
4033	DNS Security Introduction and Requirements
4034	Resource Records for the DNS Security Extensions
4035	Protocol Modifications for the DNS Security Extensions

future, due to the immense expansion of the Internet, would inevitably mean extreme effort and expense. For this reason, the conception of the Domain Name System placed great importance on the careful planning of design goals, always with a look to the future.

- **The creation of a global, scalable and consistent name space**
  The created name space should function worldwide; i.e., be capable of managing millions of computers simultaneously on a global scale. To do this it was necessary that a simple, if possible intuitively understandable, scheme be developed that would allow finding (and designating) a computer in a direct way. At the same time, it was essential to be able to to efficiently detect name duplication - even if the computers involved were located on different continents.

- **Local resources must also be managed locally**
  Inside a single network the responsible network administration must itself be in

a position to designate names for network-capable devices and computers. The necessity of having to register all the allocated names at a central point should be avoided. It is also not necessary that every local network be required to know every other computer and every network by name.

- **Conception of a distributed design in order to avoid bottlenecks**
  The idea of a central management would have to be avoided because of the high administrative costs involved. The domain name system is therefore hierarchically structured and works as a distributed database application, exchanging the necessary information required for the correct function via the Internet.

- **The highest universal application possible**
  The domain name system must support numerous functions in as many applications as possible. DNS should additionally be able to support different types of protocols including addressing variations.

The name space of the domain name system is conceived as a hierarchically structured tree that starts from a shared root and branches out into individual domains. These domains can themselves contain single computers or branch off further into sub-domains. The structure therefore also resembles the construction of a file system on a computer. An especially developed syntax specifies correct names, its terminology providing information about the structure of the name space from the root to the corresponding end device.

The domain name system has a special registration system whose administrative units correspond to the structure of the name space. A central organization thereby manages the highest level of the domain and further delegates registration and administrative queries to subsidiary organizations that represent the regional registration instance.

Special importance is placed on the conception of the mechanisms responsible for name resolution, i.e., for the translation of the symbolic names into IP addresses. The procedure implemented must take into account the distributed structure of the name space and the administrative instances it contains. At the same time, it must be able to take care of millions, and even billions, of computers. Scalability was an especially important design criteria in this regard. The administrative organizations of the respective domains or the secondary sub-domains of the name space keep so-called DNS servers. DNS servers are special programs responsible for queries regarding name resolution. Name resolution can either be taken care of directly by the DNS server or the query forwarded above or below the structure of the name space.

DNS is a central component of the Internet infrastructure today. A disturbance can result in significant costs, and a falsification of DNS data can be the starting point of attacks. The main goal of DNS attacks focuses on a targeted manipulation of DNS World Wide Web participants with the intention of steering them to false websites. It is then hoped to gain access to passwords, identification data or credit card numbers. Because of this, DNS was expanded in 2005 to include additional security functions, defined in RFC 2535 as **Domain Name System Security Extensions** (**DNSSEC**). With the help of DNSSEC, the authentication and data integrity of DNS transactions can be guaranteed and the possibility of manipulation by unauthorized third parties ruled out.

The DNS Address Space

The DNS address space is hierarchically structured. It is analogous to a postal address with its division into name, street, postal code, city and country. Through its hierarchical structure a worldwide unique address identification is ensured. In a similar way, the DNS address space is subdivided into so-called **domains**. Read in reverse a postal address provides a simple directive for sorting letters: first in countries, which are then subdivided into cities, and further into a multitude of streets where the individual members of the postal network live.

The Domain Name System analogously divides the address space in the Internet conceptually first in several hundred **top-level domains** (also called first-level domains), which are each assigned a large number of computer addresses branching off from the **root domain**. Every top-level domain is divided in further **domains** and these in **sub-domains** (cf. Fig. 9.7).

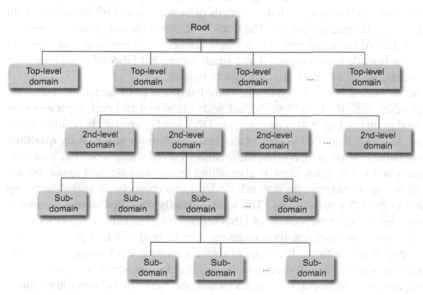

Fig. 9.7 Hierarchical division of the DNS address space.

Broken down this way, the entire DNS address space of the Internet can be represented in a tree form. The subdivision of the DNS name space is structured from the most general to the most specific. Using the terminology for graphs, which includes the tree structure as a special type, the following correspondences in the DNS name space can be identified:

- **Node**
  Every node is a structural element within the DNS name space, extending from the root nodes of the tree to the leaf nodes.

- **Root**
  The root conceptually represents the end of the DNS name structure, i.e., there is no further structure above the root. In the DNS, the root domain contains the entire DNS structure.

- **Branch**
  A branch represents a coherent substructure of the DNS name space, consisting of a domain and all of the sub-domains it contains.

- **Leaf**
  Leaf nodes form the root at the opposite end of the DNS name structure. Below the leaf nodes there are no further substructures.

A network end device in the Internet, participating in the DNS, is always located in one of the leaf nodes of the tree that stretches out over the DNS name space. Addressing proceeds in accordance with the path along which the relevant device can be reached from the root. Every node in the tree is distinguished by its name identifier, the **DNS label**, which names the corresponding domain or sub-domain. A DNS label can theoretically have a length of between 0 and 63 characters, with lengths of 1–20 characters normal. The DNS label allows the use of letters, numbers and hyphens. At the same time, DNS labels are not "case sensitive," meaning no distinction is made between capital and small letters. No DNS label may be used twice in its domain, i.e., each must be locally unique.

In order to name a network end device completely, the DNS labels are connected to each other along the path from the leaf node to the root and each separated with a decimal point (cf. Fig. 9.8). The complete DNS label character chain, identifying the end device, is identified as the correct and full **domain name (fully qualified domain name)**. A complete domain name cannot exceed a length of 255 characters. A correct and full domain name is also called an absolute domain name, because it allows the corresponding network device to be uniquely identified from any location in the DNS name space. The term **partially qualified domain name** is similarly used to describe a chain of DNS labels along the path in the DNS name space that does not reach the root, meaning is to be read within a particular context. In accordance with the hierarchical formation of the DNS name space, there exists a further hierarchically structured decentralized administration responsible for the assignment of the respective domain names: the **Network Information Center (NIC)**. Within every domain, or sub-domain, the responsible administration units must guarantee the unique state of the name identifier. The **IANA (Internet Assigned Numbers Authority)** is the highest authority for the assignment of names and numbers in the Internet. The administration of general top-level domains is transferred by the IANA to **ICANN (Internet Corporation for Assigned Names and Numbers)**. ICANN coordinates, monitors and delegates the registration tasks to subordinate organizations.

Top-level domains can have two different forms:

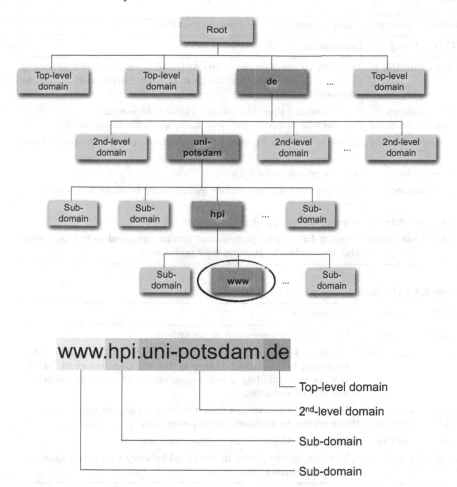

Fig. 9.8 Example of a domain name.

- Generic designations (**Generic Top-Level Domains, gTLD**), which identify different types of institutions and organizations. gTLDs have DNS labels with three or more letters. A distinction is made between two groups of gTLDs:

  - **unsponsored Top-Level Domains (uTLD):**
    These top-level domains work directly in accordance with the guidelines published by ICANN for the global Internet community (cf. Tab. 9.2).

  - **sponsored Top-Level Domains (sTLD):**
    These top-level domains are proposed and financed by private companies or organizations. They establish the user guidelines themselves for participation in the corresponding TLD (cf. Tab. 9.3).

Table 9.2 Examples of unsponsored Top-Level Domains (uTLDs).

TLD	Meaing	Expaination
.arpa	ARPANET	Originally intended for use in the ARPANET, now used as *Address and Routing Parameter Area* and designated by the IANA as infrastructure domain
.biz	business	For commercial use, theoretically available to everyone
.com	commercial	Originally for company use only, today available to everyone
.info	information	Originally intended for information providers, today available to everyone
.name	name	Only for private individuals
.net	network	Originally intended for network management facilities, today available to everyone
.org	organization	Intended for non-profit organizations
.pro	professional	Intended for specific professional groups (qualified skilled personnel who are certified to practice their profession)

Table 9.3 Examples of sponsored Top-Level Domains (sTLDs).

TLD	Meaning	Explanation
.aero	aeronautics	Use limited to aeronautic organizations
.edu	educational	Use limited to educational institutions, since 2001 limited to those confirmed by accredited agency of the US Department of Education. Notwithstanding a few exceptions, used solely by American universities and colleges.
.gov	government	Use limited to American government agencies and authorities.
.int	international	Use limited to multinational organizations
.mil	military	Use limited to American military organizations
.travel	travel	Intended for companies in the travel industry e.g., travel agencies, airline companies, etc.

- Country codes (**country code Top-Level Domains ccTLD**) provide a top-level domain for every country. There are more than 200 ccTLDs. Besides representing each country, there are also several ccTLDs for independent territories, which are usually geographically separated from their mother country. The codes for the country top-level domains are defined in ISO 3166. Here are just a few examples:

  - .at: Austria,
  - .br: Brazil,
  - .de: Germany,
  - .eu: Europe,
  - .fr: France,
  - .it: Italy,

- .uk: Great Britain,
- .za: South Africa.

Although not regulated, organizations in the USA are mostly found in a generic top-level domain. Nearly all organizations outside of the US belong to the top-level domains of their own country. A few exceptions to this rule are summarized in Fig. 9.9.

Exceptions and Special Cases of Country-Specific Top-Level Domains

For country-specific top-level domains, the original regulations specified that every country be assigned a two character long ISO 3166 country code. But there are also exceptions and special cases that exist, just a few of which are outlined as follows.

- Great Britain uses the TLD .uk instead of the abbreviation intended by ISO 3166 .gb
- In addition to the ccTLD .us, the US also uses the TLDs .mil, .edu und .gov
- Although not an independent state, the European Union uses the ccTLD .eu.
- The ccTLD .dd was intended for the former German Democratic Republic, but subsequently used by the universities of Jena and Dresden internally.
- The ccTLD .cs was used in the past by Czechoslovakia and .yu by the former Yugoslavia. After the disintegration into single states, the original codes were replaced by new ones.
- The ccTLD .za for Zaire was removed in 2004 and replaced with .cd, the code for the Democratic Republic of Congo.
- Besides the ccTLD .ru, the abbreviation .su, for the former Soviet Union, continues to be used in Russia.
- Smaller countries often market their domains with a liberal procurement policy and strong advertising. Already in 1998, Tonga became one of the first countries to put its domain .to on the market, thereby allowing the possibility for names such as come.to or go.to The case of the small Pacific Island group, Tuvalu became well-known. Their TLD .tv is especially popular with media (television) companies. With profits from the sale of their TLD, the construction of an IT infrastructure for state-run institutions could be financed as well as admission fees to the United Nations. In a similar way, the ccTLD .fm of the Federated States of Micronesia is used in the broadcasting sector, as is the Isle of Man's .im used for instant messaging services, .tk (Tokelau) for telecommunication companies, .dj (Djibouti) for disc jockeys or .ag (Antigua and Barbados) for "Aktiengesellschaften" (corporations, abbreviated AG in German).

Further reading:

Mueller, M.: Ruling the root: Internet governance and the taming of cyberspace, The MIT Press, Cambridge, MA, USA (2009)

Fig. 9.9 Exceptions and special cases of country-specific top-level domains.

The top-level domains are assigned centrally from ICANN (Internet Corporation for Assigned Names and Numbers).

The management of a top-level domain and allocation of domains below the top-level is carried out by the responsible registration authority ((Domain Name Registry).

For example, DENIC[1] (Deutsches Network Information Center) is responsible for the top-level domains .de. It generally applies that a new domain can only be set up with the approval of the next higher domain. In this way, every organization that has a domain can be completely free to set up a hierarchy itself within this domain, made up of secondary sub-domains. A company can, for instance, create sub-domains for every department so that the company structure is reflected in the DNS address space. For every domain, there exists a tabular classification of the domain names allocated in that particular domain, potentially also together with the corresponding IP addresses. This distributed data base is also designated a **DNS Name Database**. Just as the entire DNS name space is hierarchically subdivided, on the corresponding administrative computers - the so-called **DNS servers** - there are local DNS name databases in the form of so-called DNS resource records.

DNS Resource Records

Every single domain, regardless of whether it is a top-level domain or just on one computer, has access to a set of so-called **resource records**. Besides basis resource records, which contain only the IP address of one computer, there exist a large number of resource records containing various information, such as the name of the DNS server responsible for this domain, name aliases or additional descriptive information about the computing platform and operating system (cf. Fig. 9.10). If a DNS server receives an query with a domain name, it gives the requesting DNS client the number of resource records associated with this domain name. A resource record always consists of a single line of text containing five fields (cf. Fig. 9.10):

- **Name**
  The complete, valid domain name of the computer/node to which this particular resource record belongs. The length of this field is variable. As a rule there exist many entries for a single domain and every copy of the DNS database contains information about many different domains. The order in which the entries are written for a domain is irrelevant. If a query is received, all entries that apply to the domain are sent back in response.

- **Type**
  2-byte long type specification of the resource record in numerical form. This field gives an indication of the data format of the resource record and its purpose.

- **Class**
  2-byte long coded specification referring to the category of the entry (Internet or non-Internet). For the Internet - the normal class for DNS entries - the abbreviation "IN" is used.

- **Time to Live (TTL)**
  4-byte long specification of the length of time in seconds that is valid for an

---

[1] http://www.denic.de/

entry. Statistical entries contain, e.g., the value 86400 (length of time of a day in seconds). Volatile entries, in contrast, contain a lower value, e.g., 60 (1 minute).

- **RDATA**
  Additional specific resource records information, e.g., an IP address or a computer name. This field has a variable length.

- **RDLENGTH**
  2-byte long length specification for the RDATA field.

DNS Resource Records

There exist a whole series of different DNS resource records:

- **SOA** (Start of Authority): The SOA resource record indicates the beginning of a DNS zone and supplies important information about it. It contains the names of the primary information sources of the name server, the email address of the administrator in question, a unique serial number as well as multiple flags and timeout information.

- **A** (Address): The A resource record supplies the 32 bit IPv4 address for the specified computer.

- **AAAA** (Address): The AAAA resource record supplies the 128 bit IPv6 address for the specified computer.

- **MX** (Mail Exchange): The MX resource record supplies the name of the domain accepted by emails for the domain specified.

- **NS** (Name Server): The NS resource record supplies the name of the name server responsible for that domain. Every DNS zone must have at least one NS resource record that references the responsible primary DNS server. The primary DNS server must likewise have a valid A resource record.

- **CNAME** (Canonical Name): Allows the definition of an alias, which can be used in place of the DNS name. This resource record is often used to conceal changes in the internal structure of the DNS zone from the outside. This means that while an unchanging alias is used externally, it is still possible to change the assignment on a computer internally.

- **PTR** (Pointer): The resource record assigns a name to an IPv4 address.

- **HINFO** (Host Information): The resource record defines the computer concealed behind the resource record. The information can describe the implemented operating system or the CPU of the computer.

- **TXT** (Text): Free text allowing the possibility for a domain to specify further information.

Example:
```
pc063.hpi.uni-potsdam.de 86400 IN A      172.16.31.31
pc063.hpi.uni-potsdam.de 86400 IN AAAA   2001:0638:0807:020d::4bfd
pc063.hpi.uni-potsdam.de 86400 IN HINFO  PC LINUX
pc063.hpi.uni-potsdam.de 86400 IN TXT    Dr. H. Sack
pc063.hpi.uni-potsdam.de 86400 IN CNAME  hs1.hpi.uni-potsdam.de
```

Fig. 9.10 DNS - resource record

DNS Name Servers

Theoretically, a single computer containing a table with all DNS resource records could centrally serve the entire Internet. However, with the number of existing Internet hosts - which in the meantime amount to a good billion - this type of a central name server would be permanently overloaded. This is why from the very beginning a decentralized solution was implemented for DNS. The entire DNS name space is divided into numerous non-overlapping **DNS zones**. These zones always comprise entire branches of the DNS address space. There is a name server at the root of the subtree tree representing the DNS authority for the relevant zone. Application programs are designated **name servers** who answer queries about the DNS name space. Normally, the computer on which these programs run is also called a name server. A distinction is made between authoritative and non-authoritative name servers. An **authoritative name server** is responsible for a DNS zone. Its information about the zone can be considered certain. For every zone there exists at least one authoritative server - the **primary name server**. They are each listed in the SOA resource record. In order to assure a certain degree of redundancy and to share as equally as possible the accrued load of queries, authoritative name servers are usually operated as server clusters. The zone data is identical on one or more of the subordinate secondary name servers, which get their information from the primary name servers. A **non-authoritative name server** obtains its information from other name servers via one zone. Therefore, its information is not considered secured.

Primary name servers at the top of the DNS address hierarchy, i.e., in the top-level domains, are called **root servers**. A root server does not have the resource records of all computers set up in the top-level domains, but only information about how the name servers of the top-level domains can be reached. These then provide access to further information necessary for address assignment. For example, the root server of the .de domain does not know every single computer at Hasso Plattner Institute, but it can provide information about reaching the name server who can competently handle queries to the domain hpi-web.de. All DNS name servers are linked with each other to form a coherent overall system. This means every name server has the information necessary to reach a root server. It also has the information to reach those name servers responsible for the DNS address space in the hierarchy level below that particular name server.

So that a name server is able to find information about other parts of the name space located outside of its own DNS zones, it can use the following strategies:

- **Delegation**
  Normally, parts of the DNS name space of a domain are outsourced to sub-domains that each have their own name servers. The domain name server knows the responsible name server for the sub-domain from its zone data. It delegates queries to this sub-domain at one of the name servers located there.

- **Forwarding**
  If the requested DNS name space is located outside of its own domain, the query is forwarded to one of the previously established name servers.

- **Root server resolution**
  If forwarding cannot take place because the corresponding name server has not been previously established or cannot be reached, then the root servers are interviewed. Names and IP addresses of the root servers are statically stored on the name server. At this time, 13 root servers (server A to M) exist. For reasons of efficiency the root servers do not return a complete name resolution, rather determine it iteratively, meaning they answer with a reference to a responsible name server who takes over the further name resolution.

DNS name resolution

To implement name resolution, the respective application forwards the query to a so-called **resolver**. Resolvers are simply structured programs that are installed at the computer of a DNS participant. Their task consists of accessing information from name servers. As a DNS client, the resolver forms a bridge between an application and the name server. To this end, the resolver takes over the application query supplementing it, if necessary, to form a complete domain name. It then transmits it to an assigned name server. The working method of a resolver can proceed iteratively or recursively:

- **Iterative name resolution**
  In iterative name resolution, the resolver receives an answer from a name server with either the resource record desired or a reference to another name server designated as the next to be queried. In this way, the resolver works its way from name server to name server through the DNS name space until it gets the information it wants. This is then delivered to the calling program.

- **Recursive name resolution**
  In recursive name resolution, the resolver sends a recursive query to the name server assigned to it. If the desired information is not in the area of responsibility of the queried name server, then the name server must contact other name servers for as long as it takes until it gets the answer it is looking for.

Resolvers implemented on normal Internet end devices work solely according to the principle of recursive name resolution. In contrast, resolvers on name servers work according to the iterative method. Popular programs for processing name resolution are e.g., nslookup[2] or dig.
Fig. 9.11 shows an example of name resolution with the help of nslookup and dig. In both cases, the complete domain name www.hpi-net.de is requested and in each case two different resource records of type A with the associated IPv4 addresses are delivered. With dig the requested name server with the IPv4 address 192.168.2.1 can be identified as well.
DNS queries from the resolver to the name server are normally carried out via UDP port 53. The allowed length of the DNS UDP data packet is 512 bytes. Exceptions to this rule are dealt with in RFC 2671 (Extension Mechanisms for DNS, EDNS).

---

[2] http://network-tools.com/nslook/

Example of Name Resolution with nslookup and dig

```
> nslookup −q=any www.hpi−net.de

Server:          192.168.2.1
Address:         192.168.2.1#53

Non−authoritative answer:
Name:            www.hpi−net.de
Address:         80.156.86.78
Name:            www.hpi−net.de
Address:         62.157.140.133


> dig hpi−net.de

; <<>> DiG 9.6.0−APPLE−P2 <<>> hpi−net.de
;; global options: +cmd
;; Got answer:
;; −>>HEADER<<− opcode:QUERY, status:NOERROR, id:60465
;; flags:qr rd ra; QUERY:1, ANSWER:2, AUTHORITY:0, ADDITIONAL:0

;; QUESTION SECTION:
;hpi−net.de.                        IN       A

;; ANSWER SECTION:
hpi−net.de.                0        IN       A       80.156.86.78
hpi−net.de.                0        IN       A       62.157.140.133

;; Query time: 72 msec
;; SERVER: 192.168.2.1#53(192.168.2.1)
;; WHEN: Sat May  7 12:59:21 2011
;; MSG SIZE  rcvd: 60
```

Fig. 9.11 Example of name resolution with nslookup and dig.

If EDNS is not implemented, then excessively long answers are simply cut off after the maximum length has been reached. These are made identifiable by the setting of a truncated flag. In the case when the querying resolver is not satisfied with the truncated answer it can repeat the query via TCP port 53. It is then not subject to prescribed length restrictions (cf. Fig. 9.12).

If name resolution takes place on the user's side because of a query from a local application program, generally both recursive and iterative name resolution are involved in the process. Fig. 9.13 shows the principle procedure of the DNS client/server interaction in the translation of a domain address into an IP address, described as follows:

1. The client asks the resolver for a name resolution of the domain name
   www.hpi.uni-potsdam.de.

Extension Mechanisms for DNS (EDNS)

The term EDNS (Extended DNS) describes different extensions of the domain name system. These involve the transport of DNS information via the UDP protocol. As time went on, DNS, which was developed in the 80s, was equipped with a number of additional functions. The original DNS mechanisms to control the message format, via flags, return codes and label types, were not longer sufficient to cover all occurring functions and situations. In the same way, the length restrictions of the DNS data packets sent via UDP of 512 bytes led to problems - excessively long information was simply cut off.

EDNS expanded the DNS message format, as defined in RFC 2671. Besides the already existing DNS message types, a special pseudo resource record was introduced: the **OPT resource record**. The introduction of a OPT resource record is sufficient to identify a DNS message as an EDNS message. 16 additional flags for control and an 8 byte extension of the answer code is possible. The entire length of the UDP data packets used for transport and a version number are also indicated in the OPT resource record.

An OPT resource record issued with the help of the DNS utility dig:

```
;; OPT PSEUDOSECTION:
; EDNS: version: 0, flags: do; udp: 4096
```

EDNS is likewise a requirement for operation of the DNS Security Extensions (DNSS-EC). In practical application the use of overly long EDNS messages can lead to problems as the original length restrictions of 512 bytes need not be adhered to. Firewalls are based on a maximum length of the DNS message of 512 bytes and therefore often do not allow longer DNS messages to pass through.

Further reading:

Vixie, P.: Extension Mechanisms for DNS (EDNS0), Request for Comments 2671, Internet Engineering Task Force (1999)

Fig. 9.12 Extension mechanisms for DNS (EDNS).

2. The resolver (DNS client) contacts the local DNS server and requests an IP address www.hpi.uni-potsdam.de.

3. The local server checks whether the sought-after name is located in the sub-domain for which it has DNS authority. If this is the case, it translates the requested name into the associated IP address, supplementing the answer to the requesting computer to include the sought-after IP address. But if the queried name server is not capable of resolving the requested name completely, it checks the type of interaction the requesting resolver wishes. If the resolver demands a complete address translation (recursive resolution), the name server contacts further name servers who are capable of resolving the address completely. It then transmits the recursively determined IP address to the requesting resolver. The local DNS server sends the request as an iterative query further to a root DNS server as the current request cannot be answered locally.

4. The root DNS server is also not able to resolve the requested name and sends the address of the asked for top-level domain name .de back to the authorized DNS server.

5. The local DNS server contacts the DNS server for.de, as provided by the root server, and asks for the www.hpi.uni-potsdam.de IP address.

6. The DNS .de server is not able to resolve the name requested and therefore sends the address of the asked for domain name uni-potsdam.de back to the authorized DNS server.

7. The local DNS server contacts the DNS server given in the answer for uni-potsdam.de and asks for the IP address of the original request www.hpi.uni-potsdam.de.

8. The DNS server for uni-potsdam.de is also not able to resolve the requested name and sends the address for the requested domain name hpi.uni-potsdam.de back to the authorized DNS server.

9. The local DNS server contacts the DNS server given in the answer for hpi.uni-potsdam.de and asks for the IP address of the original request www.hpi.uni-potsdam.de.

10. The DNS server responsible for hpi.uni-potsdam.de is able to resolve the requested name www.hpi.uni-potsdam.de and sends back the associated IP address.

11. The answer of the target system's DNS server is passed back to the local DNS client (resolver).

12. The resolver gives the IP address back to the originally requesting application.

13. Now the client's actual request can be sent to www.hpi.uni-potsdam.de by way of the IP address it has received.

So that this procedure functions, every Internet-connected computer must at least have access to the address of a local name server. It must be indicated when the network software is configured as it cannot be determined automatically. Furthermore, every name server must have the address of the name server one level above it in the DNS address hierarchy, or of the root DNS server.

DNS Caching

The outlay involved in DNS name resolution can grow excessively if it is always necessary to involve the associated root server in the resolution of a non-local domain name. Name servers use a **caching mechanism** for this reason to keep down the cost and effort of name resolution. In DNS name resolution the well-known computer science principle of **locality (locality of reference)** can also be used to a great advantage. For example, it is highly probable that in retrieving a website by means of a browser, further websites of the same web server will be requested next by the user. If it would be necessary to carry out a new DNS name resolution in every case the result would be enormous redundancy. It is possible to avoid this by caching already completed DNS name resolutions. All non-local IP addresses determined by way of recursive address resolution are stored intermediately in the local name server as well as in the requesting resolver (DNS client) for a certain period of time. They are then available in the case of a new request. However, the lifetime of these non-local IP addresses in the cache is limited. Changes in foreign address space

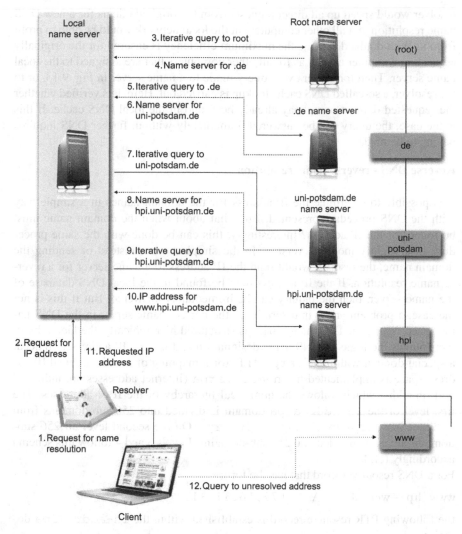

Fig. 9.13 Procedure of the DNS client/server for the resolution of a domain name into an IP address.

are not automatically passed on to the local name server. The DNS resource record therefore contains a field indicating the lifetime of the entry in question. After its lifetime expires the entry is discarded, and if there is a new query it must be recursively determined. The redundant caching of the DNS entry in both the local name server and in the requesting resolver makes sense. If storage were only to take place at the local name server with every new DNS name resolution of the same name, it would be necessary to carry out a new communication between the resolver and the name server. The end result would be the creation of an unnecessary load on network resources and a loss of valuable time. Even if a caching exclusively at the resolver would speed up all direct requests from the original clients for a new DNS name resolution, not all other computers in the local network would be able to profit from it. With doubled caching the maximum efficiency is ensured for the originally requesting computer as well as for all other computers that are assigned to the local name server. Therefore, after every query made to a name server in Fig. 9.13, or to the resolver, a so-called DNS cache lookup is carried out. Here, it is verified whether the requested domain name may already be found in the local DNS cache. If this is the case, the query can be answered immediately without further DNS inquires having to be made.

Reverse DNS - reverse name resolution

It is possible to determine the IP address for the domain names in a simple way with the DNS procedure presented. But what about when the domain name must be found for the IP address? Interestingly, this can be done with the same procedure in a slightly modified version. In the simplest case, instead of sending the domain name, the resolver would send the IP address to a name server for a reverse name resolution. If the IP address may be found in the local DNS database of the name server, then the inquiry can be immediately answered. But if this is not the case, a problem arises in determining the correct name server in the DNS name space. While the DNS name space is structured hierarchically, the hierarchy in fact follows the logic of DNS name designation and not the IP addresses. Indeed, a special domain within the `.arpa` TLD for a mapping of the hierarchical IP address space is implemented in `in-addr.arpa` (Internet addresses are indicated with in-addr), which follows the numerical hierarchy of the IPv4-addresses. The first level of the `in-addr.arpa` domain is divided into 256 sub-domains from `0.in-addr.arpa` to `255.in-addr.arpa` On the second level all 256 sub-domains are again divided into 256 sub-domains. Levels 3 and 4 follow this schema accordingly (cf. Fig. 9.14).

For a DNS resource record that has the form

www.hpi−web.de   A   172.16.31.31.

the following PTR resource record is established within the `in-addr.arpa` domain:

31.31.16.172.in−addr.arpa   PTR   www.hpi−web.de

In contrast to the A resource record, the IP address is given first and then a domain name. Now a reverse DNS name resolution of an IP address can be carried out with the resolver requesting a normal DNS name resolution for
31.31.16.172.in−addr.arpa.
The original IP address must in this case merely be turned around as DNS name resolution always proceeds from the most general to the most specific.

DNS message format

The DNS name resolution proceeds according to the client/server principle. The DNS client places a request for name resolution and the DNS server returns the

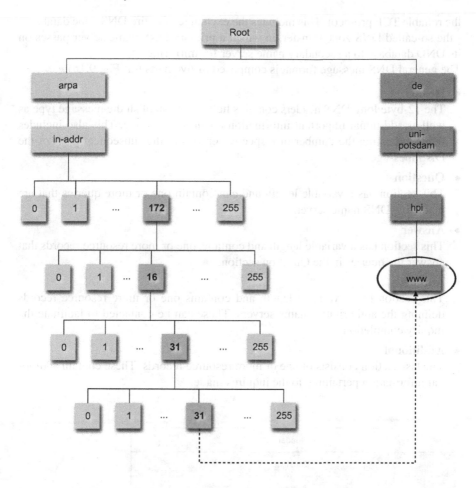

Fig. 9.14 The in-addr.arpa domain for reverse DNS name resolution.

corresponding answer. This is also the case in recursive name resolution in which a queried name server initially contacts further name servers before answering a client request. DNS works together with the transport protocols UDP and TCP. n general, the DNS client query is implemented via UPD as less administrative overhead is involved and the process is faster. As already mentioned, DNS UDP messages are limited to a length of 512 bytes. If they are longer, then they are simply cut off, unless EDNS is used in place of DNS. UDP only implements an unreliable service. This means that the DNS client cannot be sure the query it has sent actually reaches the DNS server. For this reason, the DNS client must keep track of its queries and repeat them after a specific cycle has been completed. Normally, there is also a fixed waiting period of 5 seconds to prevent the possibility of excessive DNS data traffic. However, several DNS administrative operations require implementation of the reliable TCP protocol. This includes the exchange of entire DNS zone databases - the so-called DNS zone transfer, in which a primary DNS name server passes on its DND database to a secondary name server for mirroring.

The general DNS message format is composed of five parts (cf. Fig. 9.15):

- **Header**
  The 12-byte long DNS headers contains fields that establish the message type as well as additional important information about the message. This also includes fields indicating the number of respective entries in the subsequent parts of the DNS message.

- **Question**
  This section has a variable length and can contain one or more queries that are made to a DNS name server.

- **Answer**
  This section has a variable length and contains one or more resource records that answer the queries in the Question section.

- **Authority**
  This section has a variable length and contains one or more resource records defining the authoritative name servers. These can be contacted to facilitate the inquiry completion.

- **Additional**
  The last section consists of one or more resource records. These contain additional information pertaining to the inquires made.

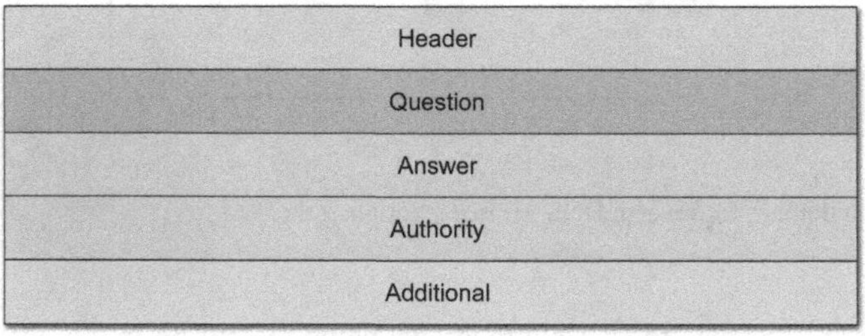

| Header |
| Question |
| Answer |
| Authority |
| Additional |

Fig. 9.15 General DNS message format.

The header of the DNS message consists of the following fields (cf. Fig. 9.16):

- **ID**
  16-bit long identification field that assigns the requesting DNS client. In this way a request or the answer to a certain query can be identified.

- **Flags and Codes Section**
  This 16-bit long bit sequence contains flags for control of the DNS queries as well as the answer codes delivered in return by the server:

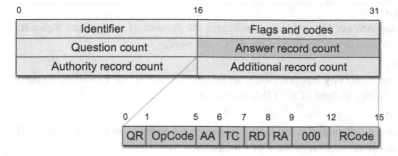

Fig. 9.16 Format of the DNS header.

- **QR**
  Query/Response flag, a distinction is made between the DNS query (QR=0) and server answer (QR=1).

- **OpCode**
  4-bit long code to establish the type of DNS query. This field is taken over unchanged in the DNS answer.

- **AA**
  Authoritative Answer flag indicates whether the server's transmitted answer is from an authoritative name server (AA=1) or not.

- **TC**
  Truncation flag indicates whether a messages transmitted via UDP should be shortened for reasons of exceeding the maximum allowed length of 512 bytes.

- **RD**
  Recursion Desired flag indicates whether the name server's query about name resolution should proceed in a recursive (RD=1) or in an iterative way (RD=0).

- **RA**
  Recursion Available flag indicates whether the name server queried about name resolution supports a recursive name resolution (RA=1).

- **Z**
  3-bit long reserved area that is set to 0.

- **RCode**
  4-bit long Response Code indicates whether the server response could be successfully answered or not. If the response code is zero (RCode=0), this means that it was possible to find a correct answer. Otherwise, the response code gives information about the type of error that occurred.

- **QCount**
  2-byte long Question Count indicates the number of inquires in the question section of the DNS message.

- **ANCount**

  2-byte long Answer Record Count indicates the number of transmitted Resource Records in the answer section of the DNS message.

- **AUCount**

  2-byte long Authority Record Count indicates the number of Resource Records in the Authority section of the DNS message.

- **ARCount**

  2-byte long Additional Record Count indicates the number of Resource Records in the Additional section of the DNS message.

Excursus 14: Secure DNS – Domain Name System Security Extensions

Just as other Internet subsystems, the domain name system is at the mercy of attacks and manipulation attempts because of its widespread use. DNS information is public, therefore the focus of attacks is not on violating confidentiality but rather on harming the integrity of DNS transmitted information and its availability.

The simplest variation of a DNS attack consists of manipulating a name server's answer when it responds to a request, i.e., intentionally tampering with the assignment of domain names and IP addresses (**DNS spoofing**). The attacker disguises itself as the name server and answers queries from the DNS clients in the place of the real DNS server. The attacker can accomplish this by paralyzing the name server with a so-called **Denial of Service** attack (**DoS**). The name server's processing is then so overburdened that the attacker is able to send the requesting client a DNS answer before the actual name server is in a position to do so. In this way, incorrect or faulty entries are generated in the DNS cache of the requesting client. The attacker is then in a position to redirect the planned data traffic to another IP address without the client noticing it. Identifying this type of DNS cache manipulation is especially difficult today as change in the practice of the dynamic assignment of IP addresses is relatively frequent. The generation of erroneous entries in the client's DNS cache is known as **cache poisoning**. If in recursive name resolution a DNS server itself makes further DNS requests to name servers in other DNS zones, cache poisoning has even more serious consequences. This is because all further requests by this DNS server are also first answered from this cache and therefore passed on further.

To carry out an attack via DNS spoofing, the attacker must know that its falsified DNS answer is only then accepted by a DNS client if the (randomly determined) ID of the query and the answer sent back match each other. The answer must also be directed to the correct IP address and port number of the query. An attacker might proceed as follows:

1. To find out the source port of the name server under attack, the attacker sends it a DNS query, provoking a DNS query from its own DNS server. As most name servers use a fixed source port for DNS queries, it can be assumed that the source port determined will also be used in the future.

2. A large number of DNS queries with identical contents and different sender addresses are now generated and sent to the name server under attack. This can be done with the help of IP spoofing, i.e., the sending of IP data packets with false IP addresses or with the help of so-called bot networks. In spite of having identical contents, queries with different IP addresses are treated as independent queries and most of the time assigned their own ID. Thus, chances increase that the correct ID can be established later.

3. In order to answer the query, the name server under attack must itself send relevant queries to the authoritative name server and wait for its answer.

4. The attacker now sends a large number of DNS answers back to the attacked name server, whereby it falsely gives IP address of the authoritative name server and addresses the source port determined in (1). Every DNS answer is given a different ID, with the purpose of randomly generating a valid answer. If the attacker succeeds in sending a valid DNS answer to the attacked name server before the authoritative DNS server, it has the possibility of introducing falsified data into its DNS cache.

A further attack variation is so-called **zone stealing**. The attacker initiates a non-authorized zone transfer, whereby the entire DNS zone database is transferred. The information gleaned is subsequently used for attacks on DNS clients querying the DNS cache.

An attacker can further attempt to send a dynamic update to a name server. Normally, changes in the DNS zone database from a primary to a secondary server are then forwarded. If the authentication of the sender is not checked at this time, an attacker can make virtually any entry in the database of the attacked DNS server. All DNS clients contacting this DNS server to make a query about name resolution may subsequently receive a fake IP address in return. A computer under the control of the attacker is then reached via this address instead of the requested goal.

By interloping on the communication between the DNS client and DNS server, the attacker can read and manipulate all transmitted messages between both of these systems. This man-in-the-middle attack is also known as **DNS hijacking**.

A DoS attack on any computer can also be accomplished easily via the DNS. In DNS protocol the messages of the DNS answer are usually considerably longer than the accompanying queries. An attacker can take advantage of this and, via the sender's address of the attacked system, send out DNS queries to many different DNS servers. These send their (long) answer back to the supposed source address, in this way creating a temporary overload. This attack variation is also called **DNS amplification**.

To better guarantee the security of the domain name system and to avoid the kind of attacks that have just been described, the following technologies can be implemented:

- Transaction Signature (TSIG)
- Domain Name System Security Extensions (DNSSEC)

Transaction Signature (TSIG)

With the help of TSIG (Transaction SIGnature) the authentication of communication partners under DNS is secured and the data integrity of the DNS transactions ensured. A DNS partner (DNS client and DNS server) should thus be able to determine whether the other communication partner is who he claims to be. Additionally, it should be clear that the transmitted DNS messages were not manipulated. There are no plans to create encryption for the transmitted DNS messages - the DNS information is public and therefore it is superfluous to treat it confidentially. The main area of TSIG application is in the communication between DNS servers. Zone transfers and dynamic updates between DNS servers can likewise be secured.

To be able to implement TSIG, the DNS communication partners need a symmetrical key, which has been exchanged (manually) between the two beforehand. With the help of this key, the sender calculates a Message Authentication Code (MAC, here the standard is HMAC-MD5) for the message. It is appended as a special **TSIG resource record** on the DNS message. The receiver carries out the same MAC calculation with its own key and compares both of the codes to each other. If they match, it is clear that the DNS message is from the partner indicated and has not been manipulated along the transmission path.

The TSIG resource record is dynamically generated by the sender of a DNS message and after receipt and verification subsequently discarded. Therefore, it is neither part of the

DNS zone database nor of the DNS cache. A TSIG resource record contains the following fields:

- Name of the key, which - modeled after the DNS - is comprised of the domain names of the participating communication partners. Therefore, if a.hpi-web.de and b.hpi-web.de exchange DNS messages via TSIG, the name of the key could be a-b.hpi-web.de. It is also possible for multiple keys to be negotiated between the partners.
- Type specification, here always TSIG.
- Class, here always ANY
- Time-to-Live (TTL), here always 0.
- Length of the TSIG resource record.
- Data, i.e., calculated MAC Code and possibly further information.

TSIG is a simple mechanism that was defined in RFC 2845 and can provide security for the DNS message in the face of manipulation. With a rise in the number of communication partners, however, the exchange of the secret key becomes more problematic. This is because the exchange is done manually and is therefore not scaled.

Domain Name System Security Extensions (DNSSEC)

In contrast to TSIG, Domain Name System Security Extensions (DNSSEC) are based on an application of public key cryptography methods (asymmetrical cryptography methods). In this way, DNSSEC gets around the existing TSIG key distribution problem. The key pair used in DNSSEC, consisting of a secret and a public key, is implemented as follows:

- The secret key is deployed to digitally sign the DNS message of the sender - as a rule of the DNS server - and in this way ensure authentication and integrity.
- The public key of the sender (DNS server) is used by the receiver of the DNS message for verifying the contained domain information.

Just as in TSIG, it does not provide encryption of the actual domain information.

The original DNSSEC, defined in RFC 2535, proved unsuitable in its first version because of the complex key exchange involved. As a result, the application of DNS was delayed for several years until in 2005 a completely new version, defined in RFC 4033, was released. Sweden (.se) became the first digitally signed domain via DNSSEC in October 2005. On May 5, 2010, DNSSEC was subsequently implemented on all 13 DNS root servers. This was followed by publication of the first public root zone key on July 15, 2010. Using it, DNS messages could be validated from the root zone.

The subsequent DNSSEC innovations include the definition of four new resource records:

- **DNSKEY Resource Record**
  The public key of a name server may be stored in the publicly accessible DNS zone database and accessed by every DNS client via an appropriate DNS query. The DNS resource record consists of the following fields:

  - **Zone name**
    The zone name indicates for which domain name (DNS zone) the defined key applies.
  - **Flags**
    This 16-bit long field contains multiple control bits. These indicate for what purpose the key may be used. If the first bit is set to 0, then the key may be used for authentication. In the case where confidentiality needs to be guaranteed, i.e., if an encryption is to take place, then the second bit is set to 0. Bit 4 is for extension purposes and permanently set to 0. Bits 7 and 8 indicate the key type, with 00. If bits 13–16 are set to zero, the key can be used for the digital signing of DNS dynamic update messages. All remaining bits are always set to 0.

- **Protocol**
  The protocol field gives the intended use of the public key. Fixed at this time are 1 for TLS, 1 for email, 3 for DNSSEC, 4 for IPsec and 255 for any protocol.

- **Encryption method**
  The following specifications have been provided up to this now as an encryption method:
  1 for RSA/MD5, 2 for Diffie Hellman, 3 for DSA/SHA-1, 4 for elliptical curves, 5 for RSA/SHA-1, 6 for DSA/SHA-1/NSEC3, 7 for RSA/SHA-1/NSEC3, 8 for RSA/SHA-256 and 10 for RSA/SHA-512.

- **Key**
  Then follows the public key for zone names, given in base64 encoding.

Example of a DNSKEY resource record:

x.hpi—web.de. IN DNSKEY 257 3 1 AQOW4333ZLdOHLR . . .

The forwarding of a public key via DNS is only adequately safe if the relevant DNSKEY resource record is digitally signed by a RRSIG resource record and the DNS query is secured by DNSSEC.

- **RRSIG Resource Record**
  Virtually any resource record in the framework of DNSSEC can be digitally signed with the signature resource record. The resolver, or requesting DNS client, can check the digital signature in the RRSIG resource record with the help of the accompanying public key from the DNSKEY resource record. In this way, the authenticity of the sender and the integrity of the signed DNS resource record can be verified. A RRSIG resource record consists of the following fields:

  - **Name**
    (Domain) Name of the issuer of the signature.

  - **Time-to-Live**
    The TTL field indicates how long the entry may be kept in the DNS client's cache.

  - **Class**
    Indicates the class the signed RR belongs to (IN for Internet).

  - **RRSIG**
    Current resource record type (here, RRSIG, type 46).

  - **TYPE**
    Type of the digitally signed resource records (e.g., A, NS, SOA, MX, etc.).

  - **Encryption method**
    The following specifications have been provided for the encryption procedures to date: 1 for MD5, 2 for Diffie-Hellman and 3 for DSA.

  - **Number of name components**
    Indicates how many subcomponents make up the domain name of the issuer; serves in wildcard resolution.

  - **TTL**
    Gives the time-to-live of the signed resource record at the time of signing.

  - **End time**
    Specification of the end time point (date) up to when the digital signature is valid.

  - **Start time**
    Specification of the start time indicating from when the digital signature is valid.

  - **Key tag**
    Unique identification; to distinguish between multiple signatures.

– **Zone name**
Name of the issuer (here DNS zone) of the digital signature.

– **Signature**
Digita signature in base64 encoding.

Example of a RRSIG resource record:

```
www.hpi—web.de.
         3600       (TTL)
         IN
         RRSIG
         A                    (Type of the original RR)
         3                    (DSA)
         3                    (3 components)
         3600                 (TTL of the orginal RR)
         20110505122208       (End time)
         20110504122208       (Start time)
         22004                (Key tag)
         hpi—web.de
         GHGds12BMTLR80WnKndatr7 ...
```

- **NSEC Resource Record**
Via a Next Secure Resource Record (NSEC) all domain names in a DNSSEC-secured DNS zone are concatenated in alphabetical order. The signing of DNS entries ensures that they are not manipulated, i.e., falsified, and that they originate from the correct authoritative DNS servers. With the means used until now it has not been possible to ascertain if a DNS entry does not exist. In other words, an attacker can remove DNS data from a server answer without the DNS client noticing it. Thus, all domain names of a DNS zone are arranged in alphabetical order, with the help of a NSEC resource record, and circularly concatenated. The last entry points to the first, e.g.,

a NSEC b
b NSEC d
d NSEC a

Every NSEC Resource Record is digitally signed via a RRSIG resource record to ensure that it cannot be manipulated. When there is a request, a DNS server also sends back the accompanying NSEC resource record. In the above example, if there is request concerning the non-existing computer c, the NSEC resource record b NSEC d is sent back as well. In this way, the DNS client can be sure that the requested domain name c in fact does not exist. Additionally, the NSEC resource record lists all resource record types that are allocated to specific domain names. The NSEC resouce record is made up of the following fields:

– **Name**
(Domain) Name of the owner of a key.

– **Type**
Type of resource record, here NSEC (type 47).

– **Next Domain Name**
Domain name of the next alphabetically ordered entry in the accompanying DNS zone.

- **Type List**
  List with the type specification of all resource records that are assigned the domain name.

Example of a NSEC resource record:

`a.hpi—web.de NSEC c.hpi—web.de NS DS RRSIG NSEC`

The disadvantage of this method is that an attacker can determine which server has a DNS zone by reading all entries in a NSEC chain. This type of attack is called **DNS walking**. To prevent its occurrence a new NSEC3 resource record is introduced in which the domain names in a DNS zone are no longer given in clear text but encrypted.

- **DS Resource Record** The DS resource records (delegation signer resource records) were introduced for concatenation of DNSSEC-secured DNS zones with RFC 3658. Multiple DNS zones can be brought together this way to form a "chain of trust" and validated via a single public key. As the number of DNS zones in the DNS is not limited, a corresponding number of suitable public keys is required for DNSSEC. To simplify the administration of an unnecessarily large number of keys, the idea arose of concatenating all zones involved and of only using the uppermost DNS zone as the secure point of entry. Only for this highest DNS zone is the distribution of the public key necessary.

A DS resource record always corresponds to a NS resource record containing the name of the authoritative name server. In the DS resource record, the hash code of the secure entry point is in the DNS zone. It is digitally signed and can be validated with the public key of the corresponding DNS zone. A DS resource record is made up of the following fields:

- **Label**
  Contains the domain name of the DNS zone to be concatenated.
- **Type**
  Type of the resource record, here DS (Typ43).
- **Key Tag**
  Unique identification number.
- **Encryption method**
  The following specifications have been provided thus far as encryption methods: 3 for DSA/SHA1, 5 for RSA/SHA1, 6 for DSA-NSEC3-SHA1, 7 RSASHA1-NSEC3-SHA1, 8 for RSA/SHA-256, 10 for RSA/SHA-512, 12 for GOST R 35.10-2001.
- **Hash type**
  As hash methods 1 SHA-1 and 2 für SHA-256 are provided.
- **Hash**
  Contains the hash code for the accompanying NS resource records.

Example of a DS resource record:

```
a.hpi—web.de.      NS ns
a.hpi—web.de.      DS 1234 1 1 7DA04267EE87E802D75C5...
```

DNS resolvers on the client side are usually not capable of validating digitally signed messages themselves. These are normally simply structured programs and would be overwhelmed by the complex DNSSEC operation. Most of the time, the DNSSEC functions are available via a local recursive name server, which has a more powerful DNS resolver. A client who wants to carry out name resolution sends a query to the local name server. By setting the DNSSEC OK bit (DO bit) in the DNS header it is indicated that an authentication should be carried out. The local name server carries out the DNSSEC operation, and after a successful authentication sets the Authenticated Data Bit (AD-Bit). To set the DO bit it is necessary to have support of the DNS extension EDNS, which is also required to support the more extensive DNSSEC UDP packet sizes needed to transmit cryptographic keys and digital signatures.

Further reading:

Aitchison, R.: ProDNS and Bind, Apress, Berkeley CA, (2005)

Arends, R., Austein, R., Larson, M. , Massey, D., Rose, S.: DNS Security – Introduction and Requirements, Request for Comments 4033, Internet Engineering Task Force (2005)

Arends, R., Austein, R., Larson, M. , Massey, D., Rose, S.: Resource Records for the DNS Security Extensions, Request for Comments 4034, Internet Engineering Task Force (2005)

Vixie, P., Gudmundsson, O., Eastlake 3rd, D., Wellington, B.: Secret Key Transaction Authentication for DNS TSIG, Request for Comments 2845, Internet Engineering Task Force (2000)

DNS and IPv6

As names of valid, corresponding IP addresses are assigned by way of the domain name system, its functionality must also be ensured in IPv6. This is of particular significance since the expansion of IP addresses from 32 bits to 128 bits in IPv6 places the human memory before an unprecedented challenge. RFC 3596 defines the new resource record type **AAAA** (often referred to as **quad-A**), which, just as an A resource record in IPv4, resolves a domain name in an IPv6 address. The reverse DNS name resolution for allocation of a domain name to a given IPv6 address works in the same way as in IPv4. Only the specially required domain that reflects the hierarchy of the IP address space is converted into `ipv6.arpa`. The PTR resource record intended for reverse lookup remain unchanged, only the address coding, which reflects the hierarchy of the address space, is changed from its former 8-bit limit to a 4-bit limit.

9.2.2 Directory Services

A directory usually contains information about technical resources or about people who are available or reachable via the Internet. The function of centrally addressing these resources or people collectively is known as a **directory service**. In the operating system UNIX, simple text files, such as `/etc/passwd` or `/etc/alias` implement simple directory services. Platform-independently they provide information about users or email routing. In order to guarantee the interoperability of different applications across system and operating system borders, general directory services with standardized access protocols are required. According to the client-server principle, these can be compared, searched, generated, modified and deleted. For the implementation of a general, global directory service, ITU-T has developed its own concept and has been standardized under the designation **X.500**. X.500 organizes the information in a X.500 directory service logically in a global **Directory Information Tree** (**DIT**). The information managed over X.500 is characterized as

objects. These can be real objects, such as people or devices as well as logical objects, such as files. Objects are always designated by associated attributes or value pairs, whereby a distinction is made between single and multiple-value attributes as well as mandatory and optional attributes. An attribute value of an object determines its so-called **Relative Distinguished Name** (**RDN**), e.g., "Friedrich Schiller." This, together with the node in the directory tree (DIT) where the object is located, provides the **Distinguished Name** (**DN**), which serves to uniquely identify the object.

Fig. 9.17 shows an example illustrating a section of the global directory tree.

Below the root it splits off into subtrees for individual countries and further into locations, organization or organizational units. Each node is itself an object with its own attributes. Therefore DNs such as in the example notation in Fig. 9.17 result: cn=Friedrich Schiller o=classics c=DE", describing a person named Friedrich Schiller (common name) in the organization classics in Germany (country).

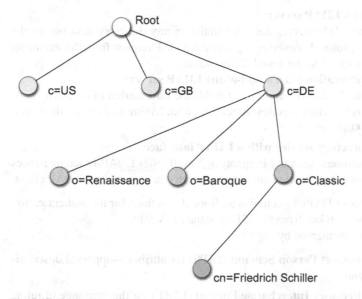

Fig. 9.17 Example for a subtree of the general, global X.500 directory information tree.

Similar to DNS, there is a distributed system involved in X.500. Here, the various X.500 servers each retain individual subtree data from the whole directory. Additionally, every X.500 server has access to information about other X.500 servers in its direct environment. In case of a client request concerning information beyond the context of its own administration name, it either contacts the other X.500 server itself or refers the client to another server.

Access to information administered via X.500 proceeds over the **Directory Access Protocol** (**DAP**, X.519). It also implements a client/server service on the basis of the ISO/OSI reference model. By means of DAP, a **Directory User Agent** (DUA) sends the relevant query to the **Directory System Agent**. Although over years, components of X.500 have been incorporated into many software systems, such as the directory service in Windows NT 5.0 that orients itself on the X.500 standard, client access to X.500 directories via DAP has proven to be quite complex. For this reason, a less complicated protocol was able to prevail over the DAP protocol: the **Lightweight Directory Access Protocol** (**LDAP**, RFC 4511). LDAP was developed at the University of Michigan (UMich) in 1993 and proposed for the first time in RFC 1487. The first server implementation also originated at the UMich and is known today as "UMich LDAP." Unlike the ITU-T protocol DAP, LDAP is not based on the entire ISO/OSI protocol stack but sets directly on top of TCP/IP. While the functionality of LDAP is limited when compared to DAP, many of the not directly supported functions are reproduced in LDAP through an adept selection of parameters. In this way, LDAP was implemented in many PCs in the early '90s and was able to gain broad application.

There are three possible variations for access to LDAP (cf. Fig. 9.18):

- **X.500 access to a LDAP server**:
  In this case, the LDAP server does not maintain any directory data but works exclusively as a protocol converter. It generates a DAP request from the incoming LDAP and forwards it to the actual X.500 server.
- **purely LDAP operation via a stand-alone LDAP server**:
  In this variation, the LDAP server itself holds the information of a naming context and can answer client queries directly without having to forward them to a dedicated X.500 server.

- **proprietary directory service with a LDAP interface**:
  Several manufacturers, such as Microsoft or Novell, offer LDAP-based interfaces to their own proprietary service directories, which can be used by a LDAP client.

The current version - LDAPv3 - contains additional functions for the authentication of clients, which were taken from the ITU-T standard X.509.

LDAP has been supplemented by

- **Lightweight Internet Person Schema** (**LIPS**) for attribute-supported descriptions of people and

- **Lightweight Directory Interchange Format** (**LDIF**) for the exchange of information between individual LDAP servers.

LDAP is used today in numerous applications and employed in different areas, for example in the address directory in Apple Address Book, IBM Lotus Notes, Microsoft Outlook, Mozilla Thunderbird or Novell Evolution. It is also employed in user administration, for example in Apple Open Directory, POSIX Accounts or Microsoft Active Directory Service. It is also used for authentication and to manage user data for the SMTP, POP and IMAP servers, as well as in a large number of mail servers, such as postfix, qmail, exim, Lotus Domino and many more.

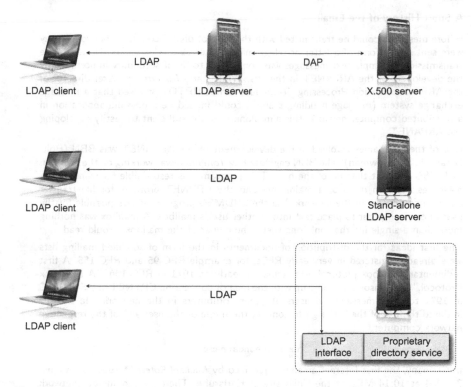

Fig. 9.18 Various way of employing LDAP.

9.3 Electronic Mail – Email

The so-called **electronic mail service** (**email**) has been one of the most popular
Internet services since its inception and retains this standing today. The first email
systems already came into being shortly after the beginning of the ARPANET. They
consisted of a file transfer program along with the convention that the first line of
every message was to start with the address of the receiver. *Ray Tomlinson* (* 1941),
who worked as an engineer at the research company BBN, a leading developer of
the ARPANET, is considered the inventor of the modern email (cf. Fig. 9.19).

The demand increased for a comfortable email protocol that would be able to send
messages to multiple recipients at the same time. It should also provide an optional
ackowledgement mechanism with which the sender would be able to determine if
the message sent had even reached the recipient. Already in 1982, the first RFCs
for today's still valid email protocol were published (RFC 821 and RFC 822). It
is interesting to note that this non-commercial proposal made by several computer
science graduate students was able to prevail over the standardized email system
recommended by ITU-T as **X.400**, and would eventually overtake it.

A Short History of the Email

Before messages could be transmitted with the help of digital computer networks, they were sent in the form of a letter or telegram and later as a telex or fax. However, the transmission of simple text messages was considered to be of secondary importance by the developers of the ARPANET in the 60s. While *Larry Roberts*, the later director of the ARPA Information Processing Techniques Office (IPTO), stressed that a message exchange system (message handling system) could indeed be a possible application in a distributed computer network, this aim alone was not sufficient to justify developing the ARPANET.

One of the companies involved in the development of the ARPANET was BBN (Bolt, Beraneck, and Newman). The BBN engineer *Ray Tomlinson* was working on the program SNDMSG there at the end of the 60's. This program was responsible for transmitting messages between users of a mainframe and the CPYNET protocol for file transfer between computers in the network. Via the SNDMSG program, it was possible for the user of one computer to place text into another user's mailbox. A mailbox was nothing more than a single file that only one user - the owner of the mailbox - could read.

The first ideas for the distribution of documents in the form of so-called mailing lists were already registered in very early RFCs, for example RFC 95 and RFC 155. A first rudimentary mailbox protocol was defined as early as 1971 in RFC 196, "A Mail Box Protocol." Tomlinson also came up with the idea of connecting CPYNET and SNDMSG in 1971 to send messages to users at other computers in the network. He thereby initiated the use of the "@" sign to connect the name of the user and of the respective network computer.

username@hostname

In Germany, the first Internet email was received by *Michael Rotert* (* 1950) on August 3, 1984 at 10:14 MEZ at the University of Karlsruhe. There were numerous network environment systems developed for message transmission at the beginning of the 80's, such as mailbox systems X.25, Novell or BTX. However, with the spread of the Internet - and with it the Internet email - these other forms were rapidly replaced by the mid-90's. The email quickly became the most popular Internet service. In 2010, 294 billion emails were being sent daily from almost 1.88 billion email users. Nearly 90% of these mails were reported to be made up of unwanted emails, better known as SPAM.

Further reading:

Roberts, L. G.: Multiple computer networks and intercomputer communication, in Proceedings of the first ACM symposium on Operating System Principles, pp 3.1–3.6., New York, NY, USA, ACM, (1967)

Internet 2010 in Numbers, Royal Pingdom Blog
http://royal.pingdom.com/2011/01/12/internet-2010-in-numbers/

Fig. 9.19 A short history of the email

9.3.1 *Message Handling Systems*

If someone were to define the term "email," it would suffice to describe it is a re-creation of the traditional mail service on the basis of a modern, digital means of communication. The email service of the Internet belongs to the family of the **Message Handling Systems** (**MHS**) (cf. Fig. 9.21).

A message handling system has two fundamental components:

E-Mail Emoticons

Communication via email - similar to every other form of written communication - is fixed on the communication channel of the written word. This means that personal interaction with the partner as expressed through gestures or facial expression is missing. To convey this additional information in writing, descriptive or explanatory words have to be used or an extra coding designed to put forth the intention of what is spoken. The latter way is applied to the communication form of the email. In this electronic form of communication - just as in the text message or Twitter - as much information as possible must be fit into the smallest space.

Thus, so-called **emoticons** have established themselves. These are used in an electronic (short) message by the author to provide additional information about his or her mood or the intention of the statement being made. In 1964, graphic artist *Harvey Ball* (1921–2001) designed the original smiley, which was used on buttons for the American insurance company State Mutual Life Assurance Company to boost employee morale. He simply drew a circle, colored it yellow, placed two dots in it and under them a half circle: the "smiley" (☺) was born.

The electronic counterpart to the smiley was first proposed on September 19, 1982 by *Scott E. Fahlman* (*1948). With the help of regular ASCII characters, it can be composed independent of support by graphic character sets. Thus, :-) stands for positive feelings and jokes and :-(for negative feelings. The order of characters presents a face tilted to the right that either look happy or sad. A large number of variations, with very different meanings, were quickly created. With their help, the author of an electronic text can express mood or personal intention.

Further reading:

Walther, J. B., D'Addario, K. P.: The impacts of emoticons on message interpretation in computer-mediated communication. Social Science Computer Review 19 (2001) pp. 324 – 347.

Fig. 9.20 Email emoticons.

- **User Agents (UA)**
  User agent is the name given to systems in which the user is able to generate and edit messages, as well as to read, send and receive messages. UAs are local application programs that via an appropriate interface provide another interface to the user to implement the messaging service. The UA is thus virtually the interface between the user and the what is actually responsible for transport: the **Message Transport System** (MTS), a subsystem of the MHS.

- **Message Transfer Agents (MTA)**
  Message transfer agents are responsible for message transport from the sender to the receiver. Normally, MTAs run as background processes that go unnoticed by the user, even as they forward a message from one system to the the next. In order that the messages reach their destination, multiple MTAs must often work with each other. Together these comprise the Message Transfer System. In the case of Internet email, they work on the basis of the **Simple Mail Transfer Protocol** (**SMTP**, RFC 821).

Continuing with the analogy of the traditional letter, the structure of an email message can best be described in terms of an envelope and a written message (cf.

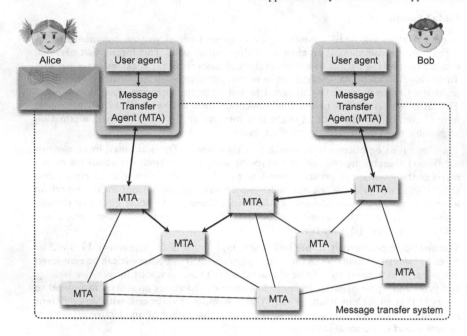

Fig. 9.21 Schematic structure of a message handling system

Fig. 9.22). The address of the receiver is noted on the envelope, as well as any information necessary for the successful delivery of the transported contents. The transported message in the envelope is made up of a message header and the actual message itself (payload). The message header contains command and control information for the user agent. Just as with standard letters, the sender must first give the message to the postal service for dispatch. The postal service then takes care of transport and delivery to the receiver, with the help of different entities. Delivery is accomplished by depositing the message in the (mail) box provided by the receiver. An email system is required to perform the following five basic functions:

- **Generation of a message (composition)**
  This function covers the generation of the actual message contents with the help of a text editor, as well as help in correct generation of the required header elements of the message, such as address details or other directives.

- **Transport of a message (transfer)**
  To be able to transfer a message successfully from the sender to the receiver there must first be a connection set up to one of the participating intermediate systems, or to the receiver over which the message can be sent directly. After successful delivery the connection is dismantled again. Connection establishment and delivery is carried out in the background without the user's direct involvement.

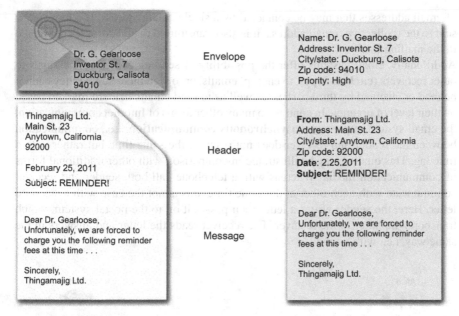

Fig. 9.22 Envelope and transported message - traditional mail and email.

- **Information about transport proceedings (reporting)**
  The sender would like to know whether its message could be successfully delivered. If this is not the case, it is important to know the reasons for failed delivery.

- **Presentation of the transported message (displaying)**
  A delivered message must be displayed at the receiver. Depending on the type of message, possible reformatting or complete format translation may be required.

- **Deposit of the transported message (disposition)**
  Once successfully delivered and displayed to the receiver, a decision has to be made as to what is to be done with the message. The possibilities range from deleting the message to storing it in a more complex archive system, or forwarding it to a further receiver.

To manage received email, most of the email systems provide the user with so-called **mailboxes**, where messages can be archived. These usually offer comfortable, content-based search possibilities that make the retrieval and handling of archived emails considerably easier. Generally, every email must indicate a receiver. The address of the receiver is made up of a user name, the sign "@" and the IP address, or the domain name of the end system where the receiver set up the email account, for example:

```
meinel@hpi.uni-potsdam.de.
```

Without the details of the unique **email address** a message cannot be successfully delivered. To make it easier for a user to simultaneously send a mail to multiple addresses, the concept of **mailing lists** was introduced. A mailing list contains a group of email addresses that may be contacted by a single email address. If a message is sent to the (collective) email address, it is also transmitted to all addresses contained in the mailing list.

Additionally, email systems offer the possibility of sending copies of messages to other receivers (**carbon copy**), to encrypt emails, or to sign them so that they cannot be read or changed by third parties, as well as to send emails that are marked based on their level of urgency. In contrast to many other forms of Internet communication, the email system implements **asynchronous communication**, i.e., communication between the sender and receiver does not occur at the same time but rather with a time lag. This quality may be illustrated in comparison with other traditional forms of communication media. Whereas with a telephone call both sender and receiver communicate with each other at the same time, this is not the case with a standard letter. Here, the sender writes a letter then passes it on to the postal system, which transports the letter to the receiver. The receiver reads the letter and answers in the same way, i.e., with a time lag.

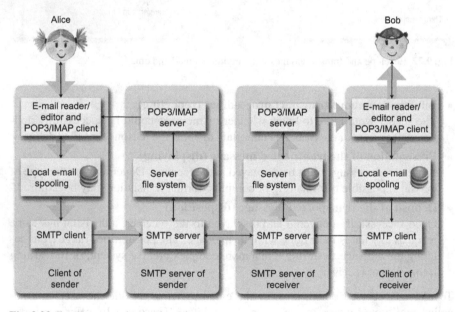

Fig. 9.23 Email communication model.

The email system implements a message handling system in the following way (cf. Fig. 9.23):

- **Client host of the sender (UA)**

  The email sender writes a message with the help of an email client program at its end device. The completed message is not immediately transmitted over the Internet. Instead, the email message is normally first kept in an intermediate storage (spooling). In this way, the user must initially not be connected to the

Internet as long as messages continue to be written. As soon as the user is finished writing, the messages are then all sent at the same time.

- **SMTP server of the sender (MTA)**
  The email protocol is implemented over the **Simple Mail Transfer Protocol (SMTP)**. When the email message of the user has been generated, then the email client connects to the Internet and sends the message to the local SMTP server of the user. This is usually operated by the Internet Service Provider (ISP) of the user. The sending of the message proceeds via SMTP.

- **SMTP server of the receiver (MTA)**
  The SMTP server of the sender transmits the message via SMTP to the SMTP server of the receiver. There, the message is put in the mailbox (inbox) of the recipient. The mailbox can be compared to intermediate storage (spooling) in the transmission of email messages, i.e., it allows the recipient to collect email messages over a longer period of time before then connecting to the Internet again to retrieve these messages.

- **Client host of the receiver (UA)**
  In certain cases it is possible that the recipient may access his mailbox directly at the local SMTP server. Most of the time, however, a special protocol is implemented for this purpose: **Post Office Protocol (POP3)** or **Internet Message Access Protocol (IMAP)**. The task of this protocol involves querying the mailbox contents of the user and transmitting the messages inside to the user so that the local email client can display them.

9.3.2 Email Message Formats

The valid Internet format for email messages was originally established in RFC 822 (today RFC 5322). Principally, an email message is made up of a primitive envelope (RFC 821, or RFC 5321), the lines of the message header, a blank space and the subsequent message body with the actual message content. While there are no requirements to observe for the format of the message body, the lines (header fields) of the message header (header section) always follows the same structure. A colon follows a special **key word** (**tag**) and an appropriate value assignment. Based on the key word, the UA of the receiver can determine the proper way to proceed with the subsequent value assignment.

The envelope for the email message to be sent is, in fact, not generated by the UA. Only first when the message is passed to the MTS with the corresponding header section, does the MTA generate the appropriate header fields and envelope necessary for further transport. The most important specifications in the message header are the address details from the sender (`From`) and recipient (`To`), followed by a `Date`, a `Subject` and further optional control directives. Table 9.4 provides an overview of the most important email header fields.

Table 9.4 Message header field of an email message, defined in RFC 5322.

Field	Content
To:	(required) Email address(es) of one or more recipient
From:	(required) Email address of the sender
Cc:	Email address of recipients who receive a copy of the message. An entry in this field signifies that this user is not being contacted directly, but that the email is being brought to the user's attention.
Bcc:	Like Cc: except that the actual receiver does not know who else receives a copy. The blind copy should protect a receiver who gets a circular email from address gathering by malicious services, e.g., so-called "spambots."
Date:	(required) Time the email was sent
Subject:	Header line of the message. In view of the increasing number of unwanted emails (spam), this field has gained increasing importance. Unsolicited email can often be detected as such in the subject line.
Reply-To:	Address(es) of one or more receivers for a possible answer to this message.
Received:	List of all MTAs who have transmitted this message
Return-Path:	Optional information indicating a path (series of email addresses), which can identify an answer back to the sender
Message-ID:	Unique identification number of the message; it is usually generated when the email is sent.
References;	Identifies further documents that are connected with this message, e.g., other email messages.
Keywords:	Key words that can be fixed by the sender to describe the contents of the email message.
X-Charset:	Set of characters used by the sender so that the recipient is able to correctly show a foreign language or special characters in the text.
X-Mailer:	An email software used to send a message.

The email standard set in RFC 5322 also allows the user to introduce several header fields. They must always be marked with the prefix "X" to distinguish them from the regular header fields. No rules apply regarding formatting for the subsequent main part of the message (message body). This is separated from the header by a blank line and it is up to the user to decide how it is to be filled and then interpreted.

The header of the email messages is evaluated during transmission in the email communication system by the instances involved. This proceeds in the following way:

- **Email creation**

  When writing the email, the sender provides several important pieces of header information, such as the receiver and subject. This is translated by the email client in the corresponding header fields.

- **Processing at the client of the sender**
  The transmitted header information from the sender is translated along with the message into the email message format. The client thereby processes the designated receiver, creating an envelop for email transmission via SMTP.
- **Processing at the SMTP server**
  SMTP servers pay hardly any attention to the respective header fields of the actual email message. There are only a few fields added during transport of the message. Newly added fields are always appended to the front of the message, to prevent changing the original sequence of the the header fields.
- **Processing at the SMTP server of the receiver**
  As soon as the email message reaches the designated recipient, the date and time of delivery may be added to the header.
- **Email client processing and read access**
  Incoming email messages are evaluated by the POP3/IMAP protocol and displayed in such a way to the receiver to enable choosing only those mails one wishes to read. Most of the time, the user is presented a list of newly incoming emails, showing the sender of a message, the subject and the date of arrival. This puts the user is in a position to decide whether the mail is worth reading or if it should be deleted.

9.3.3 MIME Standard

With the fixing of the email standard in RFC 822 at the beginning of the '80s, the message content contained in the body of the message itself was established in a plain text form in English. This meant that to define an email only 7 bit ASCII characters were available - excluding any national or other special characters, such as the German umlaut. Because of the Internet's global use, the demand quickly arose for expansion to allow for nationally-based supplements to the standard characters or even completely different character sets for languages such as Japanese or Chinese. There was a further wish to expand transmission to allow the capability to send any binary files via email, for example graphic or audio files. While RFC 822 did not fundamentally place restrictions on the message body, the problem was concentrated in a standard interpretation of the content by the UAs. A standardized solution could first be created in 1993 with the so-called **Multipurpose Internet Mail Extensions** (**MIME**). This is defined in RFC 1521 (the successor of RFC 2045 – RFC 2049). The basic principle of MIME is therefore based on the retained regulation for the email traffic defined in RFC 822. MIME only fixes a standardized message body structure and additional rules concerning how data types should be handled that are not ASCII standard. Therefore, all existing MTAs could be kept unchanged, and only the UAs had to be adapted for the display and generation of MIME-conform message formats. While a non-MIME-capable UA is able to receive a MIME-coded email, it is not able to interpret its contents correctly. Two fundamental structure types that are transferable in an email can be distinguished in MIME:

- **Simple structure (discrete media)**
  Describes MIME messages that transport a single (discrete) media type, for example a text message or an image. In these cases, only one particular variation of media coding is deployed in the message.

- **Complex structure (composite media)**
  Describes MIME messages that are made up of different types of media, such as, e.g., combined text and image information, as well as one email message embedded in another. It is possible that every single media type could be coded differently.

Five new standard header fields are defined in the MIME expansions. These describe so-called MIME entities, i.e., a MIME message or a MIME section (MIME body part)

- **MIME version**
  This header field informs the UA that the present message is a MIME-coded message, and it is also informed about the MIME version used. If this header field is not contained in an email message, the contents of the message are handled per default as an English ASCII text. As the only MIME header field, this header field refers to the entire MIME message and is not intended for individual headers of single parts of a MIME message.

- **Content description**
  This optional header contains a free description of the subsequent MIME-coded email message contents and in this way supports the operational process.

- **Content ID**
  Unique identification for message content. The format corresponds to the message ID of the standard email header. This header is optional and usually serves to identify individual sections ("body parts") of the MIME message.

- **Content transfer encoding**
  MIME allows for the transfer of virtually any binary data, it is therefore necessary to establish the way in which the data at hand is to be coded for transmission. The following types of coding are available:

  - **7 bit ASCII**: The simplest form of coding is the 7 bit US-ASCII. It corresponds to an email according to RFC 822, provided that the message does not contain more than 1,000 lines. If a content transfer encoding header is not indicated, then this variation is always adopted.

  - **8 bit ASCII**: This ASCII coding also allows for the additional integration of special national characters. The limitation of 1,000 lines of text also applies here.

  - **Base64 encoding**: With this form of coding, 24-bit long groups of binary data to be coded are sectioned into four bit sequences, each 6 bits in length (cf. Fig. 9.24). These 6 bits are encoded with the help of 7 bit US-ASCII characters (A–Z, a–z, 0–9, +/=). Binary data should always be sent in this coding.

- **Quoted printable encoding**: Coding a text with Base64 encoding - which in fact consists mainly of ASCII characters and of only a few characters that are not displayable in 7 bit US-ASCII - is not especially efficient. In this case, these individual signs can be coded better in the following way: " =" serves to introduce a character not displayable in 7 bit US-ASCII character, followed by the hexadecimal value of the corresponding character.

- **User defined**: The user can specify its own coding. However, it must bear the responsibility that the receiver is able to encode the decoded message.

- **Content type**
  In addition to the encoding of the relevant message, an identification of its type is important in order that the UA is in a position to display it correctly after a possible decoding. This type specification is particularly crucial for every kind of binary data, such as e.g. graphic files, audio files, video sequences, executable programs or mixed forms made up of different data types. The description is given in the form of a general content type and a special content subtype, e.g.,

  Contenttype : text/html.

  Optional parameter specifications can follow afterwards, such as

  Contenttype : image/jpeg;name =" testbild.jpg".

  The content type establishes whether the MIME message is composed of one discrete medium (simple structure) or of more combined together (complex structure). The MIME types are represented in Table 9.5.

An example for a header of a MIME encoded message that contains a graphic file of the JPEG type is shown in Fig. 9.25. Special attention is paid to the possibility of sending multipart messages, i.e., messages that are made up of different types of elements. A multipart message (**MIME multipart message**) is identified by the content type multipart. The individual message parts are separated from each other with a separator defined via the keyword Boundary (in the example the separator NextPart is used).
Different types of multipart MIME messages are possible:

- multipart/mixed:
  This message consists of different, non-contiguous, partial messages. These can all be of a different type (content type) and different encoding (content transfer encoding). In a multipart message of the mixed type it is possible to send texts together with multimedia data in a message.

- multipart/alternative:
  With this option it is possible to send different forms of the same message at the same time. This could be, e.g., multiple language versions of one text or different quality levels of one multimedia file.

- multipart/parallel:
  This message consists of different partial messages that must be displayed at

Base64 Encoding

The base64 encoding generally describes a method for encoding 8 bit binary data in a character sequence composed only of readable ASCII characters (A–Z, a–z, 0–9, +/=). It is specifically applied in the encoding of multi-media email contents in the MIME standard, as use of the 7 bit US-ASCII code is intended for the email message format.

For the encoding process, three bytes of the byte stream (corresponding to 24 bits) are each divided into four 6-bit long blocks. Each one of these 6-bit code blocks correlates to a number between 0 and 63

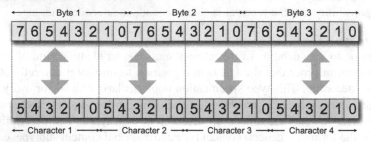

If the total number of the input bytes is not divisible by three, the text to be encoded is filled at the end with padding bytes made up of zero bits so that a number of bytes divisible by three is reached. With a n byte long binary byte stream, the space required for the base64 encoded contents is calculated in z bytes as

$$z = 4 \cdot \lfloor \frac{n+2}{3} \rfloor \ .$$

Further reading:

Josefsson, S.: The Base16, Base32, and Base64 Data Encodings, Request for Comments, 4648, Internet Engineering Task Force (2006)

Fig. 9.24 Base64 coding.

the same time. It is an option that can be implemented, e.g., for simultaneously displayed associated video and audio data.

- multipart/digest:
  This message contains a series of partial messages that all display complete email messages based on RFC 822. It is possible here to e.g., pack individual emails into a type of collection envelop and to send them together.

- multipart/encrypted:
  The message consists of two parts. The first part is not encrypted and indicates the method of how the second, encrypted part of the message is to be decrypted

Table 9.5 A selection of MIME types and subtypes.

Type/Subtype	Description	Reference
text/plain	Unformatted text according to RFC 822	RFC 2046
text/enriched	Text with formatting	RFC 1896
text/html	HTML-coded text, corresponds to the hypertext markup language used in the World Wide Web	RFC 2854
text/css	Cascading style sheet information for the World Wide Web	RFC 2318
image/jpeg	Graphic file in the JPEG format	RFC 2046
image/gif	Graphic file in the Graphics Interchange Format (GIF)	RFC 2046
image/tiff	Graphic file data in the Tagged Image File Format (TIFF)	RFC 2302
audio/basic	Basic audio data type for the MIME standard, it encodes one single audio channel as 8 bit ISDN µ-law pulse code modulation with sampling frequency of 8000 Hz	RFC 2046
audio/mpeg	MPEG standard audio (including MP3)	RFC 3003
video/mpeg	MPEG standard video data format	RFC 2046
video/dv	Digital Video (DV) video data format	RFC 3189
application/octet-stream	Unspecified binary data, e.g., an executable application program	RFC 2046
application/postscript	Print file in the postscript data format	RFC 2046
multipart/mixed	Multi-part message, whereby each part contains its own content type with its own coding	RFC 2046
multipart/alternative	Multi-part message, whereby each part contains its own content type with its own coding	RFC 2046

9.3.4 Simple Mail Transfer Protocol – SMTP

The protocol implemented for the forwarding of email messages is the **Simple Mail Transfer Protocol (SMTP)**. It was standardized in 1982 in RFC 821, and has gone through many extensions since that time (currently the last update was 2008 in RFC 5321). The sender's computer must set up a TCP connection to the receiver via the well-defined TCP port 25 in order to forward an email message. TCP port 25 always operates with a SMTP process that is capable of receiving or sending emails. SMTP servers can send messages as well as receiving them. An SMTP server that sends a message works as a logical client and the SMTP server that receives the sent message works as a logical server. To prevent misunderstandings it is easier to speak in terms of the SMTP sender and SMPT receiver. This was the original terminology defined in RFC 821 before the renaming as client and server in RFC 2821. The communication between sender and receiver via the SMTP protocol is made

From: harald@hpi.uni−potsdam.de
To: christoph@hpi.uni−potsdam.de
Date: Sun, 15 May 2011 13:28:19 −0200
MIME−Version: 1.0
Content−Type: Multipart/Mixed; Boundary="NextPart"

This is a multipart meddase in MIME format

−−NextPart
Content−Type: text/plain

Hi Christoph ,
I'm sending you a picture .
See you soon ,
Harald

−−NextPart
Content−Type: image/jpeg name="bild.jpg"
Content−Transfer−Encoding: base64

RnJhbnogamFndCBpbSBrb21wbGV0dCB2ZXJY
. . .
G9zdGVuIFRheGkgcXVlciBkdXJjaCBCYXllcm44=

Fig. 9.25 Example for a multi-part MIME coded message.

easily understandable for the human user as only plain ASCII text notifications are exchanged, based on the following simple schema (cf. Fig. 9.26):

- After the sender (client) has set up a TCP connection to port 25 it waits for the receiver (server) to begin the subsequent communication.
- The server begins communication with a line of text proving its identity. The receiver communicates to the sender whether or not it is ready to accept the sender's message.

    Server: 220 hpi-web.de SMTP READY FOR MAIL

    If the server is not ready to accept a message, the client terminates the TCP connection and attempts to take up communication at a later time.

- If the server answers that it is ready to receive email messages, the client then responds with a HELO message (abbreviation of "hello") along with proof of its identity. Since the expansion of SMTP through RFC 1425 (and later RFC 2821) ELHO - " Extended Hello" - has been used as the negotiated greeting command when the extended variation of SMTP is supported.

    Client: HELO example.de

    The server's confirmation of the connection signals the establishment of a SMTP connection between the client and the server.

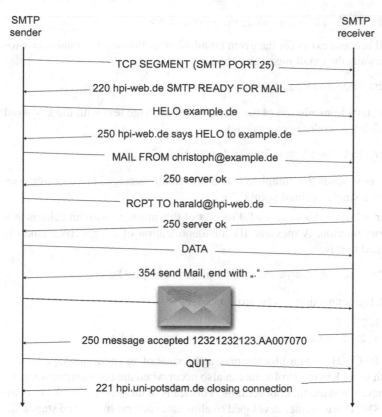

```
SMTP                                                              SMTP
sender                                                            receiver

         ———————— TCP SEGMENT (SMTP PORT 25) ————————►

         ◄———————— 220 hpi-web.de SMTP READY FOR MAIL

         ———————————— HELO example.de ————————————►

         ◄———— 250 hpi-web.de says HELO to example.de

         ———————— MAIL FROM christoph@example.de ————————►

         ◄———————————— 250 server ok

         ———————————— RCPT TO harald@hpi-web.de ————————►

         ◄———————————— 250 server ok

         ————————————————— DATA —————————————————►

         ◄———————— 354 send Mail, end with „.“
```

Fig. 9.26 Typical SMTP dialog for the transmission of an email message.

Server: 250 example.de says HELO to hpi-web.de

The client can now send one or more email messages, terminate the connection or request the server to reverse roles so that email transmission takes place in the opposite direction. In case the server supports the SMTP extensions, it then answers with ELHO and a sequence of further messages. These all begin with the code 250 and show the respectively supported SMTP extensions. If a SMTP server is contacted with a ELHO request and it does not support any SMTP extensions, it answers with an error message 500 syntax error, ...

• The client now starts the transmission of the email message with the command

    Client: MAIL FROM <christoph@example.de>

in which the sender of the email message is indicated with his email address The server acknowledges receipt.

    Server: 250 server ok

• The client next sends the email message to the email address of the receiver

Client: RCPT TO <harald@hpi-web.de>

If an email account exists for the given email address, the server signals clear-to-send and awaits the email message to be sent.

Server: 250 server ok

- The client starts transmission of the actual email message text with the key word DATA, which the sender confirms with

  Server: 354 Send mail; end with "."

- The client now sends the complete text of the email message. The last line concludes with a single decimal point (".").
- The server acknowledges successful receipt of this message with an acknowledgement confirmation. A message ID for identification of the received email is then attached to this.

  Server: 250 message accepted <message-id>

- After that, the connection can be terminated:

  Client: QUIT
  Server: 221 example.de closing connection

The original RFC 821 standard for the message format of an email message is limited to a length of 64 Kbytes. Problems can also occur when the communication partners do not have the same timeout settings. This leads to unforeseen breaks in communication. SMTP was further developed to eliminate such problems and standardized as **Extended SMTP (ESMTP)** in RFC 1425. It is possible for a sender to signal to the receiver at the beginning of communication that it wishes to use ESMTP instead of SMTP. Instead of transmitting a HELO command it will then transmit the command EHLO. The receiver's refusal of this request signals that it is only possible to communicate via SMTP. In many cases, it is also not possible to deliver emails, for example due to an incorrect email address. This non-deliverable email message is then generally returned to the sender. This procedure is known as **bouncing**. When delivery to the original sender proves impossible, the non-deliverable email is then sent to a general user - the **postmaster** (**double bounce**).

9.3.5 POP und IMAP

Until now we have looked at communication inside of an email system that runs via the SMTP protocol. A computer able to transmit and receive emails via SMTP is known as an **email gateway** (or email server) as it makes email service directly available (cf. Fig. 9.28). In companies and business, however, not necessarily every computer in the local network has this capability. In reality there are only a few computers, maybe even only one, operating as an email gateway.

Spam – Unsolicited emails

Spam is a name given to all types of unsolicited messages that are received electronically. Mass email, also called Unsolicited Bulk Email (UBE), is the best known among these types. It involves sending emails to a large number of recipients, usually with the intention of a marketing a product or service. Spam is comparable to a traditional mass mailing by means of postal distribution. Most of the time the receiver is not addressed specifically and the content is usually advertising.

The name has its origin in the brand name of an American, precooked, canned meat. SPAM has been sold since 1936 under its brand name, a combination of the words "spiced ham." During World War II, SPAM was one of the few foods that was available almost everywhere without being subject to rationing. At the beginning of the '70s, the name was connected with a skit performed on the English comedy series "Monty Python's Flying Circus" - when it became a synonym for overkill and repetition. The setting of the skit was a café that only offered meals on the menu with SPAM, resulting in the name being mentioned a total of 132 times.

The use of the term to describe widespread unsolicited bulk emails first caught on in the '90s. Today, spam has become a serious Internet problem. Almost 90% of the entire email data traffic results from spam. Counter-measures against spam may be employed in all email data traffic instances. The most commonplace today are the so-called spam filters at SMTP servers or at the user's side. Moreover, there is also an ongoing discussion about making changes to the SMTP protocol and the domain name system in order to prevent spam.

Further reading:
Zdziarski, J. A.: Ending Spam: Bayesian Content Filtering and the Art of Statistical Language Classification, No Starch Press, San Francisco, CA, USA, (2005)

Internet 2010 in Numbers, Royal Pingdom Blog
http://royal.pingdom.com/2011/01/12/internet-2010-in-numbers/

Fig. 9.27 SPAM – unsolicited emails.

So that the users of the other computers can send and receive emails it is necessary for them to communicate with the email gateway via special protocols These enable picking up received emails from the email gateway by the user as well as delivering the messages to be sent there. The **Post Office Protocol 3** (**POP3**), is a simple protocol that provides email gateway access. It evolved from the original Post Office Protocol defined in 1984 as RFC 918 and in 1996 was standardized in RFC 1939 . POP3 is based on the TCP protocol and manages the communication between email client (UA) and email gateway (cf. Fig. 9.23). Among the main tasks of the POP3 protocol are:

- Login to the email gateway,
- Authentication of the user by means of password request,
- Email retrieval from the email account of the user,
- Deletion of the retrieved email messages from the storage of the email gateway.

The protocol itself works with ASCII text notifications and is very similar to the SMTP protocol. A constant connection to the email gateway, or mail server, is not necessary with POP3. It is set up by the client as required and terminated afterwards.

Email gateways

Email gateways often serve as connecting links, joining the computers of sender and receiver with each other over the Internet. In this function, email gateways perform various tasks:

- Linking computers not directly connected to Internet email traffic.
- Linking computers that do not have RFC 822 conform email systems (e.g., X.400 email systems). In each case, email gateways undertake a format translation of the different systems.
- Processing **mailing lists**. Via mailing lists, a group of email recipients can be pooled together and addressed by a single, special group address. The email must be transmitted to all recipients contained on the mailing list when addressed to the group address of that list,. At the same time, the processing of mailing lists can lead to significant overhead. For this reason, this task is therefore often transferred to a dedicated computer, e.g., the email gateway.
- To be able to furnish all email users in a company with an email address that has a uniform suffix, the IP address of the email gateway is used. In the other case, the suffix of the email address would need to be composed of the user's computer IP address. However, this would allow conclusions to be drawn about the internals of the local network and provide a target for unauthorized intrusions. Every email directed to the address of the email gateway arrives there first. The address prefix is then evaluated, after which the email is forwarded to the actual recipient.

Further reading:

Hunt, C.: TCP/IP Network Administration O'Reilly Media, Inc., 3rd edition, (2002)

Fig. 9.28 Email gateways.

POP3's main task consists of copying the emails of the email gateway user to the local computer of the user. The user can then read and process these later at any given time.

While the POP3 server has to be installed at the same computer that manages the mailboxes of the email gateway, its counterpart, the POP3 client, is usually a component of the user's email software (e.g., Microsoft Outlook), which the POP3 server must contact to gain access to user's emails. To ensure a reliable processing of the given command via POP3, and the answer of the mail server, message transport proceeds over TCP via port 110. As soon as a TCP connection is set up over this port, the POP3 session begins. The client transmits instructions to the server and the server sends back corresponding answers, or requested email messages. The basic answers of the POP3 servers are:

- **+OK** as a positive answer to a client request, or
- **-ERR** as a negative answer, to show that an error has occurred.

The standard commands received by a POP3 server are summarized in Fig. 9.29.
A POP3 session begins with a TCP connection by the POP3 client on the user's side at the POP3 server at port 110. On the server's side, the POP3 server listens for

POP3 standard set of commands

- USER xxx for the definition of the user name, or the user account at the mail server.
- PASS xxx transmits a unencypted password as cleartext.
- STAT returns the status of the mailbox. Among other information, the number of email messages in the mailbox and their total size in bytes.
- LIST (n) returns the number and the size of the (n-th) email message.
- RETR n obtains the n-th email from the mail server.
- DELE n deletes the n-th email at the mail server.
- NOOP no special function, the mail server answers with +OK.
- RSET resets all previously transmitted DELE commands.
- QUIT ends the current POP3 session and carries out all DELE commands as indicated.

In addition to the standard commands, the following commands are optionally supported:

- APOP for a secure login at a POP3 server.
- TOP n x retrieves the email header and the first x lines of the n-th email message.
- UIDL n to display the unique ID of the n-th email message.

Further reading:

Myers, J., Rose, M.: Post Office Protocol - Version 3 Request for Comments 1939, Internet Engineering Task Force (1996)

Fig. 9.29 POP3 standard set of commands.

connection queries arriving at this port. If a corresponding TCP connection is set up, the POP3 server answers with a greeting:

Server: +OK example.de POP3 server

In the next step, the user must authenticate itself and prove that it has the authorization to query its mailbox. Authentication proceeds with the transmission of a user name, or mailbox name, and an associated secret password.

Client: USER harald@example.de
Server: +OK Please enter password
Client: PASS **********

Following successful login, the user is authenticated and can enter further commands to query its mailbox. As this unencrypted password cannot be considered secure, modern POP3 servers also maintain a secure method with the APOP command. If a POP3 server supports this variation of authentication, it transmits a timestamp uniquely identifying the POP3 session at the greeting. With this time stamp and a mutual secret key, known only to client and server, the POP3 client carries out a MD5 checksum calculation. This is sent along at the client login by means of the APOP command. The POP3 server verifies the MD5 checksum and thus authenticates the POP3 client in a secure way. Fig. 9.30 shows the exemplary procedure of a simple POP3 session.

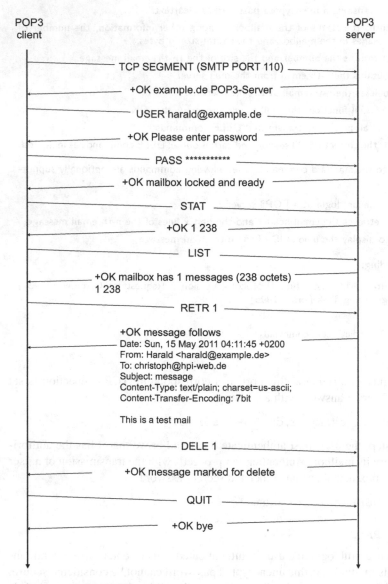

POP3
client

POP3
server

TCP SEGMENT (SMTP PORT 110)

+OK example.de POP3-Server

USER harald@example.de

+OK Please enter password

PASS ***********

+OK mailbox locked and ready

STAT

+OK 1 238

LIST

+OK mailbox has 1 messages (238 octets)
1 238

RETR 1

+OK message follows
Date: Sun, 15 May 2011 04:11:45 +0200
From: Harald <harald@example.de>
To: christoph@hpi-web.de
Subject: message
Content-Type: text/plain; charset=us-ascii;
Content-Transfer-Encoding: 7bit

This is a test mail

DELE 1

+OK message marked for delete

QUIT

+OK bye

Fig. 9.30 Typical POP3 dialog in a mailbox query.

Further possibilities are offered by the more complex **Interactive Mail Access Protocol (IMAP)**, originally defined in RFC 1730 (currently RFC 3501). IMAP was designed in 1986 with the goal of allowing access to mailboxes and messages at the user's local computer as if they were not at a remote mail server. In contrast to POP3, the emails remain at the mail server and are manged there. POP3 and IMAP were designed to correspond to the asynchronous character of email communication, i.e., the user's computer was originally not permanently connected with the Internet, but for reasons of cost and efficiency only connected when it was required. However, so that an email can be forwarded and received at any time, the operation of decoupled mail servers is necessary. These must be permanently connected to the Internet and able to process all email communication tasks as soon as they occur.

While in the case of POP3 email messages are transmitted from the mail server to the individual's computer and stored there, IMAP is designed so that emails are stored directly at the mail server. The user therefore has the advantage of contacting its mail server from virtually any computer in the Internet and, following successful authentication, gaining access to its emails. In the case of POP3, the user was always subject to the one specific computer that administers its emails with POP3. IMAP copies only the email header from the mail server, leaving up to the user the choice of email it wants to have forwarded to read and process on one of its computers. The IMAP protocol offers the following functions:

- Accessing and copying emails from a remote mail server to the local computer. While available locally, they can be administered by the mail server.

- Setting and deleting of flags for individual emails so that the user is able to track which messages have not yet been (or have already been) read.

- Managing multiple mailboxes and copying of emails from one mailbox to another. The user is thus able to set up mailboxes for different categories of messages, e.g., business and private.

- Copying information from the email header prior to a user's decision whether or not to actually copy or read the email.

- Copying sections of an email, e.g., in the case of MIME multi-part messages.

The IMAP server is operated at the mail server, while it is necessary that an IMAP client be installed at the computer of the user. The mailbox must be configured in such a way that it can create new emails via SMTP and via IMAP retrieve, research and modify emails. IMAP uses TCP as the transport protocol. The IMAP server monitors the TCP port 143 and waits for connection by the IMAP client. As soon as a TCP connection is set up via port 143, the IMAP session begins. The procedure of a typical IMAP session can be simply illustrated with the help of a finite state machine, which consists of five different states (cf. Fig. 9.31):

- **No connection**
  The IMAP session is terminated or inactive. The IMAP server waits for the client to set up a connection via TCP port 143.

- **Not authenticated**
  Initial state of an IMAP session. The IMAP server enters this state directly after accepting the TCP connection via port 143, as long as the option "preauthentication" is not chosen. To be able to transition from this state to the next, it is necessary for the IMAP client to authenticate itself at the IMAP server.

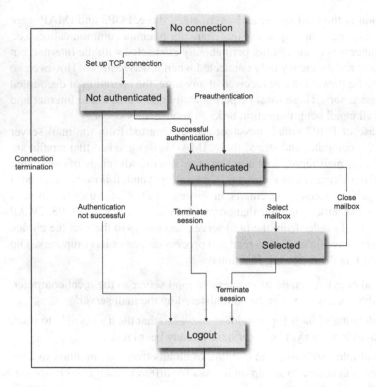

Fig. 9.31 Procedure of an IMAP session as an finite state machine.

- **Authenticated**
  In this state, the IMAP client has successfully identified itself to the IMAP server. This has either occurred by explicit authentication in the "not authenticated" state or through an authentication that has taken place already by way of another external system (preauthentication). Before further operations can take place, the IMAP client must select one of the available mailboxes at the mail server.

- **Selected**
  After a mailbox has been chosen, the IMAP client can access and manipulate the individual emails in the mailbox. As soon as the IMAP client has concluded its operation at this mailbox it can either return to the "authenticated" state to choose a new mailbox or proceed to the "logout" state and end the session.

- **Logout**
  The IMAP client can end the IMAP session from any of the other states and go into the "logout" state. The session also transitions to this state when the inactivity timer - which has been set beforehand - runs out. The server now deletes the predetermined emails and releases the reserved resources, thereby ending the session as well as the TCP connection.

The most important IMAP commands for IMAP version 4 (IMAP4) are summed up in Fig. 9.32.

9.3.6 Pretty Good Privacy – PGP

A potentially large number of computers can be involved in the transportation of an email across the Internet from the sender to the receiver. At each of the participating computers the email message can also be read, copied and even infected by unauthorized users who have gained the necessary access. Although many users are not aware of it, there is no private sphere in the Internet unless one actively provides it. A widespread method of email encryption is the method known as **Pretty Good Privacy** (**PGP**). PGP provides a complete packet for the encryption of emails as well as for authentication of the communication partner via digital signatures, and it is freely available.

If two users, Alice and Bob, want to keep their email communication confidential, the following requirements must be met:

- On obtaining possession of the exchanged email documents, an unauthorized third party should not be able to draw any conclusions based on their contents (**confidentiality**).
- When Alice sends a message to Bob, he wants to be sure that the message is in fact from Alice (**sender authentication**).
- It is also crucial for Alice and Bob that the content of the transmitted message is not changed or falsified (**data integrity**).
- When sending a message to Bob, Alice also needs to be sure that Bob is indeed the recipient of the message and no one else (**receiver authentication**).

PGP uses a public key encryption procedure (asymmetrical encryption procedure), i.e., a public key is used which makes it possible for every party to encrypt the data for a certain receiver. There is an additional private, secret key that only the receiver has. Messages sent to the receiver are encrypted with this public key and can only be decrypted with the receiver's private key. Only in this way is it possible to achieve an authentication of the participating communication partners. If both communication partners have access to the public key of the opposite party, and if both keys are certified, then the claimed identity matches the one stated in the certificate of the public key. This means that **encrypted email data traffic** can begin:

- Alice encrypts the message to be sent M with Bob's public key kp_B.
- Alice sends $kp_B(M)$ encapsulated with a corresponding MIME header also via SMTP, as if a regular email.
- Bob receives the email from Alice. He uses his private, secret key ks_B for decryption on the exempt payload section of the MIME header $kp_B(M)$ of the email.
- As $ks_B(kp_B(M)) = M$, Bob is now in possession of the secret message M.

Selection of standard IMAP4 commands

- Commands that can be issued in every state of an IMAP session
    - CAPABILITY: Query regarding the server's capabilities.
    - LOGOUT: Notification to the IMAP server that the session is ending. The IMAP session transitions to the "Logout" state.

- A selection of possible commands that can be issued in the "Not Authenticated" state:
    - LOGIN: Input of user name and password for authentication.
    - AUTHENTICATE: Selection of a specific authentication mechanism by the client.
    - STARTTLS: Selection of the transport layer security protocol for authentication.

- A selection of commands that can be issued in the "Authenticated" state:
    - SELECT: Selection of a specific mailbox.
    - EXAMINE: Selection of a specific mailbox in the read only mode. No changes can be made.
    - CREATE / DELETE / RENAME: Creation of a new mailbox, or deletion or renaming of an existing mailbox.
    - SUBSCRIBE / UNSUBSCRIBE: Addition or deletion of a mailbox to the quantity of a mail server's "active" mailboxes.
    - LIST: Request for a (partial) list of the available mailbox names at the mail server.
    - LSUB: The same as LIST, but only the names of active mailboxes will be returned.
    - STATUS: Status request of a mailbox, delivers information such as the number of emails (read and unread).
    - APPEND: Adds an email to a mailbox.

- A selection of commands that may be issued in the "Selected" state:
    - CHECK: Sets a checkpoint point in the processing of a mailbox.
    - CLOSE: Closes the mailbox currently being processed and returns the session to the "Authenticated" state.
    - EXPUNGE: Permanently removes all of the emails from the mailbox that were marked for deletion.
    - SEARCH: Searches the mailbox matching the criteria parameters.
    - FETCH: Retrieves information based on the given email parameters from the mailbox.
    - COPY: Copies a set of emails that can be specified as parameters into a specified mailbox.

Further reading:

Crispin, M.: Internet Message Access Protocol Version 4rev1 Request for Comments 3501, Internet Engineering Task Force (2003)

Fig. 9.32 Selection of standard IMAP4 commands.

The first version of the PGP software was published by *Phil Zimmerman* (*1954) in 1991. It uses a RSA algorithm for the encryption of data. Later versions use the Diffie-Hellman key exchange procedure. PGP, however, does not encrypt the entire message with an asymmetrical encryption procedure because, particularly in the transmission of multimedia data, this would be too computationally intensive. The actual message is encoded with the help of a simple, symmetrical encryption. At the same time, the symmetrical key, used by both communication partners, is exchanged beforehand via an asymmetrical encryption procedure. This process is also referred to as **hybrid encryption**. The symmetrical key encrypted this way is normally placed before every (symmetrically encrypted) email message. In this way, it is possible to encrypt a message for multiple recipients at the same time.

The PGP encryption procedure works in the following way (cf. Fig. 9.33 and Fig. 9.34):

- Alice generates a random, only to be used once, **session key** R_1.

- Alice encrypts the message to be sent via a symmetrical encryption procedure with the secret session key R_1.

- Alice encrypts the session key R_1 with Bob's public key kp_B.

- Alice concatenates the encrypted session key $kp_B(R_1)$ with the encrypted message $R_1(M)$. She sends this packet $[kp_B(R_1), R_1(M)]$ to Bob.

- Bob receives $[kp_B(R_1), R_1(M)]$, separates both parts and implements his own, secret key ks_B for the decryption of the session key on $ks_B(kp_B(R_1)) = R_1$.

- With the help of the session key R_1 Bob is now able to decrypt the encrypted message from Alice $R_1(R_1(M)) = M$.

It is possible in this way to achieve a secure encryption of the transported email. However, an authentication of both communication partners as well as the integrity of the email message must still be ensured. A digital signature and message digest can be used to do this:

- Alice, who again wants to send a message to Bob via email, first uses a hash function h on her email message M, and in this way generates a message digest (checksum) $h(M)$.

- In order to generate a digital signature from $h(M)$, Alice encrypts $h(M)$ with her private, secret key ks_A and generates $ks_A(h(M))$.

- Alice concatenates the original email message M with the digital signature $ks_A(h(M))$ and generates a packet $[M, ks_A(h(M))]$.

- At this point, Alice uses the just-described method of encryption. She generates a session key R_1, encrypts $[M, ks_A(h(M))]$ with R_1, encrypts R_1 with Bob's public key kp_B and sends the thus generated packet to Bob:

$$[kp_B(R_1), R_1([M, ks_A(h(M))])] .$$

- Bob first extracts the encrypted session key $kp_B(R_1)$ from the received packet and decrypts it with the help of his private, secret key ks_B.

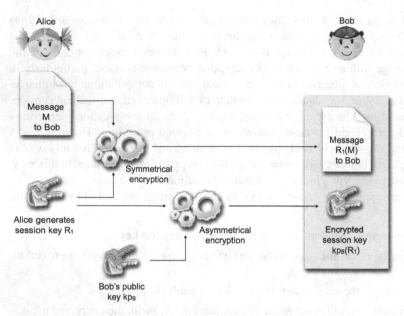

Fig. 9.33 Shared use of symmetrical and asymmetrical encryption procedure in the transmission of confidential emails with PGP encryption.

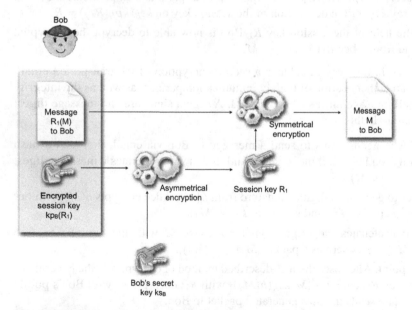

Fig. 9.34 Shared use of symmetrical and asymmetrical encryption in the transmission of confidential emails with PGP decryption.

- Using this session key R_1 Bob decrypts the second part of the packet and gets $[M, ks_A(h(M))]$.

- Bob extracts the digital signature $ks_A(h(M))$ and then, using Alice's public key kp_A for encryption, obtains the message digest $h(M)$.

- Bob uses the hash function h on the received email M' and compares both message digests. When $h(M') = h(M)$, Bob can assume that email M has been transmitted intact. In this case, because it was possible to encrypt the digital signature with Alice's public key, the identity of the sender is secured.

A secure email communication can, however, only be guaranteed when it is ensured that both Alice and Bob both have access to the "correct" public key of their counterpart. This means that a secure distribution of the public key, e.g., via a certified public key infrastructure, is the necessary requirement for sending secure emails. The PGP developed by Phil Zimmerman is a complete software packet containing all the methods and procedures required for secure email data traffic. It does not contain any independent proprietary developments but rather a working environment for known methods and procedures that are meshed with each other. PGP accordingly offers

- Encryption of emails (confidentiality),

- Authentication of the communication partner,

- Ensuring message integrity via digital signatures and

- Compression techniques for emails.

PGP is freely available for most computer platforms and operating systems as an open source. It also exists for the majority of email clients in a commercially available form, e.g, as a plugin. PGP became famous through the litigations between the US government and Phil Zimmerman. Zimmerman was accused of having violated U.S. import restrictions regarding cryptographic software with the widespread public dissemination of PGP. This claim was based on PGP's Internet availability as public domain software and its accessibility to foreign users. In the USA, cryptographic systems were considered in the realm of military defense and therefore subject to special export restrictions. Although Phil Zimmerman had developed PGP as a private citizen without the involvement of a large company, it quickly became one of the most popular encryption systems for email. Legal proceedings were dropped in 1996 after three years of litigation and Zimmerman founded the company PGP Inc, which was taken over by Network Associates, Inc. in 1997. In the meantime, PGP has been standardized by the IETF as OpenPGP in RFC 4880.

When PGP is installed, a key pair consisting of a public and a private key is generated for every user. The public key is provided to the user on the web site or via a corresponding public key server (Certificate Authority, CA). The private key, protected by a password, is stored on the user's computer. Every time the user wishes to employ its private key, it is necessary to enter the password. With PGP, the user can choose whether to just add a digital signature to his message or use encryption, or to employ both possibilities.

PGP also provides its own possibility for the certification of the public key. In this case, however, no trustworthy third party (CA) is involved. Instead, PGP certifies public keys via its own **web of trust**. This means that a user certifies each pair - consisting of a user name and matching public key - which he himself trusts. By vouching for the authenticity of other keys, one user can communicate her trust to another user via PGP.

PGP users meet each other at so-called **key signing parties**. An exchange of public keys is carried out at these events. These are certified mutually by being signed with the private key. A third possibility is the use of special **PGP public key servers**. A PGP public key server stores the public keys sent to it by a user. It distributes these to other PGP public key servers, giving the keys to any requesting party.

X.400 – Message Handling System

X.400 is an alternative message handling system on top of the ISO/OSI reference model and was standardized by the ITU-T. The same system is implemented by the ISO under the name **MOTIS** (Message Oriented Text Interchange System). The goal in developing X.400 was to secure interoperability between different systems as well as between public and private email services. As a rule, X.400 has been offered by all operators of public networks for general message exchange, thereby providing transitions to other services and networks.

Compared to the Internet email standard RFC 822, the X.400 specification is considerably more complex and, accordingly, offers more sophisticated functionality as well. Released in 1984 two years after RFC 822, the X.400 standard was then released in a revised and improved form in 1992. The X.400 system nevertheless failed to gain dominance over its simpler competitors and has been increasingly relegated to the background.

Further reading:

Betanov, C.: Introduction to X.400 Artech House, Boston, MA, USA (1993)

Fig. 9.35 X.400 message handling system.

9.4 File Transfer

Before the days of computer networks, data could only be transmitted between different computers via transportable storage media, such as magnetic tapes or floppy disks. Distributed computing with a common database was nearly impossible or could only be implemented with great inefficiency. First with the introduction of computer networks and, in particular the Internet, did the situation change dramatically. From then on, data could be electronically exchanged with remote computers in a fraction of a second and distributed applications run over different computers. In the early days of the Internet, network applications could be divided into two distinct categories. The first type was direct use of the remote computer's resources with the user or an application logging in and working at the remote computer as if it were running locally. The second category was indirect use of these resource. In this case, they were first transmitted to the local computer over the Internet and subsequently processed locally.

In order that it would not be necessary to develop separate mechanisms for every application required for data exchange, very early on **general data transfer services** were established that could be used simultaneously by many different applications. These data transfer service must be flexible enough to fulfill the various demands of the computer systems and applications involved with regard to the structuring of file names, data formats or user rights.

In this section, the Internet file transfer procedure, FTP (File Transfer Protocol) and its variations SFTP (Secure File Transfer Protocol) and TFTP (Trivial File Transfer Protocol), as well as the File System (NFS) will be introduced. Subsequently, we will look briefly at the distributed programming in the Internet via Remote Procedure Call (RPC).

9.4.1 (Secure) File Transfer Protocol – FTP and SFTP

The still widely used **File Transfer Protocol** (**FTP**) originated at the time of the ARPANET. It was already defined for the first time in 1971 in RFC 114 and since 1985 has existed as the still valid RFC 959. Just as most of the other Internet services, FTP works according to the client/server method. An FTP server makes data available that can be requested via a FTP client.

Following conversion of the ARPANET to the protocols of the TCP/IP protocol suite and the guarantee of a reliable transport service via TCP, the central task of FTP - file transfer- could be carried out in a considerably simpler way. Yet, at the same time, there remain additional tasks to be be solved, e.g., format conversion or verification of user authorization, these make FTP a relatively complex protocol. FTP provides the following functions:

- **File transfer**
  The central task of FTP, which can, however, only be successfully carried out if the following details are taken into account.

- **Formatting and presentation**
  Via FTP the client can influence the data format of the requested file. Binary data is therefore transmitted differently than plain text data, which can be coded more efficiently for transfer. The client can additionally fix whether the coding of a to-be-transferred text file should follow the ASCII or the EBCDIC standard.

- **Interactive access**
  Besides FTP being used by application programs, a direct interactive use of FTP services by the user is also possible. Following successful registration, the user can, for example, request a list of the available commands or a list of the available files before starting a file transfer.

- **Verification of user validation**

  Before an application program or an interactive user can begin a file transfer over FTP, an authorized registration must take place at the FTP server by way of user name and password. Without this information the FTP server denies file delivery.

As already mentioned, FTP service follows the client/server principle. Communication proceeds via two separate TCP connections. All of the general management and control commands between client and server are transmitted via a **control connection**, while the data to be transported is sent over the **data connection**. Both connectionS are provided on the client side as well as on the server side, in each case by their own process. FTP enables bi-directional transport, i.e., the client can request files from the server as well as sending files to the server. While the control connection remains intact for the entire session period, a separate data connection is set up for every single file transfer carried out during the FTP session.

In the **connection establishment** of a FT client with an FT server, the client starts a reliable TCP connection from any free TCP port to the TCP-port 21 of the FTP server to set up the control connection. All requesting FTP clients use the same server TCP port. They can, however, still be distinguished from one another by the TCP server. This is because a TCP connection is always identified by both connection end points. If a connection could be established successfully, a TCP connection is set up from TCP port 20 of the FTP servers to any free client TCP port (cf. Fig. 9.36) for every data transfer.

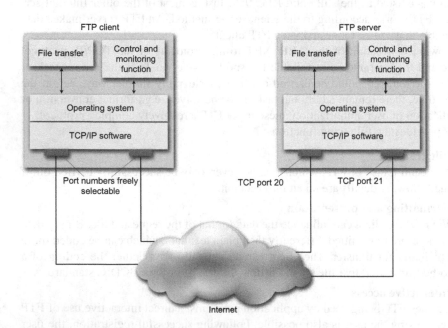

Fig. 9.36 Client/server interaction in the FTP file transfer.

There was no independent data format developed for transfer of command and control information via the control connection, instead, the TELNET protocol is implemented (cf. section 9.5.1). During a FTP session, the TCP connection for the control connection remains established throughout the whole session. In contrast, the data connection is terminated after each file transfer is completed.
FTP distinguishes two variations of connection establishment.

- **Active FTP**
  In the so-called "active mode," the FTP client opens a random port and communicates this to the FTP server together with its own IP Address over the FTP command PORT. Typically, on the client side a port above port number 1023 is chosen, while the FTP control connection on the FTP server is addressed on port 21. Fig. 9.37 shows the process of FTP connection establishment in the active mode. To begin the data connection over the server port 20, the FTP server induces a TCP connection establishment to the already known client port. The problem with this type of connection establishment is the use of the same port for a control connection as well as for a data connection on the client's side. To circumvent this situation, it is recommended that the FTP client first initiate a port change for establishment of the data connection over the PORT command.

- **Passive FTP**
  In the contrasting "passive mode" the client sends the FTP command PASV, at which the FTP server opens a port and transmits this together with its IP address to the client. Also here, on the client's side, a port above port number 1023 is used. On the FTP server's, the port that was previously transmitted to the client is addressed. This variation is implemented when the FTP server cannot establish a connection itself to the the FTP client, e.g., if the FTP client is located behind a NAT gateway that masks the IP address of the FTP client with a NAT or if a firewall shields the network of the FTP client from unauthorized external access.

From the user's point of view, the FTP client seems like an interactive application. In the simplest implementation, such as in UNIX ftp, the user is provided with a command line interface, which carries out the following tasks:

- Reading an input line,
- Interpreting an input line in order to pick out a command and the associated parameters, and
- Executing the recognized command with the given parameters.

Fig. 9.38 shows an example of an interactive FTP session in which the following commands are executed:

- ftp ipc617.hpi-web.de
  The FTP client is activated for establishment of a connection to the FTP server of the computer ipc617.hpi-web.de. The FTP server on the remote computer responds and requests authentication. In return, the user gives its name and password.

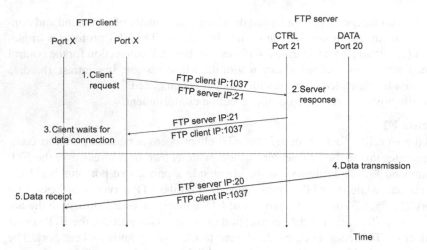

Fig. 9.37 FTP connection in the active mode.

- get /pub/public/FTP/home/sack/test.tgz
  Via the command get the user requests a file, which must be specified exactly
  in the following parameter. If the file is available, it is transmitted to the client.
- close
  The connection to the server is closed with the command close, the command
  quit terminates the client.

Example of an FTP session

```
> ftp ipc617.hpi-web.de
Connected to ipc617.hpi-web.de
220 ipc617.hpi-web.de FTP server (Version wu-2.4.2-VR16(1) ready
Name (ipc617.hpi-web.de:usera): anonymous
331 Guest login ok, send e-mail address as password
Password: harald@hpi-web.de
230 Guest login ok, access restrictions apply
ftp> get /pub/public/FTP/home/sack/test.tgz
200 PORT command ok
150 Opening ASCII mode data connection for test.tgz (1314221 bytes)
226 Transfer complete
1314221 bytes received in 13.04 seconds (1.0e+02 Kbytes/s)
ftp> close
221 Goodbye
ftp> quit
```

Fig. 9.38 An example of the file transfer protocol.

In response to a command, the FTP server always answers with a three digit status code. This is implemented by the FTP software and also shown to the user as

readable text. FTP has a simple authentication mechanism via user name and password to safeguard the access restricted files. Problematic is, however, that the user name and password are both transmitted in unencrypted form. This "cryptographic Achilles' heel" is owing to the age of the FTP protocol, which goes back to the early day of the Internet. At this time, only a closed circle of people had access to the Internet and the danger of network data traffic eavesdropping was very small. With RFC 2228 "FTP Security Extensions"RFC 2228—see File Transfer Protocol, additional cryptographic security measures were introduced in the FTP protocol for authentication and secure data transmission.

Interestingly, for safety-uncritical data the concept of the **anonymous FTP**, which does not require authentication, has prevailed. When the FTP protocol began to gain in popularity at the beginning of the 80's, the World Wide Web had not yet become a publication vehicle for public-accessible information and data. The idea of using the FTP protocol for this purpose quickly came to fruition. In the other case, it would have proven highly inefficient if each user had been forced to register in advance at the provider to access information. The involved process of registering a user account, user name and password would be particularly uneconomical in cases where access was limited to a single instance. For this reason, a collective user account was set up for FTP. Here, there is no special password necessary and the user is able to call up publicly accessible files. The following user names are available for anonymous file access via FTP.

- guest,

- anonymous,

- ftp.

Normally, no password is required for this user account. In many cases, instead of a password, the FTP server asks for the email address of the requesting client, which is used for statistical purposes. Anonymous FTP was defined in 1994 as RFC 1994. Most of the FTP servers accept both authenticated users as well as anonymous users, whereby authenticated users have more access rights than anonymous users, who can only access public areas set up for this purpose for a certain time. As soon as a data connection is established via FTP between the client and the server, data can be transmitted directly between both parties. The FTP protocol makes a distinction between three different transmissions modes.

- **Stream mode**
  The data to be transmitted is simply sent as continual data stream made up of unstructured bytes in the stream mode. The sender begins data transmission without initially sending a data header with command and control information. The FTP protocol distinguishes itself in this way from numerous other Internet protocols. These often send data in the form of discrete data blocks. The end of a data transmission is signaled solely by the sender's closing of the connection.

- **Block mode**
  In this variation of data transmission, the data to be sent is subdivided into fixed data blocks. These are sent in FTP data packets. The FTP data packet contains

a 3-byte long data header for this purpose. It contains length specifications and type information. A special algorithm makes it possible for an interrupted transmission to be carried out seamlessly.

- **Compressed mode**
  In this mode, the data to be transmitted is compressed with a simple data compression procedure, here the **run length encoding**[3]. Transmission of the compressed data proceeds analogous to the block mode.

The FTP protocol has the special quality of being able to distinguish different data types in transmission. This is necessary, for example, when a text file must be transmitted between different systems that display text data in different ways, i.e., in different coding. If the original text data were to be simply copied in the coding of system A, it could be that another text coding is used on system B. In this case, the data could not be read and interpreted, but would need to be converted separately beforehand. FTP distinguishes between the following data types.

- **ASCII**
  This data type represents text files that were coded according to the ASCII standard (American Standard Code for Information Interchange).

- **EBCDIC**
  This data type represents text files that were coded according to the EBCDIC standard (Extended Binary Coded Decimal Interchange Code), introduced by the IBM company.

- **Image**
  This type of data to be transmitted has no internal structure at all. It is sent byte by byte as binary data without an internal processing taking place.

- **Local**
  This data type was introduced to enable the connection of systems whose internal representation is not based on an 8 bit per byte display.

Today there exist a range of graphical front ends to serve FTP clients. These are located on the command line-oriented FTP client. A selection of the most important FTP commands is shown in Fig. 9.39. A FTP server can also be alternatively addressed aided by a web browser with the specifications of a corresponding Uniform Resource Locator (URL). Specifications for authentication can be indicated directly together with the URL, for example:

ftp : //mueller : 12345@ftp.download.de/dokumente/datei.txt

The user name mueller and the password 12345 are passed on to the FTP server with the domain name ftp.download.de to access the file datei.txt in the directory dokumente.

---

[3] Run length encoding and other data compression procedures are described in detail in the first volume of this trilogy, Meinel, Ch., Sack, H.: Digital Communication – Networking, Multimedia, Security, Springer (2013) in Chapter 4 , "Multimedia Data and Encoding."

A selection of important FTP commands

- `ascii` – sets the following file transfer type to the ASCII file type.
- `binary` – sets the following file transfer type to the image data type.
- `bye` – ends the current FTP session and simultaneously terminates the FTP client.
- `cd <path specification>` – changes the current path in the data directory of the FTP server.
- `close` – ends the current FTP session.
- `delete <filename>` – deletes a file from the FTP server.
- `dir` – gives a listing of the current directory on the FTP server.
- `ftp <ftp-server>` – opens a session to a FTP server.
- `get <filename1> <filename 2>` – transmits a file from the FTP server to the FTP client and stores it under the name given in `<file name2>`.
- `ls` – lists the contents of the current directory on the FTP server.
- `mkdir <directoryname>` – sets up a file directory on the FTP server.
- `mode <transfer-mode>` – sets a specific transBeyond the working of the Internet are fer mode.
- `put <filename1> <filename2>` – transfers a file from the FTP client to the FTP server and saves it under the name given in `<filename2>`.
- `type <filetype>` – sets a specific data type for data transmission.
- `user <username>` – login at a FTP server under the specified user name.

Further reading:

Postel, J, Reynolds, J.: File Transfer Protocol Request for Comments 959, Internet Engineering Task Force (1985)

Fig. 9.39 A selection of FTP commands.

As previously mentioned, in the use of FTP legitimate safety concerns arise as soon as confidential data is to be transmitted. This applies particularly to the transmitted data necessary for authentication, which in FTP is carried out unencrypted in plain text. In Fig. 9.40 different variations of safe data transfer are presented. These enable complete or partial secure user authentication as well as preserving the integrity of the transmitted messages and their encryption.

9.4.2 Trivial File Transfer Protocol – TFTP

FTP provides a general service for data transfer. As one of the most multi-faceted services within the TCP/IP protocol family, its implementation is also very complex and resource-intensive. There are, however, many applications that do not need the entire variety of functions offered by FTP or are not able to offer a complex service, such as FTP, themselves. The TCP/IP protocol family provides a second, considera-

Secure Variations of Internet Data Transmission

- **Secure FTP** The special feature of the Secure File Transfer protocol is that an otherwise unsecure FTP data transmission is tunneled in part via a secure shell (SSH, cf. Sect. 9.5). Secure FTP, the command and control operations, such as the authentication of the client at the server, as well as the exchange and listing of field directories proceed via an encrypted SSH tunnel. The actual data transfer takes place unencrypted over another randomly chosen port, negotiated beforehand by the client and server.

- **Secure Copy (SCP)**
  The secure copy protocol is an independent protocol for encrypted data transfer. SCP itself implements only the data transmission - SSH is used for authentication and connection control.

- **SSH File Transfer Protocol (SFTP)**
  The SSH file transfer protocol is a further development of SCP. Differently than the secure file transfer protocol, which only tunnels the control channel of the file transfer protocol over secure shell (SSH), SFTP was developed as an independent protocol, handling neither the authentication nor the encryption of transmitted data, rather allows these functions to be carried out by SSH.

- **FTP over SSL**
  Different than secure FTP or SFTP, FTP over SSL does not uses the secure shell for encryption of the data transmission, but for this purpose uses TLS/SSL. Thereby, both the data channel as well as the control channel of the the FTP protocol are authenticated and encrypted. FTP over SSL is specified in RFC 4217.

Further reading:

Convery, S.: Network Security Architectures, Cisco Press networking technology series, Networking Technology Series, Cisco Press (2004).

Fig. 9.40 Secure variations of Internet data transmission.

bly simpler file transfer service expressly for this purpose: the **Trivial File Transfer Protocol** (**TFTP**), which was originally defined in RFC 783 .

TFTP does not require complicated client/server interaction based on a reliable TCP transport connection, but rather relies on the considerably simpler UDP datagram of the TCP/IP transport layer. TFTP only accommodates a simple file transfer thereby forgoing authentication and authorization mechanisms. Only data can be transmitted if global access is permitted. TFTP also does not require direct user interaction. As TFTP is clearly more limited in possibilities - when compared to FTP - the TFTP software can accordingly be implemented much easier and with less space required. This reduced storage space effort also makes the implementation of TFTP beneficial in non-volatile memory (Read Only Memory, ROM) of computers that are not equipped with their own hard disk (diskless workstations). If this type of computer were started, it would be possible to load important programs for the initialization of the computer (operating system) via TFTP. Those programs in the non-volatile memory of a computer executed during system start-up are also summarized using the term **system bootstrap**.

In data transfer with TFTP, one communication partner sends the relevant divided files in blocks of 512-bytes long each. It then waits for the receiving side to acknowledge the receipt of a data block before the next one is sent. The requesting client must send a data packet to the server in advance asking for the file transfer. Here, the name of the corresponding file is given together with the direction of the transfer (from server to client or vice versa). Every block of the transmitted file is consecutively numbered - the block number contained in the header of the data packet to be sent. If a block is sent with a length of less than 512 bytes, this is a signal for the end of the file transfer. Instead of a data block transmission, it is also possible for acknowledgements or error messages to be sent, whereby an error message terminates the data transfer. If the sender of a block does not receive an acknowledgement before the end of a previously set timeout, then the data block is transmitted again.

Because in TFTP no explicit connection establishment takes place via TCP, the TFTP software must establish a connection via the connectionless and unreliable UDP protocol. To do this, the TFTP server constantly listens for requests via the UDP port 69, provided for communication via TFTP. For connection establishment the TFTP client chooses its own UDP port number beyond 1023. This is used for identifying the relevant data transfer (Transfer Identifier, TID). In contrast to FTP, the TFTP server also randomly chooses a free UDP port number beyond 1023. This is displayed by the server side TID. In this way, the client and the server side TIDs both uniquely identify a TFTP connection. This means that multiple TFTP transmissions can take place at the same time. Furthermore, because of this unique identification it is no longer necessary to identify the data transfer uniquely through a field in the TFTP data block header. Therefore, the TFTP data packet header can be kept small to maximize efficiency of the actual data traffic.

A symmetrical new transmission is a special characteristic of TFTP. This means that the communication partner sending the data, as well as the communication partner acknowledging the data, both have the possibility to transmit any data believed to have gotten lost after expiration of the timeout. As data transmission proceeds strictly sequentially - only after acknowledgement has occurred - if new data is to be sent, the TFTP protocol does not have to keep account of the already sent or acknowledged TFTP data packets. This simplifies the entire protocol. No data needs be taken into account that has arrived at the receiver in the incorrect order. Additionally, no data blocks need be reorganized in the case of delayed transmissions. Although communication via TFTP costs much more effort and proceeds in a more complicated manner than FTP, the capacity of a single data transmission is always limited to 512 bytes. While this method is considered quite robust, it can still lead to the following problems. If the acknowledgement for the data block k has not gotten lost but is only delayed, the sender re-transmits the already successfully sent data block, which is then confirmed again by the receiver. Both acknowledgements reach the sender and each triggers the transmission of data block $k + 1$. These two data blocks are acknowledged a second time, which results in a doubled transmission of data block $k + 2$ and so on and so forth. This defective multiple transmission is known as the **Sorcerer's Apprentice Syndrome**, because of the unnecessary duplication of transmitted data that results.

The TCP protocol is a very simply structured protocol, however through the years it has gone through extensions and changes that have raised its complexity. The "TFTP Option Extension" was introduced in 1995 with RFC 2347. It offered the possibility of negotiating additional data transmission options via TFTP. The TFTP Option Extension enables the negotiation of command and control parameters before the beginning of actual data transfer. It can only be implemented if both the server as well as the client support this option. For this purpose, modified read or write commands are transmitted at the start of the TFTP connection. Command and control information is sent via special option codes confirmed by the receiver with an "Option Acknowledgement" (OACK). It contains all requested options that the server supports. The following options have been defined until now.

- **Block size**
  This option makes it possible to transmit client and server data blocks that do not correspond to the fixed length of 512 bytes. In this way, the efficiency of data transmission can be increased or specific networks addressed where size restrictions from the side of the hardware or basic protocol must be taken into account.

- **Timeout interval**
  This option allows for the client and server to agree on the timeout interval length before the retransmission of previously sent data.

- **Transfer size**
  This option enables the sender of a data transfer (the client in the case of read access or the server in the case of write access) to communicate the entire data length to be transmitted before the start of actual data transfer. The receiver thus has the possibility to allocate sufficient storage space for the recording of data.

In contrast to FTP, the data transmission in TFTP takes the form of discrete data packets. There are five different types of messages in TFTP:

- Read Request (RRQ)
- Write Request (WRQ)
- Data
- Acknowledgement (ACK)
- ERROR

TFTP datapakets have the following data fields depending on their type (cf. Fig. 9.41):

- **Operation Code**:
  2-byte long field that specifies the TFTP message type.

  - 1 – Read Request
  - 2 – Write Request
  - 3 – Data
  - 4 – Acknowledgement
  - 5 – Error

- **Filename**:
Field of variable length that gives the name of the file to read or write.

- **Mode**:
Field of variable length that specifies the data transmission mode (ASCII or binary).

- **Options**:
If TFTP options are supported, this field of variable length contains the specifications of the TFTP options to be negotiated. The specifications of the supported options consist of two fields each of variable length. The first gives the option code to identify a certain option and the second the respectively selected value for this option.

- **Block**:
2-byte long field giving the number of the data block currently being sent, or confirmed.

- **Data**:
Field of variable length that is contained in the transmitted data.

- **Error code**:
2-byte long field identifying the type of error that has occurred.

- **Error message**:
Field of variable length that provides an error message in plain text.

9.4.3 Network File System – NFS

The transmission of an entire file from one computer to another is often unnecessary, because e.g., only one single line is to be appended to the file. If only part of a file should be read or changed from a remote computer, without the entire file being needed, it is possible to make use of a so-called **file access service**. The file access service allows a nearly transparent implementation of the files and resources available in the network. This means, remote files can be accessed and processed as if they were located at the local computer.

For this purpose, the TCP/IP protocol family provides the **Network File System** (**NFS**). It was originally developed by SUN Microsystems Incorporated on a client/server basis and in 1989 first defined in RFC 1094. In 1995, the NFS version 3 (NSFv3, RFC 1813) followed and in 2000/2003 NFS version 4 (NFSv4, RFC 3010, RFC 3550). NFS is also known as a **distributed file system**. NFS originally came from the UNIX world. The corresponding protocol for the Microsoft Windows operating system is known as **Server Message Block** (**SMB**). NFS services are also available on Microsoft Windows servers. In this way, UNIX computers also receive

TFTP RRQ/WRQ Message format

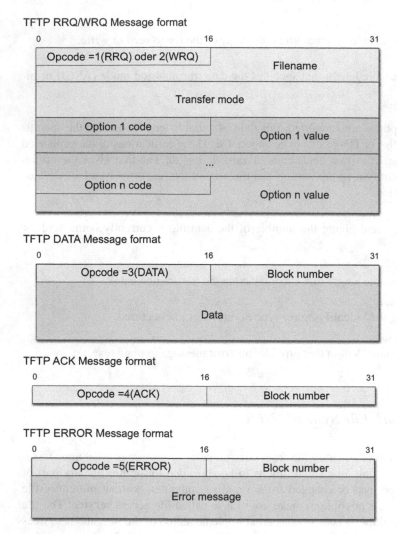

Fig. 9.41 TFTP Data packet formats.

access to their files. However, in mixed environments SMB with Samba [4] is used most of the time on the Unix page.

With NFS it is possible for an application to open a remote file, locate a specific place on the file and at that point to read, add to, or further change the data. The NFS client sends the modification data together with a corresponding request to the NFS server, where the stored file is located. The NFS server carries out the request by updating the relevant file and then sends a confirmation back to the NFS client, i.e.,

---

[4] Samba is an open source software suite that provides access to SMB services on UNIX computers.

in the implementation of NFS only the changed data is transmitted over the network and not the complete file. Moreover, NFS allows multiple access to a file. If other clients are to be kept from the file to be updated, it can be locked to foreign access. When the changes are completed, the lock is then removed by the NFS client.

In contrast to FTP, NFS is normally completely integrated into the file system of the computer and allows applications transparent access to the remote file systems. This is possible because NFS supports the regular operations of the file system, such as open, read and write. For the implementation of NFS, an individual file system is created at the client computer and linked to a remote file system at the NFS server. If an application accesses a file in this special file system, the NSF client performs the execution at the remote system. The local application works with the remote file system in the same way as if it were a local file system. Therefore, every application program allowed to access the local file system can, without hindrance, access the remote file system connected over NFS (cf. Fig. 9.42 and Fig. 9.43).

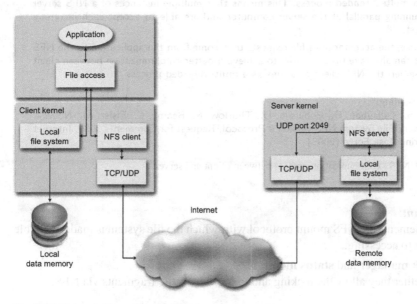

Fig. 9.42 Network file system - function

For access to the file system of a remote computer, NFS carries out a **Remote Procedure Call** (**RPC**), described in detail in the following section. A NFS system normally consists of more than just one NFS client and a NFS server. Together they implement the NFS protocol. The following additional processes are involved.

- **Port mapper**:
  Serves in assigning RPC routines to specific TCP/UDP port numbers.

A Typical Client/Server Communication Process with NFS

1. If a file is accessed by an application in a NFS file system, the access is executed in the same way as in a normal local file. On NFS, access is completely transparent for the application. The client's operating system kernel decides which components should be implemented for processing. Access to remote files are regulated via the NFS client and access to local files over the local file system of the client.

2. For access to a remote file, the NFS client sends a RPC request to the NFS server via the TCP/IP interface. Although NFS is generally associated with the UDP transport protocol, a NFS connection via TCP is also possible.

3. The NFS server receives a request from the client as an UDP datagram via the UDP port 2049 (another UDP port can also be used with the help of a port mapper).

4. The NFS server forwards the client's request to the local file system of the server who ensures access to the requested file.

5. Access to a remote file over NFS may well take some time to implement as it necessitates the NFS server contacting the local file system. So that the NFS server does not have to block any other client requests during this time, it runs as a so-called **multi-threaded** process. This means that multiple instances of a NFS server are running parallel at the server computer and are able to accept each incoming request.

6. Likewise, the acceptance of file requests that come from the application to the NFS client can also take time. In order to achieve a better synchronization between client and server, the NFS client also runs as a multi-threaded process.

Further reading:
Shepler, S., Callaghan, B., Robinson, D., Thurlow, R., Beame, C., Eisler, M., Noveck, D.: Network File System NFS version 4 Protocol, Request for Comments 3530, Internet Engineering Task Force (2003)

Fig. 9.43 NFS - communication process between client and server.

- **Mount**:
  Implements the NFS mount protocol with which the file system is made available prior to accessing.

- **Lock manager** and **status monitor**:
  Together they allow the locking and unlocking of file fragments via NFS.

9.4.4 Remote Procedure Call – RPC

In the design phase of the NFS implementation it was decided that the necessary procedures and routines would not be developed from scratch. Instead, tried-and-true routines for facilitating communication with applications that run on a foreign computer system should be applied.

For this purpose, the general mechanism of the **Remote Procedure Calls** (**RPC**) was used. It allows an operation call at a remote computer. In 1977, it was defined for the first time in RFC 707 and later in RFC1057 and in RFC 1831 (now RFC 5531).

Fig. 9.44 shows the embedding of the RPC client/server interaction in the TCP/IP reference model.

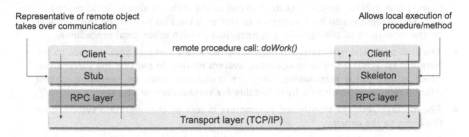

Fig. 9.44 Remote procedure call in the TCP/IP reference model.

RPC allows the implementation of a client/server-based interaction, using a special **transfer syntax** to encode client requests and server responses, defined in RFC 1014 as **external data representation (XDR)**.

Together RPC and XDR are available to the programming mechanism to program applications that can run distributed in the network. For this purpose, the programmer divides an application into a client side and a server side. These can then communicate with each other via RPC. Additionally, several operations are declared **remote** and the compiler is induced to embed RPC procedures in the code of the application. The programmer implements the desired operations on the server side and uses other RPC options to declare these as the server section of the application. If the application on the client side now calls up one of the routines declared as remote, RPC goes into action. RPC collects all necessary arguments that must be handed over to the server as parameters together with the corresponding operation call and then makes a suitable request to the server. RPC waits for the server's answer and then gives server's answer values back to the calling application. Fig. 9.45 shows a detailed representation of the RPC client/server interaction process.

The RPC mechanism shields the application as much as possible from the underlying protocol mechanisms. This also gives users who do not have in-depth insight into communication via TCP/IP the opportunity to program distributed applications. XDR, on the other hand, simplifies the transfer of data between computers with different architectures, i.e., it saves the programmer from writing conversion routines for the data formats to be transferred. XDR provides a means of displaying numerous, generally used data types, such as integer, unsigned integer, float, double, quadruple, as well as enumeration types, strings and more complex composite data types. The XDR notation is thereby oriented on the popular programing language C, which in the area of UNIX (and therefore also in TCP/IP) is widespread. In this way, XDR automates all necessary conversion mechanisms. The programmer need not call explicit XDR conversion routines to do this, rather a special XDR compiler is simply supplied with the data declarations, which must be accordingly transformed for the transmission. After that, the XDR compiler automatically generates a

The Process of a RPC Client/Server Interaction

The process of a RPC-initiated client/server interaction proceeds in the following way:

- On the client side, a local representative of the remote procedure/method (**remote object**) first calls the so-called **stub**. The call to the stub is a simple, local procedure call. The parameters to be transferred to the stub are also written by the compiler on the local stack of the operating system, just as with other local procedures.

- The stub packs the parameters to be transferred into a message that it sends to the (remote) server via a call to an operating system routine. In packing the parameters, which is also called **marshalling**, they are transferred from an operating system internal representation into a form suitable for transmission or storage.

- The message with the transferred parameters is sent by the operating system from the client to the server.

- On the server side, the transmitted message from the client stub is passed on by the port mapper to the calling procedure, the so-called **skeleton** (also called server stub). It unpacks the parameters from the received message (**demarshalling**), giving it to the (remote) procedure it has called.

- Following the procedure call, the skeleton takes the resulting parameters from the calling procedure/method and packs them into a new message. This is then sent back to the client stub the same way.

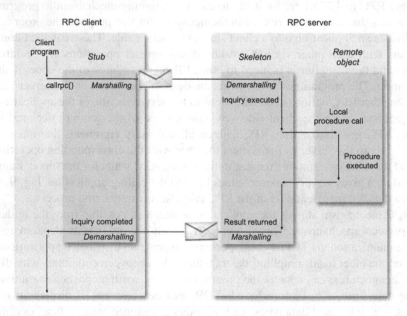

Further reading:
Thurlow, R.: RPC: Remote Procedure Call Protocol Specification Version 2, Request for Comments 5531, Internet Engineering Task Force (2009)

Fig. 9.45 The process of a RPC client/server interaction.

program that already contains the necessary XDR calls for conversion. In this way, the development of platform-independent client/server applications is made easier.

9.5 Remote Login

As previously described, there are a large number of client/server-based applications that communicate with each other via the TCP/IP protocol family. While some run without active user involvement others applications, such as FTP, offer the user an interactive interface to a remote computer. If such interactive interfaces would have to be programmed anew for every application at a remote computer and each were to follow its own protocol, the offering server would quickly be flooded with a multitude of different server processes. To prevent this, a general interactive interface is offered via which the user or an application can identify and register itself at the server computer. It creates the possibility of executing general commands there and of redirecting their output to the client computer

This **dialing in to a remote system** (**remote login**) enables the user or an application to execute commands that are in fact carried out at the remote machine. This means that programmers and application designers are not compelled to first develop specialized server applications for this purpose. The development of a general process for a remote login is a very complex matter. Normally, computers intend for the login procedure to be implemented exclusively via their own keyboards with output on the user's screen. Therefore, programming a remote login service requires access to the computer operating system. Despite the technical difficulties, there are remote login services in the meantime for nearly all operating systems. These allow users and application programs direct access to remote computers as clients. Among the most popular remote login Internet services are:

- **Telnet**:
  A standard application delivered with every UNIX implementation.

- **rlogin**:
  The remote login application of BSD UNIX Distribution. Originally intended only for remote login at Unix computers, it is available in the meantime for many operating systems.

- **rsh**:
  The remote shell (rsh), which is a variation of the rlogin service.

- **ssh**:
  The secure shell (ssh), a rsh variation, which guarantees a secure, encrypted data transmission.

9.5.1 Telnet

The TCP/IP protocol family contains a simple remote login protocol that enables the establishment of an interactive connection to a remote computer over the Internet. Called the **Telecommunications Network Protocol** (**Telnet**), it has its origins in the late 60's. It was defined in a series of RFCs, starting with RFC 15, through the 1983 published RFC 854, which is still the basic specification valid today, and on to numerous extensions up to RFC 4777.

With the help of the TCP transport protocol, Telnet establishes a connection to the remote computer in which the user's keystrokes at the local computer can be directly transferred to the remote computer as if input locally. Telnet uses the TCP port 23 at the server side to implement this. The TCP connection is maintained for the entire length of the Telnet connection. The monitoring of service quality in connection with TCP makes it possible to maintain a connection for hours, days and even weeks. Files are sent reliably and in the correct sequence with a data rate acceptable to both sender and receiver. In the opposite direction, Telnet passes the answers of the remote computer to the transmitted command back to the local computer and displays them there. Telnet offers a so-called **transparent** service. This gives the user at the local computer the impression of working directly at the remote computer and therefore completely undisturbed by the routines necessary for data transmission.

One of the difficulties the Telnet protocol must overcome is the necessity to guarantee a fundamental compatibility between the different hardware and operating systems it connects with one another. Telnet provides three basic services to the user.

* **Network Virtual Terminal (NVT)**

  The NVT provides a standard interfaces for access to remote systems. This has the advantage of hiding the details for communication of the necessary protocols, as well as the necessary conversion and calculation from the communication partners involved. The NVT works like an imaginary input and output device on which both communication partners map their local input and output, independent of which system platform is involved. The format used by the client computer is translated by the Telnet client into the NVT format, which is sent via TCP over the Internet to the server computer. There, the Telnet server receives the NVT format and translates it into the format necessary for the server computer (cf. Fig. 9.46).

  The NVT defines a virtual, character-based communication device that provides a keyboard as input device and a printer as associated output device. Data that the user enters on the keyboard, is transmitted to the server and every answer of the server is displayed via the printer. The data format used by NVT is designed very simply. The communication proceeds in 8 bit units (bytes). Thereby, NVT uses the standard 7 bit US-ASCII code for data and reserves the bit in position 8 bit (high order bit) for control commands and sequences.

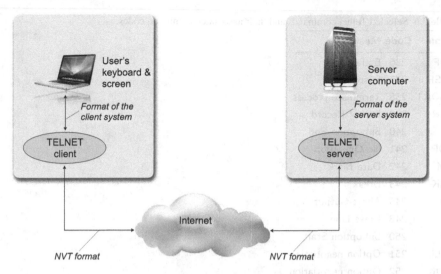

Fig. 9.46 Telnet client/server communication via the network virtual terminal.

The 7 bit US-ASCII code consists of 95 printable characters (letters, numbers or punctuation marks), which are adopted unchanged. There are also 33 so-called **control codes**, whose meaning is redefined by NVT to suit its needs. For example, the Telnet client translates an EOL control sequence (end of line), which is generated at the local computer by pressing the "RETURN" key, into the two part sequence CR-LF (Carriage Return-Line Feed). This must possibly be translated back again by the other communication partner. Additionally, NVT must also be in the position to translate control signals that are triggered by special keys on the client's keyboard (CONTROL-C causes e.g., an interruption of program execution at UNIX computers) into so-called **escape sequences**. An escape sequence contains two bytes, starting with the ASCII code for the "escape" control signal (byte 0xff (255 decimal), also called **IAC** (Interpret as Command)), followed by a byte code. This signals the control command to be carried out. Table 9.6 lists several Telnet commands and their associated control codes.

An application on the server computer that behaves erratically and needs to be brought under control by the appropriate signals from the client, could possibly block the interpretation of the running data stream. This would consequently impede the command and control sequences. In order to be able to transmit control signals in such situations as well, Telnet uses an **out-of-band** signal, that implements the special mechanism of prioritizing TCP data streams. For this purpose, the Telnet client sends a **Telnet Data Mark** (**DM**) protocol command to facilitate synchronization. This causes the TCP endpoint to set the TCP urgent bit (URG). By setting the URG bit, the regular TCP flow control is overridden and the associated data receives prioritized transmission. As soon as it notices the synchronization point in the data stream, the Telnet instance on the opposite side

Table 9.6 Selected Telnet commands and their associated command codes.

Name	Code	Meaning
EOF	236	End of File
SUSP	237	Suspend
ABORT	238	Terminate process
EOR	239	End of Record
SE	240	Suboption End
NOP	241	No Operation
DM	242	Data Mark
BRK	243	Break
AO	245	Abort Output
EL	248	Erase Line
SB	250	Suboption Start
WILL	251	Option negotiation
WONT	252	Option negotiation
DO	253	Option negotiation
DONT	254	Option negotiation

carries out a targeted search in the input buffer for Telnet control and monitoring commands, which leads directly to execution.

- **Option Negotiation**
  This service makes it possible for the involved communication partners to mutually negotiate and establish connection parameters and connection options. For the case that two mutually communicating Telnet versions do not have an identical set of available options, Telnet supplies a set of standard communication parameters with NVT. It is noteworthy that both client and server can take the initiative, the reason that the communication relationship via Telnet is called symmetrical. In case two Telnet versions communicate with each other that do not have an identical set of options, the following control commands are available to negotiate a common intersection of options:

  - WILL:
    The sender wants to activate the specified option itself.

  - DO:
    The sender wants the recipient to activate the specified option.

  - WONT:
    The sender wants to deactivate the specified option itself.

  - DONT:
    The sender wants the recipient to deactivate the specified option.

The option negotiation requires the exchange of three bytes respectively: the IAC byte (255), followed by the command control for WILL, DO, WONT or DONT, followed by a further command code identifying each specific option. Example:

    (IAC, WILL, 24)

There are more than 40 different options available over which the two communication partners can negotiate with each other. The command control named in the example, 24 says that one of the communication partners wants to set the terminal type used. Further details must then follow to specify the respective terminal type. In this case, the communication partners then negotiate about which **sub-options** to select. If the communication partner in the given example answers with

    (IAC, DO, 24)

the **sub-option negotiation** begins. The first communication partner asks the second for its terminal type:

    (IAC, SB, 24,1,IAC,SE)

„SB" indicates the beginning of the sub-option negotiation, "1" is the command code for the query regarding terminal type and "IAC SE" terminates the sub-option. The second communication partner can then respond, e.g., that the terminal type in question is a ibmpc. The respective coding of the various terminal types is specified in RFC 1091 .

    (IAC, SB, 24, 0, 'I', 'B', 'M', 'P', 'C', IAC,
    SE)

Further Telnet options are summarized in Table 9.7.

- **Symmetrical communication**
  Telnet treats both communication partners completely identically - thus symmetrically. It is not necessary for the requesting client to submit its input via a keyboard, just as the output of the answering server's results does not necessarily have to be displayed on the screen. Therefore, virtually any application program can use the services of Telnet and mutually negotiate each necessary connection option.

The procedure of a TELNET session is relatively simple (cf. Fig. 9.47).

1. The user, or application program, starts with the request to a Telnet client to establish a connection to a remote computer. The Telnet client reads the input of the user directly from the keyboard or gets it from the application program, which writes the corresponding input in the the standard input buffer.

2. Via a TCP connection, the Telnet client leads the entered command to the remote Telnet server. The Telnet server passes it on as user input to the the operating system of "its" computer for further processing.

Table 9.7 A selection of Telnet options with the corresponding RFC where they were specified.

OpCode	Name	Meaning	RFC
0	Binary Transmission	Permits the involved devices to exchange data in the 8 bit binary format instead of the 7 bit ASCII	RFC 856
1	Echo	Negotiation of a specific echo mode	RFC 857
3	Status	Query about the status of a Telnet option	RFC 858
10	Output Carriage Return	Permits the involved devices to negotiate how a line break should be handled	RFC 652
17	Extended ASCII	Permits the involved devices to use an extended ASCII coding for file transmission	RFC 698
24	Terminal Type	Permits the involved devices to negotiate a certain terminal type	RFC 1091
31	Window Size	Permits the forwarding of information on the current size of the terminal window	RFC 1073
32	Terminal Speed	Permits the forwarding of information on the current speed of the terminal	
37	Authentication	Permits the client and server to each negotiate a method for the authentication and security of data transmission	RFC 1416

3. The Telnet server passes the output generated back to the Telnet client, who displays it on "its" screen or passes it further to the requesting application program. The interface of the operating system, which allows the Telnet server to handle input from the requesting client in the same way as output over a local keyboard, is known as a **pseudo terminal**. This interface is vital when a Telnet server is to be installed.

4. Over the pseudo terminal a **login shell** is activated. Here, an interactive session at the remote computer takes place as if the respective input is entered via a normal, local terminal.

A Telnet server is capable of serving a multiple number of Telnet clients. Normally, a Telnet master server is at the server computer. With every new TCP connection a corresponding Telnet slave process is dynamically generated, which then processes the new connection. Thereby, every Telnet slave process connects a TCP stream from the Telnet client to a specific pseudo terminal in the operating system of the server computer.

9.5.2 Remote Login – rlogin

Operating systems based on BSD-UNIX have included the remote login service **rlogin** since version 4.2BSD. It was defined in RFC 1282 . Because rlogin handles remote login between Unix computers, the rlogin protocol could be designed

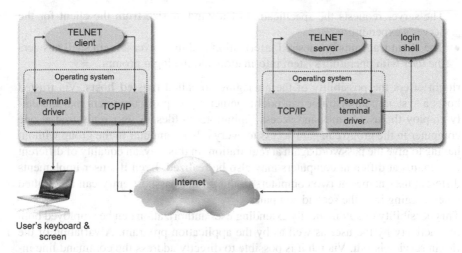

Fig. 9.47 Client/server connection via the Telnet protocol.

in a considerably simpler and less complex way than the Telnet protocol. This is because the option negotiations for mediating the respective display options could be eliminated. rlogin uses just a single TCP connection between client and server over TCP port 513. Following a successful TCP connection setup, a simple protocol starts between the client and server:

- The client sends four strings to the server:

  - the null byte 0x00,
  - the login name of the user to the client, terminated by a null byte,
  - the login name of the user to the server, terminated by a null byte and
  - the name of the user's terminal type followed by a slash (/), the transmission speed of the terminal, terminated with a null byte.

  Both login names are necessary as a user can implement different login names at the client and at the server.

- The server answers with a null byte.
- The server can then ask for the user's password. The password query and its necessary data exchange run over the regular rlogin connection without the necessity of a special protocol. If the client does not answer with the correct password before a previously established time limit has expired, the server ends the connection. The user can avoid the password query by creating a file .rhosts in his assigned directory at the server. This file contains the computer name of the client as well as the login name of the user. If a login query concerns a computer indicated there together with its respective login name, then the server omits the password query. Older rlogin implementations send the user-entered password in plain text. Their implementation is therefore not recommended for security-conscious users.

- The server requests the specification of a window size from the client for the server's expenditures.
- The client transmits the desired information and the server's login shell answers the user with operating system information and the login prompt.

rlogin offers the possibility of the managing so-called **trusted hosts**. Via trusted hosts a system administrator can pool together a group of computers that mutually employ the user names and access limitations to files. A user registered at one computer in the group can gain access to every other computer in the group without having to give the password again at registration. In this way, an equality of different user logins at different computers may also be realized. Even if a user implements different user names at two computers in the group, passport entry can be omitted when logging in at the second computer.

This possibility of automatically extending user authorization can be employed both interactively by the user as well as by the application program. A variation of the rlogin service is **rsh**. Via rsh it is possible to directly address the command line interpretor at a remote computer, without the necessity of having to run an interactive login procedure (**remote execution**). The entered command line via rsh

```
rsh -l user name IP address/domain name command
```

is simply transferred to the addressed computer, executed there and the result displayed at the output computer. As soon as the command on the remote computer is executed the session is automatically ended. If no command is given, an interactive session is initiated via rlogin at the remote system. Whether another computer is authorized to access a computer via rsh is verified based on the IP address, the source port and the user ID. Because this information is easy to counterfeit, rsh is not considered secure. The preferable alternative is discussed in the following section - the secure SSH.

9.5.3 Secure Shell – SSH

The disadvantage of the remote login services discussed so far is found in their unencrypted data transmission, which is carried out in plain text. The successor of rsh, rlogin and Telnet, **Secure Shell** (**SSH**) provides an encrypted remote login service. To sum it up, SSH offers the establishment of a secure, authenticated and encrypted connection via an insecure network.

Just as rlogin or rsh, SSH can manage access user rights for groups of computers. Moreover, the public key cryptography procedure **RSA** is used for authentication and encryption[5]. Version 2 of the SSH protocol (SSH-2) also offers DSA, Triple-DES and other encryption procedures as an alternative to RSA. Whenever data is

---

[5] RSA and other encryption procedures are presented in detail in the first volume of this trilogy, Meinel, Ch., Sack, H.: Digital Communication – Networking, Multimedia, Security, Springer (2013) in Chapter 5 "Digital Security."

to be transported between the client and server, it is automatically encrypted by ssh without the user being aware of it. The encryption provided by ssh proceeds completely transparently for the user.

Developed originally in 1995 by Finn *Tatu Ylönen* (* 1968) at the University of Helsinki as a purely UNIX application, SSH is available today from a number of system platforms as free software (OpenSSH). It is also a commercial product from the company SSH Communications Security Ltd. (founded by Tatu Ylönen). The architecture of the SSH protocol was defined in 2006 in RFC 4251. SSH is not limited to a simple terminal function in terms of features, but offers the following possibilities:

- The file transfer services SFTP and SCP offer cryptographically secure alternatives to FTP and RCP in retrieving remote procedures and objects (remote objects).

- The X Window graphic system X11 can be transported and secured over SSH. In this way, secure access is possible directly on the graphical user interface (GUI) of a remote computer.

- With the help of SSH any TCP/IP connections can be tunneled. In each case, one port from a remote server is forwarded to the client or vice verse (port forwarding). This means that an otherwise unencrypted virtual network (Virtual Network Computing, VNC) can be secured.

- An SSH client is capable of behaving in the manner of a SOCKS server[6] Thus, the client facilitates automatic access to a remote computer via a SSH tunnel and can be used to bypass a firewall.

- With the help of the SSHFS (Secure SHell FileSystem), a remote file system can be logged in (mounted) at the local computer, also by non-privileged users, via a secure shell. On the server side, SSHFS only needs a SSH server with SFTP subsystem. An authentication as well as the encryption of the transmitted data proceeds over SSH.

- It is possible to request the public key of a remote computer with the aid of "ssh-keyscan." This can help to to determine in a simple way whether the IP address or, for example, the DNS entry of a SSH server has been manipulated.

The server carries out **authentication**, by identifying itself to the client by means of a RSA certificate. In this way it is possible to detect if there has been any tampering. The client thereby has the choice of identifying itself via public key authentication with a private key, whose public key is stored at the server, or of using a normal password. The employment of a user password, however, always requires interaction of the user (as long as the password is not stored at the client computer in unencrypted form). In contrast, pubic key authentication enables the login of the client at the SSH server without user interaction. Storage of a password in unencrypted form is therefore not necessary.

---

[6] Computers behind a firewall wishing to establish connection to an external server, connect instead to a so-called SOCKS proxy. This proxy server verifies the entitlement of the requesting computer to contact the external server and forwards this request to the server (cf. also Section 9.9).

If the authentication was successful, a secret key is generated for the duration of the session. The entire data communication that follows is encrypted with it. Different encryption procedures can be employed in this process. SSH-2 uses AES with a 128 bit key length as standard procedure. Other key lengths are additionally supported by 3DES, Blowfish, Twofish, CAST, IDEA, Arcfour, SEED and AES.

9.6 Network Management

Besides the wide range of Internet services comprising the actual function of the Internet as a communication medium, we will now look more closely at just how networks are actually controlled and administered. The job of the network administrator is to detect and correct errors that can be caused by hardware or software in the network operating system. Other tasks include designing and assembling networks, configuring their components and always attempting to ensure the highest possible throughput with the greatest network availability.

Troubleshooting numbers among the most sophisticated tasks of the network administrator. Protocols can conceal possible packet loss and sporadically appearing hard and software errors can only be recognized and defined with difficulty. While the protocols in the TCP/IP protocol family have various error detection and correction mechanisms, errors always lessen network throughput and, therefore, the performance of the network. Thus, the network administrator must determine what has caused the error and fix it accordingly.

To be able to recognize hardware and software errors, the network administrator requires special tools. These can take the form of network hardware, e.g., interfaces and network analysis devices that monitor the entire network data traffic and offer a multiple number of filter and analysis functions for troubleshooting. The dedicated software also used is known as **network management software**. The different network management tasks are listed in Fig. 9.48.

The special network management software works on the application level of the TCP/IP protocol family. It has generally been designed as a client/server application. For the purpose of analysis, the network administrator establishes a connection to the server via a client on which the network components to be examined are structured. Conventional transport protocols such as TCP and UDP are employed. These exchange questions and answers based on the management protocol used. The TCP/IP protocol family provides the Dynamic Host Configuration Protocol (DHCP) and the **Simple Network Management Protocol** (**SNMP**) for administrative tasks and network control. These are discussed in detail in the following.

Network management functions

- **Fault management**
  Early detection of error conditions and the localization of error sources.

- **Configuration management**
  Query and change of the configuration of all managed system components.

- **Accounting management**
  Functions for the detection of the use of resources by a user as the basis for an accounting system.

- **Performance management**
  Monitoring functions for network services rendered, it additionally detects bottlenecks and their causes to optimize performance.

- **Security management**
  Functions for the protection of network components and services from abuse or access entry by unauthorized third parties.

Further reading:

Black, U. D.: Network Management Standards: SNMP, CMIP, TNM, MIBs and Object Libraries (2nd ed.), McGraw-Hill, Inc., New York, NY, USA (1994)

Fig. 9.48 Network management functions based on OSI-10164 or ITU-T X.700.

9.6.1 Dynamic Host Configuration Protocol – DHCP

The **Dynamic Host Configuration Protocol (DHCP)** was originally developed to facilitate the automatic assignment of IP addresses in local networks. It was standardized in 1993 as RFC 1541. In 1997, a first revision took place in what became standard RFC 2131. DHCP builds on the previously introduced BOOTP protocol (Bootstrap Protocol) in its most basic form, enabling a simple address assignment as well as the querying or provision of network configuration information.

The provision of an IP address is among the fundamental tasks of the DHCP protocol. Thereby, DHCP provides the requesting client different possibilities of address configuration:

- **Manual assignment**:
  A certain IP address is assigned to a specific device in advance by an administrator. DHCP is only used here to give the corresponding device the pre-reserved IP address.

- **Automatic assignment**:
  In this case, DHCP automatically assigns an IP address from the existing IP address pool. This is permanently assigned to the respective network device.

- **Dynamic assignment**:
  DHCP automatically assigns an IP address from the existing IP address pool. However, this address is assigned for only a limited time period. This period is either determined by the server or the end device returns the address as soon as it is no longer needed.

Compared with a manual or even static automatic address assignment, dynamic address assignment via DHCP offers the following advantages:

- **Automation**
  Every network end device receives an IP address if and when it needs it. There is no central administrator intervention necessary to decide which address should be assigned to whom.

- **Centralization**:
  All of the IP addresses available for the network are managed at a central location by a DHCP server. Thus, the network administrator can easily maintain a central overview of the currently assigned IP addresses and perform other network management task based on this.

- **Address reuse and shared use**:
  Through the limited usage time of IP addresses, it is assured that addresses are each only given to active network devices. In this way, other network devices also have the chance to use this IP address following the expiration of usage time. Thus, it is possible to manage and operate more devices in the network above and beyond the number of available IP addresses.

- **Universality and portability**:
  In BOOTP, or in manual DHCP address assignment, the identity of all network devices must be known beforehand so that a corresponding address assignment for the device can be determined via the central address administration. No advance assignment of IP addresses is however necessary for dynamic address assignment. This particularly eases the administration of mobile devices and those migrating between different networks.

- **Conflict avoidance**:
  Because all available IP addresses are administered over a central address pool, no address conflicts (competing address assignment) can occur.

Dynamic address assignment is the most popular form of DHCP address assignment. The selection of the so-called "lease time," which can be determined by the network administrator, is of great importance. Short-term assignments with a lease time in the minute range promise the greatest efficiency in a high-dynamic network environment with numerous mobile end devices signing in and out of the network. Longer lease times of up to months or years are advantageous in network environments that are primarily static. In this case, network devices should also be reachable at the same IP address for longer periods of time.

The dynamic assignment of IP addresses via DHCP makes a distinction between the following phases:

- **Allocation**:
  A client begins the cycle without having an active, assigned IP address.

- **Reallocation**:
  A client already has an active IP address that was assigned to it by the DHCP server. With the new start, the client queries the DHCP server to reconfirm the previous assignment.

- **Normal operation**:
  The IP address is assigned to the client by the DHCP server and it is thereby connected to the "DHCP lease."

- **Renewal**:
  As soon as the specific, predetermined time period of the DHCP lease has run out (expiration renewal timer T1, normally after 50% of the given lease time), the client again queries the DHCP server who had previously assigned the IP address to extend the DHCP lease and to keep the IP address.

- **Rebinding**:
  Should a renewal fail, because, for example, the DHCP was shut off (or after expiration of the rebinding timer T2 after 87.5% of the original lease time), the client tries to contact a DHCP server to confirm and extend its existing DHCP lease.

- **Release**:
  The client can return an allocated IP address at any time and thus end the DHCP lease (release).

The DHCP server plays a central role in managing the network participants in the local network. This role involves managing storage of the available IP addresses and administering their distribution. In this way, it keeps an eye on the still available IP addresses as well as on the IP addresses already assigned to the clients. Moreover, based on need, clients may also query the DHCP server regarding further parameters necessary for network operation. DHCP servers make it possible for network administrators to call up current information about the network from a central location, influence address allocation as well as to change and carry out further network management operations.

For processing client/server interaction, DHCP offers the following commands:

- DHCPDISCOVER:
  A client who does not yet have an IP address sends a broadcast request to the DHCP server or servers in the local network to get an offer for an address assignment.

- DHCPOFFER:
  The DHCP server/servers answer a DHCPDISCOVER client's request with corresponding offers.

- DHCPREQUEST:
  The client requests one of the available IP addresses in connection with further data from the DHCP server as well as an extension of the lease time.

- DHCPACK:
  Acknowledgement of the DHCP server to a DHCPREQUEST.

- DHCPNAK:
  Refusal of a DHCPREQUEST by the DHCP server.

- DHCPDECLINE:
  Refusal of an offer by a client because the IP address is already in use.

- DHCPRELEASE:
  The client releases its own address configuration so that the IP address is available again.

- DHCPINFORM:
  Query from a client about network information but not for an IP address. This is the case, for example, when the client already has a static IP address.

Fig. 9.49 and 9.50 show the schematic procedure of the DHCP duty cycle during address allocation, aided by a Finite State Machine (FSM). The DHCP-FSM is described from the perspective of the DHCP client and shows which messages are exchanged between the DHCP client and server in the course of address allocation.

The first time a new computer is started in a DHCP-equipped network, or if a computer in the network does not yet have an assigned IP address, the initial DHCP address allocation process begins. The DHCP client sends a DHCPDISCOVER message as a broadcast. This contains its own hardware address as well as a randomly generated transaction identifier. It may optionally contain a suggested IP address, a suggested lease duration as well as other parameters.

The network DHCP servers receive the DHCPDISCOVER message and check whether it could correspond to the respective request. The DHCP server answers with a DHCPOFFER message. This contains an IP address, an offered lease time, further client-specific parameters, an identification code for the relevant DHCP server and the original transaction identifier.

The DHCP client receives the DHCPOFFER messages from the DHCP server. Which strategy offer is chosen depends on the implementation. Once an offer is selected, the client sends a DHCPREQUEST message as a broadcast. Thereby, the chosen DHCP server is confirmed, while the other DHCP servers are made aware of their rejection. The message contains the DHCP server identifier providing identification of the chosen server, the IP address assigned to the client as well as optional, extra configuration parameters.

The DHCP servers receive the DHSPREQUEST message from the client. The servers rejected by the client still wait a period of time before the refused offer is assigned anew to other clients. [7] The accepted DHCP server then sends a DHCPACK message to acknowledge acceptance of the address allocation and carries out an address binding.

The client receives the DHCPACK message and normally carries out a final verification to make sure that the assigned IP address might not be in use after all. This is done with the help of an ARP query (Address Resolution Protocol). When a network participant answers the ARP query, the client rejects the assigned, already in-use IP address with a DHCPDECLINE message sent back to the server. In the other case, the client starts both lease timers T1 and T2 and uses the allocated IP address.

Besides serving its own network, a DHCP server can also operate in remote networks. For this to be possible, it is necessary that it be connected to the network

---

[7] This makes sense as there is still a possibility that the offer accepted by the client may not come to fruition, e.g., in the case of an error-related transaction abort or a timeout. In this case, after realization of the error, the client would take one of the alternate offers.

DHCP Duty Cycle as a Finite State Machine

- **INIT**: Beginning state in which the client requests an address. Final state when a lease time ends or a negotiation fails. The client sends a DHCPDISCOVER message as a broadcast to locate a DHCP server. Afterwards, the client enters the **SELECTING** state.

- **SELECTING**: The client waits for DHCPOFFER messages from DHCP servers and can select one of them. The chosen DHCP server is then contacted via DHCPREQUEST. The client enters the **REQUESTING** state.

- **REQUESTING**: The client waits for an answer from the queried DHCP server. The client then receives the DHCPACK message with IP address instructions from the DHCP server. If the IP address is not yet in use, the client takes over the associated parameters of the lease timers T1 and T2 and enters the **BOUND** state. In the other case, the address allocation is refused and the DHCP is informed with a DHCPDECLINE message. The client returns to the **INIT** state. It is also possible for the DHCP server to send back a DHCPNACK message and reject the client's address allocation wish. In this situation, the client also returns to the **INIT**.

- **INIT-REBOOT**: In the event that a client has already received a valid address but is off during the lease time, or is started new during this time, this state is entered instead of the **INIT** state. In sending a DHCPREQUEST message the client wishes to acknowledge the address it has already received. It then enters into the **REBOOTING** state.

- **REBOOTING**: The client is restarted and waits for acknowledgement of the previously allocated address by the DHCP-Server. The client receives a DHCPACK message with the DHCP server's IP address allocation. If the IP address is not already being used, the client takes over the associated parameters of the lease timers T1 and T2 enters the **BOUND** state. In the other case, the address assignment is refused and communicated to the DHCP server with a DHCPDECLINE message. The client again enters the **INIT** state. The DHCP server can also send a DHCPNACK message and refuse the client's address assignment wish. In this situation, the client likewise returns to the **INIT** state.

    BOUND: The DHCP client has a valid IP address and is in its normal state. If the first lease timer for refreshing the lease runs out, the client enters the **RENEWING** state. In the situation where the client wants to end the lease, it sends a DHCPRELEASE message and enters the **INIT** state.

- **RENEWING**: The client refreshes its DHCP lease. In doing so, it sends a DHCPRE-QUEST message in regular intervals and waits for the answer from the DHCP server. If the client receives a DHCPACK message from the DHCP server, the lease is renewed, the timers T1 and T2 are reset and the client returns to the **BOUND** state. But if the DHCP refuses a lease renewal, the client receives a DHCPNACK message and goes back to the original **INIT** state. In the case that timer T2 also runs out during the renewal phase, the lease is terminated and the client enters the **REBINDING** state.

- **REBINDING**: The client did not succeed in extending its lease at the original DHCP server and searches with a DHCP-REQUEST message for a DHCP server that will accept its extension request. Receiving a DHCPACK message from a DHCP server, the client renews its lease, timers T1 and T2 are reset and the client goes back to the normal operating mode **BOUND**. If the client receives a negative DHCPNACK message, this means that while a DHCP network server is ready to give an IP address to the client, it will only do so if the allocation process is started from the beginning. In that case, or also if the lease time runs out, the client again returns to the **INIT** state.

Fig. 9.49 DHCP duty cycle as a finite state machine.

DHCP duty cycle as a finite state machine - Part 2

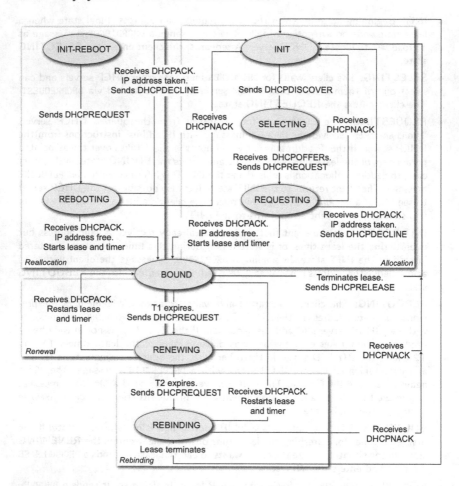

Further reading:

Droms, R.: Dynamic Host Management Protocol, Request for Comments 2131, Internet Engineering Task Force (1997)

Fig. 9.50 DHCP duty cycle as finite state machine - Part 2.

through a so-called **DHCP relay agent**, which in many cases also functions as a router. The DHCP relay agent also receives DHCP queries in the remote network and forwards them to the DHCP server. DHCP queries then reach the DHCP relay agent as a broadcast and are subsequently forwarded as a goal-oriented unicast. The IP address of the interface via which the broadcast of the DHCP relay agents is received is added to the unicast message. Based on this information, the DHCP server knows from which network segment the query comes.

The messages exchanged in the course of this process are created originally by the DHCP client. In answering the client, the DHCP server uses the message the client has sent it. If necessary, the contents of certain fields are exchanged in advance. Via an initially fixed transaction identifier (XID) from the client, all messages involved in the address allocation process are marked. Transport of the DHCP messages proceeds with help of the User Datagram Protocol (UDP). A client query is always transmitted to the server over UDP port 67. With the exception of those DHCP messages that could not be targeted to a specific IP address, the complete communication proceeds via broadcast. Because broadcast messages are considered quite inefficient, the DHCP server always attempts to first address an already known client directly via its hardware address (MAC address). UDP port 68 is used for the DHCP communication on the client side.

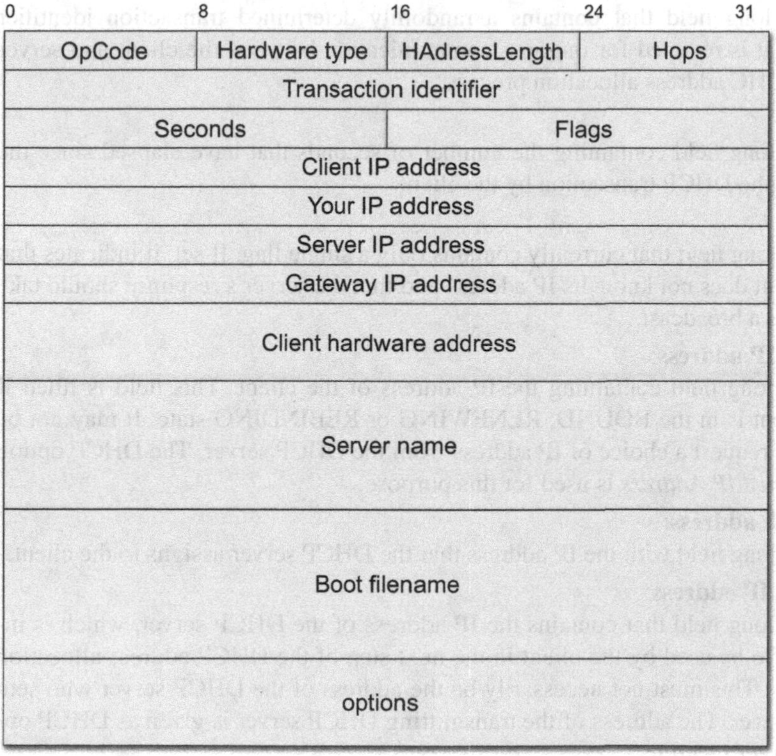

0	8	16	24	31
OpCode	Hardware type	HAdressLength		Hops
Transaction identifier				
Seconds		Flags		
Client IP address				
Your IP address				
Server IP address				
Gateway IP address				
Client hardware address				
Server name				
Boot filename				
options				

Fig. 9.51 Data format of the DHCP message.

The data format of the DHCP message is shown in Fig. 9.51 and contains the following fields:

- **Operation code**:
  1-byte long field indicating the transmission direction of a DHCP message, i.e.,

for a query from a DHCP client to a DHCP server OpCode=1 is used. An answering DHCP server uses Opcode=2 in the direction of the DHCP client.

- **Hardware type**:
  1-byte long field specifying the hardware type, analogous to implementation in the Address Resolution Protocol (ARP), e.g., Ethernet with HType=1.

- **Hardware address length**:
  1-byte long field giving the length of the hardware address in bytes, e.g., for a IEEE 802 MAC address applies Hlen=6.

- **Hops**:
  1-byte long field giving the number of DHCP relay agents passed along the data path. The initial setting of the transmitting client hop=0, and every DHCP relay agent increments the counter by the value 1.

- **Transaction identifier**:
  4-byte long field that contains a randomly determined transaction identifier (XID). It is retained for ongoing communication between the client and server in the DHC address allocation process.

- **Secs**:
  2-byte long field containing the number of seconds that have elapsed since the start of the DHCP transaction by the client.

- **Flags**:
  2-byte long field that currently contains only a single flag. If set, it indicates that the client does not know its IP address and that the server's response should take place as a broadcast.

- **Client IP address**:
  4-byte long field containing the IP address of the client. This field is filled if the client is in the BOUND, RENEWING or REBINDING state. It may not be used to request a choice of IP address from the DHCP server. The DHCP option *Requested IP Address* is used for this purpose.

- **Your IP address**:
  4-byte long field with the IP address that the DHCP server assigns to the client.

- **Server IP address**:
  4-byte long field that contains the IP address of the DHCP server, which is intended to be used by the client in the next step of the DHCP address allocation process. This must not necessarily be the address of the DHCP server who sent this answer. The address of the transmitting DHCP server is given as DHCP option *server identifier*.

- **Gateway IP address**:
  4-byte long field that contains the IP address of a DHCP relay agent. This serves in forwarding DHCP queries to remote networks. This field is not used to specify one of the routers (IP Gateway) to be used by a client. For this purpose the DHCP-Option *Router* is used.

- **Client hardware address**:
  16-byte long field containing the hardware address of the DHCP client.

- **Server name**:
  64-byte long field, which optionally has the name of the DHCP server. It can be a fully qualified domain name.

- **Boot file name**:
  128-byte long field that can be used by the DHCP client or server to request a special start directory or a special start file.

- **Options**:
  A field of variable length that may contain numerous different options necessary for the operation of the DHCP client in the network.
  A distinction is made between:

  - Manufacture-specific options required for the operation of different hardware.
  - IP-layer options, here parameters may be defined necessary for communication on the network layer.
  - Link-layer options, parameters are defined necessary for communication on the data link layer.
  - TCP-options, parameters are defined necessary for communication on the transport layer.
  - Application and service parameters necessary for control of various required applications and services.
  - DHCP extensions. These contain DHCP-specific parameters for the control and monitoring of the DHCP protocol itself. The specification of the DHCP message type numbers among these options.

DHCPv6

The new Internet protocol IPv6 has autoconfiguration possibilities available and does not need DHCP service for address allocation and for locating a router. For error-free operation, the IPv6 client needs the additional specification of a DNS server that is responsible for the resolution of domain names. Until this time there has not been a standard procedure to transmit this information to the client with the help of the IPv6 autoconfiguration.[8] For transmission of the DNS server responsible for a network participant, the **DHCPv6** protocol was specified in 2003 for IPv6, in RFC 3315. In general, DHCPv6 makes the same functionality available as the original DHCPv4 protocol. Additionally, via DHCPv6 configuration parameters for further services can be transferred, such as the Network Information Service (NIS), Session Initiation Protocol (SIP), Network Time Protocol (NTP) and other services. Moreover, it is possible for DHCPv6 to transfer information exclusively from securely authenticated network participants. Deviating from DHCPv4, the communication in DHCPv6 is carried out over the UDP ports 546 (client) and 547 (server).

---

[8] An experimental procedure for this purpose is described in RFC 5006 „IPv6 Router Advertisement Option for DNS Configuration" (2007).

9.6.2 Simple Network Management Protocol – SNMP

The Simple Network Management Protocol (SNMP) is considered the standard Internet protocol for network management. It forms the basis of most of the multi-manufacturer network management and analysis tools.

SNMP was first specified in 1990 as RFC 1157 and is currently in version SNMPv3 (RFC 3410) (cf. Fig. 9.52). SNMP allows the integration of network devices from different manufacturers within a network. All of them are uniformly managed by a single SNMP manager. SNMP messages sent from the network administrator by a client system (**manager**) to a network component, which is to examined, as the server (**agent**), all follow the standard known as **Abstract Syntax Notation.1 (ASN.1)**. While an in-depth look at ASN.1 would digress too far from the current topic, the brief example in Fig. 9.53 provides an illustration of the coding of integers with ASN.1.

The number of objects that can be administered via SNMP - the so-called "managed objects" - are described in the files known as **Management Information Base (MIB)**. These were defined in RFC 1155, RFC 1157 and RFC 1213. All variables are described in the MIB that an administered network object must provide together with the applicable operations and their meaning. Principally, the following MIB objects can be distinguished:

- **Counter**:
  For example, the number of the data packets discarded by the router because of errors in the data header or the number of detected Ethernet network interface card collisions.

- **Descriptive information**:
  For example, version and implementation of the software that runs on a DNS server, status information indicating whether a certain device functions correctly or protocol specific information, such as the routing path of data packets.

The respective contents of the objects managed by the MIB have their origin in different sources (cf. Fig. 9.54). Not only manufacturers of network components are involved, but also the network components themselves as well as their operators and users. All contribute information that is administered via MIB objects.

The MIB for TCP/IP subdivides the administered information into different categories (cf. Table 9.8) and is independent of the particular network management protocol used. The data format used for the description of the MIB objects is summarized under the heading **Structure of Management Information (SMI)**. It was defined in RFC 1155. SMI determines that, e.g., an object type counter is a positive, integer that can be counted from 0 up to $2^{32} - 1$, before starting again with the value 0.

For versions 1 and 2 of the network management protocol SNMP, all administered variables are defined in one large MIB. With the introduction of MIB2, the second version of MIB, the IETF took another path, so that for a multiple number of network objects, e.g., router, bridges and even interruption-free power supplies, MIB

A Brief History of SNMP

- **SNMPv1**:
  The first version of SNMP was developed in 1988. It was published in the three RFCs 1065, 1066 and 1067 in August 1988. These were replaced shortly afterwards by RFC 1155, 1156 and 1157. The three RFCs contain the specifications of the three Internet Standard Management Framework components: Structure of Management Information (SMI), Management Information Base (MIB) and the actual SNMP protocol. SNMPv1 gained general acceptance quickly and is still used in many places today. The main criticism of SNMPv1 is the lack of security. Client authentication proceeds in SNMPv1 solely with the help of an unencrypted "community string," i.e., a password is transmitted unsecured through the network, which means it can be intercepted and tampered with easily by unauthorized third parties.

- **SNMPsec**
  Already in 1992, a SNMP version with extra security functions was published as SNMPsec (Secure SNMP) in RFC 1351, 1352 and 1353. However, it never achieved widespread distribution.

- **SNMPv2**:
  SNMPv2 was published as RFC 1441, 1445, 1456 and 1457 in April 1993. The fundamental improvements of the new version apply in particular to improved security and confidentiality as well as to the newly introduced option of communication between multiple SNMP managers. However, just as SNMPsec, the original SNMPv2 version never found wide distribution. Different groups formed among the developers and they initiated development in various directions. Among them are:

  - **SNMPv2c**:
    SNMPv2c does not distinguish itself from SNMPv1 in the area of security but implements a few additional functions that were already included in SNMPv2. These include the GetBulk command. This facilitates reading multiple values at one time as well as communication between different SNMP managers. SNMPv2c is defined in RFC 1901, 1905 and 1906. This variation was able to survive on a broader basis and is generally synonymous today with SNMPv2.

  - **SNMPv2u**:
    An attempt was made to increase security in SNMPv2u by implementing user names. Yet, this failed to establish itself in the RFCs 1909 and 1910 specified variations of the SNMP protocol.

- **SNMPv3**:
  In SNMPv3, still the current version today, the security mechanisms were expanded considerably with the possibility of encrypting messages by means of different cryptographic procedures, including the guaranteeing of data integrity and a secure authentication. In December 2002, SNMPv3 was specified in RFCs 3410 - 3418.

A mixed form is often used in practical application, implementing different SNMP versions. These are usually made up of SNMPv1, SNMPv2c and SNMPv3.

Further reading:

Zeltserman, D.: A practical guide to SNMPv3 and network management, Prentice Hall PTR, Upper Saddle River, NJ, USA, (1999)

Fig. 9.52 A brief history of SNMP.

ASN.1 - Abstract Syntax Notation.1

In the sense of the ISO definition, ASN.1 offers a **presentation service** that mediates between different presentation types in different computer architectures. It does this by transferring them to a platform-independent presentation form. In its first version, ASN.1 was already specified in 1984 as a part of the CCITT standard X.409.

ASN.1 generally implements **Basis Encoding Rules** (**BER**) for coding. These are based on a **TLV** approach (**Type, Length, Value** - every data element to be sent is as a triple consisting of data type, length and actual value) and specify how instances of objects can be defined using the data description language ASN.1.

Integer values are also transmitted with a combination of length and value. Thereby, integers between 0 and 255 are encoded as a byte (plus one byte for the length) while integers from 256 to 32.767 need two bytes and larger integers require three or more bytes. ASN.1 encodes an integer as a pair of values:

Length **L**, followed by **L** bytes with the encoded integer.

Encoding may even exceed the specified length by one byte so that arbitrarily long integers can be encoded.

Example:

Integer Length Value (hexadecimal)		
33	01	21
24.566	02	5F F6
190.347	03	02 E7 8B

A detailed description of ASN.1 is found in the relevant ISO standard documents.

Further reading:

International Organization for Standardization: Information Processing Systems – Open Systems Interconnection – Specification of Abstract Syntax Notation One (ASN.1), International Standard 8824 (1987)

International Organization for Standardization: X.680: Information Technology Syntax Notation One (ASN.1): Specification of Basic Notes, ITU-T Recommendation X.680 (1997) / ISO/IEC 8824-1:1998

Fig. 9.53 Encoding of integers based on ASN.1.

was specified in each. In the meantime, these have been defined in more than 100 different RFCs. Table 9.9 shows several examples of MIB variables and lists their meaning.

The name of MIB objects is derived from the general ISO/ITU object identifier name space. This name space is ordered hierarchically and is globally valid, meaning that it gives all MIB objects globally unique names.

The name results from the ISO/ITU name convention based on stringing together the numerical values assigned to the nodes of the name space. At the top, the name space is divided as follows:

- **Level 1**:
  The entire name space is divided into objects that are allocated to either the ISO (1), the ITU (2) or both organizations (3).

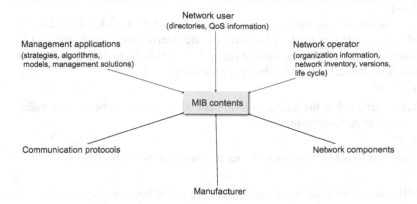

Fig. 9.54 Sources of the MIB management information.

Table 9.8 A selection of MIB2 categories.

MIB category	Information received	Number of objects
system	System information via host or router	7
interfaces	Network interfaces	23
at	Address translation (e.g., via ARP)	3
ip	IP software	38
icmp	ICMP software	26
tcp	TCP software	19
udp	UDP software	7
egp	EGP software	18
snmp	SNMP application entities	30

Table 9.9 A selection of MIB2 variables.

MIB variable	Category	Meaning
sysUpTime	system	Time since last system start
ifNumber	interfaces	Number of network interfaces
ifMtu	interfaces	MTU of a specific interface
ipDefaultTTL	ip	Given datagram lifetime
ipInReceives	ip	Number of received datagrams
ipForwDatagrams	ip	Number of forwarded datagrams
ipOutNoRoutes	ip	Number of routing errors
ipReasmOKs	ip	Number of reassembled datagrams
ipRoutingTable	ip	Contents of the IP routing tables
icmpInEchoes	icmp	Number of received ICMP echo requests
tcpRtoMin	tcp	Minimum time for new transmission
tcpMaxConn	tcp	Max. possible number of TCP connections
tcpInSegs	tcp	Number of received TCP segments
udpInDatagrams	udp	Number of received UD datagrams

- **Level 2-4**:
  MIBs are assigned to the subtree under the authority of the ISO. On level 2 this is the subtree intended by the ISO for national and international organizations (3). It is followed on level 3 by the subtree set up for the U.S. Department of Defense (6), which is succeeded by the subtree for the Internet (1).
- **Level 5**:
  The name space below the Internet subtree contains, e.g., the subtree responsible for the network management (2).
- **Level 6**:
  Here begins the subtree that starts the name space for the MIB object (1).
- **Level 7**:
  The MIB categories in the MIB name space are found here once again.

Corresponding to the numerical values, which the objects along a path in the name space are assigned, a MIB name always starts with the prefix `1.3.6.1.2.1` or, as the case may be, with the textual representation (cf. Fig. 9.55)

```
iso.org.dod.internet.mgmt.mib
```

Therefore, the MIB variable `ipInReceives` has the complete name

```
iso.org.dod.internet.mgmt.mib.ip.ipInReceives
```

or its numerical equivalent

```
1.3.6.1.2.1.4.3.
```

Besides simple variables, such as numbers, MIB variables can also follow a more complex configuration, such as tables or arrays, whose subelements are appended to the MIB name in the form of a suffix, e.g.,

```
iso.org.dod.internet.mgmt.mib.ip.ipRoutingTable.
            ipRouteEntry.field.IPdestaddr
```

which designates an IP address entry from a routing table.

Related MIB objects are grouped into so-called **MIB modules**. The MIB objects themselves are described in one of the definition languages as **SMI** (Structure of Management Information) and defined in the RFCs 1155 and RFC 1902. SMI itself is however based on the general definition language ASN.1.

To explain the SNMP protocol, several concepts from SNMP terminology need to be defined:

- **SNMP device types**: The implementation of SNMP makes it possible for the network administrator to operate special network devices to collect information from other network devices and monitor and control their operation. Two basic types of devices can thereby be distinguished:

  - **Managed Node** (**MN**)
    Term for all managed network components (also **Managed Network Entity** (**MNE**) or **Network Element** (**NE**)) that are monitored and controlled with the help of the SNMP protocol.

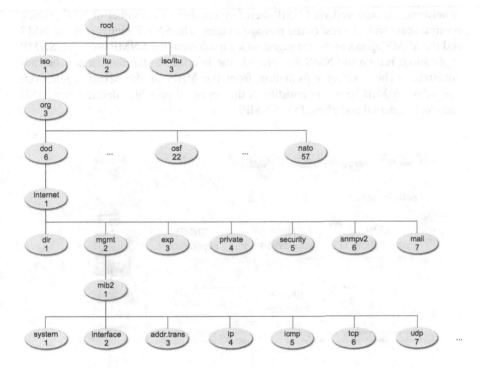

Fig. 9.55 MIB object names in the ISO/ITU object identifier name space.

- **Network Management Station (NMS)**
  Designation for a specific computer from which the network administration is operated and the regularly managed nodes can be monitored and controlled.

- **SNMP entities**
  Every device administered within a network interconnection via SNMP must have special software that implements the various functions of the SNMP protocol. Each SNMP entity consists of two primary software components whose composition determines the function of the SNMP device:

  - **Managed Node Entities (MNE):**
    The SNMP entity of a managed node consists of a **SNMP agent**, which makes the information about SNMP available for a NMS and follows its instructions. The **SNMP management information base** determines the type and amount of information provided.

  - **Network Management Station Entities (NMSE):**
    The SNMP entity of a NMS consists of a **SNMP manager**, which retrieves and collects information from the managed nodes or can send them instructions, and a **SNMP application**, which serves as network management interface for the network administrator.

A network administered via SNMP therefore consists of a number of NMS, which communicate with the rest of the managed nodes. The SNMP manager at the NMS and the SNMP agents at the managed nodes implement the SNMP protocol. SNMP applications run on the NMS and provide the interface to the human network administrator. They collect information from the MIBs of the SNMP agents (cf. Fig. 9.56). A MIB therefore constitutes the set of all variables defined in a MNE and can be queried and altered via SNMP.

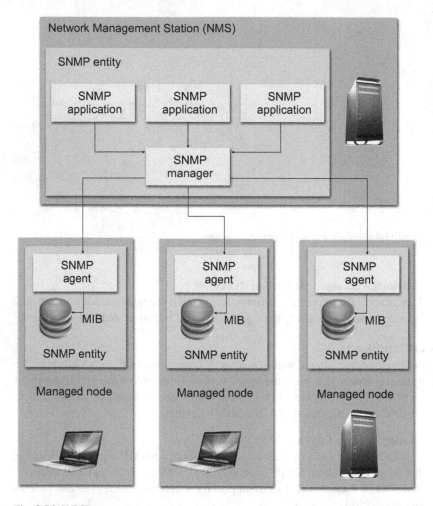

Fig. 9.56 SNMP - operational model.

Network management protocols generally have the task of regulating the communication between a client program, which serves the network administrator, and a server program, which runs on one of the network components to be checked. In contrast to what might be expected, SNMP has very few commands for this purpose. These follow a simple **fetch-store** paradigm. To determine device status, the network administrator queries the relevant component with a **fetch** command and thus "retrieves" the requested value. On the other hand, the network administrator is capable of controlling a device via a **store** command in which a MIB variable is simply assigned a certain value that must be taken into account by the specific device. All other necessary operations result virtually as a "side effect" of both of these commands. For example, an explicit command for restarting a network device does not exist. The network administrator can, however, set a variable that contains the time period until the next new start to zero and thus trigger a restart.

This type of approach was chosen because it is both robust as well as simple to implement and very flexible. The SNMP definition itself remains unchanged even if new MIB objects and MIB variables are added and additional effects can be achieved via the implementation of new value assignments. In reality, there are more than only two SNMP commands. SNMPv2 defines seven types of messages known as a **Protocol Data Unit** (**PDU**) (cf. Table 9.10 und Fig. 9.57).

Table 9.10 SNMP – possible operations

Operation	Meaning
get-request	Request for the value of a MIB variable
get-next-request	Request for a value without knowledge of its exact name
get-bulk-request	Request for a MIB variable block
response	Answer to an inquiry
set-request	Value assigned to a MIB variable
inform-request	Informing another manager (not the network administrator)
snmp2-trap	Device informs the network manager about the occurrence of an event

get-request and set-request are the two fundamental fetch and store commands. The network administrator requests data with get-request, which the relevant device then sends back via response.

With the command get-next-request the network administrator can request a sequential table entry without knowing its exact name. A trap reports an event identified by a network component, such as a failure or the restored availability of a network connection to the network administrator. The command inform-request can be used for communication between different NMS to inform them about an event or the queried status of a network component. In this way, distributed network management is also possible via SNMP.

Fundamentally speaking, SNMP can be transferred to every transport protocol because of its simple request/response architecture, yet in practical application transport always proceeds via UDP and port number 161 (trap message via port number 162), named in RFC 1906 as the preferred transport protocol for SNMP. UDP is

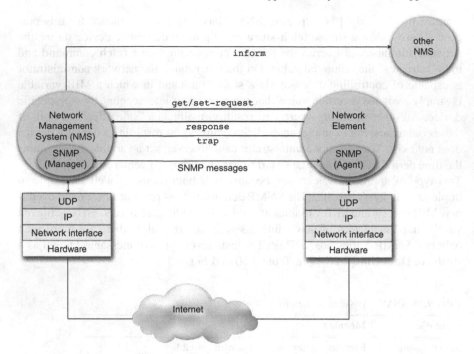

Fig. 9.57 SNMP protocol stack and SNMP operations.

an unreliable transport protocol. For this reason, there can be no guarantee that a SNMP message actually reaches a predetermined target. The SNMP data format therefore provides a request ID. It sequentially numbers the NMS for queries to an agent. The agent's corresponding answer is taken over by the request ID of the query. This results in an assignment of request and response. An additional timeout mechanism can be installed that starts a repeat transmission if a failed message has been identified. A new transmission mechanism is not explicitly required and is the responsibility of each NMS.

The generation and forwarding of messages in SNMP differs somewhat from the typical TCP/IP client/server model. There is no explicitly established client or server in SNMP as administration information can be obtained from all devices involved in the network. As previously described, the message exchange is mostly limited to simple pairs of request/ response contents. Here, NMS takes over the role of the client and the addressed SNMP agent the role of the server, even though it is not a server in the true sense of the word. SNMP traps deviate from this scheme. With the triggering of a trap, an SNMP agent sends a message to the NMS, without being requested to do so beforehand. Trap messages are not acknowledged upon receipts. This is also the reason that an answer does not follow from the side of the NMS. In contrast, an acknowledgement message is sent back to the NMS in response to a `inform request` message, which it has generated.

SNMP messages are comprised of a SNMP header, providing the data fields for control, identification and security, and a SNMP body, which contains the actual SNMP PDU. Here, the administrative information between the network devices is exchanged. PDU data format and SNMP wrapper data format are usually viewed separately. This is to distinguish the data fields for the basic functions of the SNMP protocol within the PDU from the various security additions.

A general distinction is made between the following simple PDU substructure:

- **PDU control fields**:
  A set of command and control fields that describe the PDU and transport information between one SNMP entity and another.

- **PDU variable bindings**:
  Descriptions of MIB objects in the form of a simple "binding" of a given name to a value. The name corresponds to the numerical MIB identifiers. The value at a get request remains empty and serves solely as a placeholder. A set request includes the value transferred to each addressed device.

Every PDU follows this general structure. However, the number of command and control fields or the respective variable bindings may vary (cf. Fig. 9.58).

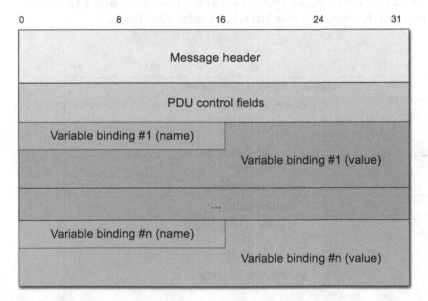

Fig. 9.58 General structure of a SNMP PDU (protocol data unit).

Representing the various SNMP versions here are the message data formats SNMPv1, SNMPv2c and SNMP3:

- **SNMPv1**: The general data format for SNMPv1 consists solely of a simple wrapper containing a short header with the following data fields (cf. Fig. 9.59):

- **Version number**:
  4-byte long field that contains the value 0 (instead of the actual version number 1).

- **Community string**:
  Data field of variable length. It identifies the respective "SNMP community" where the sender and receiver are located. This field serves as a simple (albeit inadequate) authentication.

With the exception of the `trap` PDU, all SNMPv1 PDUs have the same data format:

- **PDU type**:
  4-byte long field indicates the message type.

- **Request ID**:
  4-byte long field serves to identify the associated request/response pairs.

- **Error status**:
  4-byte long field that returns the result of a query in a `GetResponse` PDU. The value 0 indicates that the query could be carried out successfully.

- **Error index**:
  4-byte long field contains a pointer to an object that has caused an error, meaning that the associated error status has a value other than 0.

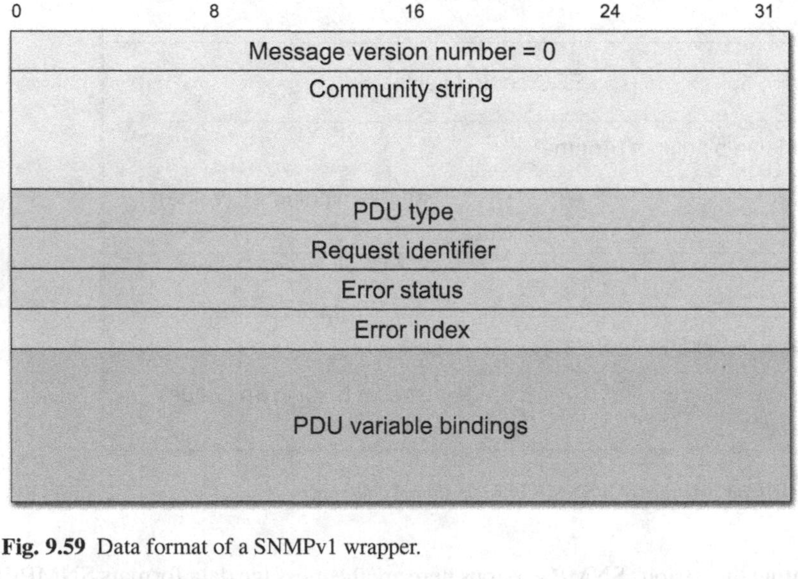

0	8	16	24	31
Message version number = 0				
Community string				
PDU type				
Request identifier				
Error status				
Error index				
PDU variable bindings				

Fig. 9.59 Data format of a SNMPv1 wrapper.

- SNMPv2c:
  The data format of the SNMPv2c wrapper corresponds to the structure of the already introduced SNMPv1 wrapper. Only here the version number has the va-

lue 1. The structure of the associated SNMPv2c PDUs also remains the same. The range of value in the PDU type field increases due to the greater number of SNMPv2 PDUs.

- **SNMPv3**:
  While the data format of the SNMPv3 wrapper follows the same concept as its predecessors, it has been expanded. It is possible to divide the fields of the header into security-critical and non-security-critical fields, whereby the security-critical fields correspond to the chosen security model. The following data fields are contained in the header (cf. Fig. 9.60):

  – **Message version number**:
    4-byte long field that gives the SNMP version. In the case of SNMPv3 the value is 3.

  – **Message identifier**:
    4-byte long field for identification of the SNMPv3 message.

  – **Maximum message size**:
    4-byte long field that gives the maximum length of a message that the sender can receive again. The minimum value is 484.

  – **Message flags**:
    1-byte long field containing command and control bits.
    Bit 1 signals the use of a secure authentication, bit 2 the use of encryption and bit 3 the return of a `Report` PDU.

  – **Message security model**:
    4-byte long field that indicates the security model implemented.

  – **Message security parameters**:
    Field of variable length that contains the associated parameters for the currently chosen security model.

  – **Context engine**:
    Field of variable length containing the ID of an application, which is to be transferred to the sent PDU for further processing.

  – **Context Name**:
    Field of variable length describing the particular context associated with a PDU.

In contrast to TCP, there is a length limitation with the simpler UDP protocol regarding the maximum allowed packet size (which is in turn dictated by the IP datagram length) The SNMP standard requires that SNMP implementations are able to transport a minimum length of 484 bytes. It is also recommended that even larger messages be carried with a length of up to of 1,472 bytes. This corresponds to the maximum amount of data that can be transported on a 1,500 bytes limited Ethernet frame as payload. If several MIBs were now to be requested at the same time via a `GetBulkRequest` operation, the prescribed maximum length for the UDP data packet could not be exceeded.

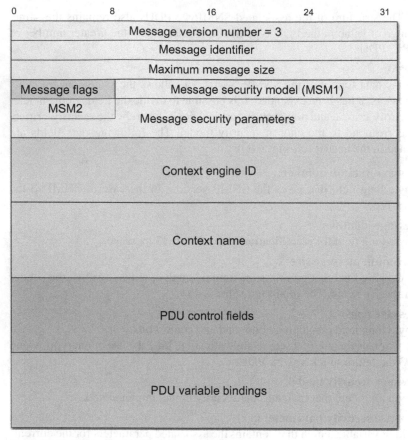

Fig. 9.60 Data format of a SNMPv3 wrapper.

SNMPv3 presents an extension of the functionality of the SNMP standards up to now. Possible SNMP applications are divided into several subunits in SNMPv3:

- **Network Management System (NMS)**

  - **Command generator**: Generates the commands `get-request`, `get-next-request`, `get-bulk-request` and `set-request` and handles the answers received to these queries

  - **Message recipient**: Receives and processes trap commands that are sent from an agent.

  - **Proxy**: Serves in forwarding queries, answers and trap commands.

- **Network Element (Agent)**

  - **Command generator**: Receives, processes and answers commands of type `get-request`, `get-next-request`, `get-bulk-request` and `set-request` with a response message.

– **Notification originator**: Generates trap commands that report any occurring events to the NMS.

In contrast to the first two versions of the SNMP specification, SNMPv3 includes additional security and administration capabilities. In particular, the introduction of new techniques for security protection were urgently needed as SNMP had been relegated to monitoring instead of active command and control, due to security concerns. An unauthorized third party who succeeded in intercepting SNMP messages and of smuggling several of its own in the management infrastructure could cause considerable harm to the network. SNMPv3 supports the following security measures:

- **Encryption**
  SNMP commands use DES (Data Encryption Standard) for the encryption of transported contents.

- **Authentication**
  SNMP combines a hash function with a secret key value to ensure both the authentication of the originator as well as the integrity of the transported data. This procedure is known as the **Hash Message Authentication Code** (**HMAC**) and is defined in RFC 2104 .

- **Protection against playback attacks**
  This is implemented via a so-called **nonce**. A nonce is a numerical value that is randomly generated and may only be used a single time in a communication. In the case of SNMP, the receiver requires that the sender embed a value in every message, based on a counter at the receiver. This counter serves as a nonce and is made up of the time that has passed since the last restart of the receiver's network management software and the total number of restarts since the last configuration of the receiver's network management software. As long as the counter in a received message only deviates from the actual value at the receiver within a certain error threshold interval then the message will be accepted.

- **Access control**
  SNMPv3 implements access control based on **View**. This indicates which network management functions may be carried out by which users. Every SNMP unit additionally implements a local configuration data store containing information regarding access rights and the associated access strategies.

9.7 Audio and Video Communication

9.7.1 Multimedia Applications on the Internet

Multimedia offers that are available via the Internet in virtually real-time have contributed significantly to the immense popularity of the medium and in large part to the data traffic on the Internet today. The constantly increasing bandwidth offers the possibility to transmit very large images as well as audio and video data in a time also acceptable for the private user. Additionally, so-called **continuous media applications**, such as Internet radio, Internet telephony, video-on-demand, video conference, teleteaching (cf. Fig. 9.69), interactive games or virtual worlds have gained increasing importance.

The continuous reproduction of a remote multimedia content, which is already carried out in real-time during the transmission process, is generally known as **streaming**, or **media streaming**, . It is achieved by buffering the transmitted data before the actual display (cf. Fig. 9.61).

Data Streams and Streaming Media

In contrast to statistical data, which is mostly stored in the database in the form of so-called tupel values, in computer science a continuous sequence of data records whose end cannot be unforeseen is known as a **data stream**. The individual data records of a data stream are of an arbitrary fixed type. They have an ordered, temporal sequence and can occur as practically unlimited. The amount of records per unit of time (data rate) can vary, which means that processing can not be carried out as a whole but only sequentially.

The concept of data streaming is widespread in modern programming languages. It goes back to the **pipes** concept, proposed by *Doug McIlroy* (* 1932), for linking macro programs. This mechanism was developed in the early 60's and implemented in the Unix operating system in 1972. Pipes involves a process data connection based on the FIFO principle.

One of the most popular data stream applications is **streaming media**. Here, audio and video data can be transmitted, received and simultaneously played via the Internet. The programs transmitted in this way are known as **livestream**. Unlike the radio, the data is not transmitted to all participants but exclusively between two (unicast) or multiple (multicast) end points.

Further reading:

Mack, S.: Streaming Media Bible, John Wiley & Sons, Inc., New York, NY, USA, 1 edition, (2002)

Fig. 9.61 Data streams and streaming media.

Compared to standard data transmission on the Internet, continuous multimedia data places other demands on the underlying communication network.

Two of these requirements are particularly critical:

● **Timing**

Continuous multimedia applications are especially delay-sensitive. If the transmission delay of individual data packets exceeds the set limit for this type of application (in Internet telephony in the millisecond range and in so-called streaming applications in the second range), the data packets are useless. Buffering the transmitted data before the actual display makes it possible to compensate for delay fluctuations (jitter).

- **Data loss**
  Continuous multimedia applications generally have a high tolerance rate with regard to data loss or data damage. While these types of errors reduce the quality
  of the reproduced multi-media data in the short term, the consumer's tolerance
  threshold is relatively high and it takes quite some time before they are recognized as a disturbance.

Continuous multimedia applications can be placed in the following categories based
on their requirements:

- **Streaming of stored audio and video data streams**
  In this case, a client requests data from a server that has previously been stored
  in its entirety there and is available in the form of a compressed file. The client is
  capable of navigating through the multimedia data during transmission, whether
  it be to fast forward, rewind or scroll. The acceptable reaction time between the
  client's inquiry and the implementation of such actions is in the area of 1 and 10
  seconds. Even when a continuous display of the transmitted files takes place, the
  conditions regarding delay in data packets during transmission are not as strict
  here as in interactive applications (video conference). **Video on demand** is e.g.,
  one of the applications of this technique. The users requests selected video files
  for transmission to his local display device.
  A basic distinction is made between **just-in-time streaming**, where the display
  already begins parallel to data transmission and is implemented via the RTP protocol (Real Time Protocol), and **HTTP streaming**. In the latter case, the multimedia data is first displayed in the WWW after complete transmission via HTTP.
  To avoid the buildup of excess waiting time, the video files to be transmitted
  with HTTP streaming are broken down in advance into small parts - so-called
  segments. If the pieces are sufficiently small, or the bandwidth large enough, the
  illusion of continual media streaming is likewise created.

- **Streaming live audio and video data streams**
  This type of continuous multimedia data transmission can best be illustrated with
  the analogy of a standard radio or live television broadcast. While being recorded, the multimedia data is transmitted directly via the Internet. The received
  data stream is then displayed with a minimum of delay, after the appropriate buffering is carried out. Buffering is necessary to make up for fluctuations caused by
  the delay of individual data packets. If the data stream is stored locally, it is also
  possible to navigate within the multimedia data (stop and rewind). A live data
  stream is generally received by many clients at the same time. To avoid an unnecessary replication of transmitted data packets and the possible network overload
  that might result, the data stream is sent as a **multicast** transmission. Restrictions
  regarding fluctuations in the delay of transmitted data packets are not as serious
  here as in the case of interactive applications (video conferences).

- **Interactive real-time audio and video**
  In this category, applications are found such as **Internet telephony** or **video
  conferences** via the Internet. Here, the user has the possibility to communicate with other users in real-time via the interactive exchange of audio and vi-

deo information. In the meantime, Internet telephony has become a wide-spread and reasonably-priced communication technique that offers all possibilities of circuit-switched telephony and even expands them (conferencing and filtering mechanisms). In video conferencing, users communicate visually and verbally over the Internet. Due to the interaction of multiple users, the time restrictions regarding delay are especially stringent, otherwise these would be distracting and affect communication considerably. Voice delays of up to 150 ms (milliseconds) are not noticeable, 150 ms to 400 ms are perceived as acceptable while delays greater than 400 ms are perceived as unpleasant.

Applications for the transmission of audio and video data are often referred to as **real-time applications** because they provide a timely transfer and display. Such a real-time signal is, however, only understandable for the communication partner if the same time conditions apply to both sender and receiver. The transmission tact must correspond exactly to the reception tact. Too big of a delay, e.g., in a telephone conversation or a video conference, is not only perceived as extremely disturbing but can also interfere with correct understanding. Communication that obeys these time restrictions is referred to as **isochronous** communication.

Yet the Internet and IP protocol it is based on is far from isochronous. Data packets can be duplicated or get lost, a minimum of delay cannot be guaranteed, and delivery is subject to strong fluctuation (**jitter**). In addition, the correct sequence of transmitted data packets delivered is not guaranteed.

So, how is this type of isochronous data traffic via the Internet even possible at all? To ensure at least virtually isochronous data traffic, the available protocols must receive additional support to circumvent the weak points mentioned.

- **Datagram duplication and order permutation**
  The individual data packets are provided with **sequence numbers** to solve this problem. In this way, duplicated data packets are immediately recognized and the original order of the data stream can be reconstructed.

- **Fluctuation and transmission delay**
  To correct an occurrence of the problem of jitter, every data packet has **time-stamp information**. This tells the receiver when the received data packet is actually to be played.

Sequence number and timestamp information make it possible for the receiver to reproduce the original signal exactly - independent of when the individual data packets in fact arrive.

To equalize fluctuation resulting from the transmission delay of individual data packets, it is necessary to additionally implement a **playback buffer** (cf. fig. 9.62) if the display of the transmitted audio and video information is to take place virtually in real-time.

At the beginning of a data transmission the receiver delays the display for the transmitted audio and video data, initially writing it in a playback buffer. When the amount of data in the playback buffer reaches a predefined threshold (**playback**

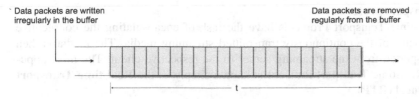

Data packets are written
irregularly in the buffer

Data packets are removed
regularly from the buffer

t

Fig. 9.62 Principle organization of a playback buffer.

point), the receiver begins the display. The threshold corresponds to the time duration **t** for a playback security specification, whereby display of the audio and video data is delayed. During display, the playback buffer is continuously being filled with incoming data packets. If these arrive at the receiver without fluctuations occurring in transmission delay, the playback buffer empties itself at one end just as quickly as it fills up at the other end. Smaller fluctuations in transmission delay can be balanced out by this type of intermediate storage and are not noticeable in playback. If a delay occurs, the buffer empties out while transmission is interrupted for up to **t** time units. As soon as the delayed data packets arrive, the playback buffer fills up again. however, a playback buffer is not capable of compensating for lost data packets. If a data packet does indeed get lost, then a break or defect occurs in the displayed data stream.

The threshold choice of **t** is critical. If a too small **t** is chosen, potentially longer fluctuations cannot be compensated. While with a too large **t** delay fluctuations can be eliminated, but this also adds to the general transmission delay. Real-time communication can be subsequently hindered or even prevented from taking place at all. On the other hand, the sound quality increases with the rising threshold **t**.

Media streaming also follows the known client/server paradigm. A client requests an audio or video file that is located at the remote server. The server could either be a normal web service (HTTP server) or a special streaming server. While the web server transmits the requested files over the standard transport protocol HTTP in its entirety, or divided into segments as HTTP streaming, streaming servers are capable of performing actual continuous media streaming.

In continuous media streaming, the media files to be transmitted are also in a special segmented form. This means, the coding of the media data does not proceed as a monolithic file, but rather in sections, divided into single segments (chunks). Each of them has its own short file header with associated command and control information.[9]

The transport of continuous streaming media data proceeds with special real-time protocols just for this purpose. These are described in detail in the following sections. According to their respective task they can be divided into the following types:

---

[9] A detailed look at the coding and compression of multimedia data appears in the first volume of this trilogy, Meinel, Ch., Sack, H.: Digital Communication – Networking, Multimedia, Security , Springer (2013) in Chapter 4 „Multimedia Data and Encoding."

- **Transport protocols**:
  Real-time Transport Protocols have the task of encapsulating the consecutive segments of the continuously transmitted streaming media. The media is then transported from the streaming server to the respective client. The most popular Real-time Transport Protocol is known simply as the **Real-time Transport Protocol (RTP)**.

- **Monitoring protocols**:
  For monitoring the real-time data transport, special monitoring protocols are implemented to exchange information about the transported audio/video data and to exchange the available network performance. The **RTP Control Protocol (RTCP)** is implemented for this.

- **Control protocols**:
  To control and direct the transport of continuous streaming media data - e.g., to halt the real-time data stream in order to continue it later or to fast forward or rewind a data stream in real-time - the **Real Time Streaming Protocol (RTSP)** is implemented.

While the HTTP protocol is based on TCP as the transport protocol, the real-time protocols RTP, RTCP and RTSP use the simpler UDP protocol, whereby RTSP also uses TCP in part (cf. Fig. 9.63).

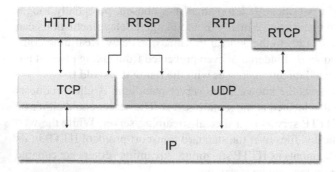

Fig. 9.63 Real-time protocols for the transport, monitoring and control of continuous streaming media data.

Nowadays help applications embedded in the web browser (Helper or Plugin) usually serve as the interactive user interface for streaming clients. They enable the playback of continuous streaming media contents in the browser. These special help applications are known as a **media player**. They are available today in a wide variety of proprietary software as well as in publicly accessible open source software.

With the help of the media player, the user requests a media data stream and can then play and navigate through it on the local computer. Among the most well-known representatives are the Windows Media Player from Microsoft, Quicktime Player from Apple, Real Player from Real Networks, but also the free VLC Media Player. Among the tasks and functions of a media player are:

- **Decompression of compressed media data streams**:
  Audio and video data is usually compressed nowadays in order to save memory storage and the necessary bandwidth for transmission. An important task of the media player is the decompressions necessary for playback.

- **Jitter elimination**:
  With the help of a suitable buffer storage, the media player ensures that the un-evenly delivered data packets of the media data stream are cached and forwarded evenly for further processing.

- **Error correction**:
  It is possible for data packets of the transmitted media data stream to get lost due to unreliable data transport via the UDP protocol. If too many data packets are affected, this loss also becomes noticeable to the user. Different strategies are implemented for error correction by media players to avoid this. These range from an explicit new request for a data packet that has gotten lost to the automatic interpolation of missing media content.

- **Graphical user interface with control elements**:
  The user interacts with the reproduced media data stream via a suitable user interface. Through it, the user can request to play new media content and navigate forward and backward.

The server side of a media streaming system provides the availability of the media data stream, and transmits it as wished based on the client's request. The server side usually consists of a combination of a traditional web server, which presents the media data stream in the World Wide web so that it can be found and requested by the user, and a special streaming server. The streaming server itself receives a request to deliver this content directly to the client (media player) - via the web server over which the user will be presented the desired media content. Otherwise, the streaming server is also responsible for reproduction and storage of the media data stream (cf. Fig. 9.64).

9.7.2 Real-time Transport Protocol – RTP

A special protocol is implemented for the transmission of digitalized audio and video signals over the Internet. Using UDP as the underlying transport protocol, it is called the **Real-time Transport Protocol (RTP)**. To satisfy the named requirements for transmission of isochronous data streams, RTP provides a sequence number for every transmitted data packet. In this way, the correct sequence can be ensured and lost data packets recognized, as well as providing a timestamp that puts the receiver in the position of being able to reproduce the delivered data packets in a timely manner.

In 1996, RTP was defined for the first time in RFC 1889 and revised in RFC 3503 in 2003. The classification of the RTP protocol in the TCP/IP reference model is not

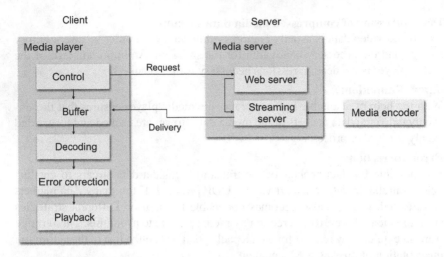

Fig. 9.64 Client/server interaction in a streaming media system.

clear-cut. On one hand, it is based on UDP as transport protocol and therefore numbers among the protocols of the application layer. While on the other hand, RTP adds a timestamp and sequence number information to the UDP data packet. It provides a general transport service that can be used by special applications. In this respect, RTP is also considered a transport protocol. RTP can therefore be designated as a transport protocol that is located in the application layer.

The RTP protocol was designed to transmit different types of multimedia information. Its basic function consists of receiving heterogeneous multi-data streams of audio, video, image and text data, multiplexing and forwarding them as RTP data packets to UDP via socket interfaces. RTP is not assigned a fixed UDP port number, instead in each case a random UDP port number greater than 1024 is chosen. Because of the implementation of the UDP protocol, there is also no possibility to ensure the timely delivery of the transmitted multimedia data over RTP, nor to give other quality of service guarantees. RTP data packets are given no special consideration by Internet routers, but treated in the same way as the rest of the data traffic unless special IP functions for the support of the service quality are activated. RTP data packets are numbered sequentially and so it is possible for a receiver to easily determine if an expected data packet has gotten lost. However, in most of the real-time applications it is not worthwhile to request a data packet that has gone missing. The available time would not be sufficient to still be able to guarantee a continuous playback. As a rule, missing data packets are therefore interpolated. For this reason, RTP is structured very simply, i.e., there are no mechanisms for flow control, no acknowledgements and there is no possibility to make a new request for a data packet.

Every RTP data packet additionally contains a timestamp. It is used to process and reproduce the received data in a timely manner. For this purpose, only the timestamp difference between the received RTP data packets is necessary. While the absolute time specifications are not important. It is also possible to combine different data streams with each other with the help of the timestamp information. For example, a video data stream can be combined with different audio data streams that provide, e.g., different playback languages. The individual fields of the data format established by RTP do not follow a fixed semantic, instead the interpretation of the transported data depends on the type-determining fields in the header of the data packet (cf. Fig. 9.65).

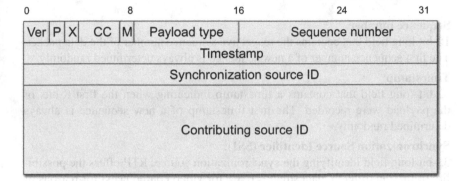

Fig. 9.65 Data format of the data packet header.

The header of the RTP data packet contains the following fields:

- **Version**:
  2-bit long field that contains the version number of the RTP protocol used (currently Ver=2).
- **Padding flag (P)**:
  1-bit long field that indicates whether the payload is supplemented with padding bits (0). Padding bits are only used if the transmitted payload is encrypted and in each case must have a fixed block length.
- **Extension flag (X)**:
  1-bit long field that stands for an optional extension of the RTP header. An optional extension of the regular RTP header can be added in several RTP applications to the area between the regular header and the payload. X-Flag=1 means this extension is being used.
- **CSRC Count (CC)**:
  4-bit long field indicating how many individual data streams are combined to make up the present data stream. It corresponds to the number of contributing source IDs added at the end of the header.
- **Marker flag (M)**:
  1-bit long field that is used by applications that must place markings in the transported payload (e.g., the beginning of a single image is marked in video data).

- **Payload type**:
  7-bit long field that defines the type of transported data. The interpretation of the remaining header field is also dependent on type, e.g.,

  - 1 – PCM audio data
  - 3 – GSM audio data
  - 8 – PCMA audio data
  - 26 – JPEG video data
  - 34 – H.263 video data

- **Sequence number**:
  16-bit long field that contains the respective sequence number of the data packet. The first sequence number of a new sequence is always determined randomly.

- **Timestamp**:
  32-bit long field that contains a timestamp indicating when the first 8 bits of the payload were recorded. The first timestamp of a new sequence is always determined randomly.

- **Synchronization Source Identifier (SSI)**:
  32-bit long field identifying the synchronization source. RTP offers the possibility of mixing different data streams (e.g., for video conferences) for reasons of efficiency and of transmitting them together. SSI identifies the sender of a data stream. If mixed data streams are involved, the SSI receives the identification of the mixer.

- **Contributing Source Identifier (CSI)**:
  32-bit long field identifying the data streams to be mixed. If a data stream that is mixed from different sources (cf. SSI) is involved, the mixer responsible for the gathering of the data streams sets the IDs of the respective original sources in these fields. Up to 15 different sources can be indicated here.

Unlike most of the other protocols that use UDP or TCP as a transport protocol, RTP uses prescribed port numbers. An individual port number is reserved for each RTP session, which the client and server must agree on beforehand. The sole stipulation dictates that RTP always implement an even port number, while the **RTCP** protocol used for the exchange of command and control information in the RTP session uses the next higher odd port number.

9.7.3 RTP Control Protocol - RTCP

The **RTP Control Protocol** (**RTCP**) is also defined with RTP in RFC 1889 (and then in RFC 3550). Via RTCP, the communication partners involved can exchange (unicast or multicast) information by means of RTP transported data as well as reports about the current performance of the underlying network infrastructures. RTCP manges feedback, the synchronization and the user interface, however does not take over any data transport itself. RTCP data packets are periodically sent for this purpose and contain sender as well as receiver reports with a statistical evaluation of the efficiency of data transmission. The specification of RTCP, or RTP, does not dictate how to proceed with this information but leaves this up it up to the respective application program using RTP, or RTCP. RTCP messages are encapsulated via UDP and always sent via the RTP port number of the succeeding port number (always odd-numbered).

Table 9.11 RTCP basic message types.

Type	Meaning
200	Sender report
201	Receiver report
202	Source description message
203	Bye message
204	Application specific message

RTCP provides five basic message types for the exchange of information about the current data transmission (cf. Table 9.11):

- **Sender report**
  The sender of a data stream periodically transmits a sender report that contains an absolute timestamp. This is necessary as the timestamp transmitted via RTP with every data packet is application-dependent. Therefore, in each case, the first is randomly chosen and the communicating application fixes the respective granularity (fineness) of the timestamp. If, however, multiple data streams are synchronized, an independent, absolute timestamp is additionally required. In a sender report there are also specifications regarding SSI (sources that are mixed together to the transmitted data streams) and the number of the sent packets/bytes in the data stream.

- **Receiver report**
  A receiver informs the sender about the quality of the data stream received with this report. For this purpose, the receiver report contains a specification of the SSI for whom the receiver report was generated. It also contains information about the number of lost data packets since the last report, the last sequence number received from a RTP data stream and a jitter value that is constantly updated in regard to deviations in the arrival time of incoming data packets. In multicast data transmissions, attention must be paid that an asymmetrical traffic load can occur between one transmitter and potentially multiple receivers. While the payload data via RTP remains the same with a rising number of receivers, the amount of exchanged RTCP information rises linearly. To avoid the overload problems caused by scaling, attention is always paid that the amount of control information with respect to the total transmitted data volume remains under 5%.

- **Source description message**
  This report contains more detailed information about the sender of the multimedia data stream, e.g., email address, name of the sender or sending application or other information in text form.

- **Bye message**
  The sender ends the data stream transmission by sending a special RTCP notification signaling this to the receivers.

- **Application specific message**
  Serves as a freely definable extension of the message type sent via RTP.

The structure of the RTCP data packet header is shown in Fig. 9.66. It contains the following fields:

- **Version**:
  2-bit long field giving the version level of the RTCP protocol (currently Ver=2).

- **Padding bits (P)**:
  1-bit long field that contains a padding bit, as in the case of the RTP data packet. The padding bit is set if one or more padding bytes are appended at the end of the RTCP data packet that do not belong to the actual payload. The last padding byte indicates the total number of appended padding bytes. Padding bytes are then appended if following protocols require fixed, predetermined block sizes. Such is the case with certain encryption procedures.

- **Report Counter (RC)**:
  5-bit long field that gives the number of reports received.

- **Packet Type (PT)**:
  8-bit long field that gives the type of the submitted report (cf. Tab. 9.11.

- **Length**:
  16-bit long field that gives the length of the RTCP data packet as the number of the 32-bit words contained therein minus 1. This includes the data packet header and the padding bytes.

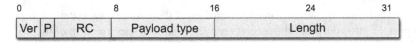

Fig. 9.66 Data format of the RTCP data packet header.

9.7.4 Real Time Streaming Protocol – RTSP

Multimedia contents have gained increasing importance on the Internet and today make up a large part of the transported data volume. Storage media is becoming more reasonable all the time, and the network bandwidth available to the private person today allows the use of broadband applications, such as video-on-demand or video conferences.

Media streaming, which functions based on the client/server principle, facilitates clients' requests for compressed multimedia data from streaming servers The transmission of the requested data proceeds segmented via the RTP protocol. Only a few seconds after the transmission has begun it is already possible for the client to start playback. Even during transmission, the client is in a position to navigate within the transmitted data to a certain extent, i.e., to stop, rewind or continue with playback. Where a live transmission is not involved, a rapid fast forward or jump to a certain point in the media data stream is possible. The necessary client/server interaction is regulated by a special protocol. This protocol was defined in 1998 in RFC 2326 **Real-Time-Streaming Protocol (RTSP)**.

Numbering among the main tasks of RTSP are:

- **Requesting a media data stream from a streaming server**
  The client first requests a media description from the server via HTTP, or another suitable protocol. If the requested media data stream involved is a data stream transmitted via multicasting, the answer contains the multicast addresses and port numbers to be used for the data stream. If the transmission to take place is a unicast only, the client provides the target address for the transmission of the media data stream.

- **Invitation to a media server to join a conference call**
  A media server can be invited to join an already running conference, either to contribute to a data stream itself or to record data streams of the conference. This type of application is used mainly in the area of teleteaching.

- **Adding media to an existing transmission**
  Especially in the case of live-streaming applications, it is an advantage for the client to already know during transmission whether other media contents are standing by that could be transmitted additionally.

A data stream provided by a streaming server can be present in different qualitative forms that each place different minimum requirements on the data connection to be bridged. A media description (presentation description), which is assigned to every form of data stream, helps the user to make the appropriate choice. RTSP provides the following commands for controlling the display of a multimedia data stream:

- SETUP
  Causes the server to reserve resources for the transmission of a requested multimedia data stream and to start the RTSP session.

- PLAY/RECORD Starts data transmission at the server after a successful SETUP has been carried out.

- PAUSE
  Causes the server to interrupt the data transmission for a short time, without the risk of the reserved resources being released.

- TEARDOWN
  Causes the server to release the reserved resources once again and ends the RTSP session.

- DESCRIBE:
  Causes the server to deliver the metadata description (presentation description) of one of the relevant resources. The presentation description usually follows in the format of the Session Description Protocol (SDP), containing the managed media data streams, their associated addresses and other parameters (cf. Fig. 9.67).

Session Description Protocol for the Transmission of Multimedia Metadata

The Session Description Protocol (SDP, RFC 4566) provides a description of the multimedia data streams. It is used for managing communication sessions, such as in Internet telephony to negotiate the implemented codecs, transport protocols and access addresses. SDP serves only in the description of the resources and does not provide mechanisms for direct negotiations of the implemented parameters.

In the following example, Harald (Origin o) provides the receiver of hpi-web.de, starting the session 1234 with the multimedia data (session name s) "HPI Video." Two data streams (media data description) m) are offered that are both sent via RTP. The implemented protocol version (v) is 0. An audio data stream in the format PCMU to port 12.000 and a video data stream in the format MPEG to port 12.001

```
v=0
o=Harald 1234 1234 IN IP4 hpi.web.de
s=HPI Video
c=IN IP4 hpi.web.de
t=0 0
m=audio 12000 RTP/AVP 97
a=rtpmap:97 PCMU/16000
m=video 12001 RTP/AVP 31
a=rtpmap:31 MPEG/180000
```

Further reading:

Handley, M., Jacobson, V., Perkins, C.: SDP: Session Description Protocol Request for Comments 4566, Internet Engineering Task Force (2006)

Fig. 9.67 Session description protocol for the transmission of multimedia metadata.

The identification of a multimedia data stream proceeds over an absolute **Uniform Resource Locator (URL)**[10]. The method given is rtsp:// followed by the address of the media server and a path specification that identifies the specific multimedia data stream. The address of the media server may follow the port number, via which the transmission of the data stream is carried out. For example:

_____

[10] URLs are used for the identification of resources in the World Wide Web (WWW). A detailed presentation of the protocols and data formats of the WWW is found in the third volume of this trilogy: Meinel, Ch., Sack, H.: Web Technologies

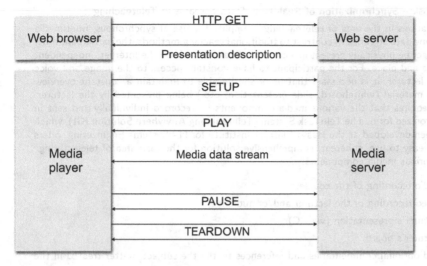

Fig. 9.68 Interaction between client and server via RTSP.

```
rtsp://www.tele-task.de:554/security-24-jan-2011.smil
```

As previously mentioned, the control of the display of the transmitted multimedia data stream - in other words, the operation of a media player - is carried out based on the operation of a VCR, with the help of the RTSP protocol. RTSP works as a so-called **out-of-band** protocol, in contrast to the actual multimedia data stream which is transmitted with RTP. RTSP implements UDP or TCP as transport protocol and is always set at port number 544. RTP and RTSP work in a similar way to FTP, which also uses a separate control canal and a data canal.

RTSP supports the synchronization of different multimedia data streams. The W3C defined the **Synchronized Multimedia Integration Language** (**SMIL**) especially for this task. SMIL is a simple markup language that enables transmitting multiple audio, video as well as text data streams and presenting them at the client. A simple example in which three audio files are simultaneously transmitted and reproduced via the protocol **rtsp** is shown in Fig. 9.70.

9.7.5 Resource Reservation and Service Quality

Because in an IP-based network there can be no guarantees provided for the much-cited **Quality of Service** (**QoS**) parameters, the IETF developed two special protocols capable of guaranteeing the advance reservations of available QoS resources. These are the **Resource Reservation Protocol** (**RSVP**, defined in 1997 in RFC 2205) and the **Common Open Policy Services** (**COPS**) protocol, defined in 2000 in RFC 2748,

TeleTask – Synchronization of Real-Time Data Streams in Teleteaching

Applications in the area of **teleteaching** exemplify the use of synchronous, multimedia real-time data streams. In contrast to traditional teaching methods, the participant must no longer be physically present in a course but can follow it on the Internet, independent of time and place. For the participant to have constant access to the image and voice of the lecturer, it is necessary that he or she be able to maintain a constant overview of the material (whiteboard, slide presentation, etc.) being presented by the lecturer. This requires that the various media components be recorded individually and sent in synchronized form. The **TeleTask** System (**Teleteaching Anywhere Solution Kit**), which has been developed at the Hasso Plattner Institute for IT- Systems Engineering, offers a very easy to use, coherent, comprehensive solution for the task area of telelecturing. The various media components:

● audio recording of the course,

● video recording of the lecturer and/or audience,

● lecturer's presentation (via PC),

● lecturer's board,

● and optional commentaries and references to the the subject matter treated in the course

are recorded separately as real-time data streams sent via the Internet and reproduced and synchronized at the participant via **SMIL** using a standard media player. Depending on the bandwidth available, the participant can choose different quality levels of the retrieved multimedia data. Moreover, it is possible to retrieve the contents later in the webportal and thus to work interactively.

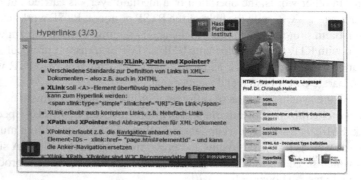

Further information:

K. Wolf, S. Linckels, Ch. Meinel: Teleteaching Anywhere Solution Kit (tele-TASK) Goes Mobile, in Proc. ACM SIGUCCS Fall Conference 2007, Orlando (Florida, USA), pp. 366 - 371 (2007)

V. Schillings, Ch. Meinel: tele-TASK - Teleteaching Anywhere Solution Kit, in Proc. SIGUCCS 2002, Providence (Rhode Island, USA), pp. 130-133 (2002)

TeleTask website: http://www.tele-task.com/

Fig. 9.69 TeleTask - synchronization of real-time data streams in a teleteaching application.

```
<SMIL>
  <BODY>
    <AUDIO SRC="rtsp://www.hpi-web.de/one.rm>
    <AUDIO SRC="rtsp://www.hpi-web.de/two.rm>
    <AUDIO SRC="rtsp://www.hpi-web.de/three.rm>
  </BODY>
</SMIL>
```

Fig. 9.70 An example of synchronized multimedia integration language.

This additional functionality cannot, however, simply be added to the application layer of the protocol stack, instead the involved network hardware must demonstrate special abilities:

- Routers must be in a position to provide guarantees that certain bandwidths are available for a reservation.

- Both of the communication partners at the end of the connection paths, or all communication partners at the leaf nodes of the multicast graph, must reach an agreement with each other about a bandwidth for transmission and every router along the transmission path must agree and comply with this requirement.

- The involved routers must monitor the entire data transfer (**traffic policing**), and if necessary intervene to control it.

- For this purpose, a router must have the appropriate queuing mechanisms available to weaken or balance out the suddenly released signal bursts that occur due to overload (**traffic shaping**).

RSVP serves to formulate and answer queries about reservations from network resources. The RSVP standard itself does not specify how the network provides the reserved bandwidth, but only allows the communicating applications to reserve the necessary connection bandwidth. Afterwards, the routers involved take over the task of actually providing the requested bandwidth. RSVP is also not included in the routing protocols category, i.e., choice of the router where the reservation takes place depends on the underlying routing protocols.

In contrast to routing protocols, RSVP manages **data flows** and does not make any routing decisions for individual datagrams, Data flows are defined by a discrete session between uniquely defined data sources and destinations. A session is defined as a unidirectional flow of datagrams to a certain receiver and a transport layer protocol. The following information uniquely specifies a session:

- Receiver's address (destination)

- Protocol identification

- Receiver's port number

RSVP supports both unidirectional unicast sessions and multicast sessions. Whereby a multicast session sends a copy of every transmitted datagram to all participants of the multicast group addresses. To begin a multicast session, the relevant receiver must first join the multicast group with the help of the Internet Group Management Protocols (IGMP). RSVP runs out before the actual transmission of a multimedia data stream takes place As soon as the data transmission is started, it is not longer possible to intervene via RSVP. An RSVP dialog can e.g., take place as follows:

- One of the communication partners sends a RSVP `path` message to determine the connection path between two end points. For this, the data packet uses the `router alert` option of the IP datagram to force all routers along the connection path to process the datagram.

- After an answer to the path request, one of the two communication partners follows with a reservation inquiry for the necessary network resources. This request contains all of the quality of service criteria to be maintained for the transmission.

- Every router along the connection path between the two communication partners must agree to this request and reserve the desired resources.

- If one of the routers along the connection path cannot deliver the desired resources, it sends a negative answer to the requesting end system via RSVP.

- When all of the involved intermediary systems agree to the reservation request, a positive RSVP message follows and the data transfer can begin.

RSVP reserves always only in one direction of the data connection, i.e., RSVP only carries out a unidirectional reservation. A unidirectional RSVP reservation does not mean, however, that the data in the opposite direction chooses the same connection path. If this is intended, it is then necessary that an RSVP reservation be carried out in the opposite direction as well.

If a router contains a RSVP request, it needs to check locally whether the requested resources can even be provided at all. This means that it needs to determine whether its downstream connection offers the necessary capacity to fill the reservation. This admission test is always carried out upon receipt of a reservation message. If this test fails, the router returns a corresponding error notification to the server. This shows that the reservation request could not be fulfilled. At the same time, this admission test is not part of the RSVP specification. RSVP assumes that the respective router will carry it out itself.

RSVP supports two different variations of reservations:

- **Distinct reservations**:
  Distinct reservations set up an individual data flow for every relevant sender of a session.

- **Shared reservations**
  Shared reservation are used together by a group of senders. It must be clear that the senders do not influence or disturb each other.

Reservation messages can be made in an explicit context or with the help of place-holders (wild cards). Accordingly, the following message reservation categories can be identified:

RSVP – A Multicast Example

Each receiver of a multicast multimedia data stream transmits a reservation message in an upstream direction via the multicast tree. Here, every receiver communicates which data rate it wants to get the data from the source. Every router that receives the reservation message aggregates it and passes it along to the next router in the upstream source direction.

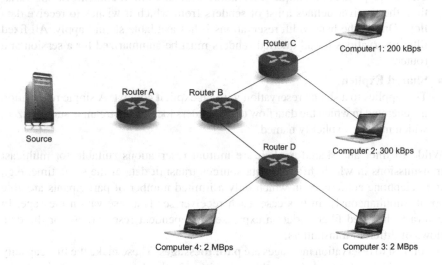

In the above example, router D requests a reservation from the upstream router B for 2 Mbps, router C requests a reservation for 300 kbps from router B. As computers 1–4 all receive the same multicast-data stream, only the the maximum of the requested bandwidth must still be reserved. Thereby, a layered coding of the data stream must be ensured so that the 200 kbps, or 300 kbps, version is contained in the 2 Mbps. Router B forwards the reservation message to the next upstream router A and requests a reservation for 2 Mbps.

In a multicast tree, every router can receive reservation information about its downstream connections. These are aggregated, the local packet scheduler carries out the appropriate reservations at the router and in each case a new reservation message is forwarded to the next upstream router

Further reading:

Braden, R., Zhang, L., Berson, S., Herzog, S., Jamin, S.: Resource ReSerVation Protocol RSVP – Version 1 Functional Specification, Request for Comments 2505, Internet Engineering Task Force (1997)

Fig. 9.71 RSVP multicast example.

- **Wildcard filter**:

  This applies to a common reservation with placeholders. In this way, the receiver expresses its wish to receive all data flow from upstream senders and that all senders are to observe its bandwidth reservation.

- **Fixed filter**:

  This corresponds to a unique reservation with an explicit context. In this variation, the receiver defines a list of senders from which it wishes to receive data flow. One of the bandwidth reservations it has available should apply. All fixed filter reservations for all queried senders must be summarized for a session at a router.

- **Shared Explicit**:

  This applies to a shared reservation with an explicit context. A single reservation is generated in which the data flow of all senders located upstream is summarized, which must be explicitly named.

Wildcard filter and shared explicit are mutual reservations suitable for multicast transmissions in which different data sources transmit data at the same time, e.g., in a telephone conference in which only a limited number of participants are able speak simultaneously. In this case, each receiver sends a reservation message. In contrast, the fixed filter variation expresses independent reservations for the data flows of different transmitters.

In contrast to reservation messages are **path messages**. These make the line capacity known, which is forwarded from the transmitters in the downstream direction to the receivers. Path messages are used for the storage of the path status in the computers along a connection path, which is itself used to forward reservation messages in the opposite direction. There are three additional variations of **error and confirmation messages**:

- **Path error messages**:

  Path error messages are triggered by path messages and travel in the upstream direction to the relevant senders. The routing of the path error messages proceeds "hop-by-hop" on the basis of the path status.

- **Reservation request error messages**:

  Reservation request error messages are triggered by reservation messages and travel in a downstream direction to the receivers. The routing of the reservation request error message proceeds "hop-by-hop" on the basis of the reservation status. The following message contents can be sent citing cause of error: admission failure, bandwidth not available, service not supported, defective flow specification and ambiguous path details.

- **Reservation request acknowledgement messages**:

  Reservation request acknowledgement messages are sent if a reservation inquiry can be answered successfully. It is likewise forwarded "hop-by-hop" by the routers in the downstream direction to the receivers.

Upon completion of a data transmission, a **teardown message** removes the reserved path and its current state. A distinction is made between **path teardown** messages, which are forwarded in the downstream direction in the same way as path messages are forwarded to the receivers, and **reservation request teardown,** which are forwarded in the opposite direction of the sending, corresponding to the reservation messages.

The data format of the RSVP messages is made up of the message header and succeeding message objects (cf. Fig. 9.72). The following data fields are in the RSVP message header:

- **Version**:
  4-bit long field that gives the protocol version (currently =1).

- **Flags**:
  4-bit long field that is to receive the guidance and control flags, is currently not used.

- **Type**:
  8-bit long field that gives the RSVP message type.

- **Checksum**:
  16-bit long field that forms a standard TCP/IP checksum for the content of the RSVP message. Zero is set for calculation of the checksum field contents.

- **Length**:
  16-bit long field that gives the length of the RSVP data packet in bytes.

- **Send TTL**:
  8-bit long field that, analogous to the IP datagram, gives "time to live" of the message.

- **Message ID**:
  32-bit long field used for the unique identification of the RSVP data packet and all of its fragments.

- **More fragments flag**:
  16-bit long field whose low order bit is set for all fragments of a messages except for the last fragment.

- **Fragment offset**:
  16-bit long field that gives the byte offset for the fragment contained in the message.

The message objects contain the following data fields:

- **Length**:
  16-bit long field giving the the entire length of the message object in bytes; must always be a multiple of 4.

- **Class-num**:
  8-bit long field giving the category (object class) of the message object. The highest value bit in the field indicates which action a network device should carry out if the category of the message object is not recognized.

  C-type:
  8-bit long field that is uniquely assigned within the message object category.

Message header

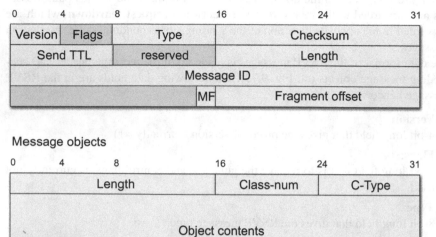

Message objects

Fig. 9.72 Data format of a RSVP message, consisting of RSVK message header and object fields.

Together with the class-num field, both of the fields, totaling 16 bits in length, can used for the unique identification of the message type.

- **Object contents**:
  Variable length field that contains the actual contents of the message object. Together with the class-num and c-type fields, the the length information specifies the structure of the contained message content.

A decision made single-handedly about resource distribution makes little sense. This is because routers follow general, fixed guidelines (**global policies**) and a decision made alone could contradict these requirements. To decide whether the query corresponds to global policies, the router must first, as a client, contact a so-called **Policy Decision Point** (**PDP**). The PDP does not forward any data traffic itself, rather it only answers queries from routers as to whether the requested resources correspond to the general guidelines. In response to the answer of the PDP, the relevant router then acts as the **Policy Enforcement Point** (**PEP**), making sure that the data traffic satisfies the fixed global policies (cf. Fig. 9.73). The protocol via which the router and PDP communicate is based on RSVP and known as **Common Open Policy Services** (**COPS**). It is defined in RFC 2748. If a router receives a RSVP query, it generates its own COPS request out of it. It does this by taking over the requested resources from the RSVP request and sending them to the PDP. This simple model, in which all decisions are outsourced to the PDP, is also known as the **outsourcing model**. Still another variation has been suggested as an alternative, and is defined in RFC 3048. It is known as the **provisional model**. Here, a PEP notifies the PDP of its own possibilities for reaching decisions. The PDP then transmits all relevant guidelines (policies) to the PEP, who based on the transmitted policies can then make decisions alone.

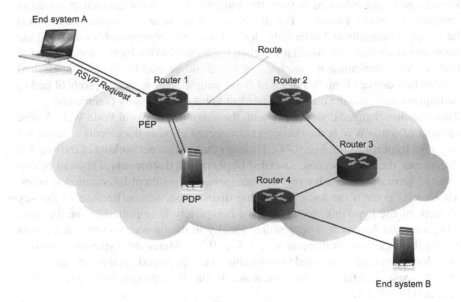

Fig. 9.73 RSVP and COPS for requesting network resources.

9.8 Other Internet Services and Applications

The email service still ranks among the most popular of Internet services today. Besides the basic types that have been discussed, there are numerous other services and applications in the application layer of the Internet that rely on the Internet transport protocols TCP and UDP. The most important of them are presented in the following.

9.8.1 World Wide Web

The popularity and global importance of today's Internet can be attributed to the development of the simple to operate user interface: the browser. An integrative interface, the browser provides unified access to a number of different Internet services, such as email or file transfer. In this way, Internet use has become simplified to a great degree, enabling it to expand as a new mass medium.

With the emergence of the World Wide Web at the beginning of the 90's, the file and service-oriented paradigm of the Internet, which had existed until that time, changed into a document-oriented paradigm. The user had previously needed special expert knowledge to get information from the Internet. This meant contacting a remote computer to initiate a session. The file system of the remote computer had to then be searched manually and in case the desired file - which presumably contained the information wanted - was found it was then transmitted to the local computer via file transfer. With transmission complete, the file could be read locally and the desired information derived from it. For all of these simple steps a special protocol had to be implemented, and the user would need to know the necessary commands.

This situation changed dramatically with the dawn of the World Wide Web. A core element of the WWW is the document connected to other documents - linked via so-called **hyperlinks** (cf. Fig. 9.74). Through the explicit networking of the original documents, these then become so-called **hypermedia documents**. They can be distributed across the entire Internet and comprised of different information sources. The crucial point is the wide geographic distribution, which, because of the key element of the hyperlink, seems to have completely disappeared to the the user. A hypermedia document is represented locally in the browser - even if it consists of distributed information resources (cf. Fig. 9.75). Moreover, hyperlinks crossing over document borders are made identifiable in the displayed documents, and with a simple mouse click the user can request and display the remote information resource concealed behind it.

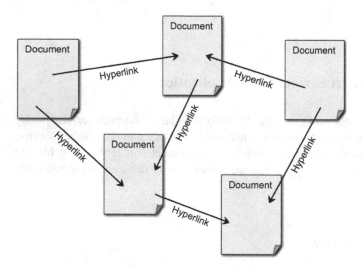

Fig. 9.74 Documents in the World Wide Web are linked with each other by way of hyperlinks.

Communication in the World Wide Web proceeds with the help of the **hypertext transfer protocol** (**HTTP**) and works according to the client/server principle. Following the documented-centered approach of WWW, the web browser (client) must

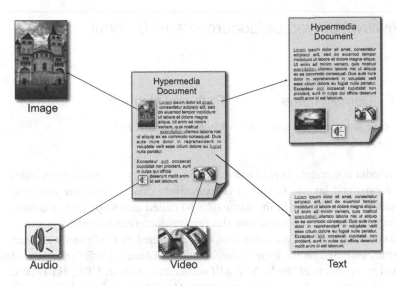

Fig. 9.75 A hypermedia document is comprised of different media components (text, graphic, audio, video).

request a document from the web server (server), which has to be uniquely identified and addressed for this purpose. To this end, in addition to the IP address, or the domain name of the web server, the document name including the path must be known in the file system of the web server. These specifications, along with the implemented protocol (http), make up the so-called **Uniform Resource Locator (URL)**. This is used in the WWW to uniquely identify information resources. Fig. 9.77 shows the schematic display of communication in the WWW via HTTP and Fig. **??** the structure of a simple URL.

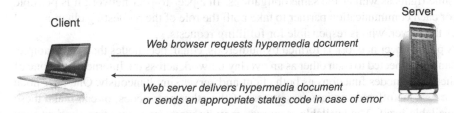

Fig. 9.76 HTTP client/server architecture.

HTTP was defined in 1997 as RFC 2068 in the version HTTP/1.1. It was replaced in 1999 with an improved version, RFC 2616. The hypermedia documents transported via HTTP are coded in a language especially designed for this purpose: the **Hypertext Markup Language (HTML)**. HTML describes the content structure

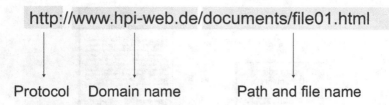

Fig. 9.77 Principle structure of a simple Uniform Resource Locator (URL).

of the hypermedia document. This is a description of the document contents independent of format and layout. Rather, it is exclusively content-structure oriented. Structure information is given in the form of a so-called **markup**. The structure-building commands are incorporated into the regular document contents via specially masked commands. These are automatically interpreted by the browser and can be issued formated based on the format specifications at hand. A brief overview of the functional components of the WWW will suffice at this time. URL, HTTP and HTML, as well as other web technologies, are the focus of the third volume in this trilogy where they are presented in detail.

9.8.2 Peer-to-Peer Applications

As we have already seen in the previous chapters, the client/server paradigm dominates most of the Internet applications today. The communication partners strictly follow the roles assigned to them, i.e., as the active component of both communication partners, the client requests information or services from the passive server. The server waits for the client requests in order to verify them and, in positive cases, to fulfill them. This stands in contrast to the **peer-to-peer** (**P2P**) communication principle. Here, each of the communication partners involved is equal, i.e., has the same rights as well as the same obligations. In a peer-to-peer network it is possible for each communication partner to take both the role of the requesting client as well as the server, who is responsible for fulfilling requests.

A peer-to-peer network is formed by numerous, equal, peer nodes that, for example, can be connected to each other as an overlay network across the Internet. Each one of these peer nodes functions as both client and server simultaneously. One peer node can use the offered services (resources) of the other peer nodes, or can make these available itself. The available resources may be memory, computing time or even information. The peer-to-peer network idea is not a new one. For example, message transfer agents of the well-known email service work according to the peer-to-peer procedure, i.e., depending on the situation they work as the client or the server. Fig. 9.78 shows the contrast between the client/server model and the peer-to-peer model.

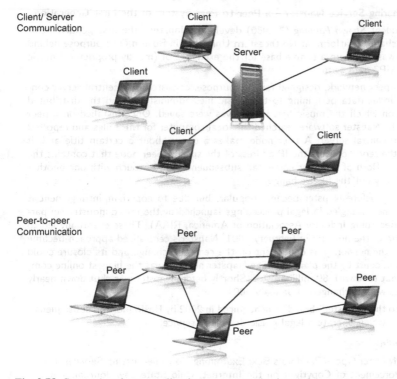

Fig. 9.78 Comparison between client/server communication and peer-to-peer communication.

Peer-to peer application first entered the public eye with the surfacing of the music exchange platform **Napster** at the end of the '90s. Along with it came the phenomena of **filesharing** (cf. Fig. 9.79). Napster is among the peer-to-peer systems of the first generation that follow an unstructured, centralized structure. In other words, a central computing node manges an index with the shared resources within the peer-to-peer system. These are then made available again to the other peers distributed across the network. The organization of a peer-to-peer system today usually provides a supplement to the already present overlay network that is on the Internet. It is a further overlay network to manage the peer-to-peer network resources and their organization and search possibilities.

Principally, a distinction is made between the following organizational forms of peer-to-peer systems:

- **Unstructured P2P systems**

  - **Centralized P2P systems**:
    In this category are peer-to-peer systems of the first generation, such as Napster. These are dependent on the management of a central server. The central server manages an index with the metadata of resources available in the entire peer-to-peer network. Not only are all of the currently participating peer

The File Sharing Service Napster – a Peer-to-Peer System of the First Generation

In 1998, student *Shawn Fanning* (* 1980) developed **Napster**., the first popular peer-to-peer file sharing platform, at Northeastern University in Boston. The purpose behind the project was to set up a simple base for the exchange of (mostly proprietary) music files in the MP3 data format.

The peer-to-peer network, designed for this purpose, consisted of a central server containing the index data pertaining to the music files administered and the distributed peer nodes on which the music files themselves were saved. Once installed on a peer computer, the Napster software searched the local computer for MP3 files and reported these to the central server. A peer node makes a query about a certain title and, if successful, the server delivers the IP address of the specific peer node that contains the requested file. Both of the peer nodes can subsequently get in touch with one another directly to carry out the file exchange.

It wasn't long before Napster became popular, but due to copyright infringement it quickly became entangled in legal proceedings launched by the music industry (in particular the Recording Industry Association of America, RIAA). These eventually led to the shutdown of the network in February 2001. Napster's centralized approach became its downfall: the network was dependent on the central exchange and its closure could be enforced in court by the plaintiffs. The Napster network was the biggest online community at that time with 80 million users. Shortly before Napster was shut down nearly 2 billion files were exchanged in January 2001.

The rights to the "Napster" brand were acquired in 2002 by Roxio Inc. and subsequently by Best Buy, which operate a legal music download service.

Further reading:

Green, M.: Napster Opens Pandora's Box: Examining How File-Sharing Services Threaten the Enforcement of Copyright on the Internet, Ohio State Law Journal, Vol. 63, No. 799. (2002)

Fig. 9.79 The file sharing service Napster - a peer-to-peer system of the first generation.

nodes contained in the index, but their resources as well. If a peer node initiates a query about a resource, the central server searches the index file for the corresponding participant who could fulfill the requirements. The server's answer contains the IP address of the relevant peer nodes who can then be contacted directly to request the desired resource. In this way, the search in the centralized peer-to-peer system still runs based on the client/server paradigm. Only the direct resource exchange takes place according to the peer-to-peer principle (cf. Fig. 9.80).

While this model offers the advantage of simple implementability as well as a faster and more efficient search for resources, on the other hand, the centralized structure increases the likelihood of censorship or interference from the side of unauthorized third parties. Such is the case with a denial-of-service attack. Moreover, the centralized model has negative scalability characteristics. This means, the server load grows as the network size increases and the possibilities of the central index storage is always restricted by size limitations.

– **Pure P2P systems**:

In this case, a central instance for the management of the peer-to-peer net-

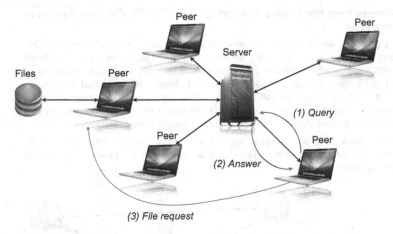

Fig. 9.80 Resource exchange process in the centralized peer-to-peer model.

work is not necessary, as in the case of e.g., Freenet or gnutella. The so-called "web of trust" networks, are a special form of pure P2P systems, where only connections to "trusted" peers are received. Peers with unknown IP addresses are simply ignored.

Gnutella was developed by *Justin Frankel* (*1978) in 2000. Just as Napster, it served as a file sharing service, however on a completely decentralized basis. Every peer node in the Gnutella network uses the same software. To still work efficiently without a central server, a peer node must know at least one of the neighboring network nodes. Peer nodes manage so-called neighborhood lists for this purpose. A peer nodes goes through its neighborhood list when starting and seeks another active peer node whose neighborhood list it can add to its own. The maximum number of active neighbors is always limited. If a user of the Gnutella network begins a query, it is initially only forwarded to its neighboring peer nodes. These forward the query from their side to their neighboring peer nodes until the requested resources are found, or a previously defined maximum number of hops is reached. Subsequently, a direct connection between the initiator of the search and the peer node supplying the resource sought for data transmission can be established (cf. Fig. 9.81).

A special advantage of the network organization in pure peer-to-peer system is its reliability. The queries themselves can be forwarded if some parts of the network are temporarily unavailable. A serious disadvantage is the long duration of the search because no central index server can answer the request. The network load is also high as the number of queries grows exponentially as the distance to the seeking peer node increases.

– **Hybrid P2P systems**:
  In hybrid peer-to-peer networks, multiple central servers (also called "supernodes") are assigned to manage the network alliance, e.g., in the case of Gnutella2, Kazaa or Morpheus. These particular nodes at the server are for the

The Pure Peer-to-Peer File Sharing Network gnutella - A Peer-to-Peer System of the First Generation

The gnutella protocol, which laid the foundation for the **gnutella** file sharing network, was developed by *Justin Frankel* (* 1978), an employee of the AOL company. On March 14, 2000 it was made available for free download on the Internet. Although he distanced himself from the project due to company pressure, Frankel's protocol had become so widespread - thanks to discussion forums and mailing lists - that it continued to be developed independent of its creator. As a result, its scope expanded considerably. By May 2005 there were more than 2.2 million Gnutella network users.

gnutella worked as a pure peer-to-peer system without a central index server, i.e., without a central operator against whom legal action could be taken in a case of copyright infringement.

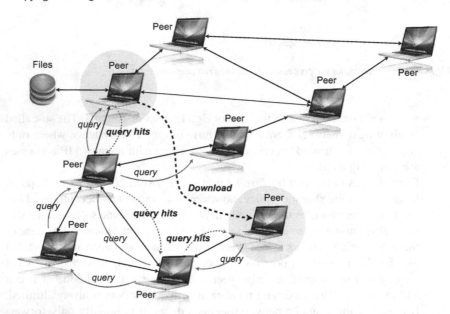

In search of resources a file query is sent to all neighboring peer nodes in gnutella. These each forward the query to their own neighbor until a maximum number of predetermined hops is reached. If the resource exists at one of the queried peer nodes, a message (query hits) is sent back along the query path to the initiating sender. The sender then requests the desired resource directly from the owner to download.

Further reading:

Ripeanu, M.: Peer-to-Peer Architecture Case Study: Gnutella Network, in Proc. of the First International Conference on Peer-to-Peer Computing (P2P '01), IEEE Computer Society, Washington, DC, USA, pp. 99–110 (2001)

Fig. 9.81 The pure peer-to-peer file sharing network gnutella - a peer-to-peer system of the first generation.

most part peer nodes that exhibit a particularly high bandwidth and are therefore preferred for this task. Supernodes maintain a purely decentralized peer-
to-peer network among themselves (cf. Fig. 9.82).

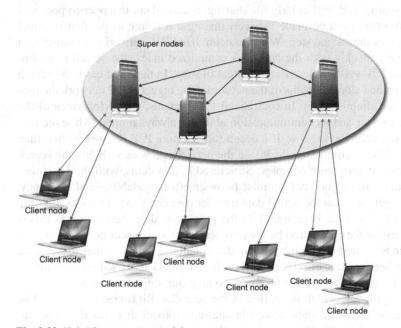

Fig. 9.82 Hybrid peer-to-peer model.

The other peer nodes are connected to the super node just as client nodes. In
this way, hybrid peer-to-peer systems combine the advantages of the centralized peer-to-peer system - in view of a low latency time in the resource search
- with the advantage of the pure peer-to-peer system in terms of scaling ability. A disadvantage is that despite having the sufficient bandwidth to fulfill the
task, not every peer node wishes to take over the role of a supernode. Furthermore, the same reliability and efficiency is not always achieved in hybrid
peer-to-peer network as in a true centralized system.

- **Structured P2P systems**:
  Decentralized systems implementing a divided hash table (Distributed Hash Table, DHT) for the management of available resources in the peer-to-peer network
  are known as structured peer-to-peer systems, e.g., the Chord system. Here, every peer has a unique key for identification. The structure of the network corresponds to a known topology. Every resource is identified using a hash value as a
  key, which results, e.g., from its name. The individual nodes of the network store
  resources whose keys fall within a particular range. This assures that if a resource
  is present in the system, it is in fact found.

Centralized and pure peer-to-peer systems are among the systems of the first generation, while decentralized systems are considered systems of the second generation. Furthermore, special peer-to-peer systems that only deliver files over non-direct connections are known as systems of the third generation.

If resource sharing and, particularly file sharing, is carried out in a peer-to-peer network, a distinction must be made between finding a resource as the first step and transferring it as the second step. While in centralized P2P networks the search can take place very quickly with the help of a centralized index, the search in decentralized, pure P2P systems needs a great deal of time. In the latter case, the search needs to be passed along all paths of the network. The bigger a P2P network the more complex its administration. In centralized P2P systems, the performance of the central index server and its communication ability is always limited, while the network itself can continue to grow. If a decentralized, pure P2P network grows, then the lengths of the connection paths inside the network grow as well and the search for resources becomes more complex. Structured P2P systems, with their distributed hash index, offer a good compromise between effort, scalability and efficiency.

We have not yet looked at the actual data transfer between two peer nodes after the searched for resource has been found. In the previously described networks this is always limited by the connection bandwidth between the two peer nodes. However, if a resource is located at multiple peer nodes simultaneously, together these could achieve a higher degree of effectiveness if each of these nodes, based on its available bandwidth, forwards a part of the resources to all requesting peer nodes.

It is just this principle which is applied in the so-called **BitTorrent** protocol. The BitTorrent protocol is a collaborative file sharing protocol that was developed in 2001 by *Bram Cohen* (*1975). It is especially suitable for the distribution of large amounts of data in a network. In contrast to other peer-to-peer technologies, BitTorrent is not based on a comprehensive network but constructs a separate distribution network for every resource to be transmitted. Unlike standard file transfer services, such as FTP, all peer nodes that have requested and downloaded a file become resource providers themselves, also when the original file transmission has not yet been completed. In this way, the available capacity can be optimally exploited and a collapse of network communication due to excessive requests to a resource owner avoided. While other file services are primarily concerned with identifying a peer node with the desired resource as fast as possible, BitTorrent's target is to quickly make a single file (resource) available to a great number of users and to reproduce it.

In order to quickly distribute a file as a **torrent**, it is divided into individual sections (usually each approx. 256 kB), for each of which a SHA-1 checksum is calculated. Additionally, with a mapping of the section in the original file and the IP address of the **trackers** this metadata is managed in a torrent file. A central server who manages all peer nodes of the **swarm** involved in the distribution of the file is identified as a tracker. Independent of the respective status of the current download in the swarm, a distinction is made between two different peer categories:

- **Seeder**:
  Peer nodes that have already received a complete version of the file to be dis-

tributed and continue to participate in distribution in the swarm are known as seeders.

- **Leecher**:
Peer nodes that have not yet received a complete version of the file to be distributed are known in the swarm as leechers.

The tracker itself is not directly involved in the distribution of the file. It only provides metadata on the file to be distributed in the form of a torrent file and manages a list of peer nodes that are currently in the swarm. A peer node can be involved in a torrent by downloading the torrent file and contacting the tracker found there. The active peer nodes of a swarm contact the tracker in regular intervals to confirm their active participation in the swarm. The parallel, distributed file transfer in a torrent is shown in in Fig. 9.83.

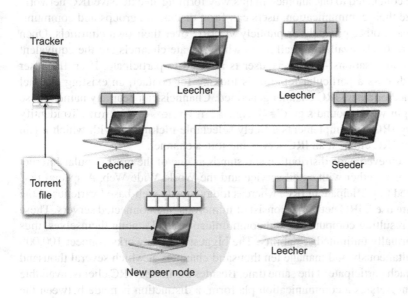

Fig. 9.83 The process of the distributed, parallel file transfer in a torrent.

The order in which the individual sections of a file are transmitted with the BitTorrent protocol is based on the principle of the largest possible guaranteed reliability. This means that the swarm must be as robust as possible when faced with retiring peer nodes. The attempt is made to handle this situation by providing as great a redundancy as possible. The degree of distribution of a file in a torrent can be determined by the frequency occurring in the rarest subsection in the swarm. To increase this frequency, the attempt is always made in a BitTorrent network to give preference to the rarest parts ("rarest first") in redistribution in order to maximize the redundancy of the torrent.

9.8.3 Internet Relay Chat – IRC

Internet Relay Chat (**IRC**) is a simple service for the exchange of text messages in real-time. It was defined in RFC 1459 in 1993. Developed in 1988 by Finnish student *Jarkko Oikarinen* (* 1967) at the University of Oulu, it was intended as a communication system for his computer-operated mailbox. IRC became hugely popular in a short time. IRC is designed as a client/server application and allows for the simultaneous participation of many clients at one server. Here, real-time communication means that the message input by the user is sent as quickly as possible from the client to the central server who provides the fastest possible display at all of the other participating clients. In contrast to the email service, which aims for asychronous communication between individuals, IRC is intended for group communication and can therefore also be seen as a kind of teleconference system. IRC servers are connected to one another, in this way forming an extensive IRC network. To organize their communication, users can form discussion groups and communicate with each other privately or publicly via IRC over their own **channels**. Open channels are freely available to all users while private channels are the equivalent of private conversations - not every user is allowed to participate. If an IRC user wishes to discuss a particular topic or is looking for contact, an existing channel can be joined or a new IRC channel generated. Channels can bear any name. These usually begin with a pound sign ("#"), e.g., #internet or #linux. To identify itself, every IRC user implements a freely selectable **nickname**, with which it can to log on its IRC client at an IRC server and join a channel.

IRC has a achieved wide distribution and stands as one of the most popular Internet user services together with email service and the World Wide Web. A special IRC client is used to participate in IRC, whereas today mostly web-based varieties of the IRC chat are used. IRC networks consist of numerous interconnected servers. These divide the resulting computation and communication load among themselves, thus creating virtually unlimited scalability. The biggest IRC networks connect 100,000 users simultaneously and manage ten thousand channels at which several thousand users can each participate at the same time. Besides the normal IRC clients, available to a human users as a communication platform, a distinction is made between the following special clients:

- **Bots**
  A bot refers to an automated client who, by virtue of its speed relative to human users, is used as a central point for information exchange and as a protective agent for individual channels.
- **Bouncer** (**BNC**):
  A bouncer refers to relay stations in the IRC network. These can connect ICR clients with IRC servers. In case a client loses its network connection, it can draw on the bouncer's ongoing data traffic recording. The session that has begun at the IRC server can be maintained, even if constant network connectivity cannot be guaranteed.

- **Services**:
  IRC network-own bots that facilitate the registration of new users and new channels are referred to as services. Via the services, messages are also forwarded to users who are not connected to the IRC network (offline users) and the management of the IRC network supported.

9.8.4 Usenet News

The Internet service called **Net News** or **Usenet News** provides a publicly available communication service. Similar to a notice board that provides an outlet for public exchange, it is divided into individual discussion groups (**newsgroups**). Users interested in a certain topic assigned to a special discussion group can register at this group via a specific client called a **newsreader**. They can then read all notices that have been posted in this discussion group, as well as posting articles of their own. Today, newsreaders are already integrated into WWW browsers most of the time. In order to play an active part in the information exchange, the user can send an article to the news server via the news client who is included in the inventory of the discussion group - either moderated or unmoderated depending on group organization. Usenet news messages implement the message format specified in RFC 822 for email service, which is supplemented with additional information in the message header (defined in RFC 1036). News messages are thus easy to transport and compatible with already existing email software.
Usenet was developed in 1979 by *Tom Truscott* and *Jim Ellis* (1956–2001) when they were students at Duke University. Usenet resembles Bulletin Board System (BBS) and is considered a precursor of today's highly popular Internet forums. The difference between Usenet news and Internet forums, or BBS, is the fact that Usenet is not dependent on a central server, but maintains a network of servers that exchange the user's messages among themselves as so-called newsfeeds. Individual users can read the Usenet news message via a local news server from their Internet provider. The discussion groups available worldwide over Usenet are organized hierarchically and regionally. Thus, the newsgroup under the name comp.lang.c refers to a discussion forum that is concerned with computer science topics (comp), especially programming languages (lang) and in particular with the programming language C (c). A purely German-language regional version of this discussion group can be found at de.comp.lang.c. Table 9.12 shows the top layer of the topic hierarchy of Usenet newsgroup.
A fundamental distinction is made between moderated discussion groups, where an editor (the **moderator**) decides which contribution gains entry into the posting inventory of the discussion group, and unmoderated discussion groups, where each user is able to carry out uncensored posting. The Usenet service is based on the **Network News Transfer Protocol (NNTP)**. It was defined in RFC 1036 and serves in the transmission of news contributions between news servers and news readers. NNTP uses TCP as a transport service. A discussion posting that is sent by the client

Table 9.12 Usenet news topic hierarchy.

Name	Topic group
comp	Computers and computer science
sci	Physics and engineering
humanities	Literature and humanities
news	Discussion and the Usenet itself
rec	Recreational activities
misc	Miscellaneous
soc	Sociability and social issues
talk	Polemics, debates and arguments
alt	Alternative branch with a wide range of topics

for release in a newsgroup to the news server must follow the format of a valid email message, based on RFC 822, i.e., it must contain an email header field, date, from, subject, message ID and newsgroup (as addressee). To access the newsreader on the news server, special commands are provided by NNTP. These are defined in RFC 977.

To get the latests posts from news groups, the client must first set up a TCP connection to the news server via port number 119. Newsreaders and news servers then begin their request/response dialog to make sure that all new posts, and only these, are transmitted to the client. This is to ensure that the client does not receive duplicates of what it already has. The most important NNTP commands are summarized in Table 9.13.

Table 9.13 Basic NNTP commands.

Command	Meaning
list	Request for a list of all news groups and posts
newsgroups DATE TIME	Request for a list of all news groups generated based on DATE/TIME
group GRP	Request for a list of all posts in GRP
newnews GRPS DATE TIME	Request for a list of all new items from the specified discussion groups
article ID	Request for a specific post
post	Sending a discussion post
ihave ID	Post ID is ready for pick up
quit	Termination of the connection

First, the client can determine the server's discussion groups via the commands list and newsgroups, before it can identify via group and newnews which individual contributions are of interest. Once a post is selected, it can be accessed via `article`. The commands `ihave` and `post` are available for information exchange in the opposite direction. With them the client can offer the server posts it has obtained from other servers or which have been written locally; `quit` terminates the connection. Although having existed for over 30 years, the popularity of the Usenet news service had been declining, despite Usenet's steadily rising data traffic with the daily news feed size exchanged in the terabyte range. Nevertheless, after 30 years of operation Duke University turned off their server in May 2010 due to low use and rising costs.

9.8.5 Other Service Offerings Via TCP/IP

A complete representation of all the Internet applications and the protocols underlying them will always be passed up by the currently ongoing developments. Therefore, the description of other protocols from the application layer is limited here to the most well-known and does not claim to be complete. Many of the application protocols from the early days of the Internet still exist, while their importance or recognition is rapidly fading. This is because their functions are now mostly hidden under the surface, (usually) carried out by web-based applications.

- **Echo**
  The **Echo Protocol**, defined in RFC 862 presents a simple client/server-based answering service, which is implemented in the Internet primarily for the testing and measuring of answering times (Round Trip Time, RTT). It is therefore suitable for testing and debugging during the development of Internet-based application programs. The echo server simply sends a requesting echo client everything back it has received from the client. The echo server is capable of transporting data over TCP as well as via UDP and is always accessible over the predefined port number 7. In Unix systems, the echo service is normally directly implemented in the network system `inetd`.

- **Discard**
  The **Discard Protocol** was defined in RFC 863. Discard defines a very simply structured client/server service whereby a discard server always deletes everything it has received from a client. It is therefore suitable for testing and error detection during the development of Internet-based application programs. The discard server can be addressed by the TCP as well as via UDP and is accessible over port number 9. The discard protocol corresponds to a TCP/IP equivalent of the Unix-own null device `/dev/null`.

- **Chargen**
  RFC 864 specifies the **Chargen Protocol** (Character Generator), which was developed in 1983 by *Jon Postel* (1943–1998). Chargen also has a client/server-based structure and can be reached over both TCP as well as via UDP. It is accessible at port number 19. If a chargen server is contacted by a client via TCP, it transmits a continuous stream of arbitrary characters back until the client ends

the connection again. Addressed over UDP, the chargen server sends back a datagram containing an arbitrary number (between 0 and 512) of randomly selected characters The Chargen Protocol is primarily used for the testing and debugging of Internet-based applications. Fig. 9.84 shows the typical process of a chargen service session.

Example for a Chargen Session

```
$ telnet localhost chargen
Trying 127.0.0.1...
Connected to localhost.
Escape character is '^]'.
!"#$%&'()*+,-./0123456789:;<=>?@ABCDEFGHIJKLMNOPQRSTUVWXYZ[\]^_`abcdefgh
"#$%&'()*+,-./0123456789:;<=>?@ABCDEFGHIJKLMNOPQRSTUVWXYZ[\]^_`abcdefghi
#$%&'()*+,-./0123456789:;<=>?@ABCDEFGHIJKLMNOPQRSTUVWXYZ[\]^_`abcdefghij
$%&'()*+,-./0123456789:;<=>?@ABCDEFGHIJKLMNOPQRSTUVWXYZ[\]^_`abcdefghijk
%&'()*+,-./0123456789:;<=>?@ABCDEFGHIJKLMNOPQRSTUVWXYZ[\]^_`abcdefghijkl
&'()*+,-./0123456789:;<=>?@ABCDEFGHIJKLMNOPQRSTUVWXYZ[\]^_`abcdefghijklm
'()*+,-./0123456789:;<=>?@ABCDEFGHIJKLMNOPQRSTUVWXYZ[\]^_`abcdefghijklmn
()*+,-./0123456789:;<=>?@ABCDEFGHIJKLMNOPQRSTUVWXYZ[\]^_`abcdefghijklmno
)*+,-./0123456789:;<=>?@ABCDEFGHIJKLMNOPQRSTUVWXYZ[\]^_`abcdefghijklmnop
*+,-./0123456789:;<=>?@ABCDEFGHIJKLMNOPQRSTUVWXYZ[\]^_`abcdefghijklmnopq
+,-./0123456789:;<=>?@ABCDEFGHIJKLMNOPQRSTUVWXYZ[\]^_`abcdefghijklmnopqr
,-./0123456789:;<=>?@ABCDEFGHIJKLMNOPQRSTUVWXYZ[\]^_`abcdefghijklmnopqrs
^]
telnet> quit
Connection closed.
```

Fig. 9.84 Example for a chargen session.

- **Time**
  The client/server-based **Time Protocol**, specified in RFC 868 in 1983, is used to determine the current time. At the request of a time client, the time server delivers the current time of the server computer as a 32-bit unsigned integer, which gives the elapsed seconds since midnight of January 1, 1900. Time can be implemented via TCP as well as UDP and is accessible at port number 37. The time protocol was replaced by the Network Time Protocol (NTP, see below).

- **Daytime**
  Just as in the case of the Time Protocol, the client/server-based **Daytime Protocol** provides the current time of the server computer in a form that is readable for the user, only that daytime also includes the date. Daytime was defined in RFC 867 and can be accessed over TCP or UDP at port number 13. The Daytime Protocol is mainly used for testing and debugging purposes in Internet-based applications.

- **Gopher, WAIS and Veronica**
  The named client/server-based services have disappeared from the Internet for the most part today. As a precursor to the World Wide Web, the **Gopher** service enabled uncomplicated access to data that was distributed worldwide. Gopher was developed in 1999 by *Mark McCahill* (*1956) along with members of his project group at the University of Minnesota and specified in RFC 1436. Similar

to a WWW server, a gopher server can be queried by a text-based gopher client, however, Gopher does not follow the concept of the hyperlink. The search tool **Veronica** (Very Easy Rodent-Oriented Net-Wide Index to Computerized Archives) is especially made for the information search in the worldwide-available Gopher information space. Another service for full text searches in databases over the Internet is **WAIS** (Wide Area Information System), specified in RFC 1625. WAIS uses the standard ANSI Z39.50 for database query, but has lost its former importance with the triumphal success of the WWW.

- **Network Time Protocol – NTP**
  A computer's current time can always be determined with the use of the Time and Daytime services. However, if multiple computers are to be uniformly synchronized to a common time, an individual protocol is necessary. The **Network Time Protocol (NTP)**, specified in 1992 in RFC 1305, provides just this kind of synchronization service. A hierarchy of reference timers is set up and, as a rule, the primary time is directly synchronized with atomic clocks. NTP implements UDP as a transport protocol, reaching a synchronization accuracy in the millisecond range. The current version NTPv4 was specified in June 2010 in RFC 5905. This makes it one of the senior Internet protocols. A service that is compatible with NTP, but less complex is the **Simple Network Time Protocol (SNTP)**. It was originally specified in RFC 1361.

- **Finger**
  Over the **Finger** Protocol, which was specified in 1991 in RFC 1288, information about one or more users can be obtained on a given computer. Finger works as a client/server application and sends ASCII encoded messages over the TCP port 79. The following information about a user can be obtained with Finger:

  - Login name,
  - Full name,
  - Name associated with the user terminal,
  - How (from where) did the user log in to the local computer,
  - When did the user log in (Login Time) and
  - How long has there been no user input (Idle Time).

  The service of the Finger Protocol is often viewed as a security risk and therefore explicitly blocked. A partial reason for this stems from the dubious fame the Finger Protocol won as one of the attack victims of the first Internet worm.

- **Whois**
  Just as the Finger Protocol, the **Whois** service is an Internet information service, which was originally specified in 1982 in RFC 812 and is currently in RFC 3912. Whois works as a client/server application over TCP port number 43. The Whois server answers a requesting client with information it has available based on the keyword supplied. Whois works as a distributed database system and administers Internet domain information as well as IP addresses and their owners. However, the possibility to identify domain owners via Whois has now ceased to exist for

reasons of data protection. Today this must proceed via the responsible Internet registries.

9.9 Security at the Application Layer – Packet Filters and Firewalls

A self-contained network (LAN) that is not connected to the global Internet is at little risk of becoming a target for attacks by unauthorized outsiders. To carry out such an attack, outsiders need a physical point of entry into the network. If physical access from the outside could be ruled out and only trustworthy users were able to enter a closed network, additional security mechanisms would hardly be necessary. At the same time, entry into the global Internet is a crucial matter of survival for companies and institutions. E-commerce, the Internet as a market place for direct sales, as well as immediate customer or prospective customer communication is imperative today. This access to a worldwide open network with its large number of questionable users, over whom the internal network administrator has no control whatsoever, presents a considerable security risk to the local network and countless headaches for any security-conscious network administrator. For this reason, the special protection of proprietary LANs - or even just private PCs - against unauthorized access is not only desirable but necessary.

Protection of this type can be carried out through the deployment of so-called **firewalls**. Firewall is the name that has been given to a network component coupled to the global Internet that conducts special protection and filtering measures to ensure the security in a LAN. The intention is to enable the LAN-internal user access to the global Internet with as little disturbance as possible. At the same time, the LAN itself should be protected from assaults by unauthorized third parties via the Internet. Independent of where the firewall is installed, a distinction is made between a **personal firewall** (also called **desktop firewall**), which is mostly installed as software on the system to be protected and an **external firewall** (also **network or hardware firewall**). This is a software or hardware solution implemented on a separate device that connects networks or network segments with each other. It monitors and filters network data traffic between the connected areas. A firewall system can be composed of different hardware and software components via which a secure network gateway can be configured corresponding to each specific requirement (cf. Fig. 9.85).

Numbering among the most important security objectives to be achieved with the use of firewall systems are:

- **Prevention of denial-of-service attacks**

  A denial-of-service attack is capable of partially or completely paralyzing the internal LAN. A competitor can e.g., gain a strategic advantage over the company under attack until the damaged is repaired. The company's communication ability is limited and it can even be disabled completely. Denial-of-service attacks are

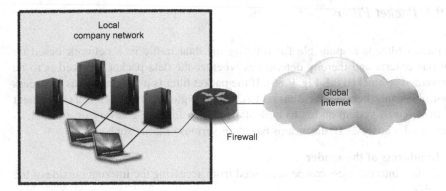

Fig. 9.85 The firewall principle.

often carried out by hackers with no economic or military motivation but simply to embarrass the company or institution

- **Prevention of intervention in data integrity**
  If an outsider succeeds in gaining access to computers in the internal LAN, it is then in a position to spy on, modify or destroy internal data. The damage resulting from such an attack is often greater than a pure denial-of-service attack, as in most cases the perpetrator could have already been operating for a long time in the network undetected. However, authorized access is usually more difficult to acquire.

- **Prevention of access to protected information**
  Even if an outsider has read-only access to the LAN-internal data, this can result in considerable damage. Companies are usually targeted with the intention of spying on trade secrets, product development plans, marketing strategies or employee plans. The military is also targeted with the intent of gaining knowledge of all types of military secrets.

If the connection of one's own LAN to the global Internet is limited to a single network component, as in the case of a firewall, then the risk zone to be monitored is focused on one single system. All other computers in the LAN remain for the most part untouched by the implemented security measures. Every access to an internal computer on the Internet must always be routed through the firewall system. Similarly, all requests from outside are filtered there, based on LAN security measures, before being forwarded to the internal LAN. For protection of the internal LAN, different access control systems can be deployed in a firewall system. While simple firewall systems may consist of just one suitably configured packet filter, more complex systems can also contain application-level gateways or circuit-level gateways. These are described in more detail in the following.

9.9.1 Packet Filter

A packet filter is responsible for filtering the data traffic in a network based on various criteria and thereby determines whether the data packet received is to be forwarded or blocked (cf. Fig. 9.86). If the packet filter is implemented as an access control instance between the internal LAN and the global Internet, it is a component of the firewall system and is to be installed at an Internet gateway or a router. A packet filter analyses IP datagrams based on various criteria, such as:

- **IP address of the sender**
  Certain internal users can be prevented from accessing the Internet outside of the LAN.

- **IP address of the receiver**
  Access to specific end systems in the global Internet (e.g., specific WWW pages) can be blocked.

- **UDP or TCP port numbers of the sender and/or the receiver**
  Many Internet services associated with UDP or TCP connections use a specific fixed port (**well-known port**). This should prevent e.g., an interactive session via TELNET being started from outside of the internal LAN. The input port 80 is blocked for all data packets originating from outside and filters out such data packets.

- **User-defined bitmasks**
  Prevents TCP being accessed from the outside, while allowing TCP connections in the opposite direction. It is therefore possible to check with a packet filter whether TCP-ACK=0 was set. With the starting of a TCP connection, the first message for the connection setup is sent with TCP-ACK=0, and all the following with TCP-ACK=1. If all data packets are filtered out that reach the internal LAN from outside containing a TCP-ACK=0, client contact from the outside can be prevented. At the same time, every connection in the opposite direction from the internal LAN to the global Internet is permitted.

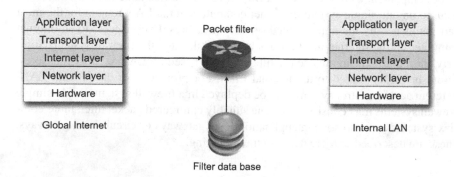

Fig. 9.86 Access control with a packet filter.

The configuration of a packet filter is carried out via a database with simple filter rules. There are basically two different approaches for the interpretation of these rules:

- Everything that is not explicitly forbidden is allowed.
- Everything that is not explicitly allowed is forbidden.

The rules in a filter database usually consists of five tuples in the form:

```
< source IP address, source TCP port, destination IP
                     address,
     destination TCP port, transport protocol>.
```

At the same time, the number of well-known ports is constantly growing, so that updating the filter base can be a problem. Besides that, dynamic port number allocation makes configuration more difficult, e.g., in the case of RPC or tunneling under a specific forbidden port via an opened port.

9.9.2 Gateways

If filtering is set on a higher layer in the TCP/IP reference model, meaning, e.g., on the transport layer or above it on the application layer, it becomes possible to implement user-based filtering. The access control systems used here are referred to as **gateways**: **application-level gateways** for filtering at the application layer, **circuit-level gateways** for filtering at the transport layer. If the access control mechanism is engaged at the transport layer, the connection between an application process and a transport connection can be identified. Such a gateway does not allow an end-to-end connection, but rather becomes itself an end point and divides the connection between the source and destination system into two separate connections. The network administrator is able to fix which types of connections are allowed and which are not. This definition can even be carried out directly on the contents of the transported data. Via an email gateway, a gateway can be fixed which, e.g., works at the application layer and monitors the complete email traffic. For example, it could be that email messages coming from a certain receiver or containing certain key words, are blocked. An email gateway can thus be used as, e.g., a **SPAM filter** to block unwanted email advertising.

9.9.3 Firewalls – Topology

The level of security achieved through the use of firewalls is not only based on the access control procedure implemented, but also to a a large degree on the choice of the firewall system location within the entire network topology. The following basic situations can be distinguished:

Fig. 9.87 Access control with an application-level gateway.

- **Border router**

  A firewall system provides the simplest level of security. The firewall software is installed at the gateway computer (router) to the Internet. If an attacker finds the weak spot at the border router, the internal LAN is open to it with no further security measures.

- **Secure subnet**

  Here, a secured subnet joins the border router of the firewall system. It consists of individual, especially protected computers, providing a connection to the global Internet over the firewall system (cf. Fig. 9.88). The actual internal LAN is for the moment completely shielded from attacks via the Internet. To get through, an attacker must first overcome the firewall system and then also a computer of the specially secured subnet.

- **Dual Home Bastion Hosts**

  Firewalls based solely on limiting routers allow connections between the internal LAN and the global Internet within the firewall system at the protocol level. A **bastion host**, in contrast, separates two networks physically. Normally, the correspondingly configured bastion host has two separate network adapters, one for coupling on the global Internet and one for the connection of the internal LAN. In this way, a physical separation of both networks is made possible. In the bastion host, the **routing** between the two network adapters must also be prevented.

  A special variation of the dual bastion host is the high-level security technology known as the **Lock-Keeper**, developed at the Hasso Plattner Institute for IT Systems Engineering in Potsdam, Germany.

Excursus 15: The Lock-Keeper Technology

The principle of physical separation is the "ultima ratio" for the securing of computer and data networks with the highest need for protection. This demands the complete isolation of the protected network from other corporate networks as well as from the open Internet. The connection to another network is not allowed at any time. Data exchange with other networks is achieved exclusively off-line using portable storage media. The data to be transported in the isolated network is subject to strict control.

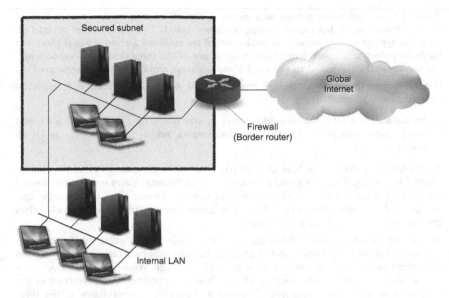

Fig. 9.88 Safeguarding the internal LAN through a border router and secure subnet.

For the technical realization of this principle there are a number of ideas that more or less all work on the implementation of a sluice principle. The data to be transferred across the network border is enclosed in a "locked chamber," examined there and transferred to the protected network. The lock principle is implemented the most consistently in the architecture of the Lock-Keeper. The Lock-Keeper consists of three independent computer components that are connected to each other via a relay-controlled switch. Either the left computer component, "OUTER," is connected with the middle computer component called the "GATE," or the GATE is connected with the right computer component, "INNER." In this way, data from an OUTER connected network, e.g., the Internet, can be transported into the network connected with INNER network, with its high security requirements, and vice versa. This can transpire without a direct physical connection between both networks existing at any time.

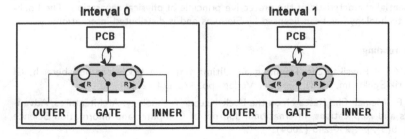

Fig. 9.89 The Lock-Keeper technology.

To meet the requirements in high-security areas, the switch mechanism in Lock-Keeper is not software-based, but rather entirely hardware-based. Therefore, hacker attacks on whatever type of control software is implemented are excluded a priori. Even if there is a hardware defect at the switch mechanism, or at one of the three computer components, the two networks connected to the Lock-Keeper always remain physically separated.

Naturally, guiding the data "though the locks," is not a simple matter. It must be ensured that

- no malware can enter the secured network by way of the data to be the sluiced, and
- the physical separation of both networks remains hidden from the user as far as possible.

As the GATE (INNER as well as OUTER) is a complete computer, (1) can be implemented at the GATE via the standard means of data filtering. Every imaginable firewall protection mechanism can be implemented or connected there. The Lock-Keeper can be operated in combination with virtually any protection system and additionally offers the "added value" of physical separation

In contrast to other security solutions, the Lock-Keeper offers guaranteed protection against all known and unknown on-line attacks. Because off-line attacks can be thwarted by appropriately integrated content filter mechanisms, there remains just a residual risk at the application level. This residual risk is, however, considerably lower than in the case of standard security solutions. Theoretically possible vulnerabilities in the filter mechanism cannot directly impact the protection performance of the overall system. The GATE in the Lock-Keeper is isolated and even in such circumstances the physical separation cannot be reversed.

Requirement (2) is more difficult. It is not possible to hide the lock operation in terms of the time needed for switching the lock system. To be fair, however, this time should not be compared with the data transmission speed in networks, but rather to the time required for the manual transport of data between physically separated networks. In the latter situation, the Lock-Keeper, with its switching cycle of only a few seconds, has the definite advantage.

However, in relation to the various network services, e.g., file transfer, email, web services, database replication or synchronization, the physical separation allows itself to be hidden from the user. Through the creation of artificial endpoints at OUTER and INNER, the services can be "dismantled" on one side, converted into sluice-ready packets and "be put back together" correctly on the other side. The user remains oblivious to these sometimes very complicated processes. Emails are received in the normal form and the web services can be offered as usual. The correct and complete disassembly on one side of the involved network protocol, and the reassembly on the other is an indispensable and essential characteristic of the protective principle of physical separation. The Lock-Keeper technology has been licensed by Siemens and is distributed internationally.

Further reading:

Meinel, Ch. : Physikalische Trennung als Ultima Ratio im Hochsicherheitsbereich, GI Informatik-Spektrum, 30(3), Springer Verlag, pp. 170-174 (2007)

Cheng, F., Meinel, Ch.: Research on the Lock-Keeper Technology: Architectures, Advancements and Applications, International Journal of Computer and Information Science (IJCIS), 5(3), pp. 236-245 (2004).

Lock-Keeper web page at Siemens AG: http://www.siemens.ch/it-solutions_en/branchen/security/kommunikation/lock_keeper.php

With the implementation of a bastion host, a so-called **Demilitarized Zone, (DMZ),** is created between the limiting router and the bastion host. For example, if the operation of WWW servers in the internal LAN is prohibited for reasons of security, it can be relocated to the DMZ. Situated behind the limiting router, but not beyond the firewall system, a WWW server is unavoidably exposed to a higher security risk. A computer inside the DMZ is therefore also known as a **sacrificial host**. While sacrificial hosts are secured as best as possible, they are still not classified as completely trustworthy by the dual-based bastion host as they constitute a prime target for an outside attacker.

Even greater security is ensured by the implementation of multiple bastion hosts in cascaded form (cf. Fig. 9.90). If an attacker succeeds in breaking into the first bastion host, there still remain several subsequent bastion hosts to get past in order to gain access to the internal network.

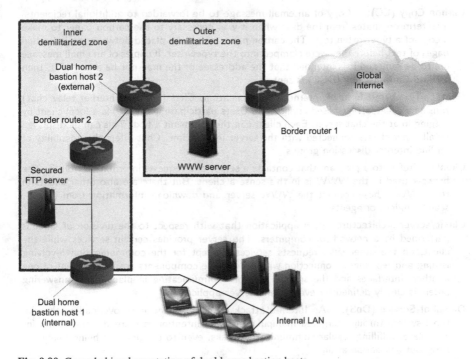

Fig. 9.90 Cascaded implementation of dual home bastion hosts.

A cascaded dual home bastion host can also be provided with a firewall system made up of individual computers, which have access to multiple network adapters. These can each be connected to subnetworks with different security requirements (**firewall with secure server network**).

9.10 Glossary

Application Programming Interface (API):   Via the API a defined interface is implemented between a programming language accessing the API, and the methods and data structures provided by the API, and a concrete hardware and software system that can be manipulated using the methods and data structures of the API libraries. An API shields the developer from so-called "low level" work, which would otherwise be necessary to access the manipulable objects from the API directly. In this way, a comfortable interface for the developer is provided.

Authentication: Serves to verify the identity of a user or the integrity of a message. In authentication, certificates of a trusted authority are used for proof of identification. For proving the integrity of a message, digital signatures are generated and sent with it.

Base64 encoding: An encoding for binary data in 7 bit US-ASCII characters, as defined in the **MIME** standard. The binary data is sectioned into 24-bit long groups, these are in turn subdivided into four groups each 6 bits long. These 6-bit long bit sequences are subsequently encoded with the help of 7 bit US-ASCII characters (A–Z, a–z, 0–9, $+/=$).

Carbon Copy (CC):   Copy of an email message to be forwarded to additional recipients. The term originates from the days when it was necessary to use carbon paper to make a copy of a typewritten text. The carbon paper had to be placed between the individual pages of text before they were clamped into the typewriter. If copies of an email message are to be sent and it is intended that the addressee of the mail not be aware of it, then a **Blind Carbon Copy (BCC)** is sent.

Chat: Text-based form of Internet communication. Chat (actually Internet relay chat) runs as a client/server-based application. Users participate in a chat via a local client by logging in at the chat server. Each client can then transmit a text. This can be displayed by all other clients connected with the server in real-time. Chat offers the possibility of on-line Internet discussion groups

Client: Refers to a program that contacts a server and requests information from it. The browser used in the WWW is in this sense a client. But there are also other clients in the WWW. These contact the WWW server and download information from it, e.g., search engines or agents.

Client/server architecture: An application that with respect to the division of labor is performed by a network of computers. The server provides certain services while the client, on the other side, requests services. Except for the communication involving issuing and replying in connection with orders, the components are independent of one another. Interfaces and the specific way of communicating in placing and answering orders is clearly defined for every client/server application.

Denial-of-Service (DoS):   An Internet attack with the intention of overloading the victim's system through targeted manipulation. The intention is to render the victim incapable of fulfilling regular communication tasks, even to the point of being unable to carry out any action at all.

Digital signature: Serves in the authentication of a document. It consists of the digital fingerprint of the document, encrypted with the private key of the originator.

Domain Name System (DNS):   Name service that can assign symbolic names to a binary IP address as a distributed database application (only the network ID is taken into account). DNS is based on TCP or UDP and organized as a client/server application. Every end system has a DNS client (resolver) that can be called by an application program to initiate name resolution. The DNS address space is organized hierarchically so that every subsection can be served by its own specific DNS server.

Electronic mail (email): The Internet service called electronic mail is a replica of standard mail based on the digital communication possibilities of the worldwide Internet. The email system is a message handling system consisting of so-called user agents (UA), which function as the user's interface for sending and receiving electronic mail, and message transfer agents, responsible for transport of the message from sender to receiver.

Email address: Users of an email system can be uniquely identified via an email address. An email address is made up of:

- User name (address prefix) and
- IP address of the email server (address suffix),

these are separated from each other by the special character "@".

Email gateway: Email gateways function as a connecting link between different email systems. They can also join computers - which are themselves not connected to the Internet - to email traffic. Their task often consists of acting as gateways to enable the LAN users a uniform email address as well as executing **mailing lists**.

File access services: Unlike a file transfer service, the file access service allows the reading or modification of a specific location in a remote file. Only the data to be changed need be transmitted across the network - and no longer the complete file. The most widespread file access service in the TCP/IP protocol family is the **NFS** service (**Network File System**), defined in RFC 1094.

File Transfer Protocol (FTP): General client/server-based application for file transfer. It is standardized in RFC 959. FTP is based on the telnet protocol and TCP as transport protocol, and it allows an interactive and comfortable file transfer with the corresponding authorization mechanisms.

File transfer services: Files from one computer can be transferred via the Internet to another computer by way of a general file transfer service. Internet file transfer services are based on the transport protocol TCP or UDP and enable file transfer between remote computers. These include the necessary format translation, directory queries and verification of user authorization. The most important representatives are **FTP** (File Transfer Protocol), **TFTP** (Trivial File Transfer Protocol) as well as the more complex **NFS** (Network File System).

Gateway: Intermediate system in the network that is able to connect individual networks into a new system. Gateways make communication possible between application programs on different end systems and are located in the application layer of the communication protocol model. They are therefore capable of translating different application protocols into each other.

Hop: Refers to a leg of the end system's journey to the next closest switching computer, or between two neighboring switching computers, or from the switching computer to an end system connected to it. The route through a network from a sender end system to a receiver end system is made up of multiple, single hops.

Internet Message Access Protocol (IMAP): A further development of the POP protocol for communication between the client application and email gateway. It is defined in RFC 1730. Via IMAP the user can log in at an email gateway and manipulate the messages assigned to the email account mailbox. This can be done without the user's client application itself providing a SMTP-based email service. In contrast to POP, not all email messages are completely transmitted by the email gateway in IMAP. Only the "subject" lines are sent by the email gateway and the user decides which email messages are to be transmitted completely.

Mailing lists: If an email message is to be sent to multiple recipients, this can be done via a mailing list. A mailing list is a file containing the email addresses of many different recipients, who can be reached at their own email address. An email gateway that receives an email message addressed to a mailing list under its local administration sends a copy of the message to all addresses on this mailing list.

Multicasting: In a multicasting transmission a source transmits simultaneously to a group of receivers. This is a 1:n communication. Multicast is often used for the transmission of real-time multimedia data.

Multimedia: If different types of media are used in displaying information, e.g., image and sound, this is referred to as a multimedia information display.

Multipurpose Internet Mail Extensions (MIME):   Expansion of the originally defined email format in RFC 822, which intended 7-bit US-ASCII encoded text to be used as the content of the email message. MIME defines additional fields for the message header of the email, with whose help different encoding and data types can be specified as content of the email message. The correct display is ensured by the respective user agent. **Base64 encoding** standard is used for the encoding of binary data.

Name service: Serves in the translation of symbolic names into IP addresses. The Internet valid IP addresses are difficult to remember. Because of this, symbolic names are used in their place. Via a name service, these symbolic names must be translated back again into binary IP addresses for Internet communication.

Network management: Collective term for all activities involving planning, configuration, monitoring, control, troubleshooting and administration of networks using appropriate hardware and software. Principally, network management can be subdivided into two areas. **Network monitoring** includes the state and behavior of end systems, intermediate systems and subnetworks. On the other hand, **network control** is responsible for correcting or improving unwanted and unfavorable network conditions. In the Internet, the **Simple Network Management Protocol (SNMP)** is used for this purpose.

Nonce: A randomly chosen numerical value that can be used in a secured communication to exclude so-called **playback attacks**. In such attacks, an unauthorized third party attempts to exert influence using a duplicate message. A specific nonce value may only be used a single time in the communication.

Post Office Protocol (POP):   A protocol, defined in RFC 1725, for the communication between client application and email gateway. Via POP, the user can log in at an email gateway and read and manipulate the messages assigned to the email account (mailbox) without it being necessary for the user's client application to itself provide a SMTP-based email service.

Port number: A 16-bit long identification for a TCP connection that is always associated with a certain application program. The port numbers 0-255 are reserved for special TCP/IP applications (**well-known ports**) and the port numbers 256-1023 for special UNIX applications. The port numbers 1024-56535 can be used for individual applications and are not subject to permanent assignment.

Pretty Good Privacy (PGP):   A system developed in 1991 by Phil Zimmerman for the secure processing of email message traffic. PGP is freely available for most hardware platforms and operating systems and offers email message encryption with symmetrical key procedures (Triple-DES, IDEA, CST), securing of the symmetrical key via an asymmetrical key procedure (RSA), as well as securing the integrity of email messages (MD5 Message Digest) and maintaining the authenticity of the communication partner (digital signatures). PGP is the most widely-used system for the secure transport of email messages.

Real-time Transport Protocol (RTP):   A specialized transport protocol, defined in RFC 1889, for the transmission of digitalized audio and video signals. RTP uses UDP as an underlying transport protocol, implementing sequence numbers and time stamp information to fulfill the requirements for a timely transport of real-time data. It does this in conjunction with the protocols **RTSP** (Real Time Streaming Protocol), for the streaming of multimedia data, and **RTCP** (RTP Control Protocol), via which the communication partners involved can exchange information about the current performance or the underlying network infrastructure. Because there is no guarantee in an IP-based network that the service parameters (**Quality of Service**) will be maintained, an advance reservations for the available network resources can be made via the two protocols: **RSVP** (Resource Reservation Protocol) and **COPS**(Common Open Policy Service).

Remote Procedure Call (RPC):   Via RPC, function calls are possible on remote computers. The necessary protocol mechanisms are defined in RFC 1057 and lay the foundation for many client/server Internet applications. The programmer of a distributed application has the possibility to program function calls to a remote server over RPC, without possessing detailed knowledge of the necessary mechanisms. For the encoding of the messages to be exchanged between the client and the server, a general transfer syntax is used, which is defined in RFC 1014 **External Data Representation** (**XDR**) . This ensures an automatic conversion between different computer architectures.

rlogin: This is a protocol defined in RFC1282 for the establishment of an interactive session (remote login) at a remote computer. Originally intended to provide a remote login service between two Unix computers, rlogin is less complex and easier to implement than Telnet. The service program **rsh** is a commonly-used variety of the rlogin service. It is a component of most Unix implementations.

Round-trip time: The reaction time of an entire network is understood as the round trip time. This is the time span required to send a signal from a source over the network to any receiver and then to send the receiver's answer over the network back to the sender.

Server: Refers to a process that is contacted by clients and sends back information to them. The computer on which a server process runs is also often referred to as a server.

Simple Mail Transfer Protocol (SMTP):   A protocol for the exchange of email messages, defined in RFC 821. As the foundation for a client/server-based email service, SMTP numbers among the most popular Internet protocols.

Simple Network Management Protocol (SNMP):   Defined in RFC 1157, SNMP is a protocol for network management in Internet-based networks. Network objects and associated parameters are described over a so-called **Management Information Base** (**MIB**) - a hierarchically organized database.

Socket: The TCP protocol provides a reliable connection between two endpoint systems. For this, so-called sockets are defined at the endpoints of the involved computers. They consist of the IP address of the computer as well as a 16-bit long port number, which uniquely defines the connection. So-called **service primitives** are provided via sockets, which allow a control and monitoring of data transmission. Sockets also associate input and output buffers each to the connections they have initiated.

Secure Shell (ssh):   An extension of the remote login service rsh, which provides a secure communication between the client and server computer. The implemented encryption procedures are RSA, DSA and TripleDES.

Streaming: A designation for the continuous playback of multimedia contents (audio and/or video) over the Internet in real-time. This means that playback already begins at the time of transmission without it being necessary to wait until it has been executed completely to start. The contents to be reproduced may already be present in the form of stored or live data, which can be immediately accessed continuously over the Internet directly after being generated. In contrast to conventional data transmission, streaming is delay-sensitive. This means that too-long delayed data packets lose their relevance. It is also error tolerant, i.e., errors and data loss are tolerated within certain limits, despite a resulting reduction in the quality of the continuous multimedia display.

Telnet: A protocol specified in RFC 854 for the establishment of an interactive session (remote login) at a remote computer. Telnet, the oldest Internet application protocol, also enables connections between different computer platforms over a joint virtual terminal interface (**Network Virtual Terminal**).

Trivial File Transfer Protocol (TFTP):   A simple variation of the considerably more complex FTP file transfer service. TFTP is defined in RFC 783. It only allows a simple file transfer without authorization possibilities. TFTP uses UPD as transport protocol and is implemented, e.g., to support the system startup procedure on computers without their own hard drive (**bootstrapping**).

X.400:   X.400 provides an email message service (Message Handling System, MHS) that was standardized by the ITU-T as **MHS** and by the ISO as **MOTIS** (Message Oriented Text Interchange System). The aim of X.400 is the interoperability of products from different manufacturers and between public and private email services. In comparison to **SMTP**, X.400 provides increased functionality, but this is paid for by its correspondingly higher complexity. For this reason, SMTP was soon able to assert itself over X.400 and replace it to a large extent.

Chapter 10
Epilogue

"What is written, is written!",
− Lord Byron (1788–1824), in "Childe Harold"

The Internet with its new communication possibilities has developed into one of the driving forces behind technical and cultural advancement in the 21st century. This volume has primarily focused on the technology of internetworking as the foundation of the global Internet, its influence in the past years on how we communicate with each other and, consequently, how our daily life has changed from the ground up.
The other two volumes in this series, "Digital Communication" and "Web Technologies" round out the spectrum of topics. The historical development and technical foundations of digital communication is described in detail as well as the World Wide Web and associated technologies built on the foundation of the Internet.

The Internet connects different computer networks of different dimensions and technologies to one world-spanning, virtual communication network. Thereby the user always has the impression that the resulting amalgamation that is the Internet is in fact a uniform and homogeneous structure. This colossal task was only possible with the help of a hierarchical modular approach in the form of a communication layer model. Here, the tasks to be addressed in the individual layers are seen as complete in themselves and an interaction between these layers can only proceed via defined interfaces. With a short historical background as the starting point, the protocols and technologies enabling internetworking to take place along the communication layer model of the TCP/IP reference model, are introduced in this book.

The TCP/IP reference model builds on the **physical layer**, which is itself not an official component of this model. If we compare simple situation parameters, for example the distance to be bridged, mobility, technical effort or expense, we see that different situations demand different technologies, which are each based on different physical communication media and infrastructures. To gain an understanding of them, the theoretical foundation of communication based on electromagnetic signals was discussed and various wired and wireless technological variations introduced. Afterwards, the focus was on the first layer of the TCP/IP reference model, the so-called **data link layer**, in the local networks (LANs) and simple wide area networks

C. Meinel and H. Sack, *Internetworking*, X.media.publishing,
DOI: 10.1007/978-3-642-35392-5_10, © Springer-Verlag Berlin Heidelberg 2013

where different technologies are located. Here we looked at **wired LAN technologies** and their most important technological examples, among them Ethernet, Token Ring, FDDI and ATM.

This was followed by an introduction of **wireless LAN technologies**. The popularity of this technology continues to rise, with its performance capability on equal footing with wired competitors. When compared to cable medium, wireless medium has other requirements as far as network communication is concerned with respect to range, reliability and, in particular, security. In addition to a presentation of the basics of wireless and mobile network technologies, the most important technological representatives, such as WLAN were discussed as well as the short-range limited technologies, e.g., Bluetooth and ZigBee. If the number of devices connected to a network grows, or also the distance between the individual communication partners, alternative technologies are implemented. These are those in the subsequently treated **wide area networks** (**WANs**). WANs can be deployed to connect local networks with each other at different locations. Of great importance in this regard are special pathfinding procedures - so-called routing algorithms. Beginning with the historical ARPANET, important technological representatives of WAN technologies were looked at up through the broadband wireless network standard WiMAX.

In order to achieve a homogeneous appearing Internet out of a plurality of individual network technologies, protocols on the **Internet layer** of the TCP/IP reference model are deployed. Above all, the Internet protocol, aided by logical addressing, makes communication possible across network borders from the source computer to the target computer. On the basis of different criteria, such as maximum throughput, reduced cost, uniform load distribution or the "best possible security," the routing process located here chooses the best path through the network. Besides giving a detailed presentation of the current IP protocol IPv4, its addressing, data formats and functionality, different routing procedures were described here. We also looked at the Internet protocol of the next generation IPv6, with its advantages but also the problems associated with its implementation, and other protocols of the Internet layer as well The Internet layer only provides an unreliable, wireless data transport service. To make reliable communication possible, protocols such as the Transport Control Protocol (TCP) are deployed in the **transport layer** above it. TCP provides a universal transport service, i.e., an explicit (per software) switched connection with assembly and dissasembly procedures, as well as secured quality criteria. Among these are error correction methods or the correct arrangement of the transmitted data packets. Data flow algorithms are responsible for a uniform load on the network infrastructure. In case of overload, the transmission volume is throttled accordingly, or if a free medium is involved its performance capacity is fully exploited. TCP therefore represents the most complex of the protocols in the TCP/IP reference model and is supplemented by the connectionless (unreliable) transport protocol User Datagram Protocol (UDP) for less demanding or time-critical transport tasks.

The real meaning and purpose of the Internet lies in the available Internet applications. These are based on the communication services of the Internet and their communication protocols. They belong to the **application layer** of the TCP/IP reference model and are made up of many protocols that have gained great popularity in data transmission in the Internet, for example HTTP (Hypertext Transfer Protocol), FTP (File Transfer Protocol) or SMTP (Simple Mail Transfer Protocol). The applications themselves, e.g., the HTML browser or the email client, are not included in this layer. They are outside of the communication model and only serve in the functionalities of the application layer. Applications such as email or WWW are the ones that influence our idea of the Internet. They are responsible for helping us forget how many technological tricks and details are necessary in getting this network of networks up and running. Additionally in the chapter covering the application layer, a detailed look was taken at the audio and and video communication in Internet-based technologies and protocols. In contrast to asynchronous email communication, synchronous audio and video communication connections are subject to strict time requirements. To implement this reliably, streaming technologies are deployed along with different real-time transport protocols.

This book is devoted to the subject of internetworking from the technical perspective of computer science. The topic "Digital Communication" naturally encompasses a great deal more and includes a number of additional areas that could not be addressed in greater detail here for lack of space. In the chapter on the physical layer here, neither the physical nor the electronic foundation of digital communication could be examined in greater detail. An introduction to cryptography and digital security as well as basic coding theory was also omitted. These are examined in detail in "Digital Communication" the first volume of the trilogy that includes "Internetworking" and "Web Technologies." The World Wide Webs (WWW) and the web technologies it is based on go beyond internetworking. The Hypertext Transfer protocol, the basic communication protocol of the WWW, and the HTML coding (Hypertext Markup Language) used there are presented in the third volume of the trilogy, as well as more advanced, dynamic and interactive web applications. Other areas examined are the new interactive technologies covered under the term Web 2.0, WWW search engines and the Semantic Web.

A brief look will now be taken at the topics addressed in the other two volumes of this series.

Volume 1 "Digital Communication," begins with a brief outline of the historical development of human communication and its technical resources as a prelude to a tour through the world of digital communication. We are led on a path from the prehistoric cave painting to today's World Wide Web.

The information theoretic implementation of communication is then discussed with an initial focus on the coding of messages. Before information can be transmitted over the communication medium from one partner to the other, it must first be brought into a suitable form for the communication medium. This means the information must first be digitally encoded and then sent in the form of a message via a digital communication medium.

Because information can exist in different medial forms, whether text, image, audio or video recording, and it is necessary to consider these forms in encoding, different encoding variations are needed for efficient technical implementation. "Efficient," in this sense initially applies to the required data memory space. The more storage space there needs to be, the longer message transmission takes. Compression procedures help to reduce this needed space by removing the redundancy of a message, i.e., those components that do not play a role in the information content of the delivered message. The Shannon criterion indicates the maximum extent of lossless message compression. This is determined by the entropy of a message (information content). If the message is to be reduced beyond this point, a lossless compression is no longer sufficient and information-bearing elements must be removed from the message (lossy data compression). The system of human perception offers approaches on how to do this. Certain limitations are set by our sensory organs and the information processing that takes place in the brain. Based on these physiological factors, it is possible to determine information components that can be left out in compression because they have little or no bearing on what is perceived. This fundamental principle forms the basis for MPEG audio coding, JPEG image coding as well as MPEG video compression.

Our previously coded messages are nowadays transmitted via digital communication channels whose scope ranges from the pico networks of one's personal space to the worldwide spanning, heterogeneous network system of the Internet. For the success of the worldwide digital communication in computer networks, all participants must follow strict rules. These are dictated in a communication protocol specifying message format as well as in rules for message exchange.

Many computer networks, and particularly the worldwide Internet, are based on the paradigm of the "open network." This means that in principle every communication participant is in the position of receiving all of the messages sent over the network. While digital communication networks share the security risks of their analog predecessors - such as the telephone network, where lines or exchange locations could be "tapped," or the postal system, where letters could be intercepted and opened undetected - the dimensions of the security problems in digital communication are in fact much more diverse. In order to guarantee a sufficient level of security and privacy in the digital communication process, secure symmetrical and asymmetrical encryption procedures with public keys have been developed. These may be implemented by everyone today for private communication and their reliability is officially guaranteed by recognized certification authorities.

Volume 3 "Web Technologies," focuses specifically on the foundation of the World Wide Web. The World Wide Web is a huge, comprehensive collection of information and data distributed across the entire world that we are able to access via the Internet, or more exactly with the help of the HTTP protocol situated on the Internet protocol and a web browser. The Internet has gained its enormous popularity and dissemination today thanks to the WWW. A large part of this success is due to the simple access interface. Owing to the browser it is also possible for laypeople to have fast and simple access to the vast range of information on the WWW.

The WWW, the world-spanning Internet, in the form it offers today for the general public has come of age. When at the end of the 1980s Tim Berners Lee and Robert Cailliau were contemplating a simple document exchange and administration system via the Internet, they never could have guessed that their idea - a byproduct of their actual work at the European Organization for Nuclear Research - CERN (Centre Européenne pour la Recherche Nucléaire) - would be the impetus for the Internet revolution. With the creation of a simple to use graphic user interface, it was possible at the beginning of the 1990s for the general population to use the Internet as a new communication and information medium. The number of computers connected to the Internet in order to use the WWW began to grow phenomenally. This development continues unabated until today.

The information itself is contained in the WWW in the form of so-called hypermedia documents. Connected with each other by so-called hyperlinks, they build an information network where the user can navigate fast and simply. So that a certain document can be found in the WWW it must be uniquely identified worldwide by an address called a **Uniform Resource Identifier (URI)**. A detailed look is taken at the URI address scheme and its subtypes URL, URN, IRI, persistent URL and openURL. Access to the user desired WWW resources, distributed across countless servers, proceeds over the **Hypertext Transfer Protocol (HTTP)**. This functions according to the client/server paradigm. A very simple and, most importantly, fast protocol, it has been continually extended since its introduction. It offers a flexible and secure infrastructure for efficient data transfer in the WWW today supplemented by components such as SSL (Secure Socket Layer) or TLS (Transport Layer Security).

As well as allowing access to hypermedia documents modern browsers also act as multifunctional clients to access alternative Internet services, such as FTP, email or video. The sometimes complex mechanisms behind the individual services and protocols remain completely hidden from the user. The user is only confronted with the simple to use graphic user interface of the browser. Along the transmission path between WWW server and the user's browser, proxy servers, caches and gateways ensure a smooth and efficient data transfer as well as considerably increasing the performance ability of the HTTP protocol. The special feature of WWW documents is the possibility of the cross-linking via so-called hyperlinks to a distributed hypermedia system. Unlike a book or a document in the traditional sense, it is possible for the user to gain direct access from one place in the document to a marked place in another document. This does not have to be even stored at the same WWW server, but can be stored at the other end of the world.

It is not only possible to link every type of text with each other in the WWW but also every type of multimedia document. This includes pictures, audio and video clips and interactive contents. The hypermedia documents of the WWW are written in a special descriptive language called **Hypertext Markup Language (HTML)**. This language makes it possible for the author to determine the structure of his or her document, i.e., the layout of the headings, paragraphs and tables, as well as to highlight certain parts of the document. This structuring is carried out via special "markups," which are recognized and by the browser as separate from the actual document and processed accordingly.

The graphic design (formatting) of a HTML documents is carried out via separate files - so-called **Cascading Style Sheets (CSS)**. Via CSS, depending on the graphic features of the respective output device (mobile telephone, computer monitor, printer etc.), different formatting can be determined for the individual structural elements of a HTML document. The desire for greater flexibility and the establishment of individualized identifying markings and a markup language led to the development of **XML**, the **Extensible Markup Language**. With the ability to establish new identifying elements, XML, as a meta-markup language, soon became the starting point for a whole series of application-specific markup languages tailored to special applications. The corresponding XML language definitions are fixed with the help of a Document Type Definition (DTD) or the data definition language schema XML. The principles of the XML, DTD and XML schema, as well as expanded object identity using XPath and XPointer, are explored in detail against this background. The reader is also introduced to the advanced hyperlink concept XLink, form processing with XForms, XML queries with XQuery and XML transformation with XSLT.

**

Up to a point it was only necessary to make pre-recorded contents available at a WWW server that the user could access with the help of a browser (static Web documents). At the same time, Content Management Systems (CMS) prevailed in the professional area. CMS enables the dynamic generation of communicated contents (**dynamic Web documents**). This means, the contents are taken from a data bank based on need and according to predefined style sheets. A CMS offers the possibility of creating style sheets independent of the frequently-changing information they offer. If there is a layout change, only the respective style sheet must be adapted, which receives the actual information content that is stored separately in a data base. Here, WWW servers work together with special server application programs. Their task consists of generating the information that is transmitted to the WWW servers and encoding the requested web documents without delay as HTML documents. These are then given to the WWW server who can deliver them to the user.

To be able to request dynamically generated web documents from WWW servers, special parameters must often be transmitted indicating the exact information to be be sent back. In this context, WWW servers offer multi-faceted possibilities for the interaction with server-side application programs, e.g.,Common Gateway Interface (CGI), Server Side Includes, Java Servlet, Enterprise JavaBeans or Web Services.

The activities on the client's side can also entail far more than the simple display of HTML pages. User involvement in the areas of user feedback and interactivity extend far beyond pure information consumption with a "genuine" participation in the WWW taking place. But in this context, HTML is no longer sufficient as a basic display language of the web document. HTML only allows a static display of information with a "simple" return channel (e.g., via so-called web forms). Early in the development of the WWW, the idea arose of transmitting programs from the WWW server to the client side where they would subsequently be executed as well as transmitting static information. Naturally there are special security regulations necessary to prevent the software transmitted in this way from causing any damage. The program code loaded by the client, e.g., a JavaScript program or a so-called Java Applet, may only run within a specially secured environment - the so-called "sandbox." Access to sensitive computing resources is not possible here. The chapter **Web Programming** provides an in-depth look at the client-side and server-side of web programming. It also explores, e.g., the Document Object Model, Java Servlets, Web Frameworks, distributed applications with Remote Procedure Call (RPC), Remote Method Invocation (RMI), Enterprise Java Beans (EJB) and introduces the world of Service-Oriented Architecture (SOA).

Today the WWW has grown into the largest information archive worldwide. Because the billions of web documents make it impossible for a user to maintain his or her bearings, the first directories and search services came into existence shortly after the launch of the WWW. Only with the development of powerful **search engines**, such as Google, has it become possible for us today to gain (usually) targeted and efficient access to the desired information. Google administers a gigantic search index that accesses the relevant web documents in a matter of seconds upon input of a search word. Based on the sheer size of the WWW this only functions with the help of automatic index compilation procedures. Using statistical methods this procedure evaluates and indexes all terms within a web document and puts the relevant documents in a ranked order with respect to a certain key word.

The WWW has also changed a great deal with respect to content since its birth in 1990. With the advent of e-commerce in the mid 1990s, the focus has shifted from the WWW as a specialized personal communication and publication medium to that of a mass communication medium. The medium of mass communication - information production and information consumption - had been strictly separate initially with only specialists being able to put their own content online in the WWW. The general public consumed the information as they did the information provided by traditional broadcast-medium commercial providers. And the WWW continued to evolve. Interactive technologies operating under the name **Web 2.0** makes it possible for laypeople as well to publish their own information content in a simple way. Blogs, chat rooms, file-sharing, social networks, tagging systems and wikis have conquered the WWW and opened up a path of genuine interaction in the digital word to a broad base of users.

The communicating resources in the WWW are generally present in the form of text documents or multimedia documents in which the information is described in normal language. The person as receiver and user of this document can understand the content (most of the time) without a problem. The situation changes completely, however, when this content is submitted for automatic processing. These dynamics occur in WWW search engines, which are often not able to deliver documents with synonymous search words (words with the same meaning). In performing their search, search engines rely solely on the occurrence of a specific string. In this way, paraphrases and synonyms cannot be found.

To make the terms behind the words of the natural language accessible for automatic processing they must be made "machine understandable," encoded and linked to the natural language document. The content semantics, i.e., the meaning of the terms and how they relate to one another, must also be described in a machine-readable, standardized format. This type of formal knowledge description is known as ontology and can be read by a program, e.g., a search engine, and processed. The tools necessary to do this have already been standardized and are the starting point for the new "**Semantic Web**." They are ontology descriptive languages with their various semantic expression, such as RDF, RDFS or OWL. Semantically annotated websites allow autonomously acting agents to collect targeted information and based on it to make independent decisions as defined by their client. They are therefore able to initiate actions independently that not only change the Web but the real world as well.

List of Persons*

Thales of Miletus,   (c. 640 BC- c. 546 BC), Greek philosopher, mathematician and astronomer (also merchant, statesman and engineer), he is generally considered to be the founder and progenitor of Greek philosophy and science. Among his many contributions was the important work in geometry (Thales Theorem). He calculated the height of the pyramids from the length of their shadows and is said to have predicted the solar eclipse of 85 BC (in all probability relying on Babylonian knowledge of the so-called "Saros Cycle," which states that an eclipse can occur every 233 lunar month). He was the first to describe the magnetic effect of static electricity produced by rubbing amber with a cloth (frictional electricity).

Harald Gormsen,   (910–987), also known as Harald Blåtand ("Bluetooth"), Danish king who united the kingdoms of Norway and Denmark and laid the foundation for the Christianization of Scandinavia. The wireless network technology "Bluetooth," specifically designed for close range and ad hoc pico networks, was named after him.

Stephen Gray,   (1666 – 1736), British physicist and amateur astronomer, he discovered that almost all material conducts electrical current, particularly copper wire.

Alessandro Volta,   (1745–1827), Italian physicist who carried out work in electrolysis and in 1800 introduced the eponymous Voltaic pile - the first reliable source of energy and the forerunner of the modern battery.

Jean-Baptiste Joseph Baron de Fourier,   (1768–1830), French mathematician and physicist who in 1822 developed the principle of trigonometric series for periodic functions in his Analytical Theory of Heat (,, *Théorie analytique de la chaleur*"). With the help of methods named after him (Fourier Analysis, Fourier Transformation), a periodic function can be represented as a superposition of sinus and cosine oscillations of varying amplitude and frequency.

André Marie Ampère,   (1775–1836), French physicist, worked on electromagnetism and developed the electromagnetic needle telegraph.

Hans Christian Oersted,   (1777–1851), Danish physicist and chemist, established the theory of electromagnetism and laid the foundation for modern electrical engineering.

---

* in chronological order

839

Carl Friedrich Gauß   (1777–1855) German mathematician famous for his pioneering work in algebra, number theory and differential geometry. Together with Wilhelm Weber, he developed one of the first electromagnetic pointer telegraphs, which linked the physics building next to the Göttingen Pauline church to the Göttingen observatory.

Michael Faraday,   (1791–1867), American physicist who in addition to discovering diamagnetism discovered the magneto optic effect and electromagnetic induction, which played a crucial role in the development of the telephone.

Samuel Morse,   (1791–1872), American portrait artist and inventor whose pioneering work in the advanced development of the telegraph led to the major breakthrough of the "writing" telegraph and the "Morse code," an encoding used in electrical telegraphy and named in his honor.

Christian Doppler,   (1803–1853), Austrian mathematician and physicist famous for the phenomena he proposed called the *Doppler effect*, which describes the change in the perceived or measured frequency of every type of wave when the source and the observer move toward or farther away from each other.

Wilhelm Weber,   (1804–1891), German physicist who together with Carl Friedrich Gauss worked on and published studies on the measurement of the earth's magnetic field and in 1833 developed the electromagnetic pointer telegraph.

John Scott Russell,   (1808–1882), British naval engineer who was the first to discover the phenomena of the soliton - wave packets that can constantly spread out unchanged for kilometers, without being influenced by other wave packets. Solitons have gained major importance in fiber optic technology; it is possible to achieve high data transmission rates over great distances by way of them - virtually without loss and interference.

Giovanni Abbate Caselli,   (1815–1892), Italian physicist who invented a copy telegraph called the Pantelegraph and had it patented in England and France in 1855. By 1865, the invention could simultaneously transmit two pictures in time-division multiplexing process line by line. In his later life Caselli was dedicated to bringing the knowledge of physics to a broader segment of the population and to this end founded the popular scientific magazine "La recreazione."

John Tyndall,   (1820–1893), British physicist, who investigated among other things light scattering in a turbid medium and was the first to explain why the sky is blue (Tyndall Effect). He made revolutionary studies in directing light along an optically transparent conductor in the form of a jet of water, thus anticipating the fiber optic cable.

James Clerk Maxwell,   (1831–1879), Scottish physicist, developed a standardized theory for electricity and magnetism. He postulated the existence of electro-magnetic waves, creating the foundation for radio technology. He proved that the light produced by electromagnetic oscillation is of a specific wave length.

Elisha Gray,   (1835–1901), American inventor who invented the telephone and had it patented at the same time as Alexander Graham Bell. Based on a court decision, it was however Bell who was later awarded the telephone patent. In 1886, Gray made the first proposal for multiple use of telegraph lines via frequency-division multiplexing.

Jean-Maurice Émile Baudot,   (1845–1903), French engineer and telecommunications pioneer, he invented the eponymous Baudot code for the encoding of letters and numbers. The unit of measure **baud** (character transmitted per second) is named after him. In 1874, he developed a telegraphy system based on synchronous time-division multiplexing with which 4-6 telegraphy signals could be transmitted over a common line.

Edouard Branly,   (1846–1940), French physicist, he discovered the possibility of converting radio waves into electrical power. He developed a detector for electromagnetic waves - the coherer, a glass tube containing metal filings that change their conductivity based on how they are influenced by electromagnetic fields and therefore can be used for detecting the same.

Alexander Graham Bell,   (1848–1922), American physiologist who, based on a decision by the US Supreme Court, is recognized as the inventor of the telephone and received the respective patents.

Oliver Heaviside,   (1850–1925), British electrical engineer, mathematician and physicist, who invented, among other things, coaxial cable, which he patented in 1880. He developed new mathematical techniques for solving differential equations and was instrumentally involved in the introduction of vectors and vector analysis.

Nikola Tesla,   (1856–1843), Serbian-American physicist, inventor and engineer famous for his pioneering inventions in the field of electromagnetism. His theoretical work and numerous patents laid the foundation for modern alternating current technology. In the field of communication technology, one of his inventions in 1900 was a frequency hopping method for remotely controlling a submarine. The SI unit of measure named after him, "Tesla," indicates the field strength of the magnetic field. The IEEE Tesla Award of the US Institute of Electrical and Electronics Engineers is named in his honor as is a moon crater.

Heinrich Hertz,   (1857–1895), German physicist, applied the theory of Maxwell and constructed devices for sending (resonator) and receiving electromagnetic waves. He thus proved the validity of Maxwell's theories. He succeeded in carrying out the first wireless message transmission. The physical measurement to describe the frequency of a wave (one cycle per second = 1 Hz) was named after him. It has been established in the international metric system since 1933.

Alexander Stephanowitsch Popow,   (1858–1906), Russian naval engineer and inventor who - based on the work of the Frenchman Branly - developed antennas and radio receivers to detect natural electrical natural phenomena such as thunderstorms. He established the first wireless Morse connections over a distance of 250 m.

Jonathan Adolf Wilhelm Zenneck,   (1871–1959), German physicist and radio pioneer, discussed frequency hopping spread spectrum for the first time in his book on radio transmission, published in 1908. This allows robust communication to be carried out on multiple channels simultaneously for multiple participants despite interference. He improved the concept of the Braum vacuum tube for the direct reception of radio signals. In 1928, he received the IEEE Medal of Honor, the highest award of the American Institute of Electrical and Electronics Engineers.

Guglielmo Marconi,   (1874–1934), Italian engineer and physicist who advanced the field of wireless communication building on the work of Hertz, Branly and Popow. He experimented with marine radio and created the first transatlantic radio link.

George Antheil,   (1900–1959), American composer, pianist, author and inventor. In 1940, he developed a secure communication system on the basis of the frequency hopping method (frequency hopping spread spectrum) with actress Hedy Lamarr. In synchronizing multiple player pianos with identical piano rolls (punch cards) the idea arose of also synchronously controlling the frequency exchange between the transmitter and the receiver. The US military chose not to implement the 1942 issued patent and it was subsequently forgotten until being used for the first time in the naval blockade of Cuba after the patent had expired.

Alec A. Reeves, (1902–1971), British engineer who in 1938 developed pulse code modulation - a procedure for the transformation of analog signals into individual pulses of constant amplitude that allow themselves to be digitally recorded and transmitted.

Hedy Lamarr, (1912–2000), Austrian-American actress and inventor who with composer George Antheil developed a secure communication system based on the frequency hopping spread spectrum. When synchronizing multiple player pianos with identical piano rolls (punch cards), the idea came up of also controlling the frequency exchange of transmitter and receiver on a synchronous basis. The 1942 issued patent (US 2292387) was not implemented by the US military and initially shelved until ultimately being put to use during the naval blockade of Cuba in 1962. By that time the patent had expired. Hedy Lamarr was awarded the EFF Pioneer Award in 1997 by the Electronic Frontier Foundation in honor of her and Antheil's invention.

James van Allen, (1914–2006), American astro-physicist and space pioneer. In 1958 he discovered the eponymous Van Allen Belt, a radiation belt of high energy cosmic particles that encircles the earth. Van was awarded the National Medal of Science by U.S. President Ronald Reagan in 1987.

Claude Elwood Shannon, (1916–2001), American mathematician who made fundamental contributions to the mathematical theory of information and coding.

Arthur C. Clarke, (1917–2008), British science fiction writer. He had already suggested the idea of world-wide radio communication based on a geostationary satellite network in 1945. His proposal became reality 19 years later and was implemented in 1963/64 when the first American communication satellites with geostationary orbits were sent into space.

Richard Bellman, (1920–1984), American mathematician who together with Lester Ford created the Bellman-Ford algorithm for the calculation of the shortest path in a network graph from a start node. Bellman is considered the inventor of dynamic programming, a special resolution procedure for mathematical optimization problems. In 1970, he received the Norbert Wiener Prize In Applied Mathematics.

Harvey Ball, (1921–2001), American commercial artist and creator of the "smiley" symbol. The smiley entered the realm of electronic communication as a so-called "emoticon." In addition to written information, emoticons are meant to provide further information about the author's mood or intention.

Delbert Ray Fulkerson, (1924–1976), American mathematician, he created the eponymous Ford-Fulkerson algorithm for the calculation of the maximum flow in a network together with mathematician Lester Ford.

Donald W. Davies, (1924–2000), British computer scientist who with Paul Baran and Leonhard Kleinrock developed the principle of packet switching as a fundamental principle of the computer network. Davis coined the term "packet switching."

Paul Baran, (1926–2011), Polish-American engineer who developed the concept of packet switching with Donald Davies and Leonhard Kleinrock as the underlying principle of computer networks.

Lester Ford,   (*1927), American mathematician. Together with Richard Bellman he developed the eponymous Bellman-Ford algorithm for determining the shortest path in a network graph and with Delbert Ray Fulkerson the Ford-Fulkerson algorithm to identify the maximum flow in a network. Both algorithms are implemented in the distance vector routing procedure.

Theodore Maiman,   (1927–2007), American physicist. In 1960, he developed the first functional laser - the ruby laser - together with his assistant Charles Asawa.

Robin K. Bullough,   (1928–2008), British physicist famous for his fundamental contributions to the theory of the optical soliton - short light impulses in the form of wave packets that can spread out in a constant and unchanged state for kilometers without being influenced by other wave packets. Solitons have achieved great importance in fiber optics technology. With them it is possible to send a high rate of data transmission over great distances that is virtually loss-free and interference-free.

Edsger Wybe Dijkstra,   (1930–2002), Dutch computer science pioneer. He became well-known for the algorithm named after him, which he developed for determining the shortest path between two nodes in a graph. Dijkstra is considered a pioneer of structured programming and was honored as a recipient of the Turing Award in 1972.

Norman Abramson,   (*1932), American engineer and computer scientist. In 1970, he developed the ALOHAnet at the University of Hawaii. This was the world's first wireless computer network and connected the main islands of Hawaii with each other using low-cost amateur radio transmitters and receivers. Abramson was awarded the IEEE Alexander Graham Bell Medal in 2007.

Doug McIlroy,   (*1932), American mathematician, engineer and programmer. He developed the pipes (pipeline) approach to linking macro programs in which data stream-oriented data processing takes place. In 1972, McIlroy's pipes were implemented in the operating system Unix.

Charles Kuen Kao,   (*1933), Chinese physicist and pioneer in fiber optic technology. In 2009, he was honored by the Royal Swedish Academy of Sciences with the Nobel Prize in Physics, together with Willard Boyle and George E. Smith, "for his groundbreaking achievements concerning the transmission of light in fibers for optical communication."

Leonard Kleinrock,   (*1934), Professor at the University of California Los Angeles. Along with Paul Baran and Donald Davies he developed the concept of packet switching and is considered to be the author of the first message sent over the Internet.

Lawrence Roberts,   (*1937), American engineer considered as one of the "fathers" of the ARPANET. In 1966, Lawrence became ARPA chief scientist and founded the Network Working Group. Under his leadership the ARPANET became the precursor of the Internet.

Robert E, Kahn,   (*1938), American engineer and member of the development team at the company BBN, which under contract for the ARPA designed the first computer communication (Interface Messenger Processor, IMP) for the ARPANET. In 1973, Kahn and Vinton Cerf began working together on the Internet protocol TCP/IP. He has also served as director of the Internet Society (ISOC). Kahn was awarded the Turing Award together with Vinton Cerf in 2004, and in 2005 both scientists received the "Presidential Medal of Freedom," the highest civilian award in the US.

Ray Tomlinson, (*1941), American engineer who in 1971 sent the first email in the world (to his own account from a computer in the next room) via the ARPANET. He was the first to use the "@" character to separate the recipient's name at the target computer.

Werner Zorn, (*1942), German computer scientist and Internet pioneer who led the project group at the University of Karlsruhe. The group, established in 1984, was the first German participant connected to the Internet. He was also founder of one of the first German ISP's: Xlink and received the German Federal Cross of Merit in 2006.

Vinton Cerf, (*1943), American mathematician and computer scientist, member of the development group for the ARPANET. Together with Robert Kahn he developed the Internet protocol **TCP/IP** in 1973, which in 1983 became the standard protocol of the Internet worldwide. Today, Cerf is vice president and Chief Internet Evangelist for Google. He and Robert E. Kahn received the Turing Award in 2004 and in 2005 the "Presidential Medal of Freedom," the highest civilian award in the US.

Jon Postel, (1943–1998), American computer scientist and Internet pioneer who as RFC editor had been responsible for the organization and publication of Internet standards since the start of the ARPANET. He also had a leading role in the IANA in the allocation and organization of Internet addresses. Postel was involved in the development of the basic Internet protocols FTP, DNS, SMTP and IP.

Charles P. Thacker, (*1943), American physicist and computer pioneer, co-inventor of the of the Ethernet LAN technology at the Palo Alto Research Center (PARC) of the Xerox company. In 1967, Thacker was involved with Butler Lampson at the University of Berkeley in the development of the first time-sharing operating system. He was project leader in the '70s at Xerox Parc for the development of the ALTO computer, the first modern personal computer, and also had a role in the development of the first laser printer. In 2009, Charles P. Thacker was honored with the ACM Turing Award.

Butler Lampson, (*1943), American computer scientist, co-inventor of Ethernet LAN technology at the Xerox company Palo Alto Research Center (PARC). In 1967, Lampson was involved in the development of the first time-sharing operating system at the University of Berkeley, together with Charles P. Thacker. With the ALTO, he realized his vision of a low cost, modern personal computer. He also played a role in the development of the first laser printer, the two-phase commit protocol and the first WYSIWIG word processing program.

Whitfield Diffie, (*1944), Cryptography expert, co-developer of the **Diffie-Hellman**-procedure named after him. This is a cryptographic procedure based on the use of a public key and therefore makes the exchange of the secret key information necessary in standard, symmetrical key exchange procedure superfluous. Diffie is politically active and committed to the rights of the individual in the cryptographically secure private sphere.

Leonard M. Adleman, (*1945), Professor of computer science at the University of Southern California, Los Angeles, co-developer of the RSA cryptographic procedure (Rivest-Shamir-Adleman, 1978) for asymmetric encryption. He invented the procedure to solve the simple Hamiltonian circuit problem and also built the first DNA computer. With Adi Shamir and Ron Rivest he received the Turing Award in 2003.

Martin Hellman,   (*1945), Cryptography expert, co-developer of the Diffie-Hellman procedure named after him. The procedure is based on the use of a public key, making the exchange of the secret key information necessary in the encryption procedure superfluous.

Robert Metcalfe,   (*1946), American engineer who developed the Ethernet LAN technology at the Palo Alto Research Center of the Xerox company. On his initiative, Ethernet became the product standard of the Digital, Intel and Xerox companies in a joint campaign to become the most widely used LAN standard today. In December 1973, he wrote the RFC 602 " *The Stockings Were Hung by the Chimney with Care*," which describes the first attack by a hacker in the still young ARPANET. He left Xerox in 1979 and founded the 3Com company, a manufacturer of network hardware and software. In 1980, he became recipient of the Grace Murray Hopper Award from the Association for Computing Machinery (ACM) for his contributions to the development of the local computer network especially the Ethernet. Metcalfe was awarded the US National Medal of Technology in 2003. He has also achieved fame for the eponymous **Metcalfe's Law**, which states that the value of a network is proportional to the square of the number of its users.

Robert Cailliau,   (*1947), Co-developer of the World Wide Web (1990). At the European Organization for Nuclear Research, CERN, Cailliau and Tim Berners Lee came up with the design for the World Wide Web as a simple hypertext-based document exchange system.

Ronald L. Rivest,   (*1947), Professor of computer science at the Massachusetts Institute of Technology (MIT), co-developer of the **RSA cryptography procedure** (Rivest-Shamir-Adleman, 1978), developer of the symmetrical encryption procedures RC2, RC4, RC5 and co-developer of RC6. Along with Adi Shamir and Leonard Adleman he was honored with the Turing Award in 2003.

Paul Mockapetris,   (*1948), American computer scientist and electrical engineer who in 1983 developed the Internet Domain Name System for mapping symbolic computer names to binary IP addresses. From 1994 to 1996, Mockapetris was chairman of the Internet Engineering Task Force (IETF), after having led several IETF working groups.

Scott E. Fahlman,   (*1948), American computer scientist, who in 1982 made the first official proposal to use emoticons in email communication as a way of providing additional information about the mood and intention of the author.

Paul Kunz,   physicist at Stanford Linear Acceleration Center. During a research stay at the Swiss nuclear research institute CERN he became familiar with the World Wide Web, and in December 1991 installed the first WWW server in the US.

John M. McQuillan,   (*1949), American computer scientist who developed the fundamentals of the link state routing procedure and was involved in the programming of the first interface message processors, the connection computer of the ARPANET.

David Reeves Boggs,   (*1950), American electrical engineer. Working together with Robert Metcalfe he developed the Ethernet LAN technology. Moreover, he was involved in the development of early prototypes of the Internet protocol, file servers, gateway and network adapter cards.

Michael Rotert,   (*1950), Professor of computer science at the University of Karlsruhe, he implemented and operated the Internet mail server germany thereby laying the foundation for the spread of the Internet and email in Germany. He set up Internet connections in Germany (1984), France (1986) and China (1990). Additionally, Rotert is a founding member of the Internet Society and of the DeNIC, as well as chairman of the eco-Association of the German Internet Industry e.V.

Ralph C. Merkle, (*1952), American computer scientist and pioneer in cryptography who together with Martin Hellman and Whitfield Diffie developed the Diffie-Hellman key exchange procedure. He also designed the block ciphers Khufu and Khafre and the cryptographic hash function SNEFRU.

Adi Shamir, (*1952), Professor at the Weizmann Institute of Science in Tel Aviv, co-developer of the RSA cryptography procedure (Rivest-Shamir-Adleman, 1978) for asymmetric encryption. He received the Turing Award in 2003 together with Leonard Adleman and Ron Rivest.

Tim O'Reilly, (*1954), Irish software developer, author and publisher. He played a decisive role in developing the Perl scripting language. With his collaborator Dale Daugherty, O'Reilly coined the term "Web 2.0".

Philip R. Zimmerman, (*1954), American software engineer and creator of the popular email encryption software, Pretty Good Privacy (PGP). Because of US export restrictions on high-strength cryptography, Zimmerman became the subject of a government investigation as this hybrid encryption software was freely available to the public for download. The investigation was however dropped in 1996 without charges being filed.

Christoph Meinel, (*1954), Director of the Hasso Plattner Institute for Software Systems Engineering at the University of Potsdam, Visiting Professor at Luxembourg International Advanced Studies in Information Technology and at the Beijing University of Technology. His work has dealt with issues involving communication complexity, e.g., as inventor of the high security network lock system, "Lock Keeper," which facilitates message exchange between physically separated networks, and as developer of the internationally implemented teleteaching system "tele-TASK." Meinel is chairman of the German IPv6 Council and one of the authors of this book.

Tim Berners Lee, (*1955), Professor at MIT and "father of the World Wide Web" (1990), he serves as director of the W3C (World Wide Web Consortium). Founded by him in 1994, the Consortium coordinates and directs the development of the WWW. He collaborated with Robert Caillieau to develop the first WWW server at the European nuclear research center CERN, thereby laying the foundation for the WWW. In 2004, he was knighted by Queen Elizabeth II as a "Knight Commander of the Order of the British Empire" (KBE) for his services to science. Tim Berners-Lee sees the future of the World Wide Web today in the Web of Data.

Mark McCahill, (*1956), American computer scientist who in 1991 at the University of Minnesota developed the Gopher system, a forerunner and early competitor of the World Wide Web. With the enormous success of the WWW, the originally widespread information service quickly lost its popularity in the late '90s.

Phil Karn, American electrical engineer and co-inventor of the Karn/Partridge algorithm, named after him, for the calculation of the network round-trip time for the retransmission of lost TCP segments. Karn contributed to the development of Internet architecture and the first implementation of the TCP protocol.

Tom Truscott, American computer scientist. As a student at Duke University, he developed the Usenet news communication service with Jim Ellis.

Jim Ellis, (1956–2001), American computer scientist who with Tom Truscott developed the Usenet new communication service as a student at Duke University.

Craig Partridge, American electrical engineer and co-inventor of the eponymous Karn/Partridge algorithm for calculating the round-trip time in retransmission of lost TCP segments. Partridge contributed to the development of the Internet with work on efficient email routing and the design of the first high-speed router.

Van Jacobson, (* 1950), American computer scientist who co-invented the eponymous Jacobson/Karels algorithm for the calculation of the network round-trip time for the retransmission of lost TCP segments. Jacobson was involved in the development of Internet architecture (TCP Header Compression, RFC 1144) and contributed, among other application: traceroute, pathchar and tcpdump. In 2001, he received the ACM SIGCOM Award for his work in developing the Internet.

Michael J. Karels, American microbiologist, co-inventor of the eponymous Jacobson/Karels algorithm for calculating network round-trip time for the retransmission of lost TCP segments.

Harald Sack, (*1965), Computer scientist and senior researcher at the Hasso Plattner Institute for Software Systems Engineering at the University of Potsdam, a founding member of the German IPv6 Council, co-founder of the video search engine Yovisto.com and one of the authors of this book. After working in the field of formal verification, his research today concentrates on multimedia retrieval, the Semantic Web, and semantic search technology.

Jarkko Oikarinen, (*1967), Finnish computer scientist who in 1988 as a student, developed a simple text-based, synchronous message exchange system for the computer mailbox he operated. This system went on to achieve huge popularity as **Internet Relay Chat (IRC)**.

Tatu Ylönen, (*1968), Finnish computer scientist who in 1995 developed the Secure Shell (SSH) protocol for secure data transmission and program execution on remote computers. He founded the company SSH Communication Services at the end of 1995 to promote and further develop SSH.

Marc Andreesen, (*1971), as a student and part-time employee of the National Center for Supercomputing Applications (NCSA) he co-authored with his colleague Eric Bina the first WWW browser with a graphical user surface called "Mosaic," (1992). After his graduation in 1994, Andreesen founded the company Netscape that became well-known due to its Web browser of the same name.

Bram Cohen, (*1975), American programmer, he developed the *BitTorrent* in 2001 for efficient file sharing in peer-to-peer networks. In contrast to other peer-to-peer networks, BitTorrent does not rely on a cross-border network but constructs a separate distribution network for every resource to be transferred.

Justin Frankel, (*1978), American software developer, who in 2001 created and published the first purely decentralized peer-to-peer file sharing network gnutella. Previously, he had developed the audio and media player *Winamp* for AOL, his employer at that time.

Shawn Fanning, (*1980), American entrepreneur who became famous through the creation and operation of his music exchange "Napster," a file sharing system, operated on the basis of the peer-to-peer file sharing principle. Between 1999 - 2001 Napster was the starting point of what became at the time the most popular Internet-based online community.

Abbreviations and Acronyms

3DES	Triple-DES
3G	3rd Generation
4G	4th Generation
A2DO	Advanced Audio Distribution Profile
AAL	ATM Adaption Layer
ACK	Acknowledgement
ACM	Association for Computing Machinery
ACL	Asynchronous Connectionless Link
ACOR	Admission Control enabled On demand Routing
ACR	Attenuation Crosstalk Ratio
ADM	Add Drop Multiplexer
ADSL	Asymmetric Digital Subscriber Line
ADU	Atomic Data Unit
AH	Authentication Header
AID	Association Identifier
AIFS	Arbitration Interframe Space
AJAX	Asynchronous JavaScript and XML
AM	Amplitude Modulation
AMA	Active Member Address
ANSI	American National Standards Institute
AODV	Ad Hoc On-Demand Distance Vector Routing
AP	Access Point
API	Application Programming Interface
APP	Atom Publishing Protocol
ARA	Ant-based Routing Algorithm
ARP	Address Resolution Protocol
ARPA	Advanced Research Project Agency
ARQ	Automatic Repeat Request
AS	Autonomes System
AS	Authentication Server

C. Meinel and H. Sack, *Internetworking*, X.media.publishing,
DOI: 10.1007/978-3-642-35392-5, © Springer-Verlag Berlin Heidelberg 2013

ASCII	American Standard Code for Information Interchange
ASK	Amplitude Shift Keying
ASM	Analog Modulation Spectrum
ASO	Address Supporting Organization
ATIM	Announcement Traffic Indicator Message
ATM	Asynchronous Transfer Mode
AuC	Authentification Center
AVRCP	Audio Video Remote Control Profile
AWDS	Ad hoc Wireless Distribution Service
B3G	Beyond 3G
BAN	Body Area Network
BASK	Binary Amplitude Shift Keying
BCD	Binary Coded Digits
BDSG	Bundesdatenschutzgesetz
BECN	Backward Explicit Congestion Notification
BGP	Border Gateway Protocol
BID	Burned-In Addresses
BIND	Berkeley Internet Name Domain
BIP	Basic Imaging Profile
B-ISDN	Broadband Integrated Service Digital Network
Bit	Binary Digit
bit	Basic Indissoluble Information Unit
BITS	Bump in the Stack
BITW	Bump in the Wire
BNC	Bayonet Neill Concelman
BNEP	Bluetooth Network Encapsulation Protocol
BPP	Basic Printing Profile
bps	Bits per Second
BPSK	Binary Phase Shift Keying
BS	Basis Station
BSC	Bit Synchronous Communication
BSC	Base Station Controller
BSR	Backup Source Routing
BSS	Basic Service Set
BSS	Base Station Subsystem
BSSID	Basic Service Set IDentifier
b/w	Black and White
BWA	Broadband WIreless Access
BT	Burst Tolerance
BTS	Base Transceiver Station
BUS	Broadcast and Unknown Server
BZT	Bundesamt für Zulassungen in der Telekommunikation
CA	Certification Authority
CAC	Connection Admission Control
CAP	Carrierless Amplitude Phase

CAPI	Common Application Interface
CBR	Constant Bit Rate
CC	Carbon Copy
CC	Creative Commons
CCIR	Comité Consultatif International des Radiocommunications
CCITT	Comité Consultatif International de Telegraphique et Telefonique
CCK	Complementary Code Keying
ccNSO	country-code Name Supporting Organization
ccTLD	country-code Top-Level Domain
CD	Compact Disc
CD-DA	Compact Disc Digital Audio
CDDI	Copper Distributed Data Interface
CDIR	Classless InterDomain Routing
CDMA	Code Division Multiple Access
CDP	Conditional DePhase Encoding
CD-ROM	Compact Disc Read Only Memory
CDTR	Cell Delay Variation Tolerance
CDV	Cell Delay Variation
CER	Cell Error Rate
CERN	Conseil Européen pour la Recherche Nucléaire
CERT	Computer Emergency Response Team
CF	Contention Free
CFP	Contention Free Period
CG	Cyclic Group
CGA	Cryptographically Generated Addresses
CGSR	Clusterhead Gateway Switch Routing Protocol
CH	Cyclic Header
CHAMP	CacHing And MultiPath routing
CHDLC	Cisco High-Level Data Link Protocol
CHAP	Cryptographic Handshake Authentication Protocol
CID	Channel IDentifier
CIP	Common ISDN Access Profile
CIR	Commited Information Rate
CLLN	Consolidated Link Layer Management
CLP	Cell Loss Priority
CLR	Cell Loss Ratio
CMR	Cell Miisinsertion Ratio
CN	Core Network
COFDM	Coded Orthogonal Frequency Division Multiplex
CP	Contention Period
CPU	Central Processing Unit
CR	Carriage Return
CRC	Cyclic Redundancy Check
CRMA	Cyclic Reservation Multiple Access
CRP	Cable Replacement Protocol

CS	Convergence Sublayer
CS	Carrier Sense
CSMA/CD	Carrier Sense Multiple Access Collision Detect
CSNet	Computer Science Network
CSS	Chirp Spread Spectrum
CTD	Cell Transfer Delay
CTP	Cordless Telephony Profile
CTS	Clear to Send
CW	Contention Window
DA	Destination Address
DAB	Digital Audio Broadcasting
DAC	Dual Attached Component
DAP	Directory Access Protocol
DARPA	Defense Advanced Research Projects Agency
DAS	Dual Attached Station
DASL	DAV Search and Locate
db	decibel
DBF	Distributed Bellman-Ford Routing Protocol
DBPSK	Differential Binary Phase Shift Keying
DCC	Digital Compac Cassette
DCCP	Datagram Congestion Control Protocol
DCE	Data Communication Equipment
DCF	Decentralized Coordination Function
DCT	Discrete Cosine Transform
DDCMP	Digital Data Communications Message Protocol
DDoS	Distributed Denial-of-Service
DECT	Digital Enhanced Cordless Telecommunications
DE-NIC	Deutscher Network Information Center
DES	Data Encryption Standard
DFN	Deutsches Forschungsnetzwerk
DFR	Direction Forward Routing
DFT	Discrete Fourier Transform
DHCP	Dynamic Host Configuration Protocol
DHT	Distributed Hash Table
DIFS	Distributed Coordination Function Interframe Space
DIGI e.V.	Deutsche Interessengemeinschaft Internet e.V.
DIN	Deutsche Industrie Norm
DLCI	Data Link Connection Identifier
DLL	Data Link Layer
DMT	Discrete Multitone
DNS	Domain Name Service
DNSO	Domain Name Support Organization
DoD	Department of Defense
DOI	Digital Object Identifier
DoS	Denial of Service

DPAM	Demand Priority Access Method
DPG	Dedicated Packet Group
dpi	dots per inch
DPSK	Differential Phase Shift Keying
DQDB	Distributed Queue Dual Bus
DQPSK	Differential Quadrature Phase Shift Keying
DPSK	Differential Phase Shift Keying
DRM	Digital Rights Management
DS	Distribution System
DSAP	Destination Service Access Point
DSDV	Destination Sequence Distance Vector Routing
DSL	Digital Subscriber Line
DSLAM	Digital Subscriber Line Access Multiplexer
DSMX	Digitaler Signalmultiplexer
DSR	Dynamic Source Routing
DS	Delegation Signer
DSS1	Digital Signalling System No. 1
DSSS	Direct Sequence Spread Spectrum
DTE	Data Terminal Equipment
DTIM	Delivery Traffic Indicator Message
DTLS	Datagram Transport Layer Security
DTR	Dedicated Token Ring
DTS	Decode Time Stamp
DTX	Discontinouse Transmission
DUA	Directory User Agent
DUAL	Diffusing Update Algorithm
DÜE	Datenübertragungseinrichtung
DUN	Dial-up Networking Profile
DVB	Digital Video Broadcasting
DVB-C	Digital Video Broadcast - Cable
DVD	Digital Versatile Disk
DVMRP	Distance Vector Multicast Routing Protocol
DXC	Digital Cross Connect
DYMO	DYnamic Manet On-demand Routing
EAP	Extensible Authentication Protocol
EAPOL	EAP Over LAN
EARN	European Academit Research Network
eBGP	exterior Border Gateway Routing Protocol
ECN	Explicit Congestion Notification
ECRIM	European Research Consortium for Informatics and Mathematics
ED	End Delimiter
EDCF	Enhanced Distribution Coordination Function
EDGE	Enhanced Data Rate for GPRS Execution
EEPROM	Electronic Erasable Programmable Read Only Memory
EFF	Electronic Frontier Foundation

EGP	Exterior Gateway Protocol
EHF	Extreme High Frequency
EIA	Electronic Industry Association
EIFS	Extended Interframe Space
EIGRP	Enhanced Interior Gateway Routing Protocol
EIR	Excess Information Rate
EIR	Equipment Identification Registry
EOB	End of Block
EOF	End of File
EOT	End of Text
ErWiN	Erweiterte Wissenschaftsnetz
ESD	End-of-Stream Delimiter
ESDP	Extended Service Discovery Profile
ESMTP	Extended Simple Mail Transfer Protocol
ESP	Encapsulating Secure Payload
ESS	Extended Service Set
ET	Exchange Termination
ETSI	European Telecommunication Standards Institute
EUI	Extended Unique Identifier
EUNet	European Unix Network
EUUG	European UNIX Users Group
FAST	Flexible Authentication via Secure Tunneling
FAXP	FAX Profile
FCC	Federal Communications Commission
FCS	Frame Check Sequence
FDC	Final Comitee Draft
FDD	Frequency Division Duplex
FDDI	Fiber Distributed Data Interface
FDMA	Frequency Division Multiple Access
FEC	Forward Error Check
FECN	Forward Explicit Congestion Notification
FFD	Full Function Devices
FHSS	Frequency Hopping Spread Spectrum
FIFO	First In First Out
FLV	Flash Video
FM	Frequency Modulation
fps	Frames per Second
FS	Frame Status
FSK	Frequency Shift Keying
FSN	Full Service Network
FTP	File Transfer Protocol
FTP	Foiled Twisted Pair
GAC	Governmental Advisory Committee
GAN	Global Area Network
GAP	Generic Access Profile

GAVDP	Generic AV Distribution Profile
GFC	General Flow Control
GFR	Guaranteed Frame Rate
GFSK	Gaussian Frequency Shift Keying
GFX	Graphical Framework Extension
GIF	Graphic Interchange Format
GMK	Group Master Key
GMSK	Gaussian Minimum Shift Keying
GNSO	Generic Name Support Organization
GNU	GNU is Not Unix
GNU-FDL	GNU Free Document License
GNU-GPL	GNU General Public License
GOEP	Generic Object Exchange Profile
GRP	General Registration Protocol
GPRS	General Paket Radio Service
GPS	Global Positioning System
GSM	Global System for Mobile Communication
GTC	Generic Token Card
GTK	Group Temporary Key
gTLD	generic Top-Level Domain
GTS	Guaranteed Time Slots
HC	Hybrid Coordinator
HCF	Hybrid Coordination Function
HCI	Host Controller Interface
HCRP	Hardcopy Cable Replacement Profile
HDB	High Density Bipolar
HDCL	High Level Data Link Protocol
HDP	Health Device Profile
HDSL	High Bit Rate DSL
HEC	Header Error Check
HF	High Frequency
HFP	Hands Free Profile
HID	Human Interface Device Profile
HiperLAN	High Performance Radio Local Area Network
HLR	Home Location Registry
HOB	Head of Bus
HPI	Hasso Plattner Institute
HRPLS	Hybrid Routing Protocol for Large Scale Mobile Ad Hoc Networks with Mobile Backbones
HSCSD	High Speed Circuit Switched Data
HSLS	Hazy Sighted Link State Routing Protocol
HSTR	High Speed Token Ring
HSP	Headset Profile
HSR	Hierarchical State Routing Protocol
HTTP	Hypertext Transfer Protocol

HTTPS	HTTP over Secure Socket Layer
HWMP	Hybrid Wireless Mesh Protocol
Hz	Hertz
IAB	Internet Architectural Board
IAC	Inquiry Access Code
IANA	Internet Assigned Number Authority
IAPP	Inter Access Point Protocol
IBSS	Independent Basic Service Set
iBGP	interior Border Gateway Routing Protocol
IC	Integrated Circuit
ICANN	Internet Corporation for Assigned Names and Numbers
ICCB	Internet Control and Configuration Board
ICCC	International Conference on Computers Communications
ICMP	Internet Control Message Protocol
ICR	Initial Cell Rate
ICT	Information and Communication Technology
ICV	Integrity Check Value
IEEE	Institute of Electrical and Electronics Engineers
IESG	Internet Engineering Steering Group
IETF	Internet Engineering Task Force
IFS	Interframe Space
IGMP	Internet Group Management Protocol
IGP	Interior Gateway Protocol
IGRP	Interior Gateway Routing Protocol
IHL	Internet Header Length
IKE	Internet Key Exchange
IMAP	Internet Message Access Protocol
IMP	Internet Message Processor
INTP	Intercom Profile
IP	Internet Protocol
IPnG	Internet Protocol – (The) Next Generation
IPsec	Internet Protocol Security
IP v6	Internet Protocol version 6
IPX	Internetworking Packet Exchange
IR	Infrared
IrDA	Infrared Data Association
IRSG	Internet Research Steering Group
IRTF	Internet Research Task Force
ISDN	Integrated Service Digital Network
IS-IS	Intermediate System to Intermediate System
ISM	Industriam, Scientific, and Medical
ISO	International Standards Organisation
ISOC	Internet Society
ISP	Internet Service Provider
ISUP	ISDN User Part

ITP	Industrial Twisted Pair
ITU	International Telecommunications Union
IV	Initialization Vector
IV-DENIC	Interessenverbund DENIC
JDC	Japanese Digital Cellular
JEITA	Japan Electronic and Information Technology Industries Association
KDC	Key Distribution Center
KEA	Key Exchange Algorithm
kHz	KiloHertz
KINK	Kerberized Internet Negotiation of Keys
L2CAP	Logical Link Control and Adaption Protocol
LAI	Location Area Identification
LAN	Local Area Network
LANE	LAN Emulation
LAP	Lower Address Part
LAPB	Link Access Procedure Balanced
LAPD	Link Access Procedure D-Channel
LAPF	Link Access Protocol for Frame Mode Bearer Service
Laser	Light Amplification by Stimulated Emission of Radiation
LC	Late Counter
LCA	Linked Cluster Architecture
LCF	Low Cost Fiber
LCP	Link Control Protocol
LDAP	Lightweight Directory Access Protocol
LDIF	Lightweight Directory Interface Format
LEAP	Lightweight EAP
LEC	LAN Emulation Client
LED	Light Emitting Diode
LECS	LAN Emulation Configuration Service
LEO	Low Earth Orbiting
LES	LAN Emulation Service
LF	Line Feed
LF	Low Frequency
LIPS	Lightweight Internet Person Schema
LLC	Logical Link Control
LLDP	Link Layer Discovery Protocol
LLTD	Link Layer Topology Discovery
LLU	Link Local Use
LMP	Link Management Protocol
LSA	Linke State Announcement/Advertisement
LSB	Least Significant Bit
LSP	Linke State Packet
LT	Line Termination
LTE	Long Term Evolution
LUT	Lookup Table

LZW	Lev Zipf Welch
MAC	Medium Access Control
MACA	Multiple Access Collision Avoidance
MAC PS	MAC Privacy Sublayer
MAC CPS	MAC Common Part Sublayer
MAN	Metropolitan Area Network
MAODDP	Mobile Ad hoc On-Demand Data Delivery Protocol
MAODV	Multirate Ad hoc On-demand Distance Vector Routing Protocol
MAP	Manufacturing Automation Protocol
MAU	Medium Attachment Unit
MCM	Multicarrier Modulation
MCR	Minimal Cell Rate
MD5	Message Digest Algorithm 5
MEO	Medium Earth Orbit
MEPA	Minimum Exposed Path to the Attack
MExE	Mobile Station Execution Environment
MF	Medium Frequency
MHS	Message Handling System
MIB	Management Information Base
MIC	Message Integrity Code
MIDI	Musical Instrument Digital Interface
MIH	Media Independent Handover
MIME	Multimedia Internet Mail Extension Format
MMF	Multimode Fiber
MMRP	Mobile Mesh Routing Protocol
MNP	Microcom Networking Protocol
MOSS	MIME Object Security Services
MOSPF	Multicast Open Shortest Path First
MS	Mobile Station
MSAU	Multi Station Access Unit
MSB	Most Significant Bit
MSC	Mobile Switching Center
MSC	Mobile Service Switching Center
MSK	Minimum-Shift Keying
MSTP	Multiple Spanning Tree Protocol
MTA	Message Transfer Agent
MTP	Message Transfer Part
MTS	Message Transfer System
MTU	Maximum Transmission Unit
MUX	Multiplexer
NAPT	Network Address Port Translation
NAPT-PT	Network Address Port Translation – Protocol Translation
NAT	Network Address Translation
NAT-PT	Network Address Translation – Protocol Translation
NAV	Network Allocation Vector

NCP	Network Control Program
NDB	Non Directional Beacon
NDP	Neighbor Discovery Protocol
NE	Node Element
NetBIOS	Network Basic Input/Output System
NEXT	Near End Crosstalk
NFC	Near Field Communication
NFS	Network File System
NIC	Network Interface Card
NIC	Network Interface Controller
NID	Namespace Identifier
NIS	Network Information Service
NIT	Network Information Table
NLPID	Network Layer Protocol ID
NMR	Noise-to-Mask Ratio
NMS	Network Management System
NNI	Network Network Interface
NRZI	Non Return to Zero Invert
NSF	National Science Foundation
NSS	Namespace Specific String
NSS	Network and Switching Subsystem
NT	Network Termination
NVT	Network Virtual Terminal
OBEX	Object Exchange
OC	Optical Carrier
ODBC	Open DataBase Connectivity
OFDM	Orthogonal Frequency Division Multiplex
OLSR	Optimized Link State Routing Protocol
OMC	Operation and Maintenance Center
ONU	Optical Network Unit
OOK	On-Off Keying
OORP	Order One Routing Protocol
OPP	Object Push Profile
OSI	Open Systems Interconnect
OSPF	Open Shortest Path First
OSS	Operation Subsystem
OUI	Organisationally Unique Identifier
PA	Preamble
PAM	Puls Amplitude Modulation
PAN	Personal Area Network
PAP	Password Authentication Protocol
PAP	Port Aggregation Protocol
PAR	Positive Acknowledgement with Retransmission
PARC	Palo Alto Research Center
PAT	Programm Association Table

PBAP	Phonebook Access Profile
PBCC	Packet Binary Convolutional Coding
PBKDFv2	Password Based Key Derivation Function version 2
PBNAC	Port Based Network Access Control
PC	Point Coordinator
PCF	Point Coordination Function
PCM	Pulse Code Modulation
PCR	Peak Cell Rate
PD	Powered Device
PDA	Personal Data Assistant
PDH	Plesiochronous Digital Hierarchy
PDP	Policy Decision Point
PDU	Protocol Data Unit
PEAP	Protected Extensible Authentication Protocol
PEP	Protocol Extension Protocol
PEP	Policy Enforcement Point
PERL	Practical Extraction and Report Language
PES	Packetized Elementary Stream
PFM	Pulse Frequency Modulation
PFR	Portable Font Resource
PHY	Physical Layer
PIFS	Point Coordination Function Interframe Space
PIM	Protocol Independent Multicast
PIM	Personal Information Management
PIN	Personal Identity Number
PING	Packet Internet Gopher
PKI	Public Key Infrastruktur
PLIP	Parallel Line Interface Protocol
PLP	Packet Layer Protocol
PMA	Parked Member Address
PMD	Physical Medium Dependent
PMK	Pairwise Master Key
PMT	Program Map Table
PMTU	Path Maximum Transfer Unit
PoE	Power over Ethernet
POTS	Plain Old Telephone Service
PPM	Puls Position Modulation
PPP	Point-to-Point Protocol
PRK	Phase Reversal Keying
PRNG	Pseudo Random Number Generator
PS	Power Save
PSE	Power Sourcing Equipment
PSK	Phase Shift Keying
PSK	Pre-Shared Key
PTI	Payload Type Identifier

PTL	Pairwise Temporary Key
PTS	Presentation Time Stamp
PUK	PIN Unblocking Key
PVC	Permanent Virtual Circuit
PWM	Pulse Width Modulation
RJ	Registered Jack
QAM	Quadrature Amplitude Modulation
QoS	Quality of Service
QPSK	Quaternary Phase Shift Keying
RAI	Remote Alarm Indication
RAM	Random Access Memory
RAN	Radio Access Network
RARP	Reverse Address Resolution Protocol
RC	Rivest Cipher (Ron's Code)
RDF	Resource Description Framework
RDS	Resolver Discovery Service
RDT	Real Data Transport
RED	Random Early Detection
RFC	Reverse Path Forwarding
RFCOMM	Radio Frequency Communications Protocol
RFD	Reduced Function Devices
RI	Ring In
RIAA	Recording Industry Association of America
RIP	Routing Information Protocol
RLE	Run Length Encoding
RLP	Radio Link Protocol
RNS	Robust Network Security
RO	Ring Out
ROM	Read Only Memory
RSA	Rivest, Shamir, Adleman encryption
RSH	Remote Shell
RSRP	Robust Secure Routing Protocol
RSTP	Rapid Spaning Tree Protocol
RSVP	Resource Reservation Protocol
RTMP	Real Time Messaging Protocol
RTS	Request to Send
RTSP	Real Time Streaming Protocol
RVSA	Remote Variant Selection Algorithm
RZ	Return-to-Zero
SA	Security Association
SA	Source Address
SAC	Single Attached Component
SACK	Selective Acknowledgement
SAD	Security Association Database
SAP	Session Announcement Protocol

SAP	Subnetwork Access Protocol
SAP	Service Access Point
SAP	SIM Access Profile
SAR	Segmentation and Reassembly
SAS	Single Attached Station
SBR	Spectral Band Replication
SCA	Software Communication Architecture
SCCP	Signal Connection Control Part
SCO	Synchronous Connection-Oriented Link
SCR	Sustained Cell Rate
SD	Starting Delimiter
SDAP	Service Discovery Application Profile
SDH	Synchronous Digital Hierarchy
SDP	Service Discovery Protocol
SDSL	Symmetric DSL
SECBR	Severly-Errored Cell Block Ratio
SEND	SEcure Neighbor Discovery
SFD	Start-of-Frame Delimiter
SFTP	Secure File Transfer Protocol
S/FTP	Screened Foiled Twisted Pair
SHF	Super High Frequency
SIFS	Short Interframe Space
SIM	Subscriber Identity Module
SIP	Session Initiation Protocol
SLA	Site Level Aggregation
SLIP	Serial Line Interface Protocol
SLPP	Simple Loop Prevention Protocol
SLRC	Station Long Retry Count
SLU	Site Local Use
SMB	Server Message Block
SMDS	Switched Multi-Megabit Metropolitan Data Service
SMF	Singlemode Fiber
SMI	Structure of Management Information
SMP	Standby Monitor Presence
SMR	Signal-to-Mask Ratio
SMR	Symbolic Music Representation
SMS	Short Message Service
SMTP	Simple Mail Transfer Protocol
SNA	System Network Architecture
SNAP	SubNet Attachment Point
SNDP	Secure Neighbor Discovery Protocol
SNMP	Simple Network Management Protocol
SNR	Signal-to-Noise Ratio
SOHO	Small Office Home Office
SONET	Synchronous Optical Network

SPAM	Spiced Ham
SPD	Secure Payload Database
S/PDIF	Sony/Philips Digital Interconnect Format
SPI	Security Parameter Index
SPI	Server Programming Interface
SPP	Serial Port Profile
SRC	Short Retry Count
SS	Subscriber Station
SS7	Signalling System No. 7
SSA	Signal Stability Adaptive Routing
SSCS	Service Specifiv Convergence Sublayer
SSD	Start-of-Stream Delimiter
SSH	Secure Shell
SSID	Service Set Identifier
SSL	Secure Socket Layer
SSP	Supplementary Special-purpose Plane
SSR	Scalable Source Routing
SSRC	Station Short Retry Count
ST	Internet Stream Protocol (Stream Transport)
STGO	Start Tag Open
sTLD	sponsored Top-Level Domains
STM	Synchronous Transfer Mode
STP	Shielded Twisted Pair
STP	Spanning Tree Protocol
STS	Station to Station Protocol
STS	Synchronous Transport Signal
SVC	Switched Virtual Circuit
SW	Short Wave
SWS	Silly WIndow Syndrome
SYNC	Synchronisation Profile
TA	Terminal Adapter
TAG	Technical Advisory Group
TAGC	Tag Close Delimiter
TBRPF	Topology Dissemination based on Reverse-Path Forwarding Routing Protocol
TBTT	Target Beacon Transmission Time
TCB	Transmission Control Block
TCM	Trellis Code Modulation
TCP	Transmission Control Protocol
TCS BIN	Telephone Control Protocol Specification – Binary
TDAC	Time Domain Aliasing Cancellation
TDD	Time Division Duplex
TDDSG	Teledienstedatenschutzgesetz
TDM	Time Division Multiplexing
TDMA	Time Division Multiple Access
TE	Terminal Equipment

TEI	Transport Error Indicator
THSS	Time Hopping Spread Spectrum
THT	Token Hold Timer
TIA	Telecommunication Industry Association
TID	Transaction Identifier
TK	Temporary Key
TKG	Telekommunikationsgesetz
TKIP	Temporal Key Integrity Protocol
TLD	Top-Level Domain
TLS	Transport Layer Security
TM	Terminal Multiplexer
TMSI	Temporary Mobile Subscriber IDentity
TORA	Temporally Ordered Routing Algorithm
TOS	Type of Service
TROLI	Token Ring Optimized Link Interface
TRT	Token Rotation Timer
TRT	Transport Relay Translation
TSC	Transport Scrambling Control
TSIG	Transaction Signature
TTG	Transmit/Receive Transition Gap
TTL	Time To Live
TTLS	Tunnelled Transport Layer Security
TTP	Timed Token Protocol
TTRT	Target Token Rotation Timer
TVC	Transient Virtual Channel
TX	Transmit
TXOP	Transmission Opportunity
UBE	Unsolicited Bulk Email
UCE	Unsolicited Commercial Email
UDP	User Datagram Protocol
UE	User Equipment
UHF	Ultra High Frequency
UMTS	Universal Mobile Telecommunication System
UNI	User Network Interface
USB	Universal Serial Bus
USIM	UMTS Subscriber Identity Module
uTLD	unsponsored Top-Level Domain
UTP	Unshielded Twisted Pair
UTRA	Universal Terrestrial Access
UTRAN	Universal Terrestrial Access Network
UV	Ultraviolet
UWB	Ultra Wide Band
VBLAN	Virtually Bridged Local Area Network
VBR	Variable Bit Rate
VC	Virtual Container

VC	Virtual Channel
VC	Virtual Circuit
VCC	Virtual Channel Connection
VCI	Virtual Channel Identifier
VDSL	Very High Bit Rate DSL
VLC	Visual Light Communication
VLF	Very Low Frequency
VLR	Visitor Location Registry
VoIP	Voice over IP
VP	Virtual Path
VPC	Virtual Path Connection
VPI	Virtual Path Identifier
VPN	Virtual Private Network
WAN	Wide Area Network
WAR	Wireless Access Revolution
WAR	Witness Aided Routing
W3C	World Wide Web Consortium
W-CDMA	Wideband Code Division Multiple Access
WD	Working Draft
WDS	Wireless Distribution System
WDMA	Wavelength Division Multiple Access
WEP	Wired Equivalence Protocol
WHAN	Wireless Home Area Network
WiMAX	Worldwide Interoperability for Microwave Access
WiN	Wissenschaftsnetzwerk
WLAN	Wireless LAN
WLL	Wireless Local Loop
WPA	Wi-Fi Protected Access
WPAN	Wireless Personal Area Network
WRAN	Wireless Regional Area Network
WRED	Weighted Random Early Detection
WRP	Wireless Routing Protocol
WUSB	Wireless Universal Serial Bus
WWAN	Wireless Wide Area Network
WWW	World Wide Web
WYSIWYG	What You See Is What You Get
XID	Transaction Identifier
XT	Crosstalk
ZED	ZigBee End Device
ZR	ZigBee Router
ZRP	Zone Routing Protocol

Bibliography

1. Abbate, J.: Inventing the Internet. MIT Press, Cambridge, MA, USA (2000)
2. Abramson, N.: THE ALOHA SYSTEM: another alternative for computer communications. In: AFIPS '70 (Fall): Proceedings of the November 17-19, 1970, fall joint computer conference, pp. 281–285. ACM, New York, NY, USA (1970)
3. Abramson, N.: Packet switching with satellites. In: AFIPS '73: Proceedings of the June 4-8, 1973, national computer conference and exposition, pp. 695–702. ACM, New York, NY, USA (1973)
4. Abramson, N.: Development of the alohanet. Information Theory, IEEE Transactions on 31(2), 119–123 (1985)
5. äger, R.J.: (Breitbandkommunikation: ATM, DQDB, Frame Relay). Addison-Wesley, Bonn (1996)
6. Aitchison, R.: ProDNS and Bind. Apress, Berkeley (CA) (2005)
7. Albrightson, B., Garcia-Luna-Aceves, J., Boyle, J.: EIGRP - a fast routing protocol based on distance vectors. In: Proc. Networld/Interop 94, pp. 192–210. Las Vegas, Nevada (1994)
8. Arends, R., Austein, R., Larson, M., Massey, D., Rose, S.: Resource Records for the DNS Security Extensions. RFC 4034 (Proposed Standard) (2005). URL http://www.ietf.org/rfc/rfc4034.txt. Updated by RFCs 4470, 6014
9. Arkko, J., Kempf, J., Zill, B., Nikander, P.: SEcure Neighbor Discovery (SEND). RFC 3971 (Proposed Standard) (2005). URL http://www.ietf.org/rfc/rfc3971.txt
10. Ash, G.R.: Dynamic Routing in Telecommunications Networks. McGraw-Hill Professional, New York, USA (1997)
11. Badach, A., Hoffmann, E.: Technik der IP-Netze - TCP/IP incl. IPv6 - Funktionsweise, Protokolle und Dienste, 2. erw. Aufl. Carl Hanser Verlag, München (2007)
12. Badach, A., Hoffmann, E., Knauer, O.: High-speed internetworking : Grundlagen, Kommunikationsstandards, Technologien der Shared und Switched LANs, 2. aufl. edn. Addison-Wesley-Longman, Bonn; Reading, Mass. [u.a.] (1997)
13. Ballardie, T., Francis, P., Crowcroft, J.: Core Based Trees (CBT). In: SIGCOMM, pp. 85–95 (1993)
14. Baran, P.: On distributed communication networks. IEEE Transactions on Communication Systems 12 (1964)
15. Baran, P.: Reliable digital communication systems using unreliable network repeater nodes, report p-1995. Tech. rep., Rand Corporation (1965)
16. Barontib, P., Pillaia, P., Chooka, V.W.C., Chessab, S., Gottab, A., Hu, Y.F.: Wireless sensor networks: A survey on the state of the art and the 802.15.4 and ZigBee standards. Computer Communications 30(7), 1655–1695 (2007)
17. Barrett, D.J., Silverman, R.E.: SSH, The Secure Shell: The Definitive Guide. O'Reilly & Associates, Inc., Sebastopol, CA, USA (2001)
18. Beck, K.: Computervermitelte Kommunikation im Internet. Oldenbourg Verlag, München (2006)
19. Bergmann, F., Gerhardt, H.J.: Taschenbuch der Telekommunikation, 2. aufl. edn. Fachbuchverlag Leipzig, Carl Hanser Verlag (2003)
20. Bertsekas, D., Gallagher, R.: Data Networks, 2nd edn. Prentice Hall, Englewood Cliffs, NJ, USA (1991)
21. Betanov, C.: Introduction to X.400 / Cemil Betanov. Artech House, Boston : (1993)
22. Bird, D.: Token ring network design. Addison-Wesley Longman Publishing Co., Inc., Boston, MA, USA (1994)
23. Black, U.: OSI: a model for computer communications standards. Prentice-Hall, Inc., Upper Saddle River, NJ, USA (1991)
24. Black, U.: Emerging communications technologies. Prentice-Hall, Inc., Upper Saddle River, NJ, USA (1994)
25. Black, U.: Emerging communications technologies (2nd ed.). Prentice-Hall, Inc., Upper Saddle River, NJ, USA (1997)

C. Meinel and H. Sack, *Internetworking*, X.media.publishing,
DOI: 10.1007/978-3-642-35392-5, © Springer-Verlag Berlin Heidelberg 2013

26. Black, U.D.: Network Management Standards: SNMP, CMIP, TMN, MIBs and Object Libraries, 2nd edn. McGraw-Hill, Inc., New York, NY, USA (1994)

27. Black, U.D.: Frame Relay Networks: Specifications and Implementations. McGraw-Hill, Inc., New York, NY, USA (1998)

28. Bless, R., Mink, S., Blaß, E.O., Conrad, M., Hof, H.J., Kutzner, K., Schöller, M.: Sichere Netzwerkkommunikation. ISBN 3-540-21845-9. Springer Verlag, Berlin, Heidelberg (2005)

29. Bless, R., Mink, S., Blaß, E.O., Conrad, M., Hof, H.J., Kutzner, K., Schöller, M.: Sichere Netzwerkkommunikation: Grundlagen, Protokolle und Architekturen. Springer, Berlin (2005)

30. Blunk, L., Vollbrecht, J.: PPP Extensible Authentication Protocol (EAP). RFC 2284 (Proposed Standard) (1998). URL http://www.ietf.org/rfc/rfc2284.txt. Obsoleted by RFC 3748, updated by RFC 2484

31. Bocker, P.: ISDN - Digitale Netze für Sprach-, Text-, Daten-, Vidoe- und Multimediakommunikation, 4. aufl. edn. Springer Verlag, Berlin, Heidelberg (2001)

32. Borisov, N., Goldberg, I., Wagner, D.: Intercepting mobile communications: the insecurity of 802.11. In: Proc. of the 7th Annual International Conference on Mobile Computing And Networking (MOBICOM 2001), pp. 180–189 (2001)

33. Borisov, N., Goldberg, I., Wagner, D.: Intercepting mobile communications: the insecurity of 802.11. In: MobiCom '01: Proceedings of the 7th annual international conference on Mobile computing and networking, pp. 180–189. ACM, New York, NY, USA (2001)

34. Borman, D., Deering, S., Hinden, R.: IPv6 Jumbograms. RFC 2675 (Proposed Standard) (1999). URL http://www.ietf.org/rfc/rfc2675.txt

35. Braden, R.: Extending TCP for Transactions – Concepts. RFC 1379 (Informational) (1992). URL http://www.ietf.org/rfc/rfc1379.txt. Updated by RFC 1644

36. Braden, R.: T/TCP – TCP Extensions for Transactions Functional Specification. RFC 1644 (Experimental) (1994). URL http://www.ietf.org/rfc/rfc1644.txt

37. Bradner, S.: The internet standards process – revision 3 (1996)

38. Bradner, S., Mankin, A.: IP: Next Generation (IPng) White Paper Solicitation. RFC 1550 (Informational) (1993). URL http://www.ietf.org/rfc/rfc1550.txt

39. Bradner, S.O., Mankin, A.: The recommendation for the ip next generation protocol. Internet RFC 1752 (1995)

40. Bradner, S.O., Mankin, A. (eds.): IPng: Internet protocol next generation. Addison Wesley Longman Publishing Co., Inc., Redwood City, CA, USA (1996)

41. Brockhaus (ed.): Der Brockhaus in einem Band. 10., vollständig überarbeitete und aktualisierte Auflage. Bibliographisches Institut F. A. Brockhaus AG, Mannheim (2005)

42. Buchanan, W.J.: The handbook of data communications and networks. Kluwer Academic Publishers, Norwell, MA, USA (2005)

43. Caicedo, C.E., Joshi, J.B., Tuladhar, S.R.: Ipv6 security challenges. Computer 42(2), 36–42 (2009). DOI http://doi.ieeecomputersociety.org/10.1109/MC.2009.54

44. Casey, L., Shelness, N.: A domain structure for distributed computer systems. SIGOPS Oper. Syst. Rev. 11(5), 101–108 (1977). DOI http://doi.acm.org/10.1145/1067625.806552. URL http://portal.acm.org/citation.cfm?id=806552

45. Cerf, V.G., Kahn, R.E.: A protocol for packet network intercommunication. IEEE Transactions on Computing COM-22, 637–648 (1974)

46. Chapman, D.B., Zwicky, E.D.: Building Internet firewalls. O'Reilly & Associates Inc., Sebastopol, CA., USA (1995)

47. Cheng, F., Meinel, C.: Research on the lock-keeper technology: Architectures, advancements and applications,. International Journal of Computer and Information Science (IJCIS) 5(3), 236–245 (2004)

48. Chu, Y.H., Rao, S.G., Zhang, H.: A case for end system multicast. In: SIGMETRICS, pp. 1–12 (2000)

49. Clark, D.: IP datagram reassembly algorithms. RFC 815 (1982). URL http://www.ietf.org/rfc/rfc815.txt

50. Clark, D.: The design philosophy of the DARPA internet protocols. In: SIGCOMM '88: Symposium proceedings on Communications architectures and protocols, pp. 106–114. ACM, New York, NY, USA (1988)
51. Comer, D.E.: Computernetzwerke und Internets. Prentice Hall, München (1998)
52. Comer, D.E.: Internetworking with TCP/IP: Principles, protocols, and architectures, 4th edn. Prentice Hall, Upper Saddle River NJ, USA (2000)
53. Conta, A., Deering, S.: Internet Control Message Protocol (ICMPv6) for the Internet Protocol Version 6 (IPv6) Specification. RFC 2463 (Draft Standard) (1998). URL http://www.ietf.org/rfc/rfc2463.txt. Obsoleted by RFC 4443
54. Conta, A., Deering, S., Gupta, M.: Internet Control Message Protocol (ICMPv6) for the Internet Protocol Version 6 (IPv6) Specification. RFC 4443 (Draft Standard) (2006). URL http://www.ietf.org/rfc/rfc4443.txt. Updated by RFC 4884
55. Cormen, T.H., Leiserson, C.E., Rivest, R.L., Stein, C.: Introduction to Algorithms. The MIT Press, Cambridge, MA, USA (2001)
56. Costa, L.H.M.K., Fdida, S., Duarte, O.C.M.B.: Hop by hop multicast routing protocol. In: SIGCOMM, pp. 249–259 (2001)
57. Costales, B., Allman, E.: Sendmail, 2nd ed. edn. O'Reilly & Associates, Inc., Sebastopol, CA, USA (1997)
58. Cox, C.: An Introduction to LTE: LTE, LTE-Advanced, SAE and 4G Mobile. John Wiley & Sons, Ltd (2012)
59. Crispin, M.: INTERNET MESSAGE ACCESS PROTOCOL - VERSION 4rev1. RFC 3501 (Proposed Standard) (2003). URL http://www.ietf.org/rfc/rfc3501.txt. Updated by RFCs 4466, 4469, 4551, 5032, 5182, 5738
60. Crow, B., Widjaja, I., Kim, J., Sakai, P.: IEEE 802.11 Wireless Local Area Networks. IEEE Communications Magazine pp. 116–126 (1997)
61. Crowcroft, J., Handley, M., Wakeman, I.: Internetworking Multimedia. Morgan Kaufman Publishers, San Francisco, CA, USA (1999)
62. D'addario, K.P., Walther, J.B.: The impacts of emoticons on message interpretation in computer-mediated communication. Social Science Computer Review **19**, 324–347 (2001)
63. Daigle, J.N.: Queueing Theory for Computer Communications. Addison-Wesley Longman Publishing Co., Inc., Boston, MA, USA (1991)
64. Davies, D.W., Barber, D.L.A.: Communication networks for computers. John Wiley, London, New York (1973)
65. Day, J.: The (un)revised osi reference model. SIGCOMM Comput. Commun. Rev. **25**(5), 39–55 (1995). DOI http://doi.acm.org/10.1145/216701.216704
66. Day, J.D., Zimmermann, H.: The OSI reference model. Proceedings of the IEEE **71**(12), 1334–1340 (1983)
67. Deering, S., Hinden, R.: Internet Protocol, Version 6 (IPv6) Specification. RFC 2460 (Draft Standard) (1998). URL http://www.ietf.org/rfc/rfc2460.txt. Updated by RFCs 5095, 5722, 5871
68. Deering, S.E., Cheriton, D.R.: Multicast routing in datagram internetworks and extended lans. ACM Trans. Comput. Syst. **8**(2), 85–110 (1990)
69. Deering, S.E., Estrin, D., Farinacci, D., Jacobson, V., Liu, C.G., Wei, L.: The PIM architecture for wide-area multicast routing. IEEE/ACM Trans. Netw. **4**(2), 153–162 (1996)
70. Dierks, T., Rescorla, E.: The Transport Layer Security (TLS) Protocol Version 1.1. RFC 4346 (Proposed Standard) (2006). URL http://www.ietf.org/rfc/rfc4346.txt. Obsoleted by RFC 5246, updated by RFCs 4366, 4680, 4681, 5746
71. Diffie, W., Oorschot, P.C., Wiener, M.J.: Authentication and authenticated key exchanges. Designs, Codes and Cryptography **2**, 107–125 (1992)
72. Dijkstra, E.W.: A note on two problems in connexion with graphs. Numerische Mathematik **1**, 269–271 (1959)
73. DIN 44302: Datenübertragung, Datenübermittlung: Begriffe. Deutsches Institut für Normierung DIN, Berlin / Köln (1979)

74. DIN 66020: Funktionelle Anforderungen an die Schnittstellen zwischen Datenendeinrichtung und Datenübertragungseinrichtungen - Teil 1: Allgemeine Anwendung. Deutsches Institut für Normierung DIN, Berlin / Köln (1999)

75. DIN 66021-9: Schnittstelle zwischen DEE und DÜE für Synchrone Übertragung bei 48000 bit/s auf Primärgruppenverbindungen. Deutsches Institut für Normierung DIN, Berlin / Köln (1983)

76. Dixon, R.C.: Spread Spectrum Systems: With Commercial Applications. John Wiley & Sons, Inc., New York, NY, USA (1994)

77. Droms, R.: Dynamic Host Configuration Protocol. RFC 2131 (Draft Standard) (1997). URL http://www.ietf.org/rfc/rfc2131.txt. Updated by RFCs 3396, 4361, 5494

78. Dube, R., Rais, C.D., Wang, K.Y., Tripathi, S.K.: Signal stability based adaptive routing (ssa) for ad-hoc mobile networks. Tech. rep., College Park, MD, USA (1996)

79. (Ed.), E.S.W.: Addressing the world: national identity and Internet country code domains. Rowman & Littlefield Publishers, Inc., Lanham (Maryland) (2003)

80. (Ed.), I.S.: Handbook on Wireless Networks and Mobile Computing. John Wiley & Sons, Inc. (2002)

81. Estrin, D., Handley, M., Helmy, A., Huang, P., Thaler, D.: A Dynamic Bootstrap Mechanism for Rendezvous-based Multicast Routing. In: INFOCOM, pp. 1090–1098 (1999)

82. Etemad, K.: Overview of mobile WiMAX technology and evolution. Communications Magazine, IEEE **46**(10), 31–40 (2008)

83. Farber, D.J., Larson, K.C.: The system architecture of the distributed computer system - The communication system. pp. pp. 21–27. Brooklyn Polytechnic Press, Brooklyn, NY, USA (1972)

84. Farmer, W.D., Newhall, E.E.: An experimental distributed switching system to handle bursty computer traffic. In: Proceedings of the first ACM symposium on Problems in the optimization of data communications systems, pp. 1–3. ACM, New York, NY, USA (1969). DOI http://doi.acm.org/10.1145/800165.805236

85. Feinler, E., Harrenstien, K., Su, Z., White, V.: DoD Internet host table specification. RFC 810 (1982). URL http://www.ietf.org/rfc/rfc810.txt. Obsoleted by RFC 952

86. Feit, S.: Local Area High Speed Networks. Macmillan Technical Publishing, Indianapolis, IN, USA (2000)

87. Fluhrer, S.R., Mantin, I., Shamir, A.: Weaknesses in the key scheduling algorithm of RC4. In: SAC '01: Revised Papers from the 8th Annual International Workshop on Selected Areas in Cryptography, pp. 1–24. Springer-Verlag, London, UK (2001)

88. Freyer, U.: Nachrichten-Übertragungstechnik, 4. Aufl. Carl Hanser Verlag, München (2000)

89. Fuglèwicz, M.: Das Internet Lesebuch - Hintergründe, Trends, Perspektiven. Buchkultur Verlagsgesellschaft m.b.H., Wien (1996)

90. Garbin, D.A., O'Connor, R.J., Pecar, J.A.: Telecommunications FactBook, 2nd edn. McGraw-Hill Professional,, Boston, MA, USA (2000)

91. Garcia-Luna-Aceves, J.J., Murthy, S.: A path-finding algorithm for loop-free routing. IEEE/ACM Trans. Netw. **5**(1), 148–160 (1997)

92. Garcia-Lunes-Aceves, J.J.: Loop-free routing using diffusing computations. IEEE/ACM Trans. Netw. **1**(1), 130–141 (1993)

93. Gast, M.: 802.11 Wireless networks: : The Definitive Guide - Creating and Administering Wireless Networks. O'Reilly & Associates, Inc., Sebastopol CA, USA (2002)

94. Gessler, R., Krause, T.: Wireless-Netzwerke für den Nahbereich - Eingebettete Funksysteme: Vergleich von standardisierten und proprietären Verfahren. Vieweg und Teubner, GWV Fachverlage GmbH, Wiesbaden (2009)

95. Göhring, H.G., Knaufels, F.J.: Token Ring: Grundlagen, Strategien, Perspektiven. Datacom Verlag, Bergheim (1993)

96. Green, M.: Napster opens pandora's box: Examining how File-Sharing services threaten the enforcement of copyright on the internet. Ohio State Law Journal **63**(799) (2002)

97. Grimm, R.: Digitale Kommunikation. Oldenbourg Verlag, München (2005)

98. Gudmundsson, O.: Delegation Signer (DS) Resource Record (RR). RFC 3658 (Proposed Standard) (2003). URL http://www.ietf.org/rfc/rfc3658.txt. Obsoleted by RFCs 4033, 4034, 4035, updated by RFC 3755

99. Haartsen, J.C.: The Bluetooth radio system. IEEE Personal Communications Magazine 7, 28–36 (2000)

100. Haaš, W.D.: Handbuch der Kommunikationsnetze: Einführung in die Grundlagen und Methoden der Kommunikationsnetze. Springer Verlag, Berlin, Heidelberg (1999)

101. Haas, Z.J., Pearlman, M.R.: ZRP: a hybrid framework for routing in ad hoc networks pp. 221–253 (2001)

102. Haaß, W.D.: Handbuch der Kommunikationsnetze: Einführung in die Grundlagen und Methoden der Kommunikationsnetze. Springer Verlag, Berlin, Heidelberg, New York (1997)

103. Häckelmann, H., Petzold, H.J., Strahringer, S.: Kommunikationssysteme. Springer, Berlin [u.a.] (2000)

104. Hafner, K., Lyon, M.: Arpa Kadabra: Die Geschichte des Internet. dPunkt Verlag, Heidelberg (1997)

105. Hahn, H., Stout, R.: The Internet Complete Reference. Osborne McGraw-Hill, Berkeley CA, USA (1994)

106. Halabi, B.: Internet Routing Architectures. Cisco Press, Indianapolis, IN (1997)

107. Hammond, N.G.L., (Hrsg.), H.H.S.: The Oxford Classical Dictionary, 2nd Ed. Oxford University Press, Oxford, UK (1992)

108. Händel, R., Huber, M.N.: Integrated Broadband Networks; An Introduction to ATM-Based Networks. Addison-Wesley Longman Publishing Co., Inc., Boston, MA, USA (1991)

109. Handley, M., Jacobson, V., Perkins, C.: SDP: Session Description Protocol. RFC 4566 (Proposed Standard) (2006). URL http://www.ietf.org/rfc/rfc4566.txt

110. Harkins, D., Carrel, D.: The Internet Key Exchange (IKE). RFC 2409 (Proposed Standard) (1998). Obsoleted by RFC 4306, updated by RFC 4109

111. Harrison, M.: The USENET handbook: a user's guide to Netnews. O'Reilly & Associates, Inc., Sebastopol, CA, USA (1995)

112. Hartley, R.V.L.: Transmission of information. Bell Syst. Tech. Journal 7, 535–563 (1928)

113. Hauben, M., Hauben, R.: Netizens: On the History and Impact of UseNet and the Internet. Wiley-IEEE Computer Society Press, Los Alamitos, CA, USA (1997)

114. Hecht, J.: Understanding fiber optics (3rd ed.). Prentice-Hall, Inc., Upper Saddle River, NJ, USA (1999)

115. Hedrick, C.L.: Routing information protocol (1988)

116. Hegering, H.G.: Open Systems Interconnection - eine kritische ürdigung. In: R. Valk (ed.) GI Jahrestagung, Informatik-Fachberichte, vol. 187, pp. 140–160. Springer Verlag, Berlin, Heidelberg (1988)

117. Heijer, P., Tolsma, R.: Data Communications. Glentop Publishers Ltd, Barnet, Herts, UK (1986)

118. Heinlein, P., Hartleben, P.: POP3 und IMAP – Mailserver mit Courier und Cyrus. Open Source Press (2007)

119. Helgert, H.J.: Integrated Services Digital Networks. Addison-Wesley Longman Publishing Co., Inc., Boston, MA, USA (1991)

120. Hinden, R., Deering, S.: IP Version 6 Addressing Architecture. RFC 4291 (Draft Standard) (2006). URL http://www.ietf.org/rfc/rfc4291.txt. Updated by RFCs 5952, 6052

121. Ho, Y.K., Liu, R.S.: A novel routing protocol for supporting qos for ad hoc mobile wireless networks. Wirel. Pers. Commun. 22(3), 359–385 (2002)

122. Hopper, A., Temple, S., Williamson, R.: Local Area Network Design. Addison-Wesley, Wokingham [u.a.] (1986)

123. Housley, R., Arbaugh, W.: Security problems in 802.11-based networks. Commun. ACM 46(5), 31–34 (2003)

124. Howes, T., Smith, M.: LDAP: programming directory-enabled applications with lightweight directory access protocol. Macmillan Publishing Co., Inc., Indianapolis, IN, USA (1997)

125. (Hrsg.), J.G.: The communications Handbook. CRC-Press, Boca Raton, FL, USA (1996)
126. (Hrsg.), J.G.: Multimedia Communications - Directions and Innovations. Academic Press, Inc., San Diego, CA, USA (1996)
127. (Hrsg.), J.L.C.: A History of Algorithms: From the Pebble to the Microchip. Springer Verlag, Berlin, Heidelberg, New York (1999)
128. Hsu, C.S., Tseng, Y.C., Sheu, J.P.: An efficient reliable broadcasting protocol for wireless mobile ad hoc networks. Ad Hoc Netw. **5**(3), 299–312 (2007)
129. Hufschmid, M.: Information und Kommunikation - Grundlagen und Verfahren der Informationsübertragung. B. G. Teubner Verlag / GWV Fachverlage GmbH, Wiesbaden (2006)
130. Hughes, L.E.: Internet E-Mail: Protocols, Standards, and Implementation. Artech House, Norwood, Massachusetts (1998)
131. Huitema, C.: Routing in the Internet. Prentice-Hall, Inc., Upper Saddle River, NJ, USA (1995)
132. Huitema, C.: IPv6: the new Internet protocol. Prentice-Hall, Inc., Upper Saddle River, NJ, USA (1996)
133. Hunt, C.: TCP/IP Network Administration, 3rd edn. O'Reilly Media, Inc. (2002)
134. Händel, R., Huber, M.N.: B-isdn network capabilities and evolution aspects. In: A. Casaca (ed.) Broadband Communications, *IFIP Transactions*, vol. C-4, pp. 19–29. North-Holland (1992)
135. IEEE: Ieee standard for information technology- telecommunications and information exchange between systems-local and metropolitan area networks-specific requirements-part 11: Wireless lan medium access control (mac) and physical layer (phy) specifications. IEEE Std 802.11-1997 pp. i –445 (1997)
136. Internet Engineering Task Force: RFC 791 Internet Protocol - DARPA Inernet Programm, Protocol Specification (1981)
137. Jeong, J., Park, S., Beloeil, L., Madanapalli, S.: IPv6 Router Advertisement Option for DNS Configuration. RFC 5006 (Experimental) (2007). URL http://www.ietf.org/rfc/rfc5006.txt. Obsoleted by RFC 6106
138. Johansen, J.: Standardisation of b-isdn. In: V.B. Iversen (ed.) Integrated Broadband Communications, *IFIP Transactions*, vol. C-18, pp. 89–93. Elsevier (1993)
139. Josefsson, S.: The Base16, Base32, and Base64 Data Encodings. RFC 4648 (Proposed Standard) (2006). URL http://www.ietf.org/rfc/rfc4648.txt
140. Kamoun, F., Kleinrock, L.: Stochastic performance evaluation of hierarchical routing for large networks. Computer Networks **3**, 337–353 (1979)
141. Kamoun, F., Kleinrock, L., Muntz, R.R.: Queueing analysis of the ordering issue in a distributed database concurrency control mechanism. In: ICDCS, pp. 13–23. IEEE Computer Society (1981)
142. Kanbach, A., Körber, A.: (ISDN - Die Technik), 3rd edn. Hüthig Buch Verlag, Heidelberg (1999)
143. Karn, P., Simpson, W.: Photuris: Session-Key Management Protocol. RFC 2522 (Experimental) (1999)
144. Karp, P.: Standardization of host mnemonics. RFC 226 (1971). URL http://www.ietf.org/rfc/rfc226.txt. Obsoleted by RFC 247
145. Kasera, S.K., Hjálmtýsson, G., Towsley, D.F., Kurose, J.F.: Scalable reliable multicast using multiple multicast channels. IEEE/ACM Trans. Netw. **8**(3), 294–310 (2000)
146. Kauffels, F.J.: Lokale Netze: Studienausgabe, 16. edn. mitp, Heidelberg (2008)
147. Kaufman, C.: Internet Key Exchange (IKEv2) Protocol. RFC 4306 (Proposed Standard) (2005). Obsoleted by RFC 5996, updated by RFC 5282
148. Kaufman, C.W., Perlman, R., Speciner, M.: Network Security: Private Communication in a Public World. Prentice-Hall, Englewood Cliffs, New Jersey (1995)
149. Kawarasaki, M., Jabbari, B.: B-isdn architecture and protocol. IEEE Journal on Selected Areas in Communications **9**(9), 1405–1415 (1991)
150. Kidwell, P.A., Ceruzzi, P.E.: Landmarks in digital computing: A Smithsonian practical history. Smithsonian Institute Press, Washington, D.C., USA (1994)

151. Kiefer, R.: Digitale Übertragung in SDH- und PDH-Netzen. Expert-Verlag, Reningen-Malmsheim (1998)
152. Kim, J.B., Suda, T., Yoshimura, M.: International standardization of b-isdn. Computer Networks and ISDN Systems **27**(1), 5–27 (1994). URL http://dblp.uni-trier.de/db/journals/cn/cn27.html#KimSY94
153. Kleinrock, L.: Information flow in large communication nets, ph.d. thesis proposal. Ph.D. thesis, Massachusetts Institute of Technology, Cambridge, MA, USA (1961). URL http://www.cs.ucla.edu/~lk/LK/Bib/REPORT/PhD/
154. Kleinrock, L.: Queueing Systems, Volume 1: Theory. John Wiley & Sons, Hoboken, NJ, USA (1975)
155. Kleinrock, L., Kamoun, F.: Hierarchical routing for large networks; performance evaluation and optimization. Computer Networks **1**, 155–174 (1977)
156. Knuth, D.E.: The Art of Computer Programming, Volume I: Fundamental Algorithms. Addison-Wesley (1968)
157. Knuth, D.E.: The Art of Computer Programming, Volume III: Sorting and Searching. Addison-Wesley (1973)
158. König, W. (ed.): Propyläen der Technikgeschichte, Bd. 1-5. Propyläen, Berlin (1990)
159. Krawczyk, H.: Skeme: a versatile secure key exchange mechanism for internet. Network and Distributed System Security, Symposium on **0**, 114 (1996)
160. Kumar, S., Radoslavov, P., Thaler, D., Alaettinoglu, C., Estrin, D., Handley, M.: The MASC/BGMP architecture for inter-domain multicast routing. In: SIGCOMM, pp. 93–104 (1998)
161. Kupris, G., Sikora, A.: ZigBee: Datenfunk mit IEEE 802.15.4 und ZigBee. Franzis Verlag GmbH, Poing (2007)
162. Kurose, J.F., Ross, K.W.: Computer Networking: A Top-Down Approach, 5th edn. Addison-Wesley Publishing Company, USA (2009)
163. Kyas, O., Campo, M.A.: IT Crackdown, Sicherheit im Internet. MITP Verlag, Bonn (2000)
164. Kyas, O., Crawford, G.: ATM Networks. Prentice Hall PTR, Upper Saddle River, NJ, USA (2002)
165. Köhler, M., Arndt, H.W., Fetzer, T.: Recht des Internet, 8th edn. C.F. Müller Verlag, Heidelberg (2008)
166. Labovitz, C., Malan, G.R., Jahanian, F.: Internet routing instability. SIGCOMM Comput. Commun. Rev. **27**(4), 115–126 (1997)
167. Leiner, B.M., Cole, R., Postel, J., Mills, D.: The darpa internet protocol suite pp. 48–53 (1992)
168. Licklider, J.C.R., Taylor, R.W.: The computer as a communication device. Science and Technology **76**, 21–31 (1968)
169. Liu, C., Albitz, P., Loukides, M.: DNS and BIND, 3rd edn. O'Reilly & Associates, Sebastopol, California (1998)
170. Macedonia, M.R., Brutzman, D.P.: MBone Provides Audio and Video Across the Internet. IEEE Computer **27**(4), 30–36 (1994)
171. Mack, S.: Streaming Media Bible, 1 edn. John Wiley & Sons, Inc., New York, NY, USA (2002)
172. Madron, T.W.: LANS: applications of IEEE/ANSI 802 standards. John Wiley & Sons, Inc., New York, NY, USA (1989)
173. Matt, H.J.: B-isdn system concept and technologies. In: ICC, pp. 1710–1714 (1986)
174. Maughan, D., Schertler, M., Schneider, M., Turner, J.: Internet Security Association and Key Management Protocol (ISAKMP). RFC 2408 (Proposed Standard) (1998). Obsoleted by RFC 4306
175. Maxwell, J.C.: A dynamical theory of the electromagnetic field. Philosophical Transactions of the Royal Society of London **155**, 459–513 (1865)
176. McQuillan, J., Richer, I., Rosen, E.: ARPANET routing algorithm improvements. Tech. rep., Cambridge (1978)
177. McQuillan, J.M., Richer, I., Rosen, E.C.: The new routing algorithm for the ARPANet. IEEE Transactions on Communications **28**(5), 711–?719 (1980)

178. Medvinsky, A., Hur, M.: Addition of Kerberos Cipher Suites to Transport Layer Security (TLS). RFC 2712 (Proposed Standard) (1999). URL http://www.ietf.org/rfc/rfc2712.txt

179. Meinel, C.: Physikalische trennung als ultima ratio im hochsicherheitsbereich. Informatik Spektrum **30**(3), 170–174 (2007)

180. Meinel, C., Sack, H.: WWW - Kommunikation, Internetworking, Web-Technologien. Springer Verlag, Heidelberg (2004)

181. Menezes, A.J., Vanstone, S.A., Oorschot, P.C.V.: Handbook of Applied Cryptography. CRC Press, Inc., Boca Raton, FL, USA (1996)

182. Metcalfe, R., Boggs, D.: Ethernet: Distributed packet switching for local computer networks. Commun. ACM **19**(7), 395–404 (1976)

183. Meyer, M.: Kommunikationstechnik - Konzepte der modernen Nachrichtenübertragung, 3rd edn. Vieweg (2008)

184. Miller, L.J.: The ISO Reference Model of Open Systems Interconnection: A first tutorial. In: ACM 81: Proceedings of the ACM '81 conference, pp. 283–288. ACM, New York, NY, USA (1981). DOI http://doi.acm.org/10.1145/800175.809901

185. Minoli, D., Alles, A.: LAN, ATM, and LAN Emulation Technologies. Artech House, Norwood MA, USA (1996)

186. Mitschke, F.: Glasfasern: Physik und Technologie. Elsevier, Spektrum Akademischer Verlag, Heidelberg (2005)

187. Mogul, J., Deering, S.: Path MTU discovery. RFC 1191 (Draft Standard) (1990). URL http://www.ietf.org/rfc/rfc1191.txt

188. Mogul, J., Kent, C., Partridge, C., McCloghrie, K.: IP MTU discovery options. RFC 1063 (1988). URL http://www.ietf.org/rfc/rfc1063.txt. Obsoleted by RFC 1191

189. Mogul, J., Postel, J.: Internet Standard Subnetting Procedure. RFC 950 (Standard) (1985). URL http://www.ietf.org/rfc/rfc950.txt

190. Myers, J., Rose, M.: Post Office Protocol - Version 3. RFC 1939 (Standard) (1996). URL http://www.ietf.org/rfc/rfc1939.txt. Updated by RFCs 1957, 2449

191. Narten, T.: Internet Routing. In: SIGCOMM '89: Symposium proceedings on Communications architectures & protocols, pp. 271–282. ACM, New York, NY, USA (1989)

192. Narten, T., Draves, R., Krishnan, S.: Privacy Extensions for Stateless Address Autoconfiguration in IPv6. RFC 4941 (Draft Standard) (2007). URL http://www.ietf.org/rfc/rfc4941.txt

193. Narten, T., Nordmark, E., Simpson, W.: Neighbor Discovery for IP Version 6 (IPv6). RFC 1970 (Proposed Standard) (1996). URL http://www.ietf.org/rfc/rfc1970.txt. Obsoleted by RFC 2461

194. Narten, T., Nordmark, E., Simpson, W., Soliman, H.: Neighbor Discovery for IP version 6 (IPv6). RFC 4861 (Draft Standard) (2007). URL http://www.ietf.org/rfc/rfc4861.txt. Updated by RFC 5942

195. Naumann, F.: Vom Abakus zum Internet : die Geschichte der Informatik. Primus-Verlag, Darmstadt (2001)

196. Neelakanta, P.S.: A Textbook on ATM Telecommunications: Principles and Implementation. CRC Press, Inc., Boca Raton, FL, USA (1999)

197. Newman, M.E.J.: The structure and function of complex networks. SIAM Review **45**(2), 167–256 (2003)

198. Nichols, K., Blake, S., Baker, F., Black, D.: Definition of the Differentiated Services Field (DS Field) in the IPv4 and IPv6 Headers. RFC 2474 (Proposed Standard) (1998). URL http://www.ietf.org/rfc/rfc2474.txt. Updated by RFCs 3168, 3260

199. Nordmark, E.: Stateless IP/ICMP Translation Algorithm (SIIT). RFC 2765 (Proposed Standard) (2000). URL http://www.ietf.org/rfc/rfc2765.txt

200. Nyquist, H.: Certain Factors Affecting Telegraph Speed. Bell System Technical Journal p. 324 (1924)

201. Ohm, J.R., Lüke, H.D.: Signalübertragung Grundlagen der digitalen und analogen Nachrichtenübertragungssysteme. Springer, Berlin; Heidelberg; New York (2007)

202. Ojanpera, T., Prasad, R. (eds.): Wideband CDMA For Third Generation Mobile Communications: Universal Personal Communications. Artech House, Inc., Norwood, MA, USA (1998)
203. O'Regan, G.: A Brief History of Computing. Springer Verlag Ldt., London, UK (2008)
204. Orman, H.: The OAKLEY Key Determination Protocol. RFC 2412 (Informational) (1998)
205. Pahlavan, K., Krishnamurthy, P.: Principles of Wireless Networks: A Unified Approach. Prentice Hall PTR, Upper Saddle River, NJ, USA (2001)
206. Paxson, V., Floyd, S.: Wide area traffic: the failure of poisson modeling. IEEE/ACM Trans. Netw. 3(3), 226–244 (1995)
207. Perkins, C., Belding-Royer, E., Das, S.: Ad hoc on-demand distance vector (aodv) routing (2003)
208. Perkins, C.E., Bhagwat, P.: Highly dynamic destination-sequenced distance-vector routing (dsdv) for mobile computers. SIGCOMM Comput. Commun. Rev. 24(4), 234–244 (1994)
209. Perlman, R.: Interconnections: Bridges and Routers, 2nd edn. Addison Wesley, Reading, Massachusetts (1999)
210. Peterson, L.L., Davie, B.S.: Computernetze - Ein modernes Lehrbuch. dPunkt Verlag, Heidelberg (2000)
211. Peterson, R.L., Ziemer, R.E., Borth, D.E.: Introduction to Spread Spectrum Communications. Prentice-Hall, Englewood Cliffs, NJ (1995)
212. Pickholtz, R.L., Schilling, D.L., Milstein, L.B.: Theory of Spread-Spectrum Communications – A Tutorial. IEEE Transactions on Communications 30(5), 855–884 (1982)
213. Picot, A., Reichwald, R., Wigand, T.R.: Die grenzenlose Unternehmung : Information, Organisation und Management ; Lehrbuch zur Unternehmensführung im Informationszeitalter, 4., vollst. überarb. u. erw. aufl. edn. Gabler, Wiesbaden (2001)
214. Pierce, J.: How far can data loops go? Communications, IEEE Transactions on 20(3), 527–530 (1972)
215. Pierce, J.R., Noll, A.M.: Signale - Die Geheimnisse der Telekommunikation. Spektrum Akademischer Verlag, Heidelberg (1990)
216. Plummer, D.: Ethernet Address Resolution Protocol: Or Converting Network Protocol Addresses to 48.bit Ethernet Address for Transmission on Ethernet Hardware. RFC 826 (Standard) (1982). URL http://www.ietf.org/rfc/rfc826.txt. Updated by RFCs 5227, 5494
217. Postel, J.: User Datagram Protocol. RFC 768 (Standard) (1980). URL http://www.ietf.org/rfc/rfc768.txt
218. Postel, J.: Transmission Control Protocol. RFC 793 (Standard) (1981). URL http://www.ietf.org/rfc/rfc793.txt. Updated by RFCs 1122, 3168
219. Postel, J., Reynolds, J.: File Transfer Protocol. RFC 959 (Standard) (1985). URL http://www.ietf.org/rfc/rfc959.txt. Updated by RFCs 2228, 2640, 2773, 3659, 5797
220. Preneel, B.: Cryptographic primitives for information authentication - state of the art. In: State of the Art in Applied Cryptography, Course on Computer Security and Industrial Cryptography - Revised Lectures, pp. 49–104. Springer-Verlag, London, UK (1998)
221. Price, R.: Further Notes and Anecdotes on Spread-Spectrum Origins. IEEE Transactions on Communications COM-31(1), 85 (1983)
222. Proebster, W.: Rechnernetze-Technik - Protokolle, Systeme, Anwendungen. Oldenbourg Wissenschaftsverlag, München (2002)
223. Ramalho, M.: Intra- and inter-domain multicast routing protocols: A survey and taxonomy. IEEE Communications Surveys and Tutorials 3(1) (2000)
224. Rech, J.: Ethernet – Technologien und Protokolle für die Computervernetzung, 2. aufl. edn. Heise Zeitschriften Verlag, GmbH & Co. KG (2007)
225. Rescorla, E., Modadugu, N.: Datagram Transport Layer Security. RFC 4347 (Proposed Standard) (2006). URL http://www.ietf.org/rfc/rfc4347.txt. Updated by RFC 5746
226. Riggert, W.: Rechnernetze, Technologien-Komponenten-Trends. Fachbuchverlag Leipzig im Carl Hanser Verlag (2001)

227. Ripeanu, M.: Peer-to-peer architecture case study: Gnutella network. In: Proceedings of the First International Conference on Peer-to-Peer Computing, P2P '01, pp. 99–110. IEEE Computer Society, Washington, DC, USA (2001). URL http://portal.acm.org/citation.cfm?id=882470.883281

228. Roberts, L.G.: Multiple computer networks and intercomputer communication. In: Proceedings of the first ACM symposium on Operating System Principles, SOSP '67, pp. 3.1–3.6. ACM, New York, NY, USA (1967). DOI http://doi.acm.org/10.1145/800001.811680. URL http://doi.acm.org/10.1145/800001.811680

229. Roberts, L.G.: Aloha packet system with and without slots and capture. SIGCOMM Comput. Commun. Rev. **5**(2), 28–42 (1975)

230. Rosenthal, D.: Internet - schöne neue Welt?: Der Report über die unsichtbaren Risiken. Orell Füssli Verlag, Zürich, Schweiz (1999)

231. Ryan, P.S.: War, Peace, or Stalemate: Wargames, Wardialing, Wardriving, and the Emerging Market for Hacker Ethics. Virginia Journal of Law & Technology **9**(7) (2004)

232. Sakane, S., Kamada, K., Thomas, M., Vilhuber, J.: Kerberized Internet Negotiation of Keys (KINK). RFC 4430 (Proposed Standard) (2006)

233. Salus, P.H.: Casting the Net: From ARPANET to Internet and Beyond... Addison-Wesley Longman Publishing Co., Inc., Boston, MA, USA (1995). Foreword By-Vinton G. Cerf

234. Saunders, S.: The McGraw-Hill high-speed LANs handbook. McGraw-Hill, Inc., Upper Saddle River, NJ, USA (1996)

235. Saunders, S.: Data Communications Gigabit Ethernet Handbook. McGraw-Hill, Inc., New York, NY, USA (1998)

236. Scherff, J.: Grundkurs Computernetze. Vieweg, Wiesbaden (2006)

237. Schiller, J.: Mobile Communications, 2 edn. Addison-Wesley (2003)

238. Schillings, V., Meinel, C.: Tele-task – tele-teaching anywhere solution kit. In: Proceedings of ACM SIGUCCS. Providence, USA (2002)

239. Schmeh, K.: Kryptografie – Verfahren, Protokolle, Infrastrukturen, pp. 199–234. dPunkt Verlag, Heidelberg (2007)

240. Schneier, B.: Applied Cryptography: Protocols, Algorithms, and Source Code in C. John Wiley & Sons, Inc., New York, NY, USA (1993)

241. Scholtz, R.A.: The origins of spread-spectrum communications. IEEE Transactions on Communications **COM-30**(5), 822–854 (1982)

242. Schwenk, J.: Sicherheit und Kryptographie im Internet. Vieweg Verlag, Wiesbaden (2002)

243. Scott, A., Chu, F., McLaughlin, D.: The soliton: A new concept in applied science. Proceedings of the IEEE **61**(10), 1443–1483 (1973)

244. Scott, C., Erwin, S.W.: Virtual Private Networks. O'Reilly & Associates, Inc., Sebastopol, CA, USA (1998)

245. Shannon, C.E.: Communication in the presence of noise. Proceedings of the IRE **37**(1), 10–21 (1949)

246. Shannon, C.E., Weaver, W.: The Mathematical Theory of Communication. University of Illinois Press, Urbana, Illinois (1949)

247. Shepler, S., Callaghan, B., Robinson, D., Thurlow, R., Beame, C., Eisler, M., Noveck, D.: Network File System (NFS) version 4 Protocol. RFC 3530 (Proposed Standard) (2003). URL http://www.ietf.org/rfc/rfc3530.txt

248. Soni, S.K., Aseri, T.C.: A review of current multicast routing protocol of mobile ad hoc network. In: ICCMS '10: Proceedings of the 2010 Second International Conference on Computer Modeling and Simulation, pp. 207–211. IEEE Computer Society, Washington, DC, USA (2010)

249. Spurgeon, C.E.: Ethernet: the definitive guide. O'Reilly & Associates, Inc., Sebastopol, CA, USA (2000)

250. Stallings, W.: Data and Computer Communications, 6th edn. Prentice-Hall, Englewood Cliffs, New Jersey (1999)

251. Stallings, W.: Wireless Communications and Networks. Prentice Hall Professional Technical Reference (2001)

252. Stamper, D.A.: Essentials of Data Communications. Benjamin Cummings, Menlo Park, CA, USA (1997)

253. Steenstrup, M. (ed.): Routing in communications networks. Prentice Hall International (UK) Ltd., Hertfordshire, UK, UK (1995)

254. Stein, E.: Taschenbuch Rechnernetze und Internet. Fachbuchverlag Leipzig, Carl Hanser Verlag, München (2001)

255. Stein, E.: Taschenbuch Rechnernetze und Internet, 3 edn. Fachbuchverlag Leipzig, Carl Hanser Verlag, München (2008)

256. Stevens, W.R.: TCP/IP Illustrated, Volume I: The Protocols. Addison-Wesley, Reading, MA (1994)

257. Stevens, W.R.: TCP/IP Illustrated, Volume 3: TCP for Transactions, HTTP, NNTP, and the UNIX Domain Protocols. Addison-Wesley Publishing Company, Reading, MA (1996)

258. Stevenson, D.: Electropolitical Correctness and High-Speed Networking, or, Why ATM is like a Nose. In: Viniotis, Y., Onvural, R. O.: Asynchronous Transfer Mode Networks. Plenum Press, New York, NY, USA (1993)

259. Stewart, J.W.: BGP4: Inter-Domain Routing in the Internet. Addison-Wesley Longman Publishing Co., Inc., Boston, MA, USA (1998)

260. Sun, J., Zhang, X.: Study of zigbee wireless mesh networks. In: HIS (2), pp. 264–267. IEEE Computer Society (2009)

261. Tanenbaum, A.S.: Computer Networks. Prentice-Hall, Inc., Upper Saddle River, NJ, USA (1996)

262. Taylor, E.: The McGraw-Hill internetworking handbook. McGraw-Hill, Inc., New York, NY, USA (1995)

263. Tews, E., Weinmann, R.P., Pyshkin, A.: Breaking 104 bit wep in less than 60 seconds. In: S. Kim, M. Yung, H.W. Lee (eds.) WISA, *Lecture Notes in Computer Science*, vol. 4867, pp. 188–202. Springer (2007)

264. Thaler, D., Ravishankar, C.V.: Distributed center-location algorithms. IEEE Journal on Selected Areas in Communications 15(3), 291–303 (1997)

265. Thurlow, R.: RPC: Remote Procedure Call Protocol Specification Version 2. RFC 5531 (Draft Standard) (2009). URL http://www.ietf.org/rfc/rfc5531.txt

266. Tseng, Y.C., hua Liao, W., Wu, S.L.: Mobile Ad Hoc Networks and Routing Protocols. In: I. Stojmenovic (ed.) Handbook on Wireless Networks and Mobile Computing, pp. 371–392. John Wiley & Sons, Inc. (2002)

267. Tseng, Y.C., Ni, S.Y., Chen, Y.S., Sheu, J.P.: The broadcast storm problem in a mobile ad hoc network. Wirel. Netw. 8(2/3), 153–167 (2002)

268. Ulaby, F.T.: Fundamentals of applied electromagnetics. Prentice-Hall, Inc., Upper Saddle River, NJ, USA (1997)

269. Vacca, J.R.: The Cabeling Handbook. Prentice Hall Professional Technical Reference, Upper Saddle River NJ, USA (1998)

270. Verma, P.K., Saltzberg, B.R.: ISDN Systems; Architecture, Technology, and Applications: Architecture, Technology, and Applicat. Prentice Hall Professional Technical Reference (1990)

271. Viterbi, A.J.: CDMA: principles of spread spectrum communication. Addison Wesley Longman Publishing Co., Inc., Redwood City, CA, USA (1995)

272. Vixie, P.: Extension Mechanisms for DNS (EDNS0). RFC 2671 (Proposed Standard) (1999). URL http://www.ietf.org/rfc/rfc2671.txt

273. Vixie, P., Gudmundsson, O., Eastlake 3rd, D., Wellington, B.: Secret Key Transaction Authentication for DNS (TSIG). RFC 2845 (Proposed Standard) (2000). URL http://www.ietf.org/rfc/rfc2845.txt. Updated by RFC 3645

274. Wei, H.Y., Ganguly, S.: Design of 802.16 wimax based radio access network. In: Proc. IEEE International Symposium on Personal, Indoor, and Mobile Radio Communications (PIMRC), pp. 1–5. IEEE (2006)

275. Weinstein, S., Ebert, P.: Data Transmission by Frequency-Division Multiplexing Using the Discrete Fourier Transform. IEEE Transactions on Communications 19(5), 628–634 (1971)

276. Wesel, E.K.: Wireless Multimedia Communications: Networking Video, Voice and Data. Addison-Wesley Longman Publishing Co., Inc., Boston, MA, USA (1997)

277. Willinger, W., Taqqu, M.S., Sherman, R., Wilson, D.V.: Self-similarity through high-variability: statistical analysis of Ethernet LAN traffic at the source level. IEEE/ACM Trans. Netw. **5**(1), 71–86 (1997)

278. Wobst, R.: Abenteuer Kryptologie. Methoden, Risiken und Nutzen der Datenverschlüsselung, 3. aufl. edn. Addison-Wesley, Pearson Education Deutschland GmbH, München (2001)

279. Wolf, K., Linckels, S., Meinel, C.: Teleteaching anywhere solution kit (tele-task) goes mobile. In: Proceedings of the 35th annual ACM SIGUCCS fall conference, SIGUCCS '07, pp. 366–371. ACM, New York, NY, USA (2007). URL http://doi.acm.org/10.1145/1294046.1294132

280. Wright, G.R., Stevens, W.R.: TCP/IP Illustrated, Volume 2: The Implementation. Addison-Wesley Publishing Company, Reading, MA (1995)

281. Yang, H.S., Yang, S., Spinney, B.A., Towning, S.: FDDI Data Link Development. Digital Technical Journal **3**, 31–41 (1991)

282. Zdziarski, J.A.: Ending Spam: Bayesian Content Filtering and the Art of Statistical Language Classification. No Starch Press, San Francisco, CA, USA (2005)

283. Zeltserman, D.: A practical guide to SNMPv3 and network management. Prentice Hall PTR, Upper Saddle River, NJ, USA (1999)

284. Zenk, A.: Leitfaden für Novell Netware : Grundlagen und Installation. Addison-Wesley, Bonn [u.a.] (1990)

285. Zhang, Y., Ryu, B.: Mobile and multicast ip services in pacs: System architecture, prototype, and performance. MONET **6**(1), 81–94 (2001)

286. Zimmermann, H.: OSI Reference Model–The ISO Model of Architecture for Open Systems Interconnection. Communications, IEEE Transactions on [legacy, pre - 1988] **28**(4), 425–432 (1980)

Index

C. Meinel and H. Sack, *Internetworking*, X.media.publishing,
DOI: 10.1007/978-3-642-35392-5, © Springer-Verlag Berlin Heidelberg 2013

Printed in the United States
by Bookmasters

Printed in the United States
By Bookmasters